杨树遗传改良研究

——张志毅林学研究文集

张志毅 著

中国林业出版社

图书在版编目(CIP)数据

杨树遗传改良研究：张志毅林学研究文集／张志毅著．—北京：中国林业出版社，2021.5
ISBN 978-7-5219-1177-0

Ⅰ.①杨… Ⅱ.①张… Ⅲ.①杨树-遗传改良-文集 Ⅳ.①S792.110.4-53

中国版本图书馆 CIP 数据核字(2021)第 099109 号

出版发行 中国林业出版社(100009 北京市西城区德内大街刘海胡同 7 号)
电话：(010)83223120
印　　刷 北京中科印刷有限公司
版　　次 2021 年 6 月第 1 版
印　　次 2021 年 6 月第 1 次印刷
开　　本 787mm×1092mm 1/16
印　　张 43　　彩插 0.75
字　　数 1100 千字
定　　价 260.00 元

本文集编委会

序　言

张志毅教授(1958.12—2011.02)是我国著名的林木遗传育种学家，为了缅怀他，在他离开我们十年之际，我们怀着无比崇敬的心情整理了他数十年学术生涯所取得的成果，汇集成《杨树遗传改良研究——张志毅林学研究文集》(以下简称《文集》)。《文集》反映了张志毅教授对林木遗传育种学科发展前沿的敏锐洞察，也体现了其对于毛白杨遗传改良研究的超前部署。张志毅教授以毛白杨优良种质资源为基础，率先在林木遗传改良领域尝试将常规杂交育种与现代分子育种进行有机融合，创建高效精准的杨树育种体系，为现代林木遗传改良开辟了一个全新的领域。文集不仅对当前杨树等林木遗传改良研究具有重要的学术价值，而且对未来林木精准分子设计育种极具借鉴意义。

张志毅教授锐意创新，勇于实践，求真务实，在林木遗传改良研究方面做出了卓越的贡献。“六五”至“十一五”期间，在国家科技攻关计划、科技支撑和“863”项目支持下，聚焦于“毛白杨良种选育”、“白杨染色体加倍技术研究及三倍体育种”及“白杨 $2n$ 花粉诱导的细胞学机理研究”等倍性育种机理和技术研究，与同事共同突破了毛白杨花粉染色体加倍技术瓶颈，建立了毛白杨三倍体育种技术体系。选育了系列毛白杨新品种，获得16个新品种权证书和3个国家级良种证书，在华北地区黄河流域大面积推广应用，产生了数十亿元的经济效益和巨大的生态与社会效益。成果荣获国家科技进步二等奖3项、部级科技进步一等奖1项、部级科技进步二等奖2项。

在国家自然科学基金项目“毛白杨遗传图谱构建和重要性状的分子标记(30170780)”、“杨树杂交育种中杂种优势分子机理的研究(30571516)”等资助下，开展了毛白杨遗传育种学基础理论研究，分别构建了首张毛白杨遗传连锁图谱和转录图谱，在毛白杨重要经济性状遗传规律、亲本起源和杂种优势理论研究方面取得了重要突破，首次提出了“响叶杨为毛白杨杂种起源母本”的新假说，并提供了叶绿体基因组的证据；率先采用转录组技术对杨树杂种优势开展了研究，从基因表达水平对杨树杂种优势的形成提出了新的见解，丰富了杂种优势育种理论，为杨树杂交育种提供了强有力的理论支撑。

在国家“948”项目“杨树控花基因分离鉴定与转化技术引进(2006-4-72)”的支持下，开展了杨树生殖生物学和基因调控研究，应用RNA干扰(RNAi)和显性负突变(dominant negative mutaion，DNM)技术在杨树开花调控研究方面取得突破，获得了一批转控花基因的毛白杨种质，经国家林业和草原局审定，其中7个转基因株系获得许可进入了中间试验阶段，研究成果为最终解决杨树花期飞絮污染问题提供了新的途径。

在国家“973”项目“分子改良树木的性状表达与鉴定(G1999016005)”、“863”项目

“速生杨树品种抗虫、抗旱耐盐转基因育种研究(2009AA10Z107)”和国家自然科学基金项目“毛白杨叶锈病病原菌诱导表达NBS型基因的分子作用机制(30872043)”等资助下，开展了杨树抗逆生理生化与分子生物学基础研究，克隆鉴定了杨树抗冻、抗锈病、耐盐基因等，在杨树中进行了抗虫、耐盐和抗冻基因的遗传转化工作，获得了转*Bt*、*CpT* I和*MdSPDS*1等抗虫、耐盐转基因杨树新种质，极大地推动了杨树抗逆基因工程育种研究工作的进展。

在国家“十一五”科技支撑项目“杨树速生丰产林培育关键技术研究与示范(2006BAD24B04)”和“863”项目“抗旱节水白杨新品种筛选与利用(2002AA2Z4011)”等项目支持下，围绕杨树光合、水分和矿质生理等方面开展了研究，探讨了栽植密度、修剪方式对杨树光合作用、水分和氮磷钾吸收的影响，提出了合理栽植密度-整枝-水分-矿质营养综合管理模型，选育了抗旱节水的白杨新品种，研究成果在杨树丰产林的生产实践中得到推广应用，并取得了显著成效。

针对杨树材性改良和木材加工利用存在的问题，在国家“863”项目“杨树纸浆材新品种选育研究(2001AA244031)”等项目支持下，开展了材性改良育种研究和木材加工工艺研究，并与纸业企业开展校企合作，选育出了纤维素长和高纤维素含量的纸浆用途新品种。同时，改进了纤维素转化乙醇的方法和工艺，研究成果在太阳纸业集团得到了推广应用，极大地降低了纸浆生产成本，显著增加了企业效益。

张志毅教授的研究涵盖了林木遗传理论、育种技术与育种实践、林木栽培理论与技术、林木加工理论与技术等林业产业化全过程，其学术见解对于从事林业教学科研和生产实践的专业人员具有重要的参考价值。

张志毅教授不仅在学术上颇有建树，而且为学校学科建设做出了重要贡献。作为林木遗传育种学科负责人，他带领学科成功申报了“林木育种国家工程实验室”，教育部重点实验室评估取得了优异成绩。引进了“长江学者”、“千人计划”等高层次人才，加强实验室和基地建设，为学科的发展奠定了坚实的基础。同时，他热衷公益，曾任北京市人大代表、北京市人大农村工作委员会副主任、北京市党外高级知识分子联谊会理事、海淀区政协委员等众多社会职务，为绿色北京和科技北京做出了突出贡献。

张志毅教授毕生执着于林业教育、科技和社会公益事业，具有崇高的奉献精神和高度的社会责任感，为我国林业教育事业、科技进步和经济建设做出了重大贡献。他的英年早逝不仅是北京林业大学和我国林木遗传育种界的重大损失，也是我国林业高等教育和科技领域的损失。

我们深切缅怀张志毅教授，同时也希望《文集》的出版能够给后辈学人以启迪，在科学探索的道路上不畏艰难，努力攻关。

北京林业大学原校长
中国工程院　　院士

2021年6月6日

前　言

张志毅教授是我国著名的林木遗传育种学家、林业教育家和社会活动家，无党派人士。他一生钟情于林木遗传改良，热爱林业教育事业，作为北京市人大代表和海淀区政协委员，积极参政议政建言献策。他成果斐然、桃李满园，对国家林业建设做出了杰出贡献。张志毅教授的英年早逝是我国林业教育界、科技界和林木良种产业界的重大损失，为缅怀张志毅教授为我国林业研究和教育做出的突出贡献，我们筹集出版了《杨树遗传改良研究——张志毅林学研究文集》(以下简称《文集》)，以此告慰九泉之下的张志毅教授。

《文集》从张志毅教授发表的200余篇论文中，选择不同时期、不同领域具有代表性论著近百篇，逾百万字汇编而成。《文集》分八部分：序言、前言、人生记事、目录、正文、附录、后记、彩插。

尹伟伦院士为《文集》作序。介绍了张志毅教授在林木遗传育种研究领域取得的实践成果和理论成就，并对他的为人处事、品格作风作出了高度评价，从中也可以领略张志毅教授教学、科研、工作及生活的不同侧面。

人生记事记录了张志毅教授学历、工作简历、行政及学术兼职等方面内容。

正文选录了张志毅教授发表的论文。内容丰富，具有较高的学术价值，包括六个方面：杨树育种技术及相关理论、分子标记技术及其应用、杨树生殖生物学与基因调控、杨树抗逆生理生化与分子生物学、杨树光合作用与水分生理及矿质营养吸收、杨树材性及木材加工。

附录部分包括：教材及国家标准的编著、研究生培养、主持和参与的科研项目、科研获奖、表彰和荣誉、社会兼职六个部分。

《文集》基本反映了张志毅教授一生的工作与学术成就，对林木遗传育种领域的科研人员和学生具有借鉴与启迪作用。张志毅教授勤勉好学、与时俱进，治学严谨、思维创新，为人谦虚谨慎、平易近人，对学生循循善诱、诲人不倦，深得师生们的尊敬与爱戴。他的优良品格值得继承和发扬，希望《文集》的出版将激励我们为我国林木遗传育种事业的发展做出更大的贡献。

编委会

2021年4月

人生记事

时 间	重要事件
1958. 12	出生于云南省昌宁县新城
1978. 09	考入北京林学院林学专业
1982. 07	毕业于北京林学院林学系，获得农学学士学位
1982. 08—1986. 03	北京林业大学任助教
1987. 04—1991. 03	北京林业大学林木遗传育种学科任讲师
1988. 03—1989. 01	赴德国哥廷根大学进修学习
1989. 09—1991. 06	北京林业大学攻读硕士研究生，获农学硕士学位
1992. 11	“全国毛白杨基因资源的收集、保存和利用的研究”获国家科技进步二等奖，排名第 13
1992. 09—1994. 06	北京林业大学攻读博士研究生，获农学博士学位
1992. 04—1995. 03	北京林业大学林木遗传育种学科任副教授，教研室主任
1996. 04—2007. 12	北京林业大学林木遗传育种学科任教授、博士生导师，系主任，生物学院学术委员会委员，国家理科生物学人才培养基地专家指导委员会委员，北京林业大学教代会副主任，教育部林木花卉遗传育种重点实验室副主任
1997. 12	“毛白杨多圃配套系列育苗技术研究”获国家科技进步二等奖，排名第 5
2004. 01	“三倍体毛白杨新品种选育研究”获国家科技进步二等奖，排名第 2
2006. 05	从美国北卡罗来纳州立大学引进“长江学者讲座教授”——李百炼教授
2008. 01—2011. 02	北京林业大学林木遗传育种学科任二级教授、博士生导师，学科负责人，生物学院学术委员会委员，国家理科生物学人才培养基地专家指导委员会委员，北京林业大学教代会副主任，教育部林木花卉遗传育种重点实验室副主任
2008. 10. 27	第 23 届国际杨树大会在北京召开，作为大会组织者参与大会的组织和筹备工作，主持了育种分会的议程
2008—2009	从美国宾夕法尼亚州立大学引进“长江学者讲座教授”、“千人计划”学者——邬荣领教授

时 间	重要事件
2008. 10. 23	成功组织申报国家发展改革委“林木育种国家工程实验室”，任林木育种国家工程实验室主任
2009. 09	作为负责人开展985优势学科创新平台—林木良种与生物学基础方向项目建设
2010. 03. 12	国家科技部曹健林副部长一行到北京林业大学调研，并考察了林木育种国家工程实验室，张志毅教授介绍了实验室建设情况和毛白杨三倍体育种成果
2010. 09. 21	带领实验室全体成员在2010年教育部林木花卉遗传育种重点实验室评估中取得了优异成绩
2011. 02. 07	林木育种国家工程实验室主任张志毅教授因病逝世

目　录

第一部分　杨树育种技术及相关理论

第二部分 分子标记技术及其应用

第三部分 杨树生殖生物学与基因调控

第四部分 杨树抗逆生理生化与分子生物学

第五部分　杨树光合作用、水分生理及矿质营养吸收

第六部分 杨树材性及木材加工

附录

后记

第一部分

杨树育种技术及相关理论

Doubling Technology of Pollen Chromosome of *Populus tomentosa* and Its Hybrid*

Abstract The big size pollen grain were obtained when the male flower buds of Chinese white poplar (*Populus tomentosa* Carr.) and its hybrid (*P. tomentosa*×*P. bolleana*) were treated with various concentrations of colchicine and by different treatment ways in this study. It was identified that the big pollen grains induced by colchicine were unreduced $2n$ pollen by measuring the DNA relative content with cytofluorimetry technology. The $2n$ pollen grain has reliable vitality and it can be used for triploid breeding by pollinating. The results show that the suitable concentration of colchicine ranged f rom 0. 1% to 0. 5% and injecting treatment is the best way. The yield of $2n$ pollen grains was higher with the treatment times from 3 to 6 in lower temperature condition. Meanwhile, the treatment effect of higher temperature (38~40℃ for two hours) was also tested and analyzed in this paper.
Key words *Populus tomentosa* Carr. , *P. tomentosa*×*P. bolleana*, colchicine, chromosomes doubling, unreduced pollen grains

1 Introduction

Polyploid breeding is a new breeding technology developed from the 1930s' (Lewis, 1980; Wright, 1976). Most researches reported that usually polyploids have 'gigantism' in their morphological organs. Furthermore, they bear such characteristics as greater growth yield, higher resistance, etc. The reason is that polyploidy plant has more chromosomes than the normal diploid in its cell nucleus that directly prompt to increase the volume of the cells. Plant polyploidy cultivars have made very high economic effect in production, especially the triploidy ones that have even greater economic values.

Although the start of the work around triploid breeding in tree species was not late, it developed rather slowly. The main reason is that the long term of flowering and bearing of tree species result in difficulties in the researches on biology and cytology. The natural European aspen (*Populus tremula*) triploid was discovered by Nilsson-Ehle and Muentzing in Sweden in 1936. Since then, other triploidy cultivars of tree species have been cultured in countries like Finland (Sarvas, 1961), the former Soviet Union (Sekawin, 1961; Mashkina, 1989), USA (Benson, 1967; Ein-

* 本文原载《北京林业大学学报》(*Journal of Beijing Forestry University*), 1997, 6(2): 9-20, 与李凤兰、朱之悌和康向阳合作发表。

spahr, 1963; 1976; Van Buijtenen, 1957; Winton, 1970), Germany (Baumeister, 1980; Weisgerber, 1979; 1980; 1983), Japan (Sasaki, 1993) and also Sweden (Johnsson, 1940), etc. The breeding efforts for tree polyploids were late in China, though. There is no report of discovering natural polyploids in the major foresting tree species, except for in very a few economic tree species. There are few researches on inducing polyploid in tree species artificially. A few putative experiments were made in the 1950s'. Some mutant plants were obtained by treating with different concentrations of colchicine to seeds and seedlings of 5 tree species by Beijing Fruit and Forest Research Institute in 1979. Sun Zhongxu et al. (1984) reported that polyploidy plants were gotten when treated with 0.2% colchicine solution for germinating seeds and seedlings of black locust (*Robinia pseudoacacia* L.). Similar experiment was done on tree species like mulberry (*Morus* spp.), Siberian elm(*Ulmus pumila* L.), etc. Researches on polyploid breeding of mulberry were relatively more. Tetraploidy plants of mulberry were obtained through colchicine treatment by Wang in 1981. Li et al. (1987) reported firstly in 1987 that the triploids were gotten through hybridizing the natural tetraploid and the diploid. Researches on doubling of mulberry were also carried out and some triploids were obtained by Yang Jinhou (1984, 1987), Yang Dequan et al. (1989). There were some reports of inducing triploids by endosperm culturing in economic tree species like sweet orange (Gao, 1988; Chen, 1991), Chinese wolfberry (Gu, 1987) and chinese date (Shi, 1985), etc. The triploid of lacquer tree was discovered by Shang (1985, 1989). Nevertheless, there is no domestic report of culturing poplar triploids by way of artificial doubling and hybridization.

Although triploidy trees exist in nature, it is hard to find them. To obtain triploids, tree breeding specialists always then utilize the natural or artificial tetraploid to hybridize with diploid. Some researchers selected $2n$ pollen grains with the unreduced chromosome from male flower of the diploidy plants and thus obtained triploids through pollinating the $2n$ pollen grains to female flowers of the normal diploid plants. But it is very difficult to get natural tetraploidy trees and natural unreduced $2n$ pollens, and it needs many years for the induced artificial tetraploidy tree to flower and bear. Large quantities of the production of triploidy trees were thus restricted. How to induce chromosome unreduced pollen in a short period becomes the key to culture triploid tree artificially.

The purpose of the study is to induce $2n$ pollen of Chinese white poplar (*Populus tomentosa*) and its hybrid (*Populus tomentosa*×*P. bolleana*) by using chemical and physical methods, to search for the optimum way, method, concentration and environmental conditions of treatment and to offer viable $2n$ pollens for artificial culturing of tripoids.

2 Materials and method

2.1 Materials

Ten male clones were selected in the experiment including 6 clones of *P. toentosa* (TJ, TL50, 5041, 5050, 3172 and 3541) collected from the clonal aboretum located at Guanxian county in Shandong province and 4 male clones of the hybrid (*Populus tomentosa*×*P. bolleana*) from the campus of China Agriculture University (the former Beijing Agriculture University). These clones disperse more pollens than the others. The sprays were collected from Jan. to Feb. of each year .

2. 2 Method

2. 2. 1 The doubling of pollen chromosomes

Two methods, the chemical and physical, were taken in the study. The chemical inducement is colchicine treatment by three ways injection, saturation and culture. The concentration of colchicine ranges from 0. 01% to 1. 00%. The injection was done by injecting the colchicine solution directly to the male flower buds in meiosis by using micro-injector. The saturation was cutting 2 or 3 cuts on the male flower bud scale by using one-side razor blade, packaging it with absorbent cotton and dropping the colchicine solution on the cotton(keep the cotton wet). Culture method was to put the male sprays directly into the colchicine solution. Orthogonal design (L9(3^4)) as carried out with treatment way (three ways as the above-mentioned) , solution concentration (0. 1% ~ 1. 0%) and treatment start time (one, two, three days after culture in water). The experiment of studying the effect of temperature conditions in the treatment environment on treatment result was done at the conditions in the greenhouse (10 ~ 20℃) and in the low-temperature cold room. The injection times ranges from one to nine.

The physical inducement method was to put the male sprays after different time of water-culture into the thermostatic container for 2 hours with the temperature controlled at 38 ~ 40℃ , relative humidity between 30% ~ 40% and then put them back in the greenhouse to water culture until pollen dispersal phase.

2. 2. 2 The identification and statistics of 2*n* pollen grains

Identification of the 2*n* pollen grains was done by morphological observation and relative DNA content test. Pollen size was measured directly under the microscope by using the dial gauge. 5 ~ 10 fields of vision of 5 ~ 10 smears for each treatment were observed and the rate was counted. The relative DNA content of the pollen grains was tested by the cytofluorimetry technology.

2. 2. 3 Separation, purification and vitality test of the 2*n* pollen grains

The 600-eye metal web was used to separate and purify the 2*n* pollen grains. Culture method and *in vivo* test method was then taken to test the vitality of the 2*n* pollen grains.

All the experiments were done 3 ~ 5 times and the original data were transformed before the variance analysis.

3 Results and analysis

3.1 The inducement of 2*n* pollen grains

Table 1 The experimental result of orthogonal design

No.	repeat				Σ
	Ⅰ	Ⅱ	Ⅲ	Ⅳ	
1	0	0	0	0	0
2	0	0	0	0	0
3	0	0	0	0	0
4	6.10	6.87	8.86	6.88	28.71
5	19.49	6.71	14.13	15.22	55.55
6	3.42	3.79	5.96	3.65	16.82
7	6.03	7.50	10.10	8.73	32.36
8	20.80	20.97	21.79	17.26	81.00
9	7.98	10.49	10.75	14.67	43.89
X_1	0	5.09	4.10	6.05	
X_2	8.42	11.38	9.14	8.15	
X_3	13.10	5.06	8.29	7.33	
R	13.10	6.32	5.04	2.10	

3.1.1 The inducement of 2*n* pollen grains by colchicines

(1) The effect of colchicine concentration and treatment way on the yield of 2*n* pollen grains

The experiment with a four-factor three level four-replication orthogonal design was made at the greenhouse condition (10~20℃). The result was shown in Table 1 and the result of variance analysis was shown in Table 2.

From Table 1, the rates of the 2*n* pollen grains from the first three groups were all zero, which means it was not feasible to induce 2*n* pollen by culturing sprays in the colchicine solution. 2*n* pollen grains were obtained from the other experiment groups with the average minimum rate 3.42% and the average maximum 21.97%. From the result of the variance analysis in Table 2, among all the treatment factors, except the material clone number, the other three factors, treatment way, colchicine concentration, and treatment start time, exist great significant differences. It indicates that the effect of the inducement of the 2*n* pollen grains was not affected by the different material clones. So in the experiment, there should be possible to obtain a certain rate of 2*n* pollen grains as long as the male flowe sprays of the selected clones bear the well-developed sufficient flower buds. The other three factors, treatment ways, colchicine concentration, and treatment start time, affected the result of the inducement of 2*n* pollen grains greatly. The multiple comparison

(q test) showed that injection is the best way of inducement, 0.5% is the most appropriate concentration, and the rate of $2n$ pollen grains is the highest when treated from the second day after the water culture.

Table 2 Variance analysis to the result of orthogonal design

Source	*df*	*SS*	*MS*	*F*
Repeat	3	14.78	4.93	1.18
Tretment	8	3294.45	411.81	98.52
Way	2	2873.23	1436.62	343.69
Concentration	2	251.53	125.77	30.39
Time	2	158.67	79.36	18.98
Clone	2	11.02	5.51	1.32
Error	24	100.25	4.18	
Total	35	34093.48		

On the basis of the or thogonal experiment, another experiment with seven levels of colchicine concentrat ions was designed. The result was shown in Table 3.

Table 3 Effect of various concentrations of colchicine on percentage of 2*n* pollen grains

Concentration	Injection			Saturation		
(%)	min.	max.	mean	min.	max.	mean
0.01	0.00	5.67	2.25	0.58	1.30	0.89
0.05	0.69	13.21	5.99	0.00	2.68	1.42
0.10	12.57	21.47	16.52	1.69	17.02	8.31
0.25	9.46	22.70	16.06	3.54	13.92	8.75
0.50	10.08	29.95	16.98	0.00	11.11	7.12
0.75	0.00	7.92	3.81	0.00	9.82	4.25
1.00	0.00	17.95	4.47	2.13	8.22	4.20

From Table 3, great significant differences in treatment effect exist among the 7 concentration levels for whatever the treatment is injection or saturation. Too high or too low of the colchicine concentration both affect the rate of the $2n$ pollen grains obtained. The variance analysis illustrated that the differences among the treatment concentrations had arrived at the great significance level ($F_{\alpha} = 8.86$, $\alpha = 0.01$). Generally, the best effect was gotten from treating with 0.10%~0.51% colchicine solution. The best treatment way was still the injection.

(2) The effect of treatment start time on the yield of $2n$ pollen grains

The male flower buds after 24 hours of water culture in the greenhouse were injected with 0.2% colchicine 1 to 9 times within two days. The yield of $2n$ pollen grains was investigated with the non-treatment taken as the control. The result was shown in Table 4.

Table 4　Effect of treatment times on yield of 2*n* pollen grains (per inflorescence)

Times	0	1	2	3	4	5	6	7	8	9
2*n* pollen(%)	0. 0	4. 7	14. 1	43. 9	57. 8	63. 5	65. 0	81. 3	83. 3	83. 8
Total pollen(mg)	42. 6	30. 1	11. 8	5. 7	5. 0	3. 4	3. 1	2. 3	2. 1	1. 9
2*n* pollen(mg)	0. 0	1. 4	1. 7	2. 5	2. 9	2. 2	2. 0	1. 9	1. 7	1. 6

As shown in Table 4, with the increase of the injection times, the rate of 2*n* pollen grains rose too, with the lowest rate (4. 7%) from treating only once and the highest rate (83. 8%) from nine times of treatment. But the average yield of pollen on each inflorescence decrease with the increase of injection times, with the highest (42. 6mg/inflorescence) from the control treatment (non – treatment), 30. 1mg/inflorescence from one time of treatment and the lowest (1. 9mg/inflorescence) from nine times of treatment. The 2*n* pollen yield of each inflorescence was computed by multiplying the rate of 2*n* pollen grains with the pollen yield of each inflorescence. From the indicator, there was no significant differences among the different treatment time, and the effect from either too frequent or too few of the treatment times were not good. The better results were obtained from 3 to 6 times of treatment and the best from 4 times of treatment, with 2. 9mg of 2*n* pollen grains on each inflorescence.

(3) The effect of environmental temperature condition on the yield of 2*n* pollen grains

It was observed that the temperature of the treatment environment influenced directly the treatment effect during the experiment. The experiment of treatment in the greenhouse in the day time and in the low temperature cold room in the night was designed accordingly. The male flowers after 24 hours of water culture were treated with injection of three colchicine concentrations. The result was shown in Table 5.

From Table 5, there was no significant differences in the treatment effect among the three concentrations. However, the effect of treating in the low temperature at night was obviously better than that in the greenhouse from beginning to end. The highest rate of 2*n* pollen grains from the former treatment arrived at more than 90% and the average more than 80%, equaling to 5 to 10 times that of the latter. The reason may result from two aspects. Firstly, the meiosis of the microsporocytes in male flower buds needs to proceed for a certain period of time under a certain low temperature condition. The higher temperature resulted in shorter meiosis time and less chance for colchicine solution and, thus worse treatment effect. On the contrary, the meiosis turned slower, the chance of colchicine action enhanced under the lower temperature and so the treatment effect turns better . Secondly, the treatments were carried out generally in the day time and stopped at night. But if the temperature at night was not low enough to cease the meiosis, parts of the microsporocyte would continue to proceed the meiosis and develop into normal pollen (1*n*). So if those sprays were moved to the low temperature condition at night, the process of meiosis will slow down relatively and the separation and development of the microsporocyte would be checked by colchicine sufficiently.

Table 5 Effect of temperature on yield of 2*n* pollen grains

(day 10~ 20℃, night 5~ 10℃)

Concentrantion (%)	2*n* pollen grains (%)			*V* (%)
	min.	max.	mean	
0. 1	70. 59	94. 34	83. 23	9. 07
0. 2	69. 12	91. 07	81. 99	16. 86
0. 5	79. 66	85. 71	83. 26	2. 76

3. 1. 2 The inducement of 2*n* pollen grains by physical method

The male flowers after culture in water in the greenhouse for various periods of time were put into the incubator, treated there for 2 hours at 38~ 40℃ and then put back to the greenhouse to be cultured in water. The rate of the big pollen grains was counted at the pollen dispersal stage. The result was shown in Table 6.

Table 6 Percentage of 2*n* pollen grains with higher temperature treatment at different time

Time cultured in water before treating(hours)	2*n* pollen grains(%)			*V* (%)
	min.	max.	mean	
24	45. 83	49. 04	47. 78	2. 37
28	50. 00	70. 00	56. 85	12. 43
32	63. 51	73. 44	66. 25	5. 57
36	60. 68	67. 61	64. 34	4. 34
40	62. 07	71. 86	65. 71	5. 30

From Table 6, differences exist in effect of the high temperature treatment among different water culture period. The treatment of high temperature after 32 hours of water culture produced the highest average rate (66. 25%) of 2*n* pollen. The result of variance analysis indicates that the differences in 2*n* pollen rate among different water culture periods arrive at the great significance level ($F_\alpha = 12.4$, $\alpha = 0.01$). At the same time, pollens from non-water-culture and within 20 hours of water culture were observed to produce nearly no 2*n* pollen grains. In conclusion, more than one day of water culture in the greenhouse before the high temperature treatment was necessary for the male flower sprays to develop a certain rate of 2*n* pollen grains. It may correlate with whether the microsporocyte had entered into the meiosis phase or not. Besides, it was found in the experiment that the total pollen grains observed in each field of vision were much less than that induced with colchicine solution. It indicated that the high temperature inducement decreased the pollen yield. The reason may be the high temperature injured the tapetum of the anther wall, but it needs to be further studied.

3. 2 The ident ificat ion of the 2*n* pollen grains

Relative DNA content was tested by the cytofluorimetry technology to determine whether or not the doubling induced big pollens are $2n$ pollen grains except for the observation of volume size and development behavior. The total DNA content of the normal pollen ($1n$) and the big pollen ($2n$) were respectively 9. 52±2. 03Au and 20. 74±2. 19 Au that showed a multiplying relationship between them, which means the relative DNA content of the big pollen was two times that of the normal pollen ($1n$). From the distribution frequency straight rectangle diagram (Figure 1) of the two kinds of pollens, they both show the normal distribution. The focal distribution area of the latter (the big pollen) located at the doubling position of that of the former (the normal $1n$ pollen). As DNA is the most principal matter composing the chromosome, the multiplying relationship of the chromosome number of the two kinds of pollens can easily be deduced from the double-relationship of the relation DNA content between them. Consequently, it may be deduced that the chromosome number of the big pollen should also be two times that of the normal pollen ($1n$) that ascertained that the induced big pollen is the chromosome unreduced $2n$ pollen grains.

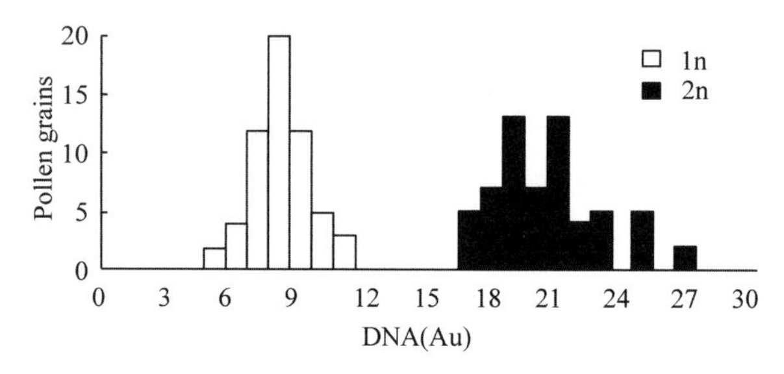

Fig. 1 The distribution of the total DNA relative content of normal ($1n$) and big ($2n$) size pollen grains

3. 3 Separation, purification and vitality of the $2n$ pollen grains

A certain rate of $2n$ pollen grains can be obtained through doubling treatment; but it is impossible to obtain 100% of pure $2n$ pollens however you improves the method or treatment temperature conditions of the environment, especially the collected pollen are actually the mixture of the normal pollen ($1n$) and $2n$ pollens treated in a big quantity. If such mixture of pollens is used to pollinate directly, the chance of fertilization of $2n$ pollens will be very low. In order to increase the chance of being fertilized, separation and purification of the $2n$ pollens will be a necessity. After many times of experiments, it was found that the method of screening with 600-eye metal web was effective, that could screen out most of the $1n$ pollen grains and the rate of the $2n$ pollen grains after the purification by screening could be raised to as high as 95% .

The vitality of $2n$ pollen grains determines its future utilization. The results of the germination test of the $2n$ pollen grains using the culture method and vivo test method show that the germination rate of the $2n$ pollen grains equals to that of the $1n$ pollen grains ranging from 16. 4% to 23. 0%. It means that the $2n$ pollen grains induced artificially are viable and can be utilized in pollination hybridization to culture triploids.

4 Conclusions

(1) A certain rate of $2n$ pollen grains can be obtained through treating the male flower buds in the meiosis of Chinese white poplar (*Populus tomentosa*) and its hybrid (*Populus tomentosa* ×*P. bolleana*) with colchicine. The pollen grain has not only larger volume size, with the diameter 37 ~48μm, over 60% than that of the control pollen grain, but also higher relative DNA content matching the double of that in the normal pollen grain ($1n$). It proves that the doubled pollen grains are indeed the $2n$ pollen grains.

(2) With the continuing improvement of the treatment method and condition in a few years of experiments, the rate of the $2n$ pollen grains is increased annually. The maximum rate of $2n$ pollen grains after some specific treatments arrives at over 90% and the average maximum rate arrives at over 70%. The rate of $2n$ pollen grains of the hybrid (*Populus tomentosa* ×*P. bolleana*) is generally higher than that of *P. tomentosa*.

(3) The main factors that affect the yield of $2n$ pollen grains are concentration of colchicine solution, way of the treatment, start time of the treatment, frequency of the treatment and the temperature condition of the treatment environment. The optimum concentration ranges from 0. 1% to 0. 5%. The better effect comes from the injection method. More $2n$ pollens grains can be obtained when the pollen mother cell of the male flower entering into the meiosis prophase is injected 3 to 6 times after one day of water culture. It is inadvisable for the environmental temperature being too high during the treatment.

(4) The simple screening method can be utilized to separate and purify the $2n$ pollen grains. The $2n$ pollen grains obtained are viable and can be used in pollination hybridization to culture triploids.

(5) Higher rate of $2n$ pollen grains is obtained when the male flower buds are treated with high temperature (38 ~ 40℃); but it is necessary to master well the meiosis time of the flower buds. That the high temperature treatment affects the yield of $2n$ pollens may contribute to the injure of the high temperature to the tapetum of the anther wall however it is left to be further investigated.

Status and Advances of Molecular Genetic Improvement of Poplar Species in China*

Abstract Poplars are among the most important deciduous tree species in China. China is replete with natural resources of poplars. Poplars have a number of good characteristics, including fast growth rate, high yield, many uses, easiness of tissue culture and small gene group that make them well suited as a model system for the application of genetic engineering in forest trees. In the last decade, much progress has been made in genetic improvement of poplar species in China. Modern biotechnology is an important tool for genetic improvement in forest trees, and its applications to genetic improvement in poplars, which covers genetic transformation, gene expression, construction of genetic linkage map, QTLs (quantitative trait loci) identification and molecular assisted selection are reviewed in this paper. At the same time, the existing problems and outlook about the application of modern biotechnology to genetic improvement in forest trees are also discussed.

Keywords poplar, gene engineering, molecular marker, genetic linkage maps, QTLs, genetic improvement

1 Introduction

Genetic improvement in forest trees based on genetic variation and genetic recombination is necessary to breed new varieties of forest trees. The effect of genetic improvement will directly depend on the breeding techniques adopted by breeders. Because of long life cycles of trees, genetic heterogeneity and most of economic traits controlled by quantitative trait loci or multiple genes, genetic improvement in forest trees by traditional breeding and selection methods have existed limitations and not enabled tree breeders to make changes in many important economic traits by design. With the development of molecular techniques that are applicable to forest tree species, genetic engineering is becoming a supplement for traditional tree breeding.

Poplars are among the most important deciduous tree species in China. There are more than 74 species, many variants and hybrids in the poplar family in China (Zhao and Chen, 1994). Fast growth rate, high yield, and its usefulness in timber industry, pulping, shelter belt and other purpose afforestation make poplar become an important tree species for afforestation, and for the

---

* 本文原载 *Forestry Studies in China*, 2002, 4(2): 1-8, 与林善枝和张谦合作发表。

same reason poplar will be the main tree species to establish million-hectare forest bases for fast growing commercial wood. Genetic engineering provides an opportunity to transfer new specific traits of interest into valuable genotypes. Modern biotechnology is an important tool for genetic improvement in poplars and its applications to genetic improvement in poplars, which covers genetic transformation, gene expression, construction of genetic linkage map, quantitative trait loci (QTLs) identification and molecular assisted selection, are reviewed in this paper. At the same time, the existing problems and outlook about the application of modern biotechnology to genetic improvement in forest trees are also discussed.

2 Genetic transformation and gene expression

2.1 Insect resistance

Poplars are susceptible to many insects and the breeding for resistance is a difficult task in long-lived cycles. Poplars are seriously damaged by many pathogens and pests, which cause heavy defoliations resulting in remarkably slower growth rates. Chemical treatments must be applied in the nursery to reduce attacks, but these treatments are expensive and potentially hazardous to the environment. Thus poplars are seriously attacked by insects in many areas and the insect resistant poplars are expected in forest production. The production of insect resistant plants is another application of genetic engineering, which has important significance in poplar improvement. In poplars, the resistance to insect has been achieved by employing two types of gene products.

2.1.1 Proteinase inhibitors

Proteinase inhibitors with small molecular weight have recently been a source of resistance to arthropod and pathogen pests for tree improvement programs. Four classes of proteinase inhibitors (cysteine, serine, metallo and aspartyl proteinase inhibitors) have been identified.

Liu et al. (1993) first reported that cowpea trypsin inhibitor (*CpT* Ⅰ) gene was cloned from their mature cowpea by RT-PCR according to the sequences of pea Bowman-Brik trypsin inhibitor in China. To study *CpT* Ⅰ gene expression in woody plants, at Beijing Forestry University, *CpT* Ⅰ gene was introduced into Chinese white poplar (*Populus tomentosa* Carr.) by *Agrobacterium tumefaciens* Conn. Assay on proteinase inhibitor activity demonstrated that leaf protein extracts of the transgenic poplar showed higher inhibition activity against trypsin than that of control plants (Hao et al., 1999). Moreover there have been no consistent differences in height and diameter growth between the transgenic and control plants. The activity of *CpT* Ⅰ was detected in the extracts of transformed bacteria. Soybean Kunitz trypsin inhibitor (*SKTI*) belongs to protease inhibitor family with insecticidal properties. Gao et al. (1998) reported that *SKTI* gene was obtained from the template cDNA which was synthesized from the total RNA of the isolated soybean (*Glycine max* L.) immature cotyledons by PCR (polymerase chain reaction). The bioassays of insects of the transgenic tobacco showed that the transgenic tobacco displayed notably resistance to the larvae of *Heliothis armigera* Hubner compared with tobacco.

2. 1. 2 Bacillus thuringiensis

Progress in engineering insect resistance in transgenic poplars has also been achieved through the use of insect toxin protein gene of *Bacillus thuringiensis* (Bt). Bt is one of the most widely used pesticides in agriculture and forestry. The high speciicity of Bt is due to the production of different δ-endotoxins, or crystal proteins. After *Bt* is eaten by larvae of susceptible insects, the crystals dissolve and the proteins (called δ - endotoxins) are released into the insects midgut, where digestive enzymes cleave the proteins into smaller toxic fragments, and display activity against certain groups of insects. In general, Bt strains devided into 4 major classes based on toxin structural similarities and activity against certain insect pests, they are: *Cry* Ⅰ with activity against lepidopterans, *Cry* Ⅱ with activity against lepidopterans and dipterans, *Cry* Ⅲ with activity against coleopterans, and *Cry* Ⅳ with activity against dipterans. So far, several different genetically engineered Bt δ-endotoxins were tested in transgenic poplar in China.

The genomic library of the plasmids from *Bacillus thuringiensis HD*-1 was firstly constructed using vector lambda Charon 28, and the Bt *HD* - 1 gene has been synthesized by Chen et al. (1987) in China. The first insect-resistant transgenic poplar (*Populus nigra* L.) plants' expression chimeric genes of *Npt* Ⅱ and 35S-Bt-Nos were obtained in 1993 by Tian et al. by *Agrobacterium* mediated transformation. Insect resistance tests showed that the transgenic plants were toxic to *Lymantria diaper* Linnacus and *Apocheimia cinerarius* Erschoff. The mortal percentages of *Lymantria diapar* Linnacus and *Apocheimia cinerarius* Erschoff were 96% and 100% respectively in 5-9d after infestation and the growth and development of survival larvae were seriously inhibited. Based on the results of southernblot of PCR producing, and the cluster analysis of their growth and insect resistance, three independent transgenic plants were selected and propagated in the nursery field. Upon field trials poplar trees derived from vegetative propagation of the original transgenic *Populus nigra* plants transformed with *Bt* gene showed strong insect-resistance to *Apocheimia cinerarius* and *Orthosia incerta* (Hu et al. 1999). Moreover the rates of damaged leaves of transgenic plants, and the amounts of pupae per square meter in the transgenic plantation soil is lower than that of the control. Up to now this is a large field test on insect resistance of transgenic poplar transformed with *Bt* gene in China. Wang et al. (1997) reported that leaves and stem segments of *Populus euramericana* were transformed with *Agrobacterium tumefaciens* harbouring a binary vectors pB 48. 214 and pB 48. 215 containing chimeric genes of *Npt* Ⅱ and 35S-δ-Bt-Nos. The three plants selected by southern blot analysis showed IPC toxicity agains tlarvae of *Lymantria diapar* Linnacus. Zheng et al. (1995, 1996) reported that the improved but genefused into express carrier pB 48. 7 was used to transform the Chinese white poplar (*Populus tomenntosa*) and hybrid poplar 741[*Populus alba* ×(*P. davidiana*×*P. simonii*)×*P. tomentosa*], respectively. According to the selection in Kanamycin-contained media and PCR test, the plantlets were tentatively proved to be transformed with the gene. Chen et al. (1995) and Rao et al. (2000) reported that the chimeric genes of partially modified But gene and *Npt* Ⅱ gene were transferred into the genome of *Populus deltoides* and poplar NL-80106(*Populus deltoides*×*P. simonii*) by *Agrobacterium tumefaciens*-

mediated transformation, respectively. The results of insect bioassay using *Lymantria diapar* Linnacus showed that the transformed plants were demonstrated to be highly resistance to the testing insects compared with the control plants.

In order to increase the insecticial activity of transgenic poplar plants and to reduce the evolution probability of insect tolerance against insect resistant transgenic plants, modification of *Bt* gene and fusion of *Bt* gene with a proteinase inhibitor gene have been made and transferred into poplar in China. Tian et al. (2000) reported that partially modified *Bt CryIAc* gene and the arrowhead proteinase inhibitor (API) gene were transferred into the genome of the hybrid poplar 741[*Populus alba*×(*P. davidiana*×*P. simonii*)×*P. tomentosa*] by *Agrobacterium* mediated transformation. Results of insect biosaaay using *Clostera anachoreta* (fabricius) indicated that the mortality of insect larvae on one transformed plant was higher than 90% in 6d after infestation and the growth of the survival larvae were seriously inhibited. This is the first report on insect-resistant transgenic hybrid poplar 741 that expresses two insecticidal protein genes. In addition, Li et al. (2000) observed that the transgenic *Populus nigra* containing *Bt* and proteinase inhibitor (*PI*) gene had toxicity to the larvae of *Lymantria dispar* L., and had enhanced toxicity to larvae by comparision with plants containing only *Bt* gene.

2.1.3 Other insect-resistant genes

Insect resistance scorpion neurotoxin *AaIT* gene inserted into a binary vector was transferred into a hybird poplar clone N2106(*P. deltoides*×*P. simonii*) growing in the southern China by Rao et al. (2000). PCR and PCR southern analysis showed that *AaIT* gene was incorporated into the genome of some recovered poplar plants. One of the transformed plants named A5 was significantly resistant to feeding by first instar larvae of *Lymantria diapar* Linnacus compared with the untransformed control plant. Moreover, a decrease in leaf consumption by larvae, a lower larval weight gain and a higher larval motality rate of *Lymantria diaper* Linnacus were observed in the transformed plants.

Antibacterial gene *Lc* I which was synthesized according to the sequence of antibacterial protein of *Bacillus* spp. was transformed into *P. nigra* and *P. euramericana* by Li et al. (1996), and more than 75% of insect mortality to *Anoplophora glabripennis* was observed in transgenic plants.

2.2 Disease resistance

A number of tree species serve as useful hosts for the viral and fungal pathogens. Poplars are also severely damaged by viral and fungal diseases. Compared with gene engineering of insect resistance, little is reported about genetic improvement in viral and fungal diseases of poplar through gene engineering in China. Rabbit defensin *NP*-1 possesses a broad resistant spectrum to pathogens. The *NP*-1 gene was transferred into *Populus tomentosa* by *Agrobacterium*-mediated transformation (Zhao et al. 1999). Antimicrobial activity test showed that the extract of transgenic plants inhibited the growth of the tested microbes. This is the first disease resistant transgenic poplar in

China.

2.3 Stress (salt, drought and freezing) resistance

Although poplars have a number of good characteristics, the limitation of inherited resistance of poplars to environment has made them grow difficultly in stress environment. It has been suggested that salt, drought and freezing resistance in woody species be altered by genetic engineering.

The first salt-resistant transgenic poplar plants were obtained in 1994 in China. Bai (1999) reported that 3 gene fragments induced and 2 gene fragments repressed by NaCl had been obtained from *P. euphratica*, and sequenced. 1-phosphate mannitol dehydrogenase gene (*mtl*-D) was cloned from *E. coli*, and was transformed into *Populus* (Balizhuangyang) via Agrobacteria (Liu et al., 2000). The results showed that several transgenic plants grew very well in 016% NaCl. Exogenous gene *Bet-A*, encoding choline dehydrogenase, was introduced into *Populus simonii*×*P. nigra* by *Agrobacterium tumefacien* (Yang et al., 2001). PCR and southern blotting analyses showed that *Bet-A* gene has been integrated into the genome of this hybrid poplar.

The 60kD antifreeze proteins (AFPs) was purified from the heat-stable proteins induced by cold acclimation at-20℃ in the branches of *Populus suaveolens* by two-dimensional electrophoresis at Beijing Forestry University, and its thermal hysteresis activity (THA) kD was 2.7℃ at 20mg · mL^{-1}, much higher than those of other antifreeze proteins observed in polar fishes and winter rye. It is appeared that AFPs with high THA in *P. suaveolens* induced by cold acclimation may play an important role in the antifreeze process in *P. suaveolens* during the period of overwintering (Lin, 2001). The gene encoding 60kD AFPs is under cloning according to its N-termina 20 amino acids of DSDLSFSNKFTVPCQDDIFL.

2.4 Wood property improvement

Lignin with important biological function is the second abundant natural product, its content is only less than cellulose in plant. But the lignin must be extracted from the cellulose fraction in making paper pulp processes in which lead to energy-requiring and cost-increasing, and produce pollution. Also, the lignin has negative effect on the digestibility of silage grass. So there is considerable interesting to reduce the lignin content by genetic engineering to improve the pulping property of wood. The Chinese white poplar is a special local tree species as an important resource for paper pulping in China. Wei et al. (2001) reported that cDNA encoding caffeoyl CoA O-methyl transferase (CCoAOMT) from Chinese white poplar (*Populus tomentosa* Carr.) was cloned by RT2PCR and sequenced, and its antisense expression of CCoAOMT cDNA in *P. tremula*×*P. alba* caused a decrease of 17.9% in Klason lignin content as compared with that of untransformed poplar and exerted basically no influence on growth of the transgenic plants, indicating that the antisense repression of CCoAOMT is an efficient way to reduce lignin content for improving pulping property in engineering tree species.

3 Applications of molecular markers to poplar breeding

In recent decades, many different techniques have been developed to detect molecular markers based on DNA since the invention of PCR technology in 1985 (Saiki et al.). Application of molecular markers to tree breeding is dependent on the advantages and limitations of forest genetics. Molecular marker technologies have been used to: the generation of genetic linkages maps, early selection of individuals with specific characteristics within larger progenies, efficient selection of parents for new breeding programs, efficient trait introgression, genetic mapping of simple or complex traits.

3.1 Construction of genetic linkage map

Genetic linkage maps may provide the basis for exploring genomes, QTL identification of some important traits and molecular marker-assisted early selection, and they have been constructed for various tree species (Cervera et al. , 2000). DNA markers used to construct genetic linkage maps in forestry include random amplified polymorphic DNAs (RAPDs), restriction fragment length polymorphisms (RELPs), amplified fragment length polymorphisms (AFLPs), expressed sequence tag polymorphisms (ESTPs), and simple sequence repeats (SSRs).

In China, Li et al. (1996) firstly used RAPD fingerprints to estimate the taxonomic and phylogenetic relation among *Aigeiros*, *Tacamahaca* and *Leuce*. Yi et al. (1998) firstly used AFLP to construct genetic linkage maps of *Populus deltoides*, and used AFLP fingerprints to identify 42 clones. In 2002, another AFLP genetic linkage map are constructed at Beijing Forestry University based on (*Populus tomentosa*×*P. bolleana*)× *P. tomentosa*. 51 pairs of AFLP primers generated 808 polymorphic fragments, among which 655 segregated in a 1 : 1 ratio. These markers were grouped to produce a paternal *P. tomentosa* map with 218 markers in 19 major linkage groups covering 2683 centiMorgan (cM), and a maternal *Populus tomentosa* ×*P. bollean* map with 144 markers in 19 major linkage groups covering 1956cm. This is the first AFLP genetic linkage map for Chinese white poplar (Zhang, 2002).

The first RAPD genetic linkage map constructed in 1998 by Su et al. is based on a 3 generation *P.* deltoides×*P. cathayana* pedigree. The published map is composed of 110 RAPD markers, and the total distance contained within 20 linkage groups is 1899.4cm. Average spacing between the markers is 17.27cm. This map should facilitate the identification of markers that "tag" genes for pest and disease resistance and other traits in poplar. Since then, other two RAPD genetic linkage maps were published in China. In 1999, RAPD genetic linkage maps of the parents of a *Populus adenopoda*×*P. alba* F_1 family were constructed by (Yi et al. , 1999), and a total of 333 segregating loci [326(1:1), 7(3:1)] were identified. Among the 326(1:1) segregating loci (238 from *P. adenopoda* and 88 from *P. alba*), 36 loci (26 in *P. adenopoda* and 10 in *P. alba*) were found to be distorted from the normal 1:1 ratio. Altogether 290 loci segregation ratio 1:1 was used to construct parent specific linkage maps, 211 for *P. alba* and 78 for *P. adenopoda*. The resulting linkage maps consisted of 1890marker loci in 20 groups, 6 triples and 16 pairs for *P. alba*, which cover the map distance about 2402. 4cm, and 41 linked marker loci for *P. adenopoda* which cover map distance about 479. 4cM. In 2000, RAPD genetic linkage map of the parents of a *P. deltoides*×*P. euramericana* F_1 family was constructed by Zhang et al. , and a total of 229 segregating loci were identified. Among the 229 loci, 15 loci distorted from the normal 1∶1 ratio. The 214 markers formed 19 main linkage groups, 6 triples and 14 pairs. The resulting linkage map of *P. deltoides*×*P. euramericana* spanned 1914. 2cm (73. 62% coverage of genome length) with an average distance of 14. 8cm between markers.

3. 2 QTLs identification and molecular assisted selection

QTLs identification and molecular assisted selection are another two important applications of molecular marker technology. Many important traits of forest trees such as growth, height, diameter, stem volume, wood property, resistance to environment and so on are complex inherited traits. These traits usually belong to quantitative traits, and are controlled by QTLs or multiple genes. Manipulation of quantitative traits, more difficult than that of simple inherited traits, has always been challenging for traditional breeding. QTLs identification according to genetic linkage map may provide opportunities for genetic improvement in forest trees.

In China, the first QTLs identification of poplar was made in 1999. QTLs associated with growth and phenology quantitative traits (height, diameter and top closure) within 80 F_2 seedlings of a cross between a *P. deltoides* female and a *P. cathayana* male by using RAPD markers were detected by Li (Li et al. , 1999). Single factor analysis of variance found that 7, 6 and 3 markers were respectively associated with height, diameter and top closure, which jointly explained 45. 94%, 41. 17% and 19. 13% of the total phenotypic variance. Detection of the significance of interaction between putative QTLs by two-way analysis of variance showed that two interactions of three markers were associated with height, and each explained 4. 57% and 5. 05% of the phenotypic variance respectively. One of the three markers was associated with height, diameter and top closure, one was associated with diameter and the other with top closure. Both quantitative genetics method and molecular marker method were used to identify marker loci associated with 5 quantitative leaf traits in a F_2-population from a cross of *P. deltoides* and *P. cathayana* (Su et al. , 2000a). The results of single-factor analysis of variance revealed that 10, 10, 4, 9 and 12 markers were significantly associated with leaf length, leaf width, ratio of length to width of leaf, petiole length and leaf area respectively, which accounted for 66. 23%, 61, 82%, 32. 86%, 59. 67% and 81. 79% of the total phenotypic variation respectively. Two−factor analysis of variance showed that 4, 2, 1, 5 and 7 markers were significant interactions among these markers associated with leaf length, leaf width, ratio of length to width of leaf, petiole length and leaf area respectively, most of marker loci associated with these quantitative traits were on the 4th, 12th, 15th and 17th line of linkage groups. Recently, QTLs of 14 traits, including stem height, height increment, basal diameter, stem volume, sylleptic branch angle, sylleptic branch numbers, leafblade length, leafblade width, the ratio of length to width of leafblade, leaf areas, leaf vein numbers, leaf petiole length, spring bud flush day and cellulose content were mapped using the F_1 pseudo-testcross strategy and AFLP genetic linkage map in China. The analysis of variance showed that 616 markers were associated with traits, of which 247 were located in the genetic linkage map. 80 QTLs associated with these traits were identified at LOD (Logarithm of Odds)> 1.0 by interval mapping, and a total of 12 traits were detected at LOD > 2.0 (Zhang, 2002). In order to investigate the genetic control of resistance to *Alternaria alternata* (Ala), the susceptibility of a three generation *P. deltoides*×*P. cathayana* hybrid poplar pedigree, comprising F_1 and F_2 progenies to Ala was tested in both the greenhouse and the field (Su et al., 2000b), suggesting that the Ala resistance be determined by a single recessive gene for *P. deltoides*. In addition, the analysis of approximately 4200 selectively amplified DNA fragments using RAPD markers, in combination with bulked segregant analysis found that two markers (RPH12-6 and RPH12-4) were linked to Ala resistance, Ala was mapped on group 3, and 3.60cm from RPH12-6 or RPH12-4, which may provide the basis for molecular marker assisted selection and early identification of disease resistance varieties.

4 Problems and prospects

Although gene engineering has been widely used in the genetic improvement in poplar and much progress has been made in genetic transformation, gene expression, construction of genetic linkage map, QTLs identification and molecular assisted selection of poplar in China, the problems about the application of modern biotechnology to genetic improvement in poplar are still existed.

4.1 Limitations of resistant gene engineering in poplar

Although gene transfer can enhance insect resistance of poplar, current methods for enhancing insect resistance include insertion of toxin gene for the bacterium *Bacillus thuringiensis* and transfer of proteinase inhibitor genes from other plant species. Moreover, the obtained transgenic poplar plants are mostly transformed with individual gene, and these individual gene effects are generally specific for a limited number of insect taxa. For example, *Bt* is toxic only to certain lepidopteran defoliators. The greatest limitations to the current use of genetic engineering to improve resistance of poplar are: insufficient knowledge of the molecular biology of insect development, insect pathogenesis and tree defenses against insects. In addition, the main environmental risk associated with the use of engineered trees is that insects may counter evolve to overcome their resistance due to long life cycles of tree. That is to say, the continuous exposure of insects to products of insecticide genes expression in engineered trees might in some cases promote the counter evolution of resistance against insecticide by insects, which is particularly observed in the stand of monoculture plantation. For these cases, the use of multiple resistance genes must be considered. Such as the fusion of a proteinase inhibitor and a *Bt* gene may provide a significant advantage on the enhancement of insect resistance and may enhance resistance to multiple pests. Also, by adding multiple genes that confer different kinds of insect resistance, the evolution probability of insect tolerance against insect resistant transgenic plants can be reduced. Such as a proteinase inhibitor and a *Bt* gene might provide a mosaic environment where insect populations must adapt to a heterogeneous tree population, insect counter evolution may be slowed or thwarted.

Knowledge about the molecular mechanism of resistance and the molecular genetic bases of these traits is seriously lacked, little progress is now being observed in the areas of drought, salt and freezing resistance compared with insect resistance.

4.2 Scarcity of gene for genetic improvement

Up to now, only a few resistance genes obtained from microbiology and some herbaceous plants have been successfully introduced into poplar, reports concerning the isolation and identification of resistance genes from woody plants are small. Moreover, small effect of these individual genes on the resistance enhancement was observed in the obtained transgenic poplars because most of economic traits in trees are controlled by quantitative trait loci or multiple genes, and genes used for genetic improvement in trees resistance are mostly individual genes, may not be key genes. The isolating and cloning resistance genes from trees have thus become increasingly important to improve resistance. At the same time, the fusion of different kinds of resistant genes must be considered for enhancing stress resistance.

4.3 Low density of genetic linkage maps constructed

Genetic linkage maps have been constructed in a few poplar species in China, however the density of the constructed genetic linkage maps was lower. The effects of QTLs identification and molecular assisted selection of many important traits may rely largely on the density of linkage maps. High density linkage maps may not only help to identify candidate genes controlling important traits and to increase the efficiency of QTL mapping, but also provide important information for the early selection of superior genotypes. Thus, for the increase of the availability for linkage maps, the density of constructed linkage maps needs to be elevated.

4.4 Developments in the areas of regeneration, genetic transformation and gene expression

Success with genetic engineering is often limited by the non availability of suitable regenerating systems. Although *Agrobacterium tumefaciens* have been successfully used in the transformation of different poplar species, its transformation efficiency is relative low. Moreover, the large and repetitive genomes possessed by trees provide further obstacles to molecular detect of transgenic plants. In addition, neither the effectiveness nor the risks of engineering resistant trees can be predicted because of the novelty of genetic engineering and the long life cycles of tree. The above mentioned factors may be the main limitations in the advance of molecular genetic improvement in poplar to a great extent. In addition, a major problem regarding the stability of gene transformation and expression is how to regulate integration and expression of foreign genes, particularly under field conditions. To obtain high frequency of regeneration and transformation, and to improve efficiency of gene expression, parallel research and developmental efforts are being undertaken, which is important for commercialization of poplar.

Although molecular techniques, genetic linkage map, QTLs identification and so on have not been practically applied to poplar improvement, recent developments in molecular biology have opened new vistas for broadening gene pools and have provided impetus to poplar improvement. Genetic engineering methods complement traditional breeding efforts by increasing the diversity of genes and germplasm available for incorporation into desirable poplar species of commercial interest, and shortening the rotation required for the production of new varieties. Thus, It is considered that applications of molecular biotechnology to genetic improvement in poplar species have become a reality in the future.

Establishment of High Frequency Regeneration System of *Populus tomentosa* *

Abstract The establishment of high frequency regeneration system is a foundation for *Agrobacterium*-mediated genetic transformation. In this work, several important factors influencing the efficiency of regeneration of plants, such as concentration of plant growth regulators, leaf explant orientation, leaf growth sequence and leaf segment, were studied. The results indicated that the differentiation rate of adventitious shoots was 90% on basal MS medium only supplemented with 1.5mg · L^{-1} BA (6-benzyladenine) and reached the highest level (95%) when 1.0mg · L^{-1} BA and 0.3mg · L^{-1}NAA (naphthalene acetic acid) were added to MS medium. 90% of differentiation rate of adventitious roots were obtained when 0.3mg · L^{-1} NAA was only added to MS medium. It was also found that more adventitious shoots were regenerated from the lower segment of leaf (with petiole) than the other segments, the number of adventitious shoots decreased from top to base of leaf growth sequence and the percentage of adventitious shoot induction with adaxial side downward was higher than that with adaxial side upward.
Keywords triploid, *Populus tomentosa*, leaf explant, regeneration

1 Introduction

Populus tomentosa has been highly valued and considered as the best tree species with fast growth and high wood quality among native poplar species in China (Zhu et al. 1997). But the distribution of *P. tomentosa* was confined by its limitation of insect-resistance, salt-resistance and cold-resistance etc. Since *P. tomentosa* was regarded as a natural host of *Agrobacterium tumefaciens* (Parsons, 1986), *Agrobacterium*-mediated transformation of *P. tomentosa* has been broadly studied for the improvement of resistance to the environment and the economically valuable traits that are difficult to be achieved by conventional breeding.

One of the important steps of genetic transformation is to establish high frequency regeneration system in the experiment. So far, many *Populus* species were successfully regenerated through adventitious shoot induction from cultured explants including node, internode, leaf, and root segment (Agrawal et al., 1991; Chun 1990; Coleman et al., 1990). But regeneration ability was varied in different species, even in different clones of the same species (Fan et al., 2001). In this paper, leaves of triploid *P. tomentosa* were used as explants, the optimal medium was

* 本文原载 *Forestry Studies in China*, 2002, 4(2): 48-51, 与杜宁霞、李云、于海武和林善枝合作发表。

screened out and the effects of leaf explant orientation, leaf growth sequence and leaf segment on the adventitious shoot induction were studied, which provided a foundation for *Agrobacterium*-mediated genetic transformation of *P. tomentosa*.

2 Materials and methods

2.1 Plant materials and culture conditions

Triploid *P. tomentosa* was obtained from the College of Biological Science and Biotechnology, Beijing Forestry University. Sterile plantlets which propagated from buds were subcultured on basal MS medium (Murashige and Skoog, 1962) supplemented with 0.3mg · L^{-1} BA and 0.1mg · L^{-1} NAA for 30d at 25℃ with illumination of cool fluorescent light for 16 h per day. Leaf materials were trimmed from sterile plantlet with 67 leaves for experiment.

2.2 Adventitious shoots regenerated from leaves

2.2.1 Effects of plant growth regulators and their concentrations on adventitious shoot induction

Leaf explants were carefully cut transversely from leaf edge to midrib, and each leaf was excised to 2 wounds (Fig. 1-B), and then they were cultured on different kinds of media: MS supplemented with 0.05 to 0.3mg · L^{-1} NAA, MS supplemented with 0.5 to 2.5mg · L^{-1}BA and MS supplemented with 1.0mg · L^{-1}BA and 0.1 to 0.3mg · L^{-1} NAA (20 replication per experiment), respectively.

2.2.2 Effects of leaf growth sequences and leaf segment on adventitious shoot induction

The first leaf to the fifth leaf (Fig. 1-A) of plantlet were cut from top to base (20 relication per experiment), each leaf was cut transversely across the midrib into three segments: upper, middle and lower (with petiole) segment (Fig. 1-C), and then each segment was cultured on medium M_1(MS+1.0mg · L^{-1} BA+0.3mg · L^{-1}NAA + 6g · L^{-1} agar + 30g · L^{-1} sucrose, pH 5.8-6.2).

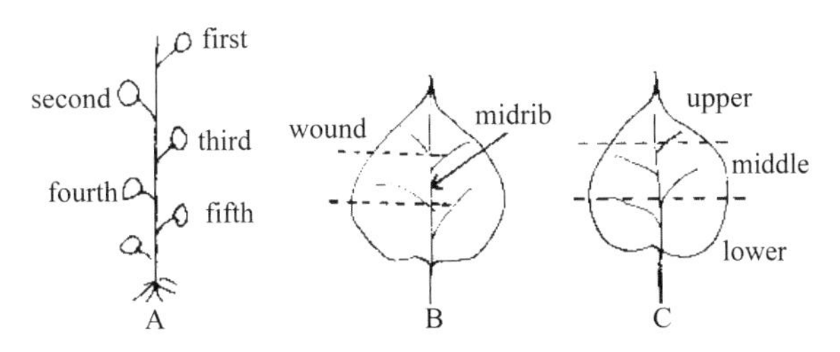

Fig. 1 Types of leaf explant

A. From top to base of plantlet, B. Leaf was cut transversely from leaf edge to midrib, C. Leaf was cut transversely across the midrib into three segments

2.2.3 Effects of leaf explant orientation on adventitious shoot induction

60 lower (with petiole) segments were used for the experiment of the effects of leaf explant o-

rientation on adventitious shoot induction. Half of them were placed on medium M_1 with the adaxial side upward (abaxial side of leaf explant contacted with medium), the other with the adaxial side downward (adaxial side of leaf explant contacted with medium).

3 Results and discussion

3.1 Effects of BA concentrations on adventitious shoot induction

From Fig. 2, it can be seen that the differentiation rate of adventitious shoots correlated with the alteration of plant growth regulator concentrations. The differentiation rate of adventitious shoots increased with the rising of BA concentration when it was at a range of 0.5~1.5mg · L^{-1}, and reached its peak (90%) when BA concentration was 1.5mg · L^{-1}. Then the differentiation rate declined gradually with the rising of BA concentration when it was ata range of 1.5~1.5mg · L^{-1}, meanwhile, the degree of vitrification increased. Shi et al. (2001) reported the same phenomena of regeneration of *Betula platyphylla* by leaf explant. BA has been used extensively in recent years, and seems to be the favorite cytokinin of plant tissue cultures, especially in the adventitious shoot induction. This was further affirmed by the experiment.

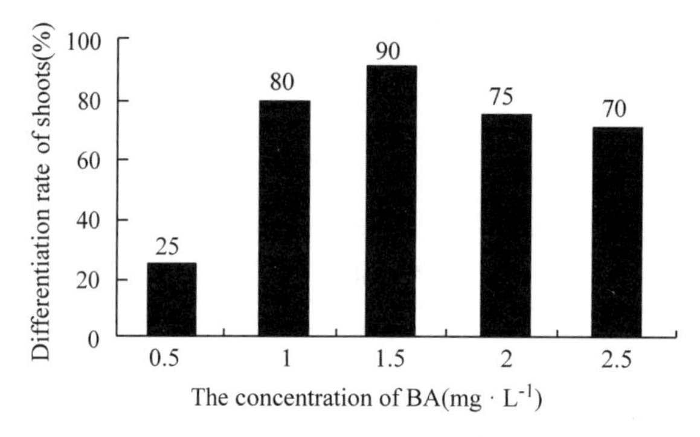

Fig. 2 Effects of BA concentrations on differentiation of shoots

3.2 Effects of NAA concentrations on adventitious shoot induction and adventitious root formation

In this experiments, callus was induced from wound surface of leaves when they were cultured on MS medium supplemented with 0.05~0.1mg · L^{-1}NAA. About 5~10 adventitious roots were formed from callus in each leaf after 15d, while the differentiation rate of adventitious shoot was always zero.

Fig. 3 showed that neither adventitious roots nor adventitious shoots were generated when NAA concentration was 0.05mg · L^{-1}, then the differentiation rate of adventitious roots increased with the rising of NAA concentration when it was at a range of 0.05~0.3mg · L^{-1}, and reached its peak (90%) when NAA concentration was 0.3mg · L^{-1}.

It was reported that adventitious roots of poplar readily formed adventitious shoots when they were cultured on modified MS (MMS) medium (Zhan 1997), in which the ammonium nitrate and potassium nitrate were reduced to 412mg · L^{-1} and 950mg · L^{-1}, respectively. Over 50% of adventitious roots developed into adventitious shoots when they were cultured on MMS medium supplemented with 0. 6mg · L^{-1} Zeatin in the experiment. So adventitious roots can potentially be used as explants in *Agrobacterium*-mediated genetic transformation of *P. tomentosa*.

3.3 Effects of combinations of BA with NAA on adventitious shoot induction

The differentiation rate of adventitious shoot increased with NAA concentration rising at a range of 0. 1~0. 3mg · L^{-1} when BA concentration was 1. 0mg · L^{-1}, which showed that maximum differentiation rate (95%) was obtained when leaves were cultured on MS medium supplemented with 1. 0mg · L^{-1} 6-BA and 0. 3mg · L^{-1} NAA.

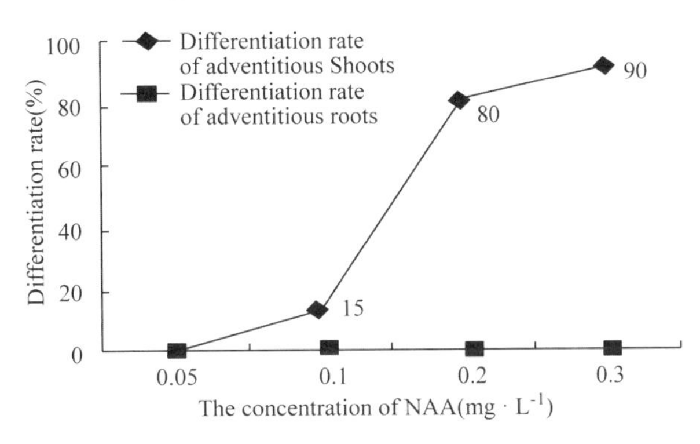

Fig. 3 Effects of NAA concentrations on differentiation of adventitious shoots and formation of adventitious roots

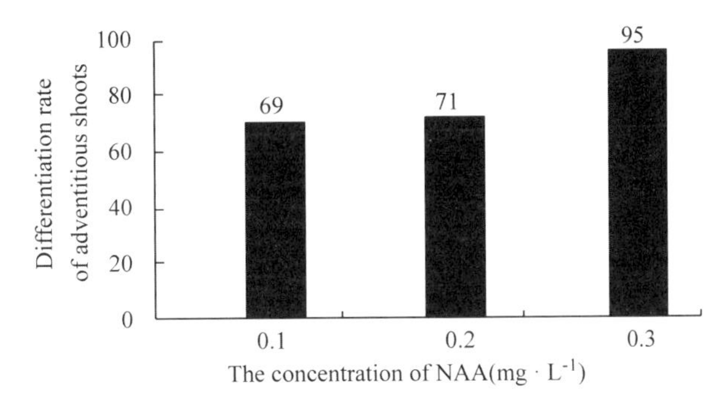

Fig. 4 Effects of combinations of BA with NAA on differentiation of adventitious shoots

In above experiment on effects of BA concentrations on differentiation rate of adventitious shoot, the differentiation rate of adventitious shoot was 80% when 1. 0mg · L^{-1} BA was only added into MS medium, while it was as high as 95% when 1. 0mg · L^{-1} BA supplemented with 0. 3mg · L^{-1} NAA were added into MS medium(Fig. 4). It was concluded that NAA could aid to promote adventitious shoots formation. Kim et al. (1994a) also reported that the combination of BA and NAA greatly affect adventitious shoot induction from leaf explant cultures of *P. davidiana*.

3.4 Effects of leaf growth sequence and leaf segment on adventitious shoot induction

As shown in Table 1, the percentages of adventitious shoot induction from the upper, middle and lower segment of leaves were 10%, 16% and 72%, respectively. The differentiation rate of adventitious shoots of lower (with petiole) segment was the highest, up to 72%, which indicated that there was significant difference in differentiation rate of adventitious shoot for different leaf segment.

Further analysis indicated that the total numbers of adventitious shoots decreased from the first leaf to the fifth leaf, which demonstrated that younger leaves were better than mature leaves in regeneration ability, which may result from the aging of leaves or discrepancy in level of endogenous hormone of leaf from top to base of whole plantlet.

3.5 Effects of leaf explant orientation on adventitious shoot induction

The differentiation rates of leaves with adaxial side upward and adaxial side downward were 85% and 95%, respectively. The average numbers of differentiation of adventitious shoots with adaxial side upward and downward leaves were 3.1 and 3.5, respectively (Table 2).

Table 1　Effects of leaf growth sequence and leaf segment on the differentiation of adventitious shoots

Leaf sequence	Leaf sequence			Total
	Upper	Middle	Lover	
First leaf	3.20	4.20	18.20	25.60
Second leaf	2.20	2.20	17.20	21.60
Third leaf	3.20	4.20	16.20	23.60
Forth leaf	2.20	3.20	12.20	17.60
Fifth leaf	0.20	3.20	11.20	14.60
Average/%	10	16	72	

Table 2　Effects of leaf explant orientation on differentiation of adventitious shoots

Leaf orientation	Rate of differ entiantion/%	Number of differentiationper leaflet
Upward	85	3.1
Downwand	95	3.5

Effects of leaf explant orientation on adventitious shoot induction were observed in shoot-producing culture of several woody species (McClelland et al. 1990). The number of adventitious shoots induced from the leaves with adaxial side downward was twice as much as that of the leaves with adaxial side upward (McClelland et al. 1990). It was also found that the percentage of adventitious shoot induction with adaxial side downward was higher than that with adaxial side upward in the experiment. Further analysis indicated that most of adventitious shoots were formed from petiole with adaxial side downward, and most of adventitious shoots were induced from wound surface of leaf with adaxial side upward.

四种抗微管物质诱导毛新杨 2*n* 花粉粒的研究*

摘　要　该文揭示了4种抗微管物质，秋水仙碱、戊炔草胺、安磺灵和氟乐灵都有能力诱导毛新杨 2*n* 花粉。秋水仙碱虽然诱导 2*n* 花粉比率最高，但花粉量最少。另外3种除草剂，诱导 2*n* 花粉效率比用秋水仙碱处理的结果要低，但花粉量较大，其中，用戊炔草胺处理效果最好。其最佳处理为注射花芽4次，药品浓度200μmol/L，这不仅能使2n花粉比率达到84.4%，花粉量达到0.70mg/序，而且注射处理10个花芽的费用仅为0.084元，远远低于用秋水仙碱处理同样数量花芽的花费。该文证明了在所采用的这4种抗微管物质中，戊炔草胺是诱导杨树花粉染色体加倍效果最好的药剂。

关键词　2n 花粉，秋水仙碱，戊炔草胺，安磺灵，氟乐灵

白杨三倍体速生、质优，尤其适合纸浆材生产，利用物理或化学诱导的白杨 2*n* 花粉(张志毅等，1992)与二倍体雌株杂交，是迅速获取白杨三倍体的有效途径之一。特别是利用秋水仙碱进行的白杨 2*n* 花粉诱导，在掌握适宜的处理时期的前提下，可以获得一定量的高比率 2*n* 花粉(康向阳等，1999，1996)。我国利用该方法已经成功地获得一定数量的白杨三倍体。

然而，由于秋水仙碱与植物微管蛋白的亲和性较低(Hansen et al.，1998)，在诱导植物染色体加倍时需要较高的浓度，不但费用较高，而且还对被处理的植物材料产生毒害作用(Hansen et al.，1998)，如果开始处理的减数分裂时期掌握不合适，则不是获得的 2*n* 花粉比率较低，就是因药害及处理的机械伤害造成花序干枯，影响花粉的产量(康向阳等，1999)。

戊炔草胺(Pronamide)、安磺灵(Oryzalin)和氟乐灵(Trifluralin)这3种除草剂对植物微管蛋白有很强的结合力，能有效阻止植物微管的聚合，诱导染色体加倍(Bouvier et al.，1994；Hansen et al.，1996；Wan et al.，1991；Ramulu et al.，1988)。本文以花粉量多，结实力强的毛新杨(*Populus tomentosa* ×*Populus bolleana*)为材料，进一步就几种除草剂诱导毛新杨 2*n* 花粉的实际效果进行了研究，期望能够找到一种高效、低毒且费用低廉的诱导剂用于花粉染色体加倍乃至三倍体育种。

1　材料与方法

1.1　材料

毛新杨雌雄花枝采自于中国农业大学，银腺杨(*Populus alba* ×*P. glandulosa*)雌花枝采自于山东省冠县苗圃。每年大约在1月中下旬，采集花芽饱满的健壮花枝，用塑料包扎严

* 本文原载《北京林业大学学报》，2002，24(1)：12-15，与黄权军和康向阳合作发表。

密，低温贮藏，毛新杨雄花枝分批置于温室内（10~25℃）进行加倍处理，毛新杨和银腺杨的雌花枝随后放入温室水培进行授粉杂交。

1.2　方法

1.2.1　花粉加倍方法

采用秋水仙碱、戊炔草胺、安磺灵和氟乐灵溶液进行 2*n* 花粉诱导。处理方法均为注射法，即用微量注射器将药剂注入已在温室内水培一定时间的毛新杨雄花芽中，每天注射 3 次，每次 30~40μL，两次处理时间间隔为 5~7h。如果次日仍需注射，则于当晚将处理花枝移入 0~4℃的室内，以减缓减数分裂过程，保持处理效用的连续性。

秋水仙碱、戊炔草胺、安磺灵和氟乐灵的处理浓度各不相同。秋水仙碱的处理浓度为 0.5%，即 12531μmol/L（张志毅等，1992）。戊炔草胺、安磺灵和氟乐灵溶解在 5%的二甲基亚砜（DMSO）水溶液里，并且低温保存。戊炔草胺的处理浓度为 50，100，200，300，400，500μmol/L；安磺灵的处理浓度为 100，250，300，400，500μmol/L；氟乐灵的处理浓度为 20，50，100，200，300，500μmol/L。并且还共同做了一个对照试验，用来观察对照和处理之间的区别。每个处理水平的处理次数均为 4 次，每次处理重复 3 次。

1.2.2　减数分裂过程观察

定期取水培花芽，卡诺液固定 2h 左右，立即用醋酸洋红涂片观察。

1.2.3　2*n* 花粉的统计

白杨的 2*n* 花粉粒直径远远大于正常的单倍体花粉粒，约在 37μm 以上，其 DNA 物质基本上为单倍体花粉的两倍（张志毅等，1992）。由于加倍所获得的大花粉与对照之间对比明显，易于判别，因此本研究中将每个处理所获得的花粉混合均匀，制成 3~5 个临时涂片，每片观察 5 个视野，统计混合花粉中的大花粉粒数，并计算大花粉粒平均比率。

2　结果

由于诱导白杨花粉染色体加倍的适宜处理时期为细线末期到终变期之间（康向阳等，1999），故各处理的开始处理时期均于细线末期进行。在配制 3 种除草剂溶液时，当它们的浓度达到 500μmol/L 时，都已经出现了部分沉淀，其溶液实际为饱和液，具体浓度要比 500μmol/L 偏低一些，不过在本实验中仍当作 500μmol/L 看待。诱导 2*n* 花粉的最佳处理次数一般为 3~5 次（康向阳等，1996）。经处理雄花芽，镜检花粉粒见图 1，其实验具体数据统计结果见表 1。

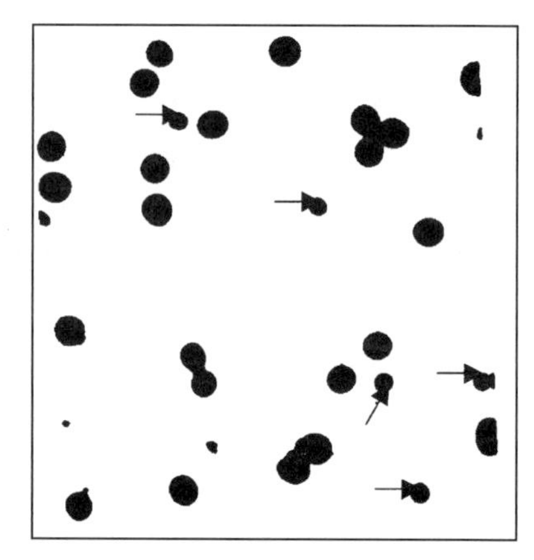

图 1　显微镜下观察得到的毛新杨混合花粉（其中→所指代表 1*n* 花粉）

从表 1 可以看出，诱导毛新杨 2*n* 花粉比率最高的仍是秋水仙碱，采用 12531μmol/L（即 0.5%）浓度的秋水仙碱诱导毛新杨 2*n* 花粉的比率最高，可达 96.7%，最低比率仍可达到 88.6%，总平均达 93.0%，比另外三种药剂的诱导比率都高，并且在它的4次处理水平之间，2*n*花粉的得率差别不大。但是从花粉量来看，用秋水仙碱处理，花粉量总平均仅为0.10mg/序，明显低于其他3种除草剂的处理水平，从而显示出秋水仙碱虽然能够诱导高比率的2*n*花粉，但是对被处理的植物材料产生了很大的毒害，大大降低了花粉产量。在具体的实验过程中同样可以看到，在用12531μmol/L（即0.5%）的秋水仙碱对毛新杨雄花芽处理以后，很多花芽极度卷缩，没有产生任何花粉，从而进一步证明了秋水仙碱会对被处理的植物材料产生大量毒害。

另外3种除草剂，戊炔草胺、安磺灵和氟乐灵诱导毛新杨2*n*花粉的比率和产生的花粉量各不相同，其中效果最好的是戊炔草胺，其诱导毛新杨2*n*花粉的比率最高可达到85.2%，已经接近秋水仙碱的诱导比率，并且经戊炔草胺处理所得到的花粉量要比经秋水仙碱处理所得到的花粉量高得多，总平均最高能达0.83mg/序。从结果可以看出，戊炔草胺在浓度为200μmol/L时，诱导毛新杨2*n*花粉的比率总平均为70.9%，花粉量为0.72mg/序，当浓度为300μmol/L时，结果分别为74.8%、0.22mg/序，虽然浓度有一定的增加，但花粉量锐减。随着其浓度的升高，2*n*花粉的得率增加不明显，即使其浓度达到饱和液500μmol/L时，2*n*花粉的诱导比率总平均也只能达到77.6%，并且其花粉量也很低，总平均仅为0.06mg/序。从高效和经济角度这两方面来考虑，可以认为戊炔草胺诱导毛新杨2*n*花粉时，在注射4次的条件下，最好的处理浓度为200μmol/L。在这时，既能得到高比率的2*n*花粉，又能产生较大的花粉量。

用安磺灵和氟乐灵诱导毛新杨2*n*花粉时，虽然花粉量比经秋水仙碱处理所得到的花粉量要高一些，但2*n*花粉的得率较低，安磺灵的诱导比率总平均最高可达24.0%，氟乐灵的诱导比率总平均最高可达34.8%，比经秋水仙碱和戊炔草胺的诱导比率要低得多。虽然安磺灵在诱导马铃薯、玉米、小麦等农作物的染色体加倍时，诱导效果比戊炔草胺要好一些，氟乐灵在浓度很低的情况下，也能使一些农作物的诱导比率很高，然而在诱导毛新杨2*n*花粉时，却没有达到预期的效果，具体原因还有待今后进一步的研究。

表1　4种药剂诱导毛新杨2*n*花粉的比率(%)及花粉产量(mg/序)

处理药剂	处理浓度(μmol/L)	处理Ⅰ		处理Ⅱ		处理Ⅲ		处理Ⅳ		总平均	
		比率	产量	比率	产量	比率	产量	比率	产量	比率	产量
对照(CK)		0	0	0	0	0	0	0	0	0	0
秋水仙碱	12531	88.6	0.10	96.3	0.10	96.7	0.08	90.5	0.12	93.0	0.1
戊炔草胺	50	53.2	0.89	45.8	0.76	—	—	—	—	49.5	0.83
	100	50.5	0.77	42.9	0.70	32.0	0.82	44.2	0.79	42.4	0.77
	200	84.4	0.70	54.3	0.65	78.8	0.77	67.1	0.75	70.9	0.72
	300	83.4	0.22	62.5	0.15	78.3	0.30	74.8	0.18	74.8	0.22
	400	83.4	0.10	75.7	0.09	70.6	0.08	80.2	0.09	77.5	0.09
	500	85.2	0.05	72.5	0.04	76.0	0.05	76.8	0.08	77.6	0.06

（续）

处理药剂	处理浓度 (μmol/L)	处理Ⅰ		处理Ⅱ		处理Ⅲ		处理Ⅳ		总平均	
		比率	产量	比率	产量	比率	产量	比率	产量	比率	产量
安磺灵	100	18.8	0. 50	9. 0	0. 48	14. 8	0.53	22. 3	0. 45	16. 2	0. 49
	250	22. 6	0. 26	10. 0	0. 24	14. 3	0. 24	28. 1	0. 22	18. 8	0. 24
	300	25. 4	0. 18	10. 9	0. 15	15. 3	0. 13	27. 4	0. 20	19. 8	0. 17
	400	29. 4	0. 10	17. 4	0. 12	19. 5	0. 09	29. 5	0. 11	24. 0	0. 11
	500	31. 4	0. 05	18. 1	0. 07	22. 5	0. 06	—	—	24. 0	0. 06
氟乐灵	20	9. 0	0. 35	12. 5	0. 30	13. 9	0. 29	12. 3	0. 40	11. 9	0. 34
	50	12. 2	0. 20	15. 1	0. 22	21. 7	0. 19	14. 5	0. 20	15. 9	0. 20
	100	10. 0	0. 11	19. 9	0. 15	21. 7	0. 10	18. 2	0. 09	17. 5	0. 11
	200	18. 1	0. 08	21. 6	0. 11	24. 0	0. 08	22. 5	0. 09	21. 6	0. 09
	300	24. 7	0. 05	36. 3	0. 09	34. 7	0. 08	28. 6	0. 06	31. 1	0. 07
	500	25. 7	0. 04	44. 4	0. 06	—	—	34. 4	0. 05	34. 8	0. 05

3 结论与讨论

秋水仙碱、戊炔草胺、安磺灵和氟乐灵在诱导毛新杨 2*n* 花粉时，诱导比率最高的是秋水仙碱，但是其产生的花粉量相当低，明显低于其他 3 种除草剂诱导产生的花粉量。在另外 3 种除草剂中，只有戊炔草胺能诱导产生高比率的 2*n* 花粉，并且能产生较多的花粉量。它在注射 4 次的情况下，最佳处理浓度为 200μmol/L，在这时，戊炔草胺能诱导产生高达 84. 4%的 2*n* 花粉，同时能产生 0. 70mg/序的花粉量。安磺灵和氟乐灵均不如秋水仙碱和戊炔草胺的诱导比率，这 2 种药剂暂时还不适合用于诱导毛新杨 2*n* 花粉。

从上面的结论可以看出，秋水仙碱诱导大花粉比率高，但产生的花粉量少，从而证明对被处理的植物材料毒害大；另外 3 种除草剂诱导产生的花粉量均高于秋水仙碱，从而证明对被处理的植物材料毒害低。戊炔草胺、安磺灵和氟乐灵这 3 种除草剂在低浓度的情况下能使花粉染色体加倍，是由于这 3 种药剂是抗微管的除草剂，对植物微管蛋白有很强的结合力，而对动物微管蛋白的结合力则较低（Wan et al. , 1991）。并且许多研究发现，秋水仙碱对植物微管蛋白的亲和性要低于对动物微管蛋白的亲和性（Wan et al. , 1991），刚好和另外 3 种除草剂的性质相反。因此这 3 种除草剂一般在 μmol/L 的浓度下就能诱导植物染色体加倍，而秋水仙碱则需要比这 3 种药剂高几十倍甚至上百倍的浓度(Wan et al. , 1991)。因此，用低浓度的除草剂和用高浓度的秋水仙碱在解聚微管、阻止纺锤体形成、诱导染色体加倍方面相当。这 4 种药剂对微管蛋白的亲和性可以从它们对白鼠的致死中量看出。秋水仙碱对白鼠急性口服致死中量 LD_{50} = 6. 1mg/kg，戊炔草胺 LD_{50} = 8350mg/kg，安磺灵 LD_{50} = 10000mg/kg，氟乐灵 LD_{50} > 10000mg/kg（林郁，1989）。因此，对动物蛋白亲和性低而对植物蛋白亲和性高的除草剂，在诱导植物染色体加倍时，只需要较低的浓度，并对被处理的植物材料毒害小。

用戊炔草胺、安磺灵和氟乐灵诱导白杨 2*n* 花粉，它们在具体注射花芽时的费用和经秋水仙碱处理所用的费用不相同。当 3 种除草剂浓度为 500μmol/L，即为饱和液时，与采用秋水仙碱处理的费用相比较，并且每个花芽每次注射按最大量计算，即每个花芽每次注射 40μL，在这种情况下，秋水仙碱每注射 10 个花芽的费用达 7.6 元，而注射戊炔草胺、安磺灵和氟乐灵的费用却分别为 0.21，0.25，0.24 元，均低于采用秋水仙碱的费用的 1/30。如果用戊炔草胺的最佳处理浓度 200μmol/L 与之比较，每个花芽每次注射也为 40μL，则每 10 个花芽的费用仅为 0.084 元，低于采用秋水仙碱的费用的 1/90。这样明显节约了药品，极大程度降低了试验经费。因此，在今后的试验中，可以大量采用这种低毒、低成本、高效率的药剂用于多倍体的诱导。

用上述抗微管物质诱导产生的花粉分别与毛新杨和银腺杨雌株授粉杂交后，产生了大量的杂种苗。在这些杂种苗中，究竟能有多少株三倍体植株，需待下一步进行镜检。如果单从育种角度考虑，需要生长到一定阶段后，在苗期表现出一定的差异来，再从这些杂种苗中选择那些生长相对高大，叶面积巨大的植株进行染色体检测，以确定出苗期优良的三倍体植株，然后再进行田间生长和其他性状的测定，最终选出优良三倍体新品种。因此，这 4 种抗微管物质诱导产生 2*n* 花粉粒授粉后究竟能获得多少三倍体植株，哪种药物诱导产生的三倍体比率最高，还需要对这批杂种苗进一步实验检测。

2*n* Pollen of *Populus tomentosa*×*P. bolleana* Induced by Four Antimicrotubule Agents

Abstrct Four antimicrotubule agents, colchicine, pronamide, oryzalin, and trifluralin, were evaluated for their ability on inducing 2*n* pollen of *P. tomentosa* ×*P. bolleana*. Colchicine was found to be the most effective on inducing 2*n* pollen among the four inducers, but the weight of pollen was the least. Although the other herbicides could not induce the rate of 2*n* pollen as high as colchicine do, more pollen was produced. Pronamide was the best. With 4 times of injection at a concentration of 200μmol/L, the percentage of 2*n* pollen could reach 84.4%, and the weight of pollen per catkin reached 0.70mg. The expense on 10 flowers' treatment by pronamide was only ￥0.084, much lower than that by colchicine. This study demonstrates that pronamide could be one of the best choices in four antimicrotubule agents for chromosome doubling of *Populus*.

Key words 2*n* pollen; colchicines; pronamide; oryzalin; trifluralin

毛白杨种子胚根发育生根性状的遗传分析*

毛白杨是我国特有的白杨派乡土树种，由于其种内有性生殖困难，败育现象严重，限制了对其杂交后代个体性状分离的研究(林惠斌等，1988；陈耀华，1983；朱大保，1990；张志毅等，1992；张志毅等，2000)。本研究首次报道了控制毛白杨种子胚根发育生根性状是质量性状，受一对完全显性基因控制。

1 材料与方法

母本毛白杨 5082，采于山东冠县全国毛白杨花枝标本园，父本截叶毛白杨，采于陕西林科院院内，切枝水培控制授粉杂交实验。杂交后 25d 收集毛白杨种子。种子用 0.1% 生汞消毒后，点播在 1/2MS 培养基上，25℃光照培养，培养 15d，调查发芽率，并以幼苗有、无可见的根分化为标准统计生根情况。

2 结论

本研究由于采用的方法确保种子的萌发、生长条件的一致性，所以保证了调查生根特性的准确性，幼苗根系发育情况如图 1 所示。实验共获得毛白杨种子 3193 粒，种子发芽率为 48.74%，萌发幼苗中生根幼苗共有 1179 棵，不生根幼苗为 376 棵，　生根幼苗:不生根

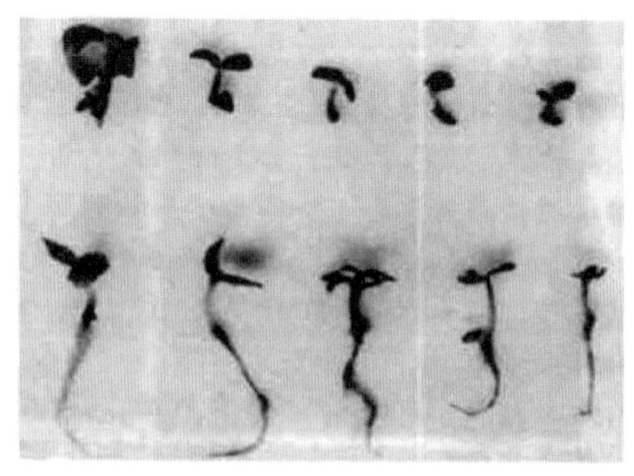

图 1　毛白杨幼苗生根情况(上为不生根幼苗，下为生根幼苗)

Fig. 1　Rooting and non rooting seedlings of *Populus tomentosa* Carr.

(Top: non rooting seedlings; Bottom: rooting seedlings)

* 本文原载《北京林业大学学报》，2002，24(2)：95-96，与张德强和杨凯合作发表。

幼苗符合 3:1，$\chi^2<\chi^2_{0.05}$($\chi^2=0.557$，$\chi^2_{0.05}=3.84$)，说明该性状属于质量性状，并由单个或几个显性基因控制。

Genetic Analysis of Seed Embryonic Root Development in *Populus tomentosa* Carr.

Abstract Experiments were carried out by introspecific controlled crossing between a female *Populus tomentosa* Carr. (clone 5082) and male *P. tomentosa* Carr. var. *truncata* Y. C. Fu et C. H. Wang, var. nov. 3193 seeds were obtained and then dibbled on 1/2 MS medium *in vitro*, and cultured by 25℃ light shining in 15 days. The rate of germination was 48.74%. Among them, there were 1179 root seedlings and 376non-root seedlings. The segregation ratio of offspring (rooting: non−rooting) was 3 :1 in *Populus tomentosa* and $\chi^2<\chi^2_{0.05}$. The conclusion is that the rooting character is belonged to qualitative trait and controlled by a single complete dominant locus.

Key words *Populus tomentosa* Carr. , rooting character, dominant gene.

滇杨遗传改良策略初论*

摘　要　滇杨是我国西南地区特有乡土树种之一，也是我国乃至世界少有的分布于低纬度高海拔地区的杨属树种。但目前国内对滇杨的研究仅局限于分布和分类学方面，有关其遗传改良研究几乎为空白。本文针对滇杨目前的资源及研究现状，对其遗传改良进行了系统探讨，并在借鉴国内外杨树遗传改良成功经验的基础上，提出将常规育种与现代生物技术育种有机结合起来，形成一个系统高效、合理可行的滇杨遗传改良策略。

关键词　滇杨，遗传改良，策略

杨树(*Populus* spp.)是世界上广泛栽培的树种之一，也是营造工业用材林和生态防护林的主要树种，它具有速生优质、轮伐期短、适应性强、繁殖容易、基因组相对较小等特点，现已成为林木遗传改良研究的模式植物(Dayton et al.，1992；Jehen et al.，1994；张绮纹等，1999；张志毅等，2002)。自20世纪40年代末以来，经过科研工作者的努力，我国在杨树优良天然杂种选择、常规杂交、倍性育种方面获得了巨大的成功(徐纬英，1988；马常耕，1994；王明庥等，1992；段安安等，1997；朱之悌等，1998)。近几年来，随着分子生物学的发展，我国杨树基因工程、遗传图谱构建、重要性状的基因定位及其分子标记辅助选择育种等方面也取得了较大的进展(张志毅等，2002；王学聘等，1997；尹佟明等，1999，1998；田颖川等，2000；李金花等，1999；刘斌等，2002；苏晓华等，2000；林善枝等，2000；饶红宇等，1999)。但与杨属其他树种相比，有关滇杨(*Populus yunnanensis*)的研究只局限于分布与分类等方面(徐纬英，1988；王战等，1984；赵能等，1991a，1991b；陈冠群等，1995；余树全等，2003)，遗传改良研究几乎为空白，至今仅有云南大学李启任等进行过美洲黑杨(*P. deltoids*)与滇杨杂交亲本和杂种的同功过氧化物酶比较研究(李启任等，1994)。

滇杨是我国西南地区特有乡土树种之一，也是全国乃至世界少有的分布于低纬度高海拔地区的宝贵杨树资源。滇杨具有速生、耐寒、易无性繁殖、抗叶锈病和叶斑病等优良特性，在杨属树种中占有特殊的位置。早在18世纪就被作为高山杨树代表种引种到了澳大利亚(Pryor，1985)，而新西兰则把滇杨作为抗叶锈病和叶斑病的基因资源加以收集、保存和利用(马常耕，1994)。滇杨的开发和利用，除可为我国杨树育种提供新的基因资源外，还可为我国高原地区的造林绿化提供一个好的树种。但目前滇杨基本处于野生状态，分布比较分散、优良成片林较少，而且它属雌雄异株，其中雌株罕见，分布群体基本由雄株组成。另外，易遭蛀干虫害，且分枝多。这些因素极大地限制了滇杨的研究与利用。基于上述原因，为了充分开发、利用现有的滇杨资源，有必要制定一个科学有效的滇杨遗传

* *本文原载《西部林业科学》，2004，33(1)：44-48，与何承忠、陈宝昆、李善文和段安安合作发表。

改良策略，以提高其抗性，改良其干形和材质，扩大其栽培种植区域和用途。据此，本文在查阅大量有关杨树遗传改良资料的基础上，并借鉴国内外杨树遗传改良的成功经验，对滇杨的遗传改良策略进行了研讨，并针对滇杨现有的资源及研究现状，对今后如何有效开展滇杨遗传改良育种研究提出了一些设想(图1)。

该育种策略的独特之处是将滇杨的生物技术育种和常规育种有机地结合起来，形成一个多层次、科学高效、合理可行的滇杨遗传改良方案。在此育种程序中，首先要对现有的滇杨资源及其遗传多样性进行深入调查与研究，并收集新的滇杨资源，建立完善的基因资源库与综合性状评价体系，为滇杨遗传育种提供丰富的优良基因资源。在此基础上，一方面开展种源、家系、无性系多层次、多地点对比试验，从中选择出一批优良种源、家系和无性系，进而确定适宜杂交亲本，通过杂交育种并结合分子生物学手段开展杂种优势选择。另一方面，对部分形态特异单株进行染色体数目鉴定，并对所选择出的优良单株进行人工染色体加倍，从中筛选出一些优质多倍体滇杨品种。另外，也可以利用基因工程技术对所选择出的优良无性系进行定向遗传改良。现将滇杨遗传育种策略的主要内容及特点概述如下，以供相关研究者借鉴。

1 滇杨基因资源及其遗传多样性研究

林木遗传改良是在研究林木遗传变异的基础上，开展的遵循其遗传变异规律来改良林木的遗传组成，进而培育林木新品种的一项活动。而基因资源及其遗传多样性研究，可为林木遗传改良育种提供丰富的优良基因资源和材料。因此，要制订合理的滇杨遗传改良方案，首先要对现有的滇杨资源及其遗传多样性进行深入调查与研究。

在已有的基础上，进一步开展滇杨资源调查、收集和保存，建立完善的基因资源库。同时，从形态、细胞、生理生化和分子等多水平上对滇杨的基因资源及其遗传多样性进行深入研究，建立综合性状评价体系，对其经济性状、抗逆性和利用潜力等做出准确评价，确定其核心种质资源并保存。在此基础上，深入了解滇杨种源间、不同群体间、群体内不同个体间的变异水平，掌握变异规律，揭示不同层次变异的内在机理，为正确制订滇杨遗传改良方案，充分利用各个层次的变异提供科学依据。

2 滇杨优良种源及优良单株的选择

分布区内复杂的地形地貌，多样的气候环境条件，独特的地史条件，加上长期自然选择和生殖隔离等，致使滇杨在长期生存进化过程中，产生了丰富的不同层次的遗传变异(不同种源间的地理变异、不同群体之间的差异以及不同个体间的遗传变异)，为滇杨优良种源及优良单株的选择育种提供了丰富的材料。针对这种现状，在开展滇杨基因资源及其遗传多样性研究的同时，通过种源、家系、无性系多层次、多地点对比试验，从中选择出一批优良种源、家系和无性系，为直接选择利用、杂交育种、倍性育种以及基因工程定向改良提供优质材料。

优良种源及优良单株的选择，具有较强的可操作性和可预见性，能使滇杨的性状在一定程度上得到改良，但也存在偶然性大，费时费工和耗财等缺点。

3　滇杨杂交育种研究

在对滇杨进行种源、家系和无性系研究的基础上，利用所筛选出的一些优良无性系，确定适宜杂交组合，进而开展杂交育种研究。但由于滇杨雌株罕见，长期以来主要以无性繁殖为主，通过种内杂交进行性状改良的成效估计不会太大。因此，滇杨杂交育种可能要以派内种间杂交为主，派间杂交和种内杂交为辅来开展，才有可能获得预期效果。

我国西南地区还分布有小叶杨(*P. simonii*)、云南青杨(*P. cathayana* var. *schneiedri*)、川杨(*P. szechuanica*)、五瓣杨(*P. yuana*)、德钦杨(*P. haoana*)、缘毛杨(*P. ciliata*)、亚东杨(*P. yatungensis*)和加杨(*P. canadensis*)等青杨派树种，黑杨派树种也有引种栽培(徐纬英，1988；王战等，1984；赵能等，1991a，1991b；余树全等，2003)。这些品种具有分布广、生物型多样性、遗传变异丰富等特点，为滇杨杂交育种提供了丰富的亲本材料。因此，可以选择优良的小叶杨、云南青杨、川杨作为母本分别与滇杨进行不同方式杂交。为了获得明显的滇杨杂种优势，在进行杂交亲本选配时要注重选择具有生态学优势互补的亲本。另外，为了提高杂交育种进程与利用效率，在滇杨杂交育种过程中，可以辅助分子生物学技术进行杂种优势早期鉴定，同时也应将杂交育种、直接选择利用以及基因工程改良有机结合起来。也就是说，所获得的优势明显的杂种一方面可以通过无性繁殖直接加以利用，另一方面也可作为进行基因工程定向改良的受体材料。

虽然常规杂交育种在杨树遗传改良和生产上具有一定的作用，但也存在较大的局限性。主要表现在育种周期长，需要多个世代才能培育出具有目标性状的优良品种；杂交难度大，且结籽率低；杂交工序复杂费时，结果不确定性；易受季节、气候、地域等多种环境因素的限制，尤其是遗传改良程度有限。

4　滇杨倍性育种研究

倍性育种目前在许多农作物和果树中已被广泛应用。我国杨树倍性育种方面，由北京林业大学朱之悌等培育的三倍体毛白杨最为成功(朱之悌等，1998，1995；李云，2001；康向阳等，2002)。实践证明，三倍体育种在遗传增益、材质改良、速生、抗病等方面有非常大的优势，是杨树遗传改良的一条有效育种捷径(朱之悌等，1998，1995；Benson et al.，1967；Einspahr et al.，1968)。目前，培育三倍体杨树的主要途径有天然杂种的直接选择；筛选天然 $2n$ 花粉培育三倍体；通过理化途径诱导 $2n$ 花粉培育三倍体；通过理化途径诱导 $2n$ 雌配子培育三倍体等(李云，2001；李云等，2000，2001；张志毅等，1992)。

针对滇杨目前的资源现状，其倍性育种应以人工诱导和天然多倍体杂种选育相结合的原则来开展。一方面，在育种资源调查和收集的基础上，对部分形态特异的单株进行染色体数目鉴定，有望从中直接筛选滇杨的三倍体品种。不过天然多倍体杂种的选择要在滇杨分布的上限、下限以及气候条件波动较大的地区进行，生长在这些地区的滇杨由于受自然环境的长期影响，发生变异的可能性较大。另一方面，也可利用遗传测定所筛选出的优良无性系进行人工染色体加倍。基于滇杨雌株少见的现状，在人工培育过程中，采用通过理化途径获得 $2n$ 花粉来培育滇杨三倍体较为适宜。此外，在滇杨倍性育种过程中，还可以把染色体加倍和染色体重组结合起来培育异源三倍体，使所培育的三倍体既具有染色体的加倍效应，又具有杂交优势，这可能更有利于滇杨遗传改良效果的提高。

尽管倍性育种已在杨树的遗传改良中发挥了一定作用，但杨树天然多倍体的发生频率较低，而且选择群体相对较小，通过人工选择天然多倍体杂种的可能性有限。另外，在人工诱导培育杨树三倍体时，也存在杂交不亲和，结籽率低等问题，使得杨树三倍体育种的成功率较低，而且进展缓慢。

5 滇杨生物技术育种研究

杨树具有生长周期长，树体高大，遗传杂合性高，遗传负荷大等特点，而且许多重要经济性状属于多基因控制的数量性状，遗传机理不明。这些因素极大地限制了杨树常规育种工作的开展。由此可见，利用常规育种手段往往难以满足不同目的定向培育滇杨新品种的要求，也就是说要在短时间内培育出人们所期望的滇杨新品种是很困难的。而生物技术育种能够打破种间界限，高效定向地改良林木性状，更能使常规育种技术难以解决的一些问题有可能得到解决，使林木遗传改良实现跨跃式的发展，缩短育种周期，加快林木遗传改良步伐。因此，开展滇杨生物技术育种研究是十分必要的。

基因工程技术是在 DNA 分子水平上直接引入所需的目的基因，然后经过筛选与鉴定，从而达到改良目的。目前，植物基因工程技术已在我国杨树遗传改良方面得到了广泛应用，培育出一批具有一定抗除草剂、抗病虫、耐盐碱的转基因植株(张志毅等，2002；林善枝等，2000；杨传平等，2001；施季森等，2001)。这可为今后开展滇杨分子遗传改良育种提供依据和机会。在滇杨分子遗传改良实际操作过程中，要以种源、家系、无性系试验以及杂交育种为基础，将所获得的滇杨优良无性系、单株或优良杂种后代直接作为转基因受体，通过外源基因的导入表达，从中筛选鉴定出性状优良的转基因滇杨植株。对所获得的转基因滇杨植株，一方面通过大田检测后可以直接加以利用，或再次作为转基因的受体；另一方面也可以作为亲本与其他杨树或优良滇杨进行杂交。

林木的重要经济性状如树高、胸径、材积、材质和抗逆性等大多数属于多基因控制的数量性状，易受外界环境条件的影响，通过表型选择效果不够理想。而林木遗传连锁图谱的构建可为控制林木重要经济性状、生物胁迫和非生物胁迫的基因进行数量性状和质量性状的 QTL 定位，以及分子标记辅助选择育种奠定基础，从而使林木早期选择成为可能。因此，利用遗传连锁图谱上的分子标记对林木性状进行基因定位是林木遗传育种研究的一个重要方面，也是分子标记技术直接应用于林木遗传改良的首项成就之一。目前，我国在杨树的遗传图谱构建、重要性状的 QTLs 定位以及分子标记辅助选择育种等方面的研究也取得了一定进展。尹佟明等首次利用 AFLP 技术绘制美洲黑杨指纹图谱，并对 42 个美洲黑杨无性系进行了鉴定(尹佟明等，1998)。至今已完成了欧美杨(*P. euramericana*)、美洲黑杨、响叶杨(*P. adenopoda*)×银白杨(*P. alba*)、美洲黑杨×欧美杨、美洲黑杨×青杨(*P. cathayana*)、毛新杨(*P. tomentosa*×*P. bolleana*)×毛白杨(*P. tomentosa*)等遗传图谱的构建

(张志毅等，2002；尹佟明等，1999)，对控制杨树叶片数量和生长等性状进行了 QTLs 定位，并利用BSA 方法结合已构建的遗传连锁图对控制杨树叶锈病与叶枯病进行了分子标记及图谱定位(李金花等，1999；苏晓华等，2000，1999)。另外，我国有关杨树的原生质体再生培养、体细胞杂交、体细胞变异和突变体筛选等方面也均获得了成功(张绮纹等，1999；施季森等，2001；诸葛强等，2000)。因此，为了克服滇杨雌株少而带来的常规杂交育种的不利，在滇杨的遗传改良过程中，可以运用原生质体再生培养、体细胞杂交、体细胞变异和突变体筛选等细胞工程技术，培育滇杨优良新品种。另外，为了提高滇杨分子遗传改良效果，可以在滇杨遗传连锁图谱构建的基础上，利用分子标记技术对控制滇杨重要经济性状、生物胁迫和非生物胁迫的基因进行数量性状和质量性状的 QTL 定位，为分子标记辅助选择育种奠定基础，从而提高选择精确度和可靠性，加速滇杨育种进程，并使滇杨早期选择成为可能。

6 滇杨良种繁育技术研究

滇杨扦插极易生根。在获得滇杨优良新品种后，可应用扦插繁殖方法扩繁大量的良种苗木用于造林。但是，扦插苗质量和扦插育苗的成活率容易受繁殖材料成熟效应的影响。有成熟效应的材料繁殖的苗木，短期内就会进入有性生殖期，开花结籽，从而使生长速率降低，发挥不出优树幼年期的速生性。另外，成熟效应使插条的再生能力降低，生根变得困难，从而影响扦插育苗的成活率(朱之悌，1986)。因此，在对滇杨筛选出的优良种源、家系和杂种后代进行无性系测定的同时，对这些初繁材料不断进行平茬，使其永处幼态之中。经无性系测定有了结果，可直接采取中选无性系的材料进行扩繁育苗，为林业生产提供优质种苗。

综上所述，尽管有关滇杨遗传改良的研究至今几乎处于空白，但是我国其他杨树的优良天然杂种选择、常规杂交、倍性育种、基因工程、遗传图谱构建、重要性状的基因定位及其分子标记辅助选择育种等方面取得了较大进展，并积累了丰富的经验，这可为今后开展滇杨遗传改良育种研究提供借鉴。传统育种方法与生物技术育种各有所长，传统育种方法不是其他技术短期内可以取代的，而生物技术也是现代杨树育种中不可忽视的一项重要技术，虽然在其实际操作和应用过程中还存在着一些问题，但随着植物基因工程技术的不断完善和广泛应用，生物技术育种在林木遗传改良中的地位日趋显著。因此，在滇杨的遗传改良中，应将现代生物技术与各种传统育种方法有机地结合起来，通过天然杂种选择、派内派间杂交、倍性育种、生物技术育种等多层次育种手段的科学搭配、取长补短，进而加速滇杨遗传改良进程，培育出优质滇杨新品种，不断满足我国生态环境建设和木材生产的迫切需求。

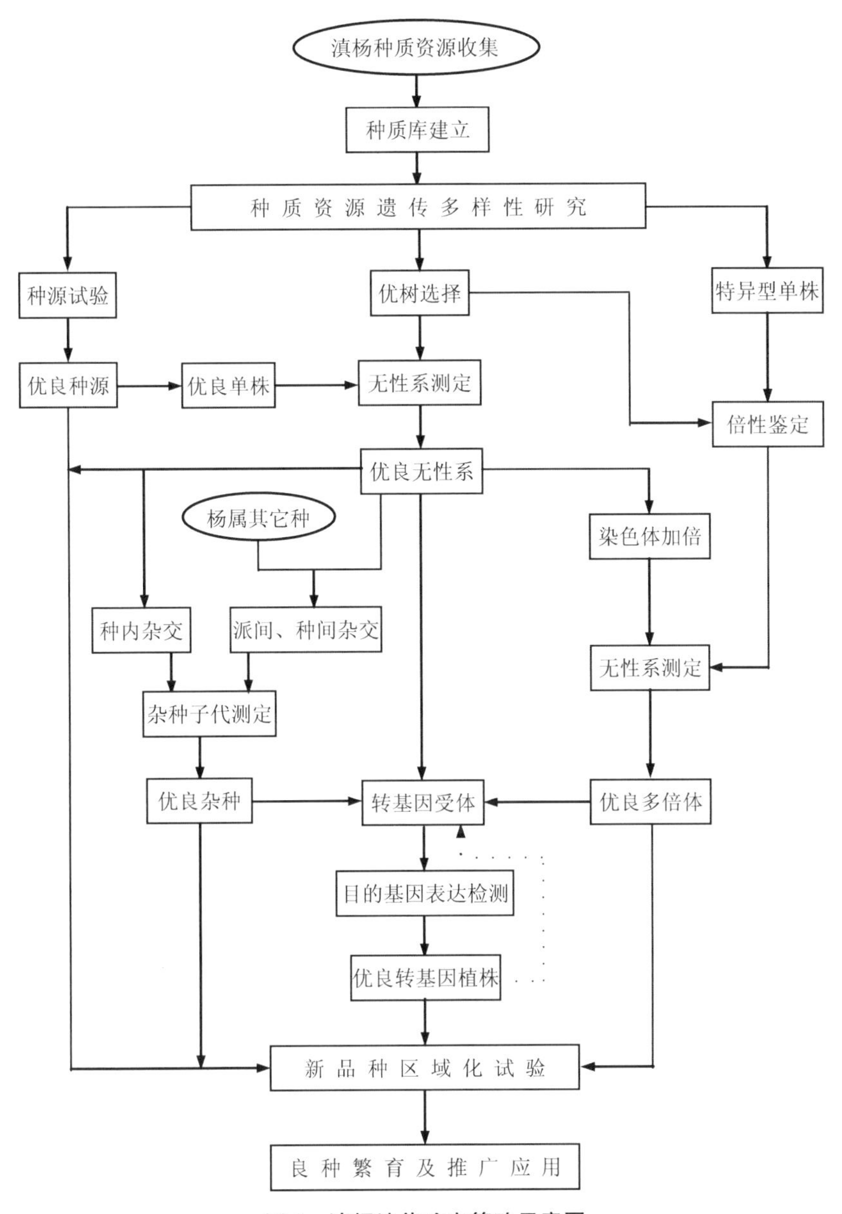

图 1　滇杨遗传改良策略示意图

Discussion on the Strategy of Genetic Improvement in *Populus yunnanensis*

Abstract: *Populus yunnanensis* is one of the unique native species in the southwest districts, and is a rare species of poplars that can distribute in low latitude but high altitude areas in our country, even in the world. But up to now, there almost has been no research on the genetic improvement in *P. yunnanensis*, except for a little reports concerning on its distribution and taxonomy. Therefore, genetic improvement is necessary to the protection and utilization of *P. yunnanensis*. In general, the effect of genetic improvement will directly depend on the breeding techniques adopted by breeders. The strategy of genetic improvement in *P. yunnanensis* is discussed in this paper, in which the reasonable combination traditional breeding methods with modern biotechnology improvement must be adopted to elevate the effect of genetic improvement, and to breed new varieties of *P. yunnanensis*.

Key words: *Populus yunnanensis*, genetic improvement, breeding, strategy

中国杨树杂交育种研究进展*

摘　要　本文对我国50余年杨属派内和派间杂交育种研究进行了综述。在众多的杨树杂交组合中，派内种间杂交以白杨派、黑杨派成就显著，派间杂交以青杨派与黑杨派杂交成果最为突出。从杂交方式看，有单交、双交、三交、回交等，以杂种作亲本进行再杂交能够获得显著杂种优势；从育种目标看，有速生、抗寒、抗旱、抗病虫、窄冠、生根等；从育种方法看，以常规人工杂交为主，将常规人工杂交与物理辐射、化学诱导等技术有机结合，能够创造出生产潜力较大的三倍体新品种。选育的杨树良种已在生产中产生了巨大的经济效益、生态效益和社会效益。文章最后对目前我国杨树杂交育种中存在的主要问题以及应对策略进行了讨论。

关键词　杨树，杂交育种

杨树是杨柳科(Salicaceae)杨属(*Populus* L.)树种的统称。由于杨树生长快、适应性强、木材用途广，在世界上栽培面积较大。中国杨树人工林面积大约667万 hm^2，约占全国人工林面积的1/5，是世界其他国家杨树人工林总面积的4倍(王世绩，1995)，在我国生态防护林和工业用材林建设中发挥着重要作用。目前，我国林业生产中推广应用的杨树良种主要是通过人工杂交培育的，杂种优势明显，具有速生、优质、抗病虫、抗逆性等特点，已产生了巨大的经济效益、生态效益和社会效益。杂交育种仍然是目前及今后更长时间培育杨树新品种的重要手段。

中国杨树杂交育种已有50余年的历史，最早是叶培忠教授于1946年在甘肃天水首次进行杨树杂交育种(叶培忠，1955)，从此以后许多林木育种工作者利用我国丰富的杨树种质资源，广泛开展了杨属派内和派间的杂交育种工作，培育出许多优良品种。本文将对这些研究内容作一综述，并对存在的主要问题以及应对策略进行必要的讨论。

1　中国杨树种质资源

杨属包括5个派：白杨派、青杨派、黑杨派、胡杨派和大叶杨派，世界共有100余种。在中国5个派杨树均有分布，共53种，特有种35种，主要分布于我国西南、西北、东北及华北地区，北纬25°~53°，东经80°~134°之间(徐纬英，1988)。

白杨派(Sect. *Leuce* Duby)有9种7个变种，如银白杨(*Populus alba* L.)、银灰杨(*P. canescens* Smith)、毛白杨(*P. tomentosa* Carr.)、山杨(*P. davidiana* Dode)、河北杨(*P. hopeiensis* Hu et Chou)、响叶杨(*P. adenopoda* Maxim)、新疆杨(*P. bolleana* Lauche)等。

* 本文原载《世界林业研究》，2004，17(2)：37-41，与李善文、何承忠和安新民合作发表。

青杨派(Sect. *Tacamahaca* Spach.) 青杨派是杨属中最大的一个派，有 34 种，21 变种，主要种如小叶杨(*P. simonii* Carr.)、大青杨(*P. ussuriensis* Kom.)、甜杨(*P. suaveolens* Fisch.)、苦杨(*P. laurifolia* Ledeb)、香杨(*P. koreana* Rehd.)、辽杨(*P. maximowiczii* Henry)等，特产种有小青杨(*P. pseudo-simonii* Kitag)、青杨(*P. cathayana* Rehd)、滇杨(*P. yunnanensis* Dode.)、青甘杨(*P. przewalskii* Maxim)、冬瓜杨(*P. purdomii* Rehd)、青毛杨(*P. Shanxiensis* C. Wang et Tung)、冒都杨(*P. qamdoensis* C. Wang et Tung)、五瓣杨(*P. yuara* C. Wang et Tung)、川杨(*P. szechuanica* Schneid.)等。

大叶杨派(Sect. *Leucoides* Spach.) 中国产 4 种 3 变型，如大叶杨(*P. lasiocarpa* Oliv.)、椅杨(*P. wilsonii* Schneid.)、灰背杨(*P. glauca* Haines.)、长序杨(*P. psudoglauca* C. Wang et Fu)等。

黑杨派(Sect. *Aigeiros* Duby) 有 3 种 2 变种，如欧洲黑杨(*P. nigra* L.)、额河杨(*P. jrtyschensis* CH. Y. Yang)、阿富汗杨(*P. afghanica* Schneid.)、钻天杨(*P. nigra*. var. *italica* Koehne.)、箭杆杨(*P. nigra* var. *thevestina* Bean.)等。

胡杨派(Sect. *Turanga* Bge.) 中国有 2 种，胡杨(*P. euphratica* Oliv.)和灰胡杨(*P. pruinosa* Schrenk)。

综上所述可知，中国具有丰富的杨树种质资源，占世界杨树种类的 50%以上，特别是在白杨派、青杨派中，许多种为中国所特有，这为开展派内及派间杂交育种提供了良好资源基础条件。

2　杂交育种

2.1　派内种间杂交

2.1.1　白杨派

叶培忠先生于 1946 年在甘肃省天水首次进行白杨派内种间杂交研究，杂交组合有河北杨×山杨、河北杨×毛白杨等(叶培忠，1955)，后来继续此项工作，选育出银毛杨(银白杨×毛白杨)、南林杨(*P.* ×'nanlin')[(河北杨×毛白杨)×响叶杨]。徐纬英等(1960)选育出毛新杨(毛白杨×新疆杨)、银山杨(银白杨×山杨)、山新杨(山杨×新疆杨)等。王绍琰等(1987)以银白杨为母本，以河北杨×山杨为父本杂交，选育出银白杨×(河北杨×山杨)优良品种，其材积超河北杨 109.8%，且抗病虫，易生根，扦插成活率 90%以上。邱明光等(1991)从河北杨×毛白杨组合中选育出河北杨×毛白杨 1 号，10 年生材积超河北杨 141%。刘培林等(1991)以银白杨为母本，山杨为父本，选育出银白杨×山杨 1333 号无性系，14 年生时比山杨快 1 倍以上；以山杨为母本，银白杨×山杨为父本，培育出山杨×银山杨 1132 号优良品种，具有速生、抗寒、抗病等特点。庞金宣等(2001)从南林杨×毛新杨组合中选出窄冠白杨 1 号、5 号(*P. leucopyramidalis*-1, 5)，从响叶杨×毛新杨组合中选出窄冠白杨 3 号、4 号(*P. leucopyramidalis*-3, 4)，从毛新杨×响叶杨组合中选出窄冠白杨 6 号(*P. leucopyramidalis*-6)，这些窄冠型杨树新品种具有冠幅窄、根系深、生长快、材质好等特点。北京林业大学对白杨派种及杂种间杂交难易程度、杂交方式进行了研究，结果表明：杂种在杂交可配性方面有着突出优势，其子代杂种优势明显；以杂种做亲本进行再度杂交是克服杂交不亲和的有效手段；选出毛新杨(*P. tomentosa*×*P. bolleana*)×银灰杨、毛新杨×毛白杨两个最佳组合(李天权等，1989)；利用毛新杨×毛白杨这一组合，采用秋水仙碱、γ射线处理花粉，选育出6个三倍体毛白杨优良无性系，这些无性系具有早期速生、材质优良等特点(朱之悌，1995)。

可见中国白杨派种间杂交育种取得了很好成绩。从杂交方式看，有单交、三交、双交等，并证明以杂种做亲本进行再度杂交是一有效方式；从育种目标看，有速生、抗寒、抗旱、抗病虫、窄冠等多个目标；从育种手段看，以常规人工杂交为主，并证明将常规人工杂交与物理辐射、化学诱导等技术有机结合，能够创造出生产潜力较大的三倍体新品种。

2.1.2 黑杨派

与白杨派相比，黑杨派种间杂交育种起步较晚。吴中伦等于1972年从意大利引进美洲黑杨南方型无性系I-69(*P. deltoides*'Lux'，又叫鲁克斯杨)，I-63(*P. deltoides* cv. 'Harvard'，又叫哈佛杨)及欧美杨无性系I-72(*P.* ×*euramericana*'San Martino'，又叫圣马丁诺杨)，在我国亚热带地区表现出较强适应性和较高生产力，一些学者利用这些无性系作亲本进行了杂交育种研究。符毓秦等(1990)以I-69杨为母本，用I-63杨、密苏里杨(*P. deltoides* var. *missouriensis* Henry)和卡罗林杨(*P. deltoides* ssp. *angulata*'Carolin')的混合花粉授粉，选育出陕林3号优良无性系，5年生单株材积超I-69杨24%，具有抗病虫特点，耐寒性、耐旱性优于亲本I-69杨。黄东森等(1991)以I-69杨为母本，欧亚黑杨混合花粉为父本，选育出中林46、中林23等优良无性系，材积超标优势30%以上。韩一凡等(1991)以I-69杨为母本，I-63杨为父本，选育出抗云斑天牛和光肩星天牛的南抗1号、2号优良品系，材积生长量超亲优势40%以上。陈鸿雕等(1995)选出鲁克斯杨×山海关杨(*P. deltoids* cv.'Shanhaiguan')、山海关杨×哈佛杨、圣马丁诺杨×山海关杨三个优良品系，简称鲁×山、山×哈、圣×山，具有抗溃疡病和灰斑病等特点，并且速生，胸径生长量超沙兰杨(*P.* ×*euramericana*'Sacrau-79')20%以上。刘月君等(1998)通过山海关杨×(美杨+哈佛杨)、鲁克斯杨×山海关杨分别选出廊坊杨1号、2号，具有明显杂种优势。

由此可知，尽管中国黑杨派基因资源较少，育种学家利用从国外引进的优良无性系作亲本，还是培育出了一批国产欧美杨、美洲黑杨杂种优良无性系；黑杨派内种间杂交以美洲黑杨改良为主，选出的无性系多数以I-69杨为母本，遗传基础相对较窄。

2.1.3 青杨派

青杨派种间杂交研究文献较少，只有徐纬英等(1960)做过小叶杨×滇杨、小叶杨×香脂杨、小叶杨×青杨、青杨×苦杨、小青杨×滇杨等杂交组合的研究工作。尽管中国具有丰富的青杨派种质资源，但种间杂交研究很少，可能有两个原因：一是青杨派种间杂交杂种优势不明显；二是青杨派内杂交，种子成熟期长，切枝水培后期蒴果营养不足，易脱落，不易得到成熟种子。

另外，大叶杨派由于分布于中部暖温带和北亚热带高山区，迄今仍处于野生状态；胡杨派种间杂交育种也未见报道。

2.2　派间杂交

2.2.1　黑杨派与青杨派

黑杨派与青杨派之间，在自然条件下容易产生天然杂种，且具有显著杂种优势。中国天然分布或广泛栽培的黑杨派与青杨派天然杂种有：二白杨(箭杆杨×小叶杨)、小钻杨(小叶杨×钻天杨)、额河杨等。另外，赤峰杨、白城杨、大关杨、泰青杨等均是从小叶杨与钻天杨天然杂种中选出的无性系，具有速生、耐寒、耐旱和耐盐碱等特点(徐纬英等，1960；赵天锡等，1994)。

表 1　黑杨派与青杨派间杂交育种

Tab. 1　Hybridization breeding between Section *Aigeiros* and Section *Tacamahaca*

杂交组合	选育目标	选育品种	选育人
钻天杨×青杨	速生、耐寒、抗病	北京杨	徐纬英等(1960，1988)
小叶杨×钻天杨	速生、耐旱、抗病	合作杨	徐纬英等(1960，1988)
小叶杨×欧洲黑杨	速生、耐旱、耐寒、抗病	小黑杨	黄东森等(1988)
赤峰杨×(欧美杨+钻天杨+青杨)	速生、耐寒、抗病虫	昭林 6 号	鹿学程等(1985)
小叶杨×美杨	速生、耐旱、耐寒	73-16、73-9	张永诚等(1990)
大关杨×钻天杨	速生、耐旱、抗病	陕林 1 号	符毓秦等(1984)
I-69×青杨	速生、耐旱、抗病	陕林 4 号	符毓秦等(1990)
I-69×小叶杨	速生、耐旱	NL-105、106、121	王明庥等(1991)
小黑杨×黑小杨	速生、耐寒	黑林 2 号	刘培林等(1993)
小青黑杨×欧洲黑杨	速生、耐寒	黑林 3 号	刘培林等(1993)
I-69×(大青杨+小叶杨)	速生、耐寒、抗病	69×大青小	张玉波等(2002)
山海关杨×塔形小叶杨	窄冠、耐盐、速生	窄冠黑青杨 6、31、69 和 70 号	庞金宣等(2001)

中国杨树育种学家开展了青杨派与黑杨派多个组合的杂交试验，获得过许多优良无性系(见表 1)。杂交亲本主要利用黑杨派的美洲黑杨(如 I-69、山海关杨)、欧洲黑杨(如钻天杨、箭杆杨)，它们具有速生、干直、抗病虫等特点；利用青杨派的小叶杨、青杨、小青杨等，它们具有耐寒、耐旱、易生根等特点。将两派杂交，选育出具有父母本优良特性的无性系；或者利用它们的杂交种，如小钻杨、小黑杨，与其亲本回交创造新品种。选育的良种由于具有速生、耐旱、耐寒、耐瘠薄等特性，因此主要在我国北方寒冷、干旱或南方高海拔地区栽培应用。

2.2.2　黑杨派与白杨派

由于黑杨派速生、易生根，而白杨派材质优、适应性强，因此许多学者想通过这两派间杂交，创造出具有两派优点的新品种。吴鸿锦等(1996)以沙兰杨为母本，毛白杨为父本，选育出沙毛杨(*P. Sacau* 79×*tomentosa* ‘Samenica’)。庞金宣等(2001)以 I-69 杨为母本，响叶杨×毛新杨为父本，培育出窄冠黑白杨。张金凤等(2000)做了 43 个黑白杨派间杂交组合，结果表明 15%的三交组合和 56%的双交组合能得到杂交苗木，用杂种〔如银腺杨(*P. alba*×*P. glandulosa*)、北京杨(*P. pyramidalis*×*P. cathayana* ‘Beijingensis’)等〕作亲本易得到黑白杨派间杂种，其扦插成活率较高，如银腺杨×中林 13 杨(*P. eur.* cv. Zhonglin 13)的杂种扦插成活率高达 83%。

2.2.3 青杨派与白杨派

关于青杨派与白杨派间杂交育种研究文献较少，仅有徐纬英等(1960)做过山杨×小叶杨、小叶杨×毛白杨、毛白杨×青杨、大青杨×毛白杨等杂交试验，结果表明青杨派与白杨派间有可配性，能获得杂种，但这些杂种的进一步研究情况未见报道。另外，有青杨派与白杨派天然杂种——青毛杨的报道(赵天锡，1994)。

2.2.4 胡杨派与其他各派

董天慈(1980)以小叶杨为母本，胡杨为父本，得到小×胡杂种，10 年生平均株高为小叶杨的 1.8 倍，为胡杨的 1.9 倍，平均胸径分别为小叶杨、胡杨的 1.9 倍，杂种优势明显。李毅等(2002)以箭杆杨为母本，用胡杨花粉和 5000 伦琴射线辐射处理的毛白杨花粉混合授粉，经强度筛选得到箭胡毛杨，经 RAPD 标记证明，该品种为胡杨和箭杆杨的杂种，它既保持了箭杆杨的窄冠、速生特点，也拥有胡杨抗旱、抗寒、耐盐碱和抗病虫特点。

综上所述可知，中国杨树杂交育种，派内以白杨派种间杂交效果突出，其次是黑杨派种间杂交；派间以青杨派与黑杨派杂交成就较大，其次是黑杨派与白杨派间杂交。

3 杂交育种存在的主要问题及对策

中国杨树杂交育种经过 50 余年历程，尽管取得许多成绩，但是还存在某些问题，如育种策略简单，无长期计划，研究树种较少，未充分利用种源、家系、无性系等水平存在的丰富遗传变异，育种目标过分追求速生，对乡土树种资源保存重视不够，未将现代生物技术尽快应用于常规杂交育种研究等。针对这些问题，应尽快采取一些相应对策。

3.1 制定长期改良计划

50 余年来，我国杨树杂交育种工作主要采用杂交-选择育种程序，重视 F_1 代无性系选择，每次杂交重新从未改良群体中随机选择亲本开始，迄今没有一个以轮回选择理论为指导的长、中、短期配套的总体计划。长期育种是指以轮回选择理论为指导，对育种亲本有计划改良，使短期育种工作随着世代前进而持续高效的螺旋式上升。意大利在杨树育种方面重视杂交亲本改良，制定了长期育种计划，其欧美杨育种策略是：首先确定美洲黑杨和欧洲黑杨最佳种源后，从中各选 120 株雌雄株组成育种群体，同时测定杂种优势和配合力。一方面，以美洲黑杨为母本，欧洲黑杨为父本，做种间杂交，取得杂种，经无性系测定后，选择杂种优势明显的植株，繁殖成无性系；另一方面，分别在美洲黑杨和欧洲黑杨育种群体内作种内控制授粉，通过配合力测定，从中挑选出雌雄各 40 株组成改良群体；对改良群体，重复上阶段的工作内容，如此循环，螺旋上升，使育种群体和杂种无性系的遗传品质不断提高(苏晓华等，1999)。我国乡土杨树遗传改良，可以借鉴该策略，并根据树种的分布范围及其特性，制订长期遗传改良计划，实行交互轮回选择为基础的遗传改良，把亲本改良作为提高杂种优势的根本。

3.2　重视多树种、多性状、多水平遗传改良

我国拥有丰富的乡土杨树资源，特别是青杨派和白杨派树种居世界之冠，它们主要分布于东北、西北、西南地区。但是近几十年来，对乡土杨树研究力度不够，也不系统。在我国杨树适生栽培区，地形复杂，气候条件各异，因此必须开展多种乡土杨树遗传改良工作，以适应不同地区生产需求，改良性状应抗逆性优先，在抗逆性基础上追求速生。

国内外许多研究表明，杨树在种源、家系、无性系等水平存在丰富遗传变异，并且这些变异显著影响杂交育种效果。Krzan(1976)做了不同种源 44 个无性系的美洲黑杨抗锈病试验，结果是南方种源和以南方种源做亲本的欧美杨抗锈病强。Cellerino(1976)从美国引种 18 个种源 52 个家系美洲黑杨种子，每一家系测 20 个无性系，内容包括对褐斑病、锈病、疮痂病及早春低温的反应，结果南方种源对上述 3 种病害是抗病的，并把它与南方型欧洲黑杨杂交，育成适应南欧的南方型欧美杨；欧洲一些学者做了许多山杨杂交，均认为用波兰种源做亲本最好。我国对青杨(杨自湘等，1995)、大青杨(苏晓华等，2001)、山杨(顾万春等，1995)等乡土树种研究表明，种源间及种源内个体间在生长、材性等方面均存在显著差异。因此，对乡土杨树遗传改良，应充分利用种源、家系、无性系这 3 个层次存在的丰富遗传变异，提高杂种优势。

3.3　注重乡土杨树育种资源收集保存

育种资源是指在选育优良品种工作种可能利用的一切材料，是选育新品种的物质基础。一个优良品种不仅应具备优良的经济性状，同时应具有较强的适应性和抗逆性。因此，选育优良品种，必须具备丰富的育种资源。如果没有丰富的育种资源做后盾，不断引进补充新的基因资源，多世代育种工作也将受到限制。为防止乡土杨树种质资源丢失，应分育种区建立育种资源保存群体，所保存群体要有代表性，不但包括生长快、材质优个体，而且包括抗病、抗虫、抗逆性个体，以满足不同育种目标需求。

3.4　将常规杂交育种与现代生物技术有机结合

在生物技术飞速发展的今天，常规杂交育种仍然是为生产提供良种的重要手段。现代生物技术与常规杂交育种有机结合，是今后杨树育种重要研究内容之一，如将转基因技术应用于杂交亲本改良，将分子标记技术应用于杂交亲本选配和杂种优势预测，将 mRNA 差异显示技术应用于杂种优势分子机理研究及杂种遗传分析，对选育的优良杂种无性系通过基因工程进一步改良等。完全有理由相信，随着生物技术与杂交育种的有机结合，传统杂交育种将再现出强大生命力。

Progress on Hybridization Breeding of Poplar in China

Abstract　The advance on intrasection and intersection cross breeding of poplar in China over the past fifties years is reviewed. Great progress has been made in Section *Leuce* and *Aigeiros*, and the best effect of intersection hybridization is the crossing between Section *Tacamahaca* and *Aigeiros*. From the view of the modes of hybridization, there are monohybrid cross, double hybrid cross, poly cross, back-cross etc. It has proved that using hybrids as parents to cross other species or hybrids is an effective and easy way to obtain heterosis. Fast growth, cold resistance, drought resistance, pest and disease resistance, narrow crown, rootage etc are the breeding goals. From the sight of breeding method, the routine artificial crossing is a main method, and the right combination of the routine artificial crossing with physical radiation and chemical inducing can creative new triploid individuals which have higher yield potential. The super clones cultivated have produced enormous benefit in practice. The problems existed and the countermeasures that should be taken in poplar hybridization breeding at present in China are also discussed in this paper.

Key words poplar, hybridization

三倍体毛白杨新无性系木材干缩性的遗传分析*

摘　要　利用9a生三倍体毛白杨测定林的9个无性系为试材，着重研究了木材干缩率的遗传学问题。结果表明：木材干缩率在无性系间的差异都达到了极显著水平，并受到强度遗传控制；木材干缩在株内纵向的变异模式是随树高增加而降低；全干体积，径向、弦向干缩率和气干体积，径向、弦向干缩率无性系重复力分别为0.89，0.95，0.84，0.67，0.84和0.68。

关键词　毛白杨，三倍体无性系，干缩率，遗传变异，重复力

木材干缩是木材加工和利用上的一大问题。木材不仅因其干缩发生尺寸和体积的缩小，尚因干缩不均而引起木材开裂、翘曲变形等缺陷(成俊卿，1985；梁世镇，1993)，同时干燥后的木材尺寸随周围环境的湿度或自身含水率的变化而变化，严重影响了木材及其制品的使用。毛白杨(*Populus tomentosa*)是我国特有的乡土树种，主要分布在黄河中下游地区，面积达100km^2。它的适应性和抗逆性强，材质优良，是我国北方重要的造林树种。由北京林业大学毛白杨研究所培育出的三倍体毛白杨新品种具有栽培周期短、超速生性、生物量大、木材白皙细密等特点，是纸浆材和胶合板材的兼性优良品种。国内较早地对火炬松(*Pinus taeda*)种源的木材干缩性能进行了研究(徐有明等，1997；1998)，但有关杨树木材干缩方面的研究较少，研究的树种主要为欧洲山杨(*Populus nigra*)、大青杨(*P. ussuriensis*)、北京杨(*P.*'Beijingensis')和毛白杨(Kariki，2001；王喜明等，2002；Cownet al.，1999)。对于三倍体毛白杨的木材品质诸如木材密度、纤维形态及热学性质等指标已有报道(邢新婷等，2000a；2002b；姚春丽等，1998；蒲俊文等，2002)，在此基础上对三倍体毛白杨的木材干缩性能进行研究，不仅可以掌握三倍体毛白杨干缩性能大小及其遗传变异规律，而且对于三倍体毛白杨的工业化加工、合理利用具有重要的现实意义。

本文利用三倍体毛白杨无性系测定林材料，较系统地研究了三倍体毛白杨木材干缩性的遗传学问题，将木材干缩这一材性差异与加工利用相结合，以期为正确制订单板材改良计划和策略提供科学理论依据。

1　材料与方法

1.1　材料

试材取自河北省晋州市国营苗圃的9a生三倍体毛白杨无性系测定林。采用完全随机区组设计，4区组10株小区，造林密度为2m×2m。测定林经过2次间伐后至采样时为4

* 本文原载《林业科学》，2004，40(1)：137-141，与刑新婷和张文杰合作发表。

株小区。该试验林在经过生长量及干形的初步评定后，初选出 L1、L2、L3、L4、L5、L6、L7、L8 和 L9 等 9 个三倍体无性系用于木材性质的研究，对照无性系 T34 为二倍体毛白杨无性系，用 CK 来表示。

1.2 取样及测定方法

对试材进行破坏性取样。每一无性系伐倒 2 株平均木并及时打枝，基部以 1.5m 为一段截取，上部均以 2m 为一段截取，分别编号，自然风干后用于木材干缩性能的测定。

试件按照 GB 1934.2—1991 规定制作，尺寸为 20mm×20mm×20mm。分别在 1.5m、5.5m 和 9.5m 处上下各 2cm 截圆盘，在圆盘上保持对应生长轮位置下截取试件。在试件各相对面的中心处，用游标卡尺分别测出弦向、径向及顺纹方向的尺寸，精确至 0.01mm，随即称重精确至 0.001g。然后将试件放入烘箱内，开始温度保持 60℃约 4h，再用(103±2)℃的温度烘 24h，取出放进干燥器的称量瓶中并盖好瓶盖。冷却至室温后称重，在各相对面的中心位置测出径向、弦向及顺纹方向的尺寸。测定完后，随即将试件浸入蒸馏水中至尺寸稳定，再测定试件径向、弦向及顺纹方向的尺寸。试样的尺寸是否达到稳定，可在浸水 20d 后选出 2~3 个试样，测定其弦向尺寸，而后每隔 3d 测量一次，如两次测量结果相差不大于 0.2mm 时即认为已达稳定尺寸。

1.3 计算方法及数据分析方法

全干试样弦向或径向干缩率(β_{max})按下式计算，精确至 0.1%。

$$\beta_{max} = [(l_{max} - l_0)/l_{max}] \times 100\%$$

式中：β_{max} 为试样径向或弦向的全干缩率,%；l_0 为试样全干时径向或弦向的尺寸，mm。

气干试样弦向或径向干缩率(β_w)按下式计算，精确至 0.1%。

$$\beta_w = [(l_{max} - l_w)/l_{max}] \times 100\%$$

式中：β_w 为试样径向或弦向的气干干缩率,%；l_{max} 为试样含水率高于纤维饱和点(即湿材)时径向或弦向的尺寸，mm；l_w 为试样气干时径向或弦向的尺寸，mm。

全干时的体积干缩率按下式计算，精确至 0.1%。

$$\beta_{Vmax} = [(V_{max} - V_0)/V_{max}] \times 100\%$$

式中：β_{Vmax} 为试样体积的全干干缩率,%；V_{max} 为试样湿材时的体积，mm^3；V_0 为试样全干时的体积，mm^3。

气干时的体积干缩率按下式计算，精确至 0.1%。

$$\beta_{Vw} = [(V_{max} - V_w)/V_{max}] \times 100\%$$

式中：β_{Vw} 为试样体积的气干干缩率,%；V_w 为试样气干时的体积，mm^3。

最后换算成含水率为 12%时的值，先将木材干缩率 4 项指标进行反正弦变换($X' = \arcsin\sqrt{x}$)，然后进行单因素方差分析和遗传参数估算(Falconer，1981)。差异干缩为弦向、径向之比值，不转换。

2　结果与分析

2.1　干缩率在无性系间的遗传变异

木材的干缩性能是衡量木材品质好坏的重要指标之一。径向干缩和弦向干缩的数值大，变异多，是影响木材利用的重要指标，而纵向干缩非常小，可以忽略不计。径向干缩率一般为3%~6%，而弦向干缩率最大，其干缩率约为径向干缩的2倍，即6%~12%(北京林业大学，1985)。三倍体毛白杨9个无性系及对照无性系的干缩性能测定结果见表1。从表1可以看出木材径向和弦向干缩率的差别，无论是全干干缩还是气干干缩都是径向小于弦向，这与木材干缩的普遍规律是一致的。全干时径向变幅为2.8%~3.5%，弦向是6.3%~8.0%，体积是10.2%~12.2%，差异干缩比值为2.0~2.7；气干时径向变幅为2.0%~2.6%，弦向为5.4%~6.5%，体积为7.8%~9.6%，差异干缩比值为2.0~2.9。本试验与柴修武(1995)的研究结果相一致。气干干缩率均值中，无性系L8最小，气干体积干缩为7.8%，气干径向干缩为2.0%，气干弦向干缩为5.0%；无性系L6干缩均值较大，气干体积干缩为9.6%，气干径向干缩为2.3%，气干弦向干缩为6.5%。对于对照无性系来说，各干缩率都较高，而三倍体中有的无性系干缩率较低，说明通过遗传改良的手段可以降低木材的干缩率。从三倍体毛白杨气干干缩均值来看，无性系L8较好，L6最差，对照无性系CK较差。

表1　三倍体毛白杨无性系干缩性能测定结果

无性系	全干干缩率均值(%)				气干干缩率均值(%)			
	体积干缩	径向干缩	弦向干缩	差异干缩	体积干缩	径向干缩	弦向干缩	差异干缩
L1	11.3	2.8	7.5	2.7	8.9	2.1	6.0	2.9
L2	10.8	2.9	6.9	2.4	8.5	2.1	5.5	2.6
L3	11.0	3.1	6.9	2.2	8.8	2.3	5.6	2.4
L4	11.2	3.3	6.9	2.1	8.6	2.4	5.4	2.3
L5	11.9	3.5	7.5	2.1	9.0	2.5	5.8	2.3
L6	12.2	3.2	8.0	2.5	9.6	2.3	6.5	2.8
L7	11.5	3.2	7.3	2.3	8.9	2.2	5.7	2.6
L8	10.2	2.9	6.3	2.2	7.8	2.0	5.0	2.0
L9	10.7	2.9	6.9	2.4	8.5	2.2	5.5	2.5
CK	11.3	3.5	6.9	2.0	9.0	2.69	5.6	2.1

无论是全干干缩率还是气干干缩率在无性系间都存在着一定的差异，方差分析结果(表2)进一步表明，各干缩率在无性系间存在显著或极显著差异，说明木材干缩率这一木材品质性状在无性系间存在较大的差异，对其开展无性系选择可能会获得较大的遗传增益，可以选择出木材密度大而干缩率又小的优良品系。

表 2　三倍体毛白杨木材干缩率的单因素方差分析

干缩率	无性系间均方	机误均方	F 值
全干体积干缩	0.7564(9)	0.0866(10)	8.7293**
全干径向干缩	0.1580(9)	0.0082(10)	19.3562**
全干弦向干缩	0.4269(9)	0.0670(10)	6.3754**
气干体积干缩	0.5588(9)	0.1828(10)	3.0574*
气干径向干缩	0.0707(9)	0.0060(10)	11.8686**
气干弦向干缩	0.3737(9)	0.1212(10)	3.0833*

注：括号内为自由度 df，“*”5%水平差异显著，“**”1%水平差异极显著，以下各表同。

2.2　干缩率的株内纵向变异

木材品质性状的变异除了种间差异、种内不同地理种源的差异、种内林分间差异、林分内单株间差异外，树木株内变异也是一种不容忽视的变异模式。大多数木材性质随树高的增加而存在一定的差异，了解株内纵向变异趋势可以为木材的合理利用提供科学依据。本试验对株内不同高度的干缩率进行了测定(表 3)，分析其结果可以看出径向干缩和弦向干缩在株内纵向变异总趋势是随着树高的增加而降低，其结果与王桂岩等(2001)对 13 种杨树的木材干缩相一致。这可能与树干下部生长轮比上部生长轮宽有关，因为生长轮宽者干缩率大而生长轮窄者则小一些。

表 3　三倍体毛白杨无性系干缩率株内纵向变异模式

无性系	取样高度(m)	全干干缩率均值(%)				气干干缩率均值(%)			
		VS	RS	TS	T/R	VS	RS	TS	T/R
L1	1.5	11.7	2.9	8.0	2.8	9.7	2.3	6.7	2.9
	5.5	11.1	2.7	7.4	2.7	8.7	2.0	6.0	3.0
	9.5	11.0	2.9	6.9	2.4	8.1	2.0	5.3	2.7
L2	1.5	11.0	2.9	7.2	2.5	8.5	2.1	5.7	2.7
	5.5	11.1	3.1	7.1	2.3	8.8	2.3	5.5	2.4
	9.5	10.3	2.8	6.5	2.3	8.0	2.0	5.2	2.6
L3	1.5	11.0	3.1	7.0	2.3	9.2	2.5	5.9	2.4
	5.5	10.8	3.0	6.9	2.3	8.8	2.3	5.7	2.5
	9.5	11.1	3.1	6.8	2.2	8.5	2.2	5.3	2.4
L4	1.5	12.0	3.6	7.5	2.1	9.4	2.8	5.8	2.1
	5.5	11.3	3.0	7.2	2.4	8.9	2.3	5.8	2.5
	9.5	9.9	3.0	6.0	2.0	7.3	2.1	4.5	2.1
L5	1.5	11.5	3.3	7.6	2.3	8.6	2.2	5.9	2.7
	5.5	12.3	3.7	7.5	2.1	9.4	2.5	5.9	2.4
	9.5	12.0	3.5	7.4	2.1	9.1	2.2	5.6	2.5
L6	1.5	12.3	3.1	8.4	2.7	9.8	2.2	6.9	3.1
	5.5	12.1	3.3	7.7	2.3	9.6	2.5	6.1	2.4
	9.5	12.1	3.3	7.8	2.4	9.4	2.3	6.3	2.7

（续）

无性系	取样高度（m）	全干干缩率均值(%)				气干干缩率均值(%)			
		VS	RS	TS	T/R	VS	RS	TS	T/R
L7	1.5	12.4	3.1	8.3	2.7	10.0	2.2	6.8	3.1
	5.5	10.6	2.9	6.7	2.3	8.4	2.2	5.3	2.4
	9.5	11.3	3.5	6.5	1.9	8.2	2.2	4.8	2.2
L8	1.5	10.7	3.1	6.6	2.1	8.6	2.4	5.5	2.3
	5.5	10.3	2.9	6.3	2.2	7.9	2.0	4.8	2.4
	9.5	9.4	2.5	6.1	2.4	6.7	1.7	4.4	2.6
L9	1.5	11.5	3.0	7.5	2.5	9.7	2.4	6.3	2.6
	5.5	10.8	3.0	6.7	2.2	8.2	2.1	5.2	2.5
	9.5	9.5	2.5	6.3	2.5	7.2	1.8	4.8	2.7
CK	1.5	11.7	3.7	7.1	1.9	9.4	2.8	5.7	2.1
	5.5	10.9	3.2	6.6	2.1	8.2	2.1	5.1	2.4

注：由于对照无性系上部很细不能作出试件，因此9.5m处没有数据。

2.3 干缩率的遗传参数估算

木材具有保守性很强的遗传性状，其遗传力愈高，则愈可以通过育种手段加以选择和控制。对三倍体毛白杨无性系干缩率遗传参数的估算见表4。在6个干缩率指标中，全干各干缩率与气干径向干缩率均属于遗传性很强的性状，其无性系重复力均在0.80以上，其他指标则相对稍低，但其值也都超过0.6，因此木材干缩率是一个呈高度遗传的木材品质性状，通过木材改良可以获得较大的遗传增益。

表4 三倍体毛白杨无性系干缩率重复力估算

干缩率	均值(%)	变异幅度(%)	无性系均方	*F* 值	重复力
全干体积干缩	11.3	9.9~12.4	0.7564	8.7293**	0.89
全干径向干缩	2.8	2.7~3.7	0.1580	19.3562**	0.95
全干弦向干缩	7.5	6.0~8.4	0.4269	6.3754**	0.84
气干体积干缩	8.9	7.3~10.0	0.5588	3.0574*	0.67
气干径向干缩	2.1	2.0~2.8	0.0707	11.8686**	0.84
气干弦向干缩	6.0	4.5~6.9	0.3737	3.0833*	0.68

3 讨论与结论

改良木材的一条有效途径就是培育出具有所希望木材性质的树木品系。大多数木材性质具有中等至较高的遗传性，因此通过遗传手段可以使其迅速地向我们需要的方向变化。木材的干缩性能也不例外，本研究的结果表明，木材的干缩性在无性系间存在较大的变异，且受到较强的遗传控制。现代育种学认为变异又是选择的基础，没有变异就没有选择，因此系统的研究三倍体毛白杨无性系的干缩变异并结合系统选择育种，有望在短期内获得生物量大且干缩率小的优良无性系。

木材变异发生于同一树种的不同地理种源、不同林分和不同的林木个体之间，甚至在同一树木的不同部位也出现较显著的差异。干缩率是衡量木材质量的一项重要指标。对于具有较高价值的产品，如家具和门窗制作，必须考虑木材的干缩性能。为制订最佳加工工艺，须测定每个树种的树内、树间干缩率差异情况。本研究表明木材干缩在株内纵向的变异呈随树高增加而下降的变异模式，对株内变异模式的了解有利于合理利用木材。但木材的干缩是木材的固有性质，干缩率的变异不仅与树种有关，同时还与木材内部早晚材比例、纤维角度的变化及纹孔多少等诸多因素有关。因此，在进行良种繁育时，应注重培育干形通直、分枝特性良好的无性系，以减少由于应力木多而引起干缩率大的机会。

由于三倍体毛白杨新无性系速生丰产，生长到 8~10a 胸径可达 30cm 以上，木材细密，与椴木性能相似，是制造人造板的好材料。Cown 等(1999)对 3 种杨树的木材质量进行了比较，发现毛白杨木材密度较高，力学性能指标以毛白杨较高，心材含量最低，干缩率较低，但含中度应拉木。因此要合理利用三倍体毛白杨必须首先掌握其材性特点，但目前木材材性的研究现状侧重于测定不同树种的材性差异，而将材性与加工利用结合起来的研究较少。因此为了提高三倍体毛白杨的综合利用水平，一方面要根据材性在无性系间及株内的变异特点，通过林木定向选育来努力达到；另一方面要根据其材性特点(应拉木、湿心材含量等)，从加工利用工艺方面设法克服其诸多不利因素，从而最终实现林—工一体化。

总之，三倍体毛白杨无性系的干缩率在无性系间存在显著差异，且受到较强的遗传控制，重复力均在 0.6 以上；株内纵向变异模式为随树高增加而降低的趋势。

Genetic Analysis of Shrinkage of New Triploid Clones of *Populus tomentosa*

Abstract The wood samples of 9 triploid clones of *Populus tomentosa* were cut from 9-year-old clonal stand. The oven-dried and air-dried samples were used to study their genetic variation in shrinkage. The results indicated that there were significant differences between clones and all the coefficients of shrinkage that were found to be under strong genetic control. The vertical variation pattern within tree of shrinkage was downward with the tree height. The clonal repeatability's in shrinkage of oven-dried and air-dried timber in volume, radial and tangential direction were respectively 0.89, 0.95, 0.84, 0.67, 0.84 and 0.68.

Key words *Populus tomentosa*, triploid clone, shrinkage, genetic variation, repeatability

QTL Analysis of Leaf Morphology and Spring Bud Flush in (*Populus tomentosa*×*P. bolleana*)×*P. tomentosa* *

Abstract One hundred and twenty progeny of *Populus* were derived from a cross between female clone "TB01" (*Populus tomentosa*×*P. bolleana*) and the male clone "LM50" (*P. tomentosa*). This population was used to detect quantitative trait loci (QTLs) affecting leaf morphology and spring bud flush. A total of 393 AFLP markers were identified and used to construct a parental specific genetic map using pseudo-test-cross mapping strategy. The total genome length corresponding to 3265. 1cm for the clone "LM50" map and 1992cm for the clone "TB01" map with 4~30 markers per linkage group was obtained. Fourteen QTLs controlling leaf morphology were identified on nine linkage groups, and 3 QTLs affecting spring bud flush were detected on three linkage groups with interval mapping software. The phenotypic variance explained by each QTL ranged from 7. 6% (on TBLG14) to 15. 8% (on TLG9). Co-localization of QTLs controlling correlated traits such as leafblade length, leafblade width and leafblade area were mainly found on linkage groups TLG2 and TLG11 in the genetic map of clone "LM50", and on linkage group TBLG1 in the genetic map of clone "TB01". Based on the genomic regions of QTLs for leaf morphology and spring bud flush, these two traits seem to be controlled by separate genes.

Keywords (*Populus tomentosa* × *P. bolleana*) × *P. tomentosa*, QTLs mapping, leaf morphology, spring bud flush

1 Introduction

The size, morphology, anatomy and phenology of leaves are important parameters for estimating the amount of biomass, as has been reported in previous studies in woody plant species (Larson et al., 1971; Bradshaw et al., 1995; Wu et al., 1997; Castro-Diez et al., 2000). In forest trees, the magnitude of leaf area directly affects the efficiency and yield of photosynthesis.

It is, therefore, a good predictor of stem volume (Leverenz et al., 1990; Wu et al., 1997). The timing of bud flush determines when new leaves emerge and trees grow. The growing tissues may be killed by late frost if bud flush occurs too early in spring. Conversely, the growing season for trees will have to be shortened due to late initiation of bud flush in spring, which reduces competitive ability and growth potential (Frewen et al., 2000). Hence, analysis of the genetic basis for leaf morphology and the timing of bud flush are important for tree breeding programs.

---

* 本文原载《林业科学》, 2005, 41(1): 42-48, 与张德强、杨凯和李百炼合作发表。

Quantitative trait locus (QTLs) mapping is a powerful and increasingly accessible tool for characterizing the genetic basis of morphological and adaptive divergence (Tanksley, 1993). QTL maps provide both a broad outline of the genetics of evolutionary change and a first step toward the isolation and identification of the particular genes involved in phenotypic differentiation (Paterson et al., 1991; Mackay, 2001). Genetic and QTL analyses of leaf morphology and adaptive traits have been performed in the population of *Eucalyptus* nitens, *Populus trichocarpa*×*P. deltoides* and Pseudotsuga menziesii (Bradshaw et al., 1995; Wu et al., 1997; Byrne et al., 1997; Jermstad et al., 2001). Previous reports showed that leaf morphology and bud flush have been found to be under strong genetic control and some moderately large QTL effects have been detected. However, the inheritance and genetic mapping of leaf morphology and spring bud flush have not been demonstrated in *P. tomentosa*.

In this paper, we described the detection and location of QTLs controlling morphological and adaptive traits in a cross between two very close poplar varieties. The interval mapping software package MAPMAKER QTL (Lander et al., 1989) was used to generate the linkage maps and detect the QTLs. The results showed that genome-wide QTL mapping was a promising strategy for the discovery of new genes controlling morphological growth and adaptive traits in forest tree species.

2 Materials and methods

2.1 Plant materials

The QTL mapping population was founded in 2000 by interspecific backcross between "TB01" clone (*P. tomentosa*×*P. bolleana*) and "LM50" clone (*P. tomentosa*). The backcross progeny and the parents were grown at the nursery of Beijing Forestry University. 120 individuals were randomly selected from this population and used for the present QTL mapping study.

2.2 Methods

2.2.1 Linkage map construction

Separate genetic linkage maps for "LM50" and "TB01" were constructed with AFLP markers using a pseudo-test-cross mapping strategy and the Kosambi mapping function (Zhang et al., 2004). To increase the number of informative markers available for QTL mapping, additional marker loci were added to the original maps using the "place" command with MAPMAKER 3.0, with the most likely marker placement accepted. The final framework map for "LM50" consisted of 247 markers ordered on 19 major linkage groups. The linked loci spanned approximately 326511cm of the poplar genome, with an average distance of 13.2cm between adjacent markers. For "TB01", the analysis resulted in 146 loci, mapping to 19 linkage groups and covering about 1992cm of the genome with an average distance between framework loci of 13.6cm (Fig. 2).

2. 2. 2 Trait measurements

Leaf morphological traits for leafblade length (LL), leafblade width (LW) and leaf petiole length (LPL) were measured in the October of 2001 in the field. Leaf area (LA) was determined with a Li-Cor 3100 leaf area meter (Bradshaw et al., 1995). The timing of spring bud flush was determined in the 2001 by twice-weekly inspection for the fully expanded leaf on the top stem.

2. 2. 3 QTL analysis

By underlying the backcross model, based on the individual linkage maps for clone "TB01" and clone "LM50", a LOD score threshold of 2. 0 was used to declare the presence of a linked QTL in a given chromosome interval. Additive genetic effects attributed to individual QTL and the percentage of phenotypic variation explained by each QTL, were estimated with the software MAPMAKERP/QTL.

3 Results

3. 1 Phenotypic trait analysis

The two original parent lines (*P. tomentosa* and *P. bolleana*) differed significantly with respect to leaf traits, including leaf size, shape and the timing of bud flush. The parent *P. tomentosa* was phenotypically superior over parent *P. bolleana* for leafblade length, leafblade width, leafblade area and leaf petiole length. For the date of spring bud flush, *P. tomentosa* was earlier than *P. bolleana*. The F_1 (*P. tomentosa*×*P. bolleana*) clone "TB01" showed intermediacy between the two parents for all leaf traits studied. A segregating population was obtained when the F_1 was backcrossed with the parental *P. tomentosa* clone "LM50". A comparison of the phenotypic characteristics of all traits indicated that there was great genetic variation in the *P. tomentosa* BC_1 population for all traits (Table 1). The frequency distribution of phenotypes for each trait in the 120 interspecific BC_1 families are shown in Fig. 1. All traits showed an approximately normal distribution. Hence, the population was meet for QTL analysis of leaf traits in *P. tomentosa*.

Table 1 Statistics of leaf traits of leafblade length, leafblade width, leafblade area, leaf petiole length and the timing of spring bud flush in mapping population①

Trait	μ	Min.	Max.	σ	CV
LL/cm	14. 38	8. 20	22. 40	2. 86	0. 20
LW/cm	15. 47	7. 50	24. 90	3. 57	0. 23
LA/cm^2	152. 13	9. 50	332. 80	62. 12	0. 41
LPL/cm	6. 37	3. 40	12. 00	1. 62	0. 25
TSBF/d	12. 60	3. 00	24. 00	3. 16	0. 43

① LL, Leafblade length; LW, Leaf blade width; LA, Leaf area; LPL, Leaf petiole length; TSBF, Timing of spring bud flush. The same below.

The correlation between two traits was estimated by regressing the phenotypic values of one trait onto those of the second trait. The correlation coefficients among the 5 traits are presented in Table 2. Significant associations were found between leafblade length, leafblade width and leafblade area. Both leafblade length and leafblade width were positively correlated with leafblade area. These correlations were highly statistically significant ($P<0.01$; $r=0.548$ and $r=0.562$).

Table 2 Phenotypic correlations among leafblade length, leafblade width, leafblade area, leaf petiole length and the timing of spring bud flush in mapping population①

Traits	LL	LW	LA	LPL
LL				
LW	0.216*			
LA	0.548**	0.562**		
LPL	0.210	0.184	0.116	
TSBF	0.084	0.012	0.026	0.086

① * $P < 0.05$; ** $P < 0.01$.

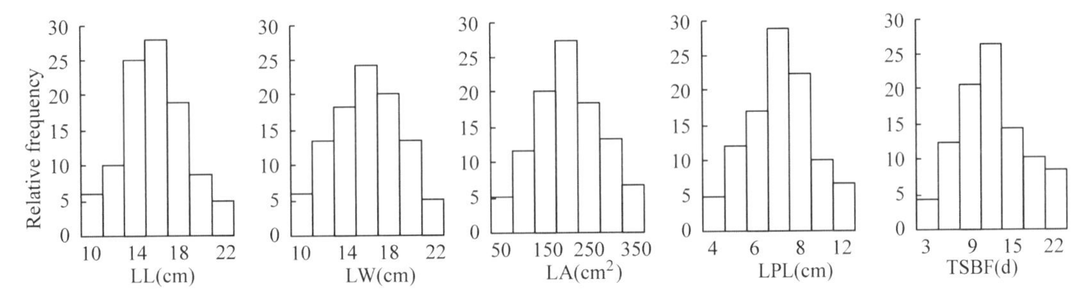

Fig. 1 Phenotypic frequency distributions for five leaf traits in mapping population

3.2 QTL detection

QTLs were detected at the threshold of LOD = 2.0 using interval mapping with MAPMAKERP/QTL, based on the separate parental linkage maps of clone "LM50" and "TB01". A total of 17 QTLs were identified for 5 leaf traits. These QTLs spanned nine linkage groups on the clone "LM50" map and three linkage groups on the clone "TB01". The proportion of phenotypic variation explained by each of these genomic regions varied from 7.6% (TBLG14) to 15.8% (TLG9). Table 3 summarized the number and direction of effect of QTL identified, together with their confidence intervals and the percentage of the additive genetic variation explained by each QTL for all traits.

3.2.1 Leaf morphology traits

Fourteen putative QTLs associated with leaf morphology, including leafblade length, leafblade width, leaf area and leaf petiole length were identified, with ten on the "LM50" map and four on the "TB01" one. Between two and five statistically significant QTLs were detected for each of these traits (Table 3) . These individual QTLs explained from 7.6% (on TBLG14) to 15.3%

(on TLG11) of the total phenotypic variance and increased the phenotypic absolute values from 32.60cm^2 up to 48.91cm^2. Seven QTLs had positive impact on phenotypic values and another 7 had negative effects.

Table 3 Putative QTLs detected for five leaf traits by interval mapping via Map Maker/QTL in mapping population of 120 progeny from clone"TB01"and clone"LM50"in poplar

	Traits/QTL	Descendance of QTL	Linkage group	Interval①	Position② (cM)	LOD	Variance③ (%)	Additive effect
LL	LLTLG2	Clone"LM50"	LG2	E63M39167r-E63M39166r	10. 1	2.60	12.90	-2.08
	LLTLG11	Clone"LM50"	TLG11	E44M50127-E63M3671	7.5	2.34	8.60	+1.96
	LLTBLG1	Clone"TB01"	TBLG1	E63M46271-E35M52249	10.8	2.46	9.42	+2.04
						Total	30.92	
LW	LWTLG2	Clone"LM50"	TLG2	E63M39167r-E63M39166r	9.8	2.40	8.72	-2.02
	LWTLG11	Clone"LM50"	TLG11	E44M50127-E63M3671	11.4	2.38	8.50	+2.07
						Total	17.22	
LA	LATLG2	Clone"LM50"	TLG2	E63M39166r-E63M40489	0.0	2.93	11.6	-41.98
	LATLG11	Clone"LM50"	TLG11	E44M50127-E63M3671	10.0	2.78	15.3	+48.91
	LATBLG1	Clone"TB01"	TBLG1	E63M46271-E35M52249	15.8	2.00	7.8	+34.30
	LATBLG14	Clone"TB01"	TBLG14	E35M50220-E60M34338	16.2	2.04	7.6	+32.60
						Total	42.3	
LPL	LPLTLG3	Clone"LM50"	TLG3	E33M32241-E61M4152r	22.3	2.1	8.2	-1.06
	LPLTLG4	Clone"LM50"	TLG4	E33M65107-E33M66166	36.5	2.3	8.6	-1.12
	LPLTLG12	Clone"LM50"	TLG12	E33M82205r-E44M40163	0.9	2.2	8.4	+1.08
	LPLTLG19	Clone"LM50"	TLG19	E60M33227r-E3340307	6.7	2.0	8.0	-1.02
	LPLTBLG17	Clone"TB01"	TBLG17	E63M52396-E60M3267	10.8	2.6	9.5	-1.32
						Total	42.7	
TSBF	TSBFTLG8	Clone"LM50"	TLG8	E60M33270r-E33M44290r	6.9	2.1	8.8	-2.84
	SBFTLG9	Clone"LM50"	TLG9	E33M79228-E44M40324	26.2	2.7	15.8	+4.51
	TSBFTLG16	Clone"LM50"	TLG16	E44M5093-E63M43178	3.0	3.7	14.5	+4.40
						Total	39.1	

① Interval between the two flanking markers (cM) where a QTL is located; ② QTL position from the first marker (cM); ③ henotypic variation explained by each QTL.

Co-localization of QTLs controlling correlated traits such as leafblade length, leafblade width and leafblade area were mainly found on linkage group TLG2 and TLG11 in the genetic map of clone "LM50", and on linkage group TBLG1 in the linkage map of clone "TB01".

3.2.2 Spring bud flush

Three putative QTLs affecting the timing of bud flush were identified. One QTL each fell into linkage group TLG8, TLG9 and TLG16, respectively, on the paternal map for clone"LM50". The clone "LM50" alleles increased the number of days to bud flush at TSBFTLG9 and TSBFTLG16 and delayed it at the TSBFTLG8 loci. These individual QTLs explained 8. 8% to 15. 8% of the total phenotypic variation, and had a positive phenotypic effect of increasing the timing of bud flush by 4. 51 and 4. 40 days and a negative effect of increasing the number of days to bud flush by 2. 84 days. Total phenotypic variation explained by all three putative QTLs was 39. 1%. The largest effect was associated with TSBFTLG9, which had additive effects of 4. 51 days that accounted for 15. 8% of the phenotypic variance.

4 Discussion

Many of the growth, development and adaptation traits in forest trees exhibit continuous phenotypic variation and are inherited quantitatively. The use of molecular markers enables one to identify the map QTLs that are involved in the variation of such complex traits. A QTL mapping strategy could dissect quantitative genetic variation into several loci at the molecular level. Over the past decade, tremendous efforts were invested by many research groups worldwide in mapping a wide spectrum of quantitative genetic variation in nearly all important animals and agricultural crops onto their genome regions. In the case of forest tree species, considerable research efforts were focused on economically important traits such as stem growth, vegetative propagation, wood quality, adaptation and disease resistance. However, surprisingly little is known about the genetic basis of continuous variation for leaf traits in *P. tomentosa*. In order to elucidate this problem, we first developed the QTL analysis of leaf morphology and the timing of bud flush traits in an interspecific *P. tomentosa* BC1 population.

4.1 QTL analysis of leaf traits through pseudo-test-cross mapping strategy

Grattapaglia and Sederoff (1994) put forward the mapping strategy of pseudo-test-cross through test-cross configuration and mapped the quantitative trait loci controlling growth and wood quality traits in *Eucalyptus grandis* (Grattapaglia et al., 1996). In the present study, quantitative trait loci affecting leaf morphology and the timing of bud flush traits were detected with genetic maps of clone "LM50" and "TB01". In this paper, we identified 17 putative QTLs which individually explained between 7. 6% and 15. 8% of the total genetic variation for five leaf traits in *P. tomentosa* BC1 population (Table 3). These results are consistent with classical quantitative genetics theory, that is, leaf morphology and spring bud flush in forest trees appear to be quantitative traits controlled by multiple genes. Estimates of the minimum number of QTLs involved, and their genome location, magnitude and direction of effect were obtained (Table 3).

The 17 QTLs found in our study were dispersed over nine linkage groups of "LM50" map and 3 linkage groups of "TB01". Considerable variation was observed in the number of QTLs detected per linkage group. For each of the linkage groups TLG3, TLG4, TLG8, TLG9, TLG12, TLG16, TLG19, TBLG14 and TBLG17, only one QTL was detected for all leaf traits. For TLG2, TLG11 and TBLG1, two or more QTLs spanning their genomic regions were detected (Table 3). For each

QTL, the direction (positive or negative) of the allele's effect on the target trait was determined. For nine out of the 17 QTLs detected in our study, a positive additive effect on phenotypic value was observed. Of these 9 trait-improving QTLs, 6 (51%) had alleles derived from the clone "LM50", and trait-improving QTL alleles from the clone "LM50" were detected for all leaf traits. Based on the genomic regions of QTLs for leaf morphology and spring bud flush, these two traits seem to be controlled by separate genes.

Correlated traits often have QTLs mapping to the same genomic regions (Veldboom et al., 1994; Xiao et al., 1996), as was observed in the present study. In our study, three QTLs located at the linkage groups TLG2 (E63M39167r-E63M39166r), TLG11 (E44M50127-E63M3671) and TBLG1 (E63M46271-E35M52249). These QTLs were detected for the positive correlated leaf traits leafblade length and leafblade area, leafblade width and leafblade area, respectively. Trait correlations may result from either tight linkage of several genes controlling these traits or the pleiotropic effect of a single gene (Bradshaw et al., 1995; Agrama, 1996).

4.2 Strategy for improving leaf traits using QTL information in *P. tomentosa*

The identification of QTLs affecting leaf traits in *P. tomentosa* is an important step in the use of molecular markers for both understanding the genetic factors that determine these traits and genetically improving economically important forest trees. When a genetic map is available from earlier generations such as a first backcross, detection of significant marker-trait associations based on such maps will be useful for marker-assisted selection (MAS) (Tanksley et al., 1996). Our results may provide useful information for *P. tomentosa* breeding. The parental lines showed marked differences for all leaf traits in the *P. tomentosa* backcross population, and the QTLs detected explained from 7.6% to 15.8% of the phenotypic variation by single QTL for leaf traits. Previous studies showed that MAS efficiency depended primarily on the heritability of the trait, on the population size and on the proportion of phenotypic variation associated with markers (Whittaker et al., 1995; Laurence et al., 1998). Compared with a previous QTL study done on the F_2 progeny of *P. trichocarpa*×*P. deltoides*, where from 21.2% to 36.1% of phenotypic variation for leaf area was explained by a single QTL (Wu et al., 1997), the power to detect QTLs in our study was probably reduced due to our use of dominant molecular markers. Use of dominant markers leads to lower LOD scores and also to an under-estimation of the magnitude of explained variation if multiple alleles were segregated at a locus in offspring of out-breeding species (Eck et al., 1994).

The QTL detection with interval mapping strategy requires a large segregation population and effective MAS should have small distance between markers and QTLs (Laurence et al., 1998). In practice, however, population size cannot usually be very large in forest tree studies. In the present study, the population consisted of only 120 progeny. MAS is usually more effective when the distance between markers and QTLs is small (Laurence et al., 1998). Thus, it is important to note that putative QTLs detected in a small sample population and close to a marker sometimes show the same effect as a putative QTL detected in a large population and located at a longer distance from a marker (Lander et al., 1989). Nevertheless, small population size decreases the overall sensitivity of detecting QTLs. Consequently, only major QTLs are usually detected, and minor ones often remain unrevealed. In our study, several markers tightly linked to major QTLs were found. Detection of these markers should provide an opportunity to select promising genotypes at early seedling stages through MAS. However, further studies are necessary to confirm the effects and ability of these QTLs in different environments and years before we can recommend their sustained usage in the *P. tomentosa* breeding program.

毛新杨×毛白杨叶片表型和春季萌芽时间 QTL 分析

摘　要　以毛新杨无性系 TB01×毛白杨无性系 LM50 的回交子代 120 株个体为作图群体，对控制叶片表型性状如叶长、叶宽、叶面积和叶柄长以及春季萌芽时间共 5 个性状的数量性状位点(quantitative trait loci, QTLs) 进行分析。运用 AFLP 标记技术结合拟测交作图策略构建了含有 393 个 AFLP 标记的毛白杨及毛新杨的遗传连锁框架图。毛白杨遗传图上共含有 247 个 AFLP 标记位点，连锁位点覆盖毛白杨基因组总长约 326.11cM，而毛新杨遗传连锁图上共含有 146 个标记位点，连锁位点覆盖毛新杨基因组总长约 1992cM，这些连锁图的每一连锁群上含有的标记数为 4~30 个。在此基础上，利用区间作图软件共检测到控制叶片表型性状的 QTLs 14 个，位于 9 个连锁群上；而对于春季萌芽时间共检测到 3 个 QTLs，分别位于 3 个不同的连锁群上。在检测到的 17 个 QTLs 中，每一 QTL 可解释表型变异的 7.6%~5.8%。此外，发现控制叶长、叶宽和叶面积等相关性状的 QTLs 位于相同的基因组区域，这些 QTLs 主要位于毛白杨遗传连锁图的 TLG2 和 TLG11 以及毛新杨连锁图的 TBLG1 连锁群上。据控制叶片表型和春季萌芽时间的 QTLs 所处的基因组区域，可推测叶片表型和春季萌芽时间这两类性状是由各自相应的基因控制。

关键词　毛新杨×毛白杨，数量性状定位，叶片表型，春季萌芽时间

我国青杨派杨树基因资源及其遗传育种研究进展*

摘　要　青杨派是我国杨属中最大的一个派，共34个种，21个变种和4个变型，分布最为广泛。本文介绍了我国青杨派杨树基因资源概况，综述了青杨派杨树遗传育种研究现状，展望了青杨派杨树基因资源开发和利用前景。

关键词　青杨派，基因资源，遗传育种

基因资源(gene resources)是林木育种的物质基础，是决定育种效果的关键。育种工作成效的大小，很大程度上取决于所掌握的基因资源以及有科学依据地选择利用这些资源(王明庥，2000)。遗传育种发展史也证明，育种工作的重大突破，往往是由于新的基因资源的发现和有效利用才得以实现的。

我国拥有十分丰富的杨树基因资源，共56个种，50个变种及类型，特产35种(徐纬英，1998；王战和方振富，1984)。充分挖掘、开发和利用这些杨树基因资源，对我国杨树新品种的培育具有重要意义。为此，本文介绍了我国杨属最大派——青杨派基因资源概况，综述了其遗传育种研究现状，在此基础上，展望了青杨派杨树基因资源开发和利用前景。

1　我国青杨派杨树基因资源概况

青杨派(Section *Tacamahaca* Spach.)是我国杨属(*Populus* L.)中最大的一个派，共34个种，21个变种，4个变型，分布最为广泛，几乎遍布全国各地(徐纬英，1998；王战和方振富，1984)。但有两个富集区，一是在青藏高原周边山区，另一个是在东北地区(马常耕，1994)。其主要种有小叶杨(*P. simonii* Carr.)、大青杨(*P. ussuriensis* Kom.)、甜杨(*P. suaveolens* Fisch.)、苦杨(*P. laurifolia* Ledeb.)、香杨(*P. koreana* Rehd.)、辽杨(*P. maximowiczii* Henry)等。特产种有小青杨(*P. pseudo-simionii* Kitag.)、青杨(*P. cathayana* Rehd.)、滇杨(*P. yunnanensis* Dode.)、青甘杨(*P. przewalskii* Maxim.)、冬瓜杨(*P. purdomii* Rehd.)、青毛杨(*P. shanxiensis* C. Wang et Tung)、昌都杨(*P. qamdoensis* C. Wang et Tung)、五瓣杨(*P. yuana* C. Wang et Tung)和川杨(*P. szechuanica* Schneid.)等。青杨派各树种基因资源概况简述于表1。

可见，我国青杨派树种大多分布于地形多变、气候复杂的地区，具有丰富的基因资源。据研究，分布于地形地貌复杂，气候环境多样，冰川活动、地史条件独特的川西高原

* 本文原载《西北林学院学报》，2005，20(2)：124~129，与何承忠、张有慧、冯夏莲和李善文合作发表。

的青杨派树种，具有三大地理类型：喜光、喜温、耐旱的滇杨、小叶杨、三脉杨、方杨（*P. fangiana*）等河谷类型；暖温带到寒温带过渡的长序杨（*P. pseudoglauca*）、乡城杨、瘦叶杨（*P. kangdingensis* var. *lancifolia*）、冬瓜杨、川杨等中间类型；环境低温、干燥的康定杨、西藏杨（*P. kangdingensis* var. *tibetica*）、青海杨（*P. przeusalskii*）等高原宽谷类型（刘军，1997）。该区更具有基因资源丰富、特有基因资源多、水平分布区域性强、垂直分布具有替代性、适应气候环境类型多样、多种区系成分共存、速生性强、古杨树资源多等特点（余树全等，2003）。表1同时也表明，我国对青杨派杨树基因资源的认识和研究还非常有限，大部分资源基本处于野生状态，其中蕴藏的珍稀基因资源有待于进一步发掘、研究、保护和利用。

2 我国青杨派杨树遗传育种研究现状

为了很好地开发和利用我国丰富的青杨派树种资源，研究者运用常规方法和生物技术对小叶杨、青杨、大青杨等树种的基因资源、遗传变异规律及其利用等开展了研究。

小叶杨分布广泛，栽培历史悠久，种内遗传变异十分丰富（王战和方振富，1984）。同时，小叶杨有很强的抗旱、耐盐碱性能，杂交亲和力强，一直被育种学者视为培育新品种的最佳亲本之一（王燕等，2000）。徐纬英等（1960）对小叶杨×滇杨、小叶杨×香脂杨、小叶杨×青杨、山杨（*P. davidiana*）×小叶杨、小叶杨×毛白杨（*P. tomentosa*）杂交组合进行了研究。而小钻杨（*P.* ×*xiaozhuanica*）、泰青杨、二白杨、大关杨（*P. dakauensis*）、赤峰杨（*P.* ‘Chifengensis’）、白城杨（*P.* ‘Baichengensis’）、群众杨（*P.* ‘Popularis’）、合作杨（*P.* ‘Opera’）、小黑杨（*P.* ‘Xiaohei’）、I-69×（大青+小）、73-16、73-9、NL-105、NL-106、NL-121、小×胡等均是以小叶杨为亲本，从其天然杂种或人工杂种中选育出的优良无性系（徐纬英，1988；1960；赵天锡和陈章水，1994；徐纬英和董永昌，1984；张玉波等，2002；张永城等，1990；王明庥等，1991；董天慈，1990）。庞金宣等（2001）从山海关杨（*P. deltoids* ‘Shanhaiguan’）×塔形小叶杨组合中，选出窄冠黑青杨6、31、69和70号。选育的这些优良品种为我国杨树造林良种化做出了重大贡献。在当今开展杨树抗性育种工作中，小叶杨因其优良的特性，已成为重要的育种资源。为了更好地研究和开发利用小叶杨基因资源，我国在科尔沁沙地建立了小叶杨基因库（王燕等，2000）。

青杨是我国特产的杨树种，由于分布广，且易产生天然杂种，因此，种内变异很大，是我国杨树育种极好的遗传资源（杨自湘等，1996a）。杨自湘等（1995）对38个试验点采集的500份样品的16个叶片特征值进行遗传变异研究表明，区分产地间叶形差异有94.2%的准确性，而区分产地内单株间叶形差异有63.3%的准确性。不同产地间青杨木材基本密度、纤维长和导管长的差异显著，纤维宽差异不显著，各产地内单株间上述4个木材性状差异不显著；木材基本密度的地理变异随纬度的增高而增加，产地间差异大于产地内差异；纤维、导管性状与产地纬度呈微弱负相关；纤维长与导管长呈极显著正相关（杨自湘等，1996a）。不同产地青杨抗寒性表现为产地内单株间有差异，产地间抗寒能力与纬度变化呈显著正相关（杨自湘等，1996b）。青杨与美洲黑杨（*P. deltoids*）杂交的 F_1 代叶片和生长性状变异研究表明，父本青杨叶片形态在种源间和种源内均存在显著差异，F_1 代叶片和

表 1　中国青杨派杨树基因资源

树　种	生物学及生态学习性	分　布	用　途	变　种　及　变　型
小叶杨 *P. simonii*	适应性强，对气候和土壤要求不严，喜光、耐旱、抗寒，耐瘠薄或轻碱性土壤，在砂荒和黄土沟谷亦能生长，根系发达，抗风力强，生长较快	东北、华北、华中、西北及西南各省（区）均产，多生在海拔 2000m 以下，最高可达 2500m	护堤固土、绿化观赏、北方防护林和用材林的主要树种之一，供建筑、家具、火柴杆和造纸等用	宽叶小叶杨（*P. s.* var. *latifolia*）、辽东小叶杨（*P. s. liaotungensis*）、圆叶小叶杨（*P. s.* var. *rotundifolia*）、秦岭小叶杨（*P. s.* var. *tsinlingensis*）、菱叶小叶杨（*P. s.* var. *rhombifolia*）5 个变种，扎鲁小叶杨（*P. s.* f. *robusta*）、垂枝小叶杨（*P. s.* f. *pendula*）、塔形小叶杨（*P. s.* f. *fastigiata*）3 个变型
青甘杨 *P. przewalskii*		主产于青海、甘肃、内蒙古一带，多生于山麓、溪流沿岸或路边	同小叶杨	
康定杨 *P. kangdingensis*	与小叶杨近似	产于四川康定一带，生于海拔 3500m 高原河边草地		
小青杨 *P. pseudo-simionii*		产于黑龙江、吉林、辽宁、河北、陕西、山西、内蒙古、甘肃、青海及四川等省，生于海拔 2300m 以下的山坡、山沟和河流两岸	经济价值同小叶杨，木材材质较软，可作一般建筑用材	
哈青杨 *P. charbinensis*	抗寒、耐干旱、抗盐碱、抗病虫害、材质好、生长快	产于黑龙江省哈尔滨市	庭园观赏和行道树树种	厚皮哈青杨（*P. c.* var. *pachydermis*）
青　杨 *P. cathayana*	喜湿润或寒冷气候	青藏高原，其中青海最为集中，西达柴达木盆地，南界澜沧江，北界祁连山，东至甘肃南部，华北、辽宁、直到四川省都有分布，常见于 2000~2800m	家具、箱板及建筑用材，为四旁绿化及防护林树种	有两变种，宽叶青杨（*P. c.* var. *latifolia*）和长果柄青杨（*P. c.* var. *pedicellata*）
冬瓜杨 *P. purdomii*		为我国特产，产于河北、河南、陕西、甘肃、湖北及四川，生于海拔 700~2600m 间的山地或沿溪两旁	可供建筑及造纸	光皮冬瓜杨（*P. p.* var. *rockii*）

（续）

树　种	生物学及生态学习性	分　布	用　途	变　种　及　变　型
三脉青杨 *P. trinervis*		产于四川大金、冕宁至九龙一带，多生于海拔 2100~3000m		
东北杨 *P. girinensis*	速生	产于黑龙江哈尔滨市	造林或城镇绿化	楔叶东北杨（*P. g.* var. *ivaschkevitchii*）
大青杨 *P. ussuriensis*	速生、耐寒、喜光、中生偏湿树种，适于微酸性棕色森林土或山地棕壤	产于东北东部山地，分布在海拔 300~1400m	建筑、造船、造纸及火柴杆等用，是东北东部山地森林更新主要树种之一	
玉泉杨 *P. nakaii*		产于黑龙江哈尔滨市以东（玉泉）及河北（东陵）	同大青杨	
甜　杨 *P. suaveolens*		产于内蒙古东部及黑龙江（大兴安岭），多生于河流两岸	经济出材率高（80%），木材可作板材、器具及火柴杆等用	
兴安杨 *P. hsinganica*		产于内蒙古、河北及黑龙江省大兴安岭南端一带，常生于河边		
香　杨 *P. koreana*	早期速生，喜光，耐寒，喜湿	自然分布区在北纬 42°~48°、东经 126°~132°之间，集中分布于小兴安岭到长白山林区，生于海拔 400~1100m 山地缓坡中下部、沟谷或河岸	供建筑、造纸及胶合板、火柴杆用	
辽　杨 *P. maximowiczii*	速生，稍耐荫、耐寒冷，适于溪谷林内肥沃土壤	产于东北、河北、陕西、内蒙古及甘肃等省区，垂直分布多在海拔 500~2000m	供建筑、造纸、造船及制火柴杆等用，也是森林更新的主要树种之一	
黑龙江杨 *P. amurensis*		产于黑龙江省北部，内蒙扎兰屯一带有栽培	同甜杨	
二白杨 *P. gansuensis*	在水肥条件好的土地上生长迅速，抗病虫害能力强	产于甘肃省武威	沙区防护林的优良树种	
苦　杨 *P. laurifolia*		产于新疆阿尔泰和塔城地区	燃料、造纸或小农具等用；叶可做饲料	

（续）

树 种	生物学及生态学习性	分 布	用 途	变 种 及 变 型
柔毛杨 *P. pilosa*		产于新疆阿尔泰地区，分布于1600～2300m		光果柔毛杨（*P. p.* var. *leiocarpa*）
帕米杨 *P. pamirica*		产新疆西昆仑山北坡，生于海拔1800～2000m的山河沿岸		
伊犁杨 *P. iliensis*		产新疆伊梨河谷地带，生于河流两岸		
密叶杨 *P. talassica*	速生，喜光，抗寒	产于新疆天山中部至西部		
梧桐杨 *P. pseudomaximowiczii*		产于河北雾灵山和陕西关山一带，多生于海拔1000～1600m山林中		光果梧桐杨（*P. p.* f. *glabrata*）
青毛杨 *P. shanxiensis*		产于山西省西部吕梁山区黑茶山附近，生于海拔1600m山谷间向阳处		
川 杨 *P. szechuanica*		产于四川、云南、甘肃和陕西	供箱板材、民用建筑或纤维原料等用；可作为行道树种	藏川杨（*P. s.* var. *tibitica*）
滇 杨 *P. yunnanensis*	适于长江以南山区生长，较耐湿热	产于云南、贵州和四川，生于海拔1300～2700m的山地	可供火柴杆及工艺等用材，常栽培为行道树	有两变种，长果柄滇杨（*P. y.* var. *pedicellata*）和小叶滇杨（*P. y.* var. *microphylla*）
昌都杨 *P. qamdoensis*		产于西藏昌都、邦达至怒江一带，多生于海拔3400～3800m的河岸		

（续）

树　种	生物学及生态学习性	分　布	用　途	变种及变型
米林杨 *P. mainlingensis*		产西藏米林、林芝一带，生于海拔 3000~3800m 山坡、河边		
德钦杨 *P. haoana*		产于云南德钦，生于海拔 2100~3600m		有三变种，大叶德钦杨（*P. h.* var. *megaphylla*）、小果德钦杨（*P. h.* var. *microcarpa*）和大果德钦杨（*P. h.* var. *macrocarpa*）
乡城杨 *P. xiangchengensis*		产于四川乡城、稻城、平武等地，生于海拔 2050~3900m，见于河岸		
缘毛杨 *P. ciliata*		产于西藏、云南等省区	可供建筑、箱板和纤维等用材，叶可做饲料	有两变种，维西缘毛杨（*P. c.* var. *weixi*）和吉隆缘毛杨（*P. c.* var. *gyirongensis*）
长叶杨 *P. wuana*		产于西藏波密，林缘或溪边常见		
五瓣杨 *P. yuana*		产于云南中甸白地一带，生于海拔 2000m 山地、河流两岸		
亚东杨 *P. yatungensis*		产于西藏亚东至陇子一带，多生于海拔 2400~3600m 的山坡		有两变种，圆齿亚东杨（*P. y.* var. *crenata*）和毛轴亚东杨（*P. y.* var. *trichorachis*）

注：资料主要来源于参考文献（徐纬英，1998；王战和方振富，1984）。

生长性状的种源间、种源内家系间和家系内无性系间变异大；杂交 F_1 代扦插苗 1 年生生长量在种源间、种源内家系间和家系内无性系间差异极显著(李金花等，2002)。F_1 代的生根性状均存在显著或极显著的无性系差异，并具有较强的超亲优势(李火根等，1998)。而对其 F_2 代的基本材性性状遗传变异研究表明，木材基本密度与纤维长受主基因的控制，表现为"质量-数量性状"，其杂种优势明显，且在 F_2 代中继续存在；纤维宽与微纤丝角表现出较为典型的微效多基因控制的数量性状遗传特点；纤维宽与基本密度、纤维宽与纤维长宽比均为显著负相关；基本密度与纤维长和微纤丝角均为弱负相关，与纤维长宽比为弱正相关；纤维长和纤维宽、纤维长和纤维长宽比显著相关；而微纤丝角与纤维长、纤维宽和纤维长宽比之间不相关(黄秦军和苏晓华，2003)。同时，运用现代生物技术手段，进行了美洲黑杨×青杨 F_2 代木材性状的 QTLs 定位研究(黄秦军等，2004)，构建了美洲黑杨×青杨的分子连锁图谱(苏晓华等，1998)，得到了青杨+胡杨(*P. euphratica*)体细胞融合杂种愈伤组织(诸葛强等，2000)，获得了青杨转双价抗虫基因植株(王永芳等，2003)，mtlD/gutD 双价基因也成功地整合到了美洲黑杨×青杨基因组中并得到表达(樊军锋等，2002)。

大青杨作为我国东北林区特有的乡土树种，木材轻软，材质洁白，致密，耐朽力强，是造纸及胶合板材的极好原料(徐纬英，1998；王战和方振富，1984)。为了保存、利用和发展大青杨这一乡土树种，中国林科院运用常规手段对大青杨的生长、物候、抗锈病和材性等性状进行了遗传分析，利用生物技术在 DNA 分子水平上研究了大青杨天然群体的遗传多样性。结果表明，不同产地的杨树封顶期和落叶期相差很大，生长量差异极为显著，抗锈病能力差异较大，群体间及群体内个体间基本材性差异均显著；从分子水平来看，大青杨群体间变异较大，占总变异的 62. 3%，群体内变异小，只占总变异的 37. 7%(苏晓华等，2001)。9 个不同产地的大青杨抗寒性存在明显差异，随纬度升高抗寒力增强；同一产地不同单株间的抗寒力差异显著；处于同一气候区内不同产地的大青杨抗寒力差异不显著(张绮纹和苏晓华，1999)。在群体变异研究的基础上，对大青杨最佳群体进行了选择研究，从 9 个不同地理产区选择出塔河大青杨为最佳群体(张绮纹等，1993)。运用 RAPD 标记对大青杨及其近缘种的遗传变异和系统关系研究表明，马氏杨(辽杨)在系统发育中出现最早，甜杨次之，香杨再次之，大青杨最晚；大青杨与香杨亲缘关系最近(苏晓华等，2001)。

随着对青杨派杨树基因资源认识的提高，研究树种也不断扩大。北京林业大学毛白杨研究所已从耐极端低温的甜杨中诱导出与抗冻性相关的特异蛋白，开始分离克隆其相关抗冻基因(林善枝等，2000)。过去不被重视的低纬度高海拔山区的滇杨，现已引起国内杨树育种学家的注意(马常耕，1994)。为改良滇杨的适应性、材质和生长速率，李启任等(1994)用美洲黑杨优良无性系与其开展杂交，对亲本和杂种的同工过氧化物酶进行了比较研究。针对滇杨现状，何承忠等(2004)对滇杨的遗传改良策略进行了探讨。

综合上述，我国对青杨派小叶杨、青杨、大青杨等代表性树种开展了深入细致的研究，取得了较大成就。从研究手段看，有常规育种和新技术育种；从研究内容看，有遗传变异规律和杂交育种，涵盖了生长、材性、抗性等性状的遗传变异规律及天然杂种选择、人工控制授粉杂交育种、体细胞杂交、遗传连锁图谱构建、主要性状 QTLs 定位等；从育种目标看，有速生、抗旱、抗寒、抗病虫、窄冠等。这些研究为进一步开展小叶杨、青杨、大青杨等树种的育种工作奠定了坚实的基础，也为青杨派其他树种基因资源的开发利用提供了科学理论依据。

3 我国青杨派基因资源利用的问题与展望

我国集中了绝大多数世界青杨派杨树(马常耕，1995)，也包含有极为稀有而珍贵的杨树资源，如甜杨，耐寒，从我国大兴安岭一直可分布到北纬72°，在-69℃的低温下仍可正常生长；康定杨分布在海拔3500m的地段；滇杨，喜湿、耐高温、速生，在一般情况下，不加特殊措施，20年生胸径可达60cm（徐纬英，1988）。然而，育种潜力很大的青杨派树种资源目前多数处于野生状态，基因资源保存力度不够；研究树种较少，某些优良基因资源未能被发现和利用；育种策略简单，无长期计划，未充分利用种源、家系、无性系等水平存在的丰富遗传变异。这些因素限制了我国杨树育种的发展，也与我国作为杨树资源大国的地位极不相称。

为此，首先要加大对青杨派杨树基因资源的搜集、研究与保护，为杨树育种储备材料和提供遗传信息。为防止乡土杨树种质资源丢失，应分育种区建立育种资源保存群体，所保存群体要有代表性，不但包括生长快、材质优的个体，也要包括抗病、抗虫、抗逆境的个体，以满足不同育种目标的需求。其次，要重视多树种、多性状、多水平遗传改良。我国青杨派树种居世界之最，在我国从南到北，从西到东均有分布，分布区地形复杂，气候条件各异，蕴藏了丰富的乡土树种资源。但是对青杨派乡土树种研究力度不够，也不系统，如作为我国特有的亚热带温暖南方型杨树种的滇杨，是我国重要的杨树速生资源，对其研究还极为有限；而对培育适应高山区生态条件的杨树新品种极为重要的川杨，研究几乎为空白；材质优良、耐寒、耐荫的辽杨(马氏杨)，国内对其分类归属尚存在争议，但国外已经从生长、遗传变异、遗传改良和抗逆性等方面开展了大量的研究工作(Honma et al.，2002；Rajora，1988；Mohanty and Khurana，2002；Shen et al.，1999；Pell et al.，1999；Riedel，1993)。因此，扩宽研究树种，挖掘优良的自然资源，开展多种乡土杨树的遗传改良，以适应不同地区的生产需求，显得极为重要。另外，由于青杨派的某些种易遭受病虫危害，限制了其推广与应用，可利用黑杨派中美洲黑杨抗病虫性强的特点，进行派间杂交，以提高其抗性(李善文等，2004)；研究目标应以抗逆性优先，在此基础上追求速生。对青杨、大青杨等乡土树种的研究表明，种源间及种源内个体间在生长、材性等方面均存在显著差异(杨自湘等，1996；苏晓华等，2001)。因此，对青杨派乡土杨树遗传改良，应充分种用种源、家系、无性系三个层次存在的丰富遗传变异，以提高育种成效。再次，要制定长、中、短期相结合的总体改良计划，以交互轮回选择理论为指导，把亲本改良作为提高杂种优势的根本，对育种亲本有计划地改良，使短期育种工作随着世代的推进而持续高效地螺旋式上升。最后，要将传统研究方法与高新技术有机地结合，高效、准确地揭示各树种的遗传变异规律，为育种亲本的选配及优良无性系的选择提供科学依据，使我国杨树育种更有预见性、高效性和目的性。

总之，随着杨树遗传育种研究的深入和现代生物技术的发展，杨树杂交育种策略从随机杂交育种修改为从选择到杂交再到选择的模式，将群体间和群体内变异用于亲本选择

(马常耕, 1994); 杂交方式也被改进, 由简单到复杂, 从单交到复交(三交、双交), 配合应用回交; 研究水平由表型到细胞、蛋白质、DNA 等。毫无疑问, 随着现代高新技术与常规研究手段的有机结合, 对我国青杨派杨树基因资源研究力度的加大, 优良基因资源不断被发掘, 育种资源将不断丰富, 杨树育种"瓶颈"之一的基因资源贫乏将得以缓解, 我国杨树育种工作将迈上新的台阶。因此, 我国青杨派杨树基因资源的研究和开发利用具有广阔的前景。

Introduction to Gene Resources and Progress in Research on Poplars Genetic Breeding of Section *Tacamahaca* in China

Abstract　Section *Tacamahaca*, including 34 species, 21 variations and 4 forms, is the largest one among the five sections of *Populus* in China. Gene resources of Section *Tacamahaca* are introduced, progress in research on poplars genetic breeding of Section *Tacamahaca* are reviewed, and the prospect of development and utilization of gene resources in Section *Tacamahaca* is discussed in the paper.

Key words　Section *Tacamahaca*, gene resources, genetic breeding

Analysis of the Leaf Characteristics Variation in *Populus tomentosa* Carr. *

Abstract An investigation was conducted to determine the extent of variation among 9 provenances of *Populus tomentosa* using leaf characters. A total of 263 accessions were studied under field conditions in the National Gene Bank of *P. tomentosa* in 2003. All the accessions were characterized for 17 characteristics from 1-dimension constructions to 2-dimension constructions. Analysis of variance for all characteristics showed that there was significantly difference among 9 provenances and among individuals within provenance. This study revealed that the evaluated germplasm appears to have a wide genetic base and highly potentiality for further genetic improvement. And it was also indicated that plenty gene resources of *P. tomentosa* have been collected and reserved in the National Gene Bank.

Key words *Populus tomentosa*, leaf characters, variation

1 Introduction

Estimates of genetic diversity are very useful for facilitating efficient germplasm evaluation, collection and management (Rabbani et al., 1998). Many tools are now available for studying variability and the relationships among accessions including total seed protein, isozymes and various types of molecular markers. However, morphological characterization is the first step in the description and classification of the germplasm (Smith and Smith. 1989; Rabbani et al., 1998). And it is a fast and efficient way to assess genetic diversity by morphological markers (Kafkas et al., 2002). Leaf character is an important morphological characterization and has highly valuation to be analyzed for the close correlation with nutrition, physiology, ecological factors and propagation of plant (Chechowitz et al., 1990).

Chinese white poplar (*Populus tomentosa* Carr.) is one of the indigenous tree species in China and mainly distributes in the middle and lower reaches of the Yellow River, covering more than 10 provinces and occupying about one million km^2. It has many desirable characteristics such as rapid growth rate, strong adaptation to harsh site conditions, super wood quality, tall and straight trunk and handsome tree form (Zhu, 1992; Zhu and Zhang, 1997). However, no systematic study has been carried out so far to investigate the extent of genetic variation of leaf characteristics in *P. tomentosa*. The main objective of this study was to characterize and classify the leaf charac-

* 本文原载 *Forestry Studies in China*, 2005, 7(3): 51-53, 与何承忠、冯夏莲、于志水和张有慧合作发表。

teristics variation among 9 provenances of *P. tomentosa*.

2 Materials and methods

2.1 Plant materials

The tested materials were sampled from the National Gene Bank of *P. tomentosa* in Guanxian County, Shandong Province. All clones, which comprised the National Gene Bank, were selected and collected from nine provinces (municipality) of Shannxi, Shanxi, Henan, Hebei, Shandong, Gansu, Anhui, Jiangsu and Beijing, which one could be named as one provenance. The existing clonal number and sample number for each provenance was listed in table 1.

Table 1 Numbers of existing clones and sampled clones of each provenance in the National Gene Bank of *Populus tomentosa*

Provenance	Numbers of existing clones	Numbers of sampled clones	Provenance	Numbers of existing clones	Numbers of sampled clones
Beijing	22	22	Shannxi	72	30
Hebei	147	73	Gansu	6	6
Shandong	20	20	Anhui	10	10
Henan	109	50	Jiangsu	6	6
Shanxi	95	46			

2.2 Measurement of leaf characteristics

Eight pieces of mature leaves, which were selected from 4 directions in middle crown, were sampled and measured the traits such as leaf blade length (LBL), leaf blade width (LBW), petiole length (PL), length from the widest to leaf base (LWLB), leaf blade middle width (LBMW), leaf blade up quarter width (LBUW), leaf blade down quarter width (LBDW), lateral vein length (LVL), leaf-venation angle (LVA), leaf apex angle (LAA), leaf base angle (LBA) and leaf margin teeth number (LMTN). All the traits can evaluate leaf 1-dimension construction variation. Subsequently, leaf blade length/leaf blade width (LBL/LBW), leaf blade length/petiole length (LBL/PL), length from the widest to leaf base/leaf blade length (LWLB/LBL), leaf blade up quarter width/leaf blade middle width (LBUW/LBMW) and leaf blade down quarter width/leaf blade middle width (LBDW/LBMW) were calculated as the index of leaf 2-demension construction variation.

2.3 Analysis method

Seventeen recorded leaf traits were analysis by numerical taxonomic techniques using variance analysis of nest design. The linear model is

$$X_{ijk} = \mu + P_i + C/P_{j(i)} + e_{ijk}$$

where 'C' refers to numbers of all measured clones, 'μ' to total mean, 'P_i' to the effect value of the i provenance, '$C/P_{j(i)}$' to the effect value of one trait of the j clone within i provenance, and 'e_{ijk}' to sampling error.

Data were analyzed among 9 provenances and among individuals within provenance using SAS soft (version 6. 0).

3 Results

It is indicated from Table 2 that extensive variation existed in leaf characteristics among 9 provenances and among individuals within provenance of *P. tomentosa*. For example, on LBL trait, the highest mean was 10. 3cm from Gansu provenance; the lowest mean was 7. 87cm from Shannxi provenance, and on LBW trait, the highest mean was 8. 67cm from Shandong provenance, the lowest mean was 6. 68cm from Shannxi provenance. Of 17 leaf characteristics, means of 7 leaf traits were the lowest value from Shannxi provenance, but other the lowest means dispersed in the other 8 provenances. The results indicated that the leaf characteristics variation was poor within Shannxi provenance. Furthermore, analysis of variation for all characteristics showed that there was significantly difference among 9 provenances and among individuals within provenances (Table 3). The results showed that there were plenty genetic diversity on leaf characteristics in *P. tomentosa*.

Table 2 Means and standard deviation of leaf traits among 9 provenances of *Populus tomentosa*

Provenance	LBL(cm)	LBW(cm)	PL(cm)	LWLB(cm)	LBMW(cm)	LBUW(cm)
Beijing	8. 82±1. 2	7. 35±1. 2	6. 38±1. 2	2. 93±0. 7	6. 58±1. 0	3. 75±0. 7
Hebei	9. 64±1. 6	8. 36±1. 6	6. 63±1. 2	3. 12±0. 8	7. 43±1. 4	4. 15±1. 0
Shandong	9. 68±1. 3	8. 67±1. 5	7. 35±1. 3	3. 55±0. 8	7. 87±1. 4	4. 88±0. 9
Henan	8. 77±1. 6	7. 60±1. 5	6. 40±1. 1	3. 04±0. 6	6. 86±1. 3	4. 05±1. 0
Shanxi	9. 70±1. 6	8. 42±1. 6	6. 99±1. 2	3. 17±0. 8	7. 46±1. 4	4. 26±1. 0
Shannxi	7. 87±1. 2	6. 68±1. 0	5. 72±1. 2	2. 74±0. 6	5. 95±1. 0	3. 44±0. 8
Gansu	10. 3±1. 5	8. 60±1. 5	6. 95±0. 9	3. 34±0. 8	7. 58±1. 4	4. 43±1. 3
Anhui	8. 89±1. 3	7. 82±1. 5	6. 39±0. 8	3. 08±0. 8	7. 05±1. 4	4. 22±1. 1
Jiangsu	8. 44±0. 9	7. 30±0. 8	6. 16±1. 0	2. 63±0. 5	6. 58±0. 7	3. 90±0. 5
Beijing	6. 99±1. 1	4. 71±0. 9	38. 95±4. 9	32. 78±6. 4	172. 90±21. 0	19. 65±2. 7
Hebei	8. 0±1. 5	5. 52±1. 1	42. 27±6. 1	33. 01±7. 7	157. 05±24. 7	18. 8±3. 3
Shandong	7. 78±1. 4	5. 69±1. 0	42. 09±5. 7	37. 78±8. 3	144. 5±27. 0	19. 64±3. 7
Henan	7. 23±1. 5	5. 10±1. 1	40. 91±7. 3	35. 55±6. 8	164. 25±27. 6	20. 95±4. 7
Shanxi	8. 04±1. 5	5. 56±1. 2	42. 73±5. 7	34. 21±7. 3	151. 39±21. 5	18. 36±2. 8
Shannxi	6. 35±1. 0	4. 59±0. 8	41. 31±5. 8	34. 17±9. 2	159. 17±33. 4	24. 17±5. 1
Gansu	8. 06±1. 5	5. 72±1. 0	42. 71±5. 6	32. 60±6. 0	137. 08±26. 7	18. 5±2. 5
Anhui	7. 34±1. 4	5. 0±1. 0	37. 44±7. 3	39. 31±8. 7	152. 13±21. 2	22. 9±5. 4
Jiangsu	6. 86±0. 8	4. 81±0. 7	39. 38±4. 5	32. 92±5. 8	153. 75±16. 8	18. 71±2. 1

（续）

Provenance	LBL/LBW	LBL/PL	LWLB/LBL	LBUW/LBMW	LBDW/LBMW
Beijing	1.21±0.1	1.41±0.2	0.33±0.1	0.57±0.1	1.06±0.1
Hebei	1.16±0.1	1.47±0.2	0.32±0.1	0.56±0.1	1.08±0.1
Shandong	1.13±0.1	1.34±0.2	0.37±0.1	0.63±0.1	1.0±0.1
Henan	1.16±0.1	1.39±0.2	0.35±0.1	0.59±0.1	1.06±0.1
Shanxi	1.16±0.1	1.40±0.2	0.33±0.1	0.57±0.1	1.08±0.1
Shannxi	1.18±0.1	1.41±0.2	0.35±0.1	0.58±0.1	1.07±0.1
Gansu	1.21±0.1	1.49±0.2	0.33±0.1	0.58±0.1	1.07±0.1
Anhui	1.15±0.1	1.40±0.2	0.35±0.1	0.60±0.1	1.05±0.1
Jiangsu	1.16±0.1	1.39±0.2	0.31±0.05	0.59±0.1	1.04±0.1

Note: LBL: leaf blade length, LBW: leaf blade width, PL: petiole length, LWLB: length from the widest to leaf base, LBMW: leaf blade middle width, LBUW: leaf blade up quarter width, LBDW: leaf blade down quarter width, LVL: lateral vein length, LVA: leaf-venation angle, LAA: leaf apex angle, LBA: leaf base angle, LMTN: leaf margin teeth number. LBL/LBW: leaf blade length/leaf blade width, LBL/PL: leaf blade length/petiole length, LWLB/LBL: length from the widest to leaf base/leaf blade length, LBUW/LBMW: leaf blade up quarter width/leaf blade middle width, LBDW/LBMW: leaf blade down quarter width/leaf blade middle width.

Table 3 Variance analysis of leaf characteristics among/within provenances in *Populus tomentosa*

Leaf characteristics	MS (df)			F value	
	Among provenances	Within provenances	Error variances	Among provenances	Within provenances
LBL(cm)	107.25(8)	13.78 (254)	0.55 (1841)	7.78**	25.05**
LBW(cm)	100.27(8)	12.61(254)	0.62 (1841)	7.95**	20.24**
PL(cm)	46.42(8)	6.75(254)	0.63 (1841)	6.87**	10.71**
LWLB(cm)	10.55(8)	1.69 (254)	0.35 (1841)	6.22**	4.78**
LBMW(cm)	77.75(8)	10.02(254)	0.56 (1841)	7.76**	17.89**
LBUW(cm)	30.47(8)	4.61(254)	0.38 (1841)	6.61**	12.17**
LBDW(cm)	86.32(8)	11.68 (254)	0.62 (1841)	7.39**	18.95**
LVL(cm)	37.67(8)	4.99 (254)	0.54(1841)	7.56**	9.20**
LVA(°)	481.69 (8)	119.001(254)	26.38 (1841)	4.05**	4.51**
LAA(°)	770.34(8)	170.54(254)	41.47(1841)	4.52**	4.11**
LBA(°)	15558.0(8)	2609.79 (254)	384.62(1841)	5.96**	6.79**
LMTN	904.83(8)	74.83(254)	6.097(1841)	12.09**	12.27**
LBL/LBW	0.0899 (8)	0.029 (254)	0.0085(1841)	3.07**	3.44**
LBL/PL	0.47(8)	0.16(254)	0.03(1841)	2.89**	5.69**
LWLB/LBL	0.0505(8)	0.0066(254)	0.0034(1841)	7.65**	1.94**
LBUW/LBMW	0.095(8)	0.017 (254)	0.0049 (1841)	5.48**	3.54**
LBDW/LBMW	0.130(8)	0.0156(254)	0.0057 (1841)	8.36**	2.74**

Note: ① ** indicates significant difference at the confidence level of 0.99. ②LBL: leaf blade length, LBW: leaf blade width, PL: petiole length, LWLB: length from the widest to leaf base, LBMW: leaf blade middle width, LBUW: leaf blade up quarter width, LBDW: leaf blade down quarter width, LVL: lateral vein length, LVA: leaf-venation angle, LAA: leaf apex angle, LBA: leaf base angle, LMTN: leaf margin teeth number. LBL/LBW: leaf blade length/leaf blade width, LBL/PL: leaf blade length/petiole length, LWLB/LBL: length from the widest to leaf base/leaf blade length, LBUW/LBMW: leaf blade up quarter width/leaf blade middle width, LBDW/LBMW: leaf blade down quarter width/leaf blade middle width.

4 Discussion

P. tomentosa is considered as the best among the native poplar species in China, and has played a key role in timber production and ecological environment construction along the Yellow River (Zhu and Zhang, 1997). In order to ensure the efficient and effective utilization of the super gene resource of *P. tomentosa*, its characterization is imperative. In the present investigations of leaf traits in *P. tomentosa*, all of 17 traits exhibited extensive genetic variation. Consequently, analysis of variance showed that there was significant variation in all of 17 traits among 9 provenances and among individuals within provenances of *P. tomentosa*. The results indicated that *P. tomentosa* has a wide genetic base and highly potentiality for further genetic improvement. And it was also indicated that plenty gene resources of *P. tomentosa* have been collected and reserved in the National Gene Bank.

Progress and Strategies in Cross Breeding of Poplars in China*

Abstract The advance in intrasection and intersection cross breeding of poplars in China over the past 50 years is reviewed. Great progress has been made in Sections *Leuce* and *Aigeiros*, and satisfactory results of intersection hybridization have been achieved in the crossing between Sections *Tacamahaca* and *Aigeiros*. The modes of hybridization include single cross, double cross, triple cross, backcross, etc. It is known that using hybrids as parents to cross with other species or hybrids is an effective and easy way to obtain heterosis. Fast growth, cold and drought tolerance, pest and disease resistance, narrow crowns and rootage, etc. are breeding goals. The conventional artificial crossing is still a major breeding method, and a combination of the conventional artificial crossing with physical radiation and chemical induction can create new triploid individuals that possess higher yield potential. The super clones cultivated have already displayed enormous socioeconomic and ecological benefits in practice. Finally, the problems that investigators have to face at present are discussed as well as some strategies in poplar cross breeding in China.

Keywords poplars, cross breeding, summary

1 Introduction

Poplars are extensively planted in the world because of their fast growth, good adaptability and wide uses of wood. There are about 6.67 million hm^2 of poplar plantations in China, accounting for 20% of the total timber plantations of the country and four times as large as the total area of poplar plantations in other countries. Poplars play an important role in the ecological projects and industrial timber use in China. The improved cultivars of poplars extensively planted for afforestation at present in China are almost cultivated by artificial cross breeding, resulting in patent heterosis with characteristics of fast growth, high-quality wood, resistance of diseases and insects and tolerance of environmental stress. These cultivars have brought about enormous economic, ecological and social benefits. Cross breeding will still be an important means to produce new cultivars of poplars for a long time.

In China, cross breeding of poplars dates back 50 years. It was originally conducted by Ye (1955) in Tianshui, west China's Gansu Province. From then on, thanks to the rich germplasm

* 本文原载 *Forestry Studies in China*, 2005, 7(3): 54-60, 与李善文、罗军民、何承忠、普映山和安新民合作发表。

resources of poplars, many scientists in this field have been engaged in intrasection and intersection cross breeding of poplars, and obtained a great deal of improved cultivars in China.

2 Germplasm resources of poplars in China

The genus *Populus* includes five sections: *Leuce* Duby, *Tacamahaca* Spach., *Leucoides* Spach, *Aigeiros* Duby and *Turanga* Bge., with a total number of over 100 species in the world. In China alone, there are 53 species in all the five sections, including 35 endemic species mostly distributed in southwest, northwest, northeast and north China between 25°~53°N and 80°~134°E (Xu, 1988).

2.1 Section *Leuce* Duby

Section *Leuce* includes 9 species and 7 varieties, such as *Populus alba* L., *P. canescens* Smith, *P. tomentosa* Carr., *P. davidiana* Dode, *P. hopeiensis* Hu *et* Chou, *P. adenopoda* Maxim, *P. bolleana* Lauche.

2.2 Section *Tacamahaca* Spach.

As the largest of the five poplar sections in China, Section *Tacamahaca* consists of 34 species and 21 varieties. The main species are *P. simonii* Carr., *P. ussuriensis* Kom., *P. suaveolens* Fisch., *P. laurifolia* Ledeb, *P. maximowiczii* Henry, *P. koreana* Rehd. etc., The endemic species are *P. pseudo-simonii* Kitag, *P. cathayana* Rehd, *P. yunnanensis* Dode., *P. przewalskii* Maxim, *P. purdomii* Rehd, *P. shanxiensis* C. Wang *et* Tung, *P. qamdoensis* C. Wang et Tung, *P. yuara* C. Wang et Tung, *P. szechuanica* Schneid.

2.3 Section *Leucoides* Spach.

Four species and three forms are distributed naturally in China in Section *Leucoides*, such as *P. lasiocarpa* Oliv., *P. wilsonii* Schneid., *P. glauca* Haines., *P. psudoglauca* C. Wang *et* Fu.

2.4 Section *Aigeiros* Duby

Section *Aigeiros* is comprised of three species and two varieties like *P. nigra* L., *P. jrtyschensis* CH. Y. Yang, *P. afghanica* Schneid., *P. nigra* var. *italica* Koehne. and *P. nigra* var. *thevestina* Bean.

2.5 Section *Turanga* Bge.

Section *Turanga* includes only two species in China, i. e. *P. euphratica* Oliv. and *P. pruinosa* Schrenk. As listed above, China is abundant in natural resources of poplars, which make up more than 50% of the total poplar species in the world. Many are endemic in China, especially those in Sections *Leuce* and *Tacamahaca*. All these are advantageous for intrasection and intersection cross breeding of poplars in China.

3 Intrasection cross breeding

3.1 Section *Leuce* Duby

Great achievements have been made concerning the interspecies cross breeding in Section *Leuce*. Ye (1955) studied initially intrasection cross breeding in Section *Leuce*, matching cross combinations of *P. hopeiensis* ×*P. tomentosa* and *P. hopeiensis* ×*P. davidiana*. Since then, similar studies have been continued and *P. alba* ×*P. tomentosa* and *P.* ×'nanlin' ((*P. hopeiensis* × *P. tomentosa*)×*P. adenopoda*) have been selected out. *P. tomentosa* ×*P. bolleana*, *P. alba* ×*P. davidiana*, and *P. davidiana* ×*P. bolleana* were screened out. By using *P. alba as* female parent and *P. hopeiensis* ×*P. davidiana* as male parent, the elite hybrid of *P. alba* ×(*P. hopeiensis* ×*P. davidiana*) was bred by Wang et al. (1987). As a result, the new variety had 9.8% more volume than *P. hopeiensis*, and was noted for its good resistance to pests and diseases, easiness to root and over 90% of cutting survival rate. *P. hopeiensis* ×*P. tomentosa* clone 1, whose 10-year-old volume was 141% as much as that of *P. hopeiensis*, was selected by Qiu and Weng (1991). Hybrid clone 1333, which grew twice as fast as *P. davidiana* when it was 14 years old, was screened out from the combination of *P. alba* ×*P. davidiana* (Liu and Zhao 1991). *P. davidiana*×(*P. alba* ×*P. davidiana*) hybrid clone 1132, characterized by rapid growth and tolerance to cold and diseases, was also bred by them. *P. leucopyramidalis*-1 and *P. leucopyramidalis*-5 were selected out from *P.* ×'nanlin'×(*P. tomentosa* ×*P. bolleana*), *P. leucopyramidalis*-3 and *P. leucopyramidalis*-4 from *P. adenopoda*×(*P. tomentosa* ×*P. bolleana*) and *P. leucopyramidalis*-6 from (*P. tomentosa* ×*P. bolleana*)×*P. adenopoda*. All of the *P. leucopyramidalis* clones were commonly characterized by their narrow crowns, deep roots, fast growth and high-quality lumber (Pang et al., 2001). Researchers at Beijing Forestry University have studied the crossability and cross modes of interspecies and interhybrids in Section *Leuce*. The results showed that crossability of hybrids had outstanding advantages and heterosis of the progenies was obvious. It was also indicated that using hybrids as parents to cross with other species or hybrids was an effective means to overcome crossing incompatibility. Meanwhile, the best combinations of (*P. tomentosa* × *P. bolleana*)×*P. canescens* and (*P. tomentosa*×*P. bolleana*)×*P. tomentosa* were selected (Li and Zhu, 1989). By using the combination of (*P. tomentosa*×*P. bolleana*)×*P. tomentosa* and treating pollen with colchicines and γ-ray, six elite clones of triploid *P. tomentosa*, which had a fast growth at early stage and excellent wood quality, were bred (Zhu et al., 1995). (*P. tomentosa* ×*P. bolleana*) × (*P. alba* ×*P. glandulosa*), (*P. alba* ×*P. tomentosa*) × (*P. alba* × *P. glandulosa*) and (*P. alba* ×*P. glandulosa*)×*P. tomentosa* 'Lumao 50' were superior cross combinations, which had higher growth and larger leaves (Li, 2004).

The modes of hybridization include single cross, double cross, three cross, backcross, etc. It is known that using hybrids as parents to cross with other species or hybrids is an effective and easy way to obtain heterosis. Advantages like fast growth, cold and drought tolerance, pest and disease resistance, narrow crowns and easy rooting ability are our goals in breeding. As to breeding methods, the traditional artificial crossing is a major method, and the combination of artificial crossing with physical radiation and chemical induction can create new triploid individuals with higher yield potential.

3.2 Section Aigeiros Duby

Interspecific cross breeding in Section *Aigeiros* lagged behind that in Section *Leuce*. *P. deltoides* 'Lux', *P. deltoides* 'Harvard' and *P.* ×*euramericana* 'San Martino' were introduced into China from Italy (Zhao and Chen, 1994), and showed better adaptability and higher productivity in subtropical regions in China. *P. deltoides* 'Lux' was polycrossed with the mixed pollen of *P. deltoides* 'Harvard', *P. deltoides* var. *missouriensis* and *P. deltoides* ssp. *angulata* 'Carolin', thus developing *P.* ×'shanlin-3' clone, whose volume of 5-year-old single tree exceeded that of *P. deltoids* 'Lux' by 24% with better cold tolerance, drought hardiness and insect and disease resistance than *P. deltoids* 'Lux' (Fu et al., 1990). The elite clones of *P.* × *euramericana* 'zhonglin-46' and *P.* ×*euramericana* 'zhonglin-23' were selected from the combination of *P. deltoides* 'Lux'×*P. nigra*, whose volume was 30% more than their parents (Huang et al., 1991). Two elite clones, *P.* ×'Nankang-1' and *P.* ×'Nankang-2' with resistance to *Anoplophora glabripennis* and *Batocera horsfieldi*, whose single tree volume was 40% more than their parents, were screened out from the combination of *P. deltoides* 'Lux'×*P. deltoides* 'Harvard' (Han et al., 1991). Three elite clones, *Populus* ×*liaoningensis* (*P. deltoides* 'Lux'×*P. deltoides* 'Shanhaiguan'), *P.* ×*liaohenica* (*P. deltoides* 'Shanhaiguan'×*P. deltoides* 'Harvard') and *P.* ×*gaixianensis* (*P.* ×*euramericana* 'San Martino'×*P. deltoides* 'Shanhaiguan'), which were known for their fast growth and good resistance to *Dothiorella gregaria* Sacc. And *Coryneum populinum* Bres. with diameters at breast-height 20% larger than *P.* ×*euramericana* 'sacrau-79', were selected out by Chen et al. (1995). *P.* ×'Langfang-1' and *P.* ×'Langfang-2' with obvious heterosis, were screened out from the combinations of *P. deltoides* 'Shanhaiguan'×(*P. pyramidalis* Bork. + *P. deltoides* 'Harvard') and *P. deltoides* 'Lux'×*P. deltoids* 'Shanhaiguan' (Liu et al., 1998). *P. deltoids* 'Lux'×*P. deltoides* 'S3244' and *P. deltoides* 'Lux' ×*P. deltoides* 'Harvard' proved to be superior cross combinations (Li et al., 2003).

As shown above, although there are less gene resources of Section *Aigeiros* in China, a passel of improved hybrid clones of *P.* ×*euramericana* and *P. deltoides* have successfully been bred by Chinese researchers using alien elite clones as parents. Hybridization in Section *Aigeiros* gave priority to the improvement of *P. deltoides*. The genetic basis of the improved clones was relatively narrow because *P. deltoides* 'Lux' was mostly used as female parents.

3.3 Section *Tacamahaca* Spach.

There is much less literature about interspecific crossing in Section *Tacamahaca*. Xu et al.

(1966) studied the combinations of *P. simonii* ×*P. yunnanensis*, *P. simonii* ×*P. koreana*, *P. simonii* ×*P. cathayana*, *P. cathayana* ×*P. laurifolia* and *P. pseudo-simonii* ×*P. yunnanensis*. The reasons of few reports on interspecific crossing lie probably in two aspects. One is the indistinct heterosis in Section *Tacamahaca*. The other is the difficulty to get mature seeds of hybrids in Section *Tacamahaca* because of the long growing period of seeds and easy falloff of capsules owing to insufficient nutrition after cutting-branch water culture.

Besides, many species in Section *Leucoides* are still in a wild state because of their vast distribution in temperate zones of the central and sub-alpine areas of north subtropics in China. The study on interspecific hybridization is not reported in Section *Turanga*.

4 Intersection cross breeding

4.1 Cross breeding between Section *Aigeiros* and Section *Tacamahaca*

It was easy to obtain natural hybrids by matching Section *Aigeiros* with *Tacamahaca* under natural condition and the heterosis was distinct. The hybrids naturally distributed or extensively planted in China are *P. gansuensis* (*P. nigra* var. *thevestina* ×*P. simonii*), *P.* ×*xiaozuanica* (*P. simonii* ×*P. nigra* var. *italica*), etc. In addition, *P.* 'Chifengensis', *P. dakuanensis*, *P.* 'Baichengensis', *P.* ×*xiaozuanica* 'Balizhuangyang', etc. are clones selected from the natural hybrids of *P. simonii* ×*P. nigra* var. *italica*, and characterized by their fast growth as well as tolerance to cold, drought and saline-alkali conditions.

Many elite clones have been obtained from combinations made by Chinese scientists among species in Section *Aigeiros* and *Tacamahaca* (Table 1). *P. deltoids* (such as *P. deltoides* 'Lux' and *P. deltoids* 'Shanhaiguan') and *P. nigra* (such as *P. nigra* var. *italica* and *P. nigra* var. *thevestina*) in Section *Aigeiros* are featured by their fast growth, straight trunk and good resistance to insects and diseases, while *P. simonii*, *P. cathayana* and *P. pseudo-simonii* in Section *Tacamahaca* have shown good tolerance to cold and drought and easiness to root. The above parental species in the two sections were selected to breed clone series that have good characteristics of their parents, and the hybrids such as *P. simonii* ×*P. nigra* var. *italica* and *P. simonii* ×*P. nigra* were used to backcross with their parents to create new varieties as well. Owing to their traits of fast growth, tolerance to drought, cold and barrenness, the improved varieties have mainly been planted in cold and arid regions in north China, or high-altitude areas in south China.

4.2 Cross breeding between section *Aigeiros* and section *Leuce*

Considering the traits of fast growth and easiness to root in Section *Aigeiros* and the features of top-quality wood and wide adaptability in Section *Leuce*, many scholars have studied the cross breeding between the two Section to create new varieties that share all the good characteristics of their parents. By using *P.* ×*euramericana* 'Sacrau-79' as female parent and *P. tomentosa* as male parent, *P.* 'Sacrau-79'×*tomentosa* 'Sa-menica' was bred (Wu et al., 1996). Narrow-crowned clones were selected from the combination of *P. deltoides* 'Lux'×(*P. adenopoda* ×(*P. tomentosa*×*P. bolleana*)) (Pang et al., 2001). Zhang et al. (1990) studied 43 combinations in the two Sections, and the results showed that hybrids were produced from 15% of triple-cross and 56% of double-cross combinations. The result indicated that it was easy to produce hybrids of the two Sections by using the hybrids such as *P. alba* ×*P. glandulosa* and *P. pyramidalis* ×*P. cathayana* 'Beijingensis' as parents, and the hybrids had a higher cutting survival rate. The cutting survival rate of the hybrid produced from (*P. alba* ×*P. glandulosa*)×*P. eur.* 'Zhonglin 13', for example, was over 83%.

Table 1 Crossing breeding between Section *Aigeiros* and Section *Tacamahaca*

Cross combination	Goals of breeding	Developed variety	Reference
P. nigra var. *italica* × *P. cathayana*	Fast growth, tolerance to cold, and resistance to diseases	*P.* ×'beijingensis'	Xu et al. (1966, 1988)
P. simonii × *P. nigra* var. *italica*	Fast growth, drought tolerance and disease resistance	*P.* ×'Opera'	Xu et al. (1966, 1988)
P. simonii × *P. nigra*	Fast growth, tolerance to cold and drought, and resistance to diseases	*P. simonii* × *P. nigra*	Huang et al. (1991)
P. 'Chifengensis'× (*P.* ×*euramericana*+*P. nigra* var. *italica* + *P. cathayana*)	Fast growth, tolerance to cold, and resistance to insects and diseases	*P.* 'zhaolin-6'	Lu et al. (1985)
P. simonii ×*P. deltoides*	Fast growth, tolerance to cold and drought	*P.* 73-16, *P.* 73-9	Zhang et al. (1990)
P. dakauensis × *P. nigra* var. *italica*	Fast growth, drought tolerance and disease resistance	*P.* 'shanlin-1'	Fu et al. (1984)
P. deltoides 'Lux'×	Fast growth, drought tolerance and disease resistance	*P.* 'shanlin-2'	Fu et al. (1990)
P. cathayana *P. deltoides* 'Lux'× *P. simonii*	Fast growth, drought tolerance	*P.* 'nanlin - 105', *P.* 'nanlin-106', *P.* 'nanlin-121'	Wang et al. (1991)
(*P. simonii* × *P. nigra*)× (*P. nigra* ×*P. simonii*)	Fast growth, tolerance to cold	*P.* 'heilin-2'	Liu et al. (1993)
(*P. pseudo-simonii* × *P. nigra*)×*P. nigra*	Fast growth, tolerance to cold	*P.* 'heilin-3'	Liu et al. (1993)
P. deltoides 'Lux'× (*P. ussuriensis*+*P. simonii*)	Fast growth, tolerance to cold, and disease resistance	*P. deltoides* 'Lux'× (*P. ussuriensis*+ *P. simonii*)	Zhang et al. (2002)
P. deltoides 'Shanhaiguan'× *P. simonii* f. *fastigiata*	Narrow crown, tolerance to saline, and fast growth	Narrow-crowned *P.* 'heiqing-6', *P.* 'heiqing-31', *P.* 'heiqing-69', and *P.* 'heiqing-70'	Pang et al. (2001)

4.3 Cross breeding between Section *Tacamahaca* and Section *Leuce*

The literature on the cross breeding between Section *Tacamahaca* and Section *Leuce* is also sparse. Only the combinations of *P. davidiana* ×*P. simonii*, *P. simonii* ×*P. tomentosa*, *P. tomentosa* ×*P. cathayana* and *P. ussuriensis* ×*P. tomentosa* were studied by Xu et al. (1966), and the results showed that crossability was possible between the two sections and hybrids could be produced. The continuous studies on these hybrids are rarely reported. However, natural hybrids between the two Sections were found (Zhao and Chen, 1994).

4.4 Cross breeding among section *Turanga* and other sections

With *P. simonii* as female parent and *P. euphratica* as male parent, a hybrid was bred, which in its age of 10 years was 1.8 times as tall as *P. simonii*, 1.9 times as tall as *P. euphratica*, and the diameter at breast-height was 1.9 times that of its parents (Dong, 1980). These indicated an obvious heterosis of the hybrid. By using *P. nigra* var. *thevestina* as female parent and mixing the pollen of *P. euphratica* with treated pollen of *P. tomentosa* by 5000 roentgen radia tion, a cultivar was selected out. Attested by a RAPD marker, the cultivar was the hybrid of *P. euphratica* and *P. nigra* var. *thevestina*; it kept the characteristics of narrow crowns and fast growth of *P. nigra* var. *thevestina*, and held the characteristics of tolerance to drought, cold and saline-alkali conditions, and resistance to insects and diseases of *P. euphratica* at well (Li et al., 2002).

5 Main problems and strategies in cross breeding

Although many achievements have been made in cross breeding of poplars in China over the past 50 years, there still exist some problems. The problems mainly include the simple breeding strategies, too few tree species studied, lack of long-term plans, failure to make full use of rich genetic variations in various levels of provenance, family and clone, overemphasis on fast growth as breeding purpose, little attention to the preservation of indigenous tree species resources, slow application of modern biotechnology to conventional hybridization breeding studies, etc. In the light of all these problems, certain strategies should be correspondingly studied, and appropriate measures should be taken as soon as possible.

5.1 A long-term improvement plan

Over the past 50 years, in the procedures of hybridization and selection, mainly adopted in cross breeding of poplars, great importance was attached to selection from F_1 clones, and every time, cross breeding started from a random selection of parents in original populations without improvement. So far, there has not been any overall plan covering a complete system of long-term, medium-term and shortterm cross breeding directed by the theory of circulation selection. Long-term breeding means that parents are improved in a planned way under the guidance of the theory of circulation selection, thus making short-term breeding spirally rise efficiently in a sustainable way with the advance of generations. Improvement of parents was given priority in cross breeding of poplars and a long-term plan was worked out for breeding in Italy. Their strategies in *P. ×euramericana* improvement demanded that the best provenance of *P. deltoides* and *P. nigra* was first found out; 120 trees per sex and per species were selected to establish the breeding population, and the heterosis and combining ability of the selected trees were measured at the same time. On the one hand, interspecific hybrids were produced by using *P. deltoids* as female parent and *P. nigra* as male parent, and then the super hybrids were selected and propagated into clones after clone tests. On the other hand, intraspecific controlled pollination was done in the breeding group of *P. deltoides* and *P. nigra* respectively, and then 40 trees per sex and per species were selected out to form an improved group after the combining ability was measured. Using the improved group as material to repeat the above work, the breeding was systematically circulated, thus improving the hybrid clones continuously and raising the genetic traits of breeding group spirally. The above strategies can be introduced in the genetic improvements of the native poplar species in China so as to make a long-term plan based on interactive circulation selection by means of emphasis on improvements of the parents in order to increase their heterosis according to the scope of the distribution areas and characteristics of the species.

5.2 Emphasis on the genetic improvements of more species, traits and levels

China is rich in native resources of poplars, especially in Sections *Leuce* and *Tacamahaca*. Most species of the two sections are mainly distributed in the Northeast, Northwest and Southwest. In the past several decades, however, studies on native poplars were neither complete nor systematic. Because the adaptive regions of poplars have complicated landforms and varied climates, different native poplars should be improved to meet the local needs of timber production. Among all traits to be improved, the emphasis should be given to stress tolerance, followed by fast growth.

Many researches showed that there were abundant genetic variations in provenances, families and clones of poplars, and these variations dominated the results of cross breeding. Krzan (1976) studied the rust resistance of 44 clones in different *P. deltoides* provenances and the results showed that the southern provenances of *P. deltoides* and these offspring varieties of *P. ×euramericana*, had strong resistance to rust. Cellerino (1976) introduced *P. deltoides* seeds from America, including 18 provenances and 52 families, and tested 20 clones per family, involving their response to brown spot, rust, corky scab and lower temperature in early spring. The results showed that the south provenances of *P. deltoides* resisted all three diseases. By using the provenance to cross with *P. nigra* of the southern type, *P. ×euramericana*, which was suitable for afforestation in South Europe, was developed. Some European scholars have carried out hybridization of *P. davidiana* and considered that it was optimal to use the Polish provenances of *P. davidiana* as parents. In China, such researches on *P. cathayana*, *P. ussuriensis* and *P. davidiana* also indicated that significant variations exist in growth and wood quality among provenances and individuals of intra-provenances. Full use of the abundant variations existing in provenances, genealogies and clones, therefore, should be made to improve native poplar species and increase their heterosis.

5.3 Collection and preservation of native poplar breeding resources

Breeding resources are all the materials that can be used to breed elite varieties. An excellent variety not only possesses good economic traits, but also has wide adaptability and strong resistance. Without plentiful breeding resources, and introducing and complementing new gene resources continuously, multi-generation breeding will be restricted. In order to prevent the native poplar resources to lose out against exotic competition and to meet the various breeding needs, protection of population breeding resources should be established for different breeding districts. The requirements include representative individuals with fast growth and high-quality wood, individuals resistant to pests and diseases as well as those with tolerance to environmental stress.

5.4 Combination of conventional breeding methods with modern biotechnologies

Today when biotechnology is developing at a fast speed, conventional breeding methods are still an important means to provide improved varieties for forest production. It is a major field of poplar breeding to combine conventional cross breeding methods with modern biotechnologies properly. These methods include the application of transgene techniques, molecular markers, mRNA differential display techniques and gene engineering technology to improve the parent material, matching of cross parents or forecast of heterosis, studies on molecular mechanism of heterosis or genetic analyses on hybrids, and further improvement of the chosen elite clones. It is undoubtedly true that conventional cross breeding of poplars will show their power with the proper combination of modern biotechnologies.

Variation Analysis of Seed and Seedling Traits of Cross Combination Progenies in *Populus*[*]

Abstract Twenty-five species and hybrids in *populus* were used as parents, and 26 cross combinations, including more than 5000 seedlings, were obtained by artificial cross breeding. The length of infructescence, number of seed per infructescence, thousand-seed weight, germination rate of seed among these cross combinations were tested. The results indicated that the cross combinational effects were significant for these traits, and demonstrated that the length of infructescence, thousand-seed weight were positively affected by female parent. In addition, seedlings height (SH), basal diameter (BD), diameter at breast height (DBH) of 17 cross combination progenies were investigated. The analysis of mean and standard deviation of the three traits showed that SH, BD, DBH had extensive variation among combinations and individuals within combination. Variance analysis and estimate of heritability indicated that the three traits had wide variation, and were controlled by heredity. It was feasible to select superior cross combinations and seedlings. Further more, the result of multiple comparison showed that *P. deltoides* 'Lux'×*P. deltoides* 'D324', *P. ussuriensis* 'No. 4'× *P. deltoides* 'T66', *P. ussuriensis* 'No. 4'× *P. deltoides* 'T26', *P. deltoides* 'Lux'× *P. ussuriensis* 'No. 3', (*P. tomentosa* ×*P. bolleana*)×(*P. alba*×*P. glandulosa*), (*P. alba*×*P. tomentosa*)×(*P. alba*×*P. glandulosa*), and (*P. alba*×*P. glandulosa* 'No. 2')×*P. tomentosa* 'Lumao 50' were superior cross combinations with higher growth. Finally, 123 elite seedlings were selected for further test.

Key words poplar, cross combination, hybrids, variation analysis

1 Introduction

Poplars are extensively planted in the world because of their fast growth, better adaptability and use for timber production. They are also the important tree species used ecological protective in China. Poplars play a key role in the projects of fast-growing and high-yield and conversion of cropland to forestland, which are carrying out now in China. The improved varieties of poplars, which are cultivated widely for afforestation at present in China, are almost produced by artificial cross breeding, resulting in obvious heterosis with characteristics of fast growth, high quality, resistance to diseases and insects, and tolerance to environmental stress. For examples, Xu (1988) selected out *P.* ×'Beijingensis', *P.* 'Opera' from the combinations of *P. nigra.* var. *italic* ×*P. cathayana* and *P. simonii*×*P. nigra* var. *italic*, respectively. By using *P. deltoides* 'Lux' as fe-

* 本文原载 *Forestry Studies in China*, 2005, 7(3): 44-48, 与何承忠、安新民、于志水和李百炼合作发表。

male parent and *P. nigra* as male parent, the elite hybrids of *P. ×euramericana* 'Zhonglin-46' and *P. ×euramericana* 'Zhonglin-23', whose volume was 30% more than those of their parents, were bred by Huang et al. (1991). *P.* ×'Langfang-1' and *P.* ×'Langfang-2' with obvious heterosis, were screened from the combinations of *P. deltoides* 'Shanhaiguan' ×(*P. pyramidalis* Bork. + *P. deltoides* 'Harvard') and *P. deltoides* 'Lux' ×*P. deltoides* 'Shanhaiguan'(Liu et al. 1998). *P. deltoides* 'Lux' ×*P. deltoides* 'S3244' and *P. deltoides* 'Lux' ×*P. deltoides* 'Harvard' were proved to be superior cross combinations(Li et al., 2003). By using the combination of (*P. tomentosa*×*P. bolleana*)×*P. tomentosa* and treating pollen with colchicines and γ-ray, 6 elite clones of triploid *P. tomentosa*, whose growth at early stage is fast and wood quality is excellent, were bred(Zhu et al., 1995). These varieties have displayed distinguished economic, ecological and social benefits. Cross breeding is still an important means for the production of new poplar cultivars for a long time(Li et al., 2004).

In this study, Using twenty-five species and hybrids in *populus* as parents, gaining hybrid seedlings by cutting-branch water culture, we analyzed variations of the length of infructescence, number of seed per infructescence, thousand-seed weight, germination rate of seed among these cross combinations and growth traits of progenies, attempting to find out superior combinations, and provide theoretical foundation and materials for plus seedlings selection and clone test.

2 Materials and methods

2.1 Experimental site description

The experimental site of seedlings test was in Houbajia Tree Nursery of Beijing Forestry University (39°34′ N, 116°29′ E), It has meadow soil with medium loam texture, annual precipitation of 631mm, annual mean temperature of 12.2 centigrade, extreme minimum temperature of-19.6 centigrade, and a frost-free period of 210d.

2.2 Materials

The survey of experimental materials is summarized in Table 1.

2.3 Methods

A cross method of cutting-branch and water-culture was used. The healthy flower branches were cut and cultured in water in a greenhouse, at 25℃ with a humidity 60%. It ought to be taken into consideration that the male flower branches must be cultured earlier than female ones. When male inflorescence maturated, pollen was collected and stored at 4℃. Artificial pollination was done and repeated three times as the stigma brightened up. Before and after artificial pollination, female flowers were covered by bags to prevent cross pollination, because many different cross combinations were done at same site simultaneously.

Table 1 Experimental materials and its sources

Section	Parents	Collection sites	Collection dates
Leuce	*P. alba*×*P. glandulosa* 'No. 1'	Guan Xian County, Shandong Province	2002. 1. 18
	P. alba×*P. glandulosa* 'No. 2'	Guan Xian County, Shandong Province	2002. 1. 18
	P. alba×*P. glandulosa* 'No. 5'	Guan Xian County, Shandong Province	2002. 1. 18
	P. tomentosa 'Lumao 50'	Guan Xian County, Shandong Province	2001. 1. 18
	P. bolleana	Research Institute of Botany, CSA	2002. 2. 27
	P. tomentosa ×*P. bolleana*	China Agriculture University	2002. 1. 20
	P. alba×*P. glandulosa*	Yangling Town, Shannxi Province	2002. 1. 22
	P. alba	Beijing Botanic Garden	2002. 2. 27
	P. alba×*P. tomentosa*	Yi Xian County, Hebei Province	2002. 1. 23
	P. alba×*P. bolleana*	Baicheng, Jilin Province	2002. 3. 20
	P. tomentosa 'No. 5082'	Beijing Forestry University	2002. 1. 20
Aigeiros	*P. deltoides* 'Lux'	Linyi City, Shandong Province	2002. 1. 22.
	P. deltoides 'D324'	Linyi City, Shandong Province	2002. 1. 22
	P. deltoides 'T26'	Linyi City, Shandong Province	2002. 1. 22
	P. deltoides 'T66'	Linyi City, Shandong Province	2002. 1. 22
	P. ×*euramericana* 'I102'	Linyi City, Shandong Province	2002. 1. 22
	P. deltoides 'Zhonghe 1'	Heze City, Shandong Province	2002. 1. 20
	P. nigra var. *italica*	Beijing Botanic Garden	2002. 2. 27
Tacamahaca	*P. simonii*	Beijing Botanic Garden	2002. 2. 27
	P. szechuanica 'No. 1'	Hanyuan City, Sichuan Province	2002. 1. 23
	P. szechuanica 'No. 2'	Hanyuan City, Sichuan Province	2002. 2. 26
	P. ussuriensis 'No. 3'	Lushuihe Town, Jilin Province	2002. 2. 26
	P. ussuriensis 'No. 4'	Lushuihe Town, Jilin Province	2002. 2. 26
	P. ussuriensis 'No. 5'	Lushuihe Town, Jilin Province	2002. 2. 26

A completely random block design was adopted for seedling test, with 4 replications, forty seedlings per block and plant spacing of 25cm×100cm.

Four female branches were randomly sampled from every cross combination and length of three infructescences per flower branch were measured. The seed number of all infructescences was investigated and the average number of seeds per infructescence calculated. Three infructescences were sampled randomly from every cross combination and the thousand-seed weight was measured. Four hundred seeds were collected randomly from every cross combination and sown in a flowerpot with 100 seeds each, and the number of seed germination per replication was investigated three days after seeding. After the defoliation, seedlings (1-year-old stem and 2-year-old root) height (SH), basal diameter (BD), diameter at breast height (DBH) of every cross combination were investigated.

2.4 Data analysis

Length of infructescence, seed number per infructescence, thousand-seed weight and germination rate of every cross combination were calculated. The mean values and standard deviations (δ) of SH, BD, DBH were obtained, followed by the variance analysis and multiple contrast of the three traits. Data were analyzed using statistical methods of SAS Institute Inc.

3 Results and analysis

3.1 Length of infructescence

Mean values and standard deviations of infructescence length are presented in Table 2. The infructescence length of every cross combination in Section *Leuce* ranges from 5.3 to 12.9cm, with the longest one falling in the combination of (*P. alba*×*P. glandulosa* 'No. 2')×*P. tomentosa* 'Lumao 50' and the shortest in the combination of (*P. alba* ×*P. bolleana*)×*P. bolleana*. The infructescence length ranges from 7.6 to 14.7cm among the cross combinations in Sections *Aigeiros* and *Tacamahaca*. The combination of *P. ussuriensis* 'No. 4'×*P. deltoides* 'Zhonghe 1' has the longest infructescence, while the combination of *P. simonii* ×*P. deltoides* 'T26' has the shortest infructescence. The maximum value is 1.9 times the minimum one. These measurements indicate that the variance of the infructescence length is quite marked among these studied cross combinations. It can also be derived from the data (Table 2) that infructescence length is affected strongly

Table 2 The infructescence length and standard deviation of cross combinations

Section	Cross combination	Mean of infructescence length (cm)
Leuce	(*P. alba*×*P. glandulosa* 'No. 1')×*P. bolleana*	12.2±1.4
	(*P. alba*×*P. glandulosa* 'No. 2')×*P. bolleana*	12.3±1.5
	(*P. alba*×*P. glandulosa* 'No. 2')×*P. tomentosa* 'Lumao 50'	12.9±1.9
	(*P. tomentosa* ×*P. bolleana*) ×(*P. alba*×*P. glandulosa*)	9.6±1.6
	(*P. tomentosa* ×*P. bolleana*)×*P. bolleana*	7.9±0.8
	(*P. tomentosa* ×*P. bolleana*)×*P. tomentosa* 'Lumao 50'	6.4±1.2
	(*P. alba*×*P. tomentosa*)×*P. tomentosa* 'Lumao 50'	8.3±1.7
	(*P. alba*×*P. tomentosa*)×(*P. alba*×*P. glandulosa*)	9.2±2.3
	(*P. alba*×*P. tomentosa*)×*P. bolleana*	8.1±1.8
	P. alba ×*P. bolleana*	5.6±0.8
	(*P. alba* ×*P. bolleana*)×(*P. alba*×*P. glandulosa* 'No. 2')	5.7±0.6
	(*P. alba* ×*P. bolleana*)×*P. tomentosa* 'Lumao 50'	5.4±0.5
	(*P. alba* ×*P. bolleana*)×*P. bolleana*	5.3±0.5

（续）

Section	Cross combination	Mean of infructescence length (cm)
Tacamahaca and *Aigeiros*	*P. deltoides* 'Lux'×*P. nigra.* var. *italica*	10.3±1.3
	P. deltoides 'Lux'×*P. deltoides* 'D324'	9.2±1.4
	P. deltoides 'Lux'×*P. ussuriensis*	9.6±1.4
	P. deltoides 'Lux'×*P. simonii*	9.4±1.3
	P. deltoides 'Lux'×*P. szechuanica* 'No. 2'	9.7±1.5
	P. ussuriensis 'No. 4'×*P. deltoides* 'Zhonghe 1'	14.7±1.8
	P. ussuriensis 'No. 4'×*P. deltoides* 'T66'	14.4±2.1
	P. ussuriensis 'No. 4'×*P. deltoides* 'T26'	13.8±1.6
	P. ussuriensis 'No. 4'×*P. szechuanica* 'No. 2'	13.3±1.8
	P. simonii ×*P. nigra.* var. *italica*	9.0±1.6
	P. simonii ×*P. deltoides* 'T66'	9.0±0.9
	P. simonii ×*P. deltoides* 'D324'	8.1±1.1
	P. simonii ×*P. deltoides* 'Zhonghe 1'	7.8±0.8
	P. simonii ×*P.* ×*euramericana* 'I102'	8.3±1.0
	P. simonii ×*P. deltoides* 'T26'	7.6±0.8
	P. simonii ×(*P. simonii* ×*P. nigra*)	8.5±0.7
	P. simonii ×*P. szechuanica* 'No. 1'	8.4±0.8

by female parent. If the same female parent is used, the difference of infructescence length is insignificant, such as three cross combinations using *P. alba*×*P. glandulosa* 'No. 1' and *P. alba*×*P. glandulosa* 'No. 2' as female parents, infructescence length was 12.2, 12.3, 12.9cm individually. Similar conclusions can be drawn from these data of cross combinations used *P. alba*×*P. tomentosa*, *P. deltoides* 'Lux', *P. ussuriensis* 'No. 4', *P. simonii* respectively as female parents.

3.2 Seed number per infructescence

The number of seed per infructescence of every cross combination is listed in Table 3. Among the cross combinations in Section *Leuce*, (*P. alba*×*P. glandulosa* 'No. 2')×*P. tomentosa* 'Lumao 50', (*P. alba*×*P. glandulosa* 'No. 1')×*P. bolleana*, (*P. alba*×*P. glandulosa* 'No. 2')×*P. bolleana* had more seeds: 436, 430 and 425 seeds per infructescence respectively. On the other hand, (*P. tomentosa* ×*P. bolleana*)×*P. tomentosa* 'Lumao 50', (*P. tomentosa* ×*P. bolleana*)×*P. bolleana* and (*P. tomentosa* ×*P. bolleana*) ×(*P. alba*×*P. glandulosa*) yielded fewer seeds than those mentioned above. (*P. alba*×*P. tomentosa*)×(*P. alba*×*P. glandulosa*) produced the least, only 72 seeds. More seeds from the cross combinations in Sections *Aigeiros* and *Tacamahaca*, *P. deltoides* 'Lux'×*P. deltoides* 'D324', *P. deltoides* 'Lux'×*P. ussuriensis*, *P. deltoides* 'Lux'×*P. szechuanica* 'No. 2', *P. deltoides* 'Lux'×*P. nigra.* var. *italica* and *P. simonii* ×*P. szechuanica* 'No. 1' were harvested, the number of seeds are 257, 241, 193, 168 and 139 respectively, *P. simonii* ×*P. deltoides* 'T66', *P. simonii* ×*P. deltoides* 'Zhonghe 1' and *P.*

simonii ×*P. deltoides* 'D324' produced relatively less seeds, which were 15, 13 and 10 respectively. Accordingly, the variance of seed number per infructescence is remarkable, which indicates that the cross-fertility of every combination is very different.

Table 3 The number of seeds per infructescence of cross combinations

Section	Combination	Number of Infructescence	Total number of seed	Seed number per infructescence
	(*P. alba*×*P. glandulosa* 'No. 2')×*P. tomentosa* 'Lumao 50'	27	11782	436
	(*P. alba*×*P. glandulosa* 'No. 1')×*P. bolleana*	51	21914	430
	(*P. alba*×*P. glandulosa* 'No. 2')×*P. bolleana*	11	4672	425
	(*P. tomentosa* ×*P. bolleana*)×*P. tomentosa* 'Lumao 50'	5	1180	236
	(*P. tomentosa* ×*P. bolleana*)×*P. bolleana*	146	31902	219
	(*P. tomentosa* ×*P. bolleana*) ×(*P. alba*×*P. glandulosa*)	24	4158	174
Leuce	*P. alba* ×*P. bolleana*	31	4151	134
	(*P. alba* ×*P. bolleana*)×(*P. alba*×*P. glandulosa* 'No. 2')	15	1943	130
	(*P. alba*×*P. tomentosa*)×*P. bolleana*	56	6955	124
	(*P. alba*×*P. tomentosa*)×*P. tomentosa* 'Lumao 50'	56	5501	98
	(*P. alba* ×*P. bolleana*)×*P. tomentosa* 'Lumao 50'	5	474	95
	(*P. alba* ×*P. bolleana*)×*P. bolleana*	17	1489	87
	(*P. alba*×*P. tomentosa*)×(*P. alba*×*P. glandulosa*)	56	4056	72
	P. deltoides 'Lux'×*P. deltoides* 'D324'	4	1027	257
	P. deltoides 'Lux'×*P. ussuriensis*	10	2414	241
	P. deltoides 'Lux'×*P. szechuanica* 'No. 2'	5	967	193
	P. deltoides 'Lux'×*P. nigra*. var. *italica*	13	2184	168
Tacamahaca and *Aigeiros*	*P. simonii* ×*P. szechuanica* 'No. 1'	10	1378	139
	P. simonii ×*P. nigra*. var. *italica*	11	1091	99
	P. deltoides 'Lux'×*P. simonii*	6	332	55
	P. simonii ×*P. deltoides* 'T66'	4	59	15
	P. simonii ×*P. deltoides* 'Zhonghe 1'	10	134	13
	P. simonii ×*P. deltoides* 'D324'	9	87	10

3.3 Thousand-seed weight

The thousand-seed weights are listed in Table 4. Among the cross combinations in Section *Leuce*, the thousand-seed weight ranged from 0.1410 to 0.3588g, the maximum is 2.5 times the minimum. The thousand-seed weights of combinations of (*P. alba*×*P. tomentosa*)×(*P. alba*×*P. glandulosa*), (*P. tomentosa* ×*P. bolleana*) ×(*P. alba*×*P. glandulosa*), (*P. alba*×*P. tomentosa*)×*P. bolleana*, (*P. alba*×*P. tomentosa*)×*P. tomentosa* 'Lumao 50' and (*P. tomentosa* ×*P.*

bolleana)×*P. tomentosa* 'Lumao 50' are greater than others. The thousand-seed weight ranged from 0.1080 to 0.4067 g among cross combinations in Section *Aigeiros* and *Tacamahaca* and these thousand-seed weights, such as combinations of *P. simonii* ×*P. nigra*. var. *italica*, *P. ussuriensis* 'No. 4'×*P. deltoides* 'T66', *P. simonii* ×*P. deltoides* 'Zhonghe 1', *P. simonii* ×*P. deltoides* 'D324', *P. simonii* ×*P. szechuanica* 'No. 1' and *P. ussuriensis* 'No. 4' ×*P. szechuanica* 'No. 2', are heavier than others. These results demonstrate that the range of thousand-seed weight is great among all of cross combinations in the present study. Additionally, when *P. alba*×*P. tomentosa*, *P. tomentosa* ×*P. bolleana*, *P. simonii* and *P. ussuriensis* 'No. 4' were used as female parents in cross combinations, the thousand-seed weight were heavier, which showed that the thousand-seed weight was controlled mainly by female parent.

Table 4 Thousand-seed weight of cross combinations

Combination	Thousand-seed weight (g)
(*P. alba*×*P. tomentosa*)×(*P. alba*×*P. glandulosa*)	0.3588
(*P. tomentosa* ×*P. bolleana*) ×(*P. alba*×*P. glandulosa*)	0.3587
(*P. alba*×*P. tomentosa*)×*P. bolleana*	0.3543
(*P. alba*×*P. tomentosa*)×*P. tomentosa* 'Lumao 50'	0.3343
(*P. tomentosa* ×*P. bolleana*)×*P. tomentosa* 'Lumao 50'	0.3258
(*P. tomentosa* ×*P. bolleana*)×*P. bolleana*	0.2802
(*P. alba* ×*P. bolleana*)×(*P. alba*×*P. glandulosa* 'No. 2')	0.2517
P. alba ×*P. bolleana*	0.2473
(*P. alba* ×*P. bolleana*)×*P. tomentosa* 'Lumao 50'	0.2383
(*P. alba*×*P. glandulosa* 'No. 2')×*P. bolleana*	0.2150
(*P. alba*×*P. glandulosa* 'No. 2')×*P. tomentosa* 'Lumao 50'	0.2137
(*P. alba*×*P. glandulosa* 'No. 1')×*P. bolleana*	0.2100
(*P. alba* ×*P. bolleana*)×*P. bolleana*	0.1638
P. tomentosa 'No. 5082'×*P. bolleana*	0.1453
P. tomentosa 'No. 5082'×(*P. alba*×*P. glandulosa* 'No. 2')	0.1410
P. simonii ×*P. nigra*. var. *italica*	0.4067
P. ussuriensis 'No. 4'×*P. deltoides* 'T66'	0.3270
P. simonii ×*P. deltoides* 'Zhonghe 1'	0.3042
P. simonii ×*P. deltoides* 'D324'	0.2870
P. simonii ×*P. szechuanica* 'No. 1'	0.2858
P. ussuriensis 'No. 4'×*P. szechuanica* 'No. 2'	0.2660
P. deltoides 'Lux'×*P. ussuriensis* 'No. 3'	0.2105
P. deltoides 'Lux'×*P. simonii*	0.2085
P. simonii ×*P. deltoides* 'T66'	0.1627
P. deltoides 'Lux'×*P. nigra*. var. *italica*	0.1391
P. deltoides 'Lux'×*P. szechuanica* 'No. 2'	0.1258
P. deltoides 'Lux'×*P. deltoides* 'D324'	0.1080

3.4 Seed germination rate

Seed germination rates per cross combination are shown in Table 5. Among the combinations in Section *Leuce*, germination rates varied between 66.9% and 96.3%, and seed germination rates of these combinations (*P. alba*×*P. glandulosa* 'No. 1')×*P. bolleana*, (*P. alba*×*P. tomentosa*)×(*P. alba*×*P. glandulosa*) and (*P. tomentosa* ×*P. bolleana*) ×(*P. alba*×*P. glandulosa*), were higher than others. Among the combinations in Section *Aigeiros* and *Tacamahaca*, seed germination rates of *P. deltoides* 'Lux' ×*P. ussuriensis* 'No. 3' and *P. deltoides* 'Lux' ×*P. deltoides* 'D324' were 30.5% and 29.3% respectively. The seed germination rate of *P. simonii* ×*P. deltoides* 'T66' was 0. These results indicate a large variance of seed germination rates. In addition, table 5 showed that seed germination rate of cross combinations in Section *Leuce* were higher than those in Section *Aigeiros* and *Tacamahaca*.

Table 5　Seed germination rates of cross combinations

Combination	Germination rate (%)
(*P. alba*×*P. glandulosa* 'No. 1')×*P. bolleana*	96.3
(*P. alba*×*P. tomentosa*)×(*P. alba*×*P. glandulosa*)	94.9
(*P. tomentosa* ×*P. bolleana*) ×(*P. alba*×*P. glandulosa*)	89.7
(*P. alba*×*P. tomentosa*)×*P. bolleana*	88.2
P. alba ×*P. bolleana*	87.9
(*P. alba*×*P. tomentosa*)×*P. tomentosa* 'Lumao 50'	84.1
(*P. alba* ×*P. bolleana*)×(*P. alba*×*P. glandulosa* 'No. 2')	82.8
(*P. tomentosa* ×*P. bolleana*)×*P. bolleana*	82.6
(*P. alba*×*P. glandulosa* 'No. 2')×*P. tomentosa* 'Lumao 50'	81.1
(*P. alba*×*P. glandulosa* 'No. 2')×*P. bolleana*	72.5
(*P. tomentosa* ×*P. bolleana*)×*P. tomentosa* 'Lumao 50'	70.7
(*P. alba* ×*P. bolleana*)×*P. bolleana*	66.9
P. deltoides 'Lux' ×*P. ussuriensis* 'No. 3'	30.5
P. deltoides 'Lux' ×*P. deltoides* 'D324'	29.3
P. ussuriensis 'No. 4' ×*P. szechuanica* 'No. 2'	18.0
P. deltoides 'Lux' ×*P. ussuriensis* 'No. 3'	13.5
P. ussuriensis 'No. 4' ×*P. deltoides* 'T26'	12.8
P. deltoides 'Lux' ×*P. szechuanica* 'No. 2'	10.4
P. deltoides 'Lux' ×*P. nigra* var. *italica*	7.7
P. ussuriensis 'No. 4' ×*P. deltoides* 'T66'	5.3
P. deltoides 'Lux' ×*P. simonii*	3.6
P. simonii ×*P. deltoides* 'Zhonghe 1'	2.8
P. simonii ×*P. deltoides* 'D324'	1.2
P. deltoides 'Lux' ×*P. szechuanica* 'No. 2'	0.6
P. simonii ×*P. deltoides* 'T66'	0.0

3.5 Seedling height, basal diameter and DBH

3.5.1 Mean value and standard deviation of growth trait

Mean value and standard deviation of seedlings height (SH), basal diameter (BD) and diameter at breast height (DBH) of every cross combination were listed in Table 6. SH varied from 2.2 to 3.7 m, the maximum was 1.7 times the minimum. BD ranged from 1.7 to 3.6cm among these cross combinations. DBH varied from 0.8 to 2.2cm. Additionally, the standard deviation of SH, BD and DBH varied from 0.5 to 0.9 m, from 0.3 to 1.0cm and from 0.2 to 0.8cm respectively. The standard deviations of SH of *P. ussuriensis* 'No. 4'×*P. deltoides* 'T26', (*P. alba*×*P. glandulosa* 'No. 2')×*P. tomentosa* 'Lumao 50', (*P. tomentosa* ×*P. bolleana*) ×(*P. alba*×*P. glandulosa*) and (*P. alba*×*P. tomentosa*)×(*P. alba*×*P. glandulosa*) were more than those of other combinations. The standard deviation of BD of *P. ussuriensis* 'No. 4'×*P. deltoides* 'T26', *P. deltoides* 'Lux'×*P. ussuriensis* 'No. 3', (*P. alba*×*P. glandulosa* 'No. 2')×*P. tomentosa* 'Lumao 50', (*P. tomentosa* ×*P. bolleana*) ×(*P. alba*×*P. glandulosa*), and (*P. tomentosa* × *P. bolleana*)×*P. bolleana* are large. The same result in DBH is also found in the combinations of *P. ussuriensis* 'No. 4'×*P. deltoides* 'T26', *P. deltoides* 'Lux'×*P. deltoides* 'D324', *P. deltoides* 'Lux'×*P. ussuriensis* 'No. 3', (*P. alba*×*P. glandulosa* 'No. 2')×*P. tomentosa* 'Lumao 50', (*P. tomentosa* ×*P. bolleana*) ×(*P. alba*×*P. glandulosa*), (*P. alba*×*P. tomentosa*)×(*P. alba*×*P. glandulosa*), (*P. tomentosa* ×*P. bolleana*)×*P. bolleana*, (*P. alba*×*P. tomentosa*)×*P. tomentosa* 'Lumao 50', (*P. alba*×*P. glandulosa* 'No. 1')×*P. bolleana*, and (*P. alba*×*P. glandulosa* 'No. 2')×*P. bolleana*. The results indicate that the growth variation is greater among individuals within combinations. Generally, it can be deduced that variation among combinations and among individuals within combinations is extensive, and it is feasible to select elite cross combinations and outstanding seedlings.

Table 6 The growth and the standard deviation of cross combinations

Cross combination	SH	BD	DBH
	$H\pm\delta$(m)	$D_0\pm\delta$(cm)	$D_{1.3}\pm\delta$(cm)
(*P. tomentosa* ×*P. bolleana*)×*P. bolleana*	2.5±0.7	2.5±0.9	1.0±0.6
(*P. alba*×*P. tomentosa*)×*P. tomentosa* 'Lumao 50'	2.6±0.8	1.9±0.7	1.0±0.6
(*P. alba*×*P. glandulosa* 'No. 1')×*P. bolleana*	3.3±0.8	2.4±0.8	1.5±0.6
(*P. alba*×*P. glandulosa* 'No. 2')×*P. tomentosa* 'Lumao 50'	3.3±0.9	2.6±0.9	1.6±0.6
(*P. tomentosa* ×*P. bolleana*)×(*P. alba*×*P. glandulosa*)	3.6±0.9	2.5±0.9	1.6±0.6
(*P. alba*×*P. tomentosa*)×(*P. alba*×*P. glandulosa*)	3.5±0.9	2.6±0.8	1.6±0.6
(*P. alba*×*P. tomentosa*)×*P. bolleana*	3.1±0.8	2.2±0.7	1.3±0.5
(*P. tomentosa* ×*P. bolleana*)×*P. tomentosa* 'Lumao 50'	3.0±0.7	2.5±0.8	1.3±0.5
(*P. alba*×*P. glandulosa* 'No. 2')×*P. bolleana*	3.2±0.7	2.5±0.8	1.5±0.6
P. alba ×*P. bolleana*	2.9±0.6	2.1±0.6	1.2±0.5
(*P. alba* ×*P. bolleana*)×*P. bolleana*	2.6±0.6	2.0±0.6	1.1±0.5

（续）

Cross combination	SH	BD	DBH
	$H±\delta$(m)	$D_0±\delta$(cm)	$D_{1.3}±\delta$(cm)
(*P. alba* ×*P. bolleana*)×(*P. alba*×*P. glandulosa* 'No. 2')	2.8±0.6	2.2±0.6	1.2±0.4
P. deltoides 'Lux'×*P. ussuriensis* 'No. 3'	3.5±0.8	2.9±1.0	1.8±0.6
P. deltoides 'Lux'×*P. deltoides* 'D324'	3.7±0.8	3.1±0.9	2.2±0.7
P. ussuriensis 'No. 4'×*P. deltoides* 'T66'	3.4±0.6	3.6±0.8	1.9±0.5
P. ussuriensis 'No. 4'×*P. szechuanica* 'No. 2'	2.2±0.5	1.7±0.3	0.8±0.2
P. ussuriensis 'No. 4'×*P. deltoides* 'T26'	3.3±1.0	2.9±1.3	1.8±0.8

3.5.2　Variance analysis and estimate of heritability

Variance analysis and estimate of heritabilities of SH, BD and DBH are shown in Table 7. The variations within the three traits were extremely significant among combinations, their cross combination heritabilities are 0.954, 0.934 and 0.968, and the individual heritabilities were 0.396, 0.312 and 0.492 respectively which indicated that variation of seedling growth among combinations and individual within combination are strongly or modestly controlled by heredity.

Table 7　Variance analysis, heritability of growth traits of cross combination and individual

Traits	*F* value	Pr>*F*	Cross combination heritability	Individual heritability
Height	21.63**	0.0001	0.954	0.396
BD	15.06**	0.0001	0.934	0.312
DBH	31.18**	0.0001	0.968	0.492

Notes: ** significant at 1% level.

3.5.3　Multiple comparison of growth traits

Based on variance analysis, multiple comparisons of SH, BD and DBG were carried out and results were presented in Table 8. For SH, the combinations of *P. deltoides* 'Lux'×*P. deltoides* 'D324', (*P. tomentosa* ×*P. bolleana*)×(*P. alba*×*P. glandulosa*), *P. ussuriensis* 'No. 4'×*P. deltoides* 'T66', (*P. alba*×*P. tomentosa*)×(*P. alba*×*P. glandulosa*), *P. deltoides* 'Lux'×*P. ussuriensis* 'No. 3', which had no significant difference among them at 5% level, had greater growth and ranked among the best five combinations. Comparatively, combinations of *P. ussuriensis* 'No. 4'×*P. szechuanica* 'No. 2', (*P. alba* ×*P. bolleana*)×*P. bolleana*, (*P.* ×*P. tomentosa*)×*P. tomentosa* 'Lumao 50', (*P. tomentosa* ×*P. bolleana*)×*P. bolleana*, and *P. alba* ×(*P. alba*×*P. glandulosa* 'No. 5') had slower growth and were listed among the last five. The other combinations fall in the middle. The combinations, which grew fast and lined up in the first four for BD, including *P. ussuriensis* 'No. 4'×*P. deltoides* 'T66', *P. deltoides* 'Lux'×*P. deltoides* 'D324', *P. ussuriensis* 'No. 4'×*P. deltoides* 'T26', and *P. deltoides* 'Lux'×*P. ussuriensis* 'No. 3', had significant differences from the other combinations at the 5% level. On the other hand, the combinations of *P. ussuriensis* 'No. 4'×*P. szechuanica* 'No. 2', (*P. alba*×*P. tomentosa*)×*P. tomentosa* 'Lumao 50', (*P. alba* ×*P. bolleana*)×*P. bolleana*, and *P. alba* ×*P.*

bolleana grew slowly and fell among the last four combinations. As for DBH, the combinations growing fast growth comprised *P. deltoides* 'Lux'×*P. deltoides* 'D324', *P. ussuriensis* 'No. 4'× *P. deltoides* 'T66', *P. ussuriensis* 'No. 4'×*P. deltoides* 'T26' and *P. deltoides* 'Lux'×*P. ussuriensis* 'No. 3', while the combinations growing slowly were the same as those having less growth in SH. The multiple comparison results of SH, BD and DBH revealed that *P. deltoides* 'Lux'×*P. deltoides* 'D324', *P. ussuriensis* 'No. 4' × *P. deltoides* 'T66', *P. ussuriensis* 'No. 4'× *P. deltoides* 'T26', *P. deltoides* 'Lux'× *P. ussuriensis* 'No. 3' were superior cross combinations with higher growth rates, while the combinations of *P. ussuriensis* 'No. 4' ×*P. szechuanica* 'No. 2', (*P. alba* ×*P. bolleana*)×*P. bolleana*, (*P. alba*×*P. tomentosa*)×*P. tomentosa* 'Lumao 50', (*P. tomentosa* ×*P. bolleana*)×*P. bolleana*, and *P. alba* ×(*P. alba*×*P. glandulosa* 'No. 5') were inferior since they had slower growth. The other eight combinations ranged in the middle. In the combinations of *Leuce*, (*P. tomentosa* ×*P. bolleana*)×(*P. alba*×*P. glandulosa*), (*P. alba*×*P. tomentosa*)×(*P. alba*×*P. glandulosa*) and (*P. alba*×*P. glandulosa* 'No. 2')×*P. tomentosa* 'Lumao 50' grew faster than the others.

Table 8 Multiple comparison of SH, BD and DBH among 17 cross combinations

H (m)					BD (cm)					DBH (cm)				
Code	Mean	0.05 level			Code	Mean	0.05 level			Code	Mean	0.05 level		
14	3.7		A		15	3.6		A		14	2.2		A	
5	3.6	B	A		14	3.1	B			15	1.9	B		
15	3.5	B	A	C	17	3.0	B			17	1.8	B		
6	3.5	B	A	C	13	2.9	B			13	1.8	B		
13	3.5	B	A	C	4	2.6			C	5	1.6			C
3	3.4	B	D	C	6	2.6			C	4	1.6			C
17	3.3	B	D	C	9	2.5			C	6	1.6			C
4	3.3		D	C	1	2.5			C	3	1.5		D	C
9	3.2	E	D	C	8	2.5			C	9	1.5		D	C
7	3.2	E	D	F	3	2.5			C	8	1.4	E	D	
10	3.1	E	D	F	5	2.5			C	7	1.3	E		
8	3.0	E		F	7	2.2		D		10	1.2	E		F
12	2.9		G	F	12	2.2		D		12	1.2	E		F
11	2.6	H	G		10	2.1	E	D		11	1.1		G	F
2	2.6	H			11	2.0	E	D	F	2	1.0		G	
1	2.5	H			2	1.9	E		F	1	1.0		G	
16	2.2		I		16	1.7			F	16	0.7	H		

Notes: 1, (*P. tomentosa* ×*P. bolleana*)×*P. bolleana*; 2, (*P. alba*×*P. tomentosa*)×*P. tomentosa* 'Lumao 50'; 3, (*P. alba*×*P. glandulosa* 'No. 1')×*P. bolleana*; 4, (*P. alba*×*P. glandulosa* 'No. 2')×*P. tomentosa* 'Lumao 50'; 5, (*P. tomentosa* ×*P. bolleana*)×(*P. alba*×*P. glandulosa*); 6, (*P. alba*×*P. tomentosa*)×(*P. alba*×*P. glandulosa*); 7, (*P. alba*×*P. tomentosa*)×*P. bolleana*; 8, (*P. tomentosa* ×*P. bolleana*)×*P. tomentosa* 'Lumao 50'; 9, (*P. alba*×*P. glandulosa* 'No. 2')×*P.*

bolleana; 10, *P. alba* ×*P. bolleana*; 11, (*P. alba* ×*P. bolleana*)×*P. bolleana*; 12, *P. alba*×(*P. alba*×*P. glandulosa* 'No. 5'); 13, *P. deltoides* 'Lux'×*P. ussuriensis* 'No. 3'; 14, *P. deltoides* 'Lux'×*P. deltoides* 'D324'; 15, *P. ussuriensis* 'No. 4'×*P. deltoides* 'T66'; 16, *P. ussuriensis* 'No. 4'×*P. szechuanica* 'No. 2'; 17, *P. ussuriensis* 'No. 4'×*P. deltoides* 'T26'. H, $LSD_{0.05}=0.2817$; BD, $LSD_{0.05}=0.2789$; DBH, $LSD_{0.05}=0.1876$.

Table 9　Elite seedlings selection of cross combinations in *Populus*

Combination	HES (m)	HHP (m)	IPHHP (%)	NPS	TNHP	ESR (%)
(*P. tomentosa* ×*P. bolleana*)×*P. bolleana*	3.9	2.5	54.2	8	325	2.46
(*P. alba*×*P. tomentosa*)×*P. tomentosa* 'Lumao 50'	4.1	2.6	56.1	7	303	2.31
(*P. alba*×*P. glandulosa* 'No. 1')×*P. bolleana*	4.6	3.3	38.8	12	530	2.26
(*P. alba*×*P. glandulosa* 'No. 2')×*P. tomentosa* 'Lumao 50'	4.5	3.3	37.7	9	571	1.58
(*P. tomentosa* ×*P. bolleana*)×(*P. alba*×*P. glandulosa*)	4.8	3.6	35.2	15	572	2.62
(*P. alba*×*P. tomentosa*)×(*P. alba*×*P. glandulosa*)	4.9	3.5	40.1	7	274	2.55
(*P. alba*×*P. tomentosa*)×*P. bolleana*	4.4	3.1	40.2	16	538	2.97
(*P. tomentosa* ×*P. bolleana*)×*P. tomentosa* 'Lumao 50'	4.3	3.0	42.6	6	195	3.08
(*P. alba*×*P. glandulosa* 'No. 2')×*P. bolleana*	4.3	3.2	33.8	6	138	4.35
P. alba ×*P. bolleana*	4.0	2.9	40.3	13	653	2.45
P. alba×(*P. alba*×*P. glandulosa* 'No. 5')	4.2	2.8	46.0	2	167	1.2
P. deltoides 'Lux'×*P. ussuriensi* 'No. 3'	4.0	3.5	15.8	5	78	6.41
P. deltoides 'Lux'×*P. deltoides* 'D324'	4.5	3.7	21.8	6	192	3.13
P. ussuriensis 'No. 4'×*P. deltoides* 'T66'	3.8	3.4	10.8	4	70	5.71
P. ussuriensis 'No. 4'×*P. szechuanica* 'No. 2'	2.8	2.2	25.0	2	85	2.35
P. ussuriensis 'No. 4'×*P. deltoides* 'T26'	4.0	3.3	19.7	5	86	5.81
Total				123	4826	
Mean value	4.1	3.1	33.6			3.11

Notes: HES is the height of elite seedling, HHP the height of hybrid population, IPHHP the increased percent of height of hybrid population, NPS the number of elite seedlings, TNHP the total number of hybrid population and ESR the elite seedlings rate.

3.5.4　Selection of plus seedlings

Following the norm of mean values plus one standard deviation or two, 123 elite seedlings were selected out from 16 hybrid population (Table 9). The SH means of superior individuals in the combinations of (*P. alba*×*P. tomentosa*)×(*P. alba*×*P. glandulosa*), (*P. tomentosa* ×*P. bolleana*)×(*P. alba*×*P. glandulosa*), (*P. alba*×*P. glandulosa* 'No. 1')×*P. bolleana*, (*P. alba*×*P. glandulosa* 'No. 2')×*P. tomentosa* 'Lumao 50', and *P. deltoides* 'Lux'×*P. deltoides* 'D324', which had more height growth than the others, were 4.9, 4.8, 4.6m、4.5 and 4.5 m respectively. Mean values of elite seedlings in height per combination exceeded its mean value of hybrid populations from 33.8%~56.1% in Section *Leuce*, 10.8%~25.0% in Section *Aigeiros* and *Tacamahaca*. The selection rates of elite seedlings per combination varied from 1.20% to 6.41%.

4 Conclusions and discussion

The traits of infructescence length, seed number per infructescence, thousand-seed weight and germination rate had broad variation among all combinations. The range of variation of each trait was 5. 6~14. 7cm, 10~436 grains, 0. 1080~0. 3588g and 0~96. 3% respectively. Synchronously, matching same female parent with different male parents, the variation of infructescence length and thousand-seed weight were small. It demonstrated that the two traits were positively affected by female parents.

Cross-fertility means the ability of producing viable seeds through crossing. It can be measured by investigating seed number per infructescence. In Section *Leuce*, these combinations such as (*P. alba*×*P. glandulosa* 'No. 2')×*P. tomentosa* 'Lumao 50', (*P. alba*×*P. glandulosa* 'No. 2')×*P. bolleana*, (*P. alba*×*P. glandulosa* 'No. 1')×*P. bolleana*, (*P. tomentosa* ×*P. bolleana*)×*P. tomentosa* 'Lumao 50', (*P. tomentosa* ×*P. bolleana*)×*P. bolleana* had higher cross-fertility. In Section *Aigeiros* and *Tacamahaca*, the combinations such as *P. deltoides* 'Lux' ×*P. deltoides* 'D324', *P. deltoides* 'Lux'×*P. ussuriensis* 'No. 3', *P. deltoides* 'Lux'×*P. szechuanica* 'No. 2' and *P. deltoides* 'Lux'×*P. nigra*. var. *italica* also had higher cross-fertility. Furthermore, *P. alba*×*P. glandulosa* 'No. 1', *P. alba*×*P. glandulosa* 'No. 2', *P. tomentosa* × *P. bolleana* and *P. deltoides* 'Lux' were the parents with higher cross ability among the studied parents.

In this study, seed germination rate of combinations in Section *Leuce* were generally higher than those in Section *Aigeiros* and *Tacamahaca*. The seed maturity period was the main cause of the difference, because the seed maturity period of species in Section *Leuce* was about 25d, but in Section *Aigeiros* and *Tacamahaca* it was about 45d. The nutrition stored in branches was enough for seed development in Section *Leuce* during cutting-branch water culture, but it was deficient in Section *Aigeiros* and *Tacamahaca*. Tissue culture could be used to increase seed germination rate.

The mean and standard deviation showed that SH, BD, DBH had wide variation among combinations and individuals within combination, and that it was feasible to select super seedlings and elite combinations. The result of multiple comparison indicated that *P. deltoides* 'Lux'×*P. deltoides* 'D324', *P. ussuriensis* 'No. 4'× *P. deltoides* 'T66', *P. ussuriensis* 'No. 4'× *P. deltoides* 'T26', *P. deltoides* 'Lux'× *P. ussuriensis* 'No. 3', (*P. tomentosa* ×*P. bolleana*)×(*P. alba*×*P. glandulosa*), (*P. alba*×*P. tomentosa*)×(*P. alba*×*P. glandulosa*), and (*P. alba*×*P. glandulosa* 'No. 2')×*P. tomentosa* 'Lumao 50' were superior cross combinations with higher growth.

The investigation and analyses were carried out on the trait variation of 1-year-old seedlings, and the progenies need further tests at different sites in the future.

Cross Breeding of *Populus* and Its Hybrids for Cold Resistance*

Abstract *Populus tomentosa* was crossed with *P. tremuloidis*, *P. grandidentata*, *P. alba*×*P. grandidentata* and *P. alba*×*Ulmus pumila* in order to maintain its rapid growth and high wood quality and improve its resistance to cold. Two methods were used to increase the germination rate from 1.5% to 41.1% and the remaining rate from 1.7% to 44.2%. Forty crossing combinations were conducted and 2744 hybrid seedlings were obtained. MX4×*P. grandidentata* (G-1-58), MX3×*P. tremuloidis* (T-44-60), MX2×*P. tremuloidis* (1-13-87-37) and MX2×(*P. alba*×*P. grandidentata*) were regarded as superior combinations after analysis and selection. Thirty seedlings of these combinations and 11 triploid seedlings identified by counting their chromosomes were selected as super plants.
Keywords *Populus tomentosa*, cross breeding, embryo abortion, in vitro embryo culture, polyploid, cold resistance

1 Introduction

Populus tomentosa is a species with rapid growth, excellent resistance to diseases and superior quality of wood, but it is distributed only in a fan-shaped area in east and central China (Zhu and Zhang, 1997). It cannot grow well in areas north of 41°N even in better conditions of water and nutrients in northeast China because it cannot resist cold there very well.

The cold resistance of trees results from their genetic background. *P. tomentosa* lacks such resistance-related genes to make it survive under low temperatures in northeast China. To improve the resistance of *P. tomentosa*, it is necessary to change its germplasm. Genetic engineering and bio-techniques are current approaches to change the genetic background of plants, but their applications are rarely reported for cold resistance of trees. Traditional cross breeding is still an effective and practical method.

Since *P. angulata* Ait and *P. trichocarpa* Torr were crossed and a crossbred, *P. generosa* Henry was obtained in 1912, which had rapid growth, various crossing investigations in poplars have been carried out worldwide and many hybrids with high cold resistance have been obtained (Duan and Zhang, 1997). For example, the hybrid of *P. simonii* Carr. ×*P. nigracan* both grow rapidly and stand temperatures as low as-43℃ (Xu, 1982). Because of the cross abortion and

* 本文原载 *Forestry Studies in China*, 2005, 7(3): 70-76, 与张冬芳合作发表。

difficulty in reproduction, the breeding of Section *Leuce* has been far less studied than other poplar sections, especially in cold resistance breeding. However, *P. tremuloidis*, *P. grandidentata*, *P. alba*×*P. gandidentata* which grow in America have a high resistance to cold. They make possible the cross breeding of *P. tomentosa* for high cold resistance.

In addition, studies on the triploid plants of *P. tomentosa* were carried out (Einspahr et al., 1963) after Nisson-Ehle first found the triploid *P. trimuloidis*. The triploid *P. trimuloidis* grow much faster and have a better wood quality than diploid ones, and their cold resistance is as good, if not better, than that of the diploid.

This research focuses on the cross breeding of *P. tomentosa* by crossing some diploid poplars and polyploid species introduced from the United States with *P. tomentosa* and *P. tomentosa* ×*P. alba* in order to obtain some crossbreds with characteristics of rapid growth, superior quality of wood and cold resistance.

2 Materials and methods

2.1 Materials

In the distribution region of *P. tomentosa*, there are rich genetic resources and wide variation (Huang, 1992). *P. tomentosa* 'Thick branch' and *P. tomentosa* 'Long branch', two types of *P. tomentosa*, were chosen from Zhengzhou in Central China's Henan Province. Their characteristics include a straight trunk, lush growth, high resistance to pests and diseases and high bearing rate (Zhu, 1990). *P. tomentosa*×*P. bolleana* was produced in the 1950's (Zhu, 1991). The flower branches of four *P. tomentosa*×*P. bolleana* were selected from the cam-pus of the China Agricultural University and named as MX1, MX2, MX3 and MX. MX had the most branches among the four plants.

P. grandidentata, the diploid, triploid and tetraploid *P. tremuloidis* and *P. alba*×*P. grandidentata* were from Wisconsin, America in the form of pollen. *P. grandidentata* with a fine trunk form and high wood quality is distributed separately in 40°—50°N, limited to Southeast Canada and Northeast America. The clones of *P. grandidentata* (G-1-58, G-1-65 and G-5-67) came from Ontonagon County in Michigan State. *P. tremuloidis* is the most widely distributed and important pulpwood species in North America (Barnes 1969). The diploid clone of *P. tremuloidis* T-44-60 was selected in Dickinson County of Michigan State, and I-13-87-12, I-13-87-35 and I-13-87-37 in Washtenaw County of Michigan. The triploid T-38-59 and tetraploid came from Ontonagon County in Michigan.

P. alba× *P. grandidentata* is a natural hybrid and its clones I-13-87-31 and 3-10-87-1 were picked in Washtenaw County of Michigan. *P. alba*×*Ulmus pumila* is a crossbred produced by the Liaoning Poplar Research Institution in northeast China. It has a similar configuration as *P. alba* and good wood quality as *U. pumila* as well as high cold resistance. The male flower branches of *P. alba*× *U. pumila* came from Gan County of Liaoning Province.

2. 2　Methods

2. 2. 1　Cutting cross in greenhouse

It could be found under the microscope that the pollen size we received was not identical. The triploid and tetraploid pollen was separated through the sieves of 30 and 40μm respectively and then stored in the refrigerator. The mother branches collected in the winter were trimmed and cultured in water in the greenhouse at 20℃. The water was changed and the low ends of the branches were cut every two or three days. When the stigma matured, pollination was done at 9 o'clock in the morning and repeated the next day. Then the pollinated flowers were covered with paper bags. Seeds were collected after 24 to 30d.

2. 2. 2　In vitro embryo culture

The immature embryos of *P. tomentosa* at different developmental stages were cultured in vitro. The basal MS medium (sucrose 20g/L) supplemented with various concentrations of IAA and 6-BA formed a series of culture media.

2. 2. 3　Growth of seedlings

Seeds were sown in clay pots in greenhouse and shoots were implanted into nutrition pots after a few days. "Hardening off" period began after the seedlings reached 5 to 10cm high. About one week later, the seedlings were transferred to the field. The culture tubes with culture stocks were taken to the greenhouse. After they adapted to the greenhouse environment, the stocks were implanted into nutrition pots. The remaining steps were the same as those for sown seedlings.

2. 2. 4　Polyploid identification

Plants with larger stomata, longer guard cells and less stomata per unit in leaves were selected and then the number of their chromosomes was counted under the microscope.

3　Results and analysis

3. 1　Effects of in vitro culture

3. 1. 1　Rescue of the abortion of immature embryos

There are some successful examples of overcoming cross sterility by in vitro embryo culture (Chen, 1979; Li and Li, 1985). As a material plant, the average bearing rate of *P. tomentosa* is only 2. 3% mainly because of the abortion of immature embryos (Li, 1987). Embryo rescue was carried out in the technique of in vitro embryo culture.

The cotyledonal embryos of two cross combinations were cultured in four different media to compare the effects of the media. As shown in Table 1, the germination rates of the cross combinations cultured in different media vary to some extent. Although the difference of the germination rates was not very significant, the qualities of stocks, leaves and roots changed greatly. In the medium MS+0. 08mg/L 6-BA, the embryos developed into stocks with shoots, leaves and few roots that were not connected directly but divided by the callus. But this medium limited the growth of radicles. In the medium MS + 0. 05mg/L IAA + 0. 08mg/L 6-BA, shoots and leaves grew, but roots were not observed. The reason of the absence of roots might be that IAA lowered the effect of 6-BA and constricted the growth of roots. The embryos planted in the MS + 0. 02mg/L 6-BA + 0. 05mg/L IAA had rapid growth and developed good shoots and roots and no callus. They came from crossbred embryos, not from the callus and they were real hybrids without genetic changes during the in vitro culture.

The germination rate of the embryos at different stages showed that embryo abortion of *P. tomentosa* in the in-group crossing happened at immature stages. But the earlier the embryos were cultured, the lower the germination rate. Chen (1983) cultured heart-shaped and torpedo embryos and found that their germination rates were 50%~54%. Li and Li (1985) cultured ovules of poplar hybrids and the results show that the germination rate of the embryos in cotyledon was highest. In this study the heart-shaped, pre-cotyledon and nearly mature embryos were cultured. The results are shown in Table 2.

Table 1 Germination rates of different cross combinations in different media

Cross combinations	0. 08mg/L 6-BA		0. 05mg/L IAA +0. 08mg/L 6-BA		0. 02mg/L 6-BA		0. 05mg/L IAAA		Total	
	NCE	GR	NCE	GR	NCE	GR	NCE	GR	NCE	GR
‘Long branch’×*P. fr*4*n*	18	88. 8%	38	78. 9%	25	76. 0	24	83. 3%	105	81. 9%
‘Thick branch’×*P. tr*4*n*	12	86. 7%	6	100. 0%	9	88. 0%	7	100. 0%	34	91. 0%
Total	30	83. 3%	44	81. 8%	34	79. 4%	31	87. 1%	139	83. 5%

Notes: ‘Long branch’×*P. tr* 4*n* is *P. tomentosa* ‘Long branch’×*P. tremuloidis* 4n; ‘Thick branch’×*P. tr* 4*n* is *P. tomentosa* ‘Thick branch’×*P. tremuloidis* 4*n*; NCE is the number of the cultured embryos; GR is the germination rate.

Table 2 Germination rates of the embryos at different stages

Cross combinations	Developmental stage of *in vitro* embryos	NCE	GR
‘Long branch’×*P. fr* 4*n*	Heart-shaped embryos	8	25. 0%
	Pre-cotyledon embryos	106	90. 0%
	Nearly mature embryos	10	0
‘Thick branch’×*P. tr* 4*n*	Heart-shaped embryos	12	16. 7%
	Pre-cotyledon embryos	16	93. 7%
	Nearly mature embryos	13	0
Total	Heart-shaped embryos	51	58. 8%

As a crossing objective, *P. tomentosa*×*P. bolleana* has a high bearing rate (Lin, 1988). But when its heart-shaped embryos were cultured, the germination rate was not necessarily high. The reason might be that they were at an early developmental stage and that the culture environment was very different from that of their parent plants. Similarly, the germination rates of the heart-shaped embryos of *P. tomentosa* ‘Long branch’ ×*P. tremuloidis* 4*n* and *P. tomentosa* ‘Thick branch’×*P. tremuloidis* 4n were low. The germination rates of pre-cotyledon embryos reached 90. 0% and 93. 7%. The nearly mature embryos all withered because the embryo stages might have all aborted. Therefore it was very crucial to culture the embryos at proper stages.

3. 1. 2　Germination rates and remaining rates by invitro culture and sowing in greenhouse

By comparing the germination and remaining rates between in vitro embryo culture and seed sowing, the extent of embryo abortion and the effect of embryo rescue are shown in Table 3. It was found that the germination rates of the seeds were 0 except for *P. tomentosa* 'Long branch' × *P. tremuloidis* 4n (2. 2%) and *P. tomentosa* 'Long branch' ×*P. tremuloidis* 3*n* (6. 7%), and that the remaining rates were also 0 except *P. tomentosa* 'Long branch' ×*P. tremuloidis* 4*n* (10. 0%). However, the germination rates of in vitro embryo culture were significantly higher than those of seeds, indicating that the embryo abortion of *P. tomentosa* could be overcome by choosing immature embryos at proper stages to culture in appropriate media.

Table 3　Comparison of the germination and remaining rates between in vitro embryos and seeds

Combinations	Seeds		In vitro embryos	
	Germination rale(%)	Remainingrate(%)	Germination rale(%)	Remainingrate(%)
'Long branch' ×*P. tr*4*n*	2. 2	10. 0	16. 7	50. 0
'Long branch' ×*P. tr*2*n*	0	0	77. 4	38. 7
'Long branch' ×*P. tr*3*n*	0	0	8. 3	0
'Thick branch' ×*P. tr*4*n*	0	0	54. 8	23. 0
'Thick branch' ×*P. tr*2*n*	0	0	41. 5	100. 0
'Thick branch' ×*P. tr*3*n*	6. 7	0	52. 5	52. 9
Mean	1. 5	41. 9	1. 7	44. 2

3. 2　Analysis of the growth, morphology and cold-resistance of the crossbred seedlings

After the field acclimatization, the seedlings from the cross seeds and in vitro embryos were implanted into the nursery from July to September and then cut at the base during the winter. Their roots were transplanted to Hebei Forest Seedling Nursery (in Yixian County of Hebei Province). The roots were arranged with a row spacing of 1 m.

3. 2. 1　Growth analysis

The seedling height and root diameter of 1-year-old crossbred seedlings were measured. Variance analysis illustrated a significant difference among combinations (Tables 4 and 5).

Table 4　Variance analysis of the height of 1-year-old seedlings of different combinations

Source	*df*	*SS*	*MS*	*F*
Between combinations	17	439032. 00	25825. 40	8. 35**
Error	785	2427880. 00	3097. 84	
Total	802	2866912. 00		

** significant at *P*= 1%

Table 5 Variance analysis of the root diameter of different combinations

Source	*df*	*SS*	*MS*	*F*
Between combinations	17	24. 52	1. 44	6. 08**
Error	491	116. 40	0. 24	
Total	508	140. 92		

** significant at $P = 1\%$

According to their average height and root diameter, MX_2× *P. tremuloidis* and MX_4×*P. grandidentata* were selected as the best combinations based on a further multiple comparison, followed by MX_3×P. tremuloidis and MX×(*P. alba*×*P. grandidentata*). It was reported that cold resistance of different poplar species could be known from their phenological traits (Wang 1982). The phenological characteristics of the cross seedlings and *P. tomentosa* 'Yi Xian' (as comparison) were observed.

It was found that the leaves of *P. tomentosa* 'Yi Xian' expanded from April 6 to 8 and MX_4 ×*P. grandidentata*, MX_3×*P. tremuloidis* and MX_4×*P. tremuloidis* (1-13-87-35) from April 14 to 18(Table 6). The leaf expanding phases of MX_2×*P. tremuloidis*, MX×(*P. alba*×*P. grandidentata*), MX_3×*P. grandidentata* and MX_4×*P. tremuloidis* were between April 10and 12. *P. tomentosa* 'Yi Xian' stopped height growth on October 20, 35d later than crossbred seedlings.

A late leaf expansion also indicates that the end of dormancy of the seedlings, the sap flow and the sprouting of buds all happen late. It can protect the seedlings from the damage of low temperatures in the early spring and improve their cold resistance. The growth end phase is also an important indicator to measure the cold resistance of plants. To a large extent, the cold resistance of poplars depends on whether they can stop growing in time in the fall. So some superior plants with good cold resistance might be selected from the hybrids.

Table 6 Phenology of the crossbred seedlings of different combinations

Combinations	Leaf expansion phase (day/month)	Growth end phase (day/month)	Detbliation phase (day/month)
MX_2×*P. grandidentata*	13/4-17/4	22/9-12/10	20/10-30/10
MX_3×*P. grandidentata*	10/4-17/4	22/9-29/9	3/11-15/11
MX_4×*P. grandidentata*	14/4-16/4	20/9-3/10	2/11-9/11
MX_2×*P. tremuloidis*	10/4-16/4	22/9-29/9	30/10-10/11
MX_3×*P. tremuloidis*	13/4-17/4	22/9-23/10	1/11-8/11
MX_4×*P. tremuloidis*	13/4-18/4	20/9-24/9	1/11-10/11
MX×(*P. alha*×*P. grandidentata*)	9/1-14/4	25/9-28/10	3/11-11/1
P. tomentosa 'Yi Xian'	6/4-8/4	22/10-28/10	6/11-8/11

Notes: Leaf expansion phase lasts from the beginning of leaf growth to the total expansion of two leaves below the terminal bud; growth end phase is the period that the terminal bud just stops growing and the bud scale appears; leaf dropping phase means that most leaves have dropped and the rest turned yellow.

Because of plentiful rain in that spring and fall and good irrigation in the nursery, the apical growth of the seedlings did not show two peaks as usual. The rapid growth of the new top lasted from June 6 to July 16. On August 6, the growth rate of the seedlings dropped sharply. After late August or early September, the seedlings stop growing. Since there was no typical fall growth, the growth increment in the fall was regarded as the growth after the beginning of autumn.

As shown in Table 7, the GIHR of *P. tomentosa* 'Yi Xian' was much higher than that of crossbred seedlings and that of MX_3× *P. tremuloidis* seedlings was the lowest. Meanwhile, MX4× *P. grandidentata* and MX×(*P. alba*×*U. pumila*) were almost the same. The growth increment of height in the fall is closely related to dehydration. The more the top grows, the more the stem loses its water and the lower the cold resistance of seedlings. In contrast, the high growth increment of diameter in the fall means more rhythmic growth which increases the cold resistance of seedlings.

Table 7　Comparison of the growth increment in the fall between cross combinations and *P. tomentosa* 'Yi Xian'

Cross combinations	MX_4× *P. grandidentata*	MX_3× *P. tremuloidis*	MX× (*P. alha*×*P. grandidentata*)	MX_2× *P. tremuloidis*	*P. tomentosa* 'Yi Xian'
GIHR (%)	14.8	9.3	13.5	16.0	26.5
GIDR (%)	10.3	12.5	12.6	7.5	2.0

Notes: GIHR means the ratio of growth increment of height in the fall to the total increment of the year; GIDR is the ratio of growth increment of the diameter at breast height in the fall to the total increment of the year.

3.2.2　Morphological analysis

Because of different parents, the crossbred seedlings of the combinations showed significant differences in their morphology. Multiple traits of the crossbred seedlings were measured and variance analysis and multiple comparisons were conducted. There was a great difference in leaf size between any two combinations except for MX_4×*P. grandidentata* and MX_2× *P. tremuloidis*.

The leaves of the seedlings of MX_4×*P. grandidentata* were large, thick, rough and very green. These traits as well as their rapid growth made the choice of seedling plants easy from among the hybrids. The leaf size of MX_2×*P. tremuloidis* was smaller than that of MX_4× *P. grandidentata*, but larger than other combinations. As in the analysis above, the growing period of MX_4 ×*P. grandidentata* and MX_2× *P. grandidentata* was shorter than that of others, but the growth of their height and diameter was superior to others.

There were also significant differences in length of leaf stalks between combinations. The average length of leaf stalks in MX×(*P. alba*×*P. grandidentata*) was 6.53cm, much longer than that of other combinations, especially the shortest one, which was only 3.42cm. Because of double cross, there was a great difference in the combination in length of leaf stalks and the longest one reached 10cm, which was close to the length of the leaves. The number of branches in various combinations was also different. Multiple analysis revealed that the number of the branches of MX

×(*P. alba*×*U. pumila*), MX×(*P. alba*×*P. grandidentata*), MX_3×*P. grandidentata* and MX_1×(*P. alba*×*U. pumila*) was much greater than that of the others. But there was no significant difference among these four combinations. If the branches of 1-year-old seedlings are too rich, the apical dominance will be weakened and the stem shape will be impaired. If there are few or even no branches, the apical dominance will be too apparent, the growth will be too rapid, the top will have branches next year and the seedling stem will be easily crooked for its low resistance to wind. Three to five branches in 1-year-old stems are optimal. MX_3× *P. tremuloidis*, MX_4×*P. grandidentata* and MX_2× *P. tremuloidis* had the proper number of branches.

Based on the analysis above, MX_3×*P. tremuloidis*, MX_4× *P. grandidentata* and MX_2×*P. tremuloidis* were selected as the superior combinations, followed by MX×(*P. alba*×*P. grandidentata*). Because of the variation among the seedlings of MX×(*P. alba*×*P. grandidentata*), they have great selection potential. Therefore the combination was also regarded as a super one.

3.3 Selection of super plants from crossbred seedlings

Super plants were selected from the four super combinations. The selection of super seedlings was made among the 1-year-old crossbred seedling populations according to their growth rate and cold resistance. The results are shown in Table 8.

Table 8 Comparison between the super plants in the super combinations and *P. tomentosa* 'Yi Xian'

Combinations	Number of super plants	Growth stopping phase (day/month)	Defotiation phase (day/month)	Leaf expansion phase (day/month)	Growing days	Days stopping growth earfier than the control	Average height (cm)	Average root diameter (cm)	Super plants/control	
									Height (%)	Root diameter (%)
MX_4×*P. gra*	13	20/9-26/9	4/11-6/11	14/4-16/4	199-209	27-39	337.4	3.03	134	135
MX_3×*P. tre*	7	24/9-15/10	4/11-6/11	14/4-16/4	203-209	7-35	319.4	2.77	126	123
MX_2×*P. tre*	1	4/10	4/11-6/11	14/4-16/4	203	18-24	364.0	3.30	145	147
MX×(*P. atbit* ×*P. gra*)	9	25/9-20/10	6/11-8/11	10/4-12/4	25-215	2-34	313.5	2.71	124	120
P. tomentosa 'Yi Xian' (control)		22/10-28/10	6/11-8/11	6/4-7/4	214-219		251.6	2.24	100	100

3.4 Identification of hybrid polyploid

When the pollen was separated by sieving, some haploid pollen might combine with the others. The sperms of haploid pollen with strong competition might go ahead of those of the diploid and dysploid pollen and combine with the eggs of the parent plants to produce diploid zygotes. So even if polyploid pollen was used to pollinate, the seedling plants might not be polyploid at all. Therefore, it was necessary to determine whether the seedling plants were polyploid or not.

The number, the size of stomas and the size of their guard cells on the leaves of the 20 polyploid hybrid seedlings were counted and measured. The results showed that the number of the stomas per unit of 11-4, 19-1, 18-5 and B 4-3 was more than that of others and that the sizes of the stomas and the guard cells were significantly smaller than those of others (data not shown). So those plants were considered to be diploid. The remaining seedling plants were regarded as polyploid and were further identified by counting their chromosomes(Table 9).

Table 9 Number of the hybrid chromosomes

Name of the hybrid phmls	Number of chromosomes	Name of the hybrid phmls	Number of chromosomes	Name of the hybrid phmls	Number of chromosomes
12-1	42-46	11-3	41-9	B_4-1	38-45
12-2	47-51	19-2	38	B_4-2	44-52
12-3	42-45	18-1	41-50	B_4-3	43-52
12-4	48-50	18-2	42-48	B_4-4	45-50
11-1	47-50	18-3	44-51		
11-2	38	18-4	43-51		

The chromosomes of diploid poplars ($2n=38$) are very small and very difficult to separate. During observation, proper estimation had been made. Table 9 showed that plants 11-2 and 19-2 were diploid. The rest could be considered triploid, for the numbers of their chromosomes were between 38 and 57 (Einspahr and Winton, 1976).

4 Conclusions and discussion

To obtain the super hybrids of *P. tomentosa*, the crucial technology is to overcome its embryo abortion. Two methods were used: changing the mother to its hybrid *P. tomentosa*×*P. bolleana* which could bear a lot of seeds and using in vitro embryo culture which could increase the germination rate from 1.5% to 41.1% and the remaining rate from 1.7% to 44.2%.

Forty crossing combinations were made and 2744 hybrid seedlings were obtained. By observation and analysis, MX_4×*P. grandidentata* G-1-58, MX_3×*P. tremuloidis* T-44-60, MX_2×*P. tremuloidis* 1-13-87-37 and MX_2× *P. alba*×*P. grandidentata* were regarded as superior combinations. From those combinations, 30 plants were selected as super plants that had a number of good characteristics such as rapid growth, possible resistance to cold, diseases and wind, straight stem etc. But further tests are needed on their physiological indices in different areas of northeast China to study their cold resistance.

Twenty hybrids were obtained by crossing MX with triploid *P. tremuloidis* and 14 plants were triploid after identification. The triploid hybrids did not grow as well as the hybrids of MX×diploid *P. remuloidis*. The stems of some plants had too many branches and were crooked without apparent apical dominance. There were also some triploid plants which suffered from serious diseases.

The reason might be that the hybrids were allotriploid, unlike autotriploids which have better characteristics than diploids in growth and resistance. On the other hand, the potential of the triploid seedlings has not been displayed fully, so we discarded the poor hybrid triploids and selected 11 seedlings for further test.

杨树体细胞胚胎发生研究概况*

摘 要 综述了杨属树种体细胞胚胎发生研究概况，共有 13 个树种进行了体细胞胚诱导。对影响体细胞胚发生的因素，如外植体、激素、氮源等进行了综合分析，并对体细胞胚胎发生中的同步性和畸形问题进行了讨论。该文对今后杨树体细胞胚研究及其意义亦提出一些看法，认为诱导胚性愈伤组织是体细胞胚胎成功发生的关键，应逐步开展体细胞胚发生的生理、生化及分子水平机理研究；加强杨树体细胞胚发生及机理研究，不仅在其遗传改良、离体繁殖等实际应用中有重要意义，而且作为林木生物学研究的模式树种，对林木发育生物学及激素调节等理论的研究亦有重要价值。

关键词 杨树，体细胞胚胎发生，愈伤组织，组织培养

植物体细胞胚胎发生的研究，是伴随着植物的细胞和组织培养技术而发展的。自 1958 年 Sterward 和 Reinert 在胡萝卜组织培养中发现胚状体，首次证明细胞具有全能性以来，对体细胞胚胎发生的认识逐步加强，概念也越来越明确(Haccius，1978；朱澂，1978)。通过体细胞胚胎发生途径形成再生植株，在植物界离体培养中是极其普遍的，并且自然界也存在这一再生途径(David，1983)。林木体细胞胚胎发生研究，最早报道于 1965 年檀香的合子胚愈伤组织体细胞胚胎发生(Rao，1965)，直到 20 世纪 80 年代后期才得到迅速发展，尤其松柏类体细胞胚发生取得了令人瞩目的成绩，已应用于生产实践和体细胞胚发育模式的分子水平研究(Dong et al.，2000；Stasotlla，2002)。杨树是杨属树种的统称，在林木生物学理论和应用研究中，一直备受关注，体细胞胚胎发生方面也有一些报道(张立功等，1981；林静芳等，1984；巴岩磊等，1986；Michler et al.，1987，1991；Cheema，1987，1989；Park et al.，1988；王影等，1991；Chung et al.，1991；Baldursson et al.，1993；Ostry et al.，1994；杜克久等，1998；马海芸等，2002)，基本处于实验条件摸索阶段，这与该属树种在林木研究中的地位是不相称的。

1 杨树体细胞胚胎发生概述

木本植物组织培养中体细胞胚胎产生也是相当普遍的(Jain，et al.，1995；汤浩茹等，1999)。杨树最早在 1981 年从小黑杨花粉植株的茎段愈伤组织中诱导出胚状体，至今已有 7 种 6 杂种进行了体细胞胚诱导，见表 1。此外，在三倍体毛白杨(马海芸等，2002)、河北杨(林静芳等，1984)、NE299(*Populus nigra* var. *betulifolia* ×*P. trichocarpa*)(Ostry et al.，1994)、欧美杨的组织培养中也有体细胞胚的报道(Chung et al.，1991)。杨树体细胞胚胎

---

* 本文原载《东北林业大学学报》，2005，33(1)：60-63，与于志水、李云、胡崇福和王胜东合作发表。

表 1　杨树体细胞胚胎发生

种名	外植体	胚性愈伤组织诱导	体细胞胚发生及发育	体细胞胚萌发	结果	文献
小黑杨花粉植株 *P. simonii*×*P. nigra*	茎段	MS30+2, 4-D2+KT1	MS30+BA0. 1 或 0. 25+NAA1; MS30+BA0. 5+IAA0. 3; 1/2MS30+KT0. 4 或 BA0. 5	1/2MS20-30+IAA0. 5	ISE PL	张立功等, 1981
缘毛杨 *P. cilicata*	叶片	MS30+BA2; MS+2, 4-D0. 5	MS+BA0. 4+NAA0. 2; MS+2, 4-D0. 2	MS+BA0. 4+NAA0. 2	ISE PL	Cheema, 1987, 1989
银白杨×大齿杨 NC-5339 *P. alba*×*P. grandidentata*	叶片	MS30+2, 4-D5+BA0. 05	MS30+BA 或 ZT0. 05(球形胚发育)	MS+BA0. 05+IAA5	I/DSE PL	Michler, *et al.*, 1987, 1991
黑杨×辽杨 *P. nigra*×*P. maximowiczii*	叶片	MS+BA0. 1+2, 4-D0. 5	MS+BA0. 1+2, 4-D0. 5	MS30	DSE PL	Park, *et al.*, 1988, 1995
小叶杨 *P. simonii*	茎段	MS+2, 4-D1-8	MS30-50+BA0. 2~2+NAA0. 02-0. 1(球形胚发育)	1/2MS	ISE PL	王影等, 1991
741 杨 *P. alba*×(*P. davidiana*×*P. simonii*)×*P. tomentosa*	茎尖	MS30+KT0. 1+2, 4-D0. 05	MS30+KT0. 3	——	ISE	杜克久等, 1998
P. nigra; *P. simonii*; (*P. simonii*×*P. nigra*)×*P. rassica*; *P. alba*×*P. laurifolia*	茎尖、茎段、叶片	H+BA1+NAA0. 5	MS+BA0. 1+NAA0. 2	MS+BA0. 1+NAA0. 2	I/DSE PL	巴岩磊等, 1986
银白杨 *P. alba*	茎尖、嫩茎、叶柄	1/2MS+ZT0. 5+IBA0. 2	MS+BA0. 3+GA0. 3+NAA0. 1	MS+BA0. 3+GA0. 3+NAA0. 1	DSE PL	巴岩磊等, 1986
香脂杨 *P. balsamifera*; 辽杨 *P. maximowiczii*; 毛果杨 *P. trichocarpa*	花药	MS+BA1. 13	MS+BA1. 13	WPM+BA0. 54+NAA0. 001 发芽; WPM+IBA0. 25 生根	DSE PL *	Baldursson, *et al.*, 1993

注：EC 代表胚性愈伤组织，D/ISE 代表直接型或间接型体细胞胚发生，PL 代表再生植株，MS30 表示 MS +蔗糖(30 g/L)，激素单位为 mg/L，* 表示香脂杨未取得小植株。

发生主要在器官、愈伤组织、小孢子、悬浮细胞 4 类培养物中产生，可分为 3 个过程：胚性愈伤组织诱导；体细胞胚的发生和发育；胚的萌发或植株再生。部分树种依体细胞胚发育特性(Michler et al.，1991；王影等，1991；杜克久等，1998)，又可以把球形胚阶段划入第一过程。发育过程形态变化与合子胚极相似，亦经球形胚、心形胚、鱼雷胚及子叶胚阶段，一般不需要休眠成熟阶段(Cheema，1989；Michler et al.，1991)；大小一般与合子胚相差不大，黑×辽(Park et al.，1988)从心形胚发育到鱼雷胚在 0.5~1.0mm 之间，NC-5339(Michler et al.，1991)心形胚为合子胚的 2 倍，子叶期略偏大。体细胞胚植株(embling)存在变异，在 NE299(Ostry et al.，1994)体细胞变异研究中，认为茎段愈伤组织悬浮细胞来源的体细胞胚植株，形态上没有差别，但对 *Septoria musiva* 敏感性上存在一定变异；花粉胚植株多数自然加倍成二倍体(Baldursson et al.，1993)。

Sharp 等(1980)把体细胞胚胎发生途径按起源分为两个类型：直接途径与间接途径。杨树的体细胞胚发生两种途径都存在，直接型体细胞胚发生采用固体培养，一般会同时伴随愈伤组织或器官发生(巴岩磊等，1986；Park et al.，1988)；间接型体细胞胚发生中，固体培养和液体悬浮培养都有应用(张立功等，1981；Cheema，1987，1989；Michler et al.，1987，1991；王影等，1991；杜克久等，1998；巴岩磊等，1986；Baldursson et al.，1993)。1987 年 Cheema 和 Michler 最早分别从缘毛杨和白杨杂种 NC5339 的细胞悬浮培养中获得体细胞胚。胚性愈伤组织表面特征不一致，缘毛杨(Cheema，1989)叶片产生的胚性愈伤组织易碎、黄白色；小叶杨(王影等，1991)胚性愈伤组织结构致密，呈淡绿色；小黑杨茎段致密，乳黄、黄或黄绿色(张立功等，1981)；741 杨茎尖产生的黄绿色坚硬的愈伤组织为胚性愈伤组织(杜克久等，1998)。缘毛杨愈伤组织悬浮培养继代 6 次后，胚性将消失，而固体培养胚性持续时间较长些(Cheema，1989)。胚性愈伤组织表面特征的不一致性，除了与遗传型和材料类型的不同有关外，培养条件，如光强、培养基成分、温度等，对其可能也有较大的影响。现在一般认为，绝大多数植物体细胞胚起源于单细胞(崔凯荣等，2000)。741 杨起源于愈伤组织的表层单细胞或小细胞团(杜克久等，1998)，而缘毛杨、NC-5339 由悬浮细胞直接诱导体细胞胚发生(Cheema，1989；Michler et al.，1991)。由接种材料组织细胞外源的，体细胞胚或胚性愈伤组织多从皮孔、叶脉及伤口处产生(巴岩磊等，1986)，这要对其进行组织学观察，以区别器官发生。关于体细胞胚的超微结构，杨属中没有进行详细观察，一般胚性细胞与胡萝卜等大多数植物胚性细胞相似，细胞质浓，含有丰富淀粉粒(Cheema，1989)。

2　体细胞胚胎发生的诱导条件

2.1　外植体选择

植物离体形态发生只有器官发生和体细胞胚发生两种途径，而不同的植物或不同部位也以不同的形态发生形成再生植株，事实也说明了在形态发生中，遗传性与生理状态可能比激素的作用更为关键(詹祥灿，1983)。

体细胞胚发生受遗传型影响。以花药为外植体诱导中，欧洲山杨、美洲山杨及灰杨未成功，体细胞胚发生的都为青杨组树种(Baldursson et al.，1993)，741 杨甚至以体细胞胚发生形式再生植株为主(林静芳，1986)。从表1中可以看出，营养器官为外植体的体细胞胚发生主要集中在银白杨和小叶杨及它们的杂交种。同一物种，基因型的不同产生胚能力也不一样。在小叶杨体细胞胚发生研究中，发现5个细胞无性系在产胚能力上存在极显著的差异(王影等，1991)。毛果杨花药培养中，产生胚状体的植株都来自同一省份，这说明种源对体细胞胚发生也有一定影响(Baldursson et al.，1993)。同种材料类型和不同的生理状态对其诱导影响也至关重要。如毛果杨(Baldursson et al.，1993)花药诱导在不同年份效果截然不同。以叶片为诱导材料的：黑杨×辽杨(Park et al.，1988)以组培苗的针刺叶片为材料，才能诱导出胚性愈伤组织；NC-5339(Michler et al.，1991)取组培苗未展开的叶片；缘毛杨(Cheema，1989)先进行BA长时期暗培养。因此，对外植体材料进行适当的选择或前处理，有利于体细胞胚的诱导。未熟胚或与生殖组织有关的材料，如珠心、珠被、花药等，在一些植物中被认为是较理想来源的外植体，因为它们具有预胚性细胞(PEDC，pro-embryogenic-determined cells)(崔凯荣，2000)，这在杨树中并未体现，叶片、茎尖、叶柄、茎段及花药都能诱导出胚状体，但花药诱导率不高，辽杨(Baldursson et al.，1993)最高诱导率为2.0%。营养型外植体产生的体细胞胚因不受时间限制，体细胞胚植株的基因型与母株又相同，因此在实际应用中意义更大。

2.2 激素影响

杨树体细胞胚诱导中，无论生殖型还是营养型外植体，生长素或与分裂素结合是体细胞胚诱导和发育所必须的。

2,4-D是植物体细胞胚诱导中使用最广泛的生长素。杨树体细胞胚胎发生的间接途径与胡萝卜“典型系统”相似，即也可分为两个阶段，脱分化阶段和体细胞胚发生和发育阶段，第一阶段需要2,4-D，第二阶段需降低浓度或去除2,4-D。有的树种略有差别，如741、NC-5339在球形胚阶段之后才撤掉(Michler et al.，1991；杜克久等，1998)。体细胞胚直接发生类型一般不需要2,4-D(巴岩磊等，1986)，如存在，会伴随不定芽或愈伤组织出现，如NC-5339(Michler et al.，1991)、黑×辽(Park et al.，1988)。杨树组织培养中极易诱导器官发生，因此体细胞胚直接发生中可能会伴有不定芽或根的出现。体细胞胚间接发生类型中，胚性愈伤组织的诱导需要2,4-D的存在(Park et al.，1988；Michler et al.，1991；杜克久等，1998)，处理时间和浓度对胚性愈伤组织发育也有影响，浓度范围因物种和发育阶段而不同，一般在0.05~10.00mg/L之间。741杨胚性愈伤组织诱导中，2,4-D处理时间10~15d最佳，过短、过长都无体细胞胚发生或频率低，形成球形胚之后需及时撤掉，否则将阻止正常发育(杜克久等，1998)；NC-5339(Michler et al.，1991)，球形胚阶段后亦要撤离2,4-D，但直接从叶脉处产生的胚在含2,4-D上却可以发育成熟，这与黑×辽(Park et al.，1988)相同；缘毛杨(Cheema，1989)高浓度能得到悬浮较好的细胞系，低浓度形成大的细胞团。在胚萌发及生根过程中一般不需要2,4-D。经2,4-D诱导产生的愈伤组织有胚性和非胚性之分，为了胚性的保持，最好把胚性组织从邻近非胚性组织上分离开。如NC-5339(Michler et al.，1991)，脉间会含有非胚性愈伤组织，在悬浮培养时，它们的快速生长会阻碍胚性组织的生长；小叶杨悬浮时，也需要挑选胚性愈伤组织(王影等，1991)。

杨树体细胞胚诱导中，分裂素的存在是有利的，这与多数植物是一致的(周俊彦，1983)。有的分裂素需和生长素结合使用(巴岩磊等，1986；Cheema，1987，1989；Park et al.，1988；杜克久等，1998；Baldursson et al.，1993)，质量浓度一般在0~1mg/L，不同树种对分裂素种类有选择性。在胚发育过程中，基本上都需要分裂素(缘毛杨细胞悬浮方式除外)。小叶杨胚性愈伤组织诱导时，外加较高浓度的BA会产生较多的体细胞胚(王影等，1991)；NC-5339(Michler et al.，1991)，BA或ZT存在的情况下可增加球形胚数量，ZT效果更好；741杨(郑均宝，1999)BA效果不如KT。与低浓度的生长素NAA或IAA结合，可促进胚(胚根或胚轴)的伸长(Michler et al.，1991；王影等，1991)。体细胞胚萌发或生根过程中一般不需要添加生长调节物质(Park et al.，1988；王影等，1991)。

2.3　碳源、氮源及其他外部因素

蔗糖作为碳源和渗透剂在被子植物体细胞胚发生中常被认为是最有效的(David，1983)，杨树的适宜质量分数在1%~6%之间。胚的不同发育阶段对蔗糖需求也有所不同，NC-5339在质量分数为3%时对球形胚的形成最有效，大于5%时将抑制球形胚形成(Michler et al.，1991)；小叶杨中认为在3%~5%时对胚发生最有效，大于8%时将降低体细胞胚发生频率(王影等，1991)。在胚发育的早期，需较多营养，蔗糖也相应较高，随着体细胞胚的发育，对环境要求下降，蔗糖需求也下降。

植物激素、碳源、氮源是调节体细胞胚发生的三大因素，该属大部分采用MS培养基(表1)，对氮源量及形态需求相关材料比较少，只有NC-5339对谷氨酰胺进行了浓度测定，谷氨酰胺的增加对体细胞胚形成并非必须的，在20~40μmol/L下能增加从叶盘产生体细胞胚的数量，但会抑制胚性愈伤组织的形成(Michler et al.，1991)。黑×辽在悬浮培养时加入少量椰子乳、麦芽膏，有利于提高体细胞胚发生的数量，混合作用效果更好(Park et al.，1995)。椰子乳、麦芽膏对杨树叶片体细胞胚直接发生诱导可能也是有利的(Park et al.，1997)。不同树种及培养方式，甚至不同发育阶段对光照会有不同的要求，甚至是模糊的。如NC-5339(Michler et al.，1991)的体细胞胚发生要求绝对黑暗条件，其萌发时有一定光照是有利的，而741杨等树种在体细胞胚发生和发育过程中对光照要求不严格(巴岩磊等，1986；王影等，1990；杜克久等，1998；Baldursson et al.，1993)。采用悬浮培养一般以黑暗条件为好，如缘毛杨(Cheema，1989)、黑×辽(Park et al.，1995)。悬浮培养是历来研究者青睐的体细胞胚培养方式，但培养方式不仅影响胚性的有无，而且对光照和激素也有不同的要求。在缘毛杨细胞悬浮培养中需要降低2, 4-D的浓度，如撤掉2, 4-D会导致细胞团生根，且只在黑暗条件下诱导体细胞胚发生；固体培养却需要撤掉2, 4-D，转到含分裂素和其他生长素的培养基上，可在光照下诱导体细胞胚。影响体细胞胚发生的因素很多，杨树这方面的材料非常有限。

3　体细胞胚胎发生存在的问题

杨树体细胞胚胎发生的成功率低、重复性差，远远达不到实际应用上的要求。虽然许多树种已成功地进行了体细胞胚诱导，但对影响体细胞胚发生的许多因素了解得尚不深入，或根本不了解，文献记录有的也不够具体，存在很多问题，有必要进一步细致地研究。

不同步化和畸形是植物体细胞胚发生中最为普遍的现象。从小黑杨花粉体细胞胚植株诱导成功以来，不同步化就伴随杨树体细胞胚的发展，由于体细胞胚诱导技术上无突破，人们还无暇进行这方面的系统研究。不同步化是多数植物体细胞胚诱导中经常出现的问题，常采用选择同步法和诱导同步法，从细胞或小细胞团的水平控制其发育过程中的不同步性。选择同步法常将胚性愈伤组织进行悬浮培养，通过体积选择(分级过筛或 Ficoll 梯度离心)等物理手段选择一致的接种物(Philip，1977)；诱导同步法常采用化学方法(饥饿法、抑制法)(Kurz，1971)或物理方法(Kubek，1978)(冷处理)及渗透压(郭钟琛，1990)等方法使细胞分裂停滞在某个时期，之后，同步进入下一阶段。Warren(1977)发明的一种用玻璃珠(glass beads)过筛分离不同阶段胚的方法，从体细胞胚自身发育不同阶段控制其不同步性，对部分树种有一定意义。如 741 杨和 NC-5339，在 2, 4-D 的存在下体细胞胚可以发育到球形胚(Michler et al.，1991；杜克久等，1998)，可在此阶段进行球形胚的同步筛选。选择恰当的外植体可能有利于体细胞胚发育的一致，巴岩磊(1986)在黑杨组和青杨组杨体细胞胚诱导中，发现叶脉产生的胚状体出现较整齐，而茎段、茎尖所产生的胚状体是不同步的。不同步化在体细胞胚诱导中是个很复杂的现象，不仅可以处于不同的发育阶段，即使同一阶段，大小形状差异也可能很大，从而影响体细胞胚的质量。体细胞胚发育过程中的畸形是相当普遍的，几乎植物体细胞胚发生中所有不正常现象在杨树中都能找到，严重地影响植株的转化。杨树体细胞胚发育过程中，畸形主要表现为：次生胚发生、胚结(clusters of embryos)、短胚轴、子叶融合、多子叶、萌发困难、胚早熟、产生无叶苗、体细胞胚苗玻璃化等等(Cheema，1989；Michler et al.，1991；Baldursson et al.，1993)。此外，小黑杨等杨树胚状体存在棍棒状现象(张立功等，1981；巴岩磊等，1986)(可能是鱼雷胚阶段)。培养方式对畸形产生是有影响的，NC-5339(Michler et al.，1991)液体悬浮培养时产生玻璃化苗，鱼雷胚阶段表现早熟，而固体培养无此现象。在处理畸形中，常采用 ABA 处理、干化或渗透处理抑制畸形产生。杨树在这方面还没有进行尝试，Michler(1991)认为影响畸形及同步化的主要因素，如 ABA、ZT 和蔗糖，还需要进一步研究。

4 杨树体细胞胚研究展望

杨树因其重要的经济生态地位及特殊的生物学特性，而成为林木生理、生化及分子生物学研究中的拟南芥(Taylor，2002)。体细胞胚胎发生再生植株与器官发生相比较，具有 3 个显著特点：数量多、速度快、结构完整，不仅能彻底解决组织培养中微繁成本高等问题，而且能提供胚性细胞，让我们更为方便地进行性状的遗传改良。体细胞胚胎发生重演受精卵形态发生的特性，是细胞全能性表达最完全的一种方式，是深入研究植物细胞分化、发育的理想离体实验体系。该属树种本身的激素水平又较特殊，在同一材料上体细胞胚发生与器官发生可以同时进行(Park et al.，1988；杜克久等，1998)，这对了解植物细胞和组织培养中，体细胞胚发生与器官发生过程中的激素调节机理亦有很大价值。

充分利用杨树已有的细胞和组织培养的经验，借鉴木本植物体细胞胚研究的成果，加强其体细胞胚研究。首先应了解内外因素对体细胞胚发生的影响，优化杨树体细胞胚胎发生的步骤。杨树细胞和组织培养中，愈伤组织是较易产生的，愈伤组织必须进行到一定状态，其细胞才能形成体细胞胚，也就是说胚性愈伤组织(胚胎状态)诱导是最为关键的，可以直接进行胚性诱导，也可以先进行愈伤组织诱导，再将其转化为胚性愈伤组织或产生胚性愈伤组织。将愈伤组织转化为胚性愈伤组织可能会减少实验的繁琐，有利于同步性控制。一般情况下，任何外植体在2,4-D作用下都能产生愈伤组织，但诱导产生具有胚性的愈伤组织还很困难，因为大部分树种是以器官发生途径再生植株为主。黑杨组杨树愈伤组织多以产生芽为主，而白杨组易生根，因此诱导条件要求格外严格。胚性愈伤组织表面特征缺少一致性，愈伤组织是否具有胚性又是以形态发生结果来判定，这会相对增加实验的难度。形态发生中不同植物对激素的敏感程度(作用程度)有选择性，一般选择敏感性强的激素进行体细胞胚诱导效果可能更理想。从报道文献的体细胞胚胎发生难易程度看，杂种植株可能更容易些。杨树通过器官发生途径的组织培养技术在20世纪90年代初就已相当成熟(Ostry et al.，1991；Bajaj，1989；Confalonieri et al.，2003)，而体细胞胚发生报道还相对很少，也就是从激素调节角度来看，绝大多数树种以器官发生为主，激素已不是影响形态发生的主要限制因子。进行体细胞胚胎发生研究时，关注其他因子，如培养基成分、温度等，可能更易诱导成功；其次，在有效的体细胞胚再生系统的基础上，逐步开展体细胞胚发生的生理、生化及分子水平的机理研究。体细胞胚发生伴随着特定的遗传信息表达，这也是人们进行体细胞胚胎发生机理研究的兴趣所在。至今，胡萝卜和苜蓿已成为体细胞胚发生系统研究的模式植物，对体细胞胚发生过程中活动基因及表达的特异蛋白质进行了克隆及检测研究，也发现了与体细胞胚发生相关的调控因子及其激素和一些理化因子诱导产生的蛋白质(崔凯荣等，2000；Zimmerman，1993)；此外，对体细胞胚胎发生中的生理代谢，如多氨、程序性细胞死亡(PCD，programmed cell death)等也是人们关心的热点(崔凯荣等，2000)。松柏类植物在林木体细胞胚研究中也是较为出色的，也开始了上述大部分工作，同时也加强了体细胞胚质量提高的研究(Dong et al.，2000；Stasotlla et al.，2002，2003)，特别是ABA及PEG对体细胞胚成熟诱导的协同作用，PEG脱水处理有取代小分子糖类的趋势。杨树被公认为林木生物学研究的模式树种，从20世纪80年代初发现体细胞胚以来，已有20多年，但至今仍未取得大的进展，只有从根本上了解体细胞胚发生机理及其控制，才能真正理解植物的形态发生并加以有效的利用。

Overview on Somatic Embryogenesis of Poplar

Abstract An overview on the studies performed on 13 species or hybrids involved in somatic embryogenesis was presented. The factors influencing somatic embryogenesis including explant, hormone and nitrogen source etc. were analyzed in detail, and a discussion was given to the synchronization and abnormality in somatic embryogenesis. Some suggestions were proposed for the studies on somatic embryogenesis and its importance. The induction of embryogenic callus is the key to the success of somatic embryogenesis. Therefore, it is necessary to conduct researches on physiology and biochemistry of somatic embryogenesis and its molecular mechanism. Acceleration at studies on somatic embryogenesis and its mechanism in pop lar is of great value to genetic improvement and micropropagation; meanwhile, it will also be beneficial to the studies on developmental biology of forest tree and the mechanism of hormone regulation in morphogenesis.

Key words *Populus*, somatic embryogenesis, callus, tissue culture

Somatic Embryogenesis from Cell Suspension Cultures of Aspen Clone*

Abstract Suspension cultures initiated from callus derived from petiole explants of aspen hybrid (*Populus tremuloides*×*P. tremula*) produced somatic embryos. Callus was induced on a MS medium supplemented with 5mg/L 2, 4-D and 0. 05mg/L zeatinunder light conditions. Embryogenic calli were obtained when a subsequent subculture of calli was suspended in the same basal medium with 10mg/L 2, 4-D. The highest number of globular embryos were induced from embryogenic calli by cell suspension culturein a MS liquid medium supplemented with 10mg/L 2, 4-D. Genotype and 2, 4-D concentration were vital to the induction of embryogenic calli producing competent cells. Embryogenic calli for each genotype were heterogeneous. Green calli with gel-like consistency could yield more competent cells than light yellow embryogenic calli. However, some globular embryos broke into slices and some developed abnormally after one month of culture under the same or other hormonal conditions.

Key words *Populus*, callus, cell suspension, somatic embryogenesis

1 Introduction

Somatic embryogenesis, the process by which asexual or somatic cells are induced to form embryos in plant cell cultures, provides a valuable tool for genetic improvement and cell development biology. It also offers a fast, reliable and reproducible method for mass propagation and seems to be the most promising way to achieve rejuvenation, especially in forest trees.

Poplars, as economically important woody plants in many countries, have become model plants in forest tree biology (Taylor, 2002). Somatic embryogenesis in poplar was first reported by Zhang et al. (1981) who obtained regenerated plantlets from callus culture of pollen-derived plants of *Populus simonii*×*P. nigra*. Subsequently, somatic embryogenesis from cell suspension of *P. alba*× *P. grandidentata* and *P. ciliate* was first obtained (Cheema, 1987; Michler and Bauer, 1987).

Up to now, somatic embryogenesis or microsporic embryogenesis was successfully demonstrated in about 7 species and 7 hybrid species of *Populus*(Deutsch et al., 2004; Yu et al., 2005). However, this technique has low probability of success, low repetition and abnormal morphology and we still lack profound understanding of somatic embryogenesis of poplars. So further studies

* 本文原载 *Forestry Studies in China*, 2005, 7(3): 43-46, 与于志水、罗军民、胡崇福和何承忠合作发表。

are needed (Confalonieri et al., 2003). In the present study, we attempt to achieve somatic embryos by means of cell suspension cultures.

2 Materials and methods

2.1 Plant materials

Dormant 1-year-old branches of *P. tremuloides*× *P. tremula* (XT-Ta-34-97-S3), *P. alba*× *P. granditata*(AG-1-56) and pure aspen of *P. tremula* (Ta-7-68), were harvested from the nursery of Beijing Forestry University. Stem cuttings 40cm long were grown to sprout at 20~25℃ in water-filled containers in the greenhouse. Petioles were collected 3~4 weeks after the bud broke and surface-sterilized with 0.1% (W/V) $HgCl_2$ for about 3 min and then rinsed three times in sterile distilled water.

2.2 Callus induction

Petioles of aspen were aseptically inoculated horizontally into 100-mL Erlenmeyer flasks containing 30mL of MS (Murashige and Skoog 1962) medium supplemented with two concentrations of 2,4-D (0.5 and 5mg/L) and zeatin (ZT) (0 and 0.05mg/L) for induction of calli. Fifteen explants from three clones were cultured in one treatment. The pH of the medium with 30g/L sugar was 6.0 prior to the addition of 0.5% agar and was autoclaved for 15min at 121℃. Cultures grew at 25±2℃ under 2000Lux irradiation by fluorescent light at a 16-h photoperiod. Petioles of AG-1-56 were cultured either in the light (2000Lux) during a 16-h photoperiod or in continuous darkness.

2.3 Suspension cultures

Four-week-old greenish white, friable fragments (2mm in diameter) of calli (2~3g fresh weight) from petioles were transferred to 250-mL Erlenmeyer flasks containing 50mL of MS liquid medium supplemented with four levels of 2,4-D (0.01, 0.1, 1 and 10mg/L). Tests of 2,4-D treatment were repeated four times. They were agitated on a gyratory shaker at 80~100r/min. Cultures were maintained under 20lx at 25±2℃. Further subcultures involved the replacement of callus suspension with fresh medium of the same formulation at 14-d intervals. Single cells from suspension were observed under a microscope during the culture periods when the callus color turned green. Five mL of suspended cell samples were either transferred to the liquid medium supplemented with 0.05mg/L ZT and 0.02mg/L NAA in order to maintain suspension cell lines, or subcultured in MS liquid medium with 2,4-D (0.01, 0.1, 1 and 10mg/L) to initiate somatic embryogenesis.

2.4 Somatic embryogenesis

Five mL of free cell lines were transferred to a liquid medium supplemented with 2,4-D

(0.01, 0.1, 1and 10mg/L). Each treatment was repeated four times. The culture condition was the same as callus suspension cultures. After two weeks, the highest number of globular embryos, of about 0.5 ~ 1 mm, were initiated from cell suspension cultures. The globular embryos were washed with sterile distilled water and were transferred to a solid medium with 0.1mg/L BA and 0.02mg/L NAA, or the liquid medium containing 3 ~ 10mg/L 2,4-D.

3 Results

3.1 Callus induction

On MS medium containing either 2,4-D or ZT, the percentage of callus induction reached 100% within 30d of culture for three clones in light or darkness. The color and texture of callus of three aspen clones were alike. These callus formed on MS medium containing only 2,4-D or combined with ZT were yellowish white and coarse and callus clumps developed over 2 g per explant within 30d of culture and occurred over the entire surface of the explant. Furthermore, the calli of AG-1-56 grew rapidly within 2 weeks in the presence of 2,4-D. The concentration of 2,4-D had no visible effect on callus production. ZT in combination with 2,4-D stimulated the formation of friable, greenish white calli. We chose the calli from the combination of 0.05mg/L ZT and 5mg/L 2,4-D to perform suspension culture and produce embryogenic cells.

3.2 Suspension cultures

When transferred to a liquid medium containing only 2,4-D, the yellow calli of AG-1-56 developed rapidly into big clumps and turned brown and became hard and would die in the following subcultures. No single cells in liquid could be found. The pale greenish calli of XT-Ta-34-97-S3 would turn green with viscidity under 2,4-D (10mg/L) after a third subculture (Fig. 1), or became brown and hard without viscidity on the liquid medium with 2,4-D (0.01, 0.1 and 1mg/L).

Cell suspension cultures initiated from this kind of callus produced the highest number of

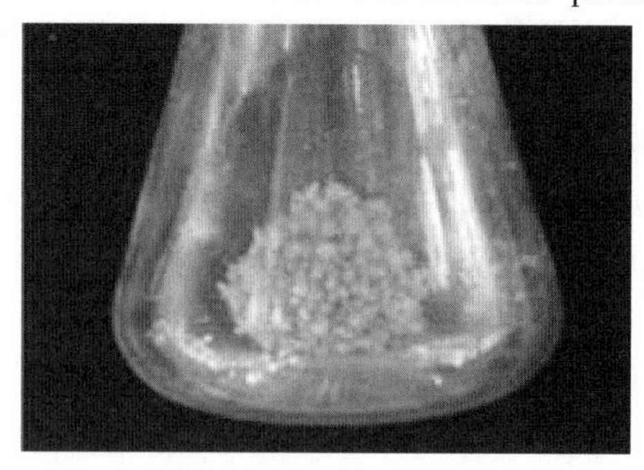

Figure 1　Embryogenic callus of *Populus tremuloides* × *P. tremula* by third suspension callus cultures

globular embryos, when 2,4-D was 10mg/L. Active division of cells was observed in the combination of 0.05mg/L ZT and 0.02mg/L NAA, which could supply resuspended cell lines. After nine subcultures of embryogenic callus suspension in liquid medium, embryogenic potential was lost, and the green embryogenic callus became brown, hard and non-viscid. The callus would produce new and yellow calli during the subculture. The free cells could also be produced from the resuspension of the new calli, but the dividing cells were few. Calli of Ta-7-68 also turned into green non-viscidly at a high concentration of 2,4-D (10mg/L). Moreover, the free cells were fewer and easy to break up. The calli at a low concentration of 2,4-D were the same as those of XT-Ta-34-97-S3. It could be concluded that genotype and 2,4-D concentration were crucial to the induction of embryogenic calli.

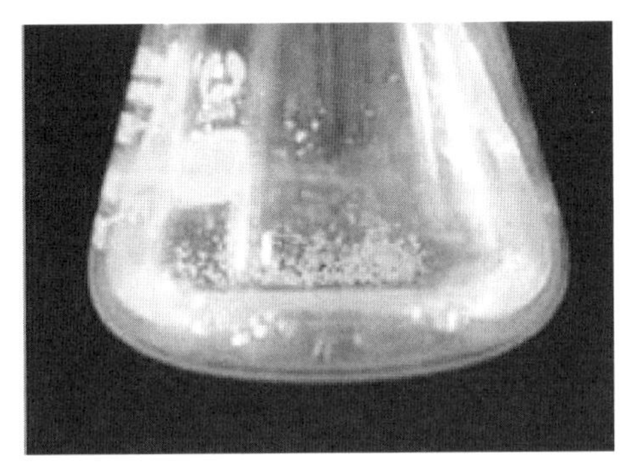

Figure 2 Globular embryos (0.5～1 mm in diameter) after two weeks of cell suspension

3.3 Somatic embryogenesis

After two weeks, cell lines from callus suspension of XT-Ta-34-97-S3 produced globular embryos in the liquid MS medium with 10mg/L 2,4-D (Fig. 2). Heart-shaped and torpedo-shaped embryos developed on the same medium after four weeks of culture, but with abnormality or vitrification and most globular embryos broke into slices (Figs. 3 and 4). Cell lines from cell suspension could also yield globular embryos within only 2 d on the same medium supplemented with 3～10mg/L 2,4-D, but it easily broke into pieces. In contrast, cell lines from calli of Ta-7-68 only produced white floccule within 3d of suspension cultures.

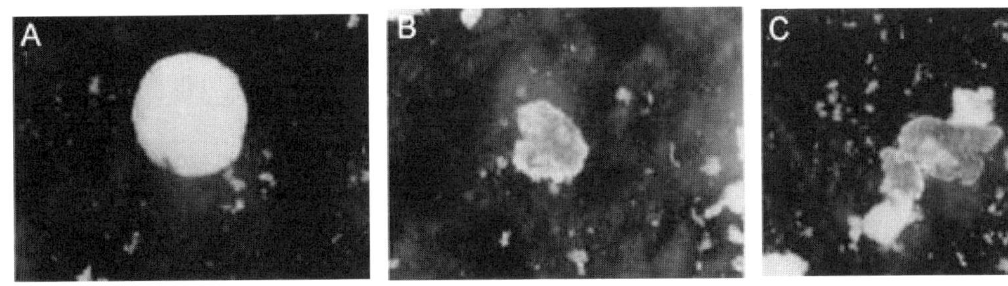

Figure 3 Embryos with abnormality and vitrification after 4 weeks of cell suspension

A: globular; B: heart-shaped; C: torpedo-shaped

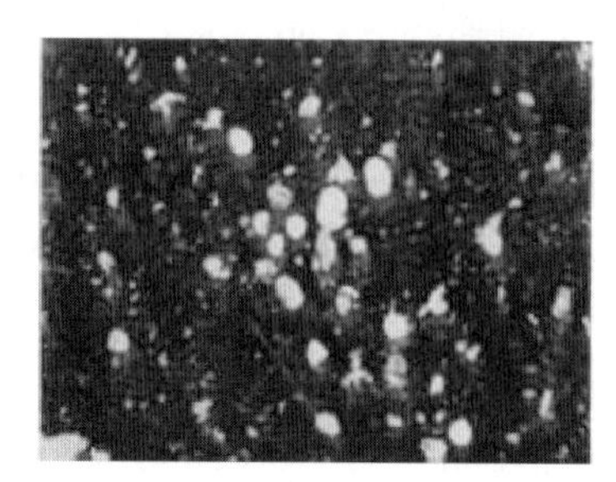

Figure 4　Dispersed globular somatic embryos

4　Discussion

Callus was easily formed from explants of poplars in the presence of 2, 4-D and the effect of genotype on the induction of embryogenic calli was obvious. AG-1-56 was easy to produce calli, but no embryogenic callus was collected. Calli of Ta-7-68 and XT-Ta-34-97-S3 were alike by suspension culture, but the latter was better than the former for the induction of embryogenic calli and production of competent cells.

Cytokinin was crucial to induce embryogenic callus. In previous studies, kinetin (Du et al. 1998), BA(Michler and Bauer, 1991; Park and Son, 1995) and ZT(Michler and Bauer, 1991) were all used to induce embryogenic calli or somatic embryogenesis. In our experiment, it was found that ZT was most effective for XT-Ta-34-97-S3. The calli of XT-Ta-34-97-S3 became brown and hard in the suspension culture without ZT. If the calli were induced without ZT and then performed suspension culture supplemented with ZT, the callus could also turn green but there were fewer free cells. That is to say, we could regulate embryogenic calli with cytokinins. When 0. 05mg/L ZT was replaced with 0. 5mg/L BA for XT-Ta-34-97-S3, the calli could also turn green but single cells were fewer.

Different types of explants would produce the same calli. Leaf-blades and petioles of Ta-7-68 produced the same calli in the same medium with 0. 5mg/L 2, 4-D and 0. 05mg/L ZT. The calli derived from leaf-blades also had the potential to yield embryogenic calli. Michler and Bauer (1991) reported that the developmental physiology of the leaf explants of *P. alba*×*P. granditata* (NC-5339) was important to the formation of embryogenic calli. However, calli induced from shoot tips, petioles and leaf-blades of *P. alb* ×*P. granditata* (AG-1-56) in our experiment had the same color and texture and became brown and hard when inducing embryogenic cells. The effect of genotype may be more important than the type of explants.

In previous reports, embryogenic calli were grown in the dark (Cheema, 1989; Michler and Bauer, 1991; Park and Son, 1995). However, in our tests, darkness was unnecessary for the induction of embryogenic callus of XT-Ta-34-97-S3 in accordance with *P. alba*×(*P. davidiana*×*P. simonii*)×*P. tomentosa* (Du et al. , 1998). At the same time, light had no clear effect on the callus induction for the three aspen petioles.

During the callus suspension of XT-Ta-34-97-S3, calli gradually turned green. The calli had viscidity and produced the highest number of free cells by the third subculture, and its characteristics were similar to those of *P. alba*×(*P. davidiana*× *P. simonii*)×*P. tomentosa* (Du et al. 1998) and *P. maximowiczii* (Stoehr and Zsuffa, 1990). Viscidity of the embryogenic calli would disappear gradually during the subculture. No single cell could be observed in nine subcultures. By suspension culture, the green embryogenic calli could develop into yellowish white calli, which could also produce free cells, but the number was fewer than green ones with viscidity. In addition, the yellowish white calli lost embryogenic potential more quickly than green callus. Characteristics of embryogenic callus for each genotype were heterogeneous, but green callus with viscidity could produce the highest number of competent cells. Yellowish white calli from leaf-blades and petioles of Ta-7-68 would also turn green but without viscidity and the single cells were also very fewer. White floccule was observed by the cell suspension culture and the floccule was like pieces of globular embryos and composed of highly elongated vacuolated cells.

A high frequency of globular embryos was observed by cell suspension from embryogenic calli in MS medium with 10mg/L 2, 4-D after two weeks (Fig. 2). Embryos would break up and abnormal heart-shaped and torpedo-shaped development appeared after four weeks under the same conditions. Embryos could also be observed after two days of suspension culture from cell lines by cell suspension, but they broke up easily. The cell lines from cell suspension would lead to somatic embryogenesis at any time, but the medium supplemented with 3 ~ 10mg/L 2, 4-D was the best. On the other hand, cell suspensionfrom embryogenic calli supplemented with 0. 05mg/L ZT and 0. 02mg/L NAA would produce perfect cell lines, although inferior embryos were observed by cell subculture. So phytohormone determined morphogenesis and the developmental physiology of cells was also important. Removal of 2, 4-D from the culture medium was necessary for embryo development for most *Populus* species (Cheema, 1989; Michler and Bauer, 1991, Du et al., 1998). In the present study, development of globular embryos was not observed in the medium with 0. 1mg/L BA and 0. 02mg/L NAA due to the low-quality embryos.

中国杨树资源与杂交育种研究现状及发展对策*

摘　要　中国是杨树种质资源十分丰富的国家，包含五大派50余种。该文在分析我国杨树资源的基础上，对我国50余年杨杂交育种研究进行了综述。在众多的杨树杂交组合中，派内种间杂交以白杨派、黑杨派成就显著，派间杂交以青杨派与黑杨派杂交成果最勾突出。从杂交方式看，有单交、双交、三交、回交等。以杂种作亲本进行再杂交能够获得显著杂种优势；从育种目标看，有速生、抗寒、抗旱、抗病虫、窄冠、生根等多个目标；从育种方法看，以常规人工杂交为主，将常规人工杂交与物理辐射、化学诱导等技术有机结合．能够创造出生产潜力较大的三倍体新品种。选育的杨树良种已在生产中产生了巨大的经济效益。牛态效益和社会效益。文章最后对我国杨树资源保护、杂交育种及杂种优势创制与利用的发展对策略进行了讨论。

关键词　杨树，杂交育种，综述

杨树是杨柳科(Salicaceae)杨属(*Populus* L.)树种的统称。由于杨树生长快、适应性强、木材用途广，在世界上栽培面积较大。中国杨树人工林面积大约667万hm^2，约占全国人工林面积的1/5，是世界其他国家杨树人工林总面积的4倍，在我国生态防护林和工业用材林建设中发挥着重要的作用。目前，我国林业生产中推广应用的杨树良种主要是通过人工杂交培育的，杂种优势明显，具有速生、优质、抗病虫、抗逆性等特点，已产生了巨大的经济效益、生态效益和社会效益。杂交育种仍然是目前及今后更长时间培育杨树新品种的重要手段。中国杨树杂交育种已有50余年的历史。最早是叶培忠教授于1946年在甘肃天水首次进行杨树杂交育种，从此以后许多林木育种工作者利用我国丰富的杨树种质资源，广泛开展了杨属派内和派间的杂交育种工作，培育出许多优良品种。本文将对这些研究内容作一综述，并对存在的主要问题以及应对策略进行必要的讨论。

1　中国杨树种质资源概况

杨属包括5个派：白杨派、青杨派、黑杨派、胡杨派和大叶杨派，世界共有100余种。在中国5个派杨树均有分布，共53种，特有种35种，主要分布于我国的西南、西北、东北及华北地区，北纬25°～53°，东经80°～134°之间。

1.1　白杨派(Sect. *Leuce* Duly)

中国有9种7个变种，如银白杨(*Poputus alba*)、银灰杨(*P. canescens* Smith)、毛白杨

* 本文原载《河北林业科技》，2006，9(增刊)：20-24，与李善文和何占国合作发表。

(*P. tomemosa* Carr.)、山杨(*P. davidiana* Dode)、河北杨(*P. hopeiensis* Hu et Chou)、响叶杨(*P. adenopoda* Maxim)、新疆杨(*P. bolleana* Lauche)等，主要分布于中国北方。

1.2 青杨派(Sect. *Tacamahaca* Spach.)

青杨派是杨属中最大的一个派，中国有34种，21变种，主要种如小叶杨(*P. simonii* Carr.)、大青杨(*P. ussuriensis* Kom.)、甜杨(*P. suaveolens* Fisch.)、苦杨(*P. laurifolia* Ledeb)、香杨(*P. koreana* Rehd.)、辽杨(*P. maximowiczii* Henry)等，特产种有小青杨(*P. pseudo-simonii* Kitag)、青杨(*P. cathayana* Rehd)、滇杨(*P. yunnanensis* Dode.)、青甘杨(*P. przewalskii* Maxim)、冬瓜杨(*P. purdomiim*)、青毛杨(*P. Shanxiensisc* Wang et Tung)、冒都杨(*P. qamdoensis* C Wang et Tung)、五瓣杨(*P. yuara* C Wang et Tung)、川杨(*P szechuanica* Sehneid.)等，主要分布在东北、西南和西北地区。

1.3 大叶杨派(Sect. *Leucoides* Spach.)

中国产4种3变种，如大叶杨(*P. lasiocarpa* Oliv.)、椅杨(*P. wilsonii* Schneid.)、灰背杨(*P. glauca* Haines.)、长序杨(*P. psudoglauca* C Wang et Fu)等，主要分布于长江以南地区。

1.4 黑杨派(Sect. *Aigeiros* Duby)

中国有3种2变种，如欧亚黑杨(*P. agra* L)、额河杨(*P. jrtyschensis* CH. Y Yang)、阿富汗杨(*P. afghanica* Sehneid.)、钻天杨(*P. nigra* var. *italica* Koehne.)、箭杆杨(*P. nigra* var. *thevestina* Bean.)等，主产新疆。

1.5 胡杨派(Sect. *turanga* Bge.)

中国有2种，胡杨(*P. euphratica* Oliv.)和灰胡杨(*P. pruinosa* Schrenk)，分布于西北地区。

综上所述可知，中国具有丰富的杨树种质资源，占世界杨树种类的50%以上，特别是在白杨派、青杨派中，有许多种为中国所特有，并且分布非常广泛，这为开展派内及派间杂交育种，为杨树各栽培区培育适合不同目的需求的优良无性系提供了良好的资源基础条件。

2 杂交育种研究进展

2.1 派内种间杂交

2.1.1 白杨派

中国杨树杂交育种始于1946年，叶培忠先生在甘肃省天水首次进行白杨派内种间杂交研究。杂交组合有河北杨×山杨、河北杨×毛白杨等，后来继续此项工作，选育出银毛杨(银白杨×毛白杨)、南林杨(*P.* ×'Nanlin')[(河北杨×毛白杨)×响叶杨]。徐纬英等(1960)选育出毛新杨(毛白杨×新搬畅)、银山杨(银白杨×山杨)、山新杨(山杨×新疆杨)等。王绍琰等(1985)以银白杨、新疆杨、河北杨、山杨、毛白杨等为亲本组配了20多个杂交组合，进行种间、种源间的最佳配合力及杂种优势测定，结果表明，以银白杨为母本的种间组合，杂交配合力高，杂种优势显著，而以银白杨为父本的种间组合效果不佳；在银白杨为母本的组合中按最佳配合力及杂种优势显著的顺序排列为银白杨×新疆杨、银白杨×河北杨、银白杨×山杨、银白杨×毛白杨；以辽宁、北京、宁夏3个地区的银白杨与同一产地新疆杨杂交，证明辽宁银白杨效果最好，并在此组合中选育出银新杨1号、2号两个优良无性系。王绍琰等(1987)以银白杨为母本，以河北杨×山杨为父本杂交，选育出银白杨×(河北杨×山杨)优良品种，其材积超河北杨109.8%，且抗病虫，易生根，扦插成活率90%以上。邱明光等(1991)从河北杨×毛白杨组合中选育出河北杨×毛白杨1号，10a生材积超河北杨141%。刘培林等(1991)以银白杨为母本，山杨为父本，选育出银白杨×山杨1333号无性系，l4a生时比山杨快1倍以上；以山杨为母本，银白杨×山杨为父本培育出山杨×银山杨1132号优良品种，具有速生、抗寒、抗病等特点。庞金宣等(2001)从南林杨×毛新杨组合中选出窄冠白杨1号、5号(*P. leucopy* ramidalis-1, 5)。从响叶杨×毛新杨组合中选出窄冠白杨3号、4号(*P. leucopyramidalis*-3, 4)，从毛新杨×响卧杨组合中选出窄冠白杨6号(*P. leucopyramidalis*-6)，这些窄冠型杨讨新品种具有冠幅窄、根系深、生长快、材质好等特点。北京林业大学对白杨派种及杂种间杂交难易程度、杂交方式进行了研究，结果表明：杂种在杂交可配性方面有着突出优势，其子代杂种优势明显；以杂种做亲本进行再度杂交是克服杂交不亲和的有效手段；选出毛新杨(*P. tomentosa*×*P. bolleana*)×银灰杨、毛新杨×毛白杨两个最佳组合；利用毛新杨×毛白杨这一组合，采用秋水仙碱、射线处理花粉，选育出6个三倍体毛白杨优良无性系，这些无性系具有早期速生、材质优良等特点。综上所述，中国白杨派种间杂交育种已取得一定成绩。从杂交方式看，有单交、三交、双交等，并证明以杂种做亲本进行再度杂交是一有效方式；从育种目标看，有速生、抗寒、抗旱、抗病虫、窄冠等多个目标；从育种手段看，以常规人工杂交为主，并注明将常规人工杂交与物理辐射、化学诱导等技术有机结合，能够创造出生产潜力较大的三倍体新品种。

2.1.2　黑杨派

与白杨派相比，黑杨派种间杂交育种起步较晚。吴中伦等于1972年从意大利引进美洲黑杨南方型无性系I-69(*deltoids*‘Lux’，又叫鲁克斯杨)，I-63(*P. del*‘Harvard’，又叫哈佛杨)及欧美杨无性系I-72(*P.* ×euramericana‘San Martino’，又叫圣马丁诺杨)，在我国亚热带地区表现出较强适应性和较高生产力，一些学者利用这些无性系作亲本进行了杂交育种研究。符毓秦等(1990)以I-69杨为母本，用I-63杨、密苏里杨(*P del.* var. *missouriensis* Henry)和卡罗林(*P. del* ssp. *angulata* Carolin，的混合花粉授粉，选育出陕林3号优良无性系，5a生单株材积超I-69杨24%，具有抗病虫特点，耐寒性、耐旱性优于亲本I-69杨。黄东森等(1991)以I-69杨为母本，欧亚黑杨混合花粉为父本，选育出中林46、中林23等优良无性系，材积超标优势30%以上。韩一凡等(1992)以I-69杨为母本，I-63杨为父本，选育出抗云斑天牛和光肩星天牛的南抗1号、2号优良品系，材积生长量超亲优势40%以上。陈鸿雕等(1995)选出鲁克斯杨×山海关杨(*P del*‘Shanhaiguan’)、山海关杨×哈佛杨、圣马丁诺杨×山海关杨三个优良品系，简称鲁×山、山×哈、圣×山，具有抗溃疡病和灰斑病等特点，并且速生，胸径生长量超过沙兰杨(*P.* ×*eur*‘Sacrau-79’) 20%以上。刘月君等通过山海关杨×(美杨+哈佛杨)、鲁克斯杨×山海关杨分别选出廊坊杨 1 号、2 号，具有明显杂种优势。

由上述分析可知，尽管中国黑杨派基因资源较少，育种学家利用从国外引进的优良无性系作亲本. 还是培育出了一批国产欧美杨、美洲黑杨杂种优良无性系；黑杨派内种间杂交以美洲黑杨改良为主，选出的无性系多数以 I-69 杨为母本，遗传基础较窄。

2.1.3 青杨派

青杨派种间杂交研究文献较少，只有徐纬英等(1960)做过小叶杨×滇杨、小叶杨×香脂杨、小叶杨×青杨、青杨×苦杨、小青扬×滇杨等杂交组合的研究工作；小叶杨×香脂杨杂种在河北张家口 4a 生树高 4.61m，胸径 2.98cm，接近北京杨的生长量，但没有大面积推广。尽管中国具有丰富的青杨派种质资源，但种间杂交研究很少，可能有两个原因：一是青杨派种间杂交杂种优势不显著；二是青杨派杂交，种子成熟期长，切枝水培后期蒴果营养不足，易脱落，不易得到成熟种子。另外，大叶杨派由于分布于中部暖温带和北亚热带高山区，迄今仍处于野生状态；有关胡杨派种间杂交育种也未见报道。

2.2 派间杂交

2.2.1 黑杨派与青杨派

黑杨派与青杨派之间，在自然条件下容易产生天然杂种，且具有显著杂种优势。中国天然分布或广泛栽培的黑杨派与青杨派天然杂种有：二白杨(箭杆杨×小叶杨)、小钻杨(小叶杨×钻天杨)、额河杨等。另外，赤峰杨、白城杨、大关杨、泰青杨等均是从小叶杨与钻天杨天然杂种中选出的无性系，具有速生、耐寒、耐旱和耐盐碱等特点。

中国杨树育种学家开展了青杨派与黑杨派多个组合的杂交试验，获得过许多优良无性系(见表 1)。杂交亲本主要利用黑杨派的美洲黑杨(如 I-69、山海关杨)、欧洲黑杨(如钻天杨、箭杆杨)，它们具有速生、干直、抗病虫等特点；利用青杨派的小叶杨、青杨、小青杨等，它们具有耐寒、耐旱、易生根等特点。将两派杂交，选育出具有父母本优良特性的无性系；或者利用它们的杂交种，如小钻杨、小黑杨，与其亲本回交创造新品种。选育的良种由于具有速生、耐旱、耐寒、耐瘠薄等特性，因此主要在我国北方寒冷、干旱或南方高海拔地区栽培应用。

另外，苏晓华等(1999)以山海关杨作为母本或父本，以同一产地的甜杨、大青杨不同个体做父本或母本杂交，研究了不同个体差异对杂交育种效果的影响后指出：同种不同个体间不仅存在表型及基因型差异，而且杂交效应也很明显，表现在杂交的亲和性和对子代的影响上；表型优良的植株做亲本其杂交后代并不一定优良，而表型不好的植株作亲本却可能得到优良后代，最好通过配合力测定选择亲本。

表 1　黑杨派与青杨派间杂交育种

杂交组合	选育目标	选育品种	选育人
钻天杨×青杨	速生、耐寒、抗病	北京杨	徐纬英等(1960，1988)
小叶杨×(美杨×旱柳)	速生、耐旱、耐盐碱	群众杨	徐纬英等(1960，1988)
小叶杨×钻天杨	速生、耐旱、抗病	合作杨	徐纬英等(1960，1988)
小叶杨×欧洲黑杨	速生、耐旱、耐寒、抗病	小黑杨	黄东森(1988)
赤峰杨×(欧美杨+钻天杨+青杨)	速生、耐寒、抗病虫	昭林 6 号	鹿学程等(1985)
小钻杨×欧洲黑杨	速生、耐寒、耐旱、抗病虫	白林杨 1 号	金志明等(1982)
欧洲黑杨×钻天杨	速生、耐寒、耐旱、抗病虫	白林杨 2 号	金志明等(1982)
青杆杨×二白杨	速生、耐旱	箭二白杨	刘榕等(1995)
小叶杨×美杨	速生、耐旱、耐寒	73-16、73-9	张水诚等(1990)
钻天杨×青杨	速生、耐旱、耐寒	74-2	张水诚等(1990)
山海关杨×小美旱	速生、耐寒、耐盐	7501	凌朝文等(1982)
大关杨×钻天杨	速生、耐旱、抗病	陕林 1 号	符毓秦等(1984)
I-69×青杨	速生、耐旱、抗病	陕林 4 号	符毓秦等(1990)
I-69×小叶杨	速生、耐旱	NL-106、106、121	王明庥等(1991)
小黑杨×*P.* 15A	速生、耐寒	黑林 1 号	刘培林等(1993)
小黑杨×黑小杨	速生、耐寒	黑林 2 号	刘培林等(1993)
小青黑杨×欧洲黑杨	速生、耐寒	黑林 3 号	刘培林等(1993)
格尔重杨×欧洲黑杨	速生、耐寒、抗病	格尔黑×欧黑	张玉波等(2002)
I-69×(大青杨+小叶杨)	速生、耐寒、抗病	69×大青小	张玉波等(2002)
I-72×欧洲黑杨	速生、耐寒、抗病	72×欧黑	张玉波等(2002)
辽河杨×小钻杨	耐寒、速生	辽育 1 号	孙远清等(2002)
山海关杨×塔形小叶杨	窄冠、耐盐、速生	窄冠黑青杨 6 号、31 号、69 号、70 号	庞金宣等(2001)
I-69×(山海关杨×塔形小叶杨)	窄冠、速生	窄冠黑杨 1 号、2 号、11 号	庞金宣等(2001)

2.2.2　黑杨派与白杨派

由于黑杨派速生、易生根，而白杨派材质优、适应性强，因此许多学者想通过这两派间杂交，创造出具有两派优点的新品种。徐纬英等(1988)在银白杨×欧洲黑杨、毛白杨×欧美杨、I-72 杨×毛白杨组合中，授粉时用母本花粉的糖蛋白处理柱头，但幼胚仍中途败育。张绮纹等(1988)用正己烷处理银白杨×美洲黑杨、银白杨×欧洲黑杨组合的花粉和柱头，未取得结果。刘培林等(1991)报道山杨与黑杨杂交，得到种子甚少，杂种生长量较小。吴鸿锦等(1996)以沙兰杨为母本，毛白杨为父本，选育出沙毛杨(*P.* ‘Sacau79’×*tometbto* ‘Samenica’)。庞金宣等(2001)以 I-69 杨为母本，响叶杨×毛新杨为父本，培育出窄冠黑白杨。张金凤等(1998)做了 43 个黑白杨派间杂交组合，结果表明 15%的三交组合和 56%的双交组合能得到杂交苗木，用杂种[如银腺杨(*P. alba*×*P. glandulosa*)、北京杨(*P. pyramidalis*×*P. cathayana* ‘Bejingensis’等)作亲本易得到黑白杨派间杂种，其扦插成活率较高，如银腺杨×中林 13 杨(*P. eur* ‘Zhonglin 13’)的杂种扦插成活率高达 83% 。

2.2.3 青杨派与白杨派

关于青杨派与白杨派间杂交育种研究文献较少，仅有徐纬英等(1960)做过山杨×小叶杨、小叶杨×毛白杨、毛白杨×青杨、大青杨×毛白杨等杂交试验，结果表明青杨派与白杨派间有可配性，能获得杂种，但这些杂种的进一步研究情况未见报道。另外，有青杨派与白杨派天然杂种——青毛杨的报道。

2.2.4 胡杨派与其他各派

董天慈(1980)以小叶杨为母本，胡杨为父本，得到小×胡杂种，10a 生平均株高为小叶杨的 1.8 倍，为胡杨的 1.9 倍，平均胸径分别为小叶杨、胡杨的 1.9 倍，杂种优势明显。李毅等(2002)以箭杆杨为母本，用胡杨花粉和 5000 伦琴射线辐射处理的毛白杨花粉混合授粉，经强度筛选得到箭胡毛杨，经 RAPD 标记证明，该品种为胡杨和箭杆杨的杂种，它既保持了箭杆杨的窄冠、速生特点，也拥有胡杨抗旱、抗寒、耐盐碱和抗病虫特点。

综上所述可知，中国杨树杂种育种，派内以白杨派种间杂交效果突出，其次是黑杨派种间杂交；派间以青杨派与黑杨派杂交成就较大，其次是黑杨派与白杨派间杂交。

3 杨树资源利用与杂种优势创制的对策

中国杨树遗传改良(主要是杂交育种工作)经过了 50 余年历程，尽管取得许多成绩，但是还存在某些问题，如育种策略简单，缺乏长期稳定的研究计划，基础研究重视不够，研究树种较少，种源、家系、无性系等水平存在的丰富遗传变异利用不系统，育种目标较单一，过分追求速生性状，抗逆性状和木材品质性状改良薄弱，对乡土杨树种质资源保护重视不够，现代生物技术与常规育种研究结合不够紧密等。针对这些问题，应尽快采取一些重大改革措施。

3.1 制定长期改良计划

50 余年来，我国杨树杂交育种工作主要采用杂交—选择育种程序，重视 F_1 代无性系选择，每次杂交重新从未改良群体中随机选择亲本开始，迄今没有一个以轮回选择理论为指导的长、中、短期配套的总体计划。长期育种是指以轮回选择理论为指导，对育种亲本有计划改良，使短期育种工作随着世代前进而持续高效的螺旋式上升。意大利在杨树育种方面重视杂交亲本改良，制订了长期育种计划。其欧美杨育种策略是：首先确定美洲黑杨和欧洲黑杨最佳种源后，从中各选 120 株雌雄株组成育种群体，分两线工作，同时测定杂种优势和配合力。一方面，以美洲黑杨为母本，欧洲黑杨为父本，做种间杂交，取得杂种，经无性系测定后，选择杂种优势明显的植株，繁殖成无性系；另一方面，分别在美洲黑杨和欧洲黑杨育种群体内作种内控制授粉，通过配合力测定，从中挑选出雌雄各 40 株组成改良群体。对改良群体，重复上阶段的工作内容，如此循环，螺旋上升，使育种群体和杂种无性系的遗传品质不断提高。我国乡土杨树遗传改良，可以借鉴该策略，并根据树种的分布范围及其特性，制订长期遗传改良计划，实行交互轮回选择为基础的遗传改良，把亲本改良作为提高杂种优势的根本。

3.2 重视多树种、多性状、多水平和多方向遗传改良

我国拥有丰富的乡土杨树资源，特别是青杨派和白杨派树种居世界之冠，它们主要分布于东北、西北、西南地区。但是近几十年来，对乡土杨树研究力度不够，也不系统。在我国杨树适生栽堵区，地形复杂，气候条件各异，因此必须开展多种乡土杨树遗传改良工作，以适应不同地区生产需求，改良性状应抗逆性优先，在抗逆性基础上追求速生。

国内外许多研究表明，杨树在种源、家系无性系等水平存在丰富遗传变异，并且这些变异显著影响杂交育种效果。Krzan(1976)做了不同种源44个无性系的美洲黑杨抗锈病试验，结果是南方种源和以南方种源做亲本的欧美杨抗锈病强。Cellerino(1976)从美国引种18个种源52个家系美洲黑杨种子，每一家系测20个无性系，内容包括对褐斑病、锈病、疮痂病及早春低温的反应，结果南方种源对上述3种病害是抗病的，并把它与南方型欧洲黑杨杂交，育成适应南欧的南方型欧美杨；欧洲一些学者做了许多山杨杂交，均认为用波兰种源做亲本最好。我国对青杨、大青杨、山杨等乡土树种研究表明，种源间及种源内个体间在生长、材性等方面均存在显著差异。因此，对国产杨树遗传改良，应充分利用种源、家系、无性系这3个层次存在的丰富遗传变异，提高杂种优势。同时，重视不同的需求开展多方向的良种选育研究，逐步走定向目标的杨树育种和产业化生产利用的道路。

3.3 注重乡土杨树育种资源收集保存与利用

育种资源是指在选育优良品种工作种可能利用的一切材料，是选育新品种的物质基础。一个优良品种不仅应具备优良的经济性状，同时应具有较强的适应性和抗逆性。因此，选育优良品种，必须具备丰富的育种资源。如果没有丰富的育种资源做后盾，不断引进补充新的基因资源，多世代育种工作也将受到限制。为防止乡土杨树种质资源丢失，应分育种区建立育种资源保存群体，所保存群体要有代表性，不但包括生长快、材质优个体，而且包括抗病、抗虫、抗逆性个体，以满足不同育种目标需求。

3.4 将现代生物技术与常规育种技术有机结合

在生物技术飞速发展的今天，常规杂交育种仍然是为生产提供良种的重要手段。现代生物技术与常规杂交育种有机结合，是今后杨树育种重要研究内容之一，如将转基因技术应用于杂交亲本改造，将分子标记技术应用于杂交亲本选配和杂种优势预测，将mRNA差异显示技术应用于杂种优势分子机理研究及杂种遗传分析，对选育的优良杂种无性系通过基因工程进一步改良等。尤其是杨树基因组已经公布的今天，完全有理由相信，随着生物技术与杂交育种的更加紧密的结合，传统杂交育种在杨树遗传改良中将再现出强大生命力。

我国杨树育种现状及其展望*

摘　要　杨树是世界上广泛栽培的造林树种之一。我国拥有十分丰富的杨树资源和最大面积的杨树人工林，杨树遗传改良对我国尤为重要。自20世纪40年代末以来，我国在杨树引种、优良天然杂种选择、常规杂交育种与倍性育种方面取得了很大的成就。近几年来，随着分子生物学的发展，我国杨树基因工程、遗传图谱构建以及分子标记辅助选择育种等方面也取得了较大进展。本文对我国50余年杨树育种的成就进行了概述，并对我国杨树育种中存在的问题和前景进行了讨论。

关键词　杨树，引种，杂交育种，倍性育种，生物技术

杨树(*Populus* spp.)是杨柳科(Salicaceae)杨属(*Populus* L.)树种的统称。由于杨树生长快、适应性强、木材用途广，在世界上栽培面积较大。因此，杨树的育种工作倍受世界主要栽植国的重视。英国学者 Henry 于1912年首次进行了杨树人工杂交试验，此后，意大利、前苏联、美国等国也相继开展了杨树的育种工作。我国以1946年叶培忠先生在甘肃天水首次开展的河北杨(*Populus hopeiensis*)与山杨(*P. davidana*)，河北杨与毛白杨(*P. tomentosa*)杂交试验为杨树育种工作的开端，经过科研工作者近半个多世纪的持续工作，现已在杨树引种、杂交育种、倍性育种以及借助生物技术开展杨树改良等方面取得了令人瞩目的成就(叶培忠，1955；马常耕，1994；何承忠，2003；何承忠等，2005；李善文等，2004)。培育出的新品种在我国工业用材林及生态林建设中发挥了巨大作用。本文对我国50余年取得的杨树育种成就进行了概述，并对杨树育种中存在的问题和今后发展的方向进行了讨论，以供相关研究工作者参考。

1　引种

引种是一项重要的育种工作。通过引种，一则直接引入优良无性系，为生产提供新品种；二则为育种工作搜集和提供新的基因资源，为培育新品种创造物质基础。我国杨树的引种工作始于19世纪，但在1949年以前的引种工作处于零星而无计划的状态，主要引进的杨树树种有钻天杨(*P. nigra* var. *italica*)、箭杆杨(*P. nigra* var. *thevestina*)、山海关杨(*P. deltoids* 'Shanhaighan')和加拿大杨(*P.* ×*euramericana* 'Guinier')等(徐纬英，1988；张绮纹等，1999)。中华人民共和国成立以后，科研工作者根据需要，开展了大量有针对性的引种工作。50年代我国从苏联、民主德国、波兰、比利时等国引进了50个品系；60年代从日本和罗马尼亚等国引进40多个无性系；70年代从意大利、荷兰、比利时等国引进

* 本文原载《西南林学院学报》，2006，26(4)：86-89，与何承忠、安新民和李善文合作发表。

100 多个无性系；80 年代我国加入国际杨树委员会，扩大了我国杨树引种的渠道，先后从 17 个国家引进了 300 多个无性系(张绮纹等，1999)。至今，我国已有 1000 余个杨树无性系引种成功，并取得了显著的经济效益和生态效益。其中尤为突出的引种实例有从意大利引入的美洲黑杨(*P. deltoids*)南方型无性系 I-69 杨和 I-63 杨，它们在我国亚热带气候条件的地区显示出高度适应性和生产力，使我国杨树栽培区南移到湖南、湖北、浙江、江苏等省的平原地区(马常耕，1994；张忠涛等，2001)。从意大利引进的喜中温的 I-214 杨(*P.* ×*euramericana* 'I-214')，从民主德国引进的沙兰杨(*P.* ×*euramericana* 'Sacrau79')生长良好。从法国、意大利等 17 个国家引入黑杨派无性系 331 个，首次建立了我国黑杨派杨树基因库，并从中筛选出了一批综合性状表现较好的无性系(张绮纹等，1999；刘闯，1993；赵天锡等，1994)。美洲山杨(*P. tremuloides*)、大齿杨(*P. grandidentata*)及银腺杨(*P. glandulosa*)等国外白杨派无性系的引种成功，为我国白杨派乡土树种的改良提供了难得的基因资源，也为我国杨树遗传改良提供了宝贵育种材料(林惠斌等，1988)。

2 杂交育种

在我国林木杂交育种中，杨树杂交育种研究工作开展较早(张忠涛等，2001)。1946 年，叶培忠先生首次进行了杨树杂交试验(马常耕，1994)。之后，研究人员相继开展了杨树派间和派内种间杂交研究，组合达 400 余个，培育出的许多优良品种已在生产中推广应用(徐建民，1996)。在众多杂交组合中，成效较大的有青杨(*P. cathayana*)×钻天杨，培育出了北京杨(*P.* ×*beijingensis*)(徐纬英，1960)；小叶杨(*P. simonii*)×钻天杨，培育出了合作杨(*P.* 'Opera')(徐纬英，1960)；小叶杨×(钻天杨+旱柳)，培育出了群众杨(*P.* 'Popularis')(徐纬英等，1984)；箭杆杨×麻皮二白杨(*P. gansuensis* '1-MP')，培育出了箭杆二白杨(*P.* 'Thevegansuensis')(刘榕等，1995)；美洲黑杨×小叶杨，培育出了 NL-80105、NL-80106 和 NL-80121 三个新无性系(王明庥等，1992)；从 I-69×I-63 杂交组合中选育出了抗云斑天牛的南抗杨(*P. delteides* cl. 'Nankang')品系(汪万江，1997)；以 I-69 为母本，欧洲黑杨(*P. nigra*)混合花粉为父本，选育出中林 46、中林 23 等优良无性系(黄东森等，1991)；南林杨(*P.* ×'Nanlin')×毛新杨，培育出了窄冠白杨(*P. leucopyramidalis*)1 号、5 号，响叶杨(*P. adenopoda*)×毛新杨，培育出了窄冠白杨 3 号、4 号，毛新杨×响叶杨，培育出了窄冠白杨 6 号(庞金宣等，2001)。同时从小叶杨与钻天杨的天然杂种群体中选出了具有速生、抗旱、抗寒性较强的大关杨(*P. dakauensis*)(南京林产工业学院主编，1980)、赤峰杨(*P.* 'Chifengensis')(徐纬英，1988)、白城杨(*P.* 'Baichengensis')(徐纬英，1988；吉林省白城地区林科所，1979)等优良无性系。选育的这些优良品种为我国杨树造林良种化做出了重大贡献。

在杂交育种的实践过程中，杨树育种策略也在不断地改进。由于在种源及家系研究中发现种内的遗传变异较大，所以将随机杂交育种策略修改为从选择到杂交再到选择的模式，将群体间和群体内变异用于亲本选择(马常耕，1994)。20 世纪 80~90 年代借鉴于玉米育种而兴起的交互轮回选择程序，使杨树育种进入亲本改良阶段，这是迄今杨树最科学的育种程序(张延桢，1996)。在不断完善杂交育种策略的同时，发现利用杂种作亲本在杂交可配性方面有着突出优势，其子代杂种优势明显，以杂种做亲本进行再度杂交是克服杂交不亲和的有效手段(李天权等，1989)。所以杨树杂交方式也被改进，由简单到复杂，从单交到复交(三交、双交)，配合应用回交，以提高杂交育种效果。

3 倍性育种

鉴于多倍体具有巨大性、生活力旺盛、抗逆性强等优点，以及单倍体在遗传育种、植物进化和起源研究等方面所具有的独特作用，故倍性育种也是杨树育种的一个热点。相对于其他倍性多倍体，三倍体的营养生长通常是最好的(鲍文奎，1988)，另外三倍体杨树有纤维长、抗性强等优点，使其倍受重视，成为速生短周期纸浆材的首选树种之一(李云，2001)。在我国杨树倍性育种中，朱之悌等利用毛新杨(*P. tomentosa*×*P. bolleana*)×毛白杨(*P. tomentosa*)这一组合，并采用秋水仙碱、γ射线处理花粉，选育出6个三倍体毛白杨优良无性系(李云，2001；朱之悌等，1995；朱之悌等 1998，朱之悌等，1995；康向阳，2002)。该品系具有速生、优质、抗病、高效等优良特性，解决了长期以来我国北方林业，尤其是工业用材林发展缺乏优良品种的难题，已被国家林业局列为黄河中下游地区纸浆原料林和人造板原料林基地的主要栽培品种(张忠涛等，2001；周生贤，2001)。

林木多倍体品种除从自然群体中选育外，主要是通过人工培育获得。目前人工培育林木三倍体的途径主要有：①利用四倍体与二倍体杂交；②采用胚乳离体培养；③选择或诱导2n配子，与正常异性配子杂交。此3种方法中只有第三种方法比较适合杨树三倍体育种，其主要是通过筛选天然2n花粉、理化途径诱导获得2n花粉或2n雌配子，再与正常异性配子杂交培育三倍体杨树品种(朱之悌等，1995a，1995b；李云等，2001；张志毅等，1992；李云等，2000；康向阳等，2000)。

杨树单倍体的培育主要是通过花药培养法来获得，陈震古等采用激光诱导花粉方法，经过培养也获得了杨树单倍体植株(陈震古，1999；陈震古等，2000)。

4 现代生物技术育种

由于杨树生长周期长，遗传杂合性高，许多重要经济性状属于多基因控制的数量性状，遗传机理不明，利用常规育种手段往往难以满足不同目的定向培育杨树新品种的要求。因此，人们将研究重点转向现代生物技术，尤其以基因工程技术作为定向培育杨树新品种的手段，以此解决常规育种技术难于解决或根本解决不了的问题。与国外相比，我国利用生物技术进行杨树遗传改良虽然起步较晚，但进展较快。目前，抗虫、抗病、抗逆境转基因杨树植株均已获得(郑均宝等，1995；王学聘等，1997；田颖川等，1993；伍宁丰等，2000；饶红宇等，2000；郝贵霞等，1999；李明亮等，2000；Rao et al.，2001；赵世民等，1999；施季森，2000；杨传平等，2001；刘斌等，2002)。其中，抗食叶害虫转Bt基因欧洲黑杨获得国家商品化许可，转双抗虫基因(部分改造的Bt *Cry1Ac*基因和*API*基因)的白杨杂种741杨[*P. alba*×(*P. davidiana* + *P. simonii*)×*P. tomentosa*]获得国家环境释放许可(苏晓华等，2003)。改良木材品质的转基因植株正在研究之中(施季森，2000；苏晓华等，2003）。原生质体培养再生植株、体细胞杂交再生植株、体细胞变异和突变体筛选植株等均已培养成功（张绮纹等，1999；施季森，2000；陈天华，1998；诸葛强等，2000；张绮纹等，1998）。杨树遗传图谱已趋于饱和，DNA 指纹图谱已经构建（尹佟明等，1999；张德强，2002；张志毅等，2002；尹佟明等，1998）。QTLs 研究进展迅速，青杨锈病已找到紧密连锁的标记，并定位在连锁图谱上，南抗杨新品种抑制害虫的物质和与抗虫相连锁的标记已找到（张绮纹等，1999；李金花等，1999；苏晓华等，2000）。这些成果的取得为杨树定向育种开辟了新的途径。

5 问题与展望

我国杨树在常规育种及现代生物技术育种方面均取得了较大成就，选育的杨树良种在生产中已发挥出巨大的经济效益、生态效益和社会效益。但是，也存在着一些不可忽视的问题，如育种资源引进、收集力度不够，使得有利于杨树遗传改良的外源基因来源贫乏；育种策略和方法落后，手段简单，主要以引种和杂交育种为主，并且无长期计划，缺乏持续性；育种目标过分集中于追求速生性，抗逆境育种成效甚微；研究树种比较单一，对丰富的乡土树种保存和重视不够；转单价基因的生物抗性局限性大，拥有知识产权的基因贫乏等。要解决我国杨树育种中面临的问题，促进杨树育种工作的进一步发展，要加大对国外杨树优良资源的引进力度；加强常规杂交育种的基础性研究，利用分子标记等手段，开展杂交亲本选配和杂交优势预测，强化育种效果；制定长期改良计划，以多世代轮回选择理论为指导，充分利用种源、家系、无性系等多水平的遗传变异，采用三交、双交或更复杂的杂交方式，提高改良程度；充分挖掘国产杨树资源，有效地开发利用优良乡土树种，拓宽研究树种，实现育种目标的多元化；不断寻找新的抗虫基因，利用调控功能强且特异性高的启动子调控抗虫基因的表达，培育抗蛀干害虫的杨树新品种；加大杨树抗逆境（抗旱、耐盐碱）机理的研究力度，逐步建立适应各树种的基因工程转化体系，加快多价基因转化技术的研究，实现多价基因转化杨树，培育出真正意义上的抗旱、耐盐碱杨树新品种也是今后研究的方向。

总之，在生物技术迅速发展的今天，许多新的育种理论和技术不断涌现，杨树育种的许多理论与技术也得到了发展。但常规育种在很长一段时间内仍然是为生产提供良种的重要手段。完全有理由相信，随着生物技术与常规育种技术的有机结合，我国杨树育种工作将会取得新的突破。

Present Situation and Prospect to Poplar Breeding in China

Abstract Poplar is one of the most important deciduous trees in the world. China is replete with rich natural resources and the largest planted areas of poplar. The applications of genetic improvement to poplar play a key role in Chinese forestry and its industry. From the 1940's, great success in introduction, selection, cross breeding and polyploid breeding of poplar has been made in China. Recently, with the development of molecular technique and its extensive applications to genetic improvement in forest tree species, the great progresses of genetic improvement in poplar which covers genetic transformation, construction of genetic linkage map, QTL identification as well as molecular-assisted selection have also been made in China. The achievements of poplar breeding in China over the past fifties years are reviewed in this paper. Meanwhile, the existing problems and outlook about the poplar breeding are also discussed.

Key words poplar, introduction, cross breeding, polyploid breeding, biotechnology

Progress in the Study of Molecular Genetic Improvements of Poplar in China*

Abstract The poplar is one of the most economically important and intensively studied tree species owing to its wide application in the timber industry and as a model material for the study of woody plants. The natural resource of poplars in China is replete. Over the past 10 years, the application of molecular biological techniques to genetic improvements in poplar species has been widely studied in China. Recent advances in molecular genetic improvements of poplar, including cDNA library construction, gene cloning and identification, genetic engineering, gene expression, genetic linkage map construction, mapping of quantitative trait loci (QTL) and molecular-assisted selection, are reviewed in the present paper. In addition, the application of modern biotechnology to molecular improvements in the genetic traits of the poplar and some unsolved problems are discussed.

Keywords poplar, gene engineering, genetic improvement, genetic linkage maps, molecular marker, quantitative trait loci (QTL) mapping.

The genus *Populus* is composed of more than 74 species classified into five or six sections. The poplar has played a key role in forest production and the construction of ecological environments in China. Fast growth, high yield, easy propagation, and a wide range of uses make the poplar one of the important tree species for artificial forests. In addition, owing to its small genome and ease of tissue culture, the poplar represents a useful model system for the investigation of the genetics and molecular biology of woody plants (Lin et al., 2000; Lin and Zhang, 2004). However, the fact that many of the traits of poplars are controlled by multiple genes, pulsed high heterozygosity, and a long life cycle makes conventional breeding very difficult. In recent years, with the development of molecular biotechnology and the study of other forest tree species, gene engineering and molecular markers have been widely used in the genetic improvement of the poplar species. In particular, because the poplar was chosen as a model tree, the annotated entire genome sequence of *Populus trichocarpa* was released to the public for the first time in early 2004 and cDNA microarray and gene chips for poplar gene profiling have been developed and deployed (Tuskan et al., 2004), which has opened up new avenues for molecular research into the poplar.

Over the past 10 years, molecular genetic improvements of the poplar have been widely studied in China. Significant progress in research that has focused on the construction of a cDNA library and the identification of genes responsible for varied characteristics, the engineered improve-

* 本文原载 *Journal of Integrative Plant Biology*, 2006, 48(9): 1001-1007, 与林善枝、张谦和林元震合作发表。

ments in biotic or abiotic resistance and wood property, as well as rooting ability, and the application of molecular markers to breeding (genetic linkage map, quantitative trait loci (QTL) mapping and marker-assisted selection) is reviewed in the present paper. All these achievements have unfolded an attractive prospect, but there are still lots of difficulties and problems.

1 cDNA Library Construction and Gene Cloning

The Chinese white poplar (*Populus tomentosa* Carr.) is a special local tree species, planted widely in northern China, with a lower lignin content and great potential usefulness as an important resource for paper pulping in the wood-processing industry. It is therefore a good material to use to clone genes encoding some of the key enzymes involved in lignin biosynthesis. Genes encoding caffeoyl conenzyme A (CoA) O-methyltransferase (COMT) and 4-coumarate-CoA ligase (4CL) have been cloned from Chinese white poplar. Northern analysis showed that COMT and 4CL were expressed specifically in the developing secondary xylem and that their expression was coincident with lignification (Wei et al., 2001; Zhao et al., 2003). In addition, one 565-bp cDNA sequence of the phenylalanine ammonialyase gene from developing second xylem of *P. euramericana* was cloned and identified (Xue et al., 2004). A previous study has determined that *P. suaveolens* can survive at temperatures of approximately -43.5℃ in winter in Daxinganling, China, and is therefore a good material from which to clone antifreeze genes (Lin and Zhang, 2004). Recently, two antifreeze genes, namely PsG6PDH and PsAFP, encoding G6PDH and antifreeze proteins, respectively, were cloned from *P. suaveolens* and their physiological functions were identified (Lin and Zhang, 2004; Lin et al., 2005).

In order to isolate genes associated with disease resistance, two cDNA libraries, namely L45-72 and L69-72, were constructed from the leaves of the black spot disease-susceptible I-45 clone (*P. euramericana*) and black spot disease-resistant I-69 clone (*P. deltoids*) inoculated with pathogen, from which the anti-disease cDNA fragments were obtained by fluorescent differential display reverse transcriptase polymerase chain reaction (DDRT-PCR) (Zeng et al., 2004). In addition, one saltsuppressed gene and two salt-induced genes were screened from the constructed cDNA subtractive library of tender roots of 2-year-old seedlings of *P. euphratica* treated with NaCl (Zeng and Shen, 2004).

2 Molecular Improvements by Genetic Engineering

2.1 Improvement in insect resistance

Poplars are known to be hosts for many herbivorous insects and breeding for insect resistance is a difficult task because of the long-lived cycle of trees. Traditional spray application of insecticidal proteins and chemical toxins can control insect damage, but has the potential to devastate the environment. Over the past several years, genetically engineered resistant poplars relied mainly on genes encoding insecticidal proteins of *Bacillus thuringiensis* and those encoding proteinase inhibitors (Delledonne et al. , 2001).

Bacillus thuringiensis (*Bt*) gene products are the most widely used pesticides in agriculture and forestry. In China, the regeneration and transformation system suitable for transgenic poplar (*P. nigra*) expression of the *Bt* gene was first established in 1991 (Wu and Fan). From then, the *Bt* gene was transferred into the genome of *P. nigra*, *P. deltoids*, *P. euramericana*, and hybrid poplar NL-80106 (*P. deltoids* ×*P. simonii*). The results of insect bioassays showed that these transformed poplars were highly resistant to the insects tested (Tian et al. , 1993; Chen et al. , 1995; Wang et al. , 1996, 1997; Rao et al. , 2000). Hu et al. (1999 and 2001) reported that field trials of the transgenic *P. nigra* trees expressing the *Bt Cry1Ac* gene showed strong resistance to *Apocheima cinerarius* and *Orthosia incerta*, and significantly reduced larva density, damaged leaf rate, and pupa number. This was a large field test on the insect-resistance of a transgenic poplar with the *Bt* gene in China. However, transgenic *P. nigra* tress with the *Bt* gene could influence insect community structure in the field (Zhang et al. , 2004c), which was found in the studies of transgenic *P. tomentosa* with the cowpea trypsin inhibitor gene (*CpT* I), suggesting that the biological safety and ecological risk assessment of insect-resistant transgenic poplars should be further evaluated (Zhang et al. , 2002b).

A novel method to control insect pests of poplars using proteinase inhibitor genes was reported by Ryan (1990). Cowpea trypsin inhibitor, a small polypeptide belonging to the Bowman-Brik type of serine proteinase inhibitors, was shown to be a more effective antimetabolite against a wide spectrum of pest insects than other proteinase inhibitors (Zhang et al. , 2004b). In China, *CpT* I , first cloned from mature cowpea by Liu et al. (1993), was introduced into *P. tomentosa* (Hao et al. , 1999) and *P. alba* var. *pyramidalis* (Zhu et al. , 2003), respectively. Assays of proteinase inhibitor activity demonstrated that leaf protein extracts of the transgenic *P. tomentosa* showed higher inhibition activity against trypsin than that of control plants (Hao et al. , 1999). Recent studies have shown that 2-year-old transgenic *P. tomentosa* with the *CpT* I gene exhibits high resistance to pests in test fields; moreover, no significant differences in height and diameter growth were observed between transgenic and control plants, indicating that the integration and expression of the *CpT* I gene is stable in transgenic poplar and that *CpTI* gene expression has no effect on the growth of transgenic poplars (Lin et al. , 2002; Zhang et al. , 2002b, 2004b).

Antibacterial gene *LcI*, synthesized according to the sequence of the antibacterial protein of *Bacillus* spp. , was transferred into *P. detoides*, *P. nigra*, and *P. euramericana*, and more than 75% insect mortality of *Anoplophora glabripennis* was observed in these transgenic poplars (Li et al. , 1996). Wu et al. (2000) found that the introduction of a gene encoding an insect-specific neurotoxin from scorpion, namely the *AaIT* gene, into the hybird poplar clone N-106 (*P. deltoids*×*P. simonii*) resulted in a significant resistance towards *Lymantria diaper*; moreover, insects feeding on the leaves of these transgenic poplars exhibited reduced growth and increased mortality. The resistance of transgenic clones of the hybrid poplar 741 expressing the agglutinin gene to *Apriona germani* has also been tested and some transgenic clones with high resistance to *A. germani* have been obtained (Wang et al., 2002).

Recently, modification of the *Bt* gene-coding region or fusion of the *Bt* gene with a proteinase inhibitor gene was achieved and transferred into poplars in China. The improved *Bt* gene was used to transform *P. tomentosa* and hybrid poplar 741 (*P. alba*×(*P. davidiana*×*P. simonii*)×*P. tometosa*) (Zheng et al., 1995b, 1996). The first insect-resistant transgenic poplar expressing two insecticidal protein genes (a partially modified *Bt Cry1Ac* gene and the arrowhead proteinase inhibitor (*API*) gene) was obtained in 2000, and the transgenic hybrid poplar 741 was shown to be highly resistant to *Clostera anachoreta*, with a higher mortality of insect larvae and a lower growth of the surviving larvae (Tian et al., 2000). However, the sensitivity of different insect species and different instar larvae of insects to the clones of transgenic poplars varied considerably (Gao et al., 2004). Moreover, years of serial observations indicated that the quality of insect-resistance had not exhibited a rhythmical drop, and no significant differences in growth, lumber quality, and seeding character were observed between the transgenic and control plants (Yang et al., 2005). In addition, a higher toxicity to larvae was observed in transgenic *P. nigra* plants expressing both *Bt* and *API* genes compared with those expressing the single *Bt* gene (Li et al., 2000). Similar findings have been reported for the transgenic hybrid poplar NL-80106(*P. deltoids*×*P. simonii*; Rao et al., 2001). Thus, it is suggested that the expression of a modified *Bt* gene or multiple genes may increase the insecticidal activity of transgenic poplars and reduce the likelihood of the evolution of insect tolerance against insect-resistant transgenic plants.

2.2 Improvements in disease resistance

Poplars are also severely damaged by viral and fungal diseases. Compared with gene engineering of insect resistance, little has been reported regarding genetic improvements in viral and fungal diseases of poplars in China. Zhao et al. (1999) reported that rabbit defensin possesses a broadspectrum resistance against pathogens and that the extract of transgenic *P. tomentosa* with the rabbit defensin *NP-1* gene inhibited the growth of the microbes tested. To our knowledge, this is the first report of a disease-resistant transgenic poplar in China. Recently, the chitinase gene (*CH5B*) has been integrated into the *P. deltoids* genome (Meng et al., 2004).

2.3 Improvements in environmental stress (salt, drought, and freezing) resistance

Environmental stresses, such as salt, drought, and freezing, are the most important factors affecting the growth, development, productivity, and distribution of poplars. Improving the resistance of poplars to environmental stresses has been an important goal for a long time. In recent years, many transgenic crop plants with enhanced tolerance for environmental stress have been obtained, which has opened up new avenues for breeding stress-resistant poplars using transgenic technology.

In China, the first salt-resistant transgenic poplar plant with an *Escherichia coli* 1-phosphate mannitol dehydrogenase gene (*mtl-D*) was obtained in 2000(Liu et al., 2000). The results of field afforestation and saline resistance tests showed that the survival rates of transgenic poplars in the field with a soil saline content of 0.3%~0.4% was higher than that of control poplars; moreover, the transferred *mtl-D* gene was stable in transgenic poplars in the field (Yin et al., 2004). Yang et al. (2001) reported that the expression of exogenous gene *Bet-A* encoding choline dehydrogenase into hybrid poplar (*P. simonii* ×*P. nigra*) resulted in an elevation of salt resistance in transgenic plants. In addition, the antisense phospholipase Dγ gene was introduced into triploid *P. tomentosa* by *Agrobacterium tumefaciens* and the transgenic plants obtained were able to grow well in culture medium with 0.7% NaCl (Liu et al., 2002). Recently, we reported that two antifreeze genes, namely *PsG6PDH* and *PsAFP*, encoding *G6PDH* and antifreeze proteins, were introduced into triploid *P. tomentosa*; freezing resistance tests of the transgenic plants obtained are currently underway (Lin and Zhang, 2004). From the studies listed above, it can be seen that reports regarding improvements in the salt, freezing, and drought resistance of poplars by genetic engineering in China are very limited. Thus, more endeavors should be made in the cloning and identification of novel functional genes that respond to environmental stress.

2.4 Improvements in wood property and rooting ability

In China, the poplar is an important resource for paper pulping because of its fast growth rate and excellent wood quality. However, for the production of high-quality paper, lignin is an undesirable component and must be eliminated from the wood during paper pulping. In general, the removal of lignin in poplar using chemical treatments is very costly and results in considerable environmental pollution during the process of making paper pulp. So, there is considerable interest in reducing the lignin content by genetic engineering to improve the pulping property of poplar wood. Wei et al. (2001) reported that antisense expression of the *CCoAOMT* gene in transgenic hybrid poplar (*P. tremula* ×*P. alba*) resulted in a decrease in the lignin content compared with that of untransformed plants and had basically no effect on the growth of transgenic plants. Recently, a transgenic triploid *P. tomentosa* with an antisense *4CL* gene was obtained, resulting in a repressive expression of the endogenous *4CL* gene and a decrease in the lignin content in transgenic plants (Jia et al., 2004). Thus, it is suggested that the antisense expression of *CCoAOMT* or *4CL* is an efficient way in which lignin content can be reduced to improve the pulping property of engineering tree species.

Rooting difficulties with hardwood cuttings has seriously hindered the popularization of some important poplar species by using the vegetative propagation technique. Bu et al. (1991) reported that the introduction of root-inducing (Ri) plasmid rol genes into *P. tomentosa* resulted in the production of hair roots and an increased rooting ability of transformed plants in vitro and in the field (Zheng et al., 1995a). In addition, a significant elevation of rooting rate was found in transformed *P. tomentosa* with phytohormone biosynthetic genes in differentation medium (Wen et al. , 1997). Recently, expression of the chimeric genes of the GH_3 promoter-*rolB* gene-nos in *P. tomentosa* increased auxin sensitivity and improved the rooting ability of the transgenic poplars (Liang, 2003).

3 Application of Molecular Markers to Poplar Breeding

In recent decades, different techniques have been developed to detect molecular markers. In China, various types of molecular marker technologies have been applied to genetic improvements of *Populus*, such as construction of genetic linkages maps, QTL mapping of many important traits, and the early selection of individuals with specific characteristics within larger progenies.

3.1 Construction of genetic linkage map

Genetic linkage maps, based on molecular markers, are useful tools for detecting loci controlling a number of traits, map-based gene cloning, molecular marker-assisted selection and for studying genome organization and evolution in many forest tree species. The four principal types of molecular markers used to construct genetic linkage maps in forestry include random amplified polymorphic DNA (RAPD), restriction fragment length polymorphisms (RFLP), amplified fragment length polymorphisms (AFLP), and simple sequence repeats (SSR). The first poplar genetic linkage map was developed by Liu and Furnier (1993); to date, various genetic linkage maps have been constructed for more than 12 poplar species abroad (Cervera et al. , 2001). In China, the first poplar genetic linkage map established with RAPD markers is based on the two populations of *P. deltoids*×*P. cathayana* (Su et al. , 1998). Since then, much more complete genetic linkage maps have been constructed for some poplar species by combining hundreds of RAPD, AFLP, or SSR markers (Yi et al. , 1999; Zhang et al. , 2000, 2003, 2004a; Huang et al. , 2004a). Although these constructed genetic maps could be exploited as potential references for QTL analysis and genomic research of *Populus*, they tend to have a lower density and a shorter map distance compared with most genetic maps constructed previously for poplars abroad (Cervera et al. , 2001). Therefore, the genetic maps constructed for *Populus* in China must be saturated with the novel molecular marker technique.

3.2 Quantitative trait loci mapping and marker-assisted selection

Quantitative trait loci (QTL) mapping and marker-assisted selection is another important application of molecular marker technology. The fact that many important traits of poplars such as growth, height, diameter, stem volume, and wood properties, are usually controlled by multiple genes makes conventional breeding very difficult. QTL mapping according to a genetic linkage map may provide opportunities for genetic improvements in poplars.

In China, QTL associated with growth and phenological quantitative traits (height, diameter, and top closure) within 80 F_2 seedlings of *P. deltoids*×*P. cathayana* were first detected by Li et al. (1999) by using RAPD markers. It was reported that the identification of QTL associated with five quantitative leaf traits in an F_2 population of *P. deltoids*×*P. cathayana* by RAPD markers showed that 10, 10, 4, 9, and 12 markers were significantly associated with leaf length, leaf width, ratio of leaf length to width, petiole length, and leaf area, respectively; moreover, most of the marker loci associated with these quantitative traits were on the fourth, 12th, 15th, and 17th lines of the linkage groups (Su et al., 2000a). Recently, 80 QTL controlling morphological and adaptive traits were detected in the Chinese white poplar using the F_1 pseudo-testcross strategy and an AFLP genetic linkage map (Zhang et al., 2004a). The study by Huang et al. (2004b) showed that 356 AFLP markers and 12 SSR markers that were aligned in 19 major groups were mapped in 87 individuals of an F_2population of *P. deltoids*×*P. cathayana* and the eight QTL identified for wood properties, including wood density, microfiber angle, fiber length, and fiber width were over-dominant. In addition, it was found that the trait of fiber length has obvious heterosis: there may be positive effects among the genes controlling the trait of fiber length, but negative effects among the genes controlling the trait of wood basic density (Huang et al., 2004a). These studies may have laid a foundation for the improvement of the wood property of poplars.

To investigate the genetic control of resistance to *Alternaria alternata* (Ala), the susceptibility of F_1 and F_2progenies and both parents (*P. deltoides* and *P. cathayana*) to Ala was tested, and two markers mapped on group 3 by using RAPD markers were found to be linked to Ala resistance. Moreover, Ala resistance may be determined by a single recessive gene for *P. deltoids* (Su et al., 2000b). Recently, two RAPD markers linked to the resistance locus of black spot disease in *P. deltoids* were identified by using bulked segregate analysis and selective genotype linkage analysis, and the genetic distances between the two markers and the resistance locus were evaluated (Zhang et al., 2002a). These results may provide the basis for molecular marker-assisted selection and early identification of disease-resistant varieties.

4 Problems and Prospects

Considerable progress has been made in breeding transgenic poplars for insect resistance in China and some transgenic poplars have started to be released into the field. However, until now, little is known about the field evaluation of insect resistance, biological safety, and ecological risk assessment of insect-resistant transgenic poplars. It has been reported that the ecological risks associated with the use of engineered plants include the counter-evolution of insect resistance to insecticidal proteins, the potential effects on non-target organisms (humans, animals, plants, and micro-organisms), and biodiversity, as well as the escape of transgenes due to pollen dispersion, all of which have been investigated in many transgenic Bt crops (Wei et al., 2002). Thus, the biological safety of transgenic insecticidal gene poplars, long-term field evaluation, and variation patterns of insect resistance, as well as the inter-planting ratio of different cultivars in a plantation, should be further investigated before these transgenic poplars can be applied. It is known that a major concern regarding genetically engineered pest resistance is its evolutionary stability. Recently, based on experimental data of risk assessment of crops transferred with the insecticidal gene, several strategies have been suggested to delay the development of insect resistance to insecticidal proteins, including the transfer of multiple genes that confer different types of insect resistance, modification of the gene-coding region by mutation techniques, highly specific gene expression using inducible promoters, application of antisense and ribozyme technologies, and the use of specific inter-plantation with other poplars, or even with other species, to provide a refuge for insects (Wei et al., 2002; Zhang et al., 2002c). In addition, the release of foreign genes into natural or feral populations could be mitigated by producing transgenic sterile poplars using antisense or promoter suppression of specific homeotic reproductive development genes. Compared with improvements in insect resistance, little progress has been observed in the areas of drought, salt, and freezing resistance in poplars, because environmental stress responses and adaptations are very complex processes. Moreover, environmental resistance-related genes and knowledge about the molecular mechanism of the stress response and the regulation of gene expression are lacking. To provide a more effective method for transferring stress-resistance genes to poplar, further research is needed in the following areas: (i) the cloning and identification of environmental resistance-related novel genes or transcriptional factors through the investigation of functional genomics; and (ii) the determination of gene expression patterns and their roles in environmental stress responses and adaptations using RNA interference technology, which may pave the way for transgenic strategies.

Various genetic maps have been constructed for some poplar species in China; however, the density of them is lower, which would result in poor practicability of QTL in markerassisted selection. It has been reported that many co-dominant markers, such as expressed sequence tags (EST) and expressed sequence tag polymorphisms (ESTP), can provide more informative polymorphic markers for mapping, which have been developed in much more detail and complete genetic linkage maps for genomic research and map-based gene cloning in some poplar species (Cervera et al., 2001). Thus, in the future, these RAPD or AFLP genetic maps constructed for poplar must be saturated with EST and ESTP markers, resulting in a much more comprehensive genetic map, which will provide a basis for the addition of other genetic markers, genome comparative mapping, map-based gene cloning, molecular-assisted selection, and QTL controlling economically important traits in poplar.

In China, the studies mentioned above should result in new breakthroughs in molecular genetic improvements and breeding of poplar species in the future.

毛白杨叶片离体再生培养的基因型效应*

摘　要　以5个毛白杨基因型(TC121、TC152、TC332、TC510和TC970)叶片为外植体，研究毛白杨不同基因型叶片再生能力的差异，通过单因子和正交试验设计，建立并优化毛白杨叶片再生体系。结果表明，毛白杨叶片离体再生能力受基因型的影响，基因型间的叶片再生率差异显著。其中基因型TC152再生率最高，在MS+6-BA 1.0mg/L+NAA 0.05mg/L+蔗糖30g/L+琼脂6g/L培养基上可达到97.0%；基因型TC970在Ms+6-BA 0.5mg/L+NAA 0.05mg/L+蔗糖30g/L+琼脂6g/L培养基上再生率可达到94.7%；经培养基优化后，基因型TC332再生率提高至72.3%，而基因型TC510和TC121再生率仍低于50.0%。

关键词　毛白杨，基因型，再生，组织培养

毛白杨(*Populus tomentosa* Carr.)属杨柳科(Salicaceae)杨属(*Populus* L.)，是我国特有的珍贵乡土树种。具有生长速度快、抗病虫能力强等优良性状(徐纬英，1988)。近年来，随着植物基因工程的发展，毛白杨的遗传转化(郑均宝等，1996；郝贵霞等，1999)研究已成为热点。筛选出再生效率较高的毛白杨基因型，建立稳定的遗传转化体系，有利于提高毛白杨的遗传转化效率。

离体再生培养是植物大规模微体繁殖和遗传转化研究的技术保障(蔡国军，2003；CHAWLA，2004)。目前已报道了毛白杨通过叶片(DU et al.，2002)、愈伤组织(沈效东等，1996)、原生质体(王善平等，1990)等途径再生完整植株，基因型是影响植物再生体系建立的关键因素之一(Victor，2005)。近年来，先后报道了基因型对小麦(*Triticum aestivum*)(Machll et al.，1998)、水稻(*Oryza sativa*)(Hoquel et al.，2004)、报春(*Primula* spp.)(Gabriele et al.，2003)、美洲黑杨(*Populus dehoides*)(COLEMAN et al.，1989)等植物的再生率和离体再生体系建立的影响。张存旭(2004)和Lu(Lu et al.，2001)分别报道了不同基因型毛白杨茎段分化能力的差异。然而，关于基因型效应对毛白杨叶片离体再生能力的影响目前尚无报道。本研究以5个优良毛白杨基因型为材料，通过调节再生培养基中植物生长调节剂的种类、组合及浓度，建立不同基因型毛白杨的叶片离体再生体系，探讨毛白杨叶片再生培养中的基因型效应，为不同基因型毛白杨再生体系的利用和下一步的研究奠定基础。

1　材料与方法

1.1　材料

在山东省国营冠县苗圃全国毛白杨基因库中，选取编号分别为TC121、TC152、

* 本文原载《北京林业大学学报》，2007，29(5)：38-42，与姚娜、安新民、王冬梅和陶凤杰合作发表。

TC332、TC510 和 TC970 的 5 个生长旺盛、无病虫害的优良毛白杨基因型，经外植体消毒后建立无菌培养体系。

1.2 方 法

1.2.1 培养基和培养条件

根据前期对基本培养基的筛选，以 MS(Murashige et al.，1962)作为基本培养基，附加 30g/L 蔗糖和 6g/L 琼脂，pH 值 5.8。培养室温度 25~28℃，光照时间 16h/d，光照强度为 1500~2000Lux.

1.2.2 毛白杨无菌苗的培养

经前期试验筛选，毛白杨的启动培养基为 MS+6-BA 0.3mg/L+ NAA 0.1mg/L；增殖培养基为 MS+6-BA 0.5mg/L+ NAA 0.1mg/L。待毛白杨茎段腋芽萌发后，切取 1cm 长的腋芽，接种在生根培养基 1/2 MS + IBA 0.4mg/L 上。叶片再生不定芽时，选取无菌生根苗顶芽下 2~3 片叶作为外植体，于无菌条件下用手术刀割伤叶片中脉 2~3 刀，远轴面向下接种在设计的培养基上。

1.2.3 试验设计与统计方法

设计单因素试验(表 1)，讨论毛白杨基因型 TC121、TC152、TC332、TC510 和 TC970 叶片在 MS_1、MS_2 和 MS_3 培养基上再生情况；设计正交试验(李艳等，2004)(表 2)，优化基因型 TC121、TC332 和 TC510 叶片再生培养基。每个处理接种 30 块外植体，重复 3 次。

接种 6 周后统计再生不定芽高于 1cm 的外植体数，并计算叶片再生率。

再生率=(再生不定芽的外植体数/接种外植体数)×100%

所得百分数结果经反正弦转换($\theta=\arcsin(\text{百分率})^{1/2}$)后，用 SPSS 12.0 分析软件进行方差分析，采用 ISD 法进行 0.05 水平多重比较(续九如等，1995)。

表 1 毛白杨叶片再生培养基

培养基	6-BA(mg/L)	NAA(mg/L)	6-BA：NAA	接种数/片
MS_1	0.5	0.05	10：1	30
MS_2	1.0	0.05	20：1	30
MS_3	1.5	0.05	30：1	30

表 2 TC121、TC332 和 TC510 毛白杨叶片再生培养基优化正交试验因素水平表 $L_9(3^4)$

水平	因素种类及浓度(mg/L)		
	6-BA	NAA	KT
1	1.0	0.0	0.5
2	2.0	0.05	1.0
3	3.0	0.1	1.5

2 结果与分析

2.1 不同基因型毛白杨叶片在相同培养基上再生率的差异

将 5 个基因型叶片分别接种在 MS_1、MS_2、MS_3 培养基上，接种 14～18d 后，基因型 TC152 和 TC970 叶脉切口处出现大量丛生不定芽；30～35d 后，基因型 TC121、TC332 和 TC510 在叶柄和叶基部有单个不定芽再生。毛白杨叶片在相同培养基上再生率(见表 3)存在基因型间的差异。基因型 TC152 和 TC970 在附加 6-BA 和 NAA 的培养基上再生率高于 90.0%，不定芽生长健壮；而在相同条件下，TC121、TC332 和 TC510 再生率均低于 40.0%，不定芽以单芽为主，叶片在培养过程中有褐化或黄化现象。因此，选用 MS_2 作为 TC152 再生培养基，MS_1 为 TC970 再生培养基；TCI21、TC332 和 TC510 叶片再生培养基有待于进一步筛选。

表 3　5 个毛白杨基因型在相同培养基上再生情况

基因型	再生率(%)		
	MS_1	MS_2	MS_3
TC121	26.3c	15.6c	4.3d
TC152	34.4bc	97.0a	60.1a
TC332	36.6b	21.1c	9.8c
TC510	21.0c	20.4c	5.4cd
TC970	94.7a	59.3b	35.5b

注：表中同列数字后字母不同表不处理的差异显著($P<0.05$)，表 6~7 同此。

2.2 TC121、TC332 和 TC510 毛白杨叶片再生培养基的优化

鉴于基因型 TCI21、TC332 和 TC510 在附加 6-BA 和 NAA 的培养基上不能高效再生不定芽，改变植物生长调节剂的种类、组合和浓度，设计正交试验(表 2)，对以上 3 个基因型再生培养基进行优化，外植体接种 25～30d 后，基因型 TCI21、TC332 和 TC510 在叶柄和叶基部切口处有不定芽再生。3 个基因型叶片均为直接再生不定芽，不定芽以单芽为主。叶片在培养过程中有褐化或黄化现象。统计结果表明(表 4~6)：

1)TCI21 叶片在处理 1 上再生率最高，为 27.8%。NAA 对叶片再生率的抑制达到显著水平，NAA 为 0 时再生率显著高于其他水平；6-BA 和 KT 对叶片再生率影响不显著。综合考虑正交试验极差分析结果，选择 6-BA 1.0mg/L+KT 1.5mg/L 为最佳不定芽诱导组合。

2)TC332 叶片在处理 4 上再生率最高，为 72.3%。6-BA 和 NAA 浓度显著影响叶片再生率，6-BA 2.0mg/L、NAA 0mg/L 时再生率显著高于其他水平。根据正交试验极差分析结果，筛选出诱导叶片再生不定芽的最佳组合为处理 4。

3)在诱导 TC510 叶片再生不定芽的 9 个处理中，处理 1 再生率最高，为 41.1%。3 个因素对再生率的影响依次为 6-BA>KT>NAA；6-BA 对叶片再生率的影响达到显著水平，

6-BA 1.0mg/L 时再生率显著高于其他水平，当 6-BA 浓度升高至 2.0mg/L 时，叶片再生率降低，说明高浓度 6-BA 抑制 TC510 叶片再生不定芽。根据正交试验极差分析结果，得出诱导不定芽再生的最佳组合为 MS+ 6-BA 1.0mg/L + NAA 0.05mg/L + KT 1.5mg/L。

表 4　TC121、TC332 和 TC510 毛白杨叶片再生培养基优化 $L_9(3^4)$ 正交试验结果

处理		6-BA (mg/L)			NAA (mg/L)			KT (mg/L)		再生率(%)		
										TC121	TC332	TC510
1		1.0			0			0.5		27.8	47.7	41.1
2		1.0			0.05			1.0		4.4	12.2	33.3
3		1.0			0.1			1.5		6.6	5.4	35.5
4		2.0			0			1.0		21.1	72.3	4.4
5		2.0			0.05			1.5		4.4	34.4	11.0
6		2.0			0.1			0.5		3.3	4.3	3.3
7		3.0			0			1.5		20.0	40.0	4.4
8		3.0			0.05			0.5		3.3	16.6	4.4
9		3.0			0.1			1.0		5.5	3.3	3.3
K_1	58.4	76.3	111.6	85.6	137.9	63.7	52.8	79.7	62.2			
K_2	49.7	106.1	41.6	34.4	79.1	66.4	52.4	87.8	57.7			
K_3	50.1	70.4	34.4	38.2	35.8	57.5	53.1	85.3	67.7			
R	2.9	11.9	25.8	17.1	34.0	3.0	0.2	2.7	3.3			

注：各因素下的 K_l、K_2、K_3 和 R 值从左至右依次属于 TC121、TC332 和 TC510。

表 5　$L_9(3^4)$ 正交试验方差分析

变异来源	df	*SS*			*MS*			*F*		
		TC121	TC332	TC510	TC121	TC332	TC510	TC121	TC332	TC510
6-BA	2	47.89	626.02	3646.21	23.95	313.01	1823.10	2.78	14.88*	166.30*
NAA	2	1627.93	5596.34	41.17	813.96	2798.17	20.58	94.46	133.05	1.88
KT	2	0.265	56.09	0.13	28.04	25.17	0.02	1.33	2.30	
误差	20	172.35	420.64	219.25	8.62	21.03	10.96			
总和	26	10188.38	28779.43	15686.98						

注：表示差异显著(P<0.05)

表 6　$L_9(3^4)$ 正交试验多重比较表

水平	TC121	TC332		TC510
	NAA	6-BA	NAA	6-BA
1	28.5a	25.9b	37.2a	47.1a
2	11.5b	35.4a	13.9b	26.8b
3	12.7b	24.6b	11.5b	11.9c

2.3　不同毛白杨基因型在最佳培养基上再生率的差异

将毛白杨基因型TCI21、TC152、TC332、TC510和TC970叶片接种在其最佳再生培养基上，基因型间再生率存在差异(表7)。TC152和TC970再生率显著高于TCI21、TC332和TC510。TC332再生率得到显著提高。而TCI21和TC510再生率仍低于50.0%；其中，TCI21再生率最低，为28.6%。由此可见，5个毛白杨基因型在经过培养基优化后，再生率仍存在显著差异。再生率由高到低依次为：TC152、TC970、TC332 TC510、TCI21。

表7　5个毛白杨基因型在最佳培养基上再生情况

处理	基因型	最佳培养基	再生率(%)
1	TC121	MS+6-BA1.0+KT1.5	28.6d
2	TC152	MS+6-BA1.0+NAA0.05	97.0a
3	TC332	MS+6-BA2.0+KT1.0	72.3a
4	TC510	MS+6-BA1.0+NAA0.05+KT1.5	49.8c
5	TC970	MS+6-BA0.5+NAA0.05	94.7a

注：表7中所用植物生长调节剂浓度单位均为mg/L。

3　讨论

3.1　不同基因型毛白杨叶片再生培养基的差异

在叶片离体再生培养过程中，再生培养基中植物生长调节剂的种类及浓度直接关系到叶片再生率和不定芽生长状况。为建立高效离体再生体系，需要很好地协调外植体基因型与再生培养基这二者之间的相互作用。本研究通过对不同毛白杨基因型最佳再生培养基的筛选，发现5个毛白杨基因型在相同培养基上叶片再生率差异显著，其最佳叶片再生培养基在植物生长调节剂种类及浓度方面也存在较大差异。这可能是由于不同毛白杨基因型有其自身独特的遗传背景，因而对再生培养基的适应性不同。

已报道的毛白杨叶片再生培养基，在附加植物生长调节剂的种类、组合和浓度方面略有差异。杜宁霞(DU et al., 2002) 和郑均宝(郝贵霞等，1999) 发现毛白杨叶片在附加6-BA和NAA的培养基上再生率最高；而郝贵霞(蔡国军等，2003)采用附加6-BA、ZT和IAA的MS培养基为再生培养基；石超(石超等，2003)则采用附加ZT、TDZ和IBA的MS培养基作为再生培养基。本研究借鉴成功报道，采取先易后难的原则，首先选用附加不同浓度6-BA和NAA的培养基作为再生培养基；对于再生率低的基因型，通过增加细胞分裂素的种类，优化再生培养基。结果表明，5个供试毛白杨基因型最佳再生培养基不同。TC152和TC970在附加6-BA和NAA的培养基上可以高效再生不定芽。TC121、TC332和TC510对植物生长调节剂种类的需要不同：细胞分裂素对叶片再生是必需的，而NAA抑制TC121、TC332叶片再生不定芽。由此可见，不同毛白杨基因型叶片离体再生培养基存在差异，很难寻找出一个适用于多个毛白杨基因型叶片再生的培养基。

3.2 不同基因型毛白杨再生能力的差异

Colemanl et al(1989)发现，16 种不同基因型美洲黑杨茎段在相同培养基上，不定芽再生能力存在差异：有 6 种难诱导不定芽，4 种再生率达 50% 以上，其余 6 种则介于 3%～23%，Lu et al(2001) 发现三倍体毛白杨不同基因型茎段分化能力存在差异。本研究发现，无论在相同培养基还是在最佳培养基上，毛白杨基因型 TC121、TC152、TC332、TC510 和 TC970 再生能力存在差异。这可能是由于不同毛白杨基因型内源激素水平引起的差异，或不同基因型对植物生长调节剂敏感性引起的差异。

依据在附加 6-BA 和 NAA 培养基上再生率的差异，可将 5 个基因型分为两种类型：① 容易高效再生不定芽的基因型，包括 TC152 和 TC970；② 较难高效再生不定芽的基因型，包括 TC332、TC510 和 TCI21。根据这一现象，在今后有关毛白杨再生培养基筛选的研究中，可将附加 6-BA 和 NAA 的培养基作为首轮筛选培养基，期待在多个候选毛白杨基因型中，首先筛选出几个再生率较高的基因型，以提高工作效率、节约研究成本。

Li 等(2003)在烟草(*Nicotiana tabacum*)再生体系研究中发现再生率低于 80%的烟草基因型，很难通过改变培养基中的添加成分来提高再生率。在本研究中，对毛白杨再生培养基进行优化后发现 TC510 和 TCI21 再生率未见明显提高。因此，毛白杨叶片离体再生能力可能存在与烟草类似的基因型限制。可以考虑增加供试基因型和培养基种类，扩大研究范围，对毛白杨叶片再生的基因型限制及其产生原因进行更加系统的研究。

3.3 不同基因型毛白杨叶片不定芽生长状况的差异

本研究发现，在毛白杨叶片再生不定芽的过程中，不同基因型叶片不定芽再生起始时间、再生部位和不定芽生长状态不同。TC152 培养 14d 后可见不定芽从叶片伤口处发生；TC510 和 TCI21 在培养 25d 后才能再生出不定芽。5 个基因型叶柄和中脉均有较强的形态发生能力，这与诸葛强(诸葛强等，2003)报道的新疆杨(*Populus alba* var. *pyramidalis*)叶片再生不定芽情况相同。5 个基因型再生不定芽的数量与其再生率一致，再生率较高的 TC152 和 TC970 不定芽呈丛生状；而 TC510 和 TCI21 多为单个出芽。综上所述，毛白杨叶片离体再生能力存在基因型的差异。在今后的毛白杨叶片离体再生培养研究中，除了进行外植体、基本培养基、植物生长调节剂筛选外，选择易于再生和再生率较高的毛白杨基因型也是考虑的重要因素。

Effects of Genotype on *in vitro* Regeneration from the Leaves of *Populus tomentosa* Carr.

Abstract The leaves of five genotypes(TC121, TC152, TC332, TC510 and TC970) in *Populus tomentosa* Carr. were used as explants to study the differences of in vitro regeneration competences among genotypes. Single factor and orthogonal experiment designs were used to establish and optimize the leaf regeneration system of *P. tomentosa*. The results showed that the regeneration competences were affected by genotypes. an d there were obvious differences among genotypes in regeneration rates. The regeneration rate of TC 152 reached 97. 0% on the medium of MS + 6-BA 1. 0mg/L+ NAA 0. 05mg/L + sucrose 30g/L + agar 6g/L. The regeneration rate of TC970 was 94. 7% on the medium of MS + 6-BA 0. 5mg/L + NAA 0. 05mg/L + sucrose 30g/L +agar 6g/L. After medium optimization, the regeneration rate of TC332 had been improved to 72. 3%. However, the regeneration rates of TC510 an d TC121 remained below 50. 0%.

Keywords *Populus tomentosa* Carr. , genotype, regeneration, tissue culture

Isolation and Analysis of a TIR-specific Promoter from Poplar*

Abstract A 5′ flanking region of the well-conserved Toll/interleukin-1 receptor domain (TIR)-encoding sequence was isolated from the genomic DNA of *Melampsora magnusiana* Wagner resistant clones of hybrid triploid poplars [(*Populus tomentosa* ×*P. bolleana*)×*P. tomentosa*]. Sequencing results and alignment analysis show that the obtained TIR-specific promoter (named as *PtTIRp*01) was 1732 bp in length; moreover 3′ region of the *PtTIRp*01 contains a 398bp complete TIR-encoding sequence, which significantly corresponds to the 5′ composition of TIR-NBS type gene *PtDRG*02, indicating that the obtained TIR-specific promoter region consists of 747 bp long 5′ region of TIR-NBS type gene *PtDRG*02 and its upstream region of promoter (985bp). It was found that the 5′ region of TIR-NBS type gene *PtDRG*02 was characterized in the downstream region of the transcriptional start, named as 5′-untranslated region (5′ UTR), consisting of one 93bp 5′-untranslation exon, one 213bp intron and one 441 bp TIR-encoding open reading frame (ORF). In addition, several putative *cis*-acting motifs were present in the obtained TIR-specific promoter of *PtDRG*02, including one TATA box, one GC-rich, one AT-rich, one P-box, one 3-AF1 binding site, two CAAT boxes, two GT-1 motifs, three typical W-boxes, four I-boxes, and one multi-*cis*-acting fragment (MCF). The latter contains five types of regulatory elements (E4, G-box, ABRE motif, box1 and HVA1s), most of which were homologous to the *cis*-acting regulatory elements involved in the activation of defense genes in plants. Thus, it can be suggested that TIR-specific promoter might be a pathogen-inducible promoter and be necessary for the inducible expression of defense-related genes.

Key words Toll/interleukin-1 receptor domain, promoter, *Cis*-acting element, poplar

1 Introduction

During normal growth and development, plants maybe severely damaged by bacterial, viral and fungal diseases, and conventional breeding for disease resistance is a difficult task because disease resistance is generally controlled by multiple genes. In addition, traditional spray application of pesticides and chemicals can control disease damage, but has the potential to devastate the environment. In recent years, with the development of molecular biotechnology, considerable progress has been made in breeding transgenic plants for disease resistance (Campbell et al., 2002;

* 本文原载 *Forestry Studies in China*, 2007, 9(2): 95~106, 与郑会全、林善枝、张谦、张珍珍、雷杨和侯璐合作发表。

Herschbach and Kopriva, 2002; Lin et al. , 2006). Many attempts at engineering increased disease resistance of plants generally using an over-expression of a single defense under the control of constitutive promoter such as CaMV 35S, and any improved disease resistance was in many cases accompanied by the spurious activation of defense response, homology-dependent gene silencing, uncontrolled spread gene expression, reduced growth, altered development or morphology, and showing disease symptoms in the absence of pathogens (Vaucheret et al. , 1998; Chenand Chen, 2002; Fitzgerald et al. , 2004; Malnoy et al. , 2006). Thus, precise control of introduced gene expression under the control of inducible promoter is preferred in any strategy to the engineering of plants with increased and durable resistance to a spectrum of disease. Moreover, alternative pathogen inducible, tissue-specific expressing or endogenous resistance gene promoters are highly required for transgene-mediated improvements in disease resistance of plants (Herschbach and Kopriva, 2002; Mentag et al. , 2003). Although many tissue-specific promoters of plants have been characterized (Joung and Kamo, 2006), up to now, reports about pathogen inducible promoters of defense-related genes are very limited, especially in wood plants.

The genus *Populus* is comprised of more than 74 species classified into five or six sections. The poplarhas played a key role in forest production and the construction of ecological environments in China. Fast growth, high yield, easy propagation, and a wide range of uses make the poplar one of the most important tree species for artificial forests. Moreover, owing to its small genome and ease of tissue culture, the poplar represents a useful model system for the investigation of the genetics and molecular biology of wood plants (Lin et al. , 2006). However, poplars are susceptible to bacterial, viral and fungal diseases, and the fact that the disease resistance of poplars is controlled by multiple genes, pulsed high heterozygosity, and a long life cycle makes conventional breeding very difficult(Lin et al. , 2006). In recent years, with the development of molecular biotechnology and the study of other forest tree species, genetic engineering and molecular markers have been widely used in the genetic improvement of the poplar species, which has opened up new avenues for engineering poplars with increased disease resistance, but little has been reported regarding genetic improvements in bacterial, viral and fungal diseases of poplars in China.

It is well known that the Toll/interleukin-1 receptor (TIR) is an inflammation and a host defense-related domain due to the homology between the cytoplasmid regions of the mammalian interleukin-1 (IL-1) receptor and the *Drosophila melanogaster* protein Toll(Hashimoto et al. , 1988; Sims et al. , 1988). The highly conserved TIR domain is found among plants, animals, and bacteria, which could provide a critical molecular switch for the immune response (O'Neil, 2000; Aderem and Ulevitch, 2000; Takeda and Akira, 2003; Nurnberger et al. , 2004; Burch-Smith et al. , 2007). In plants, TIR domain is usually one component of multi-domain protein, and TIR-type protein are classified into three families namely: TIR-NBS-LRR, TIR-X, and TIR-NBS. Within the TIR-NBS-LRR family, the tobacco *N* (TMV resistance), flax *L6* and *M* (rust resistance), *Arabidopsis RPP5* and *RPP1*(downy mildew resistance), and *Arabidopsis RPS4*(bacterial resistance) have been characterized (Anderson et al., 1997; Noel et al., 1999; Zhang et al., 2004; Howles et al., 2005; Burch-Smith et al., 2007). It has been found that the TIR-X family of proteins lacks both the NBS domain and the LRR domain, while theTIR-NBS proteins contain much of the NBS region, but lacks the LRR domain (Meyers et al., 2002). Over the last several years, approximately 50 TIR-typegenes have been cloned from the *Populus* genus. Recently, the nucleotide sequences of TIR-NBS type disease resistance-like gene *PtDRG*02 (GenBank accession no. DQ324362) obtained from hybrid triploid poplars [(*Populus tomentosa* ×*P. bolleana*)×*P. tomentosa*] have been analyzed by us, and the similar TIR-encoding sequence (identities ranging from 87%to 100% with the known TIR-encoding sequence) was present at the 5′ coding region in addition to its conserved translation product of TIR domain of *PtDRG*02. This suggests that this well conserved 5′ TIR-encoding sequence may be one of the significant hallmarks of TIR-type genes in the *Populus* genus, which could provide an important basis for the isolation of unknown TIR-specific region such as unexploited TIR-type genes and TIR-specific promoters.

Based on the above results, the isolation and analysis of the TIR-specific promoter regions of pathogen-related gene *PtDRG*02 isolated from Chinese white poplar was reported in this study.

2 Materials and methods

2.1 Plant materials

Six *Melampsora magnusiana* Wagner resistant clones of L2, L7, L8, L9, L12 and L13, and three highly susceptible clones of L5, L11 and L25 of hybrid triploid poplars [(*P. tomentosa* ×*P. bolleana*)×*P. tomentosa*] were used as experiment materials.

2.2 PCR primers for 5′-flanking region of theTIR-encoding sequence

Three putative genetic objects, either of which contains a TIR-encoding sequence (around 400bp in length) and its corresponding upstream region (2000~3000bp in length), of a 95-kb genomic sequence mapped 0.6 centiMorgan from the *M. larici-populina* (MER) resistance locus of *P. deltoids* reported by Lescot et al. (2004), were employed as BLAST probes and subjected to the poplar genome database (JGI: http://genome.jgi-psf.org/Poptr1/Poptr1.home.html) to search for the conserved fragments upstream from the well-conserved TIR-encoding sequence. Basing on these upstream conserved fragments, as well as the downstream TIR-encoding sequence, two sets of forward/reverse primers involving Pro-U I/Pro-D I and Pro-U II/Pro-DI (Table 1) were designed with program premier 5.0 and used for the PCR amplification of 5′-flanking region of the TIR-encoding sequence

Table 1 Primers used for this study

Primer	Sequence	Category	Location
Pro-U Ⅰ	5′-CTAATACCTTCCCTTTACGCAACCAG-3′	Forward	Upstream from TIR-encoding sequence
Pro-U Ⅱ	5′-AACAAATAGTTCTCCGTGAGCATACC-3′	Forward	Upstream from TIR-encoding sequenc
Pro-D Ⅰ	5′-GCTTTTCTCCACTCCTTCACCAACTT-3′	Reverse	Correspouding to tile 3′ end of the TIR-encoding sequence
Pro-D Ⅱ	5′-TCATTOAOACACCATCTAOAAOAAOC-3′	Reverse	Correspouding to the internal region of the TIR-eneoding sequence
Pro-D Ⅲ	5′-AAGTGTGGATTCCTGCTTGGACTAAG-3′	Reverse	Correspouding to the 5′ end of the TIR-eneoding sequence

2.3 PCR procedure

Genomic DNA was extracted from present-year leaves of Chinese white poplar with a DNeasy Plant Kits. Two batches of PCR procedure were employed to isolate the 5′-flanking region of TIR-encoding sequence using the different poplar genomic DNA as a template and two sets of forward/reverse primers Pro-U I/Pro-DI and Pro-U II/Pro-D I (shown in Table 1), respectively. All the PCR systems contained 1μL (20–40 ng) genomic DNA, 1μL forward primer (10μmol/L), 1μL reverse primer (10μmol/L), 5μL 10×LA Taq polymerase buffer (Mg^{2+} Plus), 0.5μL LA Taq DNA polymerase (5U/μL), 1μL dNTP (10mmol/L), and 40.5μL sterile double water. Thermal cycling conditions were: a) predenaturing at 94℃ for 5min; b) denaturing at 94℃ for 30s, primer annealing at 60℃ for 40s and extension at 72℃ for 1min, repeating for 35 cycles; c) final extension at 72℃ for 10min. Specificity of the first round PCR products was verified by two rounds of nested-PCR amplifications using the 50-fold diluted primary products as a template. Nested primers are shown in Table 1. PCR reaction systems refer to the first round amplification. Thermal cycling conditions were the same as the first procedure except for the number of cycles decreased to 25.

2.4 DNA cloning and sequencing

All the PCR products were analyzed on 1% agarose gel with the GENE GENIUS imaging system. Specific amplification bands were extracted from the gel and purified using the MinElute Gel Extraction Kit, and then cloned into the pGEM-T easy vector. Positive clones containing the full-length sequence of target fragment designed as PtTIRp01 were finally sequenced with the ABI3730 DNA Analyzer.

2.5 Computer-assisted analysis

Sequence alignment analysis, carried out with the BLAST search program in NCBI Transcriptional start site of the obtained DNA sequence, was predicted with the online program TSSP in

Softberry. Cis-acting regulatory elements located at the promoter region were predicted by using the online program PLACE, Plant CARE, NSITE-PL and ScanWM-P (Softberry).

3 Results

3.1 Isolation of the 5′-flanking region of TIR-encoding sequence

Two batches of PCR amplification were performed to isolate the 5′-flanking region of the TIR-encoding sequence from Chinese white poplars using primer combination Pro-U I/Pro-D I and Pro-U II/Pro-D I. Of the two possible primer combinations, only the Pro-UI/Pro-D I primer combination gave one obvious specific 1700bp PCR product in the disease resistant clones of L2, L7, L8, L9, L12 and L13, but no same product found in susceptible clones L5, L11 and L25(Fig. 1A), suggesting that the Pro-U I/Pro-D I primer combination was specific, and the obtained 1700bp fragment might be associated with disease resistance. To identify the specificity of the obtained 1700bp fragment potentially corresponding to the 5′-flanking region of TIR-encoding sequence, two rounds of nested PCR amplifications were employed, and one obvious 1500bp band for nested amplification I and 1400bp band for nested amplification II were only observed in all tested disease resistant clones (Fig. 1B、1C), which consists of the expected results that the nested PCR products should be 200bp and 300bp less than the primary product (1700bp fragment) respectively, due to the primer locations at TIR-encoding sequence (Table 1). Thus, it may be concluded that the1700bp fragment contained a 1300bp 5′-flanking region of the TIR-encoding sequence (approximately 400bp in length), which potentially harbored a functional promoter hypothesized previously.

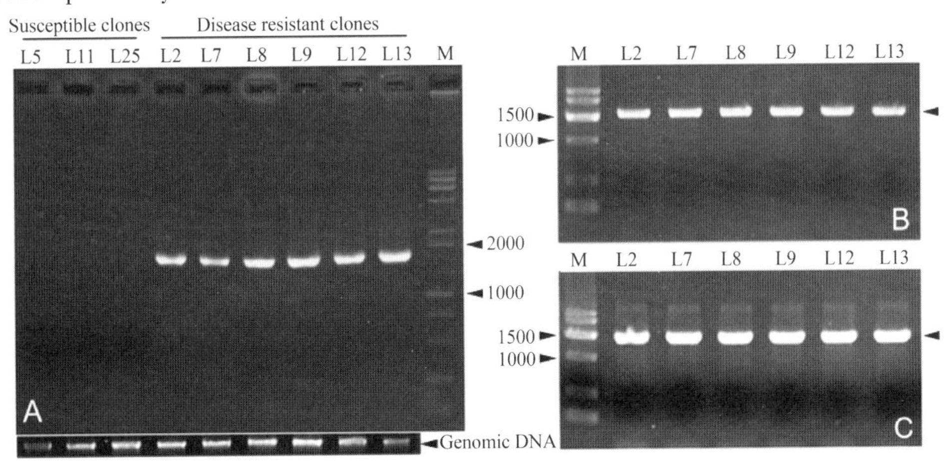

Fig. 1 PCR amplification of the 5′ flanking region of TIR-encoding sequence from hybrid triploid poplars. M: DNA marker; A: firstround PCR amplification, genomic DNA from each clone was loaded into each lane to normalize the amplification; B: nested PCR amplification I, and nested products were indicated with arrow head; C: nested PCR amplification II.

3.2 Sequence dissection of 5′-flanking region of theTIR-encoding sequence

Sequence results showed that all the length 1700bp fragments (designed as PtTIRp01) share the same nucleotide sequence except for three bases difference between clones L8 and L12, and the entire fragment was sequenced with 1732bp in length. The transcriptional start site (TSS) prediction result indicated that the base G was recognized as TSS signal and designed as +1 for sequence numbering. Align-ment analysis showed that the 3′ region of the *PtTIRp*01 sequence contains a 398 bp complete TIR-encoding sequence, which significantly corresponds to the 5′ composition of PtDRG02 gene (Gen-Bank accession no. DQ324361). In addition, the downstream region from +1 was characterized as afunctional gene that harbored a 5′-untranslated region (5′ UTR) comprising one 96bp 5′-untranslation exon (exon 1) and one 213bp intron followed by one 441bp TIR-encoding open reading frame. Upstream region 985bp (0/-985) was hypothesized to be a functional promoter specific to the TIR-encoding sequence.

3.3 Analysis of *cis*-acting regulatory elements of the TIR-specific promoter region

To identify the potential motifs responsible for thefunction of the obtained TIR-specific promoter, the *cis*-acting regulatory elements of the TIR-specificpromoter region were analyzed by using four online programs (Plant CARE, PLACE, NSITE-PL and ScanWM-P). It was found that about 985bp TIR-specific promoter region contained several putative regulatory cis-acting elements, including one TATA box (TGTCTATATA) (-32/-23), one GC-rich(-799/-793), one A/T-rich (-692/-684), one P-box(-418/-412), one 3-AF1 binding site (-306/-296), two CAAT boxes (-75/-71 and-210/-206), two GT-1 motifs(GAAAAA) (-322/-317 and-372/-366), three typical W-boxes (GTTGACT/TTGAC) (-164/-160, -549/-545 and-575/-569), four I-boxes (-275/-271, -342/-238, -494/-490 and-631/-627), and one multi-cis-acting fragment (MCF) (-788/-775) contain four types of regulatory elements of G-box, ABRE motif, box1 and HVA1s (-906/-893).

4 Discussion

In the previous study, the PtDRG02 gene obtained from hybrid triploid poplars was characterized as a TIR-NBS type gene, and its constitutive express was only observed in the leaf tissue of the disease resistant clones of hybrid triploid poplars, which indicates that *PtDRG*02 gene was closely associated with disease resistance of poplar (Zhang et al., 2006). Moreover, quantitative real time RT-PCR result has shown that the expression of *PtDRG*02 gene in leaf tissues of the tested disease resistant clones was positively induced by wounding and MeJA (Zhang et al., unpublished). The fact that a well-conserved TIR-encoding sequence was found within 18 TIR-type genes including *PtDRG*02 gene from the *Populus* genus, indicated that the well-conserved 5′ TIR-encoding of the obtained *PtDRG*02 gene could provide an important basis forthe isolation of TIR-specific promoters from poplar. In this study, a TIR-specific promoter region (designed as *PtTIRp*01) located at the upstream of the conserved TIR-encoding sequence was isolated from disease resistant clones of hybrid triploid poplars, implicating that the obtained TIR-specific promoter was potentially responsible for the resistance to *M. magnusiana* Wagner of hybrid triploid poplars. Sequencing results and alignment analysis showed that the obtained TIR-specific promoter *PtTIRp*01 was 1732 bp in length, 3′ region of the *PtTIRp*01 sequence contains a 398 bp complete TIR-encoding sequence, which significantly corresponds to the 5′ composition of TIR-NBS type gene *PtDRG*02, which indicates thatthe obtained fragment *PtTIRp*01 consists of 5′ region of TIR-NBS type gene *PtDRG*02 and its upstream region of promoter. Thus, it may be suggested that the obtained TIR-specific promoter is related to the expression of *PtDRG*02 gene. It was found that the 747bp long 5′ region of TIR-NBS type gene *PtDRG*02 was characterized in the downstream region of the transcriptional start, named as 5′-untranslated region (5′ UTR), consisting of one 93bp 5′-untranslation exon (exon 1), one 213bp intron, and one 441 bp TIR-encoding open reading frame (ORF). The presence of introns at the 5′ UTR region is a common feature found in several plant genes (García-Hernández et al., 1998; León et al., 1998; Alvarez-Buylla et al., 2000; Kim et al., 2004). It has been demonstrated that enhancer elements are generally present in the introns of the 5′ UTR region, suggesting that the introns located in the 5′ UTR region may play an important role in the regulation of gene expression by regulating mRNA stability, the efficiency of translation, or can directly impact transcriptional regulation (Maas et al., 1991; Gidekel et al., 1996; Bailey-Serres and Gallie, 1998; Plesse et al., 2001; Morello et al., 2002; Fiume et al., 2004; Wang et al., 2004; Gutiérrez-Alcalá et al., 2005). In addition, plant intron sequences were reported to enhance functional gene expression in transgenic plants, and this intron-mediated enhancement (IME) of gene expression has been reported for several genes, indicating that the first 5′-leader intron is necessary for a high-level gene expression and can autonomously promote transcription(Mascarenhas et al., 1990; Vasil et al., 1990; Maas et al., 1991; Clancy et al., 1994; Rose and Beliakoff, 2000; Kim et al., 2004; Morello et al., 2002, 2006; Gutiérrez-Alcalá et al., 2005; Kim et al., 2006; Sivamani and Qu, 2006). Thus, introns may influenceand enhance gene expression by multiple mechanisms(Le Hir et al., 2003). Recently, the pathogen-inducible wheat GstAl promoter in combination with WIR1aintron was found to have significantly resulted in the enhancement of defense-related gene expression and to have increased the pathogen resistance in transgenic plants (Altpeter et al., 2005). At present, the 213bp sequence of intron was located at the proximal 5′ region (+305/+93) of TIR-NBS type gene *PtDRG*02, and exhibited typical characteristics such as the A/T-rich (68.1%) and GT/AG consensus spice sites that are identical to other introns of many higher plants (Jung et al., 2005; Hong and H Wang, 2006), suggesting that the observed intron in the proximal 5′region of TIR-NBS gene might be required for the expression of TIR-NBS gene *PtDRG*02.

Many studies showed that the promoter regions of stress-inducible genes contain *cis*-acting regulatory elements, which are important molecular switches involved in the transcriptional regulation of a dynamic network of gene expression in response to environmental stresses (Yamaguchi-

Shinozaki and Shinozaki, 2005). In the present study, 985 bp sequence of TIR–specific promoter region (0/–985) were determined upstream of the translation start, and several putative regulatory motifs, all of which were homologous to the *cis*–acting elements involved in the activation of defense genes in plants, have been identified within the TIR–specific promoter region, indicating that TIR–specific promoter contains multiple *cis*–acting elements associated with the expression of stress–inducible genes.

A number of pathogen – induced genes and their promoters have been identified in plants (Zhang and Wang, 2005; Lee and H Wang, 2006). In general, transcriptional regulation of gene expression is largely mediated by the specific recognition of *cis*–acting promoter elements by trans–acting sequence–specific DNA – binding transcription factors (Park et al., 2006). Among the several classes of transcription factors associated with plant pathogen defensive–responses, WRKY transcription factors appear to be more prevalent in plants (Eulgem et al., 2000; Park et al., 2006). It was suggested that WRKY proteins and other transcription factors may preferentially arrange *cis*–acting elements, thereby directly binding WRKY transcription factors to the relevant target promoters (Cormack et al., 2002). To date, many WRKY proteins involved in defense against attack by pathogen have been identified in several plant species (Ryu et al., 2006), and probably have regulatory functions in plant defense responses to pathogen infection, and many defense–related genes, including pathogenesis–related(PR) genes, generally containing W–box elements in their promoter regions (Rushton et al., 1996; Eulgem et al., 1999, 2000; Dellagi et al., 2000; Hara et al., 2000; Kim et al., 2000; Du and Chen, 2000; Chen and Chen, 2002; Jung et al., 2005; Park et al., 2006; Ryu et al., 2006; Zheng et al., 2007). Thus, W–box specifically recognized by WRKY proteins is necessary for the inducible expression of *PR* gene (Rushton et al., 1996; Eulgem et al., 1999; Chen and Chen, 2002; Yamamoto et al., 2004; Jung et al., 2005). In this study, the typical W–box, with consensus sequence GTTGACT/TTGAC, is present three times and found to be directly orientated in the TIR–specific promoter, which similar to that identified in the promoters of many *PR* genes (Rushton et al., 1996; Laloi et al., 2004; Nishiuchi et al., 2004; Rocher et al., 2005; Hong and Hwang, 2006). Additionally, of the three W–boxes, spacing of two W–boxes version–549 and–575 by 18bp in length was almost identical to the 17bp of the palindromic arrangement of two W–boxes(WB and WC) in the parsley WRKY1 promoter(Eulgem et al., 1999), and exhibited a tandem repeat structure of the directly oriented two W–boxes (–549 and–575) which may resemble to that of two copies of W2–box (2×W2) in the constructed synthetic promoter identified to be the best pathogen–specific inducibility in *Arabidopsis* plants (Rushton et al., 2002). Similar findings had been reported in the pathogen-inducible promoters of *PR* genes such as *CMPGl* (Kirsch et al., 2001) and *CASAR82A* (Lee and Hwang, 2006). It has been suggested that W–box may be the binding site for transcriptional factor members of WRKY family and play an important role in regulation of the pathogen–induced defense responses, moreover the multiple W–box sites may enable cooperative binding of WRKY factors to generate a synergistic effect on transcription (Eulgem et al., 1999, 2000; Eulgem, 2006; Ryu et al., 2006; Zheng et al., 2007). Thus, in combination with the fact that the

TIR-specific promoter was obtained from disease resistant clones of poplar, it can be suggested that three typical W-boxes present in the TIR-specific promoter region may function as the key *cis*-acting regulatory elements for the regulation of the inducible expression of PR gene PtDRG02 in the poplars.

It has been reported that all the identified promoters of plants require at least one *cis*-acting regulatory element in addition to the W-box for transcriptional activation of gene expression (Yamaguchi-Shinozakiand Shinozaki, 2005). The TATA box is apparently the most conserved functional signal in eukaryotic promoters, and approximately 30% ~ 50% of all known promoters typically contain the transcriptional initiation site (TSS) and a TATA box present around −30 relative to the TSS (Shahmuradov et al., 2005). Many studies have shown that various transcription factors interact with *cis*-acting elements in promoter regions and form a transcriptional initiation complex on the TATA box, which effectively activates the RNA polymerase to start transcription of stress-responsive genes (Yamaguchi-Shinozaki and Shinozaki, 2005). In the present study, a putative typical TATA box (TGTCTATATA) was found at −23 in the TIR-specific promoter, which is consistent with the distance of TATA box from the TSS in the promoter of *PR* genes such as *PmPR*10-1.13 and *CALTPI* (Jung et al., 2005; Liu et al., 2005). Additionally, two typical CAAT boxes, located upstream of TATA box (−23), similar to those reported in other promoters of stress-induced genes (Liu et al., 2005), were identified at positions−75 and−210 of the TIR-specific promoter region. Thus, based on the fact that the TIR-specific promote was characterized by the presence of typical TATA and CAAT boxes located closely to the TSS, as has been demonstrated in many other promoters, and these consensus sequences are shown to be essential for transcriptional activation of defense-responsive genes(Jung et al., 2005; Liu et al., 2005; Lee and Hwang, 2006), it can be suggested that TATA and CAAT boxes together with W-box may be the core elements for TIR-specific promoter and be required for pathogen-responsive gene expression. Many stress-responsive genes are regulated by different signaling pathways in response to stress, and promoter regions of stress-inducible genes generally contain several different sets of *cis*-acting regulatory elements involved in stress-responsive gene expression(Yamaguchi-Shinozaki and Shinozaki, 2005). In this study, in addition to the typical W, TATA and CAAT boxes, several putative *cis*-acting elements, including G/C-rich, A/T-rich, P-box, 3-AF1 binding site, GT-1 motif, I-box, and one multi-*cis*-acting fragment (named as MCF) were identified in the isolated TIR-specific promoter region. Such *cis*-acting regulatory elements were previously shown to be involved in the activation of defense gene expression in plants. Many studies have revealed that the activation and expression of many inducible *PR* genes is not limited to pathogen infection, and may be induced and regulated by several signal molecules such as salicylic acid (SA), ethylene (ET), jasmonic acid (JA) and abscisicacid (ABA) (Rushton et al., 1996; Lebel et al., 1998; Gruner et al., 2003; Jung et al., 2003; Li et al., 2005; Lee et al., 2005; McGrath et al., 2005; Lee and Hwang, 2006; Park et al., 2006; Ryu et al., 2006). Recently, it has been reported that the promoter of *PR* gene *CALTPI* could by induced by pathogen infection and ET/SA treatments (Jung et al., 2005). The GT-1 *cis*-element, one of the many *cis*-acting elements found in plants, was first identified in the promoter of pea RBCS-3A (Green et al. , 1987), and has been characterized suggested that the GT-1 binding site of TIR-specific promoter might be necessary for *PR* gene *PtDRG*02 expression in response to biotic and abiotic stress. It has been reported that the deletion of a 33bp AT-rich sequence from the 5′ end of a pea chloroplast *GS*2 gene promoter-beta-glucuronidase (GUS) fusion resulted in a 10-fold reduction in GUS activity, suggesting that AT-rich promoter elements may be important for the transcription of plant genes (Tjaden and Coruzzi, 1994). Recently, it has been revealed that AT-rich consensus in the promoter region of the *CpC*2 and *ERD*1 genes may be critical for the responsiveness of the gene to dehydration or hormone ABA(Ditzer and Bartels, 2006; Tran et al. , 2007). Wang et al. (2007) have found that both the specific mutation of the proximal GC-rich (GC-box) region of *p*16*INK*4*a* gene promoter and siRNA-induced transcription factor SP1 silencing resulted in the complete loss of its transcription regulation, indicating that GC-rich region ofpromoter is required for SP1 function, and plays a rolein *p*16*INK*4*a* gene transcription regulation. The studiesby Fenoll et al. (1990) showed that the upstream activating sequence (UAS) of the promoter from maize streak virus (MSV) contains a GC-rich element, which was responsible for the transcriptional activation and the binding of maize nuclear factors. In our experiments, one 9 bp AT-rich (ATAATAAAA) was identified at positions -692/-684 in TIR-specific promoter region, which are present within the promoters of many defense-related genes (Tran et al. , 2007). In addition, one 7 bp GC-rich motif located at-793, with two altered bases CGCCGCG, was found to be almost similar to the typical GCC box (CGCCGCG) that is known to be bound by ethylene response factor (ERF), which were previously shown to exist in many of the promoter regions of the *PR* genes (Ohme-Takagi and Shinshi, 1995; Eulgem et al. , 1999; Ohta et al. , 2000; Hong and Hwang, 2006; Lee and Hwang, 2006). Our results thus suggested that AT-rich and GCC-like elements might be associated with the TIR-specific promoter activity. Interestingly, one 14 bp sequence (-788/-774) name as multi-*cis*-acting fragment (MCF), containing four putative *cis*-elements (G-box, ABRE motif, box 1and HVA1), has been identified within the TIR-specific promoter region. It was hypothesized that MCF may provide many potential binding site for several stress-responsive transcriptional factors such as GBF and AREB/ABF (Menkens et al. , 1995; Uno et al. , 2000; Fujita et al. , 2005). ABA-responsive elements(ABREs), with the conserved core sequence CACGTG (termed as G-box motif), are common motifs for binding bZIP (basic region leucine zipper) transcriptional factor involved in ABA-mediated regulation of transcription (Izawa et al. , 1993; Kim et al. , 1997; Ditzer and Bartels, 2006), and have been characterized in the promoters of many plant defense-related genes (Marcotte et al. , 1989; Straub et al. , not only in many light-regulated genes such as *RBCS*, *CAB*, *RCA* and *PHYA* (Zhou, 1999), but also in the pathogen-inducible promoters of the *PR*-1 (Lebelet al. , 1998), *PR*-1a (Buchel et al. , 1999), *CPr* (Lopes Cardoso et al. , 1997), *Str* (Pasquali et al. , 1999), *SAR*8. 2*b* (Song and Goodman, 2002), *SCaM*-4(Park et al. , 2004) and *CASAR*82*A* (Lee and Hwang, 2006) as well as *GIII* genes (Li et al. , 2005). Moreover, the influence of GT-1 *cis*-element of many promoters on the level of inducible gene expression has been found to be related to pathogen elicitor and to SA,

ETor JA responsiveness (Buchel et al., 1999; Park et al., 2004). In this study, GAAAAA is a GT-1 motif, represented two copies at positions-317 and-366 identified in the TIR-specific promoter region of *PR* gene *PtDRG*02, and such regulatory element is also present in the promoter region of soybean SCaM-4 which expression is rapidly increased by pathogen infection and salt treatment (Park et al., 2004). It has been suggested that the G-box identified as stress-responsive *cis*-acting element, may be responsible for the induction of the gene expression by environmental stress, JA, ABA and light (Schindler and Cashmore, 1990; Vasil et al., 1995; Xu and Timko, 2004; Qian et al., 2006). It was found that HVA1s, a low temperature and drought responsive related element, was related to stress-induced reaction (Xue, 2003) and the box 1with a sequence CGCACACGTGCGAC was shown to be involved in the response to wounding (Kawaoka et al., 1994). Thus, the cluster of G-box, ABRE motif, box 1 and HVA1 with highly overlapped with each other, present in the region from -775 to-788 of TIR-specific promoter, might be a cooperative candidate as the functional regulatory element implicated in the stress-regulated transcription of pathogen-related gene *PtDRG*02.

Additionally, a sequence element enriched with pyrimodine nucleotides (pyrimodine box, P-box) has been characterized to be necessary for high levels of hormone-regulated expression (Gubler and Jacobsen, 1992; Sutliff et al., 1993). A putative CCTTTTG motif located at -418 region of TIR-specific promoter is matched with the consensus sequence CCTTTT(named as P-box) identified in the promoters of carnation *CSDC*9 (Kim et al., 2004) and rice *OsGAEl* (Janet al., 2006). It was found to be related to the regulation of gene expression by GA induction (Morita et al., 1998; Mena et al., 2002; Jan et al., 2006). The putativeI-box, with consensus sequence GATAA, is present four times (located at -271, -238, -490 and -627) within the TIR-specific promoter region, and this element has previously been identified in many promoters, some of which are regulated by light and/or by thecircadian clock (Borello et al., 1993; Terzaghi and Cashmore, 1995; Baum et al., 1997; Agius et al., 2005; Ito et al., 2005). The 11 bp conserved sequence AAGATATATTT (at -296), termed 3-AF1 binding site is also recognized in TIR-specific promoter. To sum up, several putative *cis*-acting motifs were present in the obtained TIR-specific promoter of PtDRG02 from disease resistant clones of hybrid triploid poplars, including one TATA box, one GC-rich, one AT-rich, one P-box, one 3-AF1 binding site, two CAAT boxes, two GT-1 motifs, three typical W-boxes, four I-boxes, and one multi-*cis*-acting fragment (MCF) containing five types of regulatory elements (E4, G-box, ABRE motif, box1 and HVA1s), most of which were homologous to the *cis*-acting elements involved in the activation of defense genes in plants, indicating that the obtained TIR-specific promoter contains multiple *cis*-acting regulatory elements, which may enable cooperative binding of many transcription factors to account for gene expression in response to abiotic or biotic stresses. It may be suggested that TIR-specific promoter might be a pathogen-inducible promoter associated with the expression of defense-related genes.

杨树杂交亲本分子遗传距离与子代生长性状相关性研究*

摘　要　对杨属23个杂交组合的子代生长性状进行变异分析，并对杂交亲本作AFLP分析，结果表明：子代生长性状在组合间存在广泛变异，亲本间分子遗传距离差异较大；杂交亲本间分子遗传距离与其子代生长性状相关和回归分析表明，遗传距离在一定范围内(0.0249~0.3681)与生长性状呈线性相关，在较大范围内(0.0249~0.5314)为二次抛物线方程关系。根据亲本间分子遗传距离与子代生长量的关系，认为在杨属杂交亲本选配时，其分子遗传距离在0.19~0.36之间，可获得较理想的杂交效果。

关键词　杨树，杂交亲本，分子遗传距离，生长性状，相关性

杨树是我国重要的造林树种，在工业人工林和生态防护林建设中发挥着重要作用。目前，我国林业生产中推广应用的杨树良种多数是通过杂交选育出来的，杂种优势明显(符毓秦等，1990；黄东森等，1991；王明庥等，1991；李善文等，2004)。根据杂种优势的遗传学原理，获得杂种优势的双亲要存在遗传差异(卢庆善等，2002)。育种实践证明，在一定范围内，两亲本间遗传差异越大，其杂种优势也越大，后代分离范围就越广泛，从而获得优良个体的机会也就越多(张爱民，1994)。遗传差异评价方法很多，如数量性状遗传距离、亲缘系数、生化标记等。

随着分子生物学的不断发展，出现了多种分子标记技术，如RAPD、AFLP、SSR等，使得在分子水平研究亲本间的遗传差异成为可能(张志毅等，2002；Chauhan et al.，2004；Cervera et al.，2005)。Arcade等(1996)研究落叶松(*Larix*)发现，杂交亲本间分子遗传距离与生长性状杂种优势相关显著，但与其他数量性状的杂种优势不相关，并证明遗传距离远的亲本能产生杂种优势强的个体；Baril等(1997)用RAPD标记桉树(*Eucalyptus*)杂交亲本遗传距离发现，中等遗传距离的2个种间亲本杂交才具有正向杂种优势；李周歧等(2002)对鹅掌楸属(*Liriodendron*)种间杂交研究结果表明，亲本RAPD遗传距离与子代一年生苗高和地径家系平均生长量均表现为显著的二次曲线相关，说明利用遗传距离进行鹅掌楸亲本选配和杂种优势预测具有一定潜力。李梅等(2001)以30个杉木(*Cunninghamia lanceolata*)杂交亲本为研究材料，结果表明亲本的分子遗传距离与子代生长性状相关关系较为复杂，亲本的分子遗传变异对亲本选配有一定参考价值。

本研究选择杨属白杨派(*Leuce*)、黑杨派(*Aigeiros*)、青杨派(*Tacamahaca*)中的部分种和杂种为亲本进行杂交，调查各杂交组合2年生苗木的生长性状；对各杂交亲本进行AFLP分析，计算亲本间分子遗传距离；将生长性状与分子遗传距离进行相关分析和回归

* 本文原载《林业科学》，2008，44(5)：150-154，与李善文、于志水、何承忠、安新民和李百炼合作发表。

分析，对杨树杂交亲本间的最适分子遗传距离进行探讨，将分子标记技术与常规杂交育种相结合，为杂交亲本选配及杂种优势预测提供分子水平理论依据。

1 材料与方法

1.1 试验地概况

试验地设在北京林业大学后八家苗圃，位于北纬 39°46′，东经 116°29′，土壤为潮土，质地中壤，年降水量 503mm，年均温 12.1℃，极端最低气温-19.6℃，无霜期 210d。

1.2 杂交亲本

杂交亲本包括杨属白杨派、黑杨派和青杨派的 13 个种和杂种的 21 个无性系(表 1)。

1.3 生长性状调查

落叶后调查 2 年生苗木的苗高、地径、胸径，每重复随机调查 30 株，共 4 次重复，计算其均值。

1.4 DNA 提取及 AFLP 分析

按 Murray 等（1980）的方法提取试验材料的基因组 DNA。AFLP 分析的基本程序和 Vos 等(1995)的方法基本相同，仅对体系进行了优化，采用 *Eco*R Ⅰ/*Mse* Ⅰ 酶切组合进行基因组限制性酶切，预扩增反应选用引物组合 *Eco*R Ⅰ_{+00}/*Mse* Ⅰ_{+00}，选择性扩增反应采用引物组合 *Eco*R Ⅰ_{+3}/*Mse* Ⅰ_{+3}。PCR 扩增反应在 PE-9700PCR 仪上进行。选择性扩增产物经变性后在 6%的变性聚丙烯酰胺序列分析胶上电泳分离，条件是 95W，3000V，恒功率电泳约 90min。电泳后采用 Tixier 等(1997)的银染检测方法进行 AFLP 指纹显色反应。统计带型在相同片段位置上的谱带，按 0/1 系统纪录，有带记为 1，无带记为 0，并参照 Marker 带估计片段大小。

1.5 统计分析

采用 Nei & Li(1979)的遗传相似系数(Genetic Similarity，GS)，其计算公式：$GS=2N_{ij}/(N_i+N_j)$，式中：N_{ij} 表示样本 i 和 j 的公共带数；N_i、N_j 分别是样本 i、j 的带数；遗传距离 $D=1-GS$。将各杂交组合的子代苗高、地径、胸径均值与其亲本间的 AFLP 分子标记遗传距离进行相关分析；以遗传距离为自变量，苗高、地径、胸径为因变量，分别进行一元线性回归和二次抛物线回归分析(黄少伟等，2001)。

2 结果与分析

2.1 各杂交组合苗木生长性状均值、标准差及变异系数

23 个杂交组合的苗高、地径、胸径均值见表 1。苗高生长量最大组合有 I-69×D324、

I-69×小叶杨 2 号，均为 3.7 m，最小组合是大青杨 4 号×川杨 2 号，为 2.2m，二者相差 1.5m；地径变幅 1.7~3.6cm，最大组合是大青杨 4 号×T66 和小叶杨 3 号×欧洲黑杨 1 号，最小组合是大青杨 4 号×川杨 2 号，前者是后者的 2.1 倍；胸径变幅为 0.7~2.4cm，最大组合 I-69×小叶杨 2 号，最小组合为大青杨 4 号×川杨 2 号；苗高、地径、胸径的标准差分别为 0.4、0.5、0.4cm，变异系数分别是 12.7%、19.7%、27.2%。因此，在各杂交组合间生长性状存在广泛变异。

表 1　杨树各杂交组合的生长量及其亲本间遗传距离①

杂交组合 Cross combinations	苗高 *H*(m)	地径 D_0(cm)	胸径 $D_{1.3}$(cm)	遗传距离 *GD*
毛新杨×新疆杨（*P. tomentosa*×*P. bolleana*)×*P. bolleana*	2.5	2.5	1.0	0.0526
银毛杨×鲁毛 50(*P. alba*×*P. tomentosa*)×*P. tomentosa* cl. ‘Lumao 50’	2.6	1.9	1.0	0.1140
银腺杨 1 号×新疆杨（*P. alba*×*P. glandulosa* cl. ‘No. 1’)×*P. bolleana*	3.3	2.4	1.5	0.1684
银腺杨 2 号×鲁毛 50(*P. alba*×*P. glandulosa* cl. ‘No. 2’)×*P. tomentosa* cl. ‘Lumao 50’	3.3	2.6	1.6	0.2212
毛新杨×银腺杨 84K（*P. tomentosa*×*P. bolleana* ‘84K’)×(*P. alba*×*P. glandulosa*)	3.6	2.5	1.6	0.1819
银毛杨×银腺杨 84K（*P. alba*×*P. tomentosa*)×(*P. alba*×*P. glandulosa* ‘84K’)	3.5	2.6	1.6	0.1644
银毛杨×新疆杨（*P. alba*×*P. tomentosa*)×*P. bolleana*	3.1	2.2	1.3	0.1336
毛新杨×鲁毛 50(*P. tomentosa*×*P. bolleana*)×*P. tomentosa* cl. ‘Lumao 50’	3.0	2.5	1.3	0.1728
银腺杨 2 号×新疆杨（*P. alba*×*P. glandulosa* cl. ‘No. 2’)×*P. bolleana*	3.2	2.5	1.5	0.1896
银白杨×新疆杨 *P. alba*×*P. bolleana*	3.1	2.1	1.2	0.1589
银新杨×新疆杨（*P. alba*×*P. bolleana*)×*P. bolleana*	2.6	2.0	1.1	0.1417
银白杨杂种×银腺杨 2 号 *P. alba*×(*P. alba*×*P. glandulosa* cl. ‘No. 2’)	2.8	2.2	1.2	0.1394
银白杨杂种×鲁毛 50 *P. alba*× *P. tomentosa* cl. ‘Lumao 50’	2.9	2.6	1.2	0.1850
I-69×大青杨 3 号 *P. deltoides* cv. ‘Lux’×*P. ussuriensis* cl. ‘No. 3’	3.4	2.9	1.8	0.3463
I-69×D324 *P. deltoides* cv. ‘Lux’×*P. deltoides* cv. ‘D324’	3.7	3.1	2.2	0.0249
大青杨 4 号×T66 *P. ussuriensis* cl. ‘No. 4’×*P. deltoides* cv. ‘T66’	3.5	3.6	1.9	0.3333
大青杨 4 号×川杨 2 号 *P. ussuriensis* cl. ‘No. 4’×*P. szechuanica* cl. ‘No. 2’	2.2	1.7	0.7	0.2985
大青杨 4 号×T26 *P. ussuriensis* cl. ‘No. 4’×*P. deltoides* cv. ‘T26’	3.3	2.9	1.8	0.3246
I-69×小叶杨 2 号 *P. deltoides* cv. ‘Lux’×*P. simonii* cl. ‘No. 2’	3.7	3.4	2.4	0.3497
I-69×川杨 2 *P. deltoides* cv. ‘Lux’×*P. szechuanica* cl. ‘No. 2’	3.5	3.1	2.0	0.3681
I-69×欧洲黑杨 2 号 *P. deltoides* cv. ‘Lux’×*P. nigra* cl. ‘No. 2’	3.1	2.8	1.7	0.3512
小叶杨 3 号×中菏 1 号　*P. simonii* cl. ‘No. 3’ ×*P. deltoides* cv. ‘Zhonghe 1’	2.9	2.4	1.3	0.3255
小叶杨 3 号×欧洲黑杨 1 号　*P. simonii* cl. ‘No. 3’ × *P. nigra* cl. ‘No. 1’	3.5	3.6	2.1	0.3413
最小值 Max. value	2.2	1.7	0.7	0.0249
最大值 Min. value	3.7	3.6	2.4	0.3681

（续）

杂交组合 Cross combinations	苗高 H(m)	地径 D_0(cm)	胸径 $D_{1.3}$(cm)	遗传距离 GD
均值	3.1	2.6	1.5	0.2212
标准差	0.4	0.5	0.4	0.1045
变异系数/%	12.7	19.7	27.2	47.26

① H-Height, D_0-Basal diameter, $D_{1.3}$-Diameter at breast height, GD-Genetic distance. 下同. The same below.

2.2 亲本间遗传距离变异

根据 AFLP 分子标记结果（李善文，2004），计算 20 个杂交亲本间的遗传距离，将 23 个杂交组合的父母本间遗传距离列于表 1。亲本间遗传距离最大的组合是 I-69×川杨 2 号，为 0.3681，最小组合是 I-69×D324，为 0.0249，前者是后者的 14.8 倍；23 个杂交组合亲本间的遗传距离均值、标准差、变异系数分别是 0.2212、0.1045、47.26%。可见在各杂交组合的亲本间遗传距离也存在较大变异。

2.3 亲本间遗传距离与子代生长性状相关分析

从表 2 可见，亲本间遗传距离与子代地径、胸径的相关性均达显著水平（$P<0.05$），与子代苗高呈正相关，但未达显著水平。表明在这 23 个杂交组合中，随着父母本间遗传距离的增大，其子代地径、胸径均显著增大，苗高也有增大趋势，但不显著。

表 2　亲本间遗传距离与子代生长性状的相关系数

性状	苗高	地径	胸径
遗传距离	0.2715	0.4913*	0.4335*
Prob>r	0.2102	0.0173	0.0363

① *：显著相关。Significant correlation.

2.4 遗传距离与生长性状的一元线性回归分析

以遗传距离为自变量，苗高、地径、胸径为因变量，分别进行一元线性回归分析，表 3 列出了遗传距离与苗高、地径、胸径的一元线性回归模型拟合方差分析结果。地径、胸径的 F 值分别为 6.68、5.00（$P<0.05$），说明地径、胸径与遗传距离的一元线性模型拟合效果达显著水平；苗高的 F 值为 1.67（$P>0.05$），说明苗高拟合效果不显著。

遗传距离与苗高、地径、胸径的一元线性回归方程见表 4。对各回归方程的参数进行 t 检验表明，胸径和地径回归方程中的参数 P 值均小于 0.05，达显著水平，说明当遗传距离在一定范围内（0.0249~0.3681）变动时，随着亲本间遗传距离的增大，地径、胸径呈显著线性增加；苗高的回归方程的两个参数概率 P 值分别是 0.0001、0.2102，其中 0.2102 大于 0.05，未达显著水平。

表 3　一元线性模型方差分析

性状	变异来源	自由度	平方和	均方	F 值	$Pr > F$
苗高	模型	1	0. 2680	0. 2680	1. 67	0. 2102
	误差	21	3. 3686	0. 1604		
	总计	22	3. 6365			
地径	模型	1	1. 3773	1. 3773	6. 68*	0. 0173
	误差	21	4. 3288	0. 2061		
	总计	22	5. 7061			
胸径	模型	1	0. 7150	0. 7150	5. 00*	0. 0363
	误差	21	3. 0035	0. 1430		
	总计	22	3. 7185			

表 4　遗传距离与各性状的一元线性回归方程及其参数检验

性状	变量	参数估计值	标准误	t 值	$Pr > t$	回归方程
苗高	常数项	2. 9100	0. 1990	14. 62	0. 0001	$H=1.0558D+2.9100$
	D 项系数	1. 0558	0. 8169	1. 29	0. 2102	
地径	常数项	2. 0837	0. 2256	9. 24	0. 0001	$D_0=2.3936D+2.0837$
	D 项系数	2. 3936	0. 9260	2. 58	0. 0173	
胸径	常数项	1. 1321	0. 1879	6. 02	0. 0001	$D_{1.3}=1.7246D+1.1321$
	D 项系数	1. 7246	0. 7713	2. 24	0. 0363	

2. 5　遗传距离与生长性状的二次抛物线回归分析

遗传距离与生长性状的一元线性关系限于适度的遗传距离之内，当超过某一数值时，则生长量下降，甚至杂交不可配。根据李善文(2004，2007)研究可知，白杨派与胡杨派(*Turanga*)、黑杨派、青杨派的遗传距离分别是 0. 5314、0. 4781、0. 4732，黑杨派与胡杨派、青杨派的遗传距离各为 0. 4706、0. 3263，青杨派与胡杨派的遗传距离为 0. 4379。由于这四派间(除黑杨派与青杨派间外)按常规杂交技术，很难获得种子，因此对它们子代的生长量均赋值为 0，将这 5 个组合与前述 23 个组合置于一起，进行遗传距离与生长性状二次抛物线回归分析。由表 5 可知，苗高、地径、胸径与遗传距离的二次抛物线方程拟合效果，经方差分析检验均达极显著水平($P<0.01$)。

表 5　二次抛物线回归模型方差分析

性状	变异来源	自由度	平方和	均方	F 值	$Pr > F$
苗高	模型	2	31. 7100	15. 8500	31. 68**	0. 0001
	误差	25	12. 5111	0. 5004		
	总计	27	44. 2211			

（续）

性状	变异来源	自由度	平方和	均方	F 值	$Pr > F$
地径	模型	2	19.1477	9.5739	16.39**	0.0001
	误差	25	14.6019	0.5841		
	总计	27	33.7496			
胸径	模型	2	6.0864	3.0432	10.25**	0.0006
	误差	25	7.4236	0.2970		
	总计	27	13.5100			

表 6　遗传距离与各性状的二次抛物线回归方程及其参数检验

性状	变量	参数估计值	标准误	t 值	$Pr > t$	回归方程
苗高	常数项	1.8654	0.5329	3.5**	0.0018	$H=-38.2652D^2+15.5543D+1.8654$
	D 项系数	15.5543	4.2582	3.65**	0.0012	
	D^2 项系数	-38.2652	7.3769	-5.19**	0.0001	
地径 D_0	常数项	1.3041	0.5757	2.27*	0.0324	$D_0=-32.7796D^2+14.1792D+1.3041$
	D 项系数	14.1792	4.6003	3.08**	0.005	
	D^2 项系数	-32.7796	7.9695	-4.11**	0.0004	
胸径	常数项	0.7138	0.4105	1.74	0.0944	$D_{1.3}=-19.1048D^2+8.4335D+0.7138$
	D 项系数	8.4335	3.2801	2.57*	0.0165	
	D^2 项系数	-19.1048	5.6825	-3.36**	0.0025	

进一步对遗传距离与苗高、地径、胸径的二次抛物线回归方程的参数进行显著性 t 检验，见表 6。苗高回归方程的 3 个参数概率 P 值均小于 0.01，达极显著水平；地径回归方程的 3 个参数概率 P 值均小于 0.05，达显著水平；胸径回归方程的 3 个参数概率 P 值中有一个大于 0.05，未达显著水平。因此，杂交亲本间遗传距离与苗高的二次抛物线回归方程经 t 检验达极显著水平，与地径的回归方程达显著水平，与胸径的回归方程未达显著水平。这 3 个二次抛物线方程的极值点遗传距离分别为 0.203、0.216、0.221，表明亲本间遗传距离为 0.21 左右时，杂交子代的苗高、地径、胸径理论值最大。

3　讨论

在本研究中，相关分析与一元线性回归分析结果是一致的。23 个杂交组合的 AFLP 分子遗传距离与其子代的地径、胸径达显著相关，与苗高呈正相关但不显著；以亲本间的分子遗传距离为自变量，苗高、地径、胸径为因变量，进行一元线性回归分析，地径、胸径的回归方程，经检验达显著水平，苗高的回归方程不显著。因此，相关分析与一元线性回归结果相一致，即杂交亲本的分子遗传距离在一定范围内(0.0249~0.3681)与其子代的生长量呈正相关，随着亲本间遗传距离的增大，子代的地径、胸径生长量显著增大，苗高也有一定增大的趋势，但不显著。

杨树杂交亲本的AFLP分子遗传距离与生长性状在一定范围内呈线性相关，在较大范围内为二次抛物线方程关系。比较一元线性回归与二次抛物线回归分析可知，当亲本间分子遗传距离在0.0249~0.3681范围内变动时，遗传距离与苗高的一元线性回归方程不显著，说明苗高随着遗传距离的变化已经开始有二次抛物线趋势，但未达显著水平，当遗传距离的变幅增大(0.0249~0.5314)时，其二次抛物线关系变得明显，达到了显著水平；胸径的变化与之相反，当遗传距离变动于0.0249~0.3681时，与胸径呈显著的一元线性关系，当遗传距离增大时，这种线性关系仍然存在，因此二次抛物线关系不显著；地径的变化介于苗高和胸径之间，当遗传距离在0.0249~0.3681时，与地径呈显著的一元线性关系，当遗传距离增大时，与地径又呈显著的二次抛物线关系，这说明地径随着遗传距离的变化比苗高、胸径变化快，较敏感。由上述讨论可知，一元线性回归与二次抛物线回归分析结果是统一的。

杂交亲本间分子遗传距离与子代苗高、地径、胸径的二次抛物线方程的极值点遗传距离分别为0.203、0.216、0.221，这表明亲本间遗传距离为0.21左右时，杂交子代的苗高、地径、胸径理论值最大，分别为3.4m、2.8cm、1.6cm。根据李善文等(2007)研究可知，黑杨派与青杨派间的遗传距离为0.3263，黑杨派内美洲黑杨(*P. deltoides*)与欧洲黑杨(*P. nigra*)的遗传距离是0.3509，白杨派内美洲山杨(*P. tremuloides*)与欧洲山杨(*P. tremula*)的遗传距离是0.1978。以往许多研究表明，黑杨派与青杨派间杂交(徐纬英，1988；符毓秦等，1990；王明庥等，1991)、黑杨派内美洲黑杨×欧洲黑杨(黄东森等，1991)、白杨派内美洲山杨×欧洲山杨均为优良杂交组合(Li et al.，1993)。根据以上亲本间分子遗传距离大小与子代生长表现的关系，认为在杨属杂交亲本选配时，其分子遗传距离在0.19~0.36之间，可获得理想的杂交效果。如果对杨属派别进一步细分，则认为对于白杨派，亲本间分子遗传距离在0.20左右，对于黑杨派内种间及黑杨派和青杨派间，亲本间分子遗传距离在0.34左右，杂种优势明显。

随着分子生物学的快速发展，分子标记技术在林木研究中得到广泛应用。其中，如何将分子标记结果应用于常规育种，提高育种效果，是当代林木遗传育种研究的重要内容之一。本研究在这方面进行了尝试和探讨，应用AFLP技术对杨树杂交亲本的遗传距离进行评估，将亲本间的分子遗传距离与子代生长性状作相关和回归分析，为杂交亲本选配及杂种优势预测提供理论依据。该研究结果显示了该方法的有效性，并证明将分子标记技术应用于常规杂交育种研究是可行的。由于本文所用生长性状数据为杂交子代苗期试验的结果，进一步结论仍有待于对子代今后多年的造林试验数据开展深入分析进行验证。

Correlation between Molecular Genetic Distances among Parents and Growth Traits of Progenies in *Populus*

Abstract Variation of growth traits of progenies of twenty-three cross combinations in *Populus* was analysed, and genetic variation among parents was studied by Amplified Fragment Length Polymorphism (AFLP). The results showed that growth traits of progenies and molecular genetic distance (GD) among parents had considerable variation. Correlation and regression analysis between molecular genetic distances among parents and growth traits of progenies indicated that the relationship between GD and growth was linear equation in a certain range (0.0249~0.3681), and the relationship was quadratic parabola equation in a wide range (0.0249~0.5314). If the genetic distance among parents in *Populus* would range from 0.19 to 0.36, the ultimate effect of hybridization could be achieved based on the established relationship between GD and growth.

Key words *Populus*, crossing parents, molecular genetic distance, growth trait, correlation

Seedling Test and Genetic Analysis of White Poplar Hybrid Clones*

Abstract　Cross breeding strategies are very efficient for gaining new and superior genotypes. Ninety-eight new white poplar hybrid clones produced from 12cross combinations within the Section *Leuce* Duby were studied using genetic analysis and seedling tests. We exploited the wide variation that exists in this population and found that the differences among diameter at breast height (DBH), root collar diameter (RCD) and height (H) were statistically extremely significant. The repeatability of clones of these measured traits ranged from 0.947–0.967, which indicated that these traits were strongly controlled by genetic factors. Based on multiple comparisons, a total of 25 clones showed better performance in growth than the control cultivar. These 25 clones were from six different cross combinations, which can guarantee a larger genetic background for future new clone promotion projects. This study provides a simple overview on these clones and can guide us to carry out subsequent selection plans.

Keywords　white poplar, multi-clonal plantation, seedling test, genetic analysis

1　Introduction

Populus L. is a pioneering tree species of the forest industry, which has been planted widely in China because of its excellent properties of fast growth, adaptability and ease of propagation (Zhang et al., 1992; Li et al., 2005b). In China, there are about 6670000hm^2 of poplar plantations, four times as many as the total number in the rest of the world. This area accounts for 1/5 of all tree plantations in China (Wang, 1995). Among the cultivars of this genus, most clones were produced by cross breeding (Zhu et al., 1995; Wang and Li, 2001). Cross breeding strategy is very efficient for producing new and superior genotypes which can pass the excellent traits of two parents onto progenies and provide the progenies with the hybrid vigor, called as "heterosis". Over the long term, field tests are viewed as the best way to select clones with good performance. Although marker assisted selection developed rapidly in recent years, field tests are still the most accurate and direct way to estimate the performance of clones. No doubt, field tests need long term observations, which have become the largest drawback of this method, require large areas of land, are labor intensive and directly or indirectly increase the cost of breeding. Early-stage selection has been used to solve the problem for some time; however, the results have not been satisfactory

* 本文原载 *Forestry Studies in China*, 2008, 10(3): 149-152, 与李博、江锡兵、张有慧、李善文和安新民合作发表。

by a long shot. In most countries, some other potential questions still exist, including monoclonal plantations, aging of promoted clones and so on, which will destroy genetic diversity and threaten the stability of the bio-system. In order to overcome these problems, it is urgent that new clones are promoted and to enhance multi-clonal plantations. Therefore, our laboratory began collecting white poplar breeding material which has largely been found in northern China since the 1990s. In 2001, a series of crossbreeding experiments were carried out and 12 promising combinations were chosen from more than 30 controlled pollinations and over 3000 progenies have been harvested (Li, 2004). In 2002 and 2003, 98 elite individuals were obtained under rigid selection standards, including consideration for growth, leaf trait, stem form, etc (Li et al., 2004, 2005a). In the spring of 2004, these 98 "super" individuals had been propagated by grafting and then all 98 clones were arranged in a completely randomized block design in Guanxian County, Shandong Province. In this study, the growth traits, including height, diameter of basal area and diameter at breast height were investigated during the first growing season. This first year performance provided a simple overview of these clones and guided us to carry out the following selection plans.

2 Materials and methods

2.1 Experimental site

This clone selection trial was carried out in forestry nurseries of Guanxian County, Shandong Province.

2.2 Materials and design

A total of 98 white poplar hybrid elites were chosen from 12 cross combinations after a two-year family test. During the following spring, up to 120 individuals per elite clone were obtained by grafting. A completely randomized block design was adopted for clonal tests, with four replications, 30 seedlings per block and a plant spacing of 30cm×100cm. The details of the total cross combinations and the number of clones are presented in Table 1.

2.3 Data analysis

Growth traits including height (H), root collar diameter (RCD) and diameter at breast height (DBH) were investigated. The mean values and standard deviations of *H*, *RCD* and *DBH* were obtained, followed by an analysis of variance and multiple comparisons of these three traits. The data were analyzed using statistical methods from the SAS 8.0 software package (SAS Institute Inc.).

3 Results

3.1 Mean values and standard deviations of *RCD*, *DBH* and *H*

Mean values, standard deviations (SD), coefficients of variation (CV) of *RCD*, *DBH* and *H* are listed in Table 2. The mean values of *RCD*, *DBH* and *H* were 1.9cm, 1.2cm and 291cm respectively. The *RCD* of this population varied from 1.2 to 2.7cm and *DBH* ranged from 0.7 to 2.1 among the 98 clones. The maximum value of *H* is more than twice the minimum. Standard deviations of *RCD*, *DBH* and *H* were 0.331cm, 0.287cm and 53.1cm respectively. The coefficients of variation of *RCD*, *DBH* and *H* were 17.85%, 22.25% and 18.28%. The results indicate that the growth traits i.e., *RCD*, *DBH* and *H* varied dramatically among these 98 clones during the first growing season. The large variation existing in this population made it feasible to select outstanding clones. On the one hand, the results reflect the great diversity in this population and this variation was found just during the first year; on the other hand, this diversity further indicated that this population was established on the basis of extensive genetic variation. It satisfied our breeding aims of multi-clone selection.

3.2 Analysis of variance and estimates of repeatability

Analysis of variance and the estimation of repeatability of three traits are shown in Table 3. The variation in all three traits were statistically extremely significant among the 98 clones. The result also indicates that extensive variation exists in this population, which provides for a large room of clone selection. The repeatability of *RCD*, *DBH* and *H* were 0.962, 0.950 and 0.967 respectively, which indicates that these growth traits are strongly controlled by genetic factors.

3.3 Multiple comparisons of growth traits

Multiple comparisons of *RCD*, *DBH* and *H* were carried out among the means of the 98 clones and a control cultivar LM50. Part of the results is presented in Table 4. For *RCD*, the best 10 clones are listed in Table 4, which indicated that the No. 51 clone was the best and was significantly different from all other clones. For *H*, No. 51 still had best performance among all clones and showed significant difference with others. Generally, No. 51 had the best performance on the three growth traits. According to the result of multiple comparisons of *H* with LM50, a total of 25 clone had better growth performance than LM50. Of these, 17 clones showed significant differences with LM50. The height growth of No. 51 was 36.2% better than that of LM50.

Table 1　Cross combinations and number of clones

No	Cross combinations	Number of clones
1	(*P. tomentosa*×*P. bolleana*)×*P. bolleana*	8
2	(*P. tomentosa*×*P. bolleana*)×*P. tomentosa* 'LM50'	6
3	(*P. tomentosa*×*P. bolleana*)×(*P. alba P. glandulosa*)	14
4	(*P. alba*×*P. glandulosa*'No. 1')×*P. bolleana*	12
5	(*P. alba*×*P. glandulosa* 'No. 2')×*P. tomentosa* 'LM50'	8
6	(*P. alba*×*P. glandulosa* 'No. 2')×*P. bolleana*	6
7	(*P. alba*×*P. tomentosa*)×(*P. alba*×*P. glandtdosa*)	5
8	(*P. alba*×*P. tomentosa*)×*P. bolIeana*	16
9	(*P. alba*×*P. tomentosa*)×*P. tomentosa* 'LMSO'	7
10	*P. alba*×*P. bolleana*	12
11	(*P. alba*×*P. bolleana*)×*P. bolleana*	2
12	(*P. alba*×*P. bolleana*)×(*P. alba*×*P. glandldosa* 'No. 2')	2
Total		98

Table 2　Population characteristics

	Mean value(cm)	Maximmn (cm)	Minimtun(cm)	*SD* (cm)	*CV*(%)
RCD	1.9	2.7	1.2	0.331	1785
DBH	1.2	2.1	0.7	0.287	2225
H	291	440	190	53.1	18.28

Table 3　Analysis of variance and repeatability

	Variation resource	*F* value	*p*>*F*	Repeatability
RCD	Clone	26.10***	0.0001	0.962
DBH	Clone	20.18***	0.0001	0.950
H	Clone	29.11***	0.0001	0.967

Note: *** means sign ificaut at 1% level.

Table 4　Multiple comparisons: incomplete results

RCD(cm)			*DBH*(cm)			*H*(cm)		
No.	Mean	0.05 level	No.	Mean	0.05 level	No.	Mean	0.05 level
51	2.68	A	51	2.13	A	51	442	A
45	2.53	A B	44	1.88	B	25	402	B
33	2.50	A B	17	1.83	B C	27	397	B
74	2.48	B C	27	1.80	B C D	50	393	B C

Continued table

RCD(cm)			DBH(cm)			H(cm)		
No.	Mean	0.05 level	No.	Mean	0.05 level	No.	Mean	0.05 level
46	2.43	B C D	41	1.78	B C D	57	390	B C D
17	2.40	B C D	33	1.75	B C D E	45	385	B C D
44	2.40	B C D	74	1.75	B C D E	41	380	B C D E
25	2.38	B C D E	45	1.75	BCD E	17	367	C D E F
41	2.38	B C D E	10	1.73	BCD E F	52	362	D E F G
27	2.38	B C D E	76	1.70	BCD E F	71	355	E F G H

4 Discussion

Crossbreeding within the Section *Leuce* Duby in China began in 1946(Ye, 1955). In the following 50 years, many promising clones were selected through field testing. For example, by using *P. alba* as female parent and *P. hopeiensis* ×*P. davidiana* as male parent, the superior hybrid of *P. alba* ×(*P. hopeiensis* ×*P. davidiana*) was bred by Wang et al. (1987). This hybrid had 9.8% more volume than *P. hopeiensis* and a better cutting survival rate and resistance to diseases and pests (Wang et al., 1987). Hybrid clone 1333, which grew twice as fast as *P. davidiana* when it was 14 years old, was screened out from the combination of *P. alba* ×*P. davidiana* (Liu and Zhao, 1991). The magnitude of genetic gain through clone selection depends on variation and inheritance of desired traits (Foster et al., 1986). In our study, three traits, *RCD*, *DBH* and *H*, were chosen to reflect the growth performance of different clones, on the basis of which we carried out selection. The first growing season performance in growth showed significant differences among the 98 clones. Analyses of variance of the variables *RCD*, *DBH* and *H* indicated that variation in these three variables was extremely significant among the 98 clones. Genetic analysis indicated that these growth traits were strongly controlled by heredity. Therefore, selecting clones by these traits in this population with great variation can guarantee successful and efficient seedling tests. In these three traits, seedling height (*H*) had the highest repeatability and was viewed as the most important trait to select elite clones.

Multiple comparisons showed that 25 clones grew better than the local cultivar LM50, which was also the male tree for several cross combinations in our study (Table 1). Among the 25 clones, 17 were significantly different from LM50. More attention should be paid to these 25 clones in subsequent testing programs. Generally, about the best 10% of total clones are selected for the first phase in a clonal trail. In this study, more than 25% of the clones were selected because a conservative strategy was taken in our selection program for the first step. The selection intensity will be strengthened when field test are carried out.

Single clones of poplars have been propagated extensively. There have been disease problems and the overuse of single clones and use of large monoclonal plantations has been questioned (Stelzer and Goldfarb, 1997). Multi-clone plantation is a more efficient way to change the situation of monoclonal promotion which specifically threatens biodiversity and helps thede velopment of diseases, especially when the plantations need to be kept for a long time in order to play a role in maintaining the balance of the bio-system. In this study, the 25 genetic clones with potential for promotion have been selected from six different cross combinations, which enlarged the genetic basis of breeding population. Compared with the clones from the same cross combination, clones from different combinations showed larger variation not only in growth but in other phenotypes (data not shown). Compared with black poplars, white poplars have a better shape and a longer life cycle and can, therefore, better serve as a plantation for safeguarding the environment. We hope the clones selected can be used not only in pulp production, but in ecosystem conservation as well.

Usually, poplars have a long rotation period and different growth rhythms, so a relative long period of observations is needed to judge the clones. Observations of only one year are insufficient to draw an accurate conclusion on the performance of these clones. However, it can provide us with an overview on this population and help make a scientific plan for field selection. Clone selection is a stepwise process of evaluation. It starts with a large number of genetically different progenies and ends up with a few clones with commercial utility (Fikret and Ferit, 2004). On one hand, this population with 98 clones has been maintained for future investigations. On the other hand, these 25 clones will be propagated again and this study will be repeated next year in various locations under different environmental conditions. This strategy can save the cost of selection by shrinking the size of the original population and at the same time avoid the risk of dropping other possible good clones. These 25 clones were chosen only on the basis of one year of observation, so a repetition of this study can ensure the accuracy of our results and produce more seedlings for following field tests in different locations. In addition, during subsequent years, besides the growth traits, other traits such as morphological and physiological traits and disease resistance should be investigated.

Establishment of Cell Suspension Line of *Populus tomentosa* Carr. *

Abstract Leaves of fine *Populus tomentosa* genotype TC152 were used as explants to establish cell suspension lines. The effects of plant growth regulators on callus induction and establishment of cell suspension lines were studied. The callus induction rate was the highest on a MS solid medium supplemented with 1.0mg/L 2,4-D. A cell suspension line could be obtained by inoculating calli which were not subcultured into a MS liquid medium supplemented with 1.5mg/L 2,4-D. The best subculture medium was MS + 0.8mg/L 2,4-D + 30g/L sucrose with a subculture cycle of seven days.

Keywords *Populus tomentosa*, callus, cell suspension line

1 Introduction

Populus tomentosa Carr., a native species in northern China, is mainly used for greening of the landscape or for fiber and wood production. With the rapid developments of cell engineering, studies using suspension cells as material are gradually being conducted. A fine cell suspension line is not only a good receptor for gene transfer (Zheng et al., 2005), but also a suitable system for studying cytology, molecular biology (Chen et al., 2002), protoplast culture, production of secondary plant metabolites (Spela et al., 2005) and somatic embryogenesis. The development of the genome database of model *Populus* species and the presence of the protocols of cell suspension cultures will provide an opportunity for studying the genes about growth and development of *Populus*. On the basis of cell suspension lines, Li and Tian (2004) analyzed the expressions of genes related to spreading of cells and cellulose synthesis. Zhao et al. (2000) investigated salt tolerance of suspended cells of *P. tomentosa*. However, there are no detailed reports about the factors affecting the establishment of cell suspension lines of *P. tomentosa* until now.

TC152 an excellent genotype of *P. tomentosa*, was used as research material to study the factors affecting the establishment of cell suspension lines and the growth characteristics of cell lines. A stable cell suspension line of *P. tomentosa* was established in our study. It is the basis for research on protoplast culture and somatic embryogenesis of *P. tomentosa*.

---

* 本文原载 *Forestry Studies in China*, 2008, 10(3): 158-161, 与姚娜、安新民和杨凯合作发表。

2 Materials and methods

2.1 Plant materials

Aseptic shoots kept *in vitro* were used as explants, called TC152. The genotype was selected from the gene bank of *P. tomentosa* in the Guanxian nursery, Shandong Province of China. TC152 has many good characteristics, such as fast growth, fine wood quality, and strong resistance to stress.

2.2 Methods

2.2.1 Callus induction

New leaves of aseptic shoots were cut into 0.5cm×0.5cm as explants, which were inoculated on a callus induction media containing a solid, basal MS medium and different concentrations of BA and NAA for callus induction. Calli that were not browning were selected for subculture.

2.2.2 Suspension culture

The exuberant calli were transferred into 150mL flasks containing 80mL liquid media. The flasks were placed on a rotary shaker at a speed of 110r/min. The initial media were a MS liquid medium supplemented with different concentrations of 2,4-D.

At the early stage of cell suspension culture, the subculture cycle was 12d. During this subculture, cells were, at first, sretained for 2h. The medium of the upper layer was dumped. Cells on the lower layer were filtered to remove liquid medium. And 2.0g fresh weight cells were partitioned into 150mL flasks containing 80mL subculture medium.

2.2.3 Measurement of proliferation cells

Packed cell volume (PCV): Cells were resuspended fully, dumped into a 10mL scaled centrifuge tube and centrifuged at a speed of 1000r/min for 10min. The volume of compact cells in the centrifuge tube is referred to as PCV.

Fresh weight (FW): Suspension cells were, at first, meshed by a 60 mesh and then filtered. The weight of cells on the filter paper was the fresh weight of suspension cells.

2.2.4 Media and culture conditions

A MS (Murashige and Skoog, 1962) medium was used as a basal medium. The solid medium was a MS medium supplemented with 5 g/L agar and 30g/L sucrose. A liquid medium was also a MS medium supplemented with 30g/L sucrose. The calli and cells were cultured under a 16-h light/8-h dark regime at a constant temperature of 25℃ and an illumination of 500-1000 Lux.

2.2.5 Statistical analysis

A two-factor, three-level experiment was designed to select the best callus induction medium. Single factor experiments were designed to study the factors that affect cell suspension culture. The experiments were replicated three times. In addition, each treatment was inoculated in five flasks. The results, expressed in percentages, were transformed by the expression $\theta = \arcsin\sqrt{percentage}$. An analysis of variance was carried out and differences between the means of the treatments were determined by *LSD* tests at a 0. 05 level of significance (Xu and Huang, 1995). The analysis of variance and calculation of means was carried out with the SPSS 12. 0 statistical program.

3 Results

3. 1 Callus induction

Two kinds of plant growth regulator, 2, 4-D and 6–BA, were used to select the best callus induction medium. Each of the plant growth regulators were represented by three levels. Abundant light yellow calli could be found from cut pieces of the leaves after being inoculated for 10–15 d. Callus induction rates and growth status were observed after being inoculated for 25d. The results show that the effects of 2, 4-D, 6–BA and their interaction were all significant (Table1). The callus induction rate of the medium supplemented with 1. 0mg/L 2, 4-D is 97. 2%, which was markedly higher than that of other treatments. In this medium, the calli grew fast and the texture was friable. The media with both 2, 4-D and 6–BA had negative effects on callus induction, in which the calli were more compact and grew at a slower rate. All of the calli became brown after being cultured for more than 35d. The medium MS + 1. 0mg/L 2, 4-D + 5g/L agar + 30g/L sucrose was the optimum callus induction medium and the best subculture cycle was 25d.

3. 2 Establishment of cell suspension line

3. 2. 1 Selection of initial medium

The calli that were friable and light yellow were inoculated into liquid media supplemented with different concentrations of 2, 4-D. Suspension cells were initially generated within 15d. Morphologically, they largely consisted of cell aggregates of round to oval shaped cells. PCV was surveyed after being cultured on a shaker for 15d. The results (Table 2) indicate that the yield of cells is the highest for the medium supplemented with 1. 5mg/L 2, 4-D. The cell aggregates were easy to differentiate adventitious roots after being cultured in a medium with 0. 5mg/L 2, 4-D for 15d. When the concentration of 2, 4-D was higher than 2. 5mg/L, the suspension cells grew so fast that some of them were browning. The medium MS + 1. 5mg/L 2, 4-D + 30g/L sucrose can be used as the best initial medium for the cell suspension line.

3. 2. 2 Callus subculture times

Calli which had been subcultured 0, 4 and 8 times were inoculated into the initial medium to initiate cell suspension culture. Table 3 shows that along with the increase in the number of callus subculture times, the yields of suspension cells are regressive. In addition, browning occurred. The calli that were not subcultured too often, easily yielded fine suspension cells. Moreover, the calli that were not subcultured at all, were the best material for suspension culture.

Table 1 Effect of plant growth regulators on callus induction of *P. tomentosa*

No	2, 4-D (mg/L)	6-BA (mg/L)	Induction rate (%)	Callus stares		
				Color	Growth	Texture
1	0.5	0	56.94±3.67 c d	LY	GF	R
2	0.5	0.2	69.45±5.01 b c d	YO	GS	C
3	0.5	0.4	48.61±5.00 c d	YO	GS	C
4	1.0	0	97.22±2.78 a	LY	GF	R
5	1.0	0.2	79.16±4.17 b c	Y	OM	C
6	1.0	0.4	77.78±3.68 b c	Y	GM	C
7	1.5	0	90.28±3.68 a b	DY	GF	R
8	1.5	0.2	38.89±5.01 d	DY	GM	R
9	1.5	0.4	48.61±3.67 c d	DY	GM	R
F	34.256*	28.277*	2, 4-D×6-BA: 10.875*			

Notes: * means significant at 0.05 level. LY means light yellow, GF growing fast, R means friable, YG yellow and green, GS growing slowly, C compact, Y yellow, GM growing moderately and DY dark yellow.

3.3 Effects of 2, 4-D concentration on subculture

Approximately 2.0g cells of fresh weight were weighed from a stable cell suspension line and transferred into subculture media containing different 2, 4-D concentrations as follows: 0.2, 0.4, 0.6, 0.8, 1.0 and 1.2mg/L. Cell growth curves were derived from data obtained by means of PCV values. The error bars indicate that under strict subculture regimes the variations between cultures at any particular age are quite small. Figure 1 indicates cells in a medium, supplemented with 0.8mg/L 2, 4-D, grew best. Suspension cells can actively proliferate in this medium. The yield of the cell line increased like an "S" curve with the changing time of culture. The curves can be divided into three phases: a lagging phase (1-4d), a logarithmic phase (4-7d) and a slow phase (8-10d). The cell line proliferated slowly in the media containing 0.4mg/L 2, 4-D or less. The cells grew so fast in the media with high 2, 4-D concentrations, that some cells became old or died after being cultured for more than 7 d. The medium MS+0.8mg/L 2, 4-D+30g/L sucrose would be the best medium for cell subculture and the best subculture cycle 7 d.

Table 2 Effect of concentrations of 2, 4-D on establishment of cell suspension line

No.	2, 4-D(mg/L)	PCV(mL)	Cell growth status
1	0.5	3.37±035b	Light yellow, adventitious roots
2	1.0	4.40±0.44 ab	Light yellow
3	1.5	5.40±0.06a	Light yellow
4	2.0	4.63±0.25a	Yellow
5	2.5	4.53±0.38 a	Dark yellow, browning

Table 3 Effect of callus subculture times on establishment of cell suspension line

No.	Subculture times	Inoculation callus(g FW)	PCV(mL)	Cell browning status
1	0	2. 0	4. 20±0. 17 a	Naught
2	4	2. 0	1. 33±0. 38 b	A little
3	8	2. 0	0. 83±0. 09 b	Badly browning

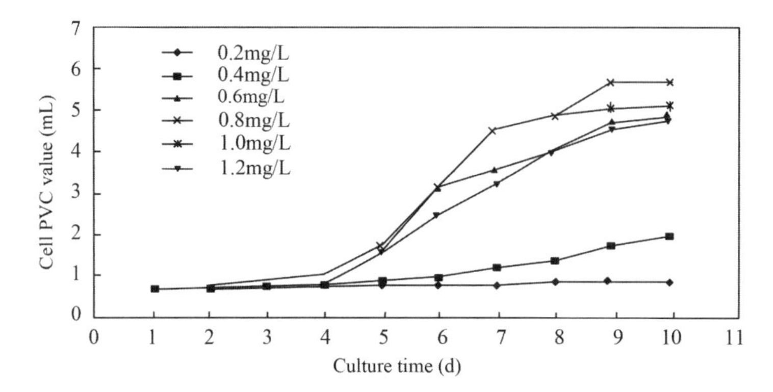

Fig. 1 Growth curves of cell suspension lines in different subculture media

4 Discussion

The category and concentration of plant growth regulators affect the induction and proliferation of calli. Fine callus can be obtained from *Populus* tissues using a media containing auxin and cytokinin, which had been reported for *P. euphratica* (Liu et al. , 2000) , *P. tomentosa* (Li and Tian, 2004) and *P. alba*×(*P. davidiana* ×*P. simonii*)×*P. tomentosa* (Du et al. , 1998). However, in our study, in order to induce callus from leaf tissues of *P. tomentosa*, the addition of 2, 4-D is essential. On the other hand, BA, a cytokinin, has a negative effect on callus induction and proliferation. The treatments of BA cause callus to become compact and to grow slowly. The result is similar to the results reported by Wang et al. (1991) about *P. simmonii* and by Lu et al. (1997) for *P. xiaohei.*

In our study, a high level of 2, 4-D concentration was needed to initiate cell suspension culture of *P. tomentosa*. However, suspension cells could proliferate in a subculture medium containing a low 2, 4-D concentration. The concentration of 2, 4-D could be decreased in the subculture in order to reduce the damage to the cells. In the next step of our study, the concentration of 2, 4-D in the subculture medium will still be reduced. Cytokinin will be added into the medium in order to promote the regeneration of cells.

The number of subculture times of callus affects the quality of suspension cells. Different plants have their best callus subculture times during the establishment of their cell suspension line. For the calli of potato, subcultured 1-3 times, it was easy to obtain suspension cells (Zhang and

Dai, 2000). However, the best material for rice suspension culture was the callus, which had been subcultured 7 times (Yin and Liu, 1994). The result in our study shows that the greater the increase in the number of subculture times, the harder to establish a suspension culture line for *P. tomentosa*. The reason may be that the ability of proliferation usually degrades when the callus is being subcultured too many times. Therefore, in order to obtain a fine cell suspension line, young, friable and coloury calli form the best initial material.

Comparison and Early Selection of New Clones in *Populus tomentosa* *

Abstract In our study, two experimental plantations, respectively, with 24 and 32 new clones of *P. tomentosa*, were established in Weixian County, Hebei Province and Wuzhi County, Henan Province using a completely randomized block design. A comparative study was conducted on the continuous 5-year-old height and diameter at breast height (DBH) of new clones in the two plantations. As well, based on genetic correlation over the years of testing of these clones, a preliminary study of early selection was carried out. Results indicate that the growth traits of the new clones in Weixian were better than those in Wuzhi. The traits show weak correlation between the two plantations. In some stands, the height, DBH and seedling volume of 5-year-old clones presented statistically significant differences among clones. In both plantations, the new clones showed over 0.6 repeatability of height, DBH and volume, as well as larger coefficients of variation (CV). The fact that these clones achieved the largest repeatability and CV in the second year suggests that these traits are highly controlled by heredity. Thus, based on the growth traits of the second year, the new clones B305, B307, B303, H75, BT18, BT17 and 21J-1 were considered suitable in Weixian. In Wuzhi, the new clones had variable repeatability and CVs in various years and their correlation of growth traits among different years was not high. We conclude that early selection of new clones was not feasible in Wuzhi.

Key words *Populus tomentosa* Carr., new clones, comparative test, early selection

1 Introduction

These days, clonal forestry is becoming popular (Ji, 2004). Chinese white poplar (*Populus tomentosa* Carr.) is one of the indigenous tree species in China, mainly distributed in the middle and lower reaches of the Yellow River. It has many desirable characteristics such as a rapid growth rate, strong adaptation to harsh site conditions, superior wood quality, a tall and straight trunk and handsome tree form (Zhu, 1992). From the 1940s, many investigators obtained superior clones through plus selection, cross breeding and so on (Pang, 1986; Jiang and Huang, 1989; Gu, 1990). At our institute of Chinese White Poplar, we also bred many new clones of *P. tomentosa*, consistent with different breeding goals from the 1980s (Zhang et al., 1992, 1997; Zhu and Zhang, 1997). The growth behavior of these new clones in different stands and years and of new clones in the same stand and year gained a broad focus of attention. With this study, we attempted

* 本文原载 *Forestry Studies in China*, 2008, 10(3): 162~167, 与郑会全、王泽亮、林善枝合作发表。

to find out superior clones for different stands. As well, early selection was studied in order to provide a basis for clonal forestry of *P. tomentosa*.

2 Experimental site and methods

2.1 Experimental site

Our experimental sites were located in Weixian County, Hebei Province and in Wuzhi County, Henan Province. Their stand profiles follow.

The topography of the forest farm in Weixian belongs to the low plains of southern Hebei, located at 36°30′N, 115°21′E. The annual average temperature is 13℃, annual accumulated temperature (≥10℃) 4302.5℃, average temperature in July 26.7℃ and average temperature in January −3.2℃. Annual average precipitation is 598.0mm. The frost-free period is 198 d.

The topography of the forest farm in Wuzhi belongs to the alluvial plain of the Yellow River, located at 35°N, 113°30′E with smooth terrain and a sandy, loamy, lightly saline and alkaline soil. The annual average temperature is 14.3℃, annual accumulated temperature (≥10℃) 4600~4700℃, average temperature in July 24~28℃ and average temperature in January −4~0℃. Annual average precipitation is 635.9mm. Its frost-free period is 210~240d.

2.2 Methods

2.2.1 Experimental design

Two experimental forests were planted in Weixian County of Heibei Province and Wuzhi County of Henan with 1-year-old cuttings in a completely randomized block design in March, 1998. The new clones used and the design plan are listed in Tables 1 and 2. From 1998 to 2002, the height and DBH were measured at the end of each year. Sapling volumes (V) were calculated with the following formula:

$V = 0.513407 \times H^{0.826956} \times (D \times 10^{-2})^{1.995375}$ ($r = 0.9872$, $s = 0.00102$, H and D represent height and diameter at breast height, DBH).

Table 1 New clones of *P. tomentosa* used in clonal test fields

Site	Source of materials	New clones
Weixian County	Breeding clones	8001,3-54-1,1313,1319
	Crossing clones	21J-1, B306,21J-17, B308,B305,B307, 51L-1, 51L-3, B304,B303,LT50, BT18, BT17, BT85,BM33,BM186,12-51, H75, H24, T18
Wuzhi Connty	Breeding clones	1313, 1316, 1319, 1327, 1521. 3541,8001,T6
	Crossing clones	B301, B302, B303, B304, B305, B306, B307, B308, B309, B310,B330, B331, B384, 12-50, 13-16, BM33, LJ, BM86, H75. LT50, TB42, YL. T18,T27

Table 2 Planting design of clonal test fields of new clones in *P. tomentosa*

Site	Density	Number of blocks	Number of clones	Number of trees in ench pot
Weixian County	3m×4m	4	24	16
WuzhiCounty	3m×4m	3	32	16

2.2.2 Analytical methods

Block means for each clone were calculated and then an analysis of varianace was conducted. For both sites, the linear model for analysis of variance was $X_{ijk} = \mu + C_i + B_j + (CB)_{ij} + e_{ijk}$, where X_{ijk} is the test value, C_i the clonal effect value, B_j the block effect value, $(CB)_{ij}$ the interaction effect between clone and block and e_{ijk} the random error. Estimation of genetic parameters of clonal repeatability was calculated according to Falconer (1981), multiple comparisons were determined with the LSD method and Spearman correlation coefficients were calculated.

3 Results

3.1 Growth comparison of new clones

3.1.1 Growth differences of new clones in Weixian

Analyses of variance indicated that the differences in height, DBH and sapling volume among 5-year-old new clones are significant in Weixian. The results of multiple comparisons of height, DBH and volume of these clones with LSD are shown in Table 3. With regard to height, DBH and volume, the R-values are 3.63m, 5.39cm and 0.0632m^3 respectively. The superior new clones are B305, B303 and B303 respectively. From the clonal trends of volume over age, we found that the accumulation rates of volume are different among the new clones tested, indicating that fast growth period are obtained by these clones in different years or ages.

3.1.2 Growth differences of new clones in Wuzhi

Similar to the results from new clones in Weixian, the differences of height, DBH and sapling volume among 5-year-old new clones are also significant in Wuzhi. But the R-value of height, DBH and volume are 3.19m, 7.77cm and 0.0535m^{3}respectively. The superior new clones are B331, B301 and B301 respectively (shown in Table 4). The clonal trends of volume over age between the two plantations are also similar. Nevertheless, all clones, except YL, were in a fast growing period during the second year after planting.

Table 3 Height, DBH and sapling volume growth, ranks and comparisons of 5-year-old new clones in *P. tomentosa* in Weixian

Clone	Height(m)	Clone	DBH(cm)	Clone	Volume(m^3)
B305	12.10	B303	15.14	B303	0.0924
BT17	11.97	BI17	14.76	BT17	0.0885
B305	11.91	B307	14.25	B307	0.0003
B306	11.69	H75	14.06	B305	0.0786
B307	11.62	21J-1	13.95	H75	0.0779
51L-1	11.51	B305	13.83	B306	0.0759
H75	11.48	B506	13.80	21J-1	0.0752
B304	11.36	B308	13.75	B308	0.0739
BT18	11.31	BTI8	13.67	BT18	0.0733
B308	11.29	B304	13.46	B304	0.0704
21J-1	11.27	LT50	13.00	LT50	0.0640
LT50	11.05	51L-1	12.43	51L-1	0.0607
51L-3	10.89	51L-3	12.38	51L-3	0.0574
21J-17	10.55	21J-17	12.14	21J-17	0.0540
TI8	10.29	1319	11.56	T18	0.0480
1313	10.22	T18	11.52	BM33	0.0448
BT85	10.16	BM33	11.37	1319	0.0433
BM33	9.96	1313	10.99	1313	0.0429
12-51	9.70	3-54-1	10.55	8001	0.0405
BM86	9.62	8001	10.54	BT85	0.0389
3-54-1	9.56	12-51	10.50	3-54-1	0.0376
8001	9.41	B385	10.48	12-51	0.0376
1319	9.02	BM86	10.13	BM86	0.0345
H24	8.47	H24	9.75	H24	0.0292
Average	10.08		12.42		0.0591
SD	0.208		0.334		0.004

3.2 Selection of new clones in *P. tomentosa*

From the results described, the new clones tested exhibited fast growth in both counties, where average values of 5-year-old sapling volumes reached 0.06 and 0.04m^3 respectively. Moreover, the growth of the same clone in the two plantations or various clones in the same plantation is different. Thus, to obtain maximum productivity, selection for superior clones should be carried out.

Table 4　Height, DBH and sapling volume growth, ranks and comparisons of 5-year-old new clones in *P. tomentosa* in Wuzhi

Clone	Height(m)	Clone	DBH(cm)	Clone	Vohume(m^3)
B331	9.67	B301	13.60	B301	0.0634
B301	9.66	B303	13.36	B303	0.0615
H75	9.65	H75	12.85	H75	0.0566
B303	9.59	B308	12.78	B308	0.0547
B304	9.51	B309	12.54	B309	0.0528
B306	9.46	B304	12.28	B304	0.0510
B309	9.45	B307	11.93	B307	0.0483
B308	9.37	1327	11.85	B306	0.0464
B307	9.32	B310	11.72	1327	0.0461
B310	9.18	T18	11.71	B331	0.0459
B330	9.15	B331	11.59	B310	0.0449
T18	9.13	B330	11.47	T18	0.0448
1327	9.03	B306	11.41	B330	0.0439
LT50	8.94	8001	11.31	BM33	0.0396
BM33	8.89	LJ	11.10	LT50	0.0396
LJ	8.87	LT50	11.09	LJ	0.0395
13-16	8.82	BM33	11.09	B305	0.0388
B305	8.79	B302	11.04	8001	0.0388
B302	8.68	B305	11.04	B302	0.0386
1316	8.66	1319	10.79	13-16	0.0339
1521	8.62	3541	10.37	1319	0.0338
3541	8.55	1313	10.35	3541	0.0333
1313	8.50	13-16	10.27	1313	0.0329
T6	8.45	1316	10.19	1316	0.0324
8001	8.34	T6	10.09	T6	0.0320
BM86	8.32	TB42	9.81	TB42	0.0302
T27	8.31	T27	9.80	BM86	0.0301
TB42	7.99	1521	9.76	1521	0.0297
B384	7.97	BM86	9.69	T27	0.0293
1319	7.88	B384	9.07	B384	0.0259
12-50	7.87	12-50	8.91	12-50	0.0237
YL	6.48	YL	5.83	YL	0.0099
Average	8.78		10.96		0.0398
SD	0.122		0.266		0.002

Table 5　Clonal repeatability and genetic coefficient of variation of new clones in *P. tomentosa* each year in Weixian

Year	Height		DBH		Volume	
	R	*GCV*	*R*	*GCV*	*R*	*GCV*
1998	0.703	0.111	0.725	0.084	0.655	0.320
1999	0.919	0.169	0.938	0.250	0.907	0.548
2000	0.817	0.120	0.882	0.199	0.853	0.429
2001	0.879	0.076	0.940	0.197	0.927	0.376
2002	0.976	0.078	0.977	0.144	0.898	0.304

3.2.1 Estimation of genetic parameters

After being dissected, clonal repeatability and genetic coefficients of variance (GCV) for each year were calculated in two plantations and shown in Tables 5 and 6. Both parameters reached or were near their maximum values in the second year after planting. Consequently, preliminary selection may be conducted according to the growth of the second year.

3.2.2 Selection of new clones

In terms of accumulation of sapling volume in 1999 and 2002, the new clones in Weixian were ranked. The best clones exhibited were B305, B307, B303, H75, BT18, BT17, 21J-1 and B303, BT17, B307, B305, H75, B306, 21J-1 respectively. The results were almost identical to those in 1999 and 2002. Based on the volumes in 1999 to those in 2002, the regression equation, $y = 6.3287x + 0.028$ ($r = 0.8944$), was obtained (y, expecting value, x, testing value). The high correlation coefficient (r) means that the growth of sapling volume is tightly linked to that of 2002 and 1999. According to the sapling volume growth, the clones B303, BT17, B307, B305, H75, B306, 21J-1 are selected as superior clones; their acquired genetic gain was 33.5%.

The same analysis was performed for the new clones in Wuzhi. The best clones were B303, 3541, B301, LT50, BM86, B310, 1319 and B301, B303, H75, B308, B309, B304, B307 in 1999 and 2002 respectively. Compared with the best clones in 1999, only clone B303 and B301 showed better in 2002, the others slowed down, confirmed by the regression equation obtained, i. e., $y = 4.3478x + 0.0216$ ($r = 0.0816$, the correlation coefficient being very low).

That is, a preliminary selection in Wuzhi cannot be made at present. How to conduct this needs to be studied further.

3.2.3 Interaction between clones and plantation sites

Two-way analyses of variance of height, DBH and volume growth of 5-year-old clones were performed (Table 7) based on the shared eight clones 1319, BM33, H75, BM86, 1313, 8001, T18, LT50 in the two sites. As a result, the growth of height and DBH is significantly different at the 1% level of significance, while sapling volumes were, statistically, not different. In addition, in regard to height, DBH and volume, the interactions between clones and plantation sites were not significantly different either.

From Tables 3 and 4, it is clear that the growth of new clones in Weixian is better than that in Wuzhi, especially the average volume growth of the saplings. Additionally, Spearman correlation coefficients of the three growth parameters were calculated between the two sites; the values were 0.243, 0.044, −0.039, also rather low. It shows that the correlation of new clones growth is not tight between the two sites. The growth of new clones in Wuzhi cannot be inferred from that in Weixian.

Table 6 Clonal repeatability and genetic coefficient of variation of new clones in *P. tomentosa* each year in Wuzhi

Year	Height		DBH		Volume	
	R	*GCV*	*R*	*GCV*	*R*	*GCV*
1998	0. 920	0. 096	0. 909	0. 146	0. 899	0. 370
1999	0. 871	0. 114	0. 904	0. 185	0. 868	0. 410
2000	0. 892	0. 074	0. 887	0. 142	0. 872	0. 298
2001	0. 871	0. 060	0. 901	0. 136	0. 896	0. 294
2002	0. 896	0. 071	0. 921	0. 129	0. 801	0. 296

Table 7 Two-way analyses of variance for growth of new clones in *P. tomentosa* in test fields

Trait	Source of variation	*df*	*MS*	*F*-value	$F_{0.05}$	$F_{0.01}$
Height	Site	1	26. 976	55. 158**	4. 080	7. 310
	Clone	7	3. 265	6. 677**	2. 250	3. 120
	Site×clone	7	0. 304	0. 621	2. 250	3. 120
	Error	40	0. 489			
	Total	55				
DBH	Site	1	2. 649	2. 189		
	Clone	7	7. 376	6. 094**		
	Site×clone	7	1. 330	1. 098		
	Error	40	1. 210			
	Total	55				
Volume	Site	1	2. 12E-03	13. 879**		
	Clone	7	1. 09E-03	7. 137**		
	Site×clone	7	1. 40E-04	0. 916		
	Error	40	1. 53E-04			
	Total	55				

4 Discussion

4. 1 Early selection of new clones in *P. tomentosa*

Previous studies showed that genetic presentation of forest trees varies with years of growing because their natural growth trend is "slow-fast-slow". From a study of 28 *Aigeiros* clones, it was found that, although unstable, the repeatability of these clones increased with age, their CVs showed the "big-small-big" trend and genetic correlation coefficients became larger than phenotypic correlation coefficients (Pan et al. , 1999). Other investigators also found that until forest trees reached a specific age, the genotypic variation and repeatibility of height and DBH did not increase significantly. After a period, these parameters became stable (Hu et al. , 2001). There are also other points of view. Ma et al. (2000) found that juvenile-mature correlation of various traits of the same clone were different. The genetic correlation of forest trees under different density differs between neighboring years (Jonsson et al. , 2000).

In our study, the results clearly show that there are extremely significant differences in height, DBH and sapling volume among 5-year-old clones on the same planting site. In both plantations, the repeatability of all tested traits were more than 0. 60, showing that they are tightly controlled by heredity; their big CVs also favor the selection for superior clones. In Weixian, the repeatability and CVs of all three traits were nearly stable, indicating that early selection is feasible. However, the two parameters were not stable in Wuzhi, showing the unstable growth of these new clones. In summary, genetic variation of the various tested new clones of *P. tomentosa* and that of the same traits of the same clone, varies with different planting sites, tested traits and tested years, regardless of whether the variable level is low or high.

4. 2 Multiple-site comparison of new clones in *P. tomentosa*

Multiple-site comparison is necessary for evaluation and popularization of new clones of forest trees. These tests could provide genetic presentation of new clones in different stands and the various genetic parameters necessary for new clone evaluation. Generally, these tests should involve various environments, stands and test sites. When there are fewer test sites, the variation of interaction is reduced and the experimental error increases (Pswarayi et al. , 1997). In addition, interactions between clones and sites differ with years of growing (Osorio et al. , 2003).

The present results show that the growth of the tested clones in Weixian is better than in Wuzhi. Moreover, the difference of interaction between 5-year-old new clones and planting site is not significant and the growth traits are not tightly correlated between the two sites. Due to only two test sites and the limited number of test years, the obtained results may just be consulted for clonal selection for each site. For other objects, such as the interaction between clones and sites, it becomes necessary to increase the number of plantations, extend testing years and make a full-scale evaluation.

第二部分

分子标记技术及其应用

毛白杨无性系同工酶基因标记的研究*

摘 要 通过对86株毛白杨优树和古树的不同组织材料的16种同工酶酶谱分析研究发现有10种酶系统遗传表达稳定。其中有8种酶系统(ACO, GOT, LAP, NDH, PER, PGM, 6-PG-DH和SKDH),共有12个基因位点,另外两种酶系统(IDH和MDH)难以确定位点,各具5种酶谱表现型。用这些酶系统可将86株毛白杨归属于79个不同遗传组成的无性系。重叠率为8.14%,平均重叠率为1.63%。由此证明全国毛白杨档案库收集的资源,具有丰富的种群代表性和可靠性。本文对用同工酶基因标记鉴别无性系所选用酶系统数目及其最佳酶系统组合也进行了讨论。

关键词 毛白杨,同工酶,基因标记,无性系鉴别

同工酶作为遗传标记,在林木遗传育种学研究的各个方面已得到广泛应用(Adams, 1983; Cheliak, 1985, 1987; Hattemer, 1978; Müller-Starck, 1976, 1988; Rudin and Lundkvest, 1977; 沈熙环, 1987; 黄敏仁, 1985; 葛颂等, 1988),其中用同工酶基因标记来鉴别树木基因型,国内外有不少报道(Bergmann, 1987; Castillo and Padro, 1987; Cheliak and Pitel, 1984b; Hattemer et al., 1990; Paule and Cheliak, 1990; Rajora, 1988; 王明庥等, 1982, 1983; 杜宗岳, 1983, 1986; 胡志昂, 1981; 钟海文等, 1982)。

毛白杨(*Populus tomentosa*)是我国特有白杨派树种。它分布广,速生,适应性和抗性强,材质优良,是华北平原人工丰产林和农田防护林的重要树种(Hattemer et al., 1990)。毛白杨起源复杂,种内变异丰富,受到广大林木育种学家重视,但是对毛白杨的同工酶研究还很少(Zhu, 1988; 杜宗岳, 1986; 胡志昂, 1981)。本文旨在借助于多种同工酶系统电泳结果,分析毛白杨优树档案库内无性系的遗传稳定性,确定其同工酶基因位点,并对众多选优无性系进行鉴定,为研究和利用毛白杨优良无性系提供理论依据。此外,对在无性系鉴别中酶系统的选择及最佳酶系统组合的确定也进行了分析讨论。

1 材料与方法

1.1 材料来源

实验中所分析86株毛白杨优树、古树材料,取自山东省冠县苗圃全国毛白杨优树档案库中(朱之涕, 1991)。它们来自其分布区的河北、北京、山东、山西、河南、陕西、甘肃7个省(自治区、直辖区)。

* 本文原载《北京林业大学学报》, 2002, 14(3): 9-18, 与朱之悌、Muller-Starck G和Hattemer H. H. 合作发表。

在86株样本中，树龄在20年以上的超过80%，100年以上的占33.7%，树龄最大的达700年。试样包括一年生根萌条及其休眠芽。实验中所用的生长芽、叶片、叶柄、嫩茎、皮和根尖等材料，均系用根萌条截成10cm长插穗扦插于有腐殖质土掺少量蛭石和石英砂作为基质的营养杯中，然后在温室中培养获得的，愈伤组织材料是利用根萌条扦插所得到小苗嫩茎、叶片和叶柄作外植体，经消毒灭菌后在MS培养基上无菌培养获得的。

1.2　酶系统

所分析的16种酶系统是：ACO，ADH，DIA，EST，GDH，GOT，IDH，LAP，MDH，NDH，PER，PEP，6-PGDH，PGI，PGM，SKDH

1.3　实验方法

同工酶电泳分离是采用淀粉凝胶水平电泳方法。酶提取缓冲液的配方如下：1.6g Tis，0.12g EDTA，1mLC_2H_6OS，4g PVP，50mg DTT，溶于100mL去离子蒸馏水中，用HCl调节pH值至7.3。

切取新鲜材料约0.1g，及时置于预先冷藏的多孔小研钵中，加入酶提取缓冲液1~2滴，然后迅速研磨匀浆，用NO.3 Whatman小滤纸片吸取酶液，并载入淀粉凝胶上电泳分离。酶的显色和染色液的配方参照Cheliak和Müller-Starck的方法(Cheliak and Pitel，1984a；Müller-Starck et al.，1988)。

1.4　结果分析方法

同工酶基因位点确定和等位基因命名参照Cheliak等和Müller-Starck的方法(Cheliak and Pitel，1984b；Müller-Starck et al.，1988)。

在利用同工酶基因型来鉴别毛白杨各无性系时，为了用数量指标来描述全部试样中无性系的重叠情况，本文采用了无性系重叠率和平均重叠率两个指标。

把同工酶基因型(遗传组成)相同的个体数(C)占所分析的总个体数(N)的百分比定义为无性系重叠率，用CO表示，即

$$CO = C/N \times 100\%。$$

而把无性系重叠率(CO)除以具有重复的同工酶基因型数(M)定义为平均重叠率，用AO表示。即

$$AO = CO/M$$

CO是指不能区分开的个体数占分析总个体数的百分比，其值范围在0~100之间。如果所分析的无性系是确定的不同无性系的话，那么$CO=0$，则表示这些无性系都能分开；若$CO=100\%$，则表示这些无性系不能被单个区分开来。AO表示在具有重叠的个体时，重叠个体的分布情况，其值范围在0~CO之间；若$AO=0$，则表示不存在重叠无性系；若$AO=CO$，则表示所重叠的个体都属于一个无性系，在有重叠无性系的情况下，AO值越小表明重叠无性系分布越均匀。

2 结果与讨论

2.1 同工酶酶谱遗传稳定性及其基因位点确定

用毛白杨各种组织器官材料，对各酶系统的最适缓冲液 pH 值进行了对比实验，结果发现酶系统 LAP，GOT，PER，EST，GDH 和 PEP 对 Ashton 和 Poulik 两种缓冲液系统均表现出较好的显色效果，无明显的差异。对其余 9 种酶系统进行不同缓冲液 pH 值比较试验，发现有 8 种酶系统的显色效果差异较大，结合不同的组织材料试验，结果列于表 1 中。

表 1 缓冲液 pH 值对不同组织材料酶显色效果的影响

酶系统	缓冲液 pH	根尖	老叶	幼叶	芽	皮	愈伤组织	酶系统	缓冲液 pH	根尖	老叶	幼叶	芽	皮	愈伤组织
	T. C. 7. 8	0	-	-	-	0	-		TC. 7. 8	-	-	-	+	-	+
	7. 6	0	-	+	+	-	+		7. 6	*	+	+	*	-	*
PGI	7. 0	0	-	+	+	-	*	ACO	7. 0	-	+	+	*	-	*
	6. 5	0	-	-	-	0	-		6. 5	+	+	-	*	+	+
	6. 2	0	-	-	-	-	0		6. 2	+	-	+	*	+	+
	T. C. 7. 8	0	0	0	-	0	-		TC. 7. 8	0	+	+	+	-	+
	7. 6	0	-	+	+	0	+		7. 6	0	+	* *	*	+	*
DIA	7. 0	0	-	-	-	0	-	PGM	7. 0	0	* *	*	* *	+	* *
	6. 5	0	-	-	-	-	0		6. 5	0	+	-	+	0	+
	6. 2	0	-	-	-	0	0		6. 2	0	-	-	*	-	-
	T. C. 7. 8	0	-	-	+	0	+		TC. 7. 8	-	-	-	+	0	0
	7. 6	0	+	+	*	+	*		7. 6	0	-	+	*	-	+
IDH	7. 0	0	+	-	+	+	+	6-PGDH	7. 0	-	-	+	*	-	+
	6. 5	+	+	-	+	-	-		6. 5	0	-	+	*	-	*
	6. 2	0	-	-	+	+	-		6. 2	+	*	+	* *	+	+
	T. C. 7. 8	-	-	-	+	-	-		TC. 7. 8	0	0	-	-	0	-
	7. 6	-	+	+	+	-	+		7. 6	0	0	+	+	-	+
MDH	7. 0	-	-	+	+	-	+	SKDH	7. 6	0	0	+	+	-	+
	6. 5	-	+	*	*	+	*		6. 5	0	0	*	* *	0	*
	6. 2	+	+	*	*	-	+		6. 2	0	-	+	+	-	+

说明：* * 表示较好，* 表示好，+表示一般，-表示差，0 表示无活性

从表 1 中可以看出不同酶系统最佳显色效果所要求的缓冲液 pH 值是不同的。如 DIA，IDH，ACO，PGM 等酶系统所要求的电泳缓冲液 pH 值要高一些，大于 7.0，而其他酶系统要求的 pH 值稍低一些，小于 7.0，同时还可以看出，不同组织器官的材料，显色效果也不尽相同。总的来看，以芽作为材料效果最好，其次是幼嫩叶片和愈伤组织的效果好，效果较差的是根尖和皮组织。

不同 pH 值的缓冲液对酶系统的显色效果的影响作用，迄今未见报道。一般是固定酶系统采用固定 pH 值的缓冲液系统。但是，由于所研究的物种是多样的，自然对电泳的条件会产生一些差异，正如对电泳缓冲液 pH 值要求不同一样。所以，当采用不同材料时，寻找最佳的电泳缓冲液 pH 值是十分关键的。

有人曾做过各种组织器官材料酶系统显色效果差异分析，如(Pitel 和 Cheliak 对 5 种针叶树的各种营养组织材料做过的实验(Pitel and Cheliak，1984)。这种差异可能是由于各种酶系统在各种组织中存在的酶活性强弱差异所致，芽和愈伤组织的酶显色效果较好，可能是由于正在生长发育的分生细胞集中的组织中，同工酶基因能充分得到表达，所以酶的活性也较强，这与欧洲山杨(Bergmann，1987)、美洲山杨(Cheliak and Pitel，1984b)、山杨和小青杨(杜宗岳，1986)等的分析结果是相似的。

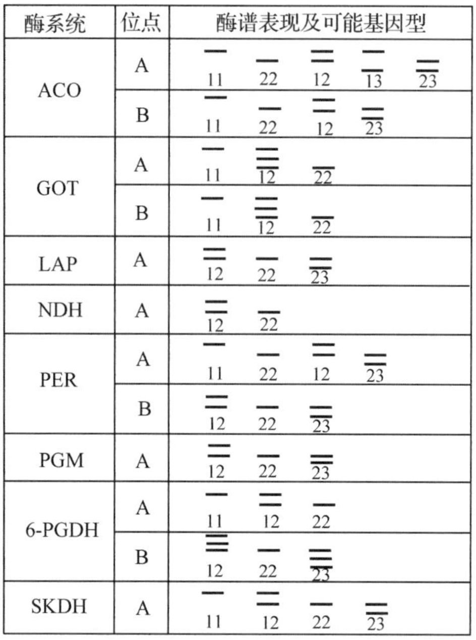

酶系统	位点	酶谱表现及可能基因型
ACO	A	11　22　12　13　23
	B	11　22　12　23
GOT	A	11　12　22
	B	11　12　22
LAP	A	12　22　23
NDH	A	12　22
PER	A	11　22　12　23
	B	12　22　23
PGM	A	12　22　23
6-PGDH	A	11　12　22
	B	12　22　23
SKDH	A	11　12　22　23

图 1　8 种酶系统酶谱表现及可能基因型

通过分析缓冲液和组织器官对酶系统的影响以后，接着对毛白杨无性系遗传稳定性进行综合分析。利用芽、叶片、叶柄、根尖、皮和组培愈伤组织等材料进行了比较实验。结果发现，毛白杨组织材料中，酶系统 GDH 和 ADH 可能不完全受遗传的控制，酶系统 EST 具有组织特异性；DIA，PEP，PGI 的显色效果比较差。最后从 16 种酶系统中选出无组织特异性、酶谱比较稳定的 10 种酶系统。

在确定毛白杨的同工酶基因位点时，由于没有毛白杨的人工控制杂交子代的同工酶分析结果，所以，参考了白杨派中其他杨树的分析结果(如 Müller-Starck 对欧洲山杨控制授粉杂交子代同工酶表现形式和 Cheliak 等对美洲山杨的同工酶电泳分析结果(Cheliak and Pitel，1984b)，并结合等电聚焦结果来综合分析毛白杨各酶系统的同工酶谱带，由此推导出其中 8 种酶系统，共具 12 个基因位点，如图 1 所示 IDH 和 MDH 酶谱因表现十分复杂，因此只能记录其不同酶谱表现类型，而不能提出直接描述遗传控制基因位点模式，电泳结果可以分别观察到 N 个酶谱类型，如图 2 所示。这与 Cheliak 对美洲山杨 MDH 系统观察到的结果是类似的(Cheliak and Pitel，1984b)。

图 2 IDH 和 MDH 的电泳谱带类型

2.2 多无性系同工酶基因型鉴别

根据基因位点标记和编码原则，将 86 株毛白杨优树、古树 10 个酶系统的遗传组成列于表 2。从表 2 最后一列可以看出，利用这 10 个酶系统分析结果，可将 86 株毛白杨归属于 79 个同工酶基因型，其中 0085 和 2201，3604 和 4143，3708、4235 和 4337，3806、4104 和 4301，3911 和 4420 等 12 株优树分别属于基因型 7，49，50，51 和 55 中，假设每株优树为一个确定的无性系的话，那么，在 86 株优树中仍有 7 株重叠在 5 个无性系中。可见，无性系的重叠率为 8.14%，而平均重叠率为 1.63%。

用同工酶作为基因标记来鉴别众多的无性系基因型，一方面要求所选的同工酶遗传稳定，另一方面则应选择较多合适的酶系统。因为可鉴别出无性系基因型的多少，取决于酶基因位点数和每个位点上等位基因的数目。在二倍体植物中，如果每个基因都有同等机会表达，并无基因连锁的话，根据计算基因型数目的通式(Cheliak and Pitel，1984b)，本文的材料能确定基因位点的 8 种酶系统的最多基因型数则多达 136048896，这对于鉴别 86 个毛白杨无性系来说是足够了。也表明我们区分的无性系是可靠的，全国毛白杨档案库中所收集保存的优树重复的材料很少，具有广泛的毛白杨种群代表性，收集的质量是很高的。等位基因最多的 ACO 和 PER，最多基因型数目为 36，等位基因最少的 NDH，最多只有 3 种基因型，即它们分别最多只能区分 36 和 3 个无性系。迄今为止，有许多研究者利用同工酶技术区分无性系。Cheliak 等曾用 8 种酶系统 13 个位点区分了 13 个美洲山杨无性系(Cheliak and Pitel，1984b)；Bergmann 用 5 种酶系统把 20 个欧洲山杨栽培无性系归属于 19 个无性系(Bergmann，1987)；Castillo 等用两种酶系统鉴别了两个欧美杨杂种无性系(Castillo and Padro，1987)；Rajora 用多酶系统鉴别辽杨和美洲杨的多个无性系(Rajora，1988)。国内在此方面也做了大量工作，由于条件所限，选用的酶系统仅一两种，能区分的无性系数目也较少(王明庥等，1983；杜宗岳，1983)。众多的实验结果都表明，只有选用较多的酶系统，才有可能把多个无性系区分开来；而选用酶系统较少的话，也只能区分较少的无性系。所以，在用同工酶鉴别无性系时，一定要根据所分析的个体数量来选用较多适合的酶系统。

表 2　86 株毛白杨优树的同工酶遗传组成

优树号	ACO		GOT		LAP	NDH	PER		PGM	6-PODH		SKDH	IDH	MDH	基因型编号
	A	B	A	B	A	A	A	B	A	A	B	A			
0001	22	22	11	12	22	22	11	22	22	22	22	13	1	1	1
0010	22	23	12	11	22	22	12	22	22	12	23	23	1	2	2
0012	12	22	12	12	22	22	11	22	22	22	22	12	3	2	3
0022	13	22	12	12	23	22	11	22	22	22	22	12	3	2	4
0034	12	22	12	12	22	22	11	23	23	22	23	12	3	1	5
0079	12	22	22	11	22	22	11	22	22	22	22	12	3	5	6
0085	12	22	11	12	22	22	22	22	22	12	22	12	3	2	7
0099	13	12	22	13	23	22	22	22	22	12	12	23	5	2	8
0101	12	22	12	12	23	22	11	23	22	22	22	12	3	4	9
1002	22	22	11	12	22	22	22	22	22	22	23	13	1	2	10
1005	12	22	12	12	33	22	11	22	22	22	12	12	3	2	11
1006	12	23	22	11	12	22	22	23	22	12	23	12	3	2	12
1201	12	22	11	12	22	22	22	22	22	22	22	12	1	2	13
1206	12	22	12	12	22	22	23	22	23	22	23	12	2	2	14
1212	12	22	12	12	22	22	22	23	22	22	22	12	3	2	15
1210	22	22	12	12	22	22	12	22	22	22	23	12	2	2	16
1232	12	22	12	12	22	22	12	22	22	22	22	13	1	2	17
1260	22	22	22	22	22	22	22	22	22	22	22	12	4	2	18
1262	22	22	12	12	22	22	12	23	22	22	23	12	4	2	10
1267	22	22	12	12	22	22	23	22	22	22	22	12	4	2	30
1272	12	22	12	12	22	22	22	22	22	22	22	12	1	6	21
1273	12	23	12	11	22	22	22	22	22	12	23	21	1	2	22
1312	22	22	12	12	22	22	12	22	22	22	22	12	4	3	23
1331	12	23	22	11	23	22	22	22	22	22	23	12	1	2	24
1338	22	22	12	12	22	22	12	23	22	23	22	12	4	2	25
1339	22	22	12	12	22	22	22	22	22	22	22	12	1	2	26
1342	12	23	12	11	23	22	22	22	12	12	23	12	1	2	27
1414	12	22	22	11	22	22	22	12	22	12	23	12	2	2	28
1501	12	23	22	11	22	22	22	22	22	13	23	12	3	2	20

（续）

优树号	ACO		GOT		LAP	NDH	PER		PGM	6-PODH		SKDH	IDH	MDH	基因型编号
	A	B	A	B	A	A	A	B	A	A	B	A			
1516	12	23	22	11	22	22	12	22	22	23	23	12	1	2	20
1701	22	22	12	12	23	22	12	22	22	22	22	12	3	1	31
1204	22	12	12	12	12	12	22	22	22	22	22	12	4	2	33
1805	22	22	12	12	22	22	12	22	23	23	22	12	1	2	33
1804	12	22	12	12	23	22	11	22	22	22	22	12	3	1	34
1910	12	12	18	12	23	22	11	23	22	22	22	12	4	1	35
1916	12	22	22	11	22	22	22	22	22	12	23	12	1	2	30
1928	12	22	12	12	72	22	11	22	22	22	22	22	3	1	37
2070	12	22	22	12	22	22	22	22	22	22	22	12	1	1	38
2104	12	22	12	12	23	22	22	22	22	22	22	12	1	2	30
2159	12	22	11	12	22	22	22	23	22	22	22	12	1	1	40
2172	12	23	22	11	12	22	22	23	22	12	23	12	1	2	41
2201	12	22	11	12	12	22	22	22	22	12	23	12	3	2	7
2608	12	22	22	12	12	22	22	22	22	22	22	12	1	2	42
2519	12	22	12	12	22	22	22	22	22	22	22	12	3	1	43
3008	12	22	11	12	22	22	22	22	22	22	22	12	3	1	44
3201	12	22	11	12	22	22	23	22	22	22	22	11	3	1	45
3302	12	12	11	12	22	22	22	22	22	22	22	12	3	2	46
3310	11	22	22	12	23	22	22	22	22	22	22	12	3	2	47
3526	11	22	11	12	12	22	00	00	22	22	22	12	1	1	48
3604	22	22	11	12	22	22	00	00	22	22	22	12	1	2	40
3708	22	22	12	12	22	22	00	00	22	22	22	12	3	2	50
3806	22	22	11	12	22	22	00	00	22	22	22	12	3	2	51
3901	12	22	12	12	22	22	00	00	22	22	22	12	3	2	52
3908	22	22	12	22	22	22	00	00	22	22	22	12	3	2	53
3909	22	22	22	12	22	22	00	00	22	22	22	12	1	2	54
3911	22	22	12	12	22	22	00	00	22	22	22	12	1	2	55
4104	22	22	11	12	22	22	00	00	22	22	22	12	3	2	51
4143	22	22	11	12	22	22	00	00	22	22	22	12	1	2	49
4150	12	22	11	12	22	22	00	00	22	32	22	12	1	2	56
4154	22	22	12	12	23	22	00	00	22	22	22	12	1	1	57
4202	22	23	12	12	22	22	00	00	22	22	22	12	1	2	53
4223	12	22	22	12	22	22	00	00	22	22	22	12	1	1	59
4235	22	22	12	12	22	22	00	00	22	22	22	12	3	2	50

（续）

优树号	ACO		GOT		LAP	NDH	PER		PGM	6-PODH		SKDH	IDH	MDH	基因型编号
	A	B	A	B	A	A	A	B	A	A	B	A			
4301	22	22	11	12	22	22	00	00	22	22	22	12	3	2	51
4302	12	23	12	12	22	22	00	00	22	22	22	12	1	2	60
4337	22	22	12	32	22	22	00	00	32	22	22	12	3	2	50
4404	22	22	12	11	22	22	00	00	22	22	22	12	1	1	51
4420	22	22	12	12	22	22	00	00	22	22	22	12	1	2	55
5103	22	23	22	11	22	22	00	00	22	22	22	12	1	2	62
5116	12	22	12	11	22	22	00	00	22	12	22	22	1	1	63
5202	22	23	13	22	12	22	00	00	12	12	22	12	1	1	64
5209	22	23	12	12	12	22	00	00	12	12	22	12	3	2	65
5214	12	23	11	12	12	22	00	00	12	11	22	22	2	2	66
5221	12	22	12	12	22	22	00	00	22	12	22	12	3	2	67
5317	22	22	12	22	12	22	00	00	22	11	22	12	1	2	68
5322	12	22	11	22	22	22	00	00	22	22	22	12	2	1	69
5323	12	22	12	12	22	22	00	00	22	11	22	13	1	2	70
5402	11	22	12	12	22	22	00	00	22	12	22	12	1	1	71
5415	12	23	11	12	22	22	00	00	22	12	22	12	1	1	72
5421	22	23	11	12	22	22	00	00	22	12	22	12	1	1	73
5425	22	22	12	12	22	22	00	00	22	12	22	12	1	2	74
6305	22	23	12	12	22	22	00	00	22	12	22	12	1	1	75
6331	12	22	12	12	22	22	00	00	22	12	22	12	1	2	76
6507	22	22	12	12	22	22	00	00	22	12	22	12	1	1	77
6510	12	23	12	12	23	22	00	00	22	12	22	12	1	2	78
6515	22	21	12	12	23	22	00	00	22	12	22	12	1	2	79

说明：PER-A，B 中的“00”示凝胶损坏了的或未分析的样本。

2.3　最佳酶系统组合选择

为了提高同工酶基因型来鉴别无性系的精确度，选用较多酶系统所得的效果好。但是，在实际工作中，由于实验条件所限，不可能选用太多的酶系统，而是应该选择适当酶系统数目的最佳组合。利用这种最佳酶系统组合，才有较大把握将一大群无性系区分开来。根据表 2 结果，对这 10 种酶系统进行不同数目的组合。不同的酶系统数目组合的无性系重叠情况见表 3。

表 3　各种酶系统组合的无性系重叠率变化

酶系统数	minOO	maxCO	$\overline{OO}$	$\overline{AO}$
1	89.53	97.67	93.49	33.99
2	63.80	95.35	94.19	16.07
3	44.10	91.36	72.70	8.58
4	27.91	86.05	60.48	5.61
5	17.44	77.91	48.47	3.94
6	18.60	61.63	39.46	3.06
7	10.47	43.84	28.84	2.47
8	9.30	34.88	22.82	2.19
9	8.14	18.60	12.79	1.82
10	—	—	8.14	1.63

从表 3 可以看出，在鉴别无性系时选用的酶系统数多少与无性系重叠率成反比关系，选用的酶系统数越多，无性系重叠率就越小，一种酶系统的 *CO* 为 93.49%。而 10 种酶系统的 *CO* 为 8.14%。从最小无性系重叠率来看，在选用 5 种以上酶系统时，最小 *CO* 值均小于 16.8%，在选用 3 种以上酶系统时，*AO* 值均小于 8.50%。说明无性系的重叠分布是比较均匀的，而不是集中重叠在少数几个无性系中。

根据各种酶系统组合的无性系重叠率值比较，将最小和最大 *CO* 值的酶系统组合列于表 4 中。

从表 4 中可看出，当选用一种酶系统时，最小 *CO* 值的酶系统是 ACO，GOT 或 PER；最大 CO 值的酶系统是 NDH，这表明选用一种酶系统时，ACO，GOT，PER 效果较好，而 NDH 的效果最差。如果选用最小 *CO* 值小于 20%的 5 种酶系统组合时，最小 *CO* 值的组合为 ACO，GOT，PER，6-PGDH 和 IDH，即为最佳组合，*CO* 为 17.44%；而最差的组合为 LAP，NDH，PGM，SKDH 和 IDH(MDH)，*CO* 为 77.91%。

在选择合适酶系统数目及其最佳组合上，林木同工酶分析中未见报道。但是在实际工作中，这是一个重要问题。因为在树木界，不同的树种的生理生化遗传特性不尽相同，各种酶的含量及表达也就会有很大的差异，所以，在具体分析某一个树种时，应该根据所分析的样本的多少、酶位点及等位基因的数目，来确定选用多少种酶系统的最佳组合，使之获得较满意的结果。

3　结论

(1)在分析测定毛白杨的 16 种酶系统中，有 10 种酶系统遗传表达稳定，其中 ACO，GOT，LAP，NDH，PER，PGM，6-PGDH 和 SKDH 8 种酶系统共受 12 个酶位点控制，IDH 和 MDH 两种酶系统难于确定位点，各具有 5 种酶谱表现类型。

(2)利用这 10 种酶系统的遗传组成可以把 86 株毛白杨优树和古树归属于 79 个无性系，有 7 株优树重复在 5 个无性系之中，无性系重叠率为 8.14%，平均重叠率 1.63%。表明全国毛白杨优树档案库所收集的基因资源具有充分的毛白杨种群代表性和可靠性。

(3)对毛白杨进行同工酶分析，以芽和组织培养愈伤组织作试样为佳，缓冲液 pH 值对酶谱显色效果有一定影响。

(4)用同工酶技术来鉴别树木无性系时，应该根据分析的具体树种、样本多少、酶位点及等位基因的数目，来综合确定选用适合的酶系统数目及其最佳的酶系统组合。

Studying Isozymes Gene Markers in Clone of *Populus tomentosa* Carr.

Abstract Isozymes inheritance and variation were studied in Chinese white poplar (*Populus tomentosa* Carr.) by means of electrophoretic technique. The material tested was sampled from clonal archive of *Populus tomentosa* in Guanxian County, Shandong Province. All plus trees and old trees in the clonal archive were selected and collected from main distribution zone of *Populus tomentosa* in China. Isozymes from six type organ tissues and separated by horizontal starch-gel eletrophoresis. The enzyme systems were genetically stable including eight systems with 12 loci and two systems with 2×5 isozyme patterns. 86 individual plus trees and old trees were classified under 79 distict clones using sets of combined genetic make-up of the ten enzyme systems. Clone overlap rate(CO) was 8. 14% and average clone overlap rate (AO) was 1. 63%. It was demonstrated that the rich gene resources were conserved in the clonal archive. The ideal numbers and perfect combination of isozyme systems during the analysis were also discussed.

Keywords *Populus tomentosa*, isozymes, gene markers, identification of clones

Segregation of AFLP Markers in A(*Populus tomentosa*× *P. bolleana*)×*P. tomentosa* Carr. BC_1 Family*

Abstract To investigate the levels of polymorphisms and Mendelian segregation ratio in clone "TB01" (*P. tomentosa*× *P. bolleana*)×clone "LM50" (*P. tomentosa*) BC_1 population at the entire genome level, amplified fragment length polymorphisms (AFLPs) analysis was conducted for both parents and 120 progenies. Forty one pairs of selective primers were used to detect 2707 bands, of which 712 (26.4%) were polymorphic. Chisquare tests were performed to examine if the observed genotypic frequencies of AFLP loci deviated from expected 1:1 Mendelian segregation ratio ($P< 0.01$) in BC1 population. Among the 712 loci 571(80.2%) fit to Mendelian 1:1 segregation ratio, corresponding to DNA polymorphisms heterozygous in one parent and a null in the other. The result shows that the AFLP markers are very suitable for finger printing and genetic mapping in the Chinese white poplar (*Populus tomentosa* Carr.).

Keywords AFLP, polymorphic loci, segregation ratio, Chinese white poplar (*Populus tomentosa* Carr.)

1 Introduction

In the past decade, various methods have been developed for the analysis of forest tree genomes at DNA level. Genome studies are enabling scientist to gain valuable insights into how the forest tree genomes are organized and are also providing a multitude of practical applications, such as variety identification by DNA fingerprinting, genetic mapping and mapping of quantitative traits loci (QTLs) which facilitates indirect selection of important economic traits such as growth and disease resistance without cumbersome screening. The four principal molecular marker techniques for identifying polymorphic loci until now have been: Restriction Fragment Length Polymorphisms (RFLPs), Random Amplified Polymorphic DNAs (RAPDs), Simple Sequence Repeats (SSRs) and Amplified Fragment Length Polymorphisms (AFLPs). Three of the four methods (RFLPs, RAPDs, and SSRs) have several inherent disadvantages. The RFLPs is a laborious technique that relies on southern blotting and detects fewer alleles, while the RAPDs is sensitive to reaction conditions and thus known to has problems of reproducibility. The SSRs provides high polymorphisms, but requires lengthy studies involving cloning and sequencing. The AFLPs overcomes

* 本文原载 *Forestry Studies in China*, 2002, 4(2): 21-26, 与张德强、杨凯和田林合作发表。

some disadvantages of these earlier techniques because the technique combines both RFLP and Polymerase Chain Reaction (PCR) techniques. This method has provided strong power of discrimination and high reproducibility. An ideal genotyping method should produce the results that are invariable from laboratory to laboratory and allows unambiguous comparative analyses and establishment of reliable databases. An increasing number of reports described the use of AFLPs analysis for forest tree genetic mapping (Margues et al., 1998; Cato et al., 1999; Remington et al., 1999; Cervera et al., 2001), medical diagnostics (Savelkoul et al., 1999) and phylogenetic studies (Naohiko et al., 1999). Although there are potentially many techniques to choose from, the choice becomes more restricted when few prior molecular studies have been carried out, as is the case for the Chinese white poplar (*Populus tomentosa* Carr.).

The genus *Populus* is composed of more than 30 species classified into five (Cervera et al., 1996) or six sections (Bradshaw et al., 2000). Poplar is one of the most beneficially and intensively studied forest tree because of its importance as a fiber or biofuel source and a model forest tree (Wu et al., 2000). The Chinese white poplar, belonging to the section *Leuce* Duby, is a member of the genus *Populus*. It is a native fast growth timber species in China and it has been cultivated more than the other poplar species by reason of its fast growth, excellent wood quality and resistance to diseases and insect pests. It is mainly distributed in the vast area of northern China and covers about one million km^2. It has played a key role in timber production and ecological environment protection along the Yellow River.

We report here an analysis of 1:1 Mendelian segregation ratio in (*P. tomentosa* × *P. bolleana*) × *P. tomentosa* population based on 120 progenies and AFLP marker technique. The analysis would facilitate the construction of an AFLP genetic map based on this segregating population.

2 Materials and methods

2.1 Plant materials

An interspecific hybrid family was founded in 2000 by crossing the female clone "TB01" (*P. tomentosa* × *P. bolleana*) from Beijing City and the male clone "LM50" (*P. tomentosa* Carr.) from Shandong Province. The generated BC1 progeny and its parents were maintained at the nursery of Beijing Forestry University, of which 120 individuals, randomly selected, were used for AFLP segregation analysis.

2.2 DNA extraction

The total genomic DNA was extracted from young leaves of BC_1 individuals and parents with methods described by Murray and Thompson (1980).

2.3 AFLP procedures

The AFLP protocol was first developed by Vos (1995) as a means of amplifying a random array of restriction fragments. The methods of Vos (1995). have been modified in this study. Total genomic DNA (0.40μg) from the parents and each BC1 individual was digested and ligased with 3.0 U *Eco*RI, 3.0 U *Mse* I and 1.5U T4 DNA ligase in 20μL reaction mixtures containing T4 ligase buffer (New England Biolab), 5 pmol *Eco*RI adapter, 50pmol *Mse*I adapter and the restriction-ligation reaction incubated at 37℃ for 4h. Pre-amplification reaction was performed with 4μL of template DNA (1∶6 solution diluted from the restriction-ligation mixture), using a pair of primers based on the sequences of the *Eco*RI and the *Mse* I adapters. Four microliters of a 1∶20 diluted pre-amplification DNA were selectively amplified using each of the *Eco*RI and *Mse*I primers with three selective nucleotides at the 3′end. All PCR reactions were performed using a 9600 Perkin Elmer thermo-cycler.

2.4 Gel electrophoresis

The selective amplification reaction products were mixed with 0.5×volume of loading buffer (98% formamide, 10mM EDTA, 0.05% bromphenol blue, and xylene cyanol), denatured and loaded on 6% denaturing polyacrylamide gel (7.5M urea) and electrophoresed at constant power (95W) in 1×TBE buffer (50mM Tris, 50mM boric acid, 1mM EDTA, pH 8.0). The PCR products were visualized using the silver-nitrate staining method as described by Tixier (1997).

2.5 Segregation analysis

Chisquare tests were performed to examine whether the observed genotypic frequencies of AFLP loci deviated from expected 1∶1 Mendelian segregation ratio ($P<0.01$). Segregation patterns in both TB01 and LuMao-50 were analyzed based on the 120 progeny.

3 Results

3.1 Primers screening

The number and sequence of selective nucleotides in system of EcoRI/Mse I primers are important factors for producing amplified DNA fragments in each AFLP reaction. A total of 200 AFLP primer combinations with the selective bases at the 3′end of each of the primers were adopted to determine the levels of polymorphism in both TB01 and LM50. Of the 200 primer pairs tested, 41 primer combinations could generate clearly visual and high polymorphic loci between the parents. The 2707 bands were detected using selected 41 primer combinations, of them 712 bands were polymorphic. The number of bands varied among different pairs of selective primers (Table 1). The combinations of EAAG/MACA gave the smallest number of bands (40), but all other primer combinations detected > 44 bands. The average number of bands per primer pair was 66, while the number of polymorphic bands and levels of polymorphisms were 17.4 and 26.4%, respectively (Table 1). This indicates that the AFLP technique could be used as a powerful tool for analysis of polymorphic loci in the Chinese white poplar.

Table 1　Segregation of AFLP markers in the clone "TB01"×clone "LM50" BC_1 family

Primer combination *Eco RI*+3/*Mse*1+3	Number of detectable fregments	Numher of polymorphic bands	Levels of polymorphisms	Size nmge of polymorphic baudy/tp	Number of 1 ratio	Nunber of skered segregation
EAAG/MAAG	50	17	34.0	68.396	15	2
EAAG/MACA	40	10	25.0	150.705	8	2
EAAG/MACT	78	23	29.5	59.518	20	3
EAAG/MAGC	48	13	27.1	51.364	13	0
EAAG/MAGG	53	21	39.6	48.702	12	9
EAAG/MAGT	70	19	27.1	56.517	17	2
EAAG/MATC	68	17	25.0	48.660	13	4
EAAG/MCAA	74	15	20.3	46.432	15	0
EAAG/MCAC	68	12	17.6	220.684	10	2
EAAG/MCGT	44	12	27.3	97.344	11	1
EAAG/MCTT	66	11	16.7	174.502	5	6
EAAG/MGAG	72	17	23.6	40.567	9	8
EAAG/MGAT	58	12	33.7	85.330	11	1
EAAG/MTAA	70	16	22.9	21.755	12	4
EAAG/MTCT	'50	8	16.0	185.555	7	1
EAAT/MATC	110	26	23.6	79.601	22	4
EAAT/MCAA	68	12	17.6	100.542	9	3
EAAT/MCAC	60	7	11.7	67.492	5	2
EACV/MCCC	68	17	25.0	57.487	13	4
EATC/MAGC	72	16	22.2	73.324	13	3
EATC/MATC	66	15	22.7	109.693	14	1
EATC/MATT	80	26	32.5	113.658	18	8
EATC/MCAT	74	18	24.3	91.650	12	6
EATC/MCTC	76	18	23.7	59.644	16	2
ECTC/MAAG	53	20	37.7	46.473	18	2
ECTC/MAAT	56	15	26.8	58.477	13	2
ECTC/MAAA	65	19	29.2	70.649	15	4
ECTC/MACG	68	19	27.9	52.473	15	4
EGAN/MAAC	70	17	24.3	99.396	14	3
ECAN/MAOC	66	18	27.3	66.317	18	0
ECAN/MAGA	76	27	35.5	49.436	17	10
ECAN/MAGT	74	23	31.1	52.440	16	7
ECAN/MATA	62	16	25.8	98.673	15	1
ECAN/MATT	68	20	29.4	58.432	18	2
ECAN/MCCC	60	21	35.0	71.961	14	7
ECAN/MCTC	64	26	40.6	60.336	18	8

(续)

Primer combination Eco RI+3/Mse1+3	Number of detectable fregments	Numher of polymorphic bands	Levels of polymorphisms	Size nmge of polymorphic baudy/tp	Number of 1 ratio	Nunber of skered segregation
ECAN/MGCA	67	17	25.4	64.650	16	1
ECAN/MAAA	73	23	31.5	40.642	20	3
ECAN/MAAT	70	21	30.0	74.827	19	2
ECAN/MAGA	60	13	21.7	178.454	11	2
Total	2707	712	1081.3	40.961	571	141
Aronge	66.0	17.4	26.4	85547.1	13.9	3.4

3.2 Segregation of AFLP markers in progeny

With modified AFLP reaction conditions, the size of the AFLP markers detected in these experiments ranged from approximately 40bp to 961 bp in this BC_1 progeny. But the bands longer than 800bp were rarely detected. Forty one pairs of AFLP primers generated 712 polymorphic fragments, among which 571 (80.2%) were segregated in a 1∶1 ratio in 120 individuals (Table 1; Fig. 1). Approximately 19.8% of the polymorphic bands displayed skewed segregation ratio. Although the number of testcross bands (e.g. 1∶1 ratio) varied among primer combinations, no linear relationship was seen between the numbers of testcross bands and selective nucleotides used in a particular primer pair.

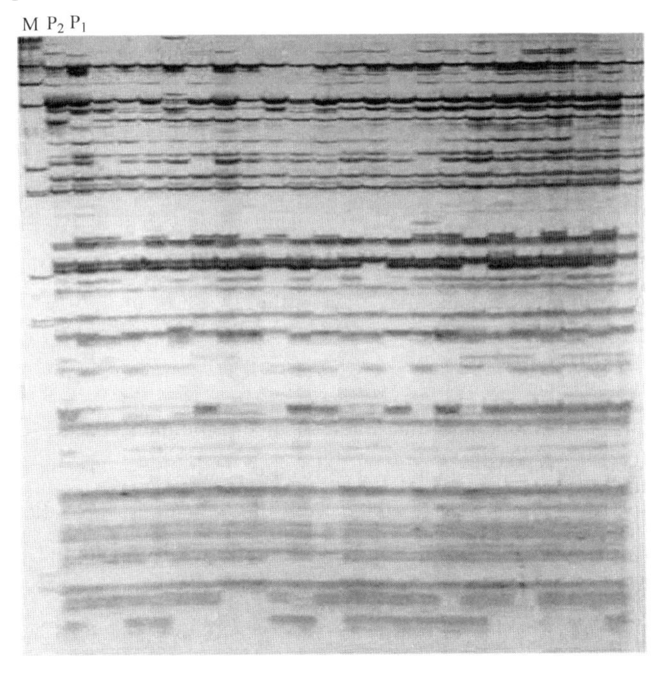

Fig. 1 Segregation of AFLP markers in the BC_1 population.

M is kb DNA ladder (M), P_2 is clone "LM50" [*P. tomentosa*], P_1 is clone "TB01" [*P. tomentosa* ×*P. bolleana*] and the other lanes are the BC_1 progeny

4 Discussion

4.1 Quality of AFLP markers

The AFLP technique is robust and reliable for detecting polymorphic loci in poplar. The method belongs to the category of selective restriction fragment amplification techniques, which are based on the ligation of adapters to genomic restriction fragments followed by a forcible PCR-based amplification with adapter-specific primers. The AFLP analysis is relatively insensitive to the amount of template DNA (≥1pg) because it has two-step PCR amplification and its results are very steady at the stringent amplified condition owing to high temperature of primer annealing. The *Eco* RI/*Mse* I primer combinations have been used to analyse levels of polymorphisms in the study. Different primer pairs which span the average-frequency restriction site can scan the entire genome. A highly informative pattern of 7 to 27 polymorphic bands is obtained, which makes it a powerful tool for the analysis of backcross population despite its high demand in skills for technical complexity. The AFLP analysis detects more polymorphic loci in virtue of mutations in the restriction sites, mutations in the sequence adjacent to the restriction sites and complementary to the selective primer extensions, and insertions or deletions within the amplified fragments than RAPDs or RFLPs in each PCR reaction. We obtain an average of 13. 9 markers in 1∶1 segregation ratio per primer pairs, approximately 5 times more than the number of RAPD markers with each pr imer combination reported for a different interspecific cross population of poplar (Yin et al. , 1999). Many studies have reported that forest trees contain many simple repeat sequences (Cato et al. , 1996; Dayanandan et al. , 1998; Brondani et al. , 1998). For example, Dayanandan et al. reported that the TC/AG SSRs were the most abundant in *Populus* tremuloides genome. In this study, we find that there are more informative AFLP markers than others when the primer combinations contain AG/GA, AT/TA, TC/CT selective bases. The results suggest that the (AG) *n*/(GA) *n*, (AT) *n*/(TA) *n*, (TC) *n*/(CT) *n* are the most frequent simple repetitive sequences in *Populus* tomentosa. Additionally, different species may have different SSRs in their genomes.

4.2 Segregation distortion

Segregation distortion of molecular markers has been commonly observed in crops (Bert et al. , 1999) , forest trees (Bradshaw et al. , 1994) and fruit trees (Lu et al. , 1998). In this study, about 19. 8% of AFLP markers deviated from the expected segregation ratio of 1∶1 in the *P. tomentosa*. Similar or lower proportions of skewed markers are presented in other populations, such as 15% in *Eucalyptus globules*×*E. tereticornis* (Marques et al. , 1998) , 15% in *P. trichocarpa*×*P. deltoids* (Bradshaw et al. , 1994) , 18% in *Quercus robur* (Barreneche et al. , 1998). Forest trees are out crossing organisms typically characterized by high genetic load. Bradshaw and Stettler found that a lethal allele in the F_2 population of *P. trichocarpa*×*P. deltoids* affecting embryonic development was the cause of deviated markers linked to it (Bradshaw et al. , 1994).

Cervera et al. reported that markers cosegregating with the *Melampsora laricipopulina* resistance gene showed a significant deviation in the *P. deltoids* because of death of susceptible trees (Cervera et al., 2001). The reasons for the segregation distortion of markers in plant are not well understood, but are believed to be related to some factors such as chromosome loss (Kuang et al., 1999), genetic isolating mechanisms (Zamir and Tadmor, 1986), presence of an allele for pollen lethality (Bradshaw and stettler, 1994), and linked lethal genes that are directly selected at either the gametic level or the zygotic level in the process of development (Hedrick and Muona, 1990). For this study, an even simpler explanation for the segregation distortion may be inbreeding depression, scoring errors or small sampling size. Further studies with larger populations and more molecular markers would be helpful to investigate the potential reason responsible for segregation distortion of molecular markers in the Chinese white poplar.

In conclusion, the AFLP technique is believed suitable for detecting marker segregation in poplar, and the results of the test cross segregation suggest that AFLP markers could be used for construction of the genetic linkage maps based on this segregating population.

Genetic Mapping in (*Populus tomentosa*×*P. bolleana*) and *P. tomentosa* Carr. Using AFLP Markers*

Abstract The AFLP genetic linkage maps for two poplar cultivars were constructed with the pseudo-test-cross mapping strategy. The hybrids were derived from an interspecific backcross between the female hybrid clone "TB01" (*Populus tomentosa*×*P. bolleana*) and the male clone 'LM50' (*P. tomentosa*). A total of 782 polymorphic fragments were obtained with a PCR-based strategy using 49 enzyme-nested (*Eco*RI/*Mse*I) primer combinations. 632 of these fragments segregated in a 1∶1 ratio ($P < 0.01$), indicating that these DNA polymorphisms are heterozygous in one parent and null in the other. The linkage analysis was performed using mapmaker ersion 3.0 with LOD 5.0 and a maximum recombination fraction (θ) of 3.0. Map distances were estimated using the Kosambi mapping function. In the framework map for "LM50" (*P. tomentosa*), 218 markers were aligned in 19 major linkage groups. The linked loci spanned along approximately 2683cM of the poplar genome, with an average distance of 12.3cM between adjacent markers. For 'TB01' (*P. tomentosa*×*P. bolleana*), the analysis revealed 144 loci, which were mapped to 19 major linkage groups and covered about 1956cM, with an average distance of 13.6cm between adjacent markers. These maps covered about 87% and 77% of the estimated genome size of parents "LM50" and "TB01", respectively. The maps developed in this study lay an important foundation for future genomics research in poplar, providing a means for localizing genes controlling economically important traits in *P. tomentosa*.

Key words Genetic linkage map, AFLP marker, *Populus tomentosa* Carr.

1 Introduction

Genetic linkage maps based on molecular markers now provide a powerful tool for detecting loci controlling a number of traits and for studying genome organization and evolution in many forest trees species (Bradshaw and Stettler, 1995; Grattapaglia et al., 1995; Devey et al., 1999; Marques et al., 1999; Sewell et al., 2002). These results in turn are the basis for map-based gene cloning and for marker-assisted selection of important traits related to faster growth, better stress adaptation or better disease resistance. The advent of robust molecular-marker systems greatly facilitates the construction of genetic linkage maps in various tree species. The four principal types of molecular markers used in mapping include: Restriction Fragment Length Polymorphisms (RFLPs), Random Amplified Polymorphic DNAs (RAPDs), Simple Sequence Repeats

* 本文原载 *Theoretical and Applied Genetics*, 2004, 108: 657-662, 与张德强、杨凯和李百炼合作发表。

(SSRs) and Amplified Fragment Length Polymorphism (AFLPs). AFLP markers are generated through a PCR-based approach that does not require either prior knowledge of sequence information or probe preparation, both of which are required for generating RFLP and SSR markers. AFLPs are more reliable and for its good reproducible compared to RAPDs, and being dominant markers, supply as much information as co-dominant markers for analysis of backcross or pseudo-test-cross populations (Staub and Serquen, 1996). Currently, the AFLP marker system has largely been used in tree genetic mapping. (Marques et al., 1998; Remington et al., 1999; Cervera et al., 2001).

The genus *Populus* is comprised of more than 30 species classified into five or six sections. Poplar is one of the most economically and intensively studied forest tree species due to its importance as a timber source and a model forest tree. The Chinese white poplar (*Populus tomentosa* Carr.), belonging to the section *Populus*, is a member of the genus *Populus*. *P. tomentosa* is a fast growing timber species native to China, with an excellent wood quality and outstanding resistance to many diseases and insects. Consequently, the Chinese have cultivated it more than any other poplar species, mainly in northern China, on an area of approximately one million km^2. However, it is difficult to generate a large segregating population of intraspecific hybrids for genetic mapping in *P. tomentosa* because its overall ability to sexually reproduce is poor, and particular individuals are often sterile. To deal with this problem, an inter-specific backcross breeding approach strategy (e.g. the hybrid *P. tomentosa*×*P. bolleana* is utilized as maternal parent and backcrossed with *P. tomentosa*) was therefore proposed by professor Zhu Zhiti (Zhu and Zhang, 1997). This strategy was then used to easily create a large segregating population by crossing (*P. tomentosa*×*P. bolleana*) with *P. tomentosa*.

Various genetic maps have been developed for genomic research in poplar to date. The first poplar genetic map was developed with RFLP and allozyme markers (Liu and Furnier, 1993). Much more detailed and complete linkage maps have also been established by combining hundreds of RAPD, RFLP, STS (sequence-tagged site), SSR and AFLP markers (Bradshaw et al., 1994; Wu et al., 2000; Cervera et al., 2001; Yin et al., 2001). However, diverse genetic maps based on distinct populations are still needed for special objectives such as QTL detection and map based cloning.

In this paper, we report an application of AFLP markers in a two-way pseudo-testcross mapping strategy (Grattapaglia et al., 1994), to map the genome of an elite clone of 'LM50' (*P. tomentosa*) and its hybrid clone 'TB01' (*P. tomentosa*×*P. bolleana*) from progeny data. This work could be exploited as a potential reference map for QTL analysis of *P. tomentosa* for important traits.

2 Materials and methods

2.1 The mapping pedigree

The inbred lines of *Populus* (696 offspring) used in this study were derived from the inter-specific backcross TB01×LM50 [TB01(*P. tomentosa* clone 3082×*P. bolleana*) and LM50(*P. tomentosa* clone 3075); *P. tomentosa* clones 3082 and 3075 belong to the same family]. One hundred and twenty progenies were randomly selected and used for genetic mapping.

2.2 AFLP procedures

The total genomic DNA was extracted from frozen young leaves of 120 inter-specific backcross hybrids and their parents by the methods described by Murray and Thompson (1980). The AFLP analysis was performed essentially as described by Vos et al. (1995). In the present study, this method was modified by Zhang et al. (2003). Products of the selective amplification reaction were detected as discrete bands on a polyacrylamide gel using electrophoresis and the silver staining method as described by Tixier et al. (1997).

2.3 Heterozygosity analysis

The average heterozygosity was defined as the ratio of bands segregating in the offspring to total number of bands observed (Cervera et al., 2001). The average heterozygosity was estimated by analyzing the 120 individuals and their parents, using 30 different AFLP primer combinations.

2.4 Data analysis and map construction

Segregation of the markers was scored only for clear, unambiguous bands having electrophoretic migration patterns that obviously represented polymorphisms between the parents. For each marker, Chi-square tests (d. f. = 1, $P<0.01$) were conducted to check deviations from expected 1∶1 Mendelian ratio. All markers deviating at the 1% significance level were excluded for the linkage analysis. Heterozygous AFLP markers present in one parent but not in the other were used to construct separate genetic linkage maps for the male clone 'LM50' parents and the female clone 'TB01' parents using the two-way pseudo-testcross strategy (Grattapaglia and Sedroff, 1994). To detect linkages in repulsion phase, the data set was duplicated and added to the original data. Both parents' genetic linkage maps were constructed using MAPMAKER version 3.0 (Lincoln et al., 1992). Markers were first grouped using a minimum LOD score of 5.0 and maximum recombination frequency (θ) of 0.30. For each linkage group, markers were ordered by using a minimum LOD score of 3.0 and a maximum θ of 0.40 using the First-Order command. The ordered marker sequences were confirmed using the Ripple command. Markers ordered with low confidence were placed again using the Try command. New markers were placed at appropriate positions of the maps with the Place command. Linkage maps were generated with the Map command. Possible errors or double crossovers were checked with the Genotype command before map construction. Map distances in centiMorgans were computed using Kosambi' s mapping function.

2.5 Genome length estimation

The recombination length of the *Populus* genome was estimated from partial linkage data using the equation

$$G_e = \sqrt{N(N-1)X/K} \quad (1)$$

with a confidence interval given by equation

$$G_e = (1 \pm 1.96/\sqrt{K}) \quad (2)$$

where *Ge* is the estimated genome length, *N* is the number of framework markers, *X* is the maximum map distance between two adjacent framework markers in centiMorgans at a certain minimum LOD score of 3.0, and *K* the number of marker pairs at the same minimum LOD score (Hulbert et al., 1988).

3 Results

3.1 Polymorphisms

The system of *Eco*RI/*Mse*I primers consisting of pre-amplification with E00/M00 and selective-amplification with E+3/M+3 was utilized for detection of DNA markers in poplar, which has a relatively small genome size (approx. 550Mbp). With modified reaction conditions, a total of 200AFLP primer combinations with the selective bases at the 3′-end of each primer were tested to determine the quality of the AFLP fingerprinting and the levels of polymorphism. Of 200 primer pairs tested, 49 were effective in revealing visually clear polymorphic loci between "LM50" and "TB01". Altogether, 3100 bands were detected using the selected 49 primer combinations, of which 782(25.2%) bands were polymorphic, with sizes ranging from approximately 40 to 961bp (note: bands longer than 800 bp were rarely detected, about 0.4%). The number of bands varied among the different pairs of selective primers (Fig. 1 and Fig. 2). The combination of EAAG/MACA and EAAG/MACC gave the smallest number of bands (40); while more than 42 bands were generally detected with other primer combinations. The average number of bands per primer pair was 63 and the number of polymorphic bands was 16 with a range of 4-27 (Fig. 1 and Fig. 2).

3.2 Segregation distortion

In our study, segregation distortion is defined as the deviation from expected Mendelian 1∶1 segregation ratio(Cervera et al., 2001). Forty-nine pairs of AFLP primers generated 782 markers, with 510(65%) heterozygous for the male "LM50" and 272(35%) heterozygous for the female "TB01" (Fig. 1 and Fig. 2). 150(19%) of the 782 AFLP markers displayed a skewed null

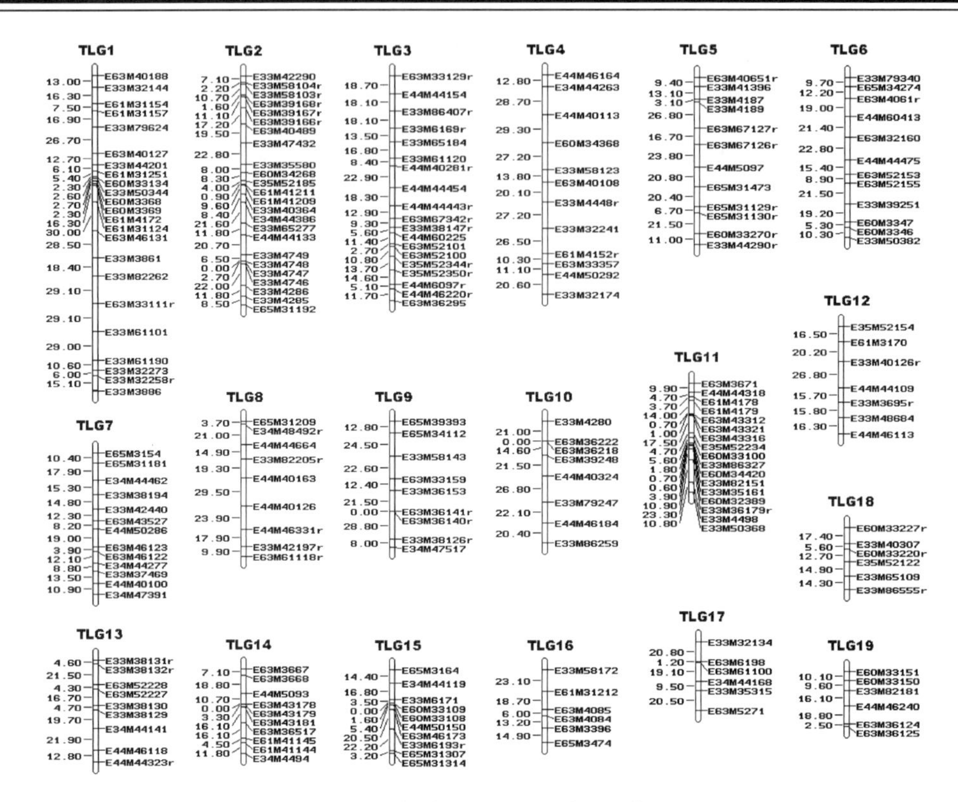

Fig. 1 AFLP genetic linkage map of *Populus tomentosa*

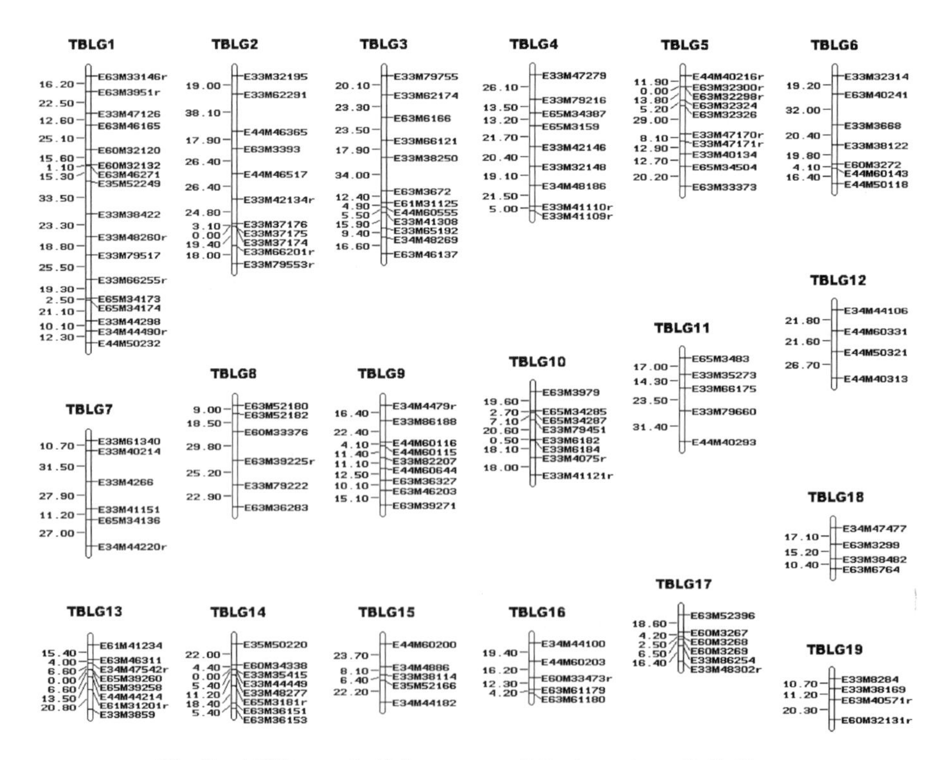

Fig. 2 AFLP genetic linkage map of *P. tomentosa*×*P. bolleana*

hypothesis of the Mendelian 1∶1 segregation ratio based on 120 progeny in the mapping population ($P < 0.01$). of these 150 markers, 114 (22%) and 36 (13%) markers were derived from "LM50" and "TB01", respectively. These deviated markers were excluded from the subsequent linkage analysis in order to avoid false linkage. As a result, 632 test-cross markers were employed for the construction of the clone "LM50" and clone "TB01" linkage maps.

3.3 Heterozygosity levels

The estimated average heterozygosity of clone "LM50" and clone "TB01" was 18.71% and 8.47%, respectively, based on the 120 progeny and 30 AFLP primer combinations.

3.4 AFLP linkage maps

The linkage analysis was based on 632 testcross AFLP markers with 396 in the male "LM50" and 236 in the female "TB01". For "LM50", 25 linkage groups, 12 triplets, 23 doublets and 74 unlinked markers were obtained at a LOD score of 5.0 and $\theta = 0.30$ using the MAPMAKER 3.0 linkage program. Nineteen of the twenty-five linkage groups were classified as "major" groups (> 50cM), which constituted the framework map for "LM50". The male map spanned a total length of 2682.8cM, and included 218 markers, with an average interval of 12.3cM between adjacent marker loci. Linkage groups ranged in length from 57.1cM (TLG19) to 326.5cM (TLG1) while the number of markers mapped in each linkage group varied from 6 (TLG16, TLG17, TLG18 and TLG19) to 24 (TLG2).

The same criteria were used to establish the linkage map of the maternal tree, "TB01". The 236 markers were assigned to 19 major groups (>40cM), including 1 triplet, 20 doublets, with 49 unlinked markers. These 19 groups in the female map covered 1956.3cM, and 144 markers were located on the map, with an average distance between marker loci of 13.6cm. The length of the groups ranged between 42.2cM (TBLG19) and 274.8cM (TBLG1) and the number of markers located in each linkage group varied from 4(TBLG12, TLG18 and TLG19) to 17 (TBLG1).

3.5 Estimated genome length

Using Equations (1) and (2), the estimated genome lengths of "LM50" and "TB01" were Ge = 3097cM, with a 95% confidence interval of 2854~3385cM, and Ge=2552cM with a 95% confidence interval of 2284~2891cM, respectively. The observed genome length based on framework maps was 2683cM for "LM50", and 1956cM for "TB01", respectively. These results showed that the linked markers used to construct our genetic maps would provide 87% coverage of the estimated genome length for "LM50" and 77% for "TB01".

4 Discussion

4.1 Characteristics of AFLPs

Unlike *P. deltoides*, *P. trichocarpa* or *P. nigra* that have been intensively studied on the molecular level, very few studies has been carried out for *P. tomentosa* on the DNA level. In this study, the AFLP marker system was successfully used in genetic mapping for *P. tomentosa*. AFLP is a PCR-based method that offers an efficient and reliable means of generating the DNA markers needed for linkage map construction (Vos et al., 1995). Small amounts of high quality template DNA are sufficient for the analysis because it involves a two-step PCR reaction with high annealing temperature. With enough primer pairs, spanning the average-frequency restriction sites, the entire genome could potentially be covered at the DNA level. In our study, forty-nine pairs of AFLP primers were used to generate 782 polymorphic fragments, among which 632 segregated in a Mendelian 1∶1 ratio, corresponding to DNA polymorphisms heterozygous in one parent and a null in the other. A highly informative pattern of 4 to 27 polymorphic bands per primer pair was obtained, providing a convenient and reliable tool for the construction of genetic maps based on an interspecific backcross population. As compared to RFLP and RAPD analyses, the labor required in detecting polymorphisms with AFLPs is considerably reduced. On average, each primer pair could produce 13 test-cross markers, approximately 5 times more than that obtained using the RAPD approach in a different interspecific cross population in section *Populus* (Yin et al., 2001). AFLPs are dominant markers but the reduced information content of dominant markers is compensated for by the use of a backcross or testcross design; many marker loci will have only two genotypes segregating among the backcross offspring: heterozygotes (Aa) and recessive homozygotes (aa). We note that the efficiency of the AFLP marker system might sometimes be affected by the constitution of the selective bases at the 3′-end of each primer. In our study, we found that more informative AFLP markers were generated when the primer combinations contained AG/GA or AT/TA.

4.2 Segregation distortion

Segregation distortion of molecular markers has commonly been observed in mapping populations of crops (Bert et al., 1999), forest trees (Bradshaw and Stettler, 1994) and fruit trees (Lu et al., 1998). In our study, 22% of AFLP markers in *P. tomentosa* and 13% in *P. tomentosa* ×*P. bolleana*, respectively, were found to deviate from the expected 1∶1 Mendelian segregation ratio. Similar or lower distorted segregation ratios were observed in previous efforts to construct genetic linkage maps using inter-or intra-specific crossing populations and AFLP markers: 16.4% in *Lolium perenne* (Bert et al., 1999), 15% in *Eucalyptus globules* and *E. tereticornis* (Marques et al., 1998), 15% in *Prunus persica* (Lu et al., 1998). Forest trees are out crossing organisms typically characterized by high genetic load. It was reported that a recessive lethal allele in the F_2 population of *P. trichocarpa*×*P. deltoides* affecting embryonic development was tightly linked to the deviated markers (Bradshaw and Stettler, 1994). Cervera et al reported that markers co-segregating with the *Melampsora larici-populina* resistance gene showed a significant deviation in *P. deltoids* due to missing data resulted from death of susceptible trees (Cervera et al., 2001). Reasons for skewed segregation ratios of molecular markers are still not well understood but are generally believed to related to genetic factors such as chromosome loss and structural rearrangements (Williams et al., 1995; Kuang et al., 1999), genetic isolating mechanisms (Zamir and Tadmor, 1986), presence of an allele for pollen lethality (Bradshaw and Stettler, 1994), or gametic selection (Zamir et al., 1982), as well as other non-biological factors such as sampling in finite mapping populations, or scoring errors (Plomion et al., 1995). The high level of segregation distortion obtained in our study indicates that *P. tomentosa*×*P. bolleana* and *P. tomentosa* have close affinity.

Because the grandparent *P. tomentosa* 3082 and the backcrossed parent *P. tomentosa* 3075 originated in the same family, as well as the limited sample scale (120 progeny), inbreeding depression and the small population size were possible reasons for segregation distortion in our study. Further studies with additional crosses, larger populations or more molecular markers would be helpful in investigating the potential reason responsible for segregation distortion of molecular markers in Chinese white poplar. Regardless of the cause for skewed segregation ratios, such skewing inevitably leads to increased difficulty in both linkage determination and recombination frequency estimation, thereby eventually affecting the map construction (Whitkus, 1998). Therefore, all markers deviating at the 1% significance level were excluded for the linkage analysis in order to minimize false linkage during map construction in our study.

4.3 Map construction

Grattapaglia and Sederoff (1994) put forward the mapping strategy of pseudo-testcross through testcross configuration in forest trees. Genetic maps have been established for two closely related poplar species, "LM50" and its hybrid "TB01" using a pseudo-testcross strategy and AFLP markers. The genetic maps of "LM50" and "TB01" were comprised of 19 major linkage groups, some minor linkage groups and some unlinked markers, respectively. Ideally, a complete genetic linkage map should contain 19 linkage groups, consistent with the 19 haploid chromosomes in *Populus*. The presence of small groups and unlinked markers in both maps indicates some vacant regions present in the maps constructed in this study. This may in part be due to the absence of neighboring markers, innate limits of using a single marker technique, or the relatively small size of the mapping population. Therefore, additional efforts will be required to map RFLPs, SSRs, ESTs, allozymes and other valuable markers in a larger mapping population to coalesce some of these linkage groups together. Furthermore, a method of bulk segregant analysis (Michelmore et al., 1991) could be used to bridge the breaches existing in these maps by screening more co-segregated markers associating with the traits in order to help coalesce the small groups with a larger linkage group.

Analysis of distributions of informative markers in our maps indicated that certain primer combinations produced more informative polymorphic markers than others for mapping. This was the case for E33M38, E63M36, E60M33 and E34M44, although many of the markers generated by E33M38 mapped to the same linkage group TLG13. Apparently, both marker screening experiments and considerable effort in searching for suitable primer combinations are needed for detection of AFLPs on the entire genome. Markers on most linkage groups are randomly distributed in these maps, with obvious gaps (≥30cm) in linkage groups of TLG1, TBLG1, TBLG2, TBLG3, TBLG6, TBLG7 or TBLG11, which could be eliminated by increasing the number of markers.

To our knowledge, this is the first report on genetic mapping in *P. tomentosa*. As compared to most previous genetic maps constructed for poplar, the population size sampled by us was larger. The genetic map of *P. tomentosa*, i. e. "LM50", was the better covered and saturated map. Many co-dominant markers such as SSR and EST have been developed for *P. trichocarpa* and *P. tremuloides*, unlike for *P. tomentosa*. Therefore, it is currently difficult to construct a genetic map of *P. tomentosa* with these co-dominant markers due to species' specialty. However, in the future these AFLP "scaffold" maps will be saturated with SSR and EST markers that are specific to *P. tomentosa*, resulting in a much more comprehensive genetic map. Such a map will provide a basis for adding other genetic markers, and ultimately the cloning major genes and QTLs controlling economically important traits in *P. tomentosa*.

毛白杨 ISSR 反应体系的建立及优化*

摘　要　为应用 ISSR 技术开展毛白杨遗传变异分析、辅助育种、品种鉴定、系统进化等研究，以$(GTG)_6$为引物，通过单因子实验分别研究了退火温度、引物浓度、dNTP 浓度、Mg^{2+}浓度、Taq DNA 聚合酶用量对毛白杨 ISSR-PCR 反应的影响，建立并优化了适宜于毛白杨 ISSR 分析的扩增体系：20μL PCR 反应体系，1×Taq DNA 酶缓冲液(10mmol/L Tris-HCl，50mmol/L KCl，0.1% Trion X-100，pH 9.0)，1.5mmol/L $MgCl_2$，1.0UTaq 酶，模板 DNA，0.2μmol/L 引物，各 0.2mmol/L 的 dNTP。引物$(GTG)_6$的最适退火温度为 61℃。

关键词　毛白杨，ISSR，PCR，条件优化

简单序列重复区间扩增多态性(inter-simple sequence repeat，ISSR)是由 Zietkiewics 等(Zietkiewics et al.，1994)提出的一种锚定 SSR 的新策略，是建立在 PCR 反应基础上的一种新型分子标记技术。ISSR 技术是利用真核生物基因组中广泛存在的简单重复序列(SSR)来设计引物，而不要求预知基因组序列信息，又可以揭示比 RFLP、RAPD、SSR 更多的多态性，结果比 RAPD 更加可靠(Hantula et al，1996；Charters et al，1996；钱韦等，2000)。与 AFLP 相比较，ISSR 技术则对模板 DNA 需用量少，实验流程简短，操作简单，成本低等优点(钱韦等，2000；Mcgregor et al，2000)。因此，近年来 ISSR 技术已被广泛应用于品种鉴定、遗传作图、基因定位、遗传多样性、进化、系统发育、分子标记辅助育种等研究方面(Arnau et al，2002；邱英雄等，2002；Cho et al，2002；Galvan et al，2003)，如采用 RAPD 和 ISSR 标记对中国疣粒野生稻(*Oryza granulata*)遗传多样性的探讨(钱韦等，2000)、应用 ISSR 标记对豇豆属(*Vigna*)种间关系的分析(Ajibade et al，2000)、利用 ISSR 技术对鹰嘴豆(*Cicer arietinum*)和杨梅(*Myrica rubra*)不同品种进行分析(Chowdhury et al，2002；邱英雄等，2002)，以及瓜类遗传图谱的构建(Yael et al，2002)等，但至今少见 ISSR 技术在毛白杨相关研究中应用的报道。

ISSR 技术虽然具有重复性高，原理简单的优点，但 PCR 条件不同，会出现不同的扩增结果。为了确保 ISSR 分析结果的可靠性和可重复性，本实验以毛白杨无性系所提取的总 DNA 为材料，比较研究了影响 ISSR-PCR 反应的多个因素，建立了毛白杨 ISSR-PCR 最佳反应体系，为 ISSR 技术应用于毛白杨的遗传变异分析、辅助育种、无性系鉴定、系统进化等研究奠定了良好的基础。

* 本文原载《北京林业大学学报》，2006，28(3)：61-65，与何承忠、安新民、杨凯和张有慧合作发表。

1 材料与方法

1.1 材料

试验所用的 8 个毛白杨无性系材料(3-50-2, 3000-2, 3-16-17, 3424, 3-53-1, 4115, 4140, 5226)采自山东省国营冠县苗圃全国毛白杨基因库。

1.2 毛白杨无性系总 DNA 的提取与检测

用 SDS 法提取毛白杨无性系总 DNA，利用分光光度计测定所提总 DNA 的含量和纯度，同时，采用琼脂糖凝胶电泳法检测提取的总 DNA 是否降解。

1.3 ISSR 反应条件及程序

以毛白杨无性系 3-50-2 为试验材料，根据预备实验的结果，以$(GTG)_6$为引物，通过单因子试验研究了引物浓度、退火温度、dNTP 浓度、Mg^{2+}浓度、Taq 酶用量对 ISSR-PCR 反应的影响。得到适合的 PCR 的基本条件是：20μL PCR 反应体系，1×Taq DNA 酶缓冲液(10mmol/L Tris-HCl，50mmol/L KCl，0.1% Trion X-100，pH 9.0)，1.5mmol/L $MgCl_2$，0.2mmol/L dNTP，模板 DNA，0.2μmol/L 引物，1.0U taq DNA 聚合酶。引物$(GTG)_6$的最适退火温度为 61℃。

用 Applied Biosystems 公司生产的 GeneAMP PCR system 9700 型 PCR 仪进行 DNA 扩增，反应程序为 94℃充分变性 5min，94℃变性 45s，61℃退火 1min；72℃延伸 1min 30s，35 个循环；最后 72℃，7 min。其中 Taq DNA 聚合酶和 dNTP 由北京天为时代公司提供，引物由上海生工公司合成。

1.4 检测方法

扩增产物用 2%琼脂糖凝胶电泳，缓冲液为 1×TAE，电压为 5 V/cm，0.5μg/mL 溴化乙锭染色，紫外光下自动凝胶成像系统观察拍照。

2 结果与分析

2.1 引物浓度对 ISSR 反应的影响

引物浓度偏高或偏低所得到的 PCR 结果均不可靠，引物浓度偏高会引起错配和非特异性产物扩增，且可增加引物之间形成二聚体的几率，浓度过低则无法测出所有 ISSR 位点(李海生等，2004；余艳等，2003；宣继萍等，2002；张志红等，2004；乔玉山等，2003)。本实验设计引物浓度梯度为：0.05、0.1、0.2、0.3、0.4、0.5、0.6、0.7μmol/L，结果发现，引物浓度为 0.05μmol/L 无扩增产物，引物浓度为 0.1μmol/L 虽然条带清晰，但条带数较少，引物浓度为 0.2~0.4μmol/L 能取得较好的扩增结果。引物浓度为 0.5~0.7μmol/L 条带亮度增加，但产生非特异性带(图 1)。所以毛白杨 ISSR 反应体系最适宜的

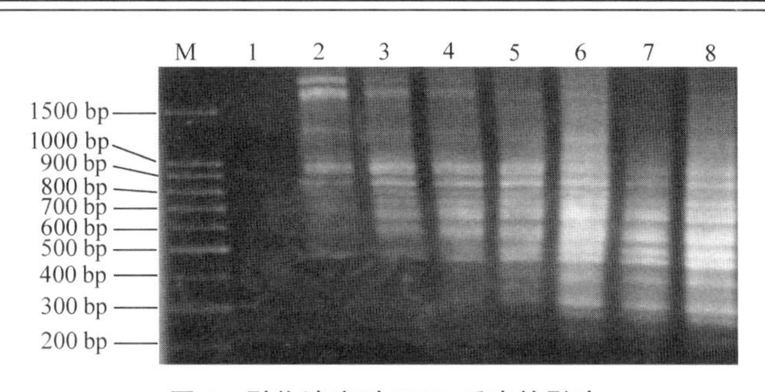

图 1 引物浓度对 ISSR 反应的影响

泳道 1~8：0.05、0.1、0.2、0.3、0.4、0.5、0.6、0.7μmol/L 引物

引物浓度为 0.2μmol/L。

2.2 退火温度对 ISSR 反应的影响

退火温度与 ISSR 指纹的稳定性至关重要。退火温度不同，产生错配的程度也不同，通常较低的温度在保证引物与模板结合稳定性的同时，也会使引物与模板之间未完全配对的一些位点间得到扩增，即产生一定的错误扩增。因此，在允许的范围内，选择较高的退火温度可减少引物和模板之间的非特异性结合，提高 PCR 反应的特异性(Lu et al，1999)。

我们设置了 57℃、58℃、59℃、60℃、61℃、62℃ 6 个退火温度。结果表明，在低于 Tm 值的退火温度下，无扩增产物(图 2a，b)或背景模糊、非特异性条带较多(图 2c，d)。退火温度过高，扩增反应受抑制、产物少或无(图 2f)。因此，毛白杨 ISSR 反应体系中引物$(GTG)_6$最适宜的退火温度为 61℃。

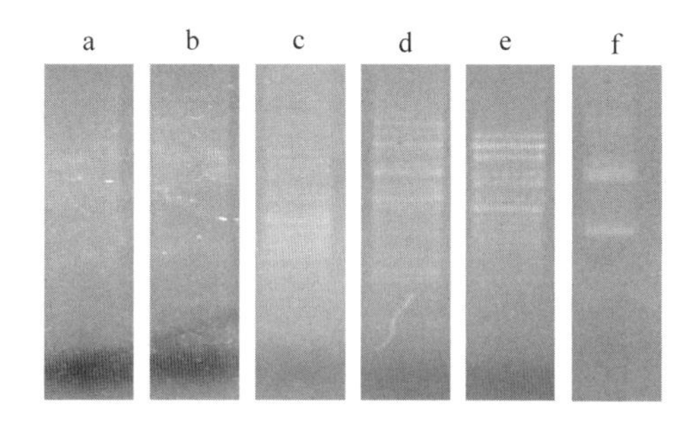

图 2 退火温度对 ISSR 反应的影响

a~f 退火温度分别为：57℃、58℃、59℃、60℃、61℃、62℃

2.3 dNTP 浓度对 ISSR 反应的影响

dNTP 是 ISSR-PCR 反应的原料，dNTP 浓度过高，会导致 PCR 错配，从而使扩增出现非特异性扩增；过低影响合成效率，甚至会因过早的消耗而使产物单链化，影响扩增效果(余艳等，2003)，为了确定最适 dNTP 的用量，我们设置了 0.04、0.12、0.16、0.20、0.24、0.28、0.36、0.44mmol/L 8 个浓度梯度。结果表明，dNTP 浓度为 0.04mmol/L 时无扩增产物出现，dNTP 浓度为 0.12~0.16 mmol/L 时扩增的条带少，dNTP 浓度 0.2~0.28mmol/L 时能取得较好的扩增结果，dNTP 浓度 0.36~0.44mmol/L 时扩增的条带数减

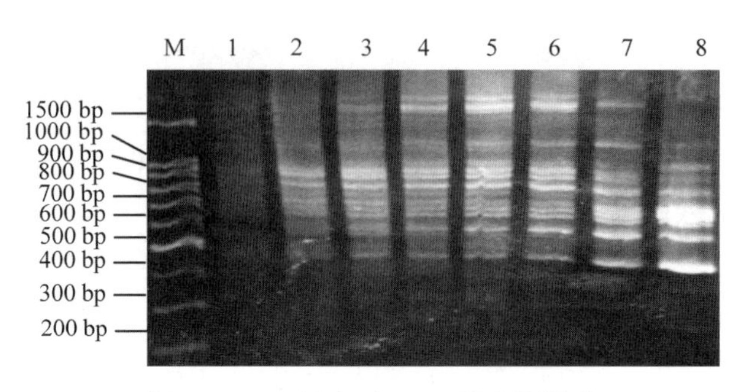

图 3　dNTP 浓度对 ISSR 反应的影响

泳道 1~8：0.04、0.12、0.16、0.20、0.24、0.28、0.36、0.44mmol/L dNTP

少(图 3)。因此，毛白杨 ISSR 反应体系最适宜的 dNTP 浓度为 0.2mmol/L。

2.4　Mg^{2+}浓度对 ISSR 反应的影响

Mg^{2+}浓度是影响 PCR 结果的重要变量之一。Taq DNA 酶是 Mg^{2+}依赖性酶，对 Mg^{2+}浓度非常敏感。引物与模板的双链杂交体的解链与退火温度受二价阳离子的影响，特别是其中的 Mg^{2+}浓度影响反应的特异性和扩增片段的产率(余艳等，2003；乔玉山等，2003)。因此，选择合适的 Mg^{2+}浓度，对 PCR 反应至关重要。

本实验设置了 Mg^{2+} 的浓度梯度比较，即 0.5、1.0、1.5、2.0、2.5、3.0、3.5、4.0mmol/L，结果表明，Mg^{2+} 浓度为 0.5mmol/L 时，无扩增产物出现；Mg^{2+} 浓度为 1.0mmol/L 时扩增条带少，不利于分析，Mg^{2+}浓度为 1.5~2.0mmol/L 时能得到清晰稳定的条带，且无非特异性扩增；Mg^{2+}浓度为 2.5~4.0mmol/L 时虽也能得到扩增条带，但具有非特异性扩增(图 4)。因此，毛白杨 ISSR 反应体系中最适宜的 Mg^{2+}浓度选用 1.5mmol/L，这一结果也与其他树种 ISSR 反应体系中的 Mg^{2+}浓度一致(Amel et al，2004；李海生等，2004；余艳等，2003；宣继萍等，2002；张志红等，2004；乔玉山等，2003)。

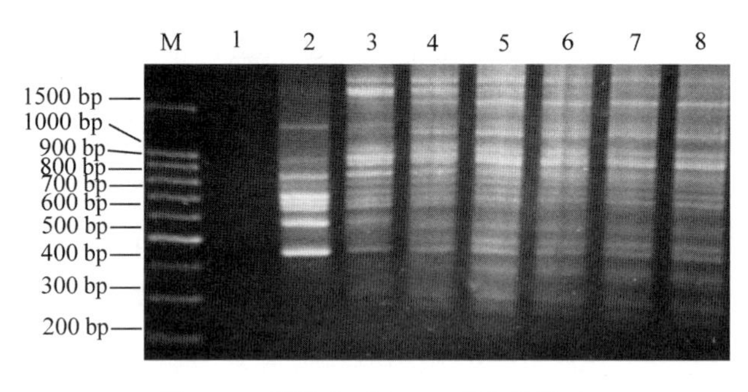

图 4　Mg^{2+}浓度对 ISSR 反应的影响

泳道 1~8：0.5、1.0、1.5、2.0、2.5、3.0、3.5、4.0mmol/L Mg^{2+}

2.5　Taq DNA 聚合酶对 ISSR 反应的影响

Taq DNA 聚合酶使用量直接影响扩增反应的成功与否，使用高浓度的 Taq DNA 聚合酶不仅提高了成本，而且也容易产生非特异扩增产物；而 Taq DNA 聚合酶浓度过低时，则会导致产物的合成效率下降。为找到合适的 Taq DNA 聚合酶单位本实验设置了 Taq DNA 聚合酶浓度梯度：0. 25、0. 5、0. 75、1. 0、1. 25、1. 75、2. 0U。结果表明：Taq DNA 聚合酶单位为 0. 25~0. 5U 时酶量过低，产物的合成效率下降，条带数量较少；Taq DNA 聚合酶单位为 0. 75~1. 0U 时可以得到清晰条带，无非特异扩增；Taq DNA 聚合酶单位为 1. 25~2. 0U 时酶量过大，虽也得到扩增产物，但出现非特异性扩增。因此，在毛白杨无性系 ISSR-PCR 实验中，Taq DNA 聚合酶的使用量为 0. 75~1. 0U 最佳，因为 Taq DNA 聚合酶为 1. 0U 时扩增产物更稳定，所以毛白杨 ISSR 反应体系最适宜的 Taq 酶用量为 1. 0U(图 5)。

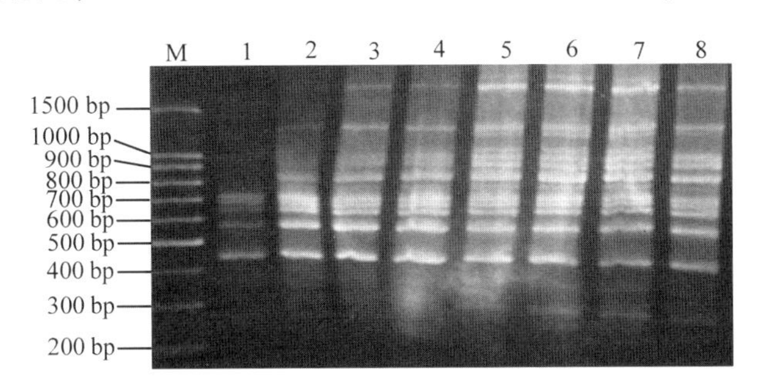

图 5　Taq 酶用量对 ISSR 反应的影响

泳道 1~8：0. 25、0. 5、0. 75、1. 0、1. 25、1. 5、1. 75、2. 0U

2. 6　优化体系

在 ISSR 反应体系中，PCR 扩增结果受模板质量浓度与纯度，Taq DNA 聚合酶用量，Mg^{2+}质量浓度，dNTP 浓度，引物浓度，退火温度，扩增反应程序等影响。为了得到较高的稳定性、可重复性和可靠性的实验结果，我们对影响因子 Taq DNA 酶用量、Mg^{2+}质量浓度、dNTP 浓度、引物浓度以及退火温度进行了梯度试验，筛选出毛白杨最适宜的 ISSR-PCR 反应体系：20μL PCR 反应体系中，1×Taq DNA 酶缓冲液(10mmol/L Tris-HCl，50mmol/L KCl，0. 1% Trion X-100，pH 9. 0)，1. 5mmol/L $MgCl_2$，0. 2mmol/L dNTP，1. 0U Taq DNA 聚合酶，模板 DNA，0. 2μmol/L 引物。利用此优化系统，以$(GTG)_6$为引物，在 61℃的退火温度下，对毛白杨 8 个无性系进行 ISSR-PCR 扩增，得到了清晰、多态性高的 ISSR 谱带，并具有较好的重复性(图 6)。

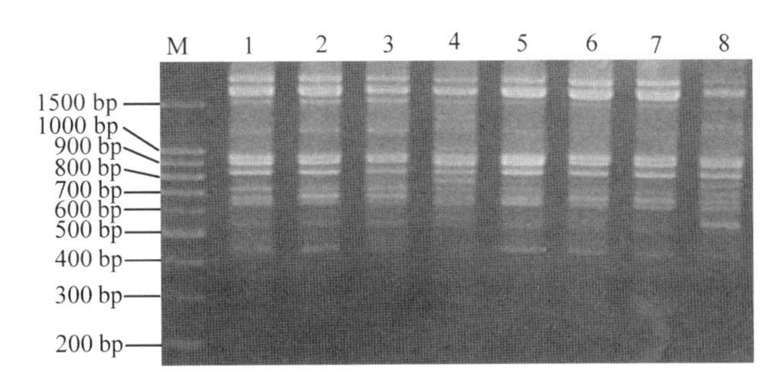

图 6　引物$(GTG)_6$对毛白杨 8 个无性系扩增的 ISSR 带型

3 讨论

ISSR标记技术和RAPD原理比较相似，所不同的是ISSR所用引物序列来源于简单重复序列区域，比RAPD引物序列长，退火温度高。乔玉山等(2003)和Jonsson等(1996)的研究认为ISSR标记比RAPD检测多态性更为灵敏，反应系统更为稳定(乔玉山等，2003；Jonddon et al，1996)。不过退火温度与ISSR指纹的稳定性至关重要。退火温度不同，产生错配的程度也不同。此外，同一引物，不同植物样品所用的退火温度可能大不相同，如余艳等用ISSR技术检测沙冬青(*Ammopiptanthus mongolicus*)时引物880和889的退火温度分别为52℃和50℃，而席嘉宾等在运用ISSR引物检测地毯草(*Axonopus compressus*)样品时发现这两个引物的最佳退火温度分别是62℃和59℃(余艳等，2003)。

与微卫星(SSR)相比，ISSR不要求已知基因组序列信息，大大减少多态分析的前期工作，引物具有广泛的通用性。尽管ISSR是一种显性遗传标记，不能检测显性纯合基因型和杂合基因型，但作为种或品种鉴定的分子标记，并不受其影响。沈永宝等(2005)在研究中发现ISSR标记能更好揭示品种间的差异，仅用2个ISSR引物就能区分13个银杏品种，鉴定效率和稳定性也明显优于RAPD标记。

在ISSR反应体系中，PCR扩增结果除了受退火温度影响外还受到很多其他因素的影响，如模板质量浓度与纯度，Taq DNA聚合酶用量，Mg^{2+}质量浓度，dNTP浓度，引物浓度，扩增反应程序等。为了得到较高的稳定性、可重复性和可靠性的实验结果，我们对影响因子Taq DNA酶用量、Mg^{2+}质量浓度、dNTP浓度、引物浓度进行了梯度试验，用筛选出的毛白杨最适宜的ISSR-PCR反应体系，以$(GTG)_6$为引物，对毛白杨8个无性系进行ISSR-PCR扩增，得到清晰、多态性高的ISSR谱带，并具有较好的重复性。当然，影响毛白杨ISSR-PCR反应产生良好清晰带型的因素很多，在具体的实验中，不同的系统构建和优化需要灵活调节，只有这样，才能够得到分析需求的谱带，得出可靠结果。

Establishment and optimization of ISSR reaction system for *Populus tomentosa* Carr.

Abstract In order to use ISSR marker to study genetic variation, assistant breeding, cultivar identification and phylogenesis of *Populus tomentosa*, the annealing temperature, concentration of primer, dNTP, Mg^{2+} and Taq DNA polymerase dosage on ISSR-PCR amplication were tested to determine their optimal levels and establish the following optimal reaction system for ISSR analysis in *Populus tomentosa*. The results showed that the optimal conditions for ISSR-PCR of *P. tomentosa* were as follows: PCR reaction volume of 20μL, 1×Taq buffer (10mmol/L Tris-HCl, 50mmol/L KCl, 0.1% Triton X-100, pH 9.0), 1.0U Taq DNA Polymerase, 0.2μmol/L primer, 0.2mmol/L dNTP, 1.5mmol/L $MgCl_2$ and 10 ng template DNA. The optimized annealing temperature was 61℃ for primer $(GTG)_6$.

Key words *Populus tomentosa*, ISSR, PCR, optimized conditions

Identification of AFLP Markers Associated with Mbryonic Root (radicle) Development Trait in *Populus tomentosa*[*]

Abstract Embryonic root (radicle) development in the mature embryo after germination is essential for the formation of the root organ in plants. In this study, seed radicle development and its association with Amplified Fragment Length Polymorphisms (AFLPs) markers were investigated based on a segregated progeny population. The progeny population was generated by intraspecific-controlled crossing between a female *Populus tomentosa* clone "5082" and a male *P. tomentosa* clone "JY". A total of 3193 seeds were obtained and sown on 1/2 Murashige and Skoog medium *in vitro*. The measurements were made at 15 to 20 days after germination. The rate of germination is 48.74% for incubated seeds. Visual inspection of the germinated seeds showed that 1179 seedlings had a normal radicle and 376 lacked a root organ. The segregation ratio of rooting versus non-rooting of seed embryos was observed to be 3:1 in *P. tomentosa*. This segregation ratio suggested that seed radicle development character is a qualitative trait and is probably controlled by a single complete dominant gene or a set of tightly linked genes. In order to identify molecular markers associated with radicle development-controlling loci, we adopted the Bulked Segregant Analysis (BSA) methodology with AFLP markers. Approximately 5600 selectively amplified DNA fragments, ranging in size from 40 to 650 nucleotides, were obtained by screening pools of rooting and non-rooting seed embryos with 78 AFLP primer combinations. Two AFLP markers (*Eco*R I $_{+GAG}$/*Mse* I $_{+AAT}$-492 and *Eco*R I $_{+GAG}$/*Mse* I $_{+CCA}$-502) were identified that were tightly linked to the radicle development-controlling locus in *P. tomentosa*. These AFLP markers can be very useful in current breeding programs and are also a firm basis for marker assisted selection (MAS) and cloning of the radicle development-controlling gene in Chinese white poplar.

Keywords Bulked segregation analysis, *Populus tomentosa* Carr., qualitative trait, dominant gene

1 Introduction

Populus tomentosa is a native species in the section *Leuce* under *Populus* in China. It is broadly distributed in the vast area of northern China and has a very long planting history (Zhu and Zhang, 1997). Many wild and cultivable *P. tomentosa* with informative genetic variation types emerged during long natural course of evolution. *P. tomentosa* has many favorable characteristics such as rapid growth, outstanding resistance to disease, and excellent wood quality (Zhang et al., 2004). It has played a key role in forest production and ecological environmental protec-

---

* 本文原载 *Silvae Genetica*, 2007, 56(1): 27-32, 与张德强和杨凯合作发表。

tion along the Yellow River. However, gamete fertility is low and incidence of sterility is high (Zhang et al., 1992). This limits research on trait segregation in *P. tomentosa* progeny. Consequently, over the past several years poplar breeders focused on vegetative propagation, which is currently the normal commercial practice. Various strategies for improving the rooting ability of cuttings were studied (Zhu and Zhang, 1986).

Previous hybrid experiments involved interspecific crossing, with all the obtained seeds then directly sown on the soil in order to observe germination (Zhang et al., 1999). As a result, few researchers were able to observe the phenomenon of seed embryonic root (radicle) development in *P. tomentosa*. In an effort to obtain clones with higher gamete fertility, Zhang et al., explored the flowering and bearing habits of clones in the arboretum of *P. tomentosa* located in Guanxian county, Shandong province. Several female and male clones with a higher ability to successfully reproduce sexually were found. The female clone "5082" was found to be highly suitable for hybrid breeding of *P. tomentosa* (Zhang et al., 1992).

Radicle development has become the subject°f considerable attention in recent years because it plays a crucial role in the life cycle of flowering plants. In most angiosperm plant species the primary root and its shoot meristems are established during embryogenesis (Natesh and Rau, 1984; Scheres et al., 1994; Laux and Jurgens, 1997; Vernoux et al., 2000). Upon germination the primary root and shoot meristems initiate post-embryonic development by producing the root organ. In order to understand the developmental mechanisms that regulate root formation, previous researchers made great efforts to study root organogenesis in *Arabidopsis thaliana* with mutant, histological or anatomical analysis strategies. Mutation analysis was mainly used for identifying and characterizing major regulatory genes; a number of genes controlling meristem and root organ identities[h]ave recently been discovered and characterized using this approach (Mayer et al., 1991; Celenza et al., 1995; Cheng et al., 1995; Kubo et al., 1999). A few genes are also known to be required for lateral root development. For example, the ALF4 locus is required for lateral root primordial initiation (Celenza et al., 1995), while the root-meristemless (Cheng et al., 1995), root meristemless1/cadmium sensitive 2 (Vernoux et al., 2000) and hobbit (Willemsen et al., 1998) genes are essential for root meristem formation in the *Arabidopsis* embryo. The cellular organization of the primary root in *Arabidopsis* is affected by a homebox gene, anthocyaninless2, and indicates that this homeobox gene is involved in both the accumulation of anthocyanin and in root development (Kubo et al., 1999). Cnops and co-workers described the genetic characterization of nine *trn* mutations that constitute two complementation groups and concluded that *TRN*1 and *TRN*2 genes are required for the maintenance of both the radial pattern of tissue differentiation in the root and for the subsequent circumferential pattern within the epidermis (Cnops et al., 2000). Surgical experiments on developing roots and shoot meristems has lead to a partial understanding of the mechanism of the patterning of cells within organs and of the patterning of lateral organs derived from the shoot meristem (Scheres et al., 1994; Van den Berg et al. 1995; Berger et al. 1998). For example, the embryonic origin of the *Arabidopsis* root and hypocotyls region has been investigated using histological techniques and the results showed that cell division in the embryo resulted in the various initials within the root promeristem (Scheres et al., 1994). The development of the post-embryonic root epidermis has been researched and three sets of initials that give rise to columella root cap cells, epidermis and lateral root-cap cells, and the cells of the cortex and endodermis were identified in *Arabidopsis thaliana* (Dolan et al., 1994). The organization and differentiation in lateral root have been described in *Arabidopsis thaliana* and these studies revealed that organization and cell differentiation in the lateral root primordial precede the visible initiation of the lateral root meristem (Malamy and Benfey, 1997). The relation between cell division and cell differentiation in *Arabidopsis* root meristems has been studied using a combination of laser ablation and anatomical techniques and the results indicated that pattern formation in the root meristems is controlled by a balance between short-range signals inhibiting differentiation and signals that reinforce cell fate decisions (Van den Berg et al., 1997). Furthermore, an auxin response element displays an important role in *Arabidopsis* root development. Sabatini et al. reported that an auxin maximum at a vascular boundary established a distal organizer in the root (Sabatini et al., 1999).

Unfortunately, the analytical techniques used in radicle development research in *Arabidopsis* are not well suited for identification of the loci controlling radical development in forest trees. Associations between molecular markers and economically important traits were first reported in the 1923 (Sax, 1923). The development of the Bulked Segregant Analysis (BSA) method (Michelmore et al., 1991) allowed tree breeders to combine molecular markers with the BSA strategy to reliably detect markers linked to economically important traits in forest trees (Cervera et al., 1996; Grattapaglia et al., 1996; Villar et al., 1996; Wilcox et al., 1996; Harkins et al., 1998; Kondo et al., 2000; Stirling et al., 2001). These research results showed that BSA is an efficient method for detecting molecular markers in specific genomic regions by using segregant populations.

In this paper, we describe the Mendelian segregation of a seed radicle development trait in a *P. tomentosa* intraspecific hybrid progeny and use the BSA method to identify the AFLP markers linked to the gene(s) controlling radicle development. For the study, a controlled hybrid experiment was carried out by intraspecifically crossing the "elite" female *P. tomentosa* clone "5082" with a *P. tomentosa* male clone "JY", sowing the resulting seed on 1/2 Murashige and Skoog medium *in vitro*, and monitoring the germination and radicle development of the seedlings. For germinated seedlings, the segregation ratio for normal vs. non-visible root organ fit to a Mendelian ratio. This segregation ratio implies that the seed radicle development characteristic in *P. tomentosa* is a qualitative trait and probably is controlled by either a single dominant gene or by several closely linked loci. BSA identified two AFLP markers (*Eco*R I_{+GAG}/*Mse* I_{+AAT}-492 and *Eco*R I_{+GAG}/*Mse* I_{+CCA}-502) that were tightly linked to the radicle development-controlling locus in *P. tomentosa*.

2 Materials and methods

2.1 Plant materials and measurements

In 2001, a controlled intraspecific cross was made between a female clone "5082" (*P. tomentosa*) (collected in the arboretum of *P. tomentosa* at Shandong province, China) and a male *P. tomentosa* Carr. var. *truncata* Y. C. Fu et C. H. Wang, var. nov clone (collected in the Shanxin province, China), which gave rise to 3193 seeds.

Seeds of *P. tomentosa* were sterilized in 0.1% hydrargyrum chloride for 30 s and then with 75% ethanol for 2min. After five washes in sterile distilled water, seeds were germinated on flask containing 0.5×Murashige and Skoog (MS) salt mixture, pH 5.8, in 0.8% agar. Flasks were incubated in a near vertical position at 22℃, 70% humidity and a cycle of 16 hr light/8 hr dark.

Measurements of phenotype were made at 15 to 20 days after germination. The rate of germination was 48.74%. We obtained a total of 1555 seedlings with 376 showing the novel phenotype (lacked a visible root). Chi-square tests (d. f. = 1, $P < 0.05$) were conducted to check the Mendelian ratio of normal rooting seedlings versus aberrant (lacking a visible root) rooting seedlings. The normal rooting seedlings were then sown on soil while the seedlings lacking a visible root were fixed in a solution of ethanol and acetic acid (75% ethanol and 25% acetic acid).

2.2 Marker assessments

Total genomic DNA was isolated from young leaves using the CTAB method (Hoisington et al., 1994). The template DNA concentration was estimated by comparing the fluorescence intensities of ethidium bromide-stained samples to those of λ-DNA standards on 0.8% agarose gel.

The AFLP protocol developed by Vos et al. (1995) for amplifying the selective restriction-ligation fragments was followed with minor modifications. Total genomic DNA (0.40μg) from each individual was digested and ligated with 3.0U *Eco*RI, 3.0U *Mse* I and 1.5 U T_4 DNA ligase in 20μL reaction mixtures containing T_4 ligase buffer (New England Biolab), 5pmol *Eco*RI adapter and 50pmol *Mse*I adapter. The restriction-ligation reaction was incubated at 37℃ overnight.

The pre-amplification reaction was performed with 4μl of template DNA (1:10 solution diluted from the restriction-ligation mixture), using a pair of primers (E00/M00) with nucleotide sequence complementary to those on the *Eco*RI and *Mse*I adapters. The reaction mixtures (20μl) consisted of 0.6 U *Taq* polymerase (Promega), 30ng E00 primer, 30 ng M00 primer, 1× PCR buffer (Promega) and 0.2m*M* each of all four dNTPs (Promega). PCR amplification consisted of 30 cycles of a 30-s denaturation at 94℃, 30-s annealing at 56℃ and a 60-s extension at 72℃ with 5-min final extension at 72℃. Four microliters of a 1:20 dilution of the final pre-amplification DNA mixture was then selectively amplified using 40 ng each of different *Eco*RI/*Mse*I primer combinations with three selective nucleotides at the 3′ end in 20μl reaction mixtures containing 0.8 U *Taq* polymerase (Biostar), 1×PCR buffer (Promega) and 0.2mM each of all four mM dNTPs (Promega) using the PCR-cycle profile described by Vos et al. (1995) with 5-min final extension at 72℃. All PCR reactions were performed using a 9600 Perkin Elmer thermo-cycler.

The selective amplification reaction products were mixed with 0.5×volume of loading buffer (98% formamide, 10mM EDTA, 0.05% bromphenol blue, and xylene cyanol), denatured at 95℃ for 10min, loaded on a 6% denaturing polyacrylamide gel (7.5 M urea) and electrophoresed at constant power (95W) in 1×TBE buffer (50mM Tris, 50mM boric acid, 1mM EDTA, pH 8.0). Products of the selective amplification reaction were detected using the silver staining method as described by Tixier et al. (1997).

2.3 Bulked segregant analysis

Bulk segregant analysis (BSA) method for detecting AFLP markers associated with loci controlling seed radicle development was performed as described by Michelmore et al., (1991). The two bulks were made by mixing equal amounts of pre-amplified DNA products from 15 normal rooting (bulk 1) and 15 non-rooting (bulk 2) seedlings, respectively. The primer combinations generating polymorphic loci between the two bulks were subsequently used to detect each individual DNA in the F_1 progeny.

Linkage analysis was performed using the formulas $r = n\text{Col}/(n\text{Col} + n\text{Ler})$ described by Mather (1938) and the genetic distances were calculated using the Kosambi mapping function with centiMorgans (cM) (Kosambi, 1944).

3 Results

3.1 Phenotypic and genetic analysis of radicle development

In 2000, different crossing combinations were designed and performed in order to establish mapping populations for construction of genetic maps in *P. tomentosa*. We adventitiously found some seedlings with non-visible root organ in the progeny derived from the intraspecific crossing between the female clone "5082" and male clone "JY" (Fig. 1). As shown in Table 1, 354 seeds were obtained in 2000. After germination, we obtained 39 seedlings with non-rooting phenotype, which accounted for 26.35% of the total seedlings. However, seedlings without a visible root organ were not obtained in progeny derived from other two interspecific crossing combinations (clone "5082"×clone "MXTB01" and clone "MXTB01"×clone "JY"). To test the reproducibility of obtaining seedlings with non-rooting phenotype in the progeny of the crossing combination between clone "5082" and clone "JY", the controlled hybrid experiments were carried out again in 2001 on a much larger scale using the same crossing materials as in 2000. As expected, the apparent phenotypic segregation for a radicle development trait appeared in the 2001 progeny. A total of 3193 seeds were obtained and the rate of germination was 48.74% (1555 seedlings) (Table 1). 1179 seedlings showed normal radicle development and 376 seedlings lacked a visible root organ. The segregation ratio of normal rooting to non-rooting seedlings was 3∶1 in *Populus tomentosa* and chi^2 = 0. 557 <$P_{0.05}$($P_{0.05}$ = 3. 84). This segregation ratio suggested that seed radicle development character is a qualitative trait and the mutation is recessive to wild type. Segregation analysis of the F_1 progeny of these seedlings indicated that a single dominant genetic locus or a set of closely linked loci was responsible for post-embryonic radicle development in *P. tomentosa*. This result implied that the genotype controlling radicle development loci is Aa in both parents (clone "5082" and clone "JY") and produced segregation ratios of 1AA : 2Aa : 1aa of genotype in the F_1 progeny. This genotype resulted in the segregation ration of 3 : 1 seedlings of normal root versus without visible root for phenotype in the F_1 progeny in *P. tomentosa*.

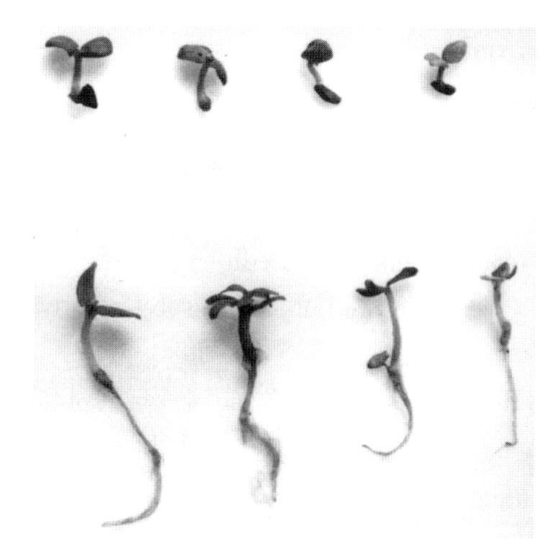

Fig. 1　Seedling appearance after seed germination in *P. tomentosa* Carr.

(Top: seedling with non-visible root organ; Bottom: seedling with normal root organ).

Table 1　Trait segregation in F_1 population derived from different crossing combinations.

Cross combinations	No. of seeds	Germination rate (%)	No. of rooting seedlings	No. of non-rooting seedlings
Clone "5082"×clone "JY"a	354	42. 10	109	39
Clone "5082"×clone "JY"b	3193	48. 74	1179	376
clone "5082"×clone "MXTB01^c"a	386	84. 50	326	0
clone "MXTB01^c"×clone "JY"a	3801	85. 20	3238	0

a: in 2000; b: in 2001; c: *P. tomentosa*×*P. bolleana*

3. 2　AFLP linked to radicle development

To expedite the identification of the AFLP markers linked to genes involved in radicle development in the post-embryo for *P. tomentosa*, screening for AFLP-polymorphisms was carried out by bulked segregant analysis (Michelmore, et al., 1991), with bulks (pools) containing equal amounts of preamplified DNA from 15 normal rooting (bulk 1) and 15 lacking visible root organ (bulk 2) seedlings, respectively. A total of 78 AFLP primer pairs were used to test bulks and approximately 5600 selectively amplified DNA fragments ranging in size from 40 to 650 nucleotides were scored. Most primer combinations showed no polymorphic loci between rooting and non-rooting bulks. However, primer pairs E_{65}/M_{34} (*Eco*R I_{+GAG}/*Mse* I_{+AAT}) and E_{65}/M_{51} (*Eco*R I_{+GAG}/*Mse* I_{+CCA}) each showed striking differences between these two bulks (Fig. 2). These two candidate AFLP markers were approximately 492bp and 502bp in size, respectively. DNA samples from 98 individuals were then amplified, scored, and the results used to analyze the linkage between the candidate AFLP markers and the radicle development-controlling allele. We identified one recombinant F_1 individual with primer combination E_{65}/M_{34}, and two for primer pair E_{65}/M_{51}. Both candidate AFLP markers thus appear to be tightly linked to the radicle development-controlling loci in *Populus tomentosa*. The linkage distance between E_{65}/M_{34-492}, E_{65}/M_{51-502} and the radicle development-controlling loci was approximately 1.02 and 2.04cM, respectively.

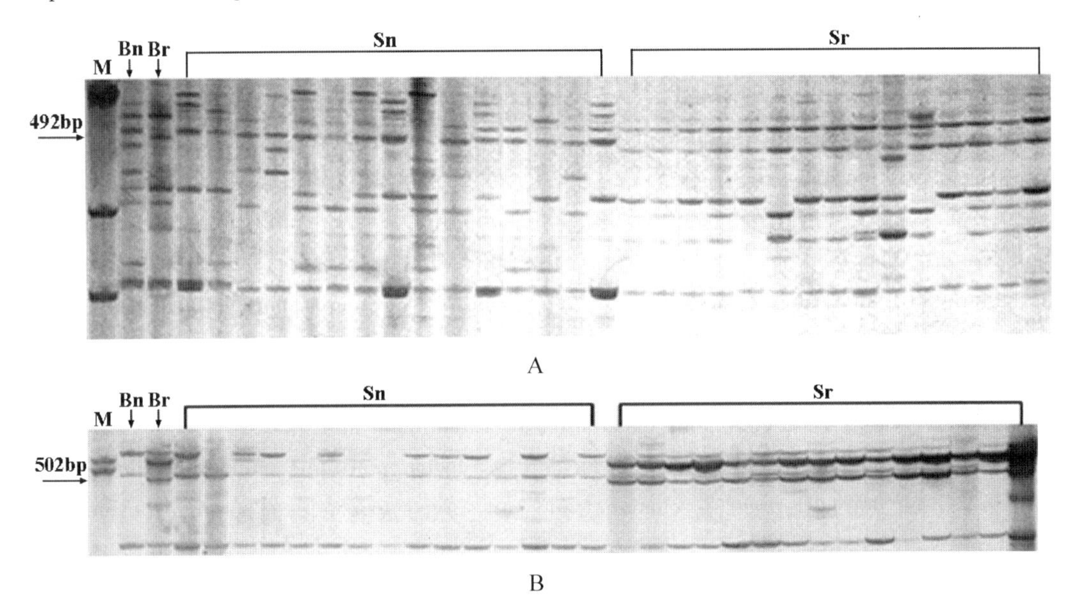

Fig. 2 AFLP markers linked to the loci of radicle development-controlling in *P. tomentosa*

(A represents marker E_{EAG}/M_{AAT}-492; B represents marker E_{EAG}/M_{CCA}-502; M represents 1kb DNA ladder (NEB Biolab), Bn represents non-visible root seedlings bulks, Br represents normal rooting seedlings bulks, Sr represents the identified markers by BSA that are present in the normal rooting seedlings, Sn represents the identified markers by BSA that are absent in the non-visible rooting seedlings.)

4 Discussion

Morphosis is species specific, indicating that it is under strong genetic control. Thus, a better understanding of the genetic basis of organ formation should eventually allow us to modify economically important traits in plants in a controlled manner. The constant activation of meristems results in the formation of new organs in plants. Previous studies have demonstrated that the primary root and its corresponding meristems are laid down during embryogenesis in angiosperms (Natesh and Rau, 1984; Scheres et al., 1994; Laux and Jurgens, 1997; Vernoux et al., 2000). They committed to post-embryonic development upon germination by producing visible root organ (lateral root and adventitious root). In our study, root organ formation in *P. tomentosa* was due to initiation of cell differentiation more than cell division. The existence of specific mutants in the root organ phenotype demonstrates that this trait is under tight genetic control. A Medelian segregation ratio (3∶1) of radicle development in F_1 progeny was observed in *P. tomentosa*. The use of a combination of Bulked Segregant Analysis (BSA) and AFLP methodology helped us identify two markers tightly linked (linkage distance ~ 1-2cM) to a radicle development trait. These results represent an important first step in the molecular analysis of radicle development trait in *P. tomentosa*.

4.1 The allelic mutation affecting radicle development

Recessive mutations in the gene regulating radicle development give rise to seedlings that contain major organs with a distinct seedling appearance that includes the normal cotyledons, hypocotyls and epicotyl, but lacks a visible root organ. The phenotypic ratio of offspring (rooting : non-rooting) was 3∶1 in *P. tomentosa* and $X < X^2_{0.05}$ ($X^2 = 0.557$, $X^2_{0.05} = 3.84$). This segregation ratio suggested that radicle development character is a qualitative trait and probably is controlled by a single complete dominant gene or by a set of closely linked genes. This result could be due to allelic mutation (from A to a) for radicle development trait in unique parental materials (clone "5082" and clone "JY") during the long course of natural evolution in the field. For example, A and a are the two alleles which might occupy one locus controlling radicle development in a diploid *P. tomentosa*. A and a are commonly considered to be dominant and recessive genes, respectively. When the individual tree has AA or Aa located in a locus controlling radicle development, the seedlings exhibit normal radicle development and can survive to complete a full life cycle in the field. Conversely, the offspring with aa (lethal recessive mutation) derived from the crossing between both parents with Aa genotype lack a functional root and die because they cannot absorb nutrient materials from soil in the field. The seedlings with a non-rooting phenotype described here are distinct from that of previously described mutants (Benfey et al., 1993; Hauser et al., 1995; Scheres et al., 1995; Willemsen et al., 1998; Kubo et al., 1999), which have no visible root organ. To our knowledge, this is the first mutant of poplar shown to lack a visible root organ in which the affected seedlings contain a naturally-occurring recessive loss of function mutation. However, such a mutant is an excellent analytical material for studying the expression and function of the relevant gene(s). Therefore, these "without visible functional root" seedlings obtained in these experiments will provide an ideal model system for unraveling the genetic basis of organ development in forest trees.

4.2 Identification of AFLP markers linked to loci controlling radicle development

In our study, a combination of Bulked Segregant Analysis and AFLP methodology was successfully employed to identify two AFLP markers, E_{65}/M_{34-492}, E_{65}/M_{51-502}, associated with a radicle development trait. Furthermore, these markers appear to be tightly linked to the loci controlling radicle development in *P. tomentosa*. Bulked Segregant Analysis is a very efficient method for identifying markers associated with monogenic qualitative traits based on a segregating population in poplar. This method allows rapid detection of markers tightly linked to the target trait segregating in the F_1 progeny, not required in the near-isogenic line, F_2 and backcross population (Michelmore et al., 1991). AFLP markers plus BSA is an excellent "combination" method for conducting linkage analysis in *P. tomentosa*. AFLP is highly reliable and reproducible due to its use of highly stringent PCR, in contrast with RAPD's problem of low reproducibility. AFLP requires no previous sequence or probe information as does RFLP and SSR. Given enough primer pairs spanning the average-frequency restriction sites, an entire genome could be scanned at the DNA level in a reasonable amount of time. In our study, only 78 *Eco*R I $_{+3}$/*Mse* I $_{+3}$ primer combinations were required. The modest number of primers used indicated that AFLP can overcome problems due to low polymorphisms between intraspecific individuals. It is also a fast technique for detecting the specific genomic region controlling organ development in poplar. Furthermore, the two AFLP markers identified here should be convertible to simple, rapid and low-cost PCR marker types like STS or SCAR. Doing so would enhance and economize the breeding programs. However, it will be necessary to isolate and clone the gene(s) controlling radicle development and perform a comparative study of the structures and function of this gene to substantiate these putative results. Identification of two molecular markers tightly linked to the target trait is an important first step toward isolation and cloning of the gene(s) controlling radicle development trait in *P. tomentosa*.

杨属部分种及杂种的 AFLP 分析*

摘　要　利用13对AFLP引物对杨属4个派19个种及杂种的47个无性系进行分析，共检测到858个标记，其中多态性标记为771个，多态性位点的百分率为89.9%，每对引物产生的多态性位点的百分率在80.7%~98.1%之间，表明杨属种间及无性系间在DNA水平上存在广泛变异。根据AFLP标记结果计算了杨属派间、派内种间、种内无性系间分子遗传距离，对它们的亲缘关系进行了定量描述。聚类分析结果表明，派间聚类与经典形态分类完全一致，派内种间及种内无性系间聚类与形态分类基本相同。最后，探讨了根据分子标记结果进行杂交亲本选配及杂种子代早期选择的可行性。

关键词　杨属，AFLP，遗传变异，亲缘关系

杨属(*Populus*)包括白杨派(*Leuce*)、黑杨派(*Aigeiros*)、青杨派(*Tacamahaca*)、胡杨派(*Turanga*)和大叶杨派(*Leucoides*)，有100余种，杨属派间、种间存在广泛遗传变异。以往研究杨属派间、种间变异，主要根据形态标记，即杨树的外部形态特征，如树高、胸径、分枝、冠形、叶形、芽形、花形、果形、叶背绒毛、叶缘锯齿等(徐纬英，1988；王明庥等，1991)；但是，形态标记数量有限、多态性差、易受环境影响等。随着分子生物学技术的发展，出现了多种分子标记技术，使从基因组水平探讨杨属派间、种间及无性系间遗传变异成为可能(张德强等，2001；张志毅等，2002；Cervera et al.，2005)。与形态标记相比，分子标记具有如下优点：直接以DNA形式表现，在植物的各个组织、器官以及不同发育时期均可检测到，不受季节、环境条件的影响，不存在表达问题；数量多，遍及整个基因组；多态性高；表现中性；有许多分子标记表现为共显性，能够鉴定出纯合基因型与杂合基因型，提供完整的遗传信息等。因此，分子标记被认为是进行遗传变异评价的理想标记(邹喻苹等，2001)。目前，广泛应用的分子标记主要有：RFLP、RAPD、AFLP、SSR等，其中AFLP是一种选择性扩增限制片段的方法，该法集RFLP与RAPD 2种方法的优点于一体，而且能提供比RFLP和RAPD更多的基因组多态性信息(Vos et al.，1995；Mukherjee et al.，2003；Faouzi et al.，2003；Zhang et al.，2004；Chauhan et al.，2004；王献等，2005)。李宽钰等(1996)应用RAPD标记对杨属白杨派、青杨派、黑杨派的20个种的28个无性系研究表明，三个派明显独立，且青杨派与黑杨派关系较进，而与白杨派则相对较远。苏晓华等(1996)对青杨派的甜杨(*P. suaveolens* Fisch.)、大青杨(*P. ussuriensis* Kom.)、香杨(*P. koreana* Rehd.)和马氏杨(*P. maximowiczii* Henry)的80个无性系进行RAPD分析得到，每个树种的全部无性系聚为一类，分子水平分类结果与经典分类一致，大青杨与香杨亲缘关系最近。Michael et al. 对西班牙的一个胡杨(*P. euphratica*

* 本文原载《林业科学》，2007，43(1)：35-41，与李善文、张有慧、安新民、何承忠和李百炼合作发表。

Oliv.)群体的257个样品进行AFLP分析，在个体间未检测到任何遗传变异，证明该群体为无性系起源。Maurizio et al.(2001)应用RAPD标记对53个银白杨(*Populus alba* L.)家系进行分析并聚类，结果明显分为3类，分别对应于3个不同的地理分布区。

本研究以杨属白杨派、黑杨派、青杨派、胡杨派中的部分种及杂种的47个无性系为试验材料，采用AFLP标记研究它们在DNA水平上的遗传变异，研究杨属无性系间、种间、派间及亲本与子代间亲缘关系；根据分子标记结果探讨杂交亲本间的亲和性和杂交子代早期选择的可行性，讨论分子标记与常规杂交育种的结合点。因此，该项研究具有重要的理论意义和实际应用价值。

1 材料与方法

1.1 试验材料及其来源

试验材料包括杨属中白杨派、黑杨派、青杨派和胡杨派的47个无性系，每个派所包括的无性系及其来源见表1。

表1 试验材料及来源[1]

编号	无性系	来源	编号	无性系	来源
1	毛新杨1 *P. tomentosa* × *P. bolleana* ‘No. 1’	中国农业大学 CAU	25	T66 *P. deltoides* T66’	山东临沂 Linyi, SD
2	毛新杨2 *P. tomentosa* × *P. bolleana* ‘No. 2’	中国农业大学 CAU	26	欧洲黑杨1 *P. nigra* ‘No. 1’	新疆阿尔泰 A’ertai, XJ
3	毛新杨3 *P. tomentosa* × *P. bolleana* ‘No. 3’	中国农业大学 CAU	27	欧洲黑杨2 *P. nigra* ‘No. 2’	新疆阿尔泰 A’ertai, XJ
4	毛新杨4 *P. tomentosa* × *P. bolleana* ‘No. 4’	中国农业大学 CAU	28	欧洲黑杨3 *P. nigra* ‘No. 3’	新疆阿尔泰 A’ertai, XJ
5	毛新杨5 *P. tomentosa* × *P. bolleana* ‘No. 5’	中国农业大学 CAU	29	T26 *P. deltoides* ‘T26’	山东临沂 Linyi, SD
6	新疆杨1 *P. bolleana* ‘No. 1’	山东冠县 Guanxian, SD	30	L35 *P.* ×*euramericana* ‘L35’	山东临沂 Linyi, SD
7	新疆杨2 *P. bolleana* ‘No. 2’	中科院植物所 PI, CAS	31	中林23 *P.* ×*euramericana* ‘Zhonglin23’	山东济南 Jinan, SD
8	新疆杨3 *P. bolleana* ‘No. 3’	北京植物园 PG, Beijing	32	中林46 *P.* ×*euramericana* ‘Zhonglin46’	山东济南 Jinan, SD
9	银毛杨 *P. alba*×*P. tomentosa*	河北易县 Yixian, HB	33	大青杨1 *P. ussuriensis* ‘No. 1’	吉林露水河 Lushuihe, JL
10	鲁毛27 *P. tomentosa* ‘LM27’	山东冠县 Guanxian, SD	34	大青杨2 *P. ussuriensis* ‘No. 2’	吉林露水河 Lushuihe, JL

（续）

编号	无性系	来源	编号	无性系	来源
11	鲁毛 50 *P. tomentosa* ‘LM50’	山东冠县 Guanxian，SD	35	大青杨 3 *P. ussuriensis* ‘No. 3’	吉林露水河 Lushuihe，JL
12	毛白杨 5082 *P. tomentosa* ‘5082’	山东冠县 Guanxian，SD	36	大青杨 4 *P. ussuriensis* ‘No. 4’	吉林露水河 Lushuihe，JL
13	银腺杨 1 *P. alba* × *P. glandulosa* ‘No. 1’	山东冠县 Guanxian，SD	37	大青杨 5 *P. ussuriensis* ‘No. 5’	吉林露水河 Lushuihe，JL
14	银腺杨 2 *P. alba* × *P. glandulosa* ‘No. 2’	山东冠县 Guanxian，SD	38	大青杨 6 *P. ussuriensis* ‘No. 6’	吉林露水河 Lushuihe，JL
15	银腺杨 4 *P. alba* × *P. glandulosa* ‘No. 4’	山东冠县 Guanxian，SD	39	滇杨 3 *P. yunnanensis* ‘No. 3’	云南耳源 Eryuan，YN
16	银腺杨 84K *P. alba* × *P. glandulosa* ‘84K’	陕西杨凌 Yangling，SX	40	小叶杨 1 *P. simonii* ‘No. 1’	吉林白城 Baicheng，JL
17	银白杨 1 号 *P. alba* ‘No. 1’	北京植物园 PG，Beijing	41	小叶杨 2 *P. simonii* ‘No. 2’	北京植物园 PG，Beijing
18	银白杨杂种 *P. alba* ‘Baicheng’	吉林白城 Baicheng，JL	42	甜杨 *P. suaveolens*	黑龙江呼中 Huzhong，HLJ
19	澳洲银白杨 *P. alba* ‘Aozhou’	北京林业大学 BJFU	43	川杨 *P. szechuanica*	四川汉源 Hanyuan，SC
20	银新杨 *P. alba*×*P. bolleana*	吉林白城 Baicheng，JL	44	滇杨 1 *P. yunnanensis* ‘No. 1’	云南耳源 Eryuan，YN
21	河北杨 *P. hopeiensis*	北京林业大学 BJFU	45	滇杨 2 *P. yunnanensis* ‘No. 2’	云南九河 Jiuhe，YN
22	山杨 *P. davidiana*	黑龙江呼中 Huzhong，HLJ	46	胡杨 1 *P. euphratica* ‘No. 1’	甘肃安西 Anxi，GS
23	I-69 *P. deltoides* ‘Lux’	山东临沂 Linyi，SD	47	胡杨 2 *P. euphratica* ‘No. 2’	甘肃安西 Anxi，GS
24	D324 *P. deltoides* ‘D324’	山东临沂 Linyi，SD			

① CAU：Chinese Agriculture University；PI，CAS：Plant Institute，Chinese Academy of Science；PG：Plant Garden；BJFU：Beijing Forestry University；HB：Hebei；SD：Shandong；SX：Shanxi；JL：Jilin；HLJ：Heilongjiang；XJ：Xinjiang；YN：Yunnan；SC：Sichuan；GS：Gansu.

1.2　DNA 提取

按 Murray 等（1980）的方法提取各试验材料的基因组 DNA。

1.3 AFLP 分析

AFLP 分析的基本程序和 Vos 等(1995)发明的方法基本相同，仅对体系进行了优化，采用 *Eco*RI/*Mse*I 酶切组合进行基因组限制性酶切，预扩增反应选用引物组合 $EcoRI_{+00}$/$MseI_{+00}$，选择性扩增反应采用引物组合 $EcoRI_{+3}$/$MseI_{+3}$。PCR 扩增反应在 PE-9700PCR 仪上进行。筛选的 13 对引物见表 2。

表 2 AFLP 引物筛选

引物 Primer	M31	M33	M34	M44	M46	M47	M60	M63
E33	+	+	+	+			+	
E34								
E44					+	+	+	
E60	+							
E63	+		+				+	
E65	+							

1.4 电泳

选择性扩增产物经变性后在 6%的变性聚丙烯酰胺序列分析胶上电泳分离，条件是 95W，3000V，恒功率电泳约 90min。电泳后采用 Tixier 等(1997)的银染检测方法进行 AFLP 指纹显色反应。Marker 为宝生物工程(大连)有限公司生产的 100bp DNA Ladder Marker。

1.5 结果记录

统计带型在相同片段位置上的谱带，按 0/1 系统纪录，有带记为 1，无带记为 0，并参照 Marker 带估计片段大小。

1.6 数据统计分析

多态性：

多态性(%)=扩增多态性片段数/总扩增片段数×100%

$$=(N_i+N_j-2N_{ij})/(N_i+N_j-N_{ij})$$

式中，N_{ij} 表示样本 i 和 j 的公共带数；N_i、N_j 分别是样本 i、j 的带数。

相似系数和遗传距离：

采用 Nei & Li(1979)的遗传相似系数(genetic similarity，GS)，其计算公式：

$$GS=2N_{ij}/(N_i+N_j)$$

式中，N_{ij}，N_i，N_j 同上；遗传距离 $D=1-GS$。

聚类分析：利用 NTSYS-pc2.1 版分析软件，采用非加权配对算术平均法（Unweighted Pair Group Method Arithmetic Averages，UPGMA)进行聚类。

2 结果与分析

2.1 AFLP 多态性分析

用 13 对 AFLP 引物组合对供试样品进行分析，共检测到 858 个标记，其中多态性标记 771 个，多态性位点的百分率为 89.9%；每对引物组合产生多态性标记数目不等，最少的是 E33/M34 检测到 37 条谱带，最多是 E44/M60，检测到 105 条谱带，平均每对引物组合扩增出 59 个多态性标记，多态性位点的百分率在 80.7%~98.1%之间(表 3)。

表 3 13 对 AFLP 引物扩增产生的多态性片段

引物组合	扩增带数	多态性带	多态性位点的百分率(%)
E33/M31	46	39	84.8
E33/M33	70	61	87.1
E33/M34	45	37	82.2
E33/M44	57	46	80.7
E33/M60	73	69	94.5
E44/M46	107	98	91.6
E44/M47	68	57	83.8
E44/M60	107	105	98.1
E60/M31	55	52	94.5
E63/M31	55	47	85.5
E63/M34	56	53	94.6
E63/M60	41	38	92.7
E65/M34	78	69	88.5

2.2 杨属种间及无性系间遗传变异

根据 13 对引物扩增结果，计算样本间的遗传距离(李善文，2004)，由此数据计算得到各派内样本间、种间遗传距离变幅及其均值(表 4)。

表 4 杨属派内样本间和种间遗传距离

派	来源	遗传距离变幅	均值
白杨派 *Leuce*	样本间 Among samples	0.0080~0.3774	0.1727
	种间 Among species	0.1036~0.3774	0.1856
黑杨派 *Aigeiros*	样本间 Among samples	0.0103~0.3621	0.1924
	种间 Among species	0.1293~0.3621	0.2435
青杨派 *Tacamahaca*	样本间 Among samples	0.0288~0.3866	0.2264
	种间 Among species	0.1306~0.3866	0.2482
胡杨派 *Turanga*	样本间 Among samples	0.0680	0.0680

2.2.1 白杨派种间和种内无性系间

各样本之间的遗传距离为0.0080~0.3774，平均遗传距离为0.1727；种间遗传距离在0.1036~0.3774，均值是0.1856。种内无性间遗传距离较小，如毛新杨的5个无性系间的遗传距离在0.0198~0.0646，新疆杨的3个无性系间遗传距离变幅为0.0080~0.0159；种间遗传距离较大，如山杨与其余样本间的遗传距离在0.2714~0.3774。

2.2.2 黑杨派种间和种内无性系间

样本间的遗传距离在0.0103~0.3621，平均遗传距离为0.1924；种间遗传距离在0.1293~0.3621，均值是0.2434。种内无性系间遗传距离较小，如美洲黑杨4个无性系间变幅为0.0249~0.1031，均值为0.0593；欧洲黑杨3个无性系间，变幅为0.0103~0.0205，均值是0.0156；3个欧美杨无性系遗传距离在0.0276~0.0767，均值是0.0568；美洲黑杨与欧洲黑杨间遗传距离较大，在0.3267~0.3621，均值为0.3509。

由上述分析又知，美洲黑杨与欧洲黑杨间遗传距离均大于美洲黑杨、欧洲黑杨和欧美杨种内无性系间遗传距离，这一遗传距离(0.3509)可能与该杂交组合产生杂种优势有密切关系，因为美洲黑杨×欧洲黑杨是世界公认的优良杂交组合，许多著名的欧美杨无性系均来自这一组合，如意大利培育的西玛杨、露易莎杨、I-72/58、I-45/51、I-214等，我国培育的中林46、中林23、中林28、中林14等(李善文等，2004)。

2.2.3 青杨派种间和种内无性系间

个体间遗传距离变动于0.0288~0.3866，平均遗传距离为0.2264；种间遗传距离变动于0.1306~0.3866，均值为0.2482。种内无性系间遗传距离较小，如6个大青杨无性系间变幅为0.0476~0.2166，均值为0.1491；3个滇杨无性系的遗传距离在0.0288~0.0641，均值为0.0476。种间遗传距离较大，如川杨与6个大青杨无性系的遗传距离变幅为0.2893~0.3345，均值为0.3021；3个滇杨无性系与6个大青杨无性系的遗传距离变动于0.2027~0.2929之间，均值为0.2397。

2.3 杨属派间遗传变异

根据47个样本间的遗传距离计算杨属派间遗传距离(表5)，其中白杨派与胡杨派间最大，为0.5314；其次是白杨派与黑杨派(0.4781)、白杨派与青杨派(0.4732)、黑杨派与胡杨派(0.4706)，这3个遗传距离较接近；再者是青杨派与胡杨派(0.4379)；派间遗传距离最小的是黑杨派与青杨派，为0.3263。

表5 杨属派间遗传距离[①]

派	白杨派	黑杨派	青杨派	胡杨派
白杨派 *Leuce*		0.4781	0.4732	0.5314
黑杨派 *Aigeiros*	0.4169~0.5457		0.3263	0.4706
青杨派 *Tacamahaca*	0.3692~0.5434	0.2788~0.4240		0.4379
胡杨派 *Turanga*	0.4390~0.5683	0.4474~0.5076	0.3930~0.4800	

①对角线下为派间遗传距离变幅，对角线上为派间遗传距离均值。

2.4　杨属种及无性系聚类分析

用各样本间遗传距离，采用非加权配对算术平均法(UPGMA)进行聚类分析，结果将47个无性系在遗传距离0.1400处划分为4类(图1)，分别是黑杨派、青杨派、白杨派、胡杨派，这与经典形态分类结果相一致。

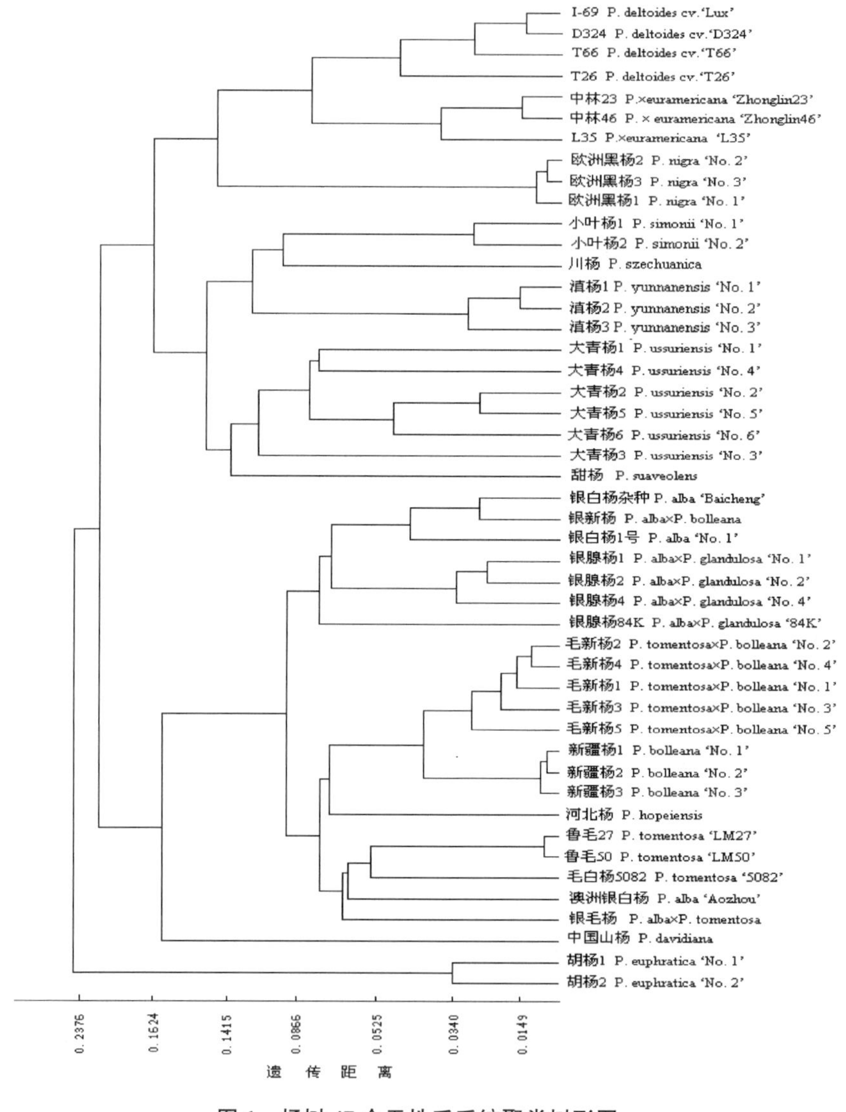

图1　杨树47个无性系系统聚类树形图

在黑杨派内，10个无性系被划分为3类：美洲黑杨、欧美杨和欧洲黑杨。在青杨派中，13个无性系聚类后可分为2类：①小叶杨、川杨、滇杨类，如果再细，每个种单独成类；②大青杨、甜杨类，大青杨6个无性系首先聚在一起，然后与甜杨聚为一类。白杨派内22个无性系聚类后可分为如下3类：①银白杨、银新杨、银腺杨类；②毛新杨、新疆杨、毛白杨、河北杨、澳洲银白杨、银毛杨类；③山杨类。胡杨派的2个无性系聚为一类。

综上分析可知，利用 AFLP 分子标记遗传距离进行聚类分析，派间聚类结果与形态分类完全一致，派内种间及种内无性系间聚类结果与形态分类也基本相同。这表明应用 AFLP 分子标记遗传距离对杨属派间、派内种间及种内无性系间的亲缘关系进行定量评价是可行的。

3 结论与讨论

利用 13 对 AFLP 引物对杨属白杨派、黑杨派、青杨派、胡杨派中 19 个种的 47 个无性系进行分析，检测到多态性位点的百分率为 89.9%，表明试验材料在分子水平存在广泛变异。聚类分析结果表明，派间聚类与经典形态分类完全一致，派内种间及种内无性系间聚类与形态分类基本相同。李宽钰等(1996)对杨属白杨派、黑杨派、青杨派的 20 个种 28 个无性系进行 RAPD 分析，筛选了 7 个引物，多态性为 79.6%，聚类分析将 28 个无性系划分 3 类，分别对应白杨派、黑杨派、青杨派。其试验材料由于未包含胡杨派，因此得到的多态性较低；除选用的试验材料外，不同的分子标记方法也可能是造成多态性差异的原因。

AFLP 分析结果与杂交试验、同工酶试验结果基本一致。杨属派间分子遗传距离的大小顺序依次为：白杨派与胡杨派(0.5314)、白杨派与黑杨派(0.4781)、白杨派与青杨派(0.4732)、胡杨派与黑杨派(0.4706)、胡杨派与青杨派(0.4379)、黑杨派与青杨派(0.3263)，这说明白杨派与胡杨派、黑杨派、青杨派，胡杨派与黑杨派、青杨派间的亲缘关系较远，这些派间进行杂交不易得到种子，而黑杨派与青杨派间亲缘关系较近，杂交容易成功。该结果与杂交试验和同工酶试验结果相一致，Li 等(1981)和 Stettler 等(1980)的杂交试验表明，白杨派与青杨派、黑杨派间杂交，花粉发育大多不正常，很难得到种子，而黑杨派与青杨派树种杂交则表现出很好的亲和性，且杂种优势明显。徐纬英(1988)以青杨派、黑杨派的种为亲本，从钻天杨×青杨(*P. nigra.* var. *italica* Koehne. ×*P. cathayana* Rehd)、小叶杨×(美杨+旱柳)[*P. simonii* Carr. ×(*P. pyramidalis*+*Salix matsudana*)]、小叶杨×钻天杨(*P. simonii* Carr. ×*P. nigra.* var. *italica* Koehne.)组合中分别选出北京杨(*P.* × *beijingensis*)、群众杨(*P.* × *popularis*)、合作杨(*P.* × *opera*)优良品系；王明庥等(1991)以美洲黑杨无性系 I-69 为母本、小叶杨为父本，选育出黑杨派和青杨派间杂种 NL-105、NL-106、NL-121 等优良无性系。胡志昂(1981)研究了杨属不同派间过氧化物同工酶，结果表明青杨派与黑杨派表现较高的同工酶谱带相似性，而白杨派与青杨派和黑杨派相似性较低。因此，应用 AFLP 标记分子遗传距离对杨属派间遗传差异进行评价，结果与常规杂交试验和同工酶分析结果相似，说明这种分析方法可行、结果可靠，可以根据亲本间分子遗传距离大小预测杂交亲本间的亲和性。

杂种优良无性系与父母本的遗传距离在不同杂交组合中存在差异。白杨杂种毛新杨的 5 个优良无性系与 3 个新疆杨无性系间的平均遗传距离是 0.0683，与 3 个毛白杨无性系间的平均遗传距离是 0.1730，这表明毛新杨杂种优良无性系与新疆杨的亲缘关系较近，其相似系数为 0.9317，与毛白杨的亲缘关系较远，相似系数为 0.8270。银毛杨杂种优良无性系与毛白杨 3 个无性系的遗传距离均值是 0. 1239，与银白杨 2 个无性系的均值是 0. 1266，这说明银毛杨杂种优良无性系与母本银白杨、父本毛白杨的亲缘关系相近，其相似系数分别是 0. 8734、0. 8761。在黑杨派中，欧美杨 3 个优良无性系与欧洲黑杨 3 个无性系的平均遗传距离是 0. 2271，与美洲黑杨 4 个无性系的遗传距离是 0. 1484，由此可知，欧美杨杂种优良无性系与美洲黑杨的亲缘关系较近，相似系数为 0. 8516，与欧洲黑杨的亲缘关系相对较远，相似系数为 0. 7729。本试验所用的毛新杨、银毛杨、欧美杨无性系分别是从毛白杨×新疆杨、银白杨×毛白杨、美洲黑杨×欧洲黑杨的杂种群体中选出的优良无性系，表明有的组合选出的优良无性系与母本相似，有的组合选出的优良无性系与父本相似，有的组合选出的优良无性系属于父母本的中间类型。形态观测结果与这一结论相一致(李善文，2004)。因此，根据亲本和子代的分子遗传距离在不同杂交组合中存在的这种特有关系，可以为其杂交子代相关性状的早期选择提供科学依据。

AFLP Analysis of Some Species and Hybrids in *Populus*

Abstract Thirteen pairs of AFLP primers were screened out and used for the analysis of forty-seven clones, which came from nineteen species and hybrids of four sections in *Populus*. Thirteen pairs of AFLP primers generated 858 fragments, 771 polymorphism fragments among them. Percentage of polymorphism fragments produced by each pair of AFLP primer was 89. 9% in average, and varied from 80. 7% to 98. 1%. All these data indicated that considerable genetic variation existed among species and clones at DNA level. Molecular genetic distances among sections, species intrasections, clones intraspecies were calculated, and the relationship among them was described quantitatively. Compared with the classical taxonomy, the result of cluster analysis among sections was consistent completely, and the results of cluster analysis among species intrasections and clones intraspecies were basically similar. Finally, the feasibility of selecting and mating parents and early selection of progenies was discussed based on the results of molecular markers.

Key words *Populus*, AFLP, genetic variation, genetic relationship

利用 AFLP 标记研究银白杨×白榆的亲子关系*

摘　要　从形态性状上看，银榆杨是银白杨×白榆的科间杂种。为了进一步从分子水平上搞清楚银榆杨与其亲本银白杨及白榆的亲缘关系，该研究采用 AnP 技术，用 20 对 *EcoR*I+*Mse* 1 引物对银榆杨、银白杨、白榆等 10 个样本进行了亲子关系分析。共获得 2040 条可统计的谱带，其中 1470 条带为多态性带，多态带百分率为 72.06%。结果表明：① 银榆杨中既含有银白杨的基因又含有白榆的基因，且出现了双亲不具有的新谱带；② 银榆杨含有的银白杨基因成分比白榆基因成分多，从聚类分析(uPGMA)结果看出，银榆杨属于偏母本型的杂种；③ 辽宁产地的银白杨是银榆杨杂种的母本得到进一步证实。在所有白榆样本中，尽管方差分析结果表明各无性系间对银榆杨杂种子代的遗传距离没有显著差异，但辽宁产地的白榆 2 号(LP-2)与银榆杨杂种各无性系的遗传距离最小，它是银榆杨父本的可能性最大；④ 银榆杨中白榆的基因在 4 个白榆样本中的存在具有普遍性。但相对于辽宁产地的白榆而言，北京的白榆与银榆杨杂种的亲缘关系较远。该文还对银榆杨杂种的形成与遗传组成进行了分析讨论。

关键词　科间杂种，银白杨，白榆，AFLP 标记，亲子关系

银白杨(*Populus alba*)(别婉丽，1990)是杨柳科杨属白杨派树种，白榆(*Ulmus pumila*)是榆科榆属树种。1979 年，别婉丽用蒙导法做了银白杨×(白榆+新疆杨死花粉)的杂交试验，并采用未成熟胚离体培养技术获得了“杂种”幼苗，在植物远缘杂交领域取得重大突破。通过形态观察，该“杂种”(暂定名银榆杨)被认为是科间杂种，并在绿化上得到了应用。

鉴别植物杂种有形态学(张金凤等，1999)、细胞学(李富生等，2004)、生化标记(张金凤等，1999；洪月云等，1994)、分子标记(李富生，2004；张德强等，2003；刘平武等，2005)等几种方法，分子标记以其高度的灵敏性被广泛应用。在众多的分子标记中，AFLP(Amplified Fragment Length Polymorphism)标记技术检测品种质量和纯度是一种十分灵敏和可靠的方法(张德强等，2000)，现在被广泛用于多样性研究(CERVERA et al.，2000) 及杂种鉴别(鹿金颖等，2005)。到目前为止还没有见到应用 AFLP 标记鉴定科间杂种真实性的报道。为了搞清楚银榆杨与银白杨及白榆的亲缘关系，从分子水平进一步证实该科间杂种的真实性，本文利用 AFLP 标记技术对银白杨×白榆的杂种与亲本的遗传关系进行了研究。

1　材料与方法

1.1　材料

1979 年，别婉丽获得银白杨×(白榆+新疆杨死花粉)的败育幼胚挽救苗 38 株，目前仅

* 本文原载《北京林业大学学报》，2007，29(2)：8-12，与杨成超、王胜东、柳志岩、何承忠和苏晓华合作发表。

存 4 株大树。当时，由于这些杂种生长一般，没有受到足够的重视。时隔 20 多年，才发现这些杂种具有较强的适应性和抗逆性。在研究档案中，银榆杨的母本银白杨采集地点记载比较清楚，在盖州市熊岳树木园内，而父本则只记载在该树木园附近采集。因此，在采样时我们采集了现存的 4 株银榆杨杂种无性系(AP-1、AP-2、AP-3 和 AP-4)，同时在盖州市熊岳树木园内采集 1 株银榆杨的母本银白杨(LA)，在树木园周围分别采集了 3 株可能的父本白榆(LP-1、LP-2 和 LP-3)，这 8 个无性系作为研究试样。另外，我们在北京采集了银白杨(BA)和白榆(BP)的样本，作为亲子鉴定的对照。共计 10 个样本(表 1)。以上试样均取休眠状态的枝条，带回实验室进行水培，获得幼嫩的叶片作为提取 DNA 的实验材料。

1.2　方法

1.2.1　模板 DNA 的制备

选取各样本枝条相同部位生长健壮的幼嫩叶片，利用 SDS 方法提取 DNA(李善文，2004)，上清液用酚—氯仿抽提，反复抽提后用乙醇沉淀水相中的 DNA。用紫外分光光度计测定 DNA 的浓度，并取少量 DNA 样品用 0.8%的 Agrose 胶检测所提 DNA 的质量。最后取适量的 DNA，稀释成 200ng/μL，置于 4℃冰箱备用。

表 1　试验材料基本情况

树种	样本采集地点	样本数	样本编号
银榆杨	辽宁省杨树研究所	4	AP-1、AP-2、AP-3、AP-4
银白杨	辽宁省盖州市熊岳树木园	1	LA
银白杨	北京市	1	BA
白榆	辽宁省盖州市熊岳树木园	3	LP-1、LP-2、LP-3
白榆	北京市	1	BP

1.2.2　AFLP 反应

本试验所选用的接头和引物由上海生工生物工程技术服务公司合成(表 2)。试验步骤基本按照 VOS(VOS et al.，1995)的方法，并略作改动(何承忠，2005)。

首先是模板 DNA 的酶切和酶切片段与接头的连接，37℃恒温酶切连接≥4h。然后，将酶切连接产物稀释 6 倍用做预扩增模板。预扩增完成后，把预扩增产物稀释 20 倍后当作选择性扩增模板 DNA，在预扩增引物的 3’ 末端加上 3 个碱基用做选择性扩增引物，对预扩增产物进行选择性扩增。选择扩增产物经变性聚丙烯酰胺凝胶电泳后，用 10%的冰醋酸脱色固定、$AgNO_3$银染、无水碳酸钠加上甲醛以及硫代硫酸钠显影、10%的冰醋酸定影、蒸馏水冲洗、自然干燥后照相。

1.2.3　数据处理

谱带统计时将具有相同迁移率的扩增片段按 0/1 系统记录，有带记为 1，无带记为 0，并参照标准 Marker 带估计扩增片段的大小。

利用 DCFA1.1 软件(张富民等，2002)将参试的 10 个样本的 AFLP 谱带统计结果转换为 POPGENE1.32 软件适用文件(NEI M，1987)，用 POPGENE1.32 软件计算多态性(P)、遗传相似系数(Gs)和遗传距离(D)，公式分别如下：

$$P=(N_i+N_j-2N_{ij})/(N_i+N_j)\times100\% \quad (1)$$

$$GS=2N_{ij}/(N_i+N_j) \quad (2)$$

$$D=1-GS \quad (3)$$

式中，N_{ij}表示样本 i 和 j 的公共带数；N_i、N_j 分别是样本 i、j 的带数。

用 SPSS12.0 软件进行方差分析。利用 NTSYS pc2.1 版分析软件，采用 Nei(NEI，1987)的遗传一致度 I 进行非加权配对算术平均聚类分析(UPGMA)。

表 2 AFLP 标记所用接头和引物

接头和引物	代号	碱基序列	接头和引物	代号	碱基序列
*Eco*R Ⅰ接头	*Eco*R Ⅰ adapter	5′-CTGGTACACTGCGTACC-3′	*Eco*R Ⅰ-65	M65	5′-GACTGCGTACCAATTCGAG-3′
		3′-CTGACGCATGGTTAA-5′	*Mse* Ⅰ-00	M00	5′-GATGAGTCCTGACTAA-3′
Mse Ⅰ接头	*Mse* Ⅰ adepler	5′-GACGATGAGTCCTGAG-3′	*Mse* Ⅰ-31	M31	5′-GATGAGTCCTGAGTAAAAA-3′
		3′-TACTCAGGACTCAT-5′	*Mse* Ⅰ-33	M33	5′-GATGAGTCCTGAGTAAAAG-3′
*Eco*R Ⅰ-00	E00	5′-GACTGCGTACCAATTC-3′	*Mse* Ⅰ-40	M40	5′-GATGAGTCCTGAGTAAAGC-3′
*Eco*R Ⅰ-33	E33	5′-GACTGCGTACCAATTCAAG-3′	*Mse* Ⅰ-44	M44	5′-GATGAGTCCTGAGTAAATG-3
*Eco*R Ⅰ-34	E34	5′-GACTGCGTACCAATTCAAT-3′	*Mse* Ⅰ-46	M46	5′-GATGAGTCCTGAGTAAATT-3′
*Eco*R Ⅰ-44	E44	5′-GACTGCGTACCAATTCAAT-3′	*Mse* Ⅰ-47	M47	5′-GATGAGTCCTGAGTAACAA-3′
*Eco*R Ⅰ-48	E48	5′-GACTGCGTACCAATTCCAC-3′	*Mse* Ⅰ-60	M60	5′-GATGAGTCCTGAGTAACTC-3′
*Eco*R Ⅰ-60	E60	5′-GACTGCGTACCAATTCCTC-3′	*Mse* Ⅰ-63	M63	5′-GATGAGTCCTGAGTAAGAA-3′
*Eco*R Ⅰ-63	E63	5′-GACTGCGTACCAATTCGAA-3′			

2 结果与分析

2.1 引物组合

用 20 对 AFLP 引物，对参试的 10 个样本做了选择性扩增，结果表明所有的引物组合都能得到 PCR 产物，但是不同的引物组合所得的扩增谱带多少、均匀程度及清晰度不同。扩增出来的有效片段长度的范围主要集中在 50~650bp。筛选出的引物对组合见表 3。

表 3 AFLP 标记试验所用引物组合

引物	M31	M33	M40	M44	M46	M47	M60	M63
E33						+	+	
E34					+	+	+	
E44	+			+		+	+	+
E48		+						
E60		+			+			+
E63				+		+		
E65	+		+			+	+	

注：“+”表示不形成引物组合。

2.2 标记多态性

通常情况下，AFLP 每一条扩增带都对应着一个 DNA 分子位点，出现多态性扩增带，说明某个或某些样品在该位点上存在变异；如果树种间扩增的谱带位点一致，则说明树种间具有相同的基因。利用亲本和子代谱带位点一致性程度，可以了解子代和“亲本”是否具有亲缘关系。

用 20 对引物进行了 AFLP 分析，共得到 2040 条可统计的谱带，其中 1470 条带为多态性带，多态带百分率为 72.06%。引物对及谱带分布见表 4。AFLP 标记结果表明，银榆杨各无性系谱带中都有与白榆和银白杨相同的谱带，只是与白榆相同的谱带数少，与银白杨相同的谱带数多。辽宁的 3 个白榆 LP-1、LP-2、LP-3 与 4 个银榆杨杂种无性系 AP-1、AP-2、AP-3、AP-4 有共同的谱带，这说明银榆杨中含有白榆的基因。此外，银榆杨各无性系具有新的特异谱带，这种新谱带是银白杨和白榆都没有的谱带，这说明银榆杨杂种不仅遗传了亲本的特征，还产生了重组变异。

表 4　20 对 AFLP 引物扩增产生的多态谱带条

引物组合	扩增总带数	差异带数	引物组合	扩增总带数	差异带数
E48/M33	94	61	E60/M63	128	75
E60/M33	100	77	E44/M63	153	119
E65/M60	94	70	E65/M40	145	113
E44/M60	92	87	E33/M47	78	70
E63/M44	54	38	E34/M47	115	98
E44/M44	72	34	E65/M31	110	65
E44/M31	98	52	E33/M60	104	78
E34/M60	106	84	E63/M47	78	62
E60/M46	144	106	E44/M47	64	45
E34/M46	141	84	E65/M47	70	55

2.3 银榆杨与银白杨、白榆间的遗传变异

根据 20 对引物的扩增结果，计算 10 个样本间的遗传相似系数和遗传距离，并以此建立遗传矩阵(表 5)。由此表数据计算得到各树种内样本间和树种间遗传距离变幅及其均值。

从银白杨、银榆杨、白榆种内样本间遗传距离变幅及均值(表 6)可以看出，每个树种内也存在多态性，银榆杨无性系间遗传距离最小，两个地理位置不同的银白杨无性系之间遗传距离最大，白榆无性系之间的遗传距离在这 3 个树种中居中。仔细研究表 5，可以看出，北京地理种源的树种与银榆杨的遗传距离较大，即亲缘关系较远(图 1)。计算银白杨、银榆杨和白榆种间遗传距离变幅及均值得出，银白杨和银榆杨的遗传距离范围在 0.2261~0.2537 之间，平均遗传距离为 0.2408。两个树种亲缘关系比较接近；而白榆和银

榆杨的遗传距离在 0. 8001～0. 8572 之间，平均遗传距离为 0. 8272；白榆和银白杨的遗传距离在 0. 8230～0. 8706 之间，平均为 0. 8475，比白榆和银榆杨的遗传距离大。说明银白杨与银榆杨的亲缘关系更近一些。

从表 5 中提取“白榆各无性系与银榆杨的遗传距离”的数据并计算平均值(表 7)，可以看出，辽宁的白榆 2 号(LP-2)与各子代的遗传距离最小，亲缘关系最近。它是银榆杨的父本的可能性最大，但在用 SPSS 12. 0 软件进行白榆各无性系对子代遗传距离的单因素方差分析后发现。各白榆无性系间对子代的遗传距离并没有显著差异(表 8)。

表 5　银白杨、白榆和银榆扬各树种及无性系间遗传距离

	BP	LJ-1	LJ-3	LJ-2	AP-1	AP-2	AP-3	AP-4	LA	BA
BP		0. 1357	0. 1484	0. 1646	0. 8499	0. 8572	0. 838	0. 5498	0. 8706	0. 8672
LP-1	0. 1357		0. 0712	0. 1126	0. 8120	0. 8202	0. 8122	0. 8435	0. 8436	0. 8375
LP-3	0. 1484	0. 0712		0. 1193	0. 8332	0. 8412	0. 8367	0. 8342	0. 8582	0. 8460
LP-2	0. 1616	0. 1126	0. 1193		0. 8001	0. 8084	0. 8011	0. 8030	0. 8335	0. 8230
AP-1	0. 8499	0. 8120	0. 8342	0. 8001		0. 0111	0. 0147	0. 0244	0. 2426	0. 2537
AP-2	0. 8572	0. 8202	0. 8412	0. 8084	0. 0114		0. 0023	0. 0147	0. 2343	0. 2466
AP-3	0. 8538	0. 8122	0. 8367	0. 8041	0. 0147	0. 0023		0. 0134	0. 2370	0. 2467
AP-4	0. 8498	0. 8135	0. 8342	0. 8030	0. 0244	0. 0142	0. 0134		0. 2261	0. 2397
LA	0. 8706	0. 8436	0. 8582	0. 8335	0. 2426	0. 2343	0. 2370	0. 2261		0. 0854
BA	0. 8672	0. 8375	0. 8460	0. 8230	0. 2517	0. 2466	0. 2467	0. 2397	0. 2854	

表 6　银白杨、白榆和银榆杨杂种的种内遗传距离变幅及均值

样本材料	来源	遗传距离变幅	遗传距离平均值
银榆杨	样本内	0. 0073～0. 0244	0. 0143
银白杨	样本内	0. 0854	0. 0854
白榆	样本内	0. 0712～0. 1646	0. 1253

表 7　白榆各无性系与银榆杨的遗传距离

子代	BP	LP-1	LP-2	LP-3
AP-1	0. 8499	0. 8120	08342	0. 8001
AP-2	0. 8572	0. 8202	0. 8412	0. 8084
AP-3	0. 8538	0. 8172	0. 8367	0. 8041
AP-4	0. 8498	0. 8135	0. 8342	0. 8030
平均值	0. 8527	0. 8157	0. 8039	0. 8366

2. 4　银榆杨、银白杨、白榆树种及无性系间的聚类分析

用各样本间遗传距离数据，采用非加权配对算术平均法(uPGMA)进行聚类分析(图 1)，结果将 l0 个样本在 0. 2 处划分为三类：分别是白榆、银榆杨、银白杨；在 0. 4 处划分

为两类：白榆，银榆杨、银白杨。从聚类图表更加明显地看出各树种之间的关系，银榆杨与母本银白杨的亲缘关系比其与白榆的亲缘关系近，也就是说，银榆杨更像母本，属于偏母本融合型的杂种。

在各个树种内部，都有多态性出现。银榆杨 4 个无性系中 AP-2 与 AP-3 的亲缘关系最近，AP-1 与这两个无性系的关系比 AP-4 与这两个无性系的关系还要近一些。与北京白榆(BP)相比，辽宁白榆(IJP)各无性系之间亲缘关系更近。

表 8　白榆各无性系与银榆杨的遗传距离的单因素方差分析

	离差平方和	自由度	均方	均方比	*Sig.*
组间	0.000	3	0.000	0.100	0.958
组内	0.006	12	0.000		
合计	0.006	15			

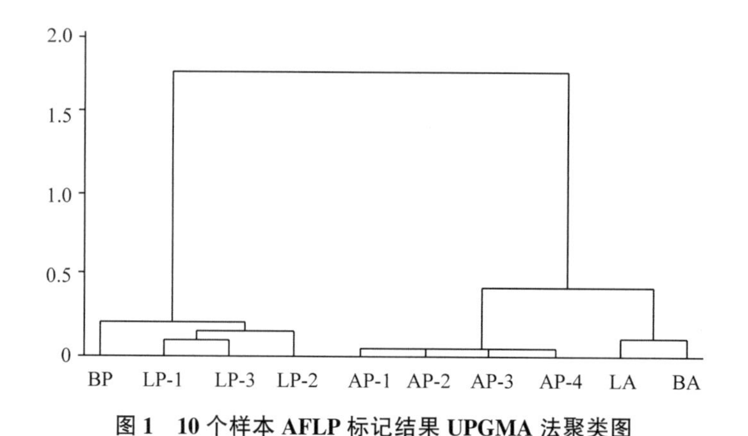

图 1　10 个样本 AFLP 标记结果 UPGMA 法聚类图

3　结论与讨论

通过用 20 对引物组合对参试的 10 个样本进行了 AFLP 标记分析，发现 AFLP 标记稳定性强，重复性好，可用于亲子分析鉴定。结果表明，在 4 个银榆杨无性系中均包含有银白杨和白榆的 DNA 片断，由此可以认为它们是银白杨与白榆的杂种。

从形态性状的观察看，银榆杨幼叶背面没有浓密的白绒毛，而银白杨幼叶有白色绒毛；银榆杨叶缘锯齿不裂，叶片大于白榆，树冠长椭圆形，侧枝多而细，呈羽状排列，似榆树分枝，与银白杨粗壮的侧枝明显不同；银榆杨多为雌雄同株，有的甚至雌雄同一花序，一般 3~4 年开花，而银白杨为雌雄异株，5~6 年开花，白榆雌雄同株，花两性，3 年开花；在花的性状上，银榆杨与白榆相似；从植株的表型上看，银榆杨与银白杨更相似一些，这一点与 AFLP 标记结果相同，也为分子标记结果做了验证。

从 AFLP 标记结果还可看出，银榆杨中含有银白杨的基因较多，白榆的基因较少，从聚类分析结果看出，银榆杨属于偏母本类型的杂种。银榆杨中含有白榆的基因较少，可以用这样的推测来解释：在组培过程中。“银白杨×白榆”杂种胚中的大部分白榆的染色体被消除了，这在农作物杂交育种中发现过类似情况(杨成超，2005)。只有少量白榆的染色体片段导入到银白杨染色体之中或者白榆的单基因渗入到银白杨染色体之中，最后成为现在的银榆杨杂种。当然这个推测还有待于运用细胞学手段和银榆杨的子代分析进一步研究证明。

各个树种内也存在多态性，银榆杨无性系间遗传距离最小，两个地理位置不同的银白杨无性系之间遗传距离最大，白榆次之；相对辽宁的亲本而言，北京地理种源的对照亲本与银榆杨的亲缘关系较远，辽宁的银白杨是银榆杨的母本这是清楚的。银榆杨中的白榆基因在这4个白榆样本中具有普遍性，虽然单因素方差分析结果表明，各白榆无性系间对子代的遗传距离并没有显著差异，但是辽宁的白榆2号(LP-2)与银榆杨各无性系的遗传距离最小，亲缘关系最近，因此推测该白榆个体为银榆杨的父本的可能性最大。AFLP标记结果表明银榆杨出现了父母本都不具有的谱带，这说明银榆杨在遗传了银白杨和白榆基因的同时发生了重组变异。

开展植物科间远缘杂交的研究不少，但成功的例子几乎没有，造成失败的原因很多。其中原因之一认为，由于所有远缘的种间杂交，胚和后代中都存在很严重的染色体异常行为，在染色体配对重组过程中，存在着双亲染色体配对交叉和异源重组重排事件(Kalloo et al.，1992)。银白杨($2n=38$)和白榆($2n=28$)的染色体数目不等，在杂交和败育胚挽救过程中，染色体的重新组合是十分复杂的，不同的染色体组合在表型上会发生什么变化和产生什么效应是下一步值得深入研究的课题。

Genetic relationship between parents and hybrid progenies of *Populus alba* L. ×*Ulmus pumila* L. using AFLP marker

Abstract The interfamilial hybrids between *Populus alba* L. ×*Ulmus pumila* L. were described by morphological characters. In order to identify the hybrids and analyse genetic relationship between the hybrids an d their parents further at the molecular level, the technique of AFLP marker was used in this study. Twenty pairs of primers were employed and 2040 clear bands were detected. Among them, 1470 bands were polymorphic and the ratio of polymorphic bands was 72.06%. The results were as the followings: 1) both gene markers of *P. alba* and *U. pumila* were exhibited on the banks of hybrids. And some new extra bands were detected in the hybrids; 2) the gene marker bands of the hybrids from *P. alba* were more than those from *U. pumila*. this showed that the hybrids were maternal progenies. And the analysis of UPGMA showed similar results; 3) *P. alba* from Liaoning station was infered as female parent of the hybrids. Since genetic distance was the least between the clone(LP-2) and the hybrids and the variance analysis indicated that there was no significant deviation among all the samples of *U. pumila*, the male parent would be the clone(LP-2), also from Liaoning station; 4) although the genes of *U. pumila* were prevalent in the hybrids. the genetic relationship between comparison clone(BP) and the hybrids was mole distant. Meanwhile, the recombination ways and genetic composition of the hybrids were also discussed in this paper.

Keywords interfamilial hybrids, *Populus alba* L., *Ulmus pumila* L., AFLP marker, genetic relationship

杨树 Genomic-SSR 与 EST-SSR 分子标记遗传差异性分析*

摘　要　利用16个杨树无性系对Genomic-SSR和EST-SSR两种SSR标记的遗传差异性进行分析。统计分析表明：Genomic-SSR平均检测到的等位基因数为4.1、Shannon指数1.0637、杂合度0.4427、期望杂合度0.5523；EST-SSR平均检测到的等位基因数为2.8、Shannon指数0.6771、杂合度0.2081、期望杂合度0.4492。聚类分析结果显示：EST-SSR相对Genomic-SSR能更精准地鉴别基因型。研究结果表明，在标记多态性及基因型鉴别等方面EST-SSR与Genomic-SSR均具有显著的遗传差异性。通过对两种分子标记遗传差异的比较分析，可为合理利用SSR标记进行物种遗传多样性等相关研究提供依据。

关键词　杨树，Genomic-SSR，EST-SSR，遗传差异性

杨树是全球广泛分布与应用的速生丰产树种，对其进行遗传改良具有重要意义(苏晓华等，2004)。目前，杨树遗传改良的最有效手段仍然是杂交育种，然而杨树杂交育种需要选择具有较大遗传差异的亲本，因此，开展杨树遗传多样性研究成为其遗传改良的重要基础(孙其信等，1996)。遗传多样性评价方法主要包括表型标记、生化标记和分子标记等，其中分子标记由于不受环境条件影响并分布在整个基因组，被普遍用于遗传多样性研究。现已广泛使用的分子标记有RAPD、RFLP、AFLP和SSR等。SSR (simple sequence repeats)由于具有共显性、多态性高和容易检测等特点，已广泛应用于遗传图谱构建、基因定位、遗传多样性、物种进化分析以及分子标记辅助选择育种等方面(Tuskan et al.，2004；Varshney et al.，2005；Yin et al.，2008)。早期SSR标记开发需要构建基因组文库，通过测序发现含有SSR结构的克隆，再根据SSR侧翼的保守序列设计特异性引物进行PCR扩增，这种基于基因组数据开发的SSR标记被称为Genomic-SSR。随着功能基因组学的发展，各个重要物种的EST(express sequence tags)序列数量迅速增长，基于EST序列开发EST-SSR成为微卫星标记新的发展方向，而且EST是来源于基因转录区域的3′或5′端测序的cDNA序列，与功能基因紧密连锁，所以利用EST序列开发SSR作为功能标记已广泛用于构建与特定功能相关的遗传连锁功能图谱。由于标记开发的来源不同，在小麦*Triticum aestivum*等植物研究中发现Genomic-SSR与EST-SSR具有一定的遗传差异性(杨新泉等，2005；常玮等，2009)。那么，在杨树中是否存在这种差异性，对于杨树遗传多样性评价以及相关研究具有非常重要的意义。因此，本研究利用16个杨树无性系对Genomic-SSR以及EST-SSR遗传差异性进行分析，为下一步综合运用不同来源的SSR标记进行相关研究奠定基础。

* 本文原载《北京林业大学学报》，2010，32(5)：1-8，与宋跃朋、江锡兵、王泽亮和李博合作发表。

1 材料与方法

1.1 材 料

分别选取白杨派的毛白杨(*Populus tomentosa* ‘LM50’)、毛新杨(*P. tomentosa*×*P. bolleana*)、银腺杨(*P. alba*×*P. glandulosa*)，青杨派的大青杨(*P. ussuriensis*)、藏川杨(*P. szechuanica*)，黑杨派的美洲黑杨(*P. deltoides*)以及部分杂交种，共16个杨树无性系(表1)。取每份材料的新鲜叶片利用CTAB法(Dellaport et al.，1983)提取基因组DNA。

表1 杨树试验材料

编号	无性系	树种	来源地	编号	无性系	树种	来源地
1	9	毛新杨×毛白杨(LM50)	本实验室(北京)	9	I69	美洲黑杨	山东省泰安市
2	10	毛新杨×毛白杨(LM50)	本实验室(北京)	10	Pu-1	大青杨	吉林省抚松县
3	13	毛新杨×毛白杨(LM50)	本实验室(北京)	11	135	美洲黑杨×大青杨	本实验室(北京)
4	46	毛新杨×毛白杨(LM50)	本实验室(北京)	12	136	美洲黑杨×大青杨	本实验室(北京)
5	51	毛新杨×毛白杨(LM50)	本实验室(北京)	13	139	美洲黑杨×大青杨	本实验室(北京)
6	LM50	毛白杨	山东省冠县	14	144	美洲黑杨×大青杨	本实验室(北京)
7	毛新杨	毛白杨×新疆杨	本实验室(北京)	15	147	美洲黑杨×大青杨	本实验室(北京)
8	银腺杨	银腺杨	山东省冠县	16	SN21	藏川杨	西藏山南地区

1.2 SSR引物设计

在NCBI下载杨树EST序列，采用EST－trimmer软件(http://pgrc.ipk－gatersleben.de)对5359条EST序列进行分析，去除载体序列、poly T或poly A并剔除长度小于100bp的EST序列。利用SSRIT软件(http://www.gramene.org/db/searches/ssrtool)搜索全部EST序列中的SSR结构。利用primer3在线设计SSR引物。引物设计参数为：引物长度18~23bp；复性温度53~56℃，(G+C)含量47.7%~56.2%。与基因组数据库进行比对后，去除受内含子影响的引物。最终选择出10对EST-SSR引物。随机挑选的10对Genomic-SSR引物序列来源于IPGC(http://www.ornl.gov/sci/ipgc/ssr_ resource.html)主页的SSR资源(表2)。

1.3 PCR扩增及电泳检测

PCR反应体积为20μL，内含DNA约为100ng，Mg^{2+}浓度为1.5mmol/L，Taq酶0.7U，dNTP浓度为0.5mmol/L，引物浓度0.5μmol/L。反应程序为：94℃预变性5min；94℃变性30s，复性30s，退火温度56~59℃(视不同引物而定)，72℃延伸40s，35个循环；72℃延伸10min；4℃保存。采用6%变性聚丙烯酰胺凝胶电泳检测PCR扩增产物，银染检测多态性。

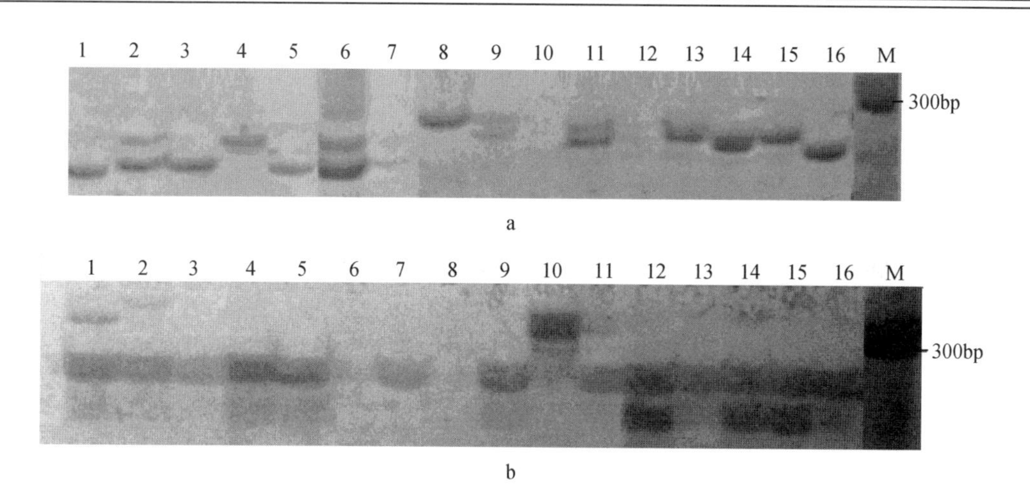

图 1　Genomic-SSR(a)与 EST-SSR(b)在 16 个杨树无性中的扩增结果

1.4　统计方法

统计电泳产生的基因型数据，并将数据转换成“0”、“1”表示的二元型数据。利用该数据计算 Shannon 多样性指数、期望杂合度、观测杂合度和相似系数，对两种 SSR 标记的遗传差异性进行分析。

Shannon 多样性指数计算公式：

$$I = -\sum_{i=1}^{n} P_i \ln P_i$$

式中，P_i 表示第 i 个条带的频率(Botstein et al.，1980)。

期望杂合度(He)计算公式：

$$He = 1 - \sum_{i=1}^{n} P_i^2$$

观测杂合度(Ho) 计算公式：

$$Ho = \text{观测杂合体数}/\text{观测总个体数}$$

应用 Nei 等(1979)的方法计算相似系数：

$$GS = 2N_{ij}/(N_i + N_j)$$

式中，N_i、N_j、N_{ij} 分别表示第 i 个材料扩增带数、第 j 个材料的扩增带数、第 i 和第 j 个材料扩增的共有带数(Nei and Li，1979)。利用 NTSYS-pc 2.1 软件按照 UPGMA 方法进行聚类分析，并绘制聚类图。

2　结果与分析

2.1　Genomic-SSR 与 EST-SSR 多态性分析

利用 10 对 Genomic-SSR 引物对 16 个杨树无性系进行扩增(图 1a)，共得到 41 个条带，引物扩增条带数在 2~5 之间，平均 4.1 条；检测的 10 个位点全部具有多态性。利用 10 对

EST-SSR 引物扩增(图 1b)得到 28 个条带，引物扩增条带数在 2~4 之间，平均 2.8 条；检测的 10 个位点同样全部具有多态性。通过计算，Genomic-SSR 的 Shannon 指数变化范围为 0.5454~1.6000，平均为 1.0646；EST-SSR 计算的 Shannon 指数变化范围为 0.1168~1.3800，平均为 0.6985。对利用两种标记计算出的 Shannon 指数进行方差分析，结果显示，Genomic-SSR 与 EST-SSR 多态性差异极显著($P<0.01$)。

2.2 Genomic-SSR 与 EST-SSR 相似系数的相关性分析

根据 41 个 Genomic-SSR 标记计算 16 个杨树无性系的相似系数，得出其值在 0.3902~0.9268 之间，平均为 0.6623，其中 10、9 与 3 号无性系相似系数最小，11 和 14 号无性系相似系数最大。由 28 个 EST-SSR 标记计算出的相似系数变化范围为 0.4117~0.9706，平均为 0.6676，其中 10 与 2、5、6、7 号无性系相似系数最小，3 与 4 无性系相似系数最大(表 3)。利用全部 69 个 SSR 标记计算出的相似系数在 0.4400~0.9333 之间，平均为 0.6647。6 与 11 号无性系相似系数最小，11 和 14 号无性系相似系数最大(表 4)。根据相似系数的比较发现，Genomic-SSR 与 EST-SSR 计算出的相同 2 个个体间的相似系数存在一定的差异性，推测可能是由于 16 个杨树无性系来源于杨树的不同派别，而两种标记在派间的鉴别能力不同而导致的。对白杨回交群体和黑杨与青杨杂交群体内的个体间相似系数进行分析发现，两种标记计算的相似系数，均表明在白杨回交群体中个体 2 与 4 号无性系的相似系数最小；在青杨和黑杨杂交群体中 12 与 14 号的相似系数最小。这表明 Genomic-SSR 与 EST-SSR 都能比较准确地反映出不同基因型之间的亲缘关系。

对 Genomic-SSR、EST-SSR 以及 Genomic-SSR+EST-SSR 计算出的相似系数进行相关性分析(表 5)。结果显示 3 部分相似系数呈显著正相关；EST-SSR 与 Genomic-SSR+EST-SSR 相关系数稍高，即两种标记计算的综合相似系数与 EST-SSR 标记更为相似，说明 EST-SSR 能更准确地揭示基因型之间的遗传关系(杨新泉等，2005)。

2.3 聚类分析

为确定两种标记对 16 个杨树材料遗传关系的鉴定准确度，利用 UPGMA 方法进行聚类分析，鉴定各供试基因型的亲缘关系(图 2~图 4)。由图 2 可知，本实验所用 Genomic-SSR 标记可以将 16 个杨树无性系区分开来，其中包括亲缘关系较近的回交以及杂交子代。在相似系数为 0.55 的水平上，将 16 个无性系区分为Ⅰ、Ⅱ两大类群：第Ⅰ类包括 LM50、毛新杨及其 5 个杂交子代和银腺杨；第Ⅱ大类包括美洲黑杨 I69、大青杨 Pu-1 及其 5 个杂交子代以及藏川杨。而 EST-SSR 则在 0.51 上水平将全部无性系区分为Ⅰ、Ⅱ两个类群(图 3)。以上分析结果说明 Genomic-SSR 和 EST-SSR 都能将不同基因型区分开来。

对图 2、3 进行分析发现：在Ⅰ大类中，Genomic-SSR 未能区分出银腺杨与其他白杨的差别，而是与毛新杨×毛白杨(LM50)的回交子代聚类到一起：EST-SSR 则准确地将银腺杨单独聚为一类。这说明 EST-SSR 比 Genomic-SSR 能更精确地鉴别出基因型的差别。根据 Genomic-SSR 和 EST-SSR 两种标记计算出的综合相似系数进行聚类分析得到图 4，发现其聚类结果与 EST-SSR 标记更为相似，表明 EST-SSR 能更准确地揭示基因型之间的遗传关系(杨新泉等，2005)。

表 2 引物信息与统计结果

引物名称	引物序列(5′-3′)	重复类型	扩增产物	扩增带数	连锁群	Shannon指数	观测杂合度	期望杂合度
ESTCU310415	F：CCCGAGTCAATCTGAGTTAGTA R：CTTGTTTATTGGAGATGGAGC	$(AAAT)_4$	122 bp	3	Ⅳ	0.8813	0	0.6133
ESTCU310401	F：GTGCAGGCAGATATTTATGGA R：GGAAGCAGCAGTTGAAGAAG	$(TGCTTC)_3$	127 bp	4	Ⅹ	1.3800	0.8125	0.7257
ESTCU310140	F：AACCGTATGAAACTTTAGGCA R：AAACCCACCCACTGTTATTG	$(TTTATA)_3$	334 bp	4	Ⅳ	0.8571	0	0.6036
ESTCU316835	F：GCAGAGGAAGCAGCAAGAG R：CCAAGTCACGGGACAGTAAAG	$(CTT)_5$	449 bp	2	Scaffold-28	0.4404	0.5000	0.4444
ESTCU306954	F：CCCCGAATATCTCGTCTT R：GGTTTGGTTGGGTTGTCT	$(CGG)_5$	522 bp	2	Ⅰ	0.2900	0	0.1244
ESTCU307135	F：AAGGTGAGGAGCAGCAGAG R：AAATCAAACCCTAAAGCACAG	$(ATC)_5$	464 bp	3	Ⅻ	0.8982	0	0.5562
ESTCU309310	F：TTATCCACCCTCCCTGTCTC R：AAAGGAAAGGTCCATCGTAAT	$(TTTG)_4$	225 bp	3	Ⅸ	1.0972	0.4285	0.6888
ESTCU307757	F：AACAATCTCGCAGCAGGAA R：GAAATGTCAGCGTTGGGTC	$(CGG)_5$	358 bp	2	Ⅰ	0.1168	0	0.2343
ESTCU307720	F：CGAGGGAATGGATGAGATG R：CTTAAATAGAGCCAGGGAAATAC	$(GAGGAA)_3$	390 bp	3	Scaffold-140	0.6018	0.2142	0.3190
ESTCU308303	F：AAGAAAGTGTAGGAGCTGGACC R：GGCAGATCATCACAACGAAAT	$(AAAT)_4$	328 bp	2	Ⅱ	0.4221	0.3750	0.3750
GenomicPMGC-93	F：ATCATGCGTTCGGCTACAGC R：CTCAAACTCCAACTGTTATAAC	$(CTT)_7$	350 bp	3	Ⅰ	0.9742	0.0625	0.5712

（续）

引物名称	引物序列(5′-3′)	重复类型	扩增产物	扩增带数	连锁群	Shannon 指数	观测杂合度	期望杂合度
GenomicPMGC-2826	F：GCTTCTTTAGCGACATGCATC R：GTCAGAACTGTGACAGTAACC	$(GA)_9$	237 bp	4	Ⅳ	0.5454	0.2000	0.3578
GenomicORPM-21	F：GGCTGCAGCACCAGAATAAT R：TGCATCCAAAATTTTCCTCTTT	$(AG)_{12}$	206 bp	2	Ⅸ	0.5982	0.4000	0.3200
GenomicORPM-344	F：GGAGATTGTCGGAGAATGGA R：TGGACGTTACGATAGGAGTGG	$(TC)_8$	229 bp	5	Ⅹ	1.6000	0.8660	0.7181
GenomicGCPM-1013	F：TGCTCCACTCAATGTCAATA R：GACGGTGATAAGAGGAACTG	$(TA)_{11}$	207 bp	4	—	0.9919	0.1818	0.3808
GenomicGCPM-1017	F：GTTTAATTCCCACGTCGTTA R：CGAATGAAGAAAAACCATTC	$(GT)_{11}$	183 bp	5	—	1.1330	0.6250	0.6960
GenomicGCPM-1019	F：CAGGTCCGTAGCACTATTTC R：GCTCAAATGGACATCAAAGT	$(TAA)_6$	208 bp	5	—	1.2267	0.5714	0.6787
GenomicGCPM-1353	F：GAAAACTGATTCCTGATTCG R：CAAGAATCAATGCATGTCTG	$(AT)_9$	150 bp	4	—	0.8016	0.3750	0.4062
GenomicGCPM-136-1	F：TATTGGCAGCAAGAAAGAAT R：TAACTTTTGACATTCCCACC	$(AGG)_6$	154 bp	5	—	1.3936	0.5625	0.7172
GenomicWPMS-16	F：CTCGTACTATTTCCGATGATGACC R：AGATTATTAGGTGGGCCAAGGACT	$(GTC)_8$	145 bp	4	Ⅶ	1.3815	0.5833	0.6779

表 3 利用 Genomic-SSR 与 Genomic-SSR+EST-SSR 标记计算的 16 个杨树无性系相似系数

编号	1	2	3	4	5	6	7	8	9	10	11	12	13	14	15	16
1	1	0.8933	0.8533	0.8533	0.8667	0.8400	0.8000	0.8133	0.5600	0.5867	0.5467	0.6133	0.5600	0.5867	0.6133	0.6000
2	0.8780	1	0.8800	0.8267	0.8400	0.8667	0.8000	0.7600	0.4800	0.5333	0.4667	0.5600	0.4533	0.4800	0.5600	0.5200
3	0.8049	0.8293	1	0.8933	0.8533	0.9067	0.8133	0.8000	0.4667	0.5200	0.4800	0.5467	0.4667	0.4933	0.5467	0.5333
4	0.8293	0.7561	0.8293	1	0.9067	0.9333	0.8133	0.7733	0.4933	0.5200	0.4533	0.5467	0.4667	0.4933	0.5200	0.5333
5	0.8780	0.8049	0.7805	0.8537	1	0.8667	0.8267	0.7867	0.5067	0.5867	0.5200	0.6400	0.5067	0.5600	0.5867	0.6267
6	0.8293	0.8049	0.8780	0.9024	0.8049	1	0.7467	0.7600	0.4533	0.4800	0.4400	0.5067	0.4533	0.4800	0.5067	0.4933
7	0.7561	0.7317	0.7561	0.7805	0.7805	0.6829	1	0.6667	0.5200	0.5733	0.5067	0.6000	0.5200	0.5733	0.5733	0.5867
8	0.8537	0.7805	0.8049	0.7805	0.8293	0.7805	0.6585	1	0.5333	0.5867	0.5467	0.6133	0.5333	0.5600	0.6133	0.6000
9	0.5366	0.4634	0.3902	0.4634	0.5122	0.4146	0.5366	0.5366	1	0.6800	0.8533	0.7067	0.8933	0.8667	0.8400	0.6933
10	0.6585	0.6341	0.5610	0.5854	0.7317	0.5366	0.7073	0.6585	0.6341	1	0.7733	0.8667	0.7333	0.7333	0.8133	0.6667
11	0.5366	0.4634	0.4390	0.4146	0.5610	0.4146	0.5366	0.5854	0.7561	0.7805	1	0.7733	0.9333	0.9333	0.9067	0.7333
12	0.6341	0.6098	0.5366	0.5610	0.7073	0.5122	0.6341	0.6829	0.6585	0.8780	0.7561	1	0.7333	0.7333	0.8133	0.7733
13	0.5366	0.4146	0.3902	0.4146	0.5122	0.4146	0.5366	0.5366	0.8537	0.6829	0.9024	0.6829	1	0.9200	0.8667	0.6933
14	0.6098	0.4878	0.4634	0.4878	0.5854	0.4878	0.6098	0.6098	0.8293	0.7561	0.9268	0.6585	0.9268	1	0.8933	0.7467
15	0.6341	0.6098	0.5366	0.5122	0.6585	0.5122	0.6341	0.6829	0.7561	0.8780	0.8537	0.8049	0.8049	0.8780	1	0.7733
16	0.6098	0.5366	0.5122	0.5366	0.6829	0.4878	0.5610	0.7073	0.6829	0.7073	0.7317	0.7805	0.6829	0.7561	0.7805	1

注:左下角为 Genomic-SSR 标记的相似系数,右上角为 Genomic-SSR+EST-SSR 标记的相似系数。

表4 利用 EST-SSR 与 Genomic-SSR+EST-SSR 标记计算的16个杨树无性系相似系数

编号	1	2	3	4	5	6	7	8	9	10	11	12	13	14	15	16
1	1	0.8933	0.8533	0.8533	0.8667	0.8400	0.8000	0.8133	0.5600	0.5867	0.5467	0.6133	0.5600	0.5867	0.6133	0.6000
2	0.9118	1	0.8800	0.8267	0.8400	0.8667	0.8000	0.7600	0.4800	0.5333	0.4667	0.5600	0.4533	0.4800	0.5600	0.5200
3	0.9118	0.9412	1	0.8933	0.8533	0.9067	0.8133	0.8000	0.4667	0.5200	0.4800	0.5467	0.4667	0.4933	0.5467	0.5333
4	0.8824	0.8529	0.9706	1	0.9067	0.9333	0.8133	0.7733	0.4933	0.5200	0.4533	0.5467	0.4667	0.4933	0.5200	0.5333
5	0.9118	0.8824	0.9412	0.9706	1	0.8667	0.8267	0.7867	0.5067	0.5867	0.5200	0.6400	0.5067	0.5600	0.5867	0.6267
6	0.8529	0.9412	0.9412	0.9706	0.9412	1	0.7467	0.7600	0.4533	0.4800	0.4400	0.5067	0.4533	0.4800	0.5067	0.4933
7	0.8529	0.8824	0.8824	0.8529	0.8824	0.8235	1	0.6667	0.5200	0.5733	0.5067	0.6000	0.5200	0.5733	0.5733	0.5867
8	0.7647	0.7353	0.7941	0.7647	0.7353	0.7353	0.6765	1	0.5333	0.5867	0.5467	0.6133	0.5333	0.5600	0.6133	0.6000
9	0.5882	0.5000	0.5588	0.5294	0.5000	0.5000	0.5000	0.5294	1	0.6800	0.8533	0.7067	0.8933	0.8667	0.8400	0.6933
10	0.5000	0.4118	0.4706	0.4412	0.4118	0.4118	0.4118	0.5000	0.7353	1	0.7733	0.8667	0.7333	0.7333	0.8133	0.6667
11	0.5588	0.4706	0.5294	0.5000	0.4706	0.4706	0.4706	0.5000	0.9706	0.7647	1	0.7733	0.9333	0.9333	0.9067	0.7333
12	0.5882	0.5000	0.5588	0.5294	0.5588	0.5000	0.5588	0.5294	0.7647	0.8529	0.7941	1	0.7333	0.7333	0.8133	0.7733
13	0.5882	0.5000	0.5588	0.5294	0.5000	0.5000	0.5000	0.5294	0.9412	0.7941	0.9706	0.8235	1	0.9200	0.8667	0.6933
14	0.5588	0.4706	0.5294	0.5000	0.5294	0.4706	0.5294	0.5000	0.9118	0.7059	0.9412	0.7041	0.9118	1	0.8933	0.7467
15	0.5882	0.5000	0.5588	0.5294	0.5000	0.5000	0.5000	0.5294	0.9412	0.7353	0.9706	0.8235	0.9412	0.9118	1	0.7733
16	0.5882	0.5000	0.5588	0.5294	0.5588	0.5000	0.6176	0.4706	0.7059	0.6176	0.7353	0.7647	0.7059	0.7353	0.7647	1

注:左下角为 EST-SSR 标记的相似系数,右上角为 Genomic-SSR+EST-SSR 标记的相似系数。

表 5　Genomic-SSR 与 EST-SSR 相似系数的相关性

标记	Genomic-SSR 相似系数	EST-SSR 相似系数
Genomic-SSR	1	
EST-SSR	0. 96**	
Genomic-SSR+EST-SSR	0. 98**	0. 99**

注：**表示在 $P<0.01$ 水平下相关显著。

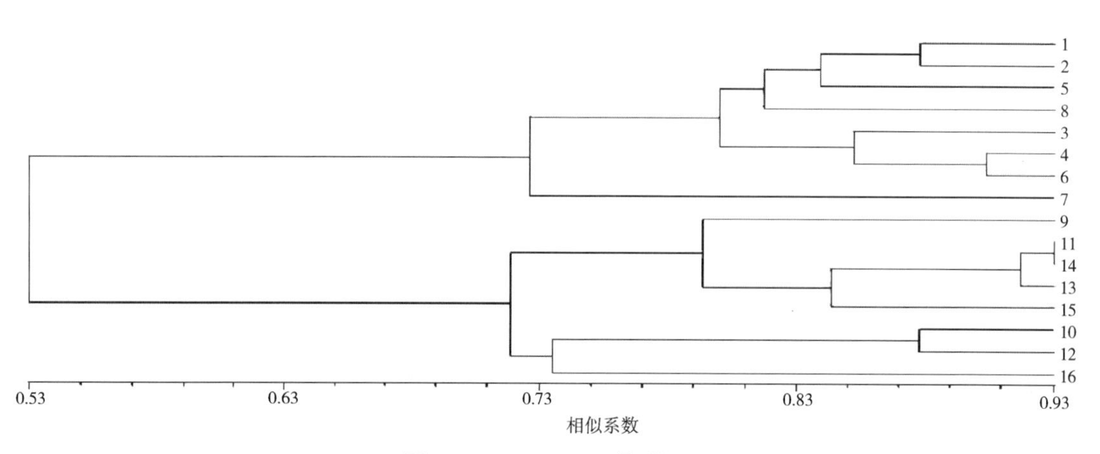

图 2　Genomic-SSR 聚类图

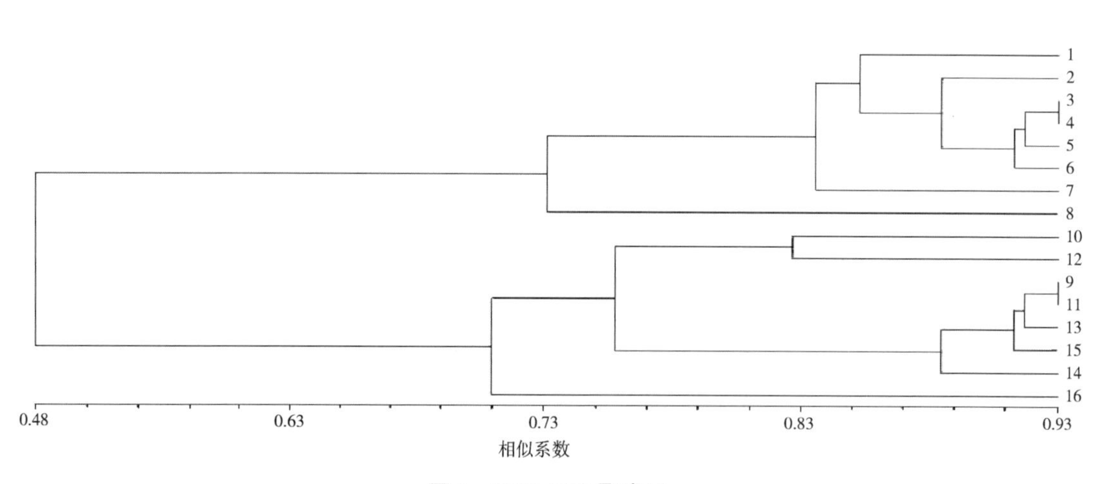

图 3　EST-SSR 聚类图

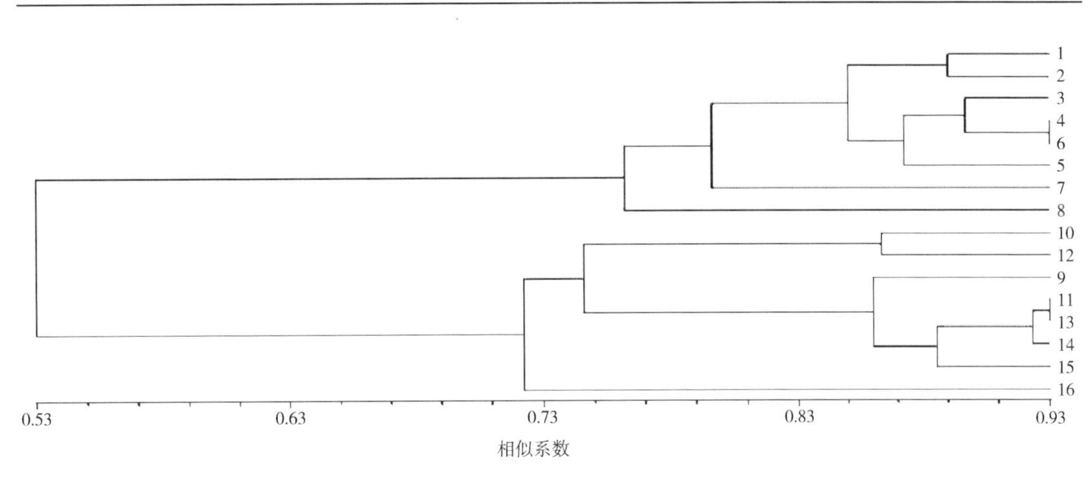

图 4 Genomic-SSR+EST-SSR 聚类图

3 结论与讨论

3.1 Genomic-SSR 与 EST-SSR 遗传差异比较

本研究利用 16 个杨树无性系对 Genomic-SSR 和 EST-SSR 两种分子标记的遗传差异性进行比较分析，结果显示：由 Genomic-SSR 计算出的 Shannon 指数、He 等数据均明显高于 EST-SSR；同时，在杨树中 EST-SSR 多态性明显低于 Genomic-SSR，具有相对较强的保守性。这可能是物种进化过程中的选择牵连效应导致 EST 序列多态性降低造成的(Maynard et al. , 1974)。对相似系数进行分析发现，对于相同个体用两种分子标记计算出的相似系数存在差异，这与小麦的相关研究结果一致(杨新泉等；2005)。以上结果证明，Genomic-SSR 与 EST-SSR 分子标记存在显著的遗传差异性。物种遗传多样性评价是种质资源保护、开发与利用的基础，客观的评价方法是其重要前提(Ellis et al. , 2007)。本研究结果显示，利用不同来源的 SSR 标记对杨树同一群体多样性的评价存在一定的差异。因此，在物种遗传多样性的研究中应利用不同来源的分子标记进行综合评价，以获得更加客观的遗传多样性评价。

3.2 EST-SSR 的特点与应用

本研究结果表明 EST-SSR 在种间保守性强，这将有利于进行连锁图谱的比较分析以及种间遗传多样性的研究(Maria et al. , 2009)。Ellis 等(2007)认为 EST-SSR 较强的保守性意味着在近邻的种间具有通用性。杨彦伶等(2008)对杨树 EST-SSR、Genomic-SSR 两种 SSR 标记在柳树中的通用性进行分析，发现 EST-SSR 的通用性为 54.2%，而 Genomic-SSR 的通用性仅为 10.4%；在结缕草(*Zoysia japonica*)、三枝九叶草(*Epimedium sagittatum*)等物种中也得到了同样的验证(赵岩等，2008；Eujayl et al. , 2001；Ayers et al. , 1997；朱之悌，2005)。因此，利用 EST-SSR 在不同种属构建的遗传图谱具有可比性，为阐述物种不同种属间的系统演化提供新的途径。通过聚类分析发现，EST-SSR 具有相对于

Genomic-SSR 更强的种间基因型鉴别能力，这与 Eujayl 在硬粒小麦中的研究结果相吻合(Eujayl et al.，2001)，并表明 EST-SSR 可能更适合于种间的遗传差异性研究。EST-SSR 是对基因转录区域变异的直接评价，与表型性状、生理生化指标、代谢特征等关联紧密，例如，水稻 *Waxy* 基因 5′不翻译区的$(CT)_n$重复与淀粉含量有关(Ayers et al.，1997)。因此，开展 EST-SSR 与上述性状的相关研究不仅为鉴定基因的功能多样性提供可能，同时作为功能标记对已有遗传连锁图谱加密将为基因的精细定位奠定基础(Hanai et al.，2010)。

3.3 杨树杂交群体的遗传多样性拓宽策略

杂交育种是目前杨树品种选育的最有效方式。根据育种经验，同一杂交组合可以选育出多个优良品种。进一步利用种内品种间杂交，可获得遗传结构比种间杂种更为稳定的杂种但会导致遗传基础变窄(朱之悌，2005)。所以，尽量利用杂交子代群体内的遗传变异，拓宽杂交群体遗传基础十分必要。本实验根据分子标记计算的毛新杨×毛白杨(LM50)回交子代 2 号与 4 号，I-69×Pu-1 杂交子代 12 号与 14 号的相似系数明显小于其他子代间相似系数的平均值，说明可以通过分子标记技术在同一系谱中选出遗传差异较大的优良后代作为交配亲本，从而拓宽杂交育种的遗传基础。例如，在杨树简单连续选择、交互连续选择育种过程中，对第二轮选择群体的表型性状进行测定，从而选出优良子代，然后利用分子标记计算遗传距离，确保入选个体间具有最大的遗传距离，以保持遗传多样性。

Genetic Differences Revealed by Genomic-SSR and EST-SSR in Poplar

Abstract Sixteen poplar clones were used to analyse the genetic differences between Genomic-SSR and EST-SSR. The statistical results showed that the average number of alleles detected by Genomic-SSR was 4.1, Shannon was 1.0637, Heterozygosity was 0.4427, Expected heterozygosity was 0.5523; while for the EST-SSR, the average number of alleles was 2.8, Shannon was 0.6771, Heterozygosity was 0.2081, Expected heterozygosity was 0.4492. Cluster analysis indicated that the EST-SSR capacity of genotypes identification was more precise than Genomic-SSR. In a word, all these results reveal that the EST-SSR and Genomic-SSR have significant genetic differences in markers polymorphism and genotype identification, and that the differences would provide theoretical basis for the rational use of SSR marker in species diversity and other related research.

Keywords poplar, Genomic-SSR, EST-SSR, genetic differences

基于ESTs序列的杨树木质部形成相关EST-SSRs标记的开发与应用*

摘　要　ESTs数据库的发展为分子标记的开发提供了宝贵的资源。本研究从NCBI下载与杨树木质部形成相关的5359条ESTs序列，在其中257条序列中检测到279个SSRs位点，检出率为5.2%。全部SSRs共分为84种重复单元，平均长度18.2bp，变幅为15~36bp，平均分布频率1/6.94kb。所有类型中三核苷酸重复单元占总SSRs的41.3%，处于主导地位。AAG占三核苷酸重复单元的比例为27.5%，是三核苷酸重复单元中的优势重复类型。利用primer3在线设计引物，通过生物信息学方法筛选出7对引物，其中6对扩增效果较好。引物扩增的多态性条带数目为3~6条，平均4.16条。引物多态信息量变化范围为0.2267~0.9845，平均0.5485。应用NTSYS软件在相似系数为0.68水平可将19个杨树无性系准确聚类。研究结果表明利用ESTs序列进行SSRs的开发是一种切实有效的方法，为EST-SSRs在杨树遗传图谱构建以及遗传多样性等方面的应用奠定了基础。

关键词　杨树，木质部形成，EST-SSRs，聚类分析，应用

简单重复序列标记(simple sequence repeats，SSRs)是共显性标记，具有多态性高、容易检测等特点，已广泛应用于遗传多样性研究以及遗传图谱构建等方面。早期SSRs标记开发需要构建基因组文库，通过测序发现含有SSRs结构的克隆并将其分离，再根据SSRs两端的保守序列设计特异性引物进行PCR扩增。然而该方法由于试验周期长、费用高等缺点制约着SSRs的开发。随着功能基因组学的发展，各个重要物种的ESTs(Express Sequence Tags)序列数量迅速增长，基于ESTs序列开发EST-SSRs成为微卫星标记新的发展方向。而且EST是来源于基因转录区域的3′或5′端的测序cDNA序列，与功能基因连锁紧密，因此利用ESTs序列开发SSRs不仅可以避免传统方法的缺点，并且可以作为功能标记用于构建与特定功能相关的遗传连锁功能图谱。目前EST-SSRs在农作物上开发应用广泛，在水稻(*Oryza sativa*)(Cho et al.，2000)、甘蔗(*Saccharum afficenarum*)(Cordeiro et al.，2001)、小麦(*Triticum aestivum*)(Eujayl et al.，2002；Gupta et al.，2003；Yu et al.，2004)、黑麦(*Secale sereale*)(Hackauf et al.，2002)、大麦(*Hordeum vulgare*)(Thiel et al.，2003)、葡萄(*Vitis vinifera*)(Decroocq et al.，2003)等中已有报道。在木本植物中仅在杏(*Prunus armeniaca*)(Decroocq et al.，2003；Xu et al.，2004)、猕猴桃(*Actinidia chinensis*)(Fraser et al.，2004)、云杉(*Picea asperata*)(Rungis et al.，2004)和茶树(*Camellia sinensis*)(金基强 等，2006)、白桦(*Betula platyphylla*)(王艳敏 等，2008)等树种中报道过。

杨树具有生长速度快、生长周期短、基因组相对较小等特点，是重要高大乔木用材生

* 本文原载《分子植物育种》，2010，8(5)：960~970，与宋跃朋、江锡兵、李博和薄文浩合作发表。

物学研究的模式树种(Taylor, 2002)。目前，大量关于杨树木材形成的研究迅速开展(李博, 2009)。其中在分离木材形成主效基因方面，分子标记技术迅速发展。利用 AFLP、SSRs 等分子标记技术进行遗传连锁作图构建和 QTL 定位成为分析木材形成主效基因的重要手段。目前利用各种分子标记技术已构建出多张杨树遗传连锁图谱，但标记的密度尚不足以将主效 QTL 分离。而利用 ESTs 序列的功能注释开发出与功能基因连锁紧密的 SSRs 标记，可以增加遗传连锁图谱的标记密度，对于杨树功能基因的精确定位具有重要意义。目前关于杨树特定功能相关的 EST-SSRs 方面的研究尚未见报道。本研究从 NCBI 下载了关于与杨树木质部形成相关的 5359 条 ESTs 序列，利用生物信息学工具进行 SSRs 的查验和分析。在此基础上根据 SSRs 两端的保守序列设计引物，从中筛选出 6 个与杨树木质部形成相关的 EST-SSRs 标记，进而利用新开发的标记对 19 个杨树无性系进行聚类分析，探讨 EST-SSRs 在杨树遗传多样性以及遗传图谱构建研究中的可行性。

1　材料与方法

1.1　植物材料以及 DNA 的提取

试验材料如表 1 所示，包括 19 个杨树无性系，分别为白杨派的毛白杨(*Populus tomentosa*‘LM50’)、毛新杨(*P. tomentosa*×*P. bolleana*)、青杨派的大青杨(*P. ussuriensis*)、藏川杨(*P. szechuanica*)、滇杨(*P. yunnanensis*)，黑杨派美洲黑杨(*P. deltoides*)以及部分杂交种。每份材料取新鲜叶片利用 CTAB 法(Dellaport et al. , 1983)提取基因组 DNA。

表 1　杨树试验材料

编号	无性系	树种	来源地	编号	无性系	树种	来源地
1	Pu-1	大青杨	吉林省露水河镇	11	滇杨	滇杨	云南省昆明市
2	Pu-5	大青杨	吉林省露水河镇	12	大青杨♂	大青杨	吉林省抚松县
3	150	大青杨×美洲黑杨	北京市	13	大青杨♀	大青杨	吉林省抚松县
4	165	大青杨×美洲黑杨	北京市	14	毛新杨	毛白杨×新疆杨	北京市
5	181	大青杨×美洲黑杨	北京市	15	LM50	毛白杨	北京市
6	LS14	藏川杨	西藏自治区拉萨地区	16	9	毛新杨×毛白杨	北京市
7	LS23	藏川杨	西藏自治区拉萨地区	17	45	毛新杨×毛白杨	北京市
8	SN08	藏川杨	西藏自治区山南地区	18	46	毛新杨×毛白杨	北京市
9	SN06	藏川杨	西藏自治区山南地区	19	51	毛新杨×毛白杨	北京市
10	LS17	藏川杨	西藏自治区拉萨地区				

1.2　ESTs 序列处理和 EST-SSRs 检索

从 NCBI 下载了 5359 条与杨树木质部形成相关的 ESTs 序列。ESTs 是对随机挑取的 cDNA 克隆的外源插入片段的一端或两端进行一次性测序产生的 cDNA 序列(Adams et al. , 1991)，序列中可能掺杂载体序列、poly T 或 poly A。因此本试验采用 EST-trimmer 软件(http://pgrc. ipk-gatersleben. de)对 5359 条 ESTs 序列进行分析，去除载体序列、poly T 或

poly A 并剔除长度小于 100bp 的 ESTs 序列。

利用 SSRIT 软件(http://www.gramene.org/db/search es/ssrtool)对全部 ESTs 序列中的 SSRs 结构进行搜索，其中参数设置为二核苷酸重复、三核苷酸重复、四核苷酸重复、五核苷酸重复、六核苷酸重复分别不得小于 8、5、4、3、3 次。

1.3 EST-SSRs 引物设计

利用 primer3 在线设计引物。引物设计参数为：引物长度 18~22bp；复性温度 52~57℃，(G+C)含量 48.9% ~57.1%，预期扩增产物 76~449bp。引物由英骏生物技术有限公司合成。

1.4 PCR 扩增及电泳检测

PCR 反应体积为 20μL，内含 DNA 约为 100ng，Mg^{2+} 浓度为 1.5mmol · L^{-1}，Taq 酶 0.7U，dNTP 浓度为 0.5mmol · L^{-1}，引物浓度 0.5μmol · L^{-1}。反应程序为：94℃ 预变性 5min；94℃ 变性 30s，复性 30s，退火温度 56~59℃(视不同引物而定)，72℃ 延伸 40s，35 个循环；72℃ 延伸 10min；4℃ 保存。采用 6% 变性聚丙烯酰胺凝胶电泳检测 PCR 扩增产物，银染检测多态性。

1.5 统计方法

对电泳检测结果进行统计，有带记录为“1”，无带记录为“0”，缺失记录为“9”。多态信息量(Polymorphism information content，PIC)计算公式为 $PIC = 1 - \sum p_i^2$ ♁，其中 p_i 表示第 i 个条带的频率(Botstein et al.，1980)。

应用 Nei 等(1979)的方法计算相似系数：

$$GS = 2N_{ij}/(N_i + N_j)$$ ，N_i、N_j、N_{ij}

分别表示第 i 个材料扩增带数、第 j 个材料的扩增带数、第 i 和第 j 个材料扩增的共有带数。利用 NTSYS-pc 2.1 软件按照 UPGMA(Unweight pair group method with arithemetic averages)方法进行聚类分析，绘制聚类图。

2 结果与分析

2.1 杨树 ESTs 分布频率及特点

在全部 5359 条 ESTs 序列中，共发现 257 条 ESTs 序列存在 279 个 SSR 位点，占全部 ESTs 序列的 4.8%，SSRs 出现频率为 5.2%。其中含有一个 SSRs 位点的 ESTs 序列有 259 条，两个 SSRs 位点的有 18 条，三个 SSRs 位点的 2 条。SSRs 长度在 15~36bp 之间，平均长度 18.2bp，平均分布频率 1/6.94kb。在检测结果中，各种重复类型出现频率以及所占 SSRs 比例各有不同，结果见表 2。其中三核苷酸重复类型共 115 个，在 ESTs 中出现频率为 2.1%，在 SSRs 中所占比例达到 41.3%，其次为二核苷酸重复出现频率为 1.0%，所占 SSRs 比例为 19.3%。其余三种重复类型出现频率分别为四核苷酸 0.6%，五核苷酸 0.8%，六核苷酸 0.7%。所占 SSRs 比例分别为 11.5%，14.7%，13.2%。检测结果表明三核苷酸重复是杨树 SSRs 中的优势类型。

在所得到的杨树 EST-SSRs 中，共发现 84 种重复单元，其中二核苷酸 3 种，三核苷酸 10 种，四核苷酸 19 种，五核苷酸 22 种，六核苷酸 30 种。二核苷酸重复中 AG/CT 出现 34 次，占所有二核苷酸重复的 63%，AT/TA 与 AC/TG 分别占 24%和 13%，GC/CG 重复类型未出现。在三核苷酸重复中 AAG/CCT 出现 31 次，占所有重复类型出现次数的 27.5%，其次是 AAT/ATT、ATC/AGT，分别出现 21 次和 14 次，且分别占所有重复类型出现次数的 18.3%、11.9%，其余类型均低于 10%(表 3)。

2.2　EST-SSRs 引物的筛选

来源于 cDNA 3′或 5′端测序产生的 ESTs 序列不包含内含子序列。因此以 ESTs 序列为模板设计引物对植物 DNA 进行扩增时，引物结合位点以及产物长度可能会受到内含子的影响。若设计的引物结合位点落在 DNA 中的内含子区域，则引物无法与 DNA 模板结合并最终导致扩增失败。若扩增产物中含有较大的内含子片段，则不易于进行电泳检测，从而影响试验的准确性与可靠性。

本研究利用杨树基因组数据库(http://genome.jgi-psf.org/Poptr1/Poptr1.home.html)的 BLAST 工具将通过分析与筛选得到的 12 条 ESTs 序列与杨树基因组数据库进行对比，发现有 3 条 ESTs 序列对应的 DNA 中含有超过 1000bp 的内含子，有 2 条无法与基因组序列匹配。因此，初步选择了 7 对引物进行合成，利用 PCR 进行筛选，其中 6 对扩增效果较好，可用于进一步的多态性分析。

表 2　EST-SSRs 序列中不同重复核苷酸数的数量及百分比

重复核苷酸数	EST-SSRs 数量(条)	全部 EST-SSRs 中所占比例(%)	SSRs 数目	全部 SSRs 中所占比例(%)	在 ESTs 中出现频率(%)	平均分布频率(1/kb)
二核苷酸	51	19.8	54	19.3	1.0	35.89
三核苷酸	106	41.3	115	41.3	2.1	16.85
四核苷酸	28	10.9	32	11.5	0.6	60.56
五核苷酸	41	15.9	41	14.7	0.8	47.27
六核苷酸	31	12.1	37	13.2	0.7	52.38
EST-SSR 总数	257	100	279	100	5.2	6.94

2.3　杨树 EST-SSRs 的多态性分析及应用

利用 6 对 EST-SSRs 引物对 19 个杨树材料的 DNA 进行扩增，共产生出 25 个扩增条带，且全部具有多态性，结果如表 4 及图 1 所示。6 对引物扩增条带数目分别为 3~6 个不等，平均 4.16 条。引物 CU310401 和 CU310140 扩增条带最多，为 6 条。引物 CU307924 和 CU316835 扩增条带最少，为 3 条。SSRs 引物的多态信息含量(PIC)为 0.2267~0.9845，平均 0.5485。其中多态信息含量最高的是引物 CU310140，为 0.9845，最低的是引物 CU310401，为 0.2267。

统计扩增条带并进行聚类分析，结果如图 2 所示，19 个杨树材料在相似系数为 0.68 的水平被分为Ⅰ、Ⅱ两个类群。第Ⅰ类群包括 10 个青杨派的无性系和 3 株大青杨与美洲黑杨杂交种无性系相似系数在 0.72~1 之间。第Ⅱ类群包含全部的 6 个白杨无性系。第Ⅰ类群在相似系数 0.72 的水平上又分为两个类群，一类由 9 个青杨组成，其中来自拉萨和山南两个种源的藏川杨在 0.91 水平上聚到一起，然后和滇杨一起在 0.72 水平上和远源的大青杨聚到一起。另一类包含 1 个大青杨和 3 个大青杨与美洲黑杨杂交种。第Ⅱ类群中，毛新杨和四个子代在 0.87 的水平先聚到一起，然后和毛白杨在 0.82 水平聚到一起。

表 3　EST-SSRs 序列中不同重复类型及其数量和百分比

重复核苷酸数	重复单元	SSRs 数量	百分比(%)
二核苷酸	AG/CT	34	63.0
	AT/TA	13	24.0
	AC/TG	7	13.0
	GC/CG	0	0
三核苷酸	AAG/CTT	31	27.5
	AAT/ATT	21	18.3
	AAC/GTT	7	6.4
	ATC/AGT	14	11.9
	ACT/ATG	10	8.2
	AGG/CCT	9	7.4
	AGC/CGT	9	7.4
	ACC/GGT	5	4.6
	ACG/CGT	2	1.8
	CCG/CGG	7	6.4

3　讨论

本研究分析了与杨树木质部形成相关的 ESTs 序列中 SSRs 分布情况，并初步探讨了 EST-SSRs 在杨树遗传多样性、遗传图谱构建、基因功能鉴定研究中的可行性。对 5359 条 ESTs 序列进行分析，发现 257 条 ESTs 序列含有 279 个 SSRs 位点，SSRs 出现频率为 5.2%，与小麦的 5.4%(Gupta et al.，2003)、香蕉(*Musa nana*)的 5.3%(王静毅等，2008)接近，高于甘蔗的 2.9%(Cordeiro et al.，2001)、水稻的 4.7%(Kantety et al.，2002)，但是远低于柑橘(*Citrus reticulata*)的 21.6%(江东等，2006)、茶树的 17.7%(金基强等，2006)等。SSRs 平均分布频率为 1/6.9kb，与大麦的 1/6.3kb(Thiel et al.，2003)接近，低于茶树的 1/2.6kb(金基强等，2006)、柑橘的 1/5.7kb(Chen et al.，2006)，而远高于小麦的 1/15.6kb(Kantety et al.，2002)、拟南芥(*Arabidopsis thaliana*)的 1/13.8kb 以及棉花(*Gossypium* spp.)的 1/20kb(Carde et al.，2000)。据 Rota 等(2005)研究结果显示，在水稻中最小 SSRs 长度标准由 12bp 增加到 30bp 时，EST-SSRs 的频率从 50%减少到 1%，同时二核苷酸重复的数量从为三核苷酸重复的 1/10 到二者基本接近，重复基元的主导类型也由 CCG 变为 AG 重复，而当 SSRs 最小长度为 40bp 时，二核苷酸重复的数量超过了三核苷酸重复。因此，以上这些差异可能来源于不同物种之间 SSRs 分布频率的真实差异，也可能是由于 ESTs 来源不同以及在检索 SSRs 位点时设置的参数不同所造成的。张新叶等(2009)以及 Carde L 等(2000)研究结果显示杨树中 EST-SSRs 的分布频率分别为 1/3.83kb 和 1/14kb，与本试验结果差距较大。本试验与前者的差异是由检索 SSR 的设置参数不同造成的，而后者设置参数与本试验相似，但结果有一定差距。推测可能是由于本研究选择的与木质部形成相关的 ESTs 序列相比杨树其他功能相关 ESTs 序列含有更丰富的 SSRs 位点。

表 4　引物信息与多态信息含量

引物名称	引物序列(5'-3')	重复类型	扩增产物	扩增带数	多态信息含量
CU310415	F：CCCGAGTCAATCTGAGTTAGTA R：CTTGTTTATTGGAGATGGAGC	$(AAAT)_4$	122bp	3	0.7459
CU310401	F：GTGCAGGCAGATATTTATGGA R：GGAAGCAGCAGTTGAAGAAG	$(TGCTTC)_3$	127bp	6	0.2267
CU310140	F：AACCGTATGAAACTTTAGGCA R：AAACCCACCCACTGTTATTG	$(TTTATA)_3$	334bp	6	0.9845
CU309310	F：TTATCCACCCTCCCTGTCTC R：AAAGGAAAGGTCCATCGTAAT	$(TTTG)_4$	76bp	3	0.4628
CU316835	F：GCAGAGGAAGCAGCAAGAG R：CCAAGTCACGGGACAGTAAAG	$(CTT)_5$	449bp	4	0.3787
CU307924	F：TTAAAGGAAGCCACAAGAAGT R：ACAAACATACCCAGAAACCAA	$(TTG)_6$	167bp	3	0.4924

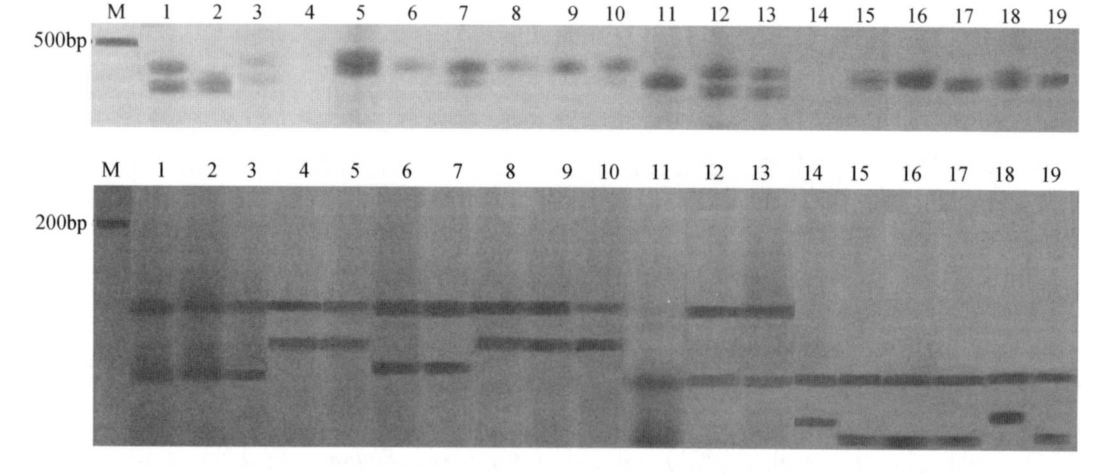

图 1　引物 CU316835(上)和 CU310401(下)在 19 个杨树无性系材料中的扩增结果

从目前研究报道来看，大多数植物的 EST-SSRs 都是以三核苷酸重复类型为主，例如，在甘蔗、小麦中三核苷酸重复类型分别为 90%、78%、33.2%(Corerio et al.，2001；陈军方等，2005）。在大麦、小麦、玉米（*Zea mays*）、黑麦、高粱（*Sorghum vulgare*）以及水稻等重要禾谷类作物的 ESTs 分析中发现，三核苷酸重复出现次数最多（54%~78%），其次为二核苷酸重复（17%~40%）（Varshney et al.，2002）。在本研究中三核苷酸重复占据所有重复类型的 41.3%，属于优势重复类型，其次是二核苷酸重复占据所有重复类型的 19.8%。在所有 10 种三核苷酸重复类型中，AAG/CTT 类型出现频率最高，占据的比例达到 27.5%，与 Gao 等（2003）关于在双子叶植物中 AAG/CTT 重复丰度很高的推测一致可能是由于密码子以三核苷酸为一功能单位，因此由插入缺失突变造成的三核苷酸位移不会给一个表达基因的阅读框造成太大影响（Metzgar et al.，2000）。二核苷酸重复类型中 AG/CT 占据 4 种二核苷酸重复类型的 63%，且没有发现 GC/CG 重复类型。研究表明，当 4 种碱基随机组合时，若 EST-SSRs 数量足够多且无偏倚性，将可能产生 4、10、33、102 和 350 种二、三、四、五、六核苷酸重复（Rota et al.，2005）。本研究中出现了所有类型的三核苷酸重复，二核苷酸重复仅 GC/CG 类型未出现，四核苷酸、五核苷酸和六核苷酸重复类型仅少量出现，这可能与杨树本身的特性相关，也可能是 EST-SSRs 数量不足所造成。此外，Temnykh 等（2001）研究表明当 SSRs 长度在 12bp 以下的 SSRs 在不同品种间显示出多态性最低，长度在 12~20bp 之间的 SSRs 多态性较高，而显示出最高多态性的是 20 及 20bp 以上的 SSRs。杨树 EST-SSRs 平均长度达 18.2bp，其中大于 20bp 的有 75 条，占全部的 26.8%，属于在品种间多态性最高的 SSRs，其余大部分属于有较高多态性的 SSRs。对 EST-SRRs 的出现频率、平均分布频率以及平均长度进行分析，结果表明杨树与木质部形成相关的 ESTs 序列中存在着多态性较高并且数量丰富的 SSRs 位点，有利于高密度遗传连锁功能图谱的构建。

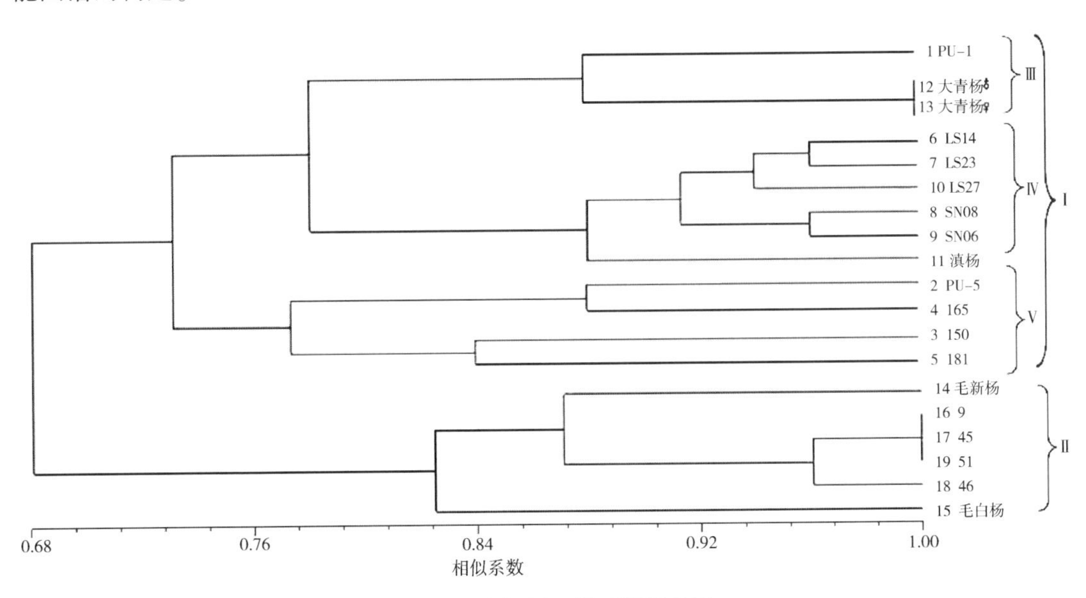

图 2 19 个杨树无性系的聚类图

EST-SSRs 是以 PCR 为基础的，因此在 ESTs 序列中检索出 SSRs 位点后，还应该考虑所含 SSRs 的 ESTs 序列在引物设计时的可操作性（李小白等，2007）。ESTs 序列通常在

150~500bp 之间，若 SSRs 位点出现在 ESTs 的两端，将导致没有足够长度的侧翼序列用于设计特异性引物。本试验利用 BLAST 工具在基因组数据库寻找 ESTs 序列的同源序列，将 ESTs 序列以基因组数据库的同源序列为模板进行延伸。通过该方法使 ESTs 序列有足够的长度用于设计特异性高的引物。并通过与基因组数据库的比对去除内含子的干扰。最终合成的 7 对引物中有 6 对能够成功扩增出产物，占全部引物的 85.7%，远高于香蕉的 65.1%（王静毅等，2008）、白桦的 59.1%（王艳敏等，2008）等。因此，可以认为这是一种去除内含子干扰的有效方法。由于 ESTs 序列大部分来源于基因 3′或 5′的 UTR（Untranslated Region），该区域相对于 DNA 其他非编码区域保守性强（Cherdsak et al.，2004；Shalu et al.，2009）。利用这种特性，没有全基因组测序的物种可以与邻近物种的全基因组序列数据库进行比对或尽量利用来自 3′端的 ESTs 序列进行设计引物也可以避免内含子的干扰，增加引物设计的准确性（Temesgen et al.，2001）。此外，EST-SSRs 在不同种间保守性强的特性有利于进行连锁图谱的比较分析以及种间遗传多样性的研究（Maria et al.，2009）。

本研究计算得出的 ESTs 序列多态信息含量在 0.2267 ~0.9845 之间，表明 EST-SSRs 的多态性变化较大。这可能是由选择牵连效应（Hitchhiking Effects）造成的（Maynard et al.，1974）。选择牵连效应对个别基因的正向选择致使其侧翼遗传多样性降低。一般来说，对维持植物生存相关基因的选择所引起的牵连效应较强。利用选择牵连效应原理，可以通过对 EST-SSRs 多态性变化的研究为杨树抗性基因的定位提供一个新思路。此外，另有研究表明，水稻 Waxy 基因 5′非翻译区域的（CT）$_n$ 重复次数与淀粉含量有关（Ayers et al.，1997）。因此，EST-SSRs 用于遗传差异研究，使得基因功能多样性的鉴定成为可能，为进一步开展杨树木质部形成相关基因转录区域的 SSRs 重复次数差异与相关基因功能变化关系的研究奠定了基础。

EST-SSRs 具有很强的基因型鉴别能力，在小麦相关研究中，使用较少的引物也可以将亲缘关系密切的不同基因型区分开（Eujayl et al.，2001）。选择不同遗传背景的材料可以检测 EST-SSRs 鉴别基因型能力的精准度，基于此本研究选择包括杨树不同亚派、派间杂交种、派内杂交种三种类型的材料。在本研究中，开发出的 6 对 EST-SSRs 引物可以准确的区分 19 个杨树材料表现出精准的基因型鉴别能力，证实 EST-SSRs 可以应用于杨树遗传多样性研究。

Information Analysis of SSRs in ESTs Related with Xyligenesis and EST-SSRs Marker Application in Poplar

Abstract Expressed sequence tags (ESTs) data provide valuable resources for development of molecular markers. In present study, a total of 5359 poplar unigene sequences related with xyligenesis were downloaded from NCBI, resulting in the identification of 279 SSRs in 257 sequences, accounting for 5.2% of the whole EST sequences. Among them, 84 SSR motifs were founded. The length of SSRs were changed between 15 ~ 36bp, the average length was 18.2bp and the average distance was 1/6.94kb. The trinucleotide repeats appeared to be the most abundant SSRs (41.3%). The rich repeats AAG were predominant, accounting for 27.5% in tri-nucleotide repeats. Seven primer pairs were designed. Six primers were validated as usable markers. These primers revealed moderate to high polymorphism information content. The polymorphic bands were ranged from 3 to 6, averagely 4. 16. The mean PIC value was 0. 5485, ranging from 0. 2267 to 0. 9845. Nineteen poplar clones were separated into two major clusters by UPGMA analysis. These results show that derive SSRs from ESTs sequence is a realistic and effective method, laying the foundation of EST-SSRs application in poplar genetic map construction and genetic diversity research.

Keywords poplar, xylogenesis, EST-SSRs, cluster analysis, application

第三部分

杨树生殖生物学与基因调控

杨树生殖生物学研究进展*

摘　要　该文介绍了杨树(*Populus* spp.)生殖生物学的研究概况，包括：①杨树有性生殖过程中雌雄生殖系统形态结构和细胞化学变化及其生物学意义；②雄性不育的机理；③杨树杂交不亲和性的表现和原因及其克服的方法，花粉-柱头识别反应；④杨树生殖工程取得的成就。

关键词　杨树，生殖生物学，有性生殖，雄性不育，杂交不亲和性，生殖工程

植物生殖生物学是在植物胚胎学的基础上发展起来的。19 世纪末，随着人们对植物性别认识的不断深入，逐渐形成了以描述植物生殖过程为主的植物胚胎学。应运而生的植物生殖生物学，不仅仅用植物学的知识和技术手段对植物生殖行为进行研究，并且已经渗透到解剖学、细胞学、生理学、生物化学、遗传学和分子生物学等多学科领域，产生了实验胚胎学、植物生殖生理学和植物生殖遗传学等分支学科。植物生殖生物学与植物育种学有着密切的关系。

杨树具有速生等特点，广泛分布于欧洲、亚洲和北美洲，一般在北纬 30°~72°之间(徐纬英，1988)。杨树为雌雄异株，偶有雌雄同株(张志毅等，1992；Boes et al.，1994)。目前林业生产上广泛应用的大多是人工培育的杂种无性系(徐纬英，1988)。杨树主要以无性繁殖为主，对杨树有性生殖研究较少。杨树生殖生物学的研究与农作物等草本植物相比，还处于较落后的水平。从 20 世纪 70 年代以来，随着生殖生物学理论和实验手段的不断丰富，杨树生殖生物学的研究进展很快，对杨树各派代表树种的整个有性生殖过程有了细致的了解(董源，1982；1984；1988；樊汝汶，1984；1983；1983；李文钿等，1988；1982；Khurana，1980；Nagaraj，1952)；对杨树受精前后的组织化学变化及其生理意义进行了探究(朱彤等，1988，1989)；从细胞化学和超微结构等方面对杂交不亲和性进行了研究(李文钿等，1986；朱彤等，1988；1990；Knox，1972；Rougier，1988；Villar，1987)；利用子房、胚和花粉离体培养成功地诱导出多种杨树植株(遗传学报，1977；王敬驹，1975；遗传学报，1976；朱湘渝，1980)。

1　有性生殖过程

1.1　花药的结构、小孢子发生与花粉的发育

杨树的花药壁分为四层，由表及里依次为——表皮层、纤维层、中层和绒毡层(董源，1982；樊汝汶，1982；李文钿，1983)。其中绒毡层对小孢子和花粉的发育有重要意义，主

* 本文原载《北京林业大学学报》，2000，22(6)：69-74，与于雪松合作发表。

要是因为它分泌胼胝质酶能分解四分体胼胝质壁使小孢子分离，合成识别蛋白释放到花粉外壁，绒毡层的降解产物可作为花粉合成DNA、RNA、蛋白质和淀粉的原料。随着小孢子的发生和花粉的发育，腺质绒毡层细胞相应发生一系列变化(樊汝汶，1982；李文钿，1983)。李文钿等(1988)在胡杨(*P. euphratica* Oliv.)花粉扫描电镜图像中发现，绒毡层与花粉壁之间有细丝相连，推测是孢粉素"桥"，并发现绒毡层的内切向壁的乌氏体，不随绒毡层细胞的退化而消失，而是有些散落在药室中，有些转移到花粉外壁上(李文钿等，1988)。

花粉母细胞的减数分裂为同时型(董源，1982；樊汝汶，1982；1983；李文钿等，1988；刘玉喜，1986)，四分体小孢子大多呈四面体形，少左右对称形(董源，1982；樊汝汶，1982；李文钿等，1988；朱彤等，1988)。董源(1988)在毛白杨(*P. tomentosa* Carr.)小孢子中发现细胞自体消化细胞质的活动极为活跃，并有多层的同心圆膜状复合系统(董源，1988)。刚形成的小孢子体积很快增大，渐渐变为圆形，细胞质增多，并开始液泡化，核增大，从中央移到边上(樊汝汶，1982)。刘玉喜等(1979)在中东杨(*P. berolinensis* Dippel.)中发现有多核和多核仁小孢子存在(刘玉喜，1986)。经过短暂的休眠期，休眠期的长短与温度条件有关，毛白杨为2~7 d(董源，1982)。休眠期结束后，小孢子进行有丝分裂，由于纺锤体的不对称，造成分裂不均等，营养细胞明显大于生殖细胞，营养细胞内质体积累大量淀粉粒(樊汝汶，1983)。樊汝汶(1983)和Rougier(1991)分别报道了美洲黑杨(*P. deltoides* Marshall)雄性生殖单位(male germ unit，MGU)在发育过程中细胞超微结构的变化。杨树成熟花粉属于22细胞型(董源，1982；樊汝汶，1982；李文钿等，1988；1982)，球型，无萌发孔，外壁雕纹有皱波状、细网状、小穴状、颗粒状等多种形状(李文钿等，1982；张绮纹，1988)。

1.2　大孢子发生和胚囊的发育

杨树一般为胚珠倒生，在发育的早期是双珠被，在大孢子母细胞减数分裂期内珠被滞育，并开始解体，最终成为单珠被(樊汝汶，1984；李文钿等，1988；1982；Khurana，1980)，但大叶杨(*P. lasiocarpa* Oliv.)的内珠被不解体，一直为双珠被(朱彤等，1989)。造孢细胞直接分化为大孢子母细胞(樊汝汶，1982；Khurana，1980)，细胞核大而明显，经减数分裂形成四分体，呈T形或直线形排列，靠珠孔端的三个大孢子退化，合点端的大孢子延伸最终发育成为八核胚囊(董源，1984；樊汝汶，1982；李文钿等，1988；1982；Khurana，1980)，在欧美杨(*P. euramericana*)和胡杨中发现有时合点端第二个大孢子成为功能孢子(樊汝汶，1982；李文钿等，1988)。卵器穿透珠心伸入珠孔，卵细胞和助细胞极性相反(樊汝汶，1982；李文钿等，1982)。杨树胚囊的发育属于蓼型(董源，1984；樊汝汶，1982；李文钿等，1988；1982)。

1.3　受精过程

杨树的花粉落到雌株的柱头上，一般几小时后开始在柱头上萌发，沿狭窄的花柱道进入子房(樊汝汶，1982；朱彤等，1989；Khurana，1980；Russel，1990；Villar，1987)，在此期间花粉管中的生殖细胞分裂成2个精子(李文钿，1982)，花粉管经珠孔进入退化的助细胞，在此处或在卵细胞与次生核之间释放其内含物，进行双受精（董源，1984；李文钿，1982；朱彤等，1989；Khurana，1980；Russel，1990）。反足细胞是“短命”的，受精后很快就退化（樊汝汶，1982；Khurana，1980）。整个受精作用要消耗大量淀粉粒（朱彤等，1989）。

关于杨树受精过程中精卵结合的详细情况报道不多，对于雌雄性细胞的融合机制，雄性细胞质在受精过程中的作用，卵细胞受精前后生理生化变化等，有待于进一步深入探讨。董源（1984）认为毛白杨的受精作用是介于有丝分裂前型和有丝分裂后型之间的中间型，精子起初整个贴附在卵核的核膜上，处于静止期，随后精子进入卵核中，先是核仁合并，接着核质融合（董源，1984）。朱彤等（1989）发现大叶杨的受精作用为有丝分裂前型。Russel（1990）对美洲黑杨受精前的精子进行了研究，发现精子在退化的助细胞发生液泡化，细胞质稀释，细胞核染色体浓缩，细胞质减少，核质分离（Russel，1990）。

1.4 胚乳和胚的发育

精核与次生核融合后，形成大的富含 DNA、RNA 和蛋白质的初生胚乳核，一般不经过休眠，直接开始初生胚乳核的分裂，比受精卵的分裂早（董源，1984；李文钿，1982；李文钿，1982；朱彤，1989；Khurana，1980；Winton，1968），但在毛白杨中发现，有时受精卵已经形成八细胞原胚，初生胚乳核尚处于休眠状态（董源，1984）。胚乳在胚的发育过程中被吸收，故成熟种子不含胚乳（樊汝汶，1982；李文钿，1982；朱彤，1989；Khurana，1980）。杨树胚乳发育类型属于核型（董源，1984；樊汝汶，1982；李文钿，1982；朱彤，1989）。精卵融合后，进入合子休眠期，核内 DNA、RNA 含量上升（朱彤，1989）。大叶杨和小叶杨（*P. simonii* Carr.）的合子休眠期分别为 6~8d（朱彤，1989）和 6~10d（李文钿，1982）。合子在休眠期并不是静止的，而是进行着十分活跃的生理过程，在大叶杨中发现，合子经历了液泡消失、合子皱缩、液泡再现、合子伸长等极性化过程（朱彤，1989）。在杨树属各派树种中大叶杨的胚胎学特征十分特殊（表 1），大叶杨胚胎发育类型为茄型，而其他各派均为柳叶菜型（樊汝汶，1982；李文钿，1982）。

1.5 有性生殖过程的组织化学

植物有性生殖过程中，除了形态和结构发生变化外，各种物质（DNA，RNA，蛋白质，淀粉粒和胼胝质）的含量和分布也发生相应的改变。中国林科院朱彤等（1988）首次报道了杨树小孢子发生过程中的组织化学变化，发现大叶杨小孢子母细胞的营养源位于表皮和纤维层，绒毡层细胞和小孢子母细胞逐渐积累多糖，绒毡层细胞始终含丰富的 DNA、RNA 和蛋白质，RNA 和蛋白质在小孢子发生过程中表现一致性的分布和含量变化（朱彤等，1988）。在小孢子发生过程中，蛋白质的作用主要是作为功能蛋白和结构蛋白参与代谢，淀粉粒则作为发育中所需的营养来源（李文钿等，1988；朱彤等，1988）。PAS 反应表明，胡杨单核花粉晚期细胞质中开始积累淀粉粒，在 22 细胞花粉的生长过程中，营养细胞内的淀粉粒日渐增多并增大，花粉散出前夕，淀粉粒和其他内含物的积累达到高峰（李文钿等，1988）。小孢子母细胞在减数分裂前期 I 开始沉积胼胝质，并逐渐加厚，到四分体后期胼胝质壁解体，释放出小孢子（李文钿等，1988）。

表 1　杨树各派代表树种胚胎学观察结果

派名	种名	心皮	种子	珠被	胚胎发生类型	胚乳	胚囊	受精所需时间
大叶杨派	大叶杨	2~3		双	茄型	核型	蓼型	6~10d
青杨派	小叶杨	2~3	4~6	单	柳叶菜型	核型	蓼型	3~6d
黑杨派	美洲黑杨	3~4	4~32	单	柳叶菜型	核型	蓼型	2~3d
	欧洲黑杨	2		单	柳叶菜型	核型	蓼型	
白杨派	毛白杨	2	1~2	单	柳叶菜型	核型	蓼型	40~48b
	响叶杨	2		单	柳叶菜型	核型	蓼型	5~7d
	美洲山杨	2		单	柳叶菜型	核型	蓼型	
	银白杨	2~4	3~4	单	柳叶菜型	核型	蓼型	
胡杨派	胡杨	3		单	柳叶菜型	核型	蓼型	

大叶杨的大孢子母细胞、四分体及功能大孢子中含不溶性多糖较少，含 RNA 和蛋白质丰富，成熟胚囊中除反足细胞外充满淀粉粒，围绕在卵核及次生核周围(朱彤等，1989)，助细胞具多糖性质的丝状器，胚囊的营养来源于子房和胎座细胞内贮存的淀粉粒，卵细胞 PAS 反应呈阴性。受精后，胚胎发育过程中，DNA、RNA 和蛋白质含量和分布发生一系列变化，游离胚乳核中 RNA 和蛋白质含量较高，不含可溶性多糖及淀粉粒，围绕在卵核及次生核周围的淀粉粒迅速消失，消耗的淀粉一方面水解作为能量的来源，另一方面为合子多糖性质的细胞壁的合成提供原料，合子 PAS 反应呈阳性(朱彤等，1989)。Label (1994)发现欧洲黑杨(*P. nigra* L.) 受精前后子房内的 ABA 含量不变(Label，1994) 。

2　雄性不育

雄性不育是由于植株不能产生正常的花药、花粉或雄配子，表现形式有：不能形成正常的小孢子发生组织，小孢子发生异常，形成不完善、不能存活的、畸形或败育的花粉，花粉不能成熟或无萌发能力，花药不开裂。花粉发育的每个时期都可能发生败育。雄性不育是植物界中一种普遍现象，这方面的研究报道很多，但大多以农作物为研究对象，对树木雄性不育机理的研究不多。樊汝汶等(1992)认为中国鹅掌楸花粉败育是遗传因子造成的，导致小孢子母细胞不能正常启动减数分裂或减数分裂异常。杨树雄性不育机理的研究主要停留在对形态结构的观察，尚未深入到分子遗传水平。目前认为，杨树雄性不育的原因主要是绒毡层发育异常。由于绒毡层是花粉粒发育时的营养来源，它的异常发育将导致绒毡层无法为花粉的发育提供适当的营养，使花粉处于饥饿状态。李文钿等(1982)在小叶杨中同时观察到绒毡层的早期败育和延迟败育，在出现早期败育的绒毡层花粉囊中可以看到败育的花粉母细胞不正常的减数分裂和败育的小孢子四分体；在绒毡层延迟退化的花粉囊中，往往有败育的粘连成片的空壳花粉和畸形的多核大花粉(李文钿，1982)。董源(1982)认为毛白杨雄性不育是由于绒毡层过度增生抑制了花粉的发育，绒毡层异常通常发生在小孢子母细胞和四分体时期。败育的四分体无核或核不明显，染色极深(董源，1982)，并认为低温能诱导雄性不育。康向阳(1996)则认为绒毡层的异常发育只是毛白杨雄性不育的一种表象，其根源在于毛白杨染色体的异源性，导致减数分裂时同源染色体不均衡分配(康向阳，1996)。纤维层内壁和侧壁加厚，有利于失水后花粉囊的裂开和散粉，纤维层的加厚异常将导致花药不开裂(李文钿，1982)。

3 种间杂交不亲和性

探索种间杂交不亲和性的原因，是植物生殖生物学中一个很重要的内容，为植物杂交育种工作的深入开展提供了条件。20 世纪 70 年代以来，国内外针对杨树种间杂交不亲和性进行了大量的研究(李文钿，1986；朱彤等，1988；Rougier，1992；Stettler，1980)，揭示了一些种间杂交不亲和性的原因。如花粉萌发率很低(Guries et al.，1976；Knox，1972)，花粉-柱头不亲和，花粉管不能进入柱头(李文钿等，1986；Gaget，1984)；花柱不亲和，花粉管无法达到子房(Gaget，1984；Guries et al.，1976)；胚囊发育迟缓，花粉管无法进入胚囊(李文钿等，1986)；杂种胚乳发育不正常导致杂种胚死亡等(李文钿等，1986)。不亲和性的表现主要有花粉及花粉管分泌物少，柱头表面和花粉管前端沉积胼胝质(Gaget，1984；Li，1991)，花粉管在柱头上表现异常，扭曲、盘绕、爬行(朱彤等，1988；Li，1991；Von Melchior，1968)，柱头细胞提前解体(朱彤等，1988)，胚乳和胚败育(李文钿等，1986；Li，1991；Von Melchior，1968)等。朱彤等(1989)认为柱头细胞的提取解体退化是雌蕊拒绝反应的一部分，表现为细胞质解体，细胞内部液泡化，膜结构逐渐消失以及细胞质电子密度增大。

花粉落到柱头上后，花粉-柱头间将发生主动识别反应(Knox，1972)，柱头一旦识别出了不亲和的花粉壁蛋白，通常在乳突细胞内迅速产生胼胝质阻止花粉管的继续侵入(Gaget，1984)。Knox(1972) 首次在杨树的种间杂交研究中提出了主动识别(positive recognition)的概念，他在美洲黑杨×银白杨的杂交中发现，银白杨花粉与美洲黑杨柱头不亲和是由于绒毡层组织合成的一种特异蛋白，授粉后被释放出来，与柱头上的特异蛋白进行相互识别，并产生一系列反应，从而决定了花粉管能否进入柱头实现受精(Knox，1972)。花粉-柱头相互作用这一识别过程涉及多种酶系的活动，在花粉外壁识别蛋白中常见酯酶、过氧化物酶、酸性磷酸酶和转移酶类等酶系活动，在柱头表膜中有非特异性酯酶、酸性磷酸酶及过氧化物酶等酶系活动(朱彤等，1990；Ashford and Knox，1980)。李文钿等(1990)首次在杨树中报道了柱头表膜 ATP 酶的活动，发现大叶杨柱头 ATP 酶与花粉-柱头相互作用有一定相关性，ATP 酶的活性随柱头可授期的到来而增加，幼嫩期和过盛花期的柱头薄膜上 ATP 酶活性低或没有(朱彤等，1990)。Villar(1993)发现亲和授粉后(欧洲黑杨×欧洲黑杨)β_2 半乳糖苷酶的活动上升，而在不亲和杂交组合中(欧洲黑杨×银白杨)，β_2 半乳糖苷酶未表现活性上升，认为 β_2 半乳糖苷酶在花粉管的异养生长中起重要作用，与花粉-柱头相互作用有关。Rougier(1988)在美洲黑杨×银白杨的不亲和杂交中发现，腺苷酸环化酶(CAMPase)的活性比亲和杂交的活性低，认为 CAMPase 与花粉在柱头上粘附、吸水及萌发有关，活性下降是配子体表型不亲和的细胞化学表现之一(Rougier，1988)。Gaude(1982)用血细胞凝集技术研究了欧美杨花粉外壁可扩散的细胞粘附分子(CAMS)，推测 CAMS 可能是一种类外源凝集素，与花粉-柱头识别反应有关(Gaude，1982)。总之，对这些酶类的研究为探讨不亲和性机理以及寻求克服方法提供了新的思路和途径。

杨树杂交中的单向不亲和性较为普遍，对于单向不亲和性的原因尚无一致意见。Villar 等(1987)认为欧洲黑杨×银白杨的单向不亲和性是由于两个种的柱头类型不同造成的，欧洲黑杨具有干柱头的特征，表面有一层薄膜，而银白杨是湿柱头，有丰富的分泌物(Villar，1987)。这与李文钿等(1995)的研究结果不同，李文钿认为杨树柱头都是湿型分泌柱头(李文钿，1995)。克服杂交不亲和性的方法很多，如花粉蒙导、用生理活性物质处理雌蕊、重复授粉、受精工程等。在杨树杂交中主要采用蒙导花粉授粉，使柱头不能识别不亲和的花粉，以克服种间杂交不亲和性，获得杂种(Gaget，1989；Stettler，1968；1976)。美国华盛顿大学的 Stettler 教授在这方面作了许多工作(Stettler，1968；1976；樊汝汶，1982)。

4 生殖工程

目前，杨树生殖工程的研究主要是以子房、胚珠和花药离体培养为主，已成功地诱导出生长正常的植株。

杨树杂交胚胎学的研究表明，杨树一些种间杂交的障碍不是发生在受精之前，而是发生在受精之后，胚在发育过程中败育(李文钿等，1986；1983；Li，1991)，因此利用子房和胚珠离体培养可以克服种间杂交胚的败育。迄今为止，利用子房和胚珠培养诱导植株成功的有：小叶杨×美杨，小叶杨×胡杨，小叶杨×大叶杨，欧洲黑杨×大叶杨，大叶杨，银白杨，欧洲黑杨，小黑杨(*P.* ×*simonigra* Chon-Lin)等(李文钿等，1985；吴克贤等，1983；Kouider，1984；Raquin and Troussard，1993)。李文钿等(1983)对不同发育阶段小叶杨×美杨杂种胚的胚珠进行离体培养，认为离体胚珠的最适接种时间是在大多数杂种胚尚未败育之前，胚龄越大，越容易成活(李文钿，1983)。

70 年代以来，相继成功地利用花药离体培养诱导出多种杨树花粉植株，有：小叶杨×美杨、小叶杨×黑杨、北京杨、中东杨、美杨、美杨×青杨、大青杨、欧洲黑杨、欧洲黑杨×小叶杨、银白杨×小叶杨、美杨×欧洲黑杨、箭杆杨(*P. thevestina* Dode)×欧洲黑杨、胡杨等(东北林学院树木育种组，1977；王敬驹，1975；黑龙江省林业研究院林业研究生，1976；朱湘渝，1980；Mofidabadi，1995)，并对花粉植株的倍性和育性进行了研究，杨树花粉植株一般是混倍体和非整倍体，随着树龄的增大，自行调整和加倍到二倍体(朱湘渝，1980；刘玉喜，1986)。刘玉喜(1986)发现 9~10 年生的小叶杨×欧洲黑杨花粉植株的花粉母细胞减数分裂正常，能产生有活力的成熟花粉，授粉后能产生饱满种子(刘玉喜，1986)。一般认为杨树花粉愈伤组织的诱导以单核期接种最为适宜，有利于花粉进行异常分裂和形成多细胞花粉(王敬驹，1975)。离体培养条件下，雄核发育以均等分裂为主，非均等分裂和多核花粉类型较少，并很快解体(王敬驹，1975)。

5 结语

纵观杨树生殖生物学研究发展情况，它已经形成了初步的理论体系，研究重点从单纯的形态描述转向对结构与功能关系的综合研究。20世纪70年代以来，应用电镜和细胞化学等生物新技术，研究探讨了杨树有性生殖过程中某些结构和生理功能的关系，诸如花粉壁和柱头表面的结构成分与花粉-柱头相互识别的关系，绒毡层细胞结构成分与雄性不育的关系等。同时，我们还应看到在杨树生殖生物学的研究中，还有许多亟待解决的问题，如雄性不育的机理，受精作用中不亲和性的机制，配子体形成过程中的遗传控制机制和配子体的基因表达等。我们相信，随着对这些问题的深入了解和认识，对杨树有性生殖过程(传粉、受精和胚胎发生等)的完全控制将会成为现实，这必将会使杨树育种工作更紧密地朝着人类所期望的方向发展。

Advances in Reproduction Biology of Poplars

Abstract In this article, the advances in reproduction biology of poplars are reviewed in four respects as follows: ①The morphological and cytochemical changes and the biological significance of female and male reproductive systems in sexual reproduction of poplars; ②The mechanism of male sterility in poplars; ③The reasons and descriptions of cross incompatiblity in poplars, the methods to overcome the crossing compatibility and the stigma pollen recognition reaction; and ④The progress made in reproductive engineering in poplars.

Key words *Populus*, reproduction biology, sexual reproduction, male sterility, cross incompati-bility, reproductive engineering

三倍体毛白杨有性生殖能力的研究*

摘　要　该文通过对 7 个毛白杨三倍体和 3 个二倍体无性系的花序量、花粉量及生命力、种子品质的比较研究和生殖器官形态的观察，发现三倍体无性系的花序量明显比二倍体的少；雄性无性系每个花序的花粉量及其发芽率在三倍体和二倍体之间差异不大；而三倍体雌性无性系的结实能力、果实形态大小和种子品质明显较二倍体的差。结果表明三倍体毛白杨的有性生殖能力较低，因此，种植三倍体毛白杨将会减少毛白杨早春散粉和飞絮对环境的污染。
关键词　毛白杨，二倍体，三倍体，有性生殖

对于毛白杨来说，长期以来人们主要关心的是它的生长和材性性状，而对毛白杨开花结实等生殖生长现象关注不够。仅有朱大保(1990) 对毛白杨有性生殖能力进行了研究(朱大报 1990)、张志毅等(1992) 对毛白杨开花结实特性进行了研究(张志毅等，1992)。造成这种状况的原因，主要是由于毛白杨败育极其严重，有性生殖能力弱，一般不依靠有性繁殖途径培育苗木，因此，毛白杨开花结实这一正常的生殖现象，对于林业生产来说意义不大。由于毛白杨早春飞絮、落花和落果严重，对城乡环境造成一定的污染，在某种程度上限制了毛白杨作为行道树和四旁绿化功能的发挥，不利于毛白杨在城乡的推广和应用。因此，选育出开花结实少且生长迅速的优良无性系是毛白杨育种工作者面临的又一新课题。三倍体毛白杨具有生长快、材质优良等特性(朱之悌，1992；1995；张志毅，1994；鹿振友等，1995)，但对其生殖生长情况知之甚少。在这种情况下，有必要对三倍体毛白杨开花结实和有性生殖能力进行研究，以揭示三倍体毛白杨无性系间生殖生长的差异，为少花甚至无花优良无性系的选育，以及今后毛白杨杂交育种工作的深入开展提供一定依据。

1　材料与方法

1.1　材料来源

试验材料来源于河北省晋州市毛白杨无性系试验林，分别于 1997 年和 1998 年两年春季对试验林进行调查取样。试验林是 1990 年春布置的，包括 7 个毛白杨三倍体无性系和 3 个毛白杨二倍体对照无性系，株行距为 3m×4m 和 3m×8m，分别为 6 株和 8 株小区，8 次和 4 次重复，设有 1 行保护行。参试的毛白杨无性系基本情况见表 1。

* 本文原载《北京林业大学学报》，2000，22(6)：1~4，与于雪松和朱之悌合作发表。

表 1 河北省晋州市毛白杨试验林各无性系基本情况

序号	无性系	倍性	性别
1	T6	2n=2X	♀
2	38	2n=2X	♂
3	39	2n=2X	♂
4	B166	2n=3X	♀
5	B180	2n=3X	♀
6	B196	2n=3X	♀
7	B173	2n=3X	♀
8	MT1	2n=3X	♂
9	B175	2n=3X	N
10	B193	2n=3X	N

注：2n=2X 为二倍体，2n=3X 为三倍体，N 为未开花。

1.2 生殖生长状况调查

花量观测是在花芽伸长但尚未展叶时进行，这时调查便于观察清楚。在小区内每个无性系随机观测 3 株，花量少则实数花序量，花量多则采取标准枝法。为了尽量排除造林密度对生长的影响，以株行距为 3m×8m 的试验林为研究对象。

1.3 生殖能力调查

(1) 花粉量调查：采用称重法，在雄花序即将散粉时，收集花粉，每个无性系选取 3 株，每株取 30~50 个花序，以 10 个花序为单位收集花粉，过筛、称重。

(2) 花粉生命力测定：采取醋酸洋红染色法、培养基法和活体授粉法。活体授粉法具体操作如下：选取生长良好的雌株，在温室中切枝水培，在适授期(柱头充分展开，发亮而有粘液)，将待测花粉授到柱头上，6h 后，每隔 1h 取样用卡诺固定液固定 2~20h 后，取少量柱头用 1NHCl 在 60℃下恒温解离处理 10~20min，水洗后用改良碱性品红染色，加盖玻片轻压，显微镜下观察拍照。

(3) 种子品质调查：在蒴果刚开裂尚未飞毛时，将果序取下，收集种子，计数、称重。在温室中播种，计算种子饱满率、千粒重以及发芽率

2 结果与分析

2.1 无性系生殖生长差异

本研究在两年中对各无性系花量进行了调查，调查结果见表 2。

从表 2 可以看出，毛白杨二倍体各无性系之间花序量差异不大，而三倍体各无性系之间花量差异极显著，有的无性系至今尚未开花，如 B175，B193。二倍体和三倍体各无性系之间花量差异极显著，毛白杨杂种三倍体花序量明显比二倍体少，三倍体无性系花序量

一般在 50~1000 个/株，花序量大多在 100~500 个/株，而二倍体无性系花序量大多在 500~2000 个/株。8 年生二倍体无性系平均花序量为 1536 个/株，而三倍体无性系平均花序量为 304 个/株，二倍体无性系的平均花序量是三倍体无性系平均花序量的 5 倍多，可以看出二倍体生殖生长明显比同龄三倍体旺盛。比较 1997 年和 1998 年两年的毛白杨无性系开花量，发现该试验林的无性系花量尚未稳定，1998 年无性系 39，B180 和 B196 的花量较 1997 年的花量有所增加，而无性系 B173 和 MT1 的花量略有减少。花量减少可能是由于大小年，基因型与环境因子互作等原因造成的。但无论如何三倍体毛白杨无性系的开花量还是明显的比二倍体毛白杨无性系的开花量要少得多。

表 2　毛白杨无性系花序量

个

性别	无性系	1997 年					1998 年				
		Ⅰ	Ⅱ	Ⅲ	Ⅳ	平均	Ⅰ	Ⅱ	Ⅲ	Ⅳ	平均
♂	38	2600	1260	2160	1500	1880	2360	1260	2100	1900	1905
	39	750	1700	482	430	840	882	2020	760	530	1048
	MT1	120	90	54	210	118	160	68	82	110	105
♀	T6	1164	1125	1865	2100	1563	1200	1240	2200	1980	1655
	B166	360	950	540	60	477	468	800	540	80	472
	B180	780	420	78	196	368	820	650	80	210	440
	B196	440	220	100	180	235	910	280	66	190	361
	B173	180	220	120	85	151	180	180	140	68	142

2.2　三倍体毛白杨有性生殖能力

2.2.1　花粉量调查与花粉生命力测定

花粉量和花粉生命力是毛白杨有性生殖能力的重要性状之一，仅仅了解开花量，难以全面地认识毛白杨三倍体生殖能力，因此有必要对毛白杨三倍体散粉量和花粉生命力进行调查。关于过去对毛白杨花粉量和花粉发芽率的研究认为不同地点和不同无性系毛白杨花粉量与花粉发芽率差异很大(张志毅等，1992)，花粉发芽率普遍偏低，但这并不是造成毛白杨杂交困难的关键因素(朱大保，1990；朱之悌，1992)。

本研究对试验林中 3 个雄性无性系的花粉量作了 3 次重复调查，无性系 38、39 和 MT1 的每个花序平均花粉量分别为 4.10，3.04 和 5.55mg，三倍体(MT1)花粉量与二倍体(38，39)花粉量差异不大，三倍体无性系 MT1 的花粉量反而比二倍体略高。但由于所调查的三倍体无性系每个花序的花粉量仅仅只是那些能正常发育的产生花粉的极少数花序，而那些不能正常发育的花序一般在散粉之前就已脱落。加之三倍体无性系 MT1 的花序量只是二倍体无性系 38，39 的花序量的 1/8 以下，所以，对于三倍体毛白杨单株或群体而言，其散粉量是非常少的。

如果是欲开展进一步杂交育种工作，保证足够数量的花粉是基本前提。但这是不够的，花粉是否有活力，能否在柱头上正常萌发并受精才是至关重要的，为此进行了花粉生命力的检测。用醋酸洋红染色法，无论是二倍体无性系(38，39)的花粉，还是三倍体无性系(MT1)的花粉，无论是新鲜花粉，还是干低温储存(0~4℃)较长时间的花粉，绝大多数的花粉都能染成红色，只有少数空瘪的花粉为无色或浅色。进而进行了花粉发芽实验，用0.5%~1.0%的琼脂培养基，加入10%~20%的蔗糖和0.01%的硼酸，25℃恒温条件下培养24h。花粉发芽率普遍极低，只有1%左右，效果很不理想。为了更好地模拟花粉生长的环境，真实地反映出花粉的生命力，进行了活体授粉实验。发现无论是三倍体无性系还是二倍体无性系的花粉均能在柱头上正常萌发，长出花粉管，花粉发芽率在40%~60%之间。表明只要能产生花粉，三倍体和二倍体无性系的花粉均有活力。

2.2.2　种子品质

毛白杨为蒴果，基生胎座，子房内一般有4个胚珠，但通常最多只有1~2个能正常发育成种子，其余败育，多数子房内的4个胚珠都败育。

表3　毛白杨无性系种子品质

无性系	结籽率(个/序)	饱满率(%)	千粒重(g)	发牙数(个)	发芽率(%)
T6	0.6	21	0.198	9	7.8
B166	0.2	0	0.100	0	0.0
B180	0.7	3	0.081	1	2.7
CK1	5.6	93	0.368	12	4.3
CK2	6.1	91	0.349	53	17.3

通过对晋州试验林中能开花结实产生种子的两个三倍体无性系和3个二倍体无性系进行调查，结果发现不同无性系结籽率存在一定的差异，各无性系种子品质见表3。最高结籽率平均可达每果序6.1粒种子(CK2)，并且饱满率、千粒重和发芽率也较高。而三倍体无性系B166结籽率最低，每果序平均只有0.2粒种子，自然状态下难以获得饱满的种子，种子无活力，发芽率为0。在花粉总体情况基本一致的条件下，无性系种子品质出现如此大的差异，说明毛白杨种子品质在很大程度上取决于雌株的育性。虽然各无性系获得了一定数量的幼苗，但由于生命力极弱，大部分幼苗未能保存下来，只有CK2获得少量幼苗存活下来，这再一次说明毛白杨有性生殖能力极低。

从表3可以看出，三倍体无性系种子品质明显比二倍体无性系的差，结实率极低，种子空瘪，仅有一层种皮，而无胚，发芽率极低，有活力的种子几乎没有。而二倍体无性系相对有较高的结实率、种子饱满、发芽率相对较高。三倍体无性系B196和B173由于花量少，落花落果严重，大多数植株在飞絮之前，果序上的小蒴果或整个果序已基本落完，未收集到种子，这也说明三倍体毛白杨高度败育。Sasaki(1993)发现日本柳杉和日本扁柏三倍体种子发芽率较低，分别不到5%和2%，育性极差(Sasaki，1993)。这些都证明了树木三倍体与一般的植物三倍体一样具有高度的不育性。

2.3　生殖器官形态观察

毛白杨不同无性系雌花及蒴果的形态和大小存在明显的差异。对无性系花器官形态进行比较，发现早期雌花柱头尚未充分展开时，花器官无明显差异，到适授期(3月10日至

3 月 15 日)柱头完全展开，三倍体与二倍体的柱头、花盘及苞片的形态、大小和颜色明显不同，比较结果见表 4。

表 4　毛白杨无性系雌花及蒴果形态特征

倍性	无性系	苞片	花盘	柱头		子房	花序	果实
				形态	颜色			
三倍数	B166 B180 B173 B196	褐色，较大前缘具不规则浅裂，绒毛细长	杯状，边缘呈浅波状	4 裂，细而长，形如角状	未成熟时浅绿色，后呈浅黄色，成熟后，近无色透明	细长	较短，4.0~4.5cm	少而大，卵形或近椭圆形
二倍体	T6	深褐色，中等，前缘具不规则浅裂，绒毛细密较长	杯状，边缘呈浅波状	2 裂，各裂片具浅裂，宽而大，似花瓣片状	同上	呈倒三角形	短而粗，4.0cm 左右	多而瘦小，呈月牙形

对成熟后期的果实(4 月 10~14 日)形态进行观察，发现三倍体的蒴果明显比二倍体的果实大，无性系 B166 的果实大约是 T6 果实的二倍，但绝大多数三倍体的蒴果内无种子，只有种毛。三倍体无性系之间果实形态和大小有明显差异，无性系 B173 和 B196 的果实明显小于 B166 和 B180。这是由于无性系 B173 和 B196 败育严重，果实中途停止发育，并出现落果现象，可见不同基因型的三倍体果实生长差异很大。

有的雌株发现有个别雄花存在，三倍体和二倍体无性系中都有雌雄同花序的现象，以二倍体居多。三倍体中只在无性系 B180 中发现了花序上有雄花共存现象，尚未在雄株中发现有雌花存在的现象。雌花序中的花药形态和颜色与雄株的花药无差异，但发育较雄株的花药迟缓，正常的雄花在 3 月 14 日之前都基本开裂撒粉，而此时雌花序上的雄花花丝尚未伸长，花药未伸出花盘。毛白杨三倍体与二倍体无性系雄花的形态和颜色无明显差异，雄花外具苞片，雄蕊着生于浅杯状花盘上，花药的颜色从早期的浅黄色，到深黄，最后呈深红色，直至开裂。三倍体无性系 MT1 具花药 7~12 个，二倍体无性系花药数一般在 7~16 个，可见三倍体无性系花药数和二倍体无性系差异不大。三倍体无性系 MT1 的苞片明显比二倍体 39 的大。与雌花相比，毛白杨无性系雄花变异不大。

3　讨论

(1)植物从营养生长进入生殖生长是在外界环境和自身内在因子的共同作用下完成的，首先形成花序分生组织，然后逐步形成花分生组织(floral meristem)，进而产生花原基，最后逐步分化为成熟的花器官(Gustafson and Savidage，1994)。外界环境(如光照、温度和水分等)和自身内在因子(如基因型和激素水平)的改变都将影响花器官的形成。由于三倍体毛白杨无性系是经过杂交得到的，除倍性增加以外，基因杂合度也很高，无性系基因型各不相同，导致三倍体无性系之间花量差异很大，有的三倍体毛白杨无性系 8 年生时尚未开花，而有的已大量开花。三倍体毛白杨开花结实极少，一方面可能是由于三倍体比二倍体多一套染色体，成为奇数倍的染色体组数，在减数分裂时染色体分配不均等，不能产生育性正常的雌雄配子，从而不育；另一方面可能是由于三倍体比二倍体多一套染色体，使体内激素水平和酶系统发生变化，不利于花原基的形成；再一方面也可能是由于三倍体营养生长过于旺盛，其生殖生长相对较弱。无论如何这些结果为今后少花毛白杨无性系的选育提供了一定的基础。在目前种植三倍体毛白杨新品种将会在很大程度上减小一般二倍体毛白杨早春飞絮或飞散花粉对环境造成的污染。

(2)由于毛白杨结籽率极低，发芽率和存活率较低，一般自然状态下难以获得实生苗，长期以来毛白杨主要通过无性繁殖。目前尚未从根本上搞清楚造成毛白杨有性生殖困难的原因，大多数的研究认为造成毛白杨结实率低的主要原因是雌株败育，而与父本花粉关系不大。朱大保(1990)通过进行不同类型毛白杨的种内杂交试验，认为毛白杨有性生殖能力低与其杂种起源密切相关，雌株的育性对毛白杨有性生殖能力起决定性作用，而雄株的花粉发芽率对结籽量并未产生显著影响(朱大保，1990)。朱之悌(1992)比较了毛白杨和毛新杨正交与反交的结籽力和种子发芽率，并观察了不同组合幼胚的发育过程，认为用毛白杨作为母本种子品质极低，毛白杨有性杂交困难主要原因在于雌株的育性差(朱之悌，1992)。进一步的工作有待运用现代分子生物学技术手段，从生殖的细胞学、遗传学和分子生物学方面去开展更深入的研究，有望弄清楚毛白杨的有性生殖困难的根本原因，根据不同的目标去培育更新、更优良的新品种。

Sexual Reproduction of Hybrid Triploids in *Populus tomentosa*

Abstract The number of inflorescence, quantity of pollen, seed quality and morphological characteristics ofreproductive organs of 3 diploids and 7 triploids in *Populus tomentosa* were investigated in this study. Inflorescence number of triploid is less than that of diploid obviously. No significant differences existed in pollen quantity per inflorescence and in pollen vitality between triploids and diploids. The fruit bearing capability, fruit size and seed quality of triploids are inferior than that of diploids. It is showed that sexual reproduction in triploids is not very capable and to plant more triploids of *Populus tomentosa* will be profitable for the environment because of less catkin and pollen.

Keywords *Populus tomentosa* Carr., diploid, triploid, sexual reproduction

Cloning and RNAi Construction of a *LEAFY* Homologous Gene from *Populus tomentosa* and Preliminary Study in Tobacco*

Abstract *PtLFY*, a *LEAFY* (*LFY*) gene, was cloned from *Populus tomentosa* (LM50) by PCR. Sequencing analysis indicated that *PtLFY* was 2629 bp long, composed of three exons and two introns and encoded 378 amino acids. The splice donor sites and the splice acceptor sites were in identical positions to the *LFY* and its homologues. The amino acid sequence inferred was 68%~99% homologous to those of *LFY* and its homologues by blast analysis in GenBank. The Southern blot analysis indicated that there was a single copy of the *PtLFY* gene in genomic DNA of male and female *P. tomentosa* (LM50 and 5082). The pBI121-Ptalfy (reverse)-intron-Ptlfy-GUS-nos was constructed using RNA interference (RNAi) technique and verified by PCR and digestion identification and transformed into tobacco. Some transgenic tobacco plants were obtained by PCR and PCR-Southern identification. The growth was generally repressed in transgenic tobacco plants compared with wild-type ones and some phenotypic differences were observed.

Key words *Populus tomentosa*, *PtLFY*, cloning, RNAi, construction, transformation

1 Introduction

Populus tomentosa Carr., a native species in Sect. *Leuce* in China, is mainly distributed in the vast area of northern China. It plays a key role in forest production and ecological projects along the Yellow River. As a model forest tree, it has also been used as a tool in the genetic improvement of forest trees.

In the near future, modification of flowering will play an important role in breeding of *P. tomentosa*. Firstly, promoting early flowering should take into consideration shorter breeding cycles of *P. tomentosa*. Additionally, interference of flowering will lessen pollen and catkin production and helps to improve urban environments. These purposes can be fulfilled by genetic engineering.

It is well known that a number of floral meristem identity genes have been isolated from *Arabidopsis thaliana* and other model plant species. For example, *LEAFY* (*LFY*) is involved in controlling the transition from an inflorescence to a floral meristem (Weigel et al. 1992). Another floral

* 本文原载《林业研究》(*Forestry Studies in China*), 2005, 7(3): 15-21, 与安新民、王冬梅、李善文和何承忠合作发表。

meristem identity gene, *APETALA*1 (*AP*1), is required for sepal and petal development and is also involved in controlling the transition mentioned above (Mandel et al., 1992). When either *LFY* or *AP*1 is expressed constitutively, transgenic *Arabidopsis* plants flowered *in vitro* in just 10d (Mandel and Yanofsky, 1995; Weigel and Nilsson, 1995). When the *LFY* gene from *Arabidopsis* is expressed constitutively in hybrid aspen (*P. tremula* ×*P. tremuloides*), it flowered *in vitro* within 7 months (Weigel and Nilsson, 1995). Overexpression of *MdMADS*5, an *APETALA*1-like gene in apple, causes early flowering in transgenic *Arabidopsis* (Kotoda et al., 2002). In *Citrus*, expression of *Arabidopsis* genes *LFY* and *AP*1 induced flowering within the first year. The shortening of the juvenile period was stable and also observed in the zygotic and nucellar-derived seedlings, which demonstrates the stability of the trait (Peña et al., 2001). In poplar, the influence of flowering genes is more complex. Over expression of *PTLF*, a *LFY* homologue from *P. trichocarpa* in several poplar species induced precocious flowering in only 2 of 19 transgenic lines, although in the flowering of *Arabidopsis* the over expression of *PTLF* was accelerated (Rottmann et al., 2000). However, little is known about antisene repression of *LFY* or its homologue in poplars. Furthermore, RNA interference (RNAi) of *LFY* in poplars has not been reported up to now. In this paper, cloning of *PtLFY*, a *LFY* homologue from *P. tomentosa* (LM50), construction of expression vector containing reverse repeat sequences like RNAi and transformation in tobacco were carried out.

2 Materials and methods

2.1 Materials and extraction

P. tomentosa trees used in the study were planted at Beijing Forestry University. DNA in young leaves of poplar including male and female plants was extracted according to the methods of Sambrook et al. (1989) and used as templates of PCR and Southern hybridization analysis. Sterile tobacco (*Nicotiana tabacum*) plant was used as transgenic materials.

2.2 Vectors and bacterial strains

The pGEM® -T easy vector was purchased from Promega Co. Vector pBI121, *E. coli* TG1 and *Agrobacteria* strains GV3101 were stored in our laboratory.

2.3 Enzymes and other reagents

Restriction enzymes and other genetic engineering tool enzymes were obtained from Promega Co. QIAquick™ Gel Extraction Kit, DIG DNA Labeling and Detection Kit were purchased from QIAGEN and Roche Co.

2.4 Cloning and sequencing of *PtLFY*

Primers used for PCR cloning of *PtLFY* and DNA probe labeling were designed according to

PTLF (U93196), a *LFY* homologue from *P. balsamifera* and synthesized by Sangon Co. (Table 1). The PCR reaction system of 50μL was composed of 1×PCR buffer (Tris 10mmol·L^{-1}, pH 8.3, KCl 50mmol·L^{-1}, $MgCl_2$ 2.5mmol·L^{-1}), genomic DNA 50ng, 1μL of 10pmol·μL^{-1}P1U/P2U/P3U and P1D/P2D/P3Da, b and a 2.5 unit Taq DNA polymerase. Thermo cycling was performed at 94℃ for 5min, then at 94℃ for 40s, 60℃ for 40s, 72℃ for 2min for 30 cycles, at 72℃ for 10min and finally kept at 4℃.

Sequencing of PCR products was done by the Sangon Co.

2.5 Southern analysis

A 1076-bp fragment amplified by P3U and P3Da (Table 1) was used as the PCR probe for DNA gel blot analysis. Tenμg of genomic DNA from male and female *P. tomentosa* were digested with restriction enzymes, *Bam*HI, *Hin*dIII and *Eco*RI, respectively, and electrophoresed on 0.8% agarose gel, then blotted onto a nylon membrane positively charged by capillary transfer with 20× SSC, hybridized with DIG-labeled DNA probe at 58℃, and washed at a high stringency (2×SSC, 0.1% SDS, at 15~25℃ for 2×5min, then 0.5×SSC, 0.1% SDS at 65~68℃ for 2×15min). Immunological detection was according to the protocol of DIG DNA Labeling and Detection Kit (Roche).

2.6 Construction of RNAi vector

PCR products amplified by P3U and P3Da, P3U and P3Db were respectively digested with *Xba*I/*Nco*I and *Bam*HI/*Nco*I. pBI121 was digested with *Xba*I/*Bam*HI. The corresponding fragments were reclaimed and purified using QIAquickTM Gel Extraction Kit (QIAGEN) and ligated with T4-DNA ligase at 16℃ overnight, then transferred into competent cells of *E. coli* TG1. Construction containing reverse repeat sequences of *PtLFY* was generated by digestion and PCR identification using primers P4U and P4D (Table 1).

Table 1 Primers used for PCR cloning of *PtLFY*, DNA probe, construction of RNAi vector and PCR identification of transformants

Primers	Oligonucleotide
P1U (upstream primer of *PtLFY* -1)	5′-ATGGATCCGGAGGCTTTCACGGC-3′
P1D (downstream primer of *PtLFY* -1)	5′-AGGGGAGAAAAATGCCCCACTAAG-3′
P2U (upstream primer of *PtLFY* -2)	5′-GTTTCTCTGAGGAGCCAGTACAGC-3′
P2D (downstream primer of *PtLFY* -2)	5′-CTGTAGGCACCAGCAGCCTA-3′
P3U (upstream primer of DNA probe)	5′-GAAGTGGCACGTGTGGCAAAAAGAA-3′
P3Da (downstream primer of DNA probe)	5′-TTCTAGACGGAGTTTGGTGGGCACA-3′(introduce *Xba*I)

(续)

Primers	Oligonucleotide
P3Db (downstream primer of DNA probe)	5′-GGATCCACGGAGTTTGGTGGGCACA-3′(introduce *Bam*HI)
P4U (upstream primer of PCR identification)	5′-GGATTGATGTGATATCTCCACTG-3′(partial sequence of 35S promoter)
P4D (downstream primer of PCR identification)	5′-CCACAGTTTTCGCGATCCAGACT-3′(partial sequence of GUS gene)

2.7 Transformation in tobacco

The construction mentioned above was introduced into the *Agrobacterium* strain GV3101 by an efficient and direct transformation method reported previously (Tzfira et al. , 1997). A single colony of *Agrobacterium* GV3101 containing target construction grows in LB with streptomycin (25mg·L^{-1}) and kanamycin (100mg·L^{-1}) at 28℃ for 24h. Leaves of sterile tobacco plant were cut into 1~2cm^2 disks and immersed into a 10×dilution of the overnight culture for 30s and the disks were dried by blotting onto the filter paper and placed on solid MS (callus medium MS containing 1mg·L^{-1} benzyladenine, 0. 1mg·L^{-1} naphtalene acetic acid, 3% sucrose and 4. 5g·L^{-1}phytoagar) light at room temperature. Then they were cultured for 2 d and transferred to CM containing 100mg·L^{-1} carbenicillin and 100mg·L^{-1} kanamycin for 7 d. Thereafter, they were transferred to fresh medium weekly until the transgenic calli were large enough to be excised (2-3mm in diameter). The excised calli were placedon solidified RMOP medium (MS containing 1mg·L^{-1}benzyladenine, 3% sucrose and 4. 5g·L^{-1} phytoagar) containing 100mg·L^{-1} carbenicillin and 100mg·L^{-1} kanamycin. The material was transferred every two weeks until transgenic shoots developed from callus, and excised shoots were placed on 1/2MS containing 1. 5% sucrose and 100mg·L^{-1} kanamycin to regenerate roots.

2.8 Molecular identification of transformed plants

Genomic DNA in young leaves of transformants was extracted and used as the template of the PCR test. The PCR reaction was according to the system mentioned above and thermocycling was performed at 94℃ for 5min, then at 94℃ for 40s, 58℃ for 40s, 72℃ for 1. 5min for 30 cycles and 72℃ for 10min. PCR products were electorophoresed on 0. 8% agarose gel, blotted onto the nylon membrane positively charged by capillary transfer with 20×SSC and then hybridized with a DIG-labeled PCR probe as described previously.

3 Results

3.1 Cloning and sequencing of *PtLFY*

Two fragments, *PtLFY*-1 and *PtLFY*-2 whose length were about 1. 4kb (Fig. 1A) and 1. 6kb (Fig. 1B), were obtained by PCR amplification, and cloned into a pGEM® -T easy vector. The recombinant plasmids verified by PCR and digestion (Fig. 1) were sequenced by the

Sangon Co. The results indicated that *PtLFY* from *P. tomentosa* was 2629 bp long and encoded 378 amino acids. By analysis using Seqaid II and SPL (search for potential splice sites) at http://dot.imgen.bcm.tmc.edu:9331/gene-finder/gf.html, 3 exons and 2 introns (1st exon 1-436; 1st intron 437-1031; 2nd exon 1032-1366; 2nd intron 1367-2034; 3rd exon 2035-2397; downstream sequence 2398-2629) (Figs. 2 and 3) were found. The donor and acceptor sites of 1st intron (596bp) were at 43 and 1032bp, the donor and acceptor sites of 2nd intron (669 bp) were at 1366 and 2035bp and the splice sites were shown by arrows. The positions of splice sites were the same as *LFY* homologue genes of other plants, but there were great differences in the length of the introns. For example, the intron of *Arabidopsis LFY* is 910bp long in the position of the 2nd intron. In addition the analysis of restriction site using Seqaid II software provided useful information in the later steps of the construction of the expression vector (Fig. 2).

The homologous genes were searched by blast analysis in the GenBank. The results indicated that the deduced amino acid of *PtLFY* was up to 99% homologous to *P. balsamiofera PTLF* (U93196), at the same time 68%~75% homologous to *Eucalyptus globulus ELF*1 (AF034806), *Malus domestica ALF*1 (AB056158) and *ALF*2 (AB056159), *Nicotiana tabacum NFL*1 (U15798) and *NFL*2 (U15799), *Antirrhinum majus FLO* (M55525) respectively. A highly conserved domain near the C-terminal of *LFY* homologues was found by comparison of amino acid sequences of *LFY* homologous genes with different species using Bioedit and Clustalx software. It suggested that *PtLFY* might perform similar functions to *LFY* in the flowering of *P. tomentosa*, although there were some differences between *PtLFY* and *LFY* homologues from other plants.

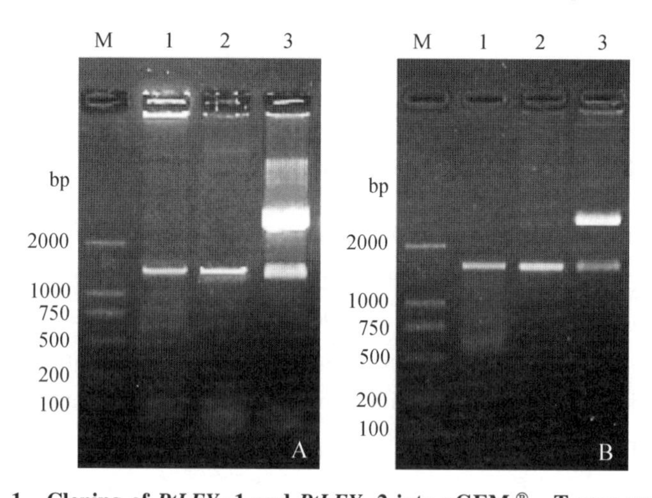

Fig. 1 Cloning of *PtLFY*-1 and *PtLFY*-2 into pGEM® -T easy vector

Lane M: DL-2000 marker; Lane 1: PCR products *PtLFY*-1/*PtLFY*-2; Lane 2: recombinant plasmid A and B identified by PCR; Lane 3: recombinant plasmid A and B digested by *Eco*RI

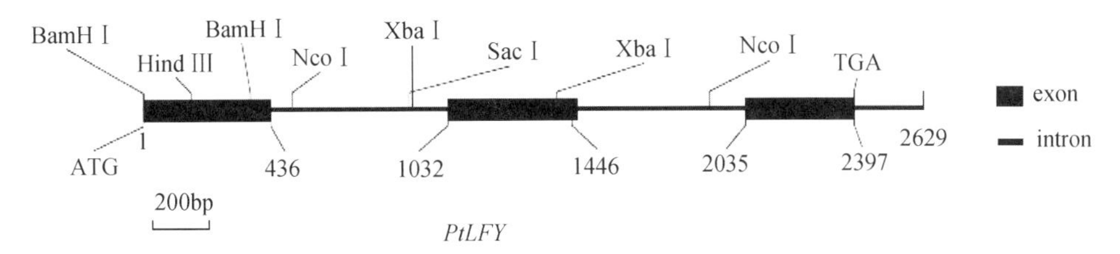

Fig. 2 Structural and restriction map of *LEAFY* homologous gene from male clone of *P. tomentosa* (*PtLFY*)

3.2 Southern blot analysis

In order to detect the copies of *PtLFY* gene in genomic DNA of *P. tomentosa*, genomic DNA from male (LM50) and female (5082) clones was digested with *Bam*HI, *Eco*RI and *Hin*d III, respectively, electrophoresed on 1.0% agarose gel and blotted onto the nylon membrane positively charged by capillary transfer with 20×SSC, hybridized with DIG-labeled DNA probe at 58℃ and washed at a high stringency. The result indicated that there was the same pattern of hybridization in genomic DNA of male and female clones of *P. tomentosa*; there was only a single copy of *PtLFY* in genomic DNA of *P. tomentosa* (Fig. 3).

3.3 Construction of expression vector containing reverse repeat sequences

Two reverse primers, P3Da and P3Db which introduced *Xba*I and *Bam*HI at 3′ end were designed and synthesized. The corresponding PCR product *PtLFYa* was digested with *Xba*I and *Nco*I, and *PtLFYb* was digested with *Bam*HI and *Nco*I. Meanwhile, the plasmid pBI121 was also digested with *Xba*I and *Bam*HI. Corresponding target fragments were ligated to the binary vector pBI121 under the CaMV 35S promoter. A RNAi vector containing antisense, sense-oriented same sequence and an intron inserted in the middle of them was generated, resulting in pBI121-Ptalfy-Ptlfy (Fig. 4). Thereafter, it was introduced into competent cells of *E. coli* TG1. The RNAi vector construction proved to be successful by digestion with *Xba*I and *Bam*HI and PCR identification using primers P4U and P4D.

3.4 Tobacco transformation and molecular identification of transformed plants

Tobacco leaf disks without pre-culture were used as acceptor materials, cocultured with *Agrobacterium* for 2d, then transferred on screening media MS with carbenicillin 250mg·$L^{-1}$and kanamycin 100mg·L^{-1}. Some resistant buds from transformed disks were found for about 7d. When resistant buds were about 1cm long, the resistant buds were cut and transformed on rooting medium MS with carbenicillin 250 mg·L^{-1} and kanamycin 200mg·$L^{-1}$to root. Some transformed plants were obtained.

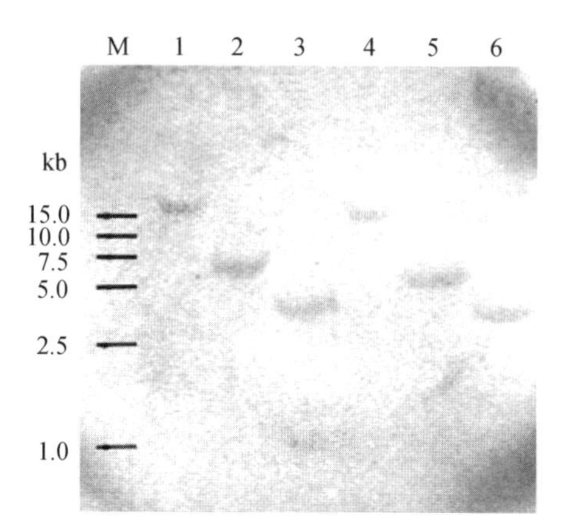

Fig. 3 Southern blot analysis of *PtLFY* in *P. tomentosa*

Lane M: DL-15000 marker; Lanes 1-3: genomic DNA from male clone *P. tomentosa* (LM50) digested with *Bam*HI, *Eco*RI and *Hin*dIII respectively; Lanes 4-6: genomic DNA of female clone (5082) of *P. tomentosa* digested with *Bam*HI, *Eco*RI and *Hin*dIII respectively.

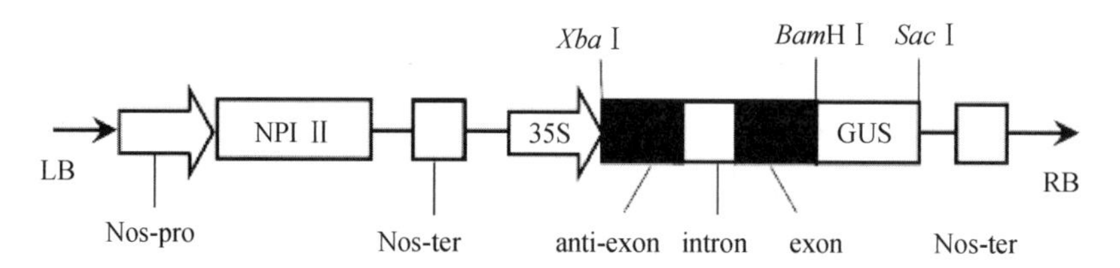

Fig. 4 Representation of the transformation vector pBI121-Ptalfy-Ptlfy

DNA in leaves of transformed and non-transformed tobacco plantlets was used as the template of PCR reaction, and the recombined plasmid as positive control. The results of PCR showed that the bands of lanes 2~7 were PCR products of transformed plants 24~27, 34 and 36 and their molecular weight, was about 1.1kb. The band of the same length in lane 9 was positive control, whereas there was no similar band in the non-transformed tobacco plants in lane 8 (Fig. 5A). It indicated that an exogenous gene had been introduced in these tobacco plants. In order to identify further transgenic tobacco plants verified by PCR tests, a PCR-Southern blot analysis was performed (Fig. 5B). The results indicated that the RNAi construction of inverted repeat Ptlfy fragments had been integrated into genomes of 6 tobacco plants.

All of these suggested that 6 transgenic tobacco plants were obtained. Their growth was generally repressed and some phenotypic differences were observed.

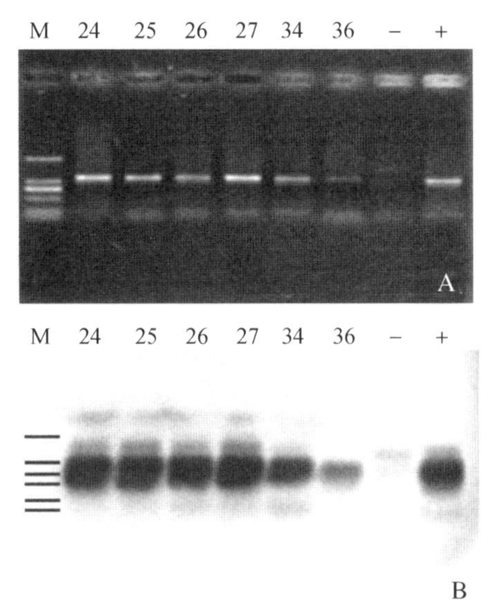

Fig. 5 Identification of the transformed tobacco plants by PCR (A) and PCR-Southern blot (B) analysis

Lane M: DL-2000 marker; Lanes 24-26, 34 and 36: transgenic plants; Lane-: non-transgenic plants; Lane +: positive control (plasmid).

4 Discussion

Floral development has been well studied in the model angiosperm *Arabidopsis thaliana*. This model plant provides a general framework for the understanding of floral gene interaction. *LFY* is a key gene in controlling the transition from vegetative meristems to inflorescence and floral meristems (Weigel and Nilsson, 1995; Yanofsky, 1995). The construction and activity of *LFY* from different plants are well conserved (Weigel et al., 1992; Weigel and Nilsson, 1995). *PtLFY*, a *LFY* homologue was isolated from *P. tomentosa* by PCR. The result of blast in GenBank indicated that the amino acid sequence coded by it had high homology (68%~99%) to *LFY* and its homologues. It was believed that *PtLFY* might have similar functions in controlling flowering because of its similarity of construction and high homology of amino acid sequences. Furthermore, the result of Southern blot analysis showed that there was one copy of *PtLFY* in the genomic DNA of *P. tomentosa*. So we considered that *PtLFY* might play an important role in floral development of *P. tomentosa*.

As an important shade tree, Chinese white poplar is distributed in a large area in China. It is well known that poplar releases much pollen and catkins during its flowering season every spring, which are environmental pollutants and harm some people. So how to lessen and prevent the emission of pollen and catkins is one of the current topics. RNAi (RNA interference) is a new biological technique and an effective method in functional analysis of genes. We designed and produced the construction of inverted repeat Ptlfy fragments to suppress *PtLFY* using this technique. Further studies and observations are under way.

*APETALA*3 Homologous Gene (*PtAP*3) Cloning from *Populus tomentosa*: A Preliminary Study on Its Sense and Anti-sense Transformation in Tobacco[①]

Abstract A pair of primers were designed according to published literature on *Populus trichocarpa* gene (PTD), and *PtAP*3, an *AP*3 homologous gene from *P. tomentosa* was isolated by PCR using genomic DNA of the male clone of *P. tomentosa* (L50) as a template. The result indicated that the sequence was 1, 813 bp (*Bam*H I and *Sac* I were introduced at the 5′ and 3′ end) including 7 extrons and 6 introns, coding 238 amino acids. It was found that there was 52%~82% homology to proteins from *Lilium regale* (AF503913), *Petunia hybrida* (AF230704), *Gerbera hybrida* (AJ009724), *Rosa rugosa* (AB055966), *Malus domestica* (AJ251116), and *P. trichocarpa* (AF057708) determined by blast analysis in the GenBank. There was a highly conserved MADS-box motif in the protein of *PtAP*3, so it was putatived to be a transcription factor. The result of Southern blot analysis indicated that there were double copies of *PtAP*3 or two members which had a high homology to each other in *P. tomentosa* (L50, male) genomic DNA, and there was single copy *PtAP*3 in *P. tomentosa* (5082, female) genomic DNA. Sense and antisense expression vectors of *PtAP*3 were constructed by PCR and restriction enzymes digestion identification, and transformed into tobacco (*Nicotiana tabacum*) by *Agrobacterium* GV3101 and LBA4404. Some transgenic tobacco plantlets were obtained by PCR identification. The results mentioned above have provided important data to understand the molecular mechanism of male flower development of *P. tomentosa*, and has contributed to the study on controlling flowering of *P. tomentosa* using genetic engineering.

Keywords *Populus tomentosa*, *APETALA*3 homologous gene, cloning, transformation

1 Introduction

Extensive studies on floral development have revealed a general model for the control of floral organ identity based on three genetic functions A, B and C (Coen and Meyerowitz, 1991; Chasan, 1991). Flowers of representative dicotyledons have a concentric arrangement of four types of organs: sepals in whorl 1, petals in whorl 2, stamens in whorl 3, and carpels in whorl 4. According to the model, the identity of these organs depends on the action in the combinations A, AB, BC, and C in whorls 1~4, respectively. A noticeable event is the antagonistic function of A and C, which indicates that when C is silenced, A expresses, and vice versa (Wang et al.,

① 本文原载《中国林业动态》(*Frontiers of Forestry in China*), 2006, 1(4): 404~412, 与王冬梅、安新民、李善文和何承忠合作发表。

2003). Usually, genes *AP*1 and *AP*2 act as function A, gene *AG* as function C, whereas *AP*3 and *PI* as function B.

The proteins contain a highly conserved DNA binding domain called the MADS domain that are encoded by the above-mentioned genes and may work as transcription factors or MADS-box genes (Yanofsky, 1990). Other two systems of MADS-box genes are: D-class genes that are confirmed to determine ovule development (Angenent, 1995; Colombo, 1995), and E-class genes that are proposed to be together with other genes required for the formation of protein compounds.

Populus tomentosa is a native tree species in China that has not been studied sufficiently compared with *Arabidopsis* and *Antirrhinum*; it is rather important to investigate its molecular mechanism of flowering and to control floral development. Floral development in poplars differs significantly from that of a typical hermaphroditic annual, the apices of the branches do not become inflorescences and the flowers are borne on axillary inflorescences or catkins, with male and female flowers found on separate trees. Instead of four concentric whorls of organs, the *Populus* flower has only two whorls, a reduced perianth cup surrounding either stamens or carpels (Sheppard, 2000). The mechanism causing the difference between *Arabidopsis* and *Populus* is still unknown. In the early of 1980s, Dong (1982, 1984) observed embryological development of *P. tomentosa*, then Zhu (1990) studied the capacity of sexual reproduction, and Zhang et al. (1992, 2000) studied flowering, fruit bearing, and sexual reproduction of hybrid triploids. Recently, Sheppard et al., (2000) have isolated PTD, an *AP*3 homologous gene, and studied its spatio-temporal expression. *LFY*, *AG*, and *AP*1 etc from *Arabidopsis*, the homologue *PTLF*, *PTAG*, and *PTAP*1 have been isolated one after the other (Rottmann, 2000, Meilan, 2001). Up to now, however the floral genes in *P. tomentosa* involved in flowering development has not been reported. To begin this study, *AP*3 homologue from *P. tomentosa* was isolated and cloned. It is a good choice for revealing the molecular mechanism of flowering in *P. tomentosa* and plays as an important base for shortening the breeding cycle and suppressing contamination of pollen and catkins.

2 Materials and methods

2.1 Plant materials

In spring, DNA in young leaves of female (5082) and male (LM50) clones were extracted as acceptor materials and Southern blot analysis from Beijing Forestry University.

DNA in young leaves of tobacco including transgenic and nontransgenic (control) plants was extracted as templates of PCR test.

2.2 Methods

2.2.1 Extraction of genomic DNA

Genomic DNA was extracted according to the methods of Wang et al. (2002)

2. 2. 2 PCR reaction

A pair of primers were designed according to published literature on *Populus trichocarpa* gene (PTD), the forward primer 5′-TTGGATCCATGGGTCGTGGAAAGA-3′ and the reverse primer 5′-AAGAGCTCTCAAGGAAGGCGAAGTT-3′. Then *PtAP3*, an *AP3* homologous gene from *P. tomentosa* was isolated by PCR using genomic DNA of the male clone of *P. tomentosa* (L50) as template. The PCR reaction system of 50μL was composed of a 10×PCR buffer (Tris 10mmol/L, pH 8. 3, KCl 50mmol/L, $MgCl_2$2. 5mmol/L), genomic DNA 10~50ng, 1μL of 10pmol/μL forward and reverse primers respectively, and 2. 5 unit Taq DNA polymerase. Thermocycling was performed at 94℃ for 5min, then at 94℃ for 40s, at 60℃ for 40s, at 72℃ for 2min for 30 cycles, at 72℃ for 10min, and finally kept at 4℃.

2. 2. 3 Recombine to pGEM-T easy vector

PCR products were reclaimed and purified using the Promega Co. QIAquick™ Gel Extraction Kit, and were ligated with T4-DNA ligase after which the products of ligation were transformed into competent cells of *Escherichia coli* TG1 and spread on a LB plate containing 50μg/mL Amp, 20μg/mL IPTG, and 20ng/mL X-gal. The positive recombinant plasmids were identified by PCR tests and digested with restriction enzymes.

2. 2. 4 DNA sequencing and analysis

The sequencing method was based on the original "Sanger" methods of dideoxy chain termination and ABI-377 DNA autosequencing system (PE Co. , USA) was used as sequencing models to take full-length DNA sequences. The result of sequencing was analyzed by blast analysis in GenBank and some software was applied to translating, comparing, and drawing.

2. 2. 5 DNA probe preparation and southern blotting analysis

The purified PCR products were labeled as blotting probe by DIG-11-dUTP for 20h in 37℃, 10μg of genomic DNA from male and female *P. tomentosa* were digested with restriction enzymes, *Bam*H I, *Eco*R I, and *Hin*d III, respectively, and electrophoresed on 0. 8% agarose gel The products were then blotted onto a nylon membrane positively charged by capillary transfer with 20× SSC, prehybridized for 30min at 42℃, and hybridized with DIG-labeled DNA probe at 55℃ for 14~16 h. They were then washed at a high stringency (2×SSC, 0. 1% SDS, at 15~25℃ for 2× 5min, then 0. 5×SSC, 0. 1% SDS at 65℃ for 2×15min). Immunological detection was carried out according to the protocol of DIG DNA Labeling and Detection Kit (Roche).

2. 2. 6 Construction and identification of sense and antisense expression vectors

The recombinant plasmid pGEM-T-*PtAP3* and pBI121 were digested with *Bam*H I/*Sac* I and *Bam*H I/*Xba* I respectively. The corresponding fragments were released, reclaimed, and purified by QIAquick™ Gel Extraction Kit (QIAGEN) and ligated in a sense and antisense-oriented manner with T4-DNA ligase, then transferred into competent cells of *E. coli* TG1 and spread on a LB plate containing 100μg/mL Kan respectively. The positive recombinant plasmids were obtained by PCR tests (the forward primer 5′-GGATTGATGT GATATCTCCACTG-3′ and the reverse primer 5′-CCACAGTTTTCGCGATCCAGACT-3′ were used to identify the antisense construct, and the reaction system was the same as thc abovc PCR), sequenced and digested with restriction enzymes.

2.2.7 Tobacco transformation and PCR tests of transgenic tobacco

pBI121-*PtAP*3 and pBI121-*Ptap*3 were introduced into *Agrobacterium* strains LBA4404 and GV3101. A single colony of *Agrobacterium* GV3101 containing target construction grew in LB with streptomycin (25mg/L) and kanamycin (100mg/L) at 28℃ for 24h. Leaves of sterile tobacco plant were cut into 1 ~ 2cm^2 disks and immersed into a 10×dilution of the overnight culture for 5min and the disks were dried by blotting onto the filter paper and placed on solid MS (callus medium MS containing 1mg/L benzyladenine, 0.1mg/L naphtalene acetic acid, 3% sucrose and 4.5g/L phytoagar). Then they were cultured for two days in dark at room temperature whereafter they were transferred to a CM containing 500mg/L carbenicillin and 100mg/L kanamycin for kan-resistant adventitious shoots to appear. Thereafter, they were transferred to fresh medium weekly until the transgenic shoots were big enough to root on 1/2MS containing 1.5% sucrose and 100mg/L kanamycin. Genomic DNA of young leaves of transgenic tobacco plants and control were extracted by cetyl-trimethylammonium bromide (CTAB)-based methods and used as a DNA template for PCR tests. The system and thermocycling conditions of PCR test were conducted according to the methods described previously.

3 Results and analysis

3.1 Isolation of *PtAP*3

A pair of primers were designed according to *P. trichocarpa AP*3 gene (PTD), and *PtAP*3, an *AP*3 homologous gene from *P. tomentosa* was isolated by PCR using genomic DNA of male clones of *P. tomentosa* (L50) as template. The corresponding fragment was obtained by PCR amplification whose length was about 1.8 kb. The results are shown in Fig. 1A. The purified PCR products was cloned into a pGEM® -T easy vector, then transferred into competent cells of *E. coli* TG1 and spread on a LB plate containing 100μg/mL Kan. Two kinds of positive recombinant plasmids (PTAP3XA-1 and PTAP3XA-2) were obtained and were amplified using the recombined plasmids as DNA template by further PCR identification. The results indicated that they were absolutely coincident with the lengths of the previous PCR products (Fig. 1B). It revealed that the length of the recombined plasmids is obviously bigger than pBS (Fig. 1C). Meanwhile, the recombinant plasmids were digested with *Bam*H I and *Sac* I; the length of corresponding fragments matched the length of the PCR products, too. It suggested that the PCR products had been cloned into the pGEM® -T easy vector successfully.

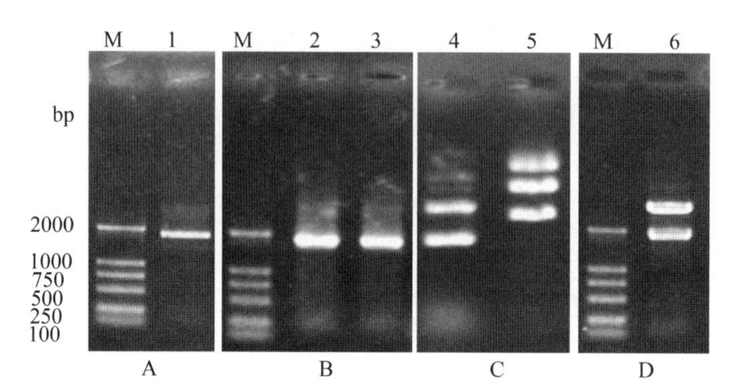

Fig. 1 Cloning of *AP*3 homologous gene from male clone of *P. tomentosa*

M: DL2000 marker; 1: PCR product of genomic DNA; 2 and 3: PCR products of PtAP3XA-1 and PtAP3XA-2; 4: The plasmid pBS; 5: The recombinant plasmid; 6: The recombinant plasmid digested with *Bam*H Ⅰ and *Sac* Ⅰ; A: PCR product of genomic DNA; B: PCR tests of the recombinant plasmid; C: Test of length of the recombinant plasmid; D: The recombinant plasmid digested with restriction endonucleases.

3.2 Sequence analysis

The recombinant plasmids, which were verified by PCR, were sequenced by the Sangon Co. The results indicated that *PtAP*3 from *P. tomentosa* was 1, 813 bp long (including the introduced 5′ and 3′ end digested sites). Combining with the rule of intron slice in eukaryotic genome DNA and the analysis of SPL & Seqaid II (search for potential splice sites) at http://dot.imgen.bcm.tmc.edu: 9331/gene-finder/gf.htm, the results indicated that there were 7 extrons and 6 introns in this gene (1st extron 9-196, 1st intron 197-331; 2nd extron 332-398, 2nd intron 399-491; 3rd extron 492-553, 3rd intron 554-664; 4th extron 665-764, 4th intron 765-1, 323; 5th extron 1, 324-1, 365, 5th intron 1, 366-1, 446; 6th extron 1, 447-1, 491, 6th intron 1, 492-1, 592; 7th extron 1, 593-1, 805) and encoded 238 amino acids (Fig. 2). There was a highly conserved MADS-box motif in the 1^{st}extron (shown by underlined section in Fig. 2), so it was accepted to be a transcription factor.

In addition, the structure and map of the restriction site of homologous AP3 from *P. tomentosa* were drawn by DNAMAN software (Fig. 3), the 1st, 2nd, 3rd, 4th extrons were distributed in the front half of the section, the 5th, 6th, 7th extrons in the back one-third section, and the 4th intron was the longest (559bp) in the sequence. Except in the introduced *Bam*H Ⅰ and *Sac* Ⅰ sites, there were *Eco*R Ⅴ, *Nde* Ⅰ, and *Xba* Ⅰ in 516bp, 719 bp, and 862bp respectively, which provided important information for further studies including preparing a probe of Southern and Northern blotting hybridization, genomic DNA function, and enzyme digestion.

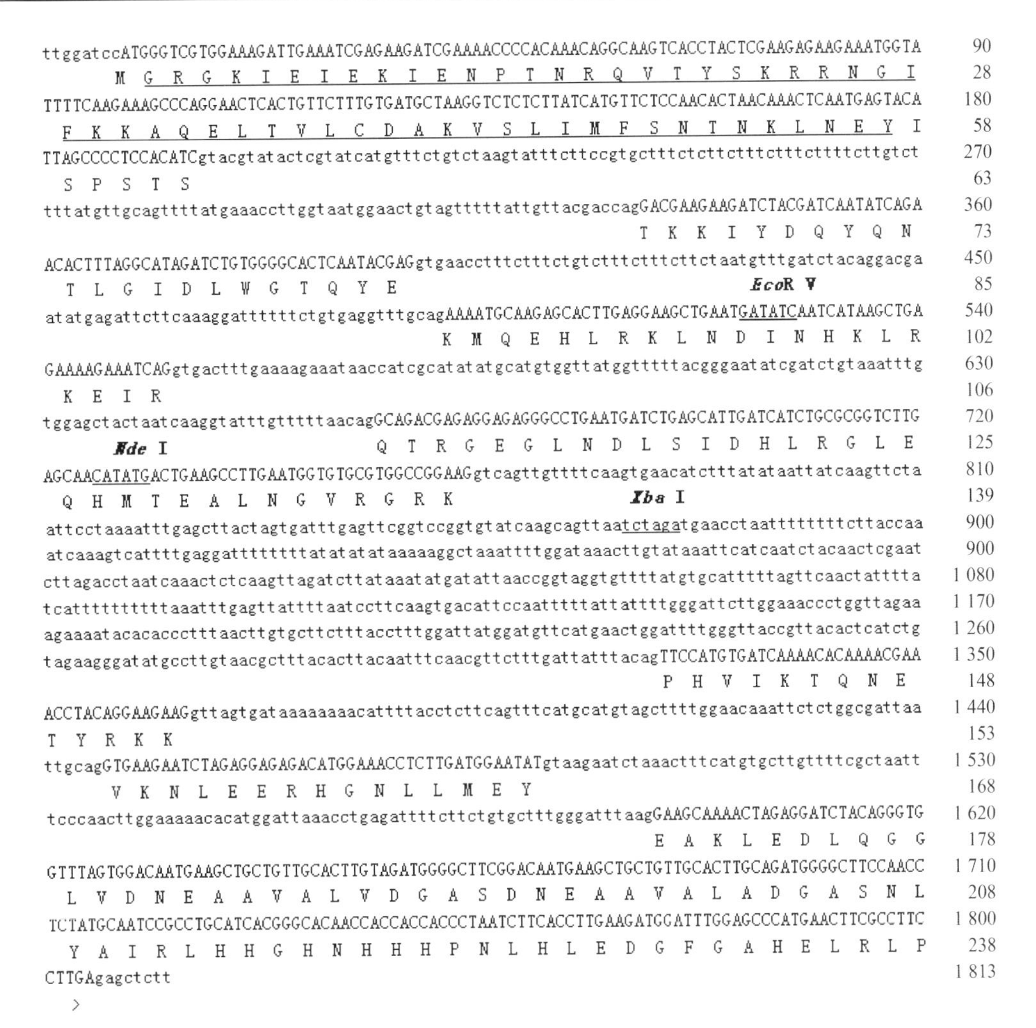

Fig. 2 The full nucleotide and deduced amino acid sequence of putative *AP3* homologous gene (*PtAP3*) from male clone of *P. tomentosa*.

The small letters belong to noncoding sequence, and the capital letter belong to coding sequence

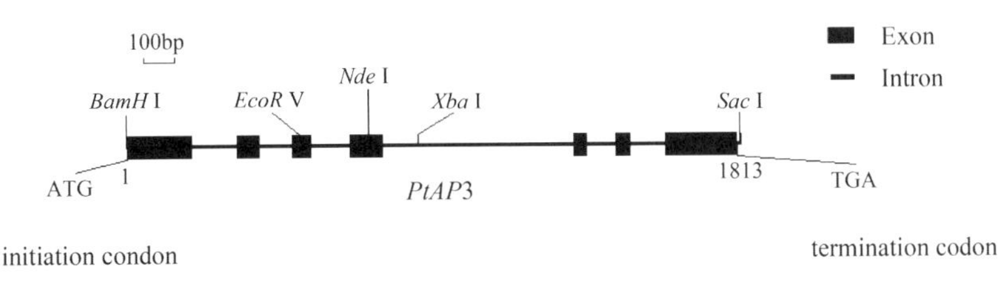

Fig. 3 Structural and restriction map of *AP3* homologous gene from male clone of *P. tomentosa* (*PtAP3*)

```
                   10         20         30         40         50         60         70
PtAP3       MGRGKIEIEK IENPTNRQVT YSKRRNGIFK KAQELTVLCD AKVSLIMFSN TNKLNEYISP STSTKKIYDQ
AF05770     MGRGKIEIKK IENPTNRQVT YSKRRNGIFK KAQELTVLCD AKVSLIMFSN TNKLNEYISP STSTKKIYDQ
AJ251116.   MGRGKIEIKL IENQTNRQVT YSKRRNGIFK KAQELTVLCD AKVSLIMLSN TNKMHEYISP TTTTKSMYDD
AB081093.   MGRGKIEIKL IENQTNRQVT YSKRRNGIFK KAQELTVLCD AKVSLIMLSN TSKMHEYISP TTTTKSMYDD
AB055966.   MGRGKIEIKL IENQTNRQVT YSKRRNGIFK KAQELTVLCD AQVSLIMQSS TDKIHEYISP TTTTKKMFDL
AJ009724.   MGRGKIEIKK IENNTNRQVT YSKRRNGIFK KAHELTVLCD AKVSLIMFSN TGKFHEYISP STTTKKMYDQ
AF230704.   MGRGKIEIKK IENSTNRQVT YSKRRNGLFK KAKELTVLCD AKICLIMLSS TRKFHEYTSP NTTTKKMIDL
AF503913.   MGRGKIEIKK IENSTNRQVT YSKRRTGIIK KATELTVLCD AEVSLLMFSS TGKLSEFCSP STDTKKIFDR
AF209729.   MGRGKIEIKK IENSTNRQVT YSKRRSGIMK KAKELTVLCD ADVSIIMFSS TGKFSEYCSP GTDTKTVFER
Clustal Co  ********:  *** ****** *****.*::* ** ******* *.:.::* *. * *: *: **  * **.: :

                   80         90        100        110        120        130        140
PtAP3       YQNTLGIDLW GTQYEKMQEH LRKLNDINHK LRKEIRQT-R GEGLNDLSID HLRGLEQHMT EALNGVRGRK
AF05770     YQNALGIDLW GTQYEKMQEH LRKLNDINHK LRQEIRQR-R GEGLNDLSID HLRGLEQHMT EALNGVRGRK
AJ251116.   YQKTMGIDLW RTHEESMKDT LWKLKEINNK LRREIRQR-L GHDLNGLSFD ELASLDDEMQ SSLDAIRQRK
AB081093.   YQKTMGIDLW RTHYESMKDT LWKLKEINNK LRREIRQR-L GHDLNGLSYD DLRSLEDKMQ SSLDAIRERK
AB055966.   YQKNLQIDLW SSHYEAMKEN LWKLKEVNNK LRRDIRQR-L GHDLNGLSYA ELQDLEETMS QSVQIIRDRK
AJ009724.   YQSTVGFDLW SSHYERMKET MKKLKDTNNK LRREIRQRVL GEDFDGLDMN DLTSLEQHMQ DSLTLVRERK
AF230704.   YQRTLGVDIW NKHYEKMQEN LNRLKDINNK LRREIRQR-T GEDMSGLNLQ ELCHLQGNVS DSLAEIRERK
AF503913.   YQQLSGINLW SAQYEKMQNT LNHLSEINRN LRKEISQR-M GEELDGLDIK DLRGLEQNLD EALKLVRHRK
AF209729.   YQQATQTNLW STQYEKMQNT LNHLKEINHN LRKEIRQR-I GEELDGMDFK ELRGLEQNLD EALKSVRARK
Clustal Co  **      ::*   : * *::   : :*.: *.: **::* *     *. :..:.    .*  *:  :   .::  :* **

                  150        160        170        180        190        200        210
PtAP3       PHVIKTQNET YRKKVKNLEE RHGNLLMEY- --EAKLEDLQ GGLVDNEAAV ALVDGASDNE AAVALADGAS
AF05770     YHVIKTQNET YRKKVKNLEE RHGNLLMEY- --EAKLEDRQ YGLV------ -------DNE AAVALANGAS
AJ251116.   YHVIKTQTET TKKKVKNLEQ RRGNMLHGYF DQEAAGEDPQ YGYEDNEG-- -------DYE SALALSNGAN
AB081093.   YHVIKTQTET TKKKVKNLEE RRGNMLHGY- --EAASENPQ YCYVDNEG-- -------DYE SALVLANGAN
AB055966.   YHVLKTQAET TRKKVKNLEE RNSNLMHGYG --APGNEDPQ YGYVDNEG-- -------DYE SAVALANGAS
AJ009724.   YHVIKTQTDT CRKRVRNLEQ RNGNLRLDYE --TIHQLDKK YDTGENEG-- -------DYE SVVAYSNGVS
AF230704.   YHVIKTQTDT CRKRVRNLEE QHGSLVHDL- --EAKSEDPT YGVVENEG-- -------HFN SAMAFANGVH
AF503913.   YHVINTQTET YKKKVKNSEE AHKNLLRDLV NREMKDENPV YGYVDEDP-- ------SNYD GGLGLANGAS
AF209729.   YHVITTQTDT YKKKVKNSQE AHKTLLHELD -------DAV YGYADEDP-- ------GNYD SSLALAHGGS
Clustal Co   **:.** :*  :*:*:* ::  . .:              :         :                : . :  :.*

                  220        230        240        250        260        270
PtAP3       NLYAIR---- ---------- ---------- -----LHHGH NHHH---PNL HLEDGFGAHE LRLP-
AF05770     NLYAFR---- ---------- ---------- -----LHHGH NHHHHL-PNL HLGDGFGAHE LRLP-
AJ251116.   NLYTF----- ---------- --------HL H-------HR NLHHG--GSS LGSSITHLHD LRLA-
AB081093.   NLYTF----- ---------- --------QL HRNSDQLHHP NLHHHR-GSS LGSSITHLHD LRLA-
AB055966.   NLYFFNRVHN NHNLDHGHGG GSLVSSITHL Q-NPNNHGNH NLENGHGGGS LISSITHLHD LRLA-
AJ009724.   NLYAF----- ---------- ---------- -----CVHPN NIPHG---AG YEL-----HD HQHTN
AF230704.   NLYAFR---- ---------- ---------L Q----TLHPN -LQN---GGG FGS-----RD LRLA-
AF503913.   HLYEFR---- ---------- ---------- ------VQPS Q-PNLH-GMG YGS-----HD LRLA-
AF209729.   NMYAYR---- ---------- ---------- ------VQPS Q-PNLH-GMS YGP-----HD LRLA-
Clustal Co  ::*                                         :            ::  : .
```

Fig. 4 Similarity of amino acid sequences coded by *AP*3 homologous gene of several plants

The amino acid sequences have been aligned by introducing gaps (—) to maximize homology (Clustal W software). Conserved MADS-box motif are underlined, and totally conserved and conservatively replaced amino acids are indicated by asterisks and dots, respectively. AF057708: *Populus trichocarpa*; AJ251116 and AB081093: *Malus domestica*; AB055966: *Rosa rugosa*; AJ009724, *Gerbera hybrida*; AF230704: *Petunia hybrida*; AF503913: *Lilium regale*; AF209729: *Hemerocallis* hybrid cultivar.

It indicated that the deduced amino acid of *PtAP*3 was up to 82% homologous to *P. trichocarpa* (AF057708); at the same time, 60%, 60%, 55%, 52% and 52% homologous to *Malus domestica*, (AJ251116, AB081093), *Rosa rugosa* (AB055966), *Lilium regale* (AF503913, AB071378), *Petunia hybrida* (AF230704), and *Gerbera hybrida* (AJ009724) respectively by blast analysis in GenBank. A highly conserved domain near the 5′ end of *AP*3 homologues was found by comparing amino acid sequences of *AP*3 homologous genes with different species (*P. tomentosa*, *P. trichocarpa*, *Malus domestica*, *Rosa rugosa*, *Lilium regale*, *Petunia hybrida*, *Gerbera hybrida*, and *Hemerocallis* hybrid cultivar (AF209729)) by Bioedit and Clustalx software

(Fig. 4). Although this region belonged to a highly conserved MADS-box motif, representing a transcription factor, it suggested that *PtAP*3 might perform as a transcription factor.

3.3 Southern blot analysis

To verify that the homologous *AP*3 gene comes from male clones of *P. tomentosa*, *PtAP*3 was labeled as probe and genomic DNA from male (LM50) and female (5082) clones was digested with *Bam*H I, *Eco*R I, and *Hin*d III, respectively, electrophoresed on 1.0% agarose gel and blotted onto the nylon membrane positively charged by capillary transfer with 20×SSC and washed at a high stringency. The result indicated that *PtAP*3 actually comes from the genome of *P. tomentosa*, the male and female clones of *P. tomentosa*, but there was remarkable difference between male and female clones—double copies or two high homologue members in male genomic DNA, whereas a single copy in female clones (Fig. 5).

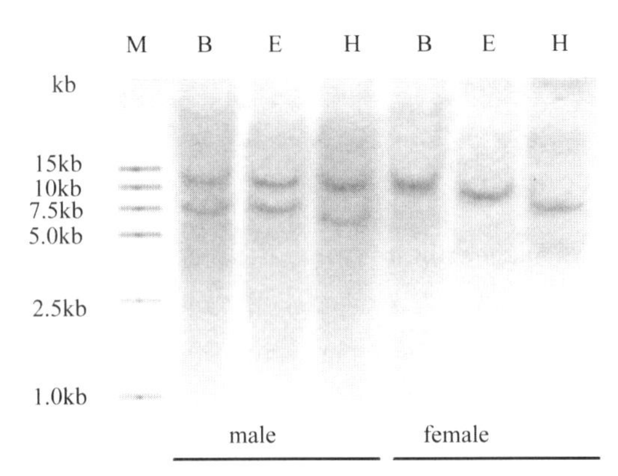

Fig. 5 Southern blot analysis of DNA from male and femaleclones of *P. tomentosa* using probe of *PtAP*3

M: DL 15000 marker; B: *Bam*H I; E: *Eco*R I; H: *Hin*d III.

3.4 Construction of sense and antisense expression vectors

To study the function of *PtAP*3 in stamen development, sense and antisense expression vectors of *PtAP*3 were constructed. The recombinant plasmids pGEM-T-*PtAP*3 and pBI121 were digested with *Bam*H I and *Sac* I, the corresponding target fragments were recycled, purified, and ligased to produce a sense-oriented and anti-sense-oriented construction.

The recombined plasmid pBI121-*PtAP*3 and pBI121-*Ptap*3 were used as DNA templates respectively, and the released fragments were also about 1.8kb and 1.1kb long (Fig. 6A, B) including partial sequences of the 35S promoter and *GUS* gene, which was about 230bp. It showed that the results of the PCR test and digestion with *Bam*H Ⅰ/*Xba* Ⅰ and *Bam*H Ⅰ/*Xba* Ⅰ were coincident, that is, the sense and anti-sense constructs were also successful.

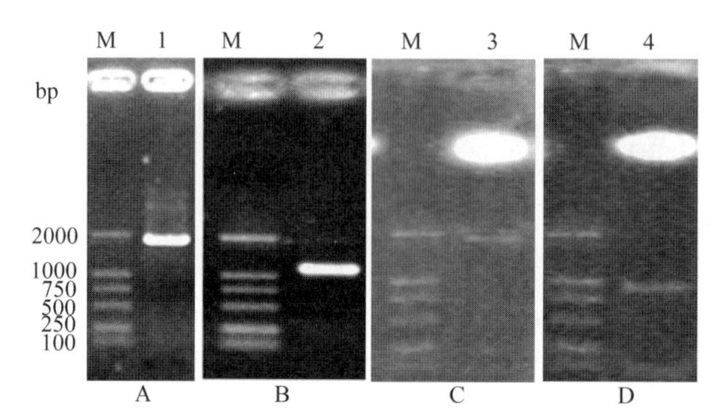

Fig. 6　The tests of PCR and digestion of both sense and anti-sense expression vectors of *PtAP3*

A: The test of PCR of pBI121-PtAP3; B: The test of PCR of pBI121-*PtAP3*; C: The test of digestion of pBI121-*PtAP3*; D: The test of digestion of pBI121-*PtAP3*. M: Marker DL2000, 1: PCR product of pBI121-*PtAP3*, 2: PCR product of pBI121-*PtAP3*; 3: pBI121-*PtAP3* digested with *Bam*H Ⅰ and *Sac* Ⅰ, 4: pBI121-*PtAP3* digested with *Bam*H Ⅰ and *Xba* Ⅰ.

3.5　Tobacco transformation

Tobacco leaf disks without preculture were used as acceptor materials, cocultured with *Agrobacterium* for two days (Fig. 7A, L: sense and R: antisense), then transferred on screening media MS with carbenicillin 250mg/L and kanamycin 100mg/L. Some resistant buds from transformed disks were found in about seven days (Fig. 7B, L: sense and R: antisense). When resistant buds were long enough, they were cut and transformed on rooting medium MS with carbenicillin 250mg/L and kanamycin 200mg/L to root (Fig. 7C: sense and D: antisense). Some transformed plants were obtained, and DNA in leaves of transformed and nontransformed tobacco plantlets was used as the template of PCR reaction, and the recombined plasmid as positive control. The results of PCR showed that one sense transformant (Fig. 8A) and three antisense transformants (Fig. 8B) had been introduced in these tobacco plants.

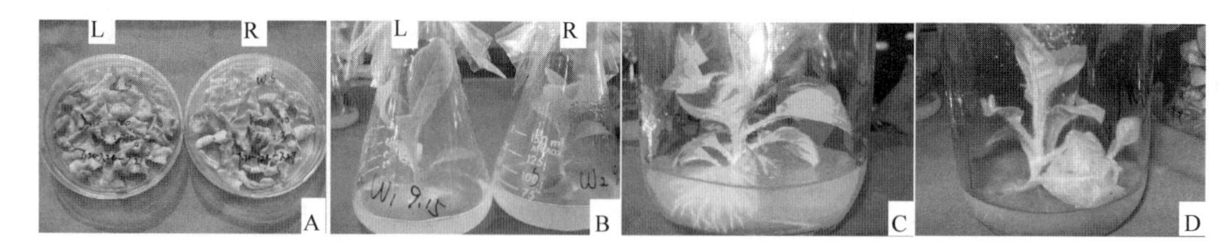

Fig. 7　Generation of Kan-resistant plantlets from tobacco leaf disks

A: The tobacco leaf disks infected by *Agrobacterium*; B: Kan resistant shoots from tobacco leaf disks; C: Culture of regenerated shoots (sense transformation) on generation medium; D: Culture of regenerated shoots (anti-sense transformation) on generation medium.

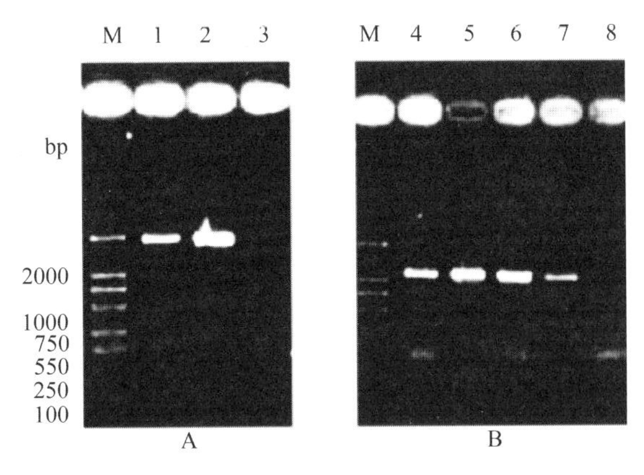

Fig. 8 Transgenic tobacco plantlets verified by PCR

A: The sense transgenic plant verified by PCR, B: The anti-sense transgenic tobaccos verified by PCR. M. DL2000 Marker, 1: The sense transgenic plant; 2: The positive control (pBI121-*PtAP*3); 3: The nontransgenic tobacco; 4~6: The anti-sense transgenic tobaccos; 7: The positive control (pBI121-*PtAP*3); 8: The nontransgenic tobacco.

4 Discussion

So far *AP*3 homologous gene has been well isolated in many herbs, such as, *Arabidopsis thaliana*, *Antrirrhinum majus*, *Petunie hybrida*, *Solanum tuberosum* etc., whereas only *Malus domestica* and *P. trichocarpa* for woody species. *PtAP*3, an *AP*3 homologue, was isolated from male clones of *P. tomentosa* by PCR in this research, indicating that homologous *AP*3 gene exist widely in higher plants. The prominent trait of the *AP*3 homologue is with conservative MADS-box motif. It is believed that MADS-box might have functions of DNA integration and dimer formation while participating in transcriptional regulation, therefore the proteins that were coded by the genes belong to transcription factors. Conserved MIKC structure with the highly conserved DNA-binding MADS domain at the amino terminus exists widely in many plants (Alvarez-Buylla, 2000). The moderately conserved K domain in the downstream of MADS-box genes, the MADS, and K domains are linked to one another by a weakly conserved I domain, as well as a carboxyl-terminal (C) region. It indicates that the I region (which follows the MADS box) have been characterized as a key molecular determinant for the selective formation of DNA-binding dimers. The DNA-binding specificity is mediated to a large degree by AP3 protein located in I region and K region (Riechmann, 1996a; 1996b). The microarray analysis suggests that *AP*3 and *PI* regulate a relatively small number of genes, implying that many genes used in petal and stamen development are not tissue specific and likely have roles in other processes as well (Moriyah, 2003).

*PtAP*3 is highly homologous to other plants, especially higher in conserved MADS-box region, so it is implicated to be a transcription factor. *APETALA*3 (*AP*3) in *Arabidopsis* is one of the genes that confer a B-class function. In accordance with their roles in specifying petal and stamen identity, the expression of *AP*3 or *PI* and *AG* are maintained throughout the development of petals and stamens (Honma, 2000; Pelaz, 2001). The floral homeotic genes *APETALA*3 (*AP*3) and *PISTILLATA* (*PI*) have been proved crucial for understanding the molecular mechanisms that lead to petal and stamen formation in this research. *AP*3 and *PI* expression is necessary and sufficient to regulate basic morphology of petals and stamens (Moriyah, 2003; Kramer, 2003). However, the result of the Southern blot analysis indicates that there are obvious differences; double copies of *PtAP*3 or two members have high homology to each other in male clones (L50) in *P. tomentosa*, whereas this research indicated a single copy in *PtAP*3 in *P. tomentosa* (5082, female) genomic DNA in female clones (5082). Meilan et al. (2001) have also verified that *PtAP*3 was expressed at high levels both in male flower primordia and female flower primordia.

The interrelated researches indicate that *P. tomentosa* is abundant in resource of male clones; however, abortive pollen are easier to happen (Zhang, 1992), and breeding cycle is rather long with a long juvenile phase of about 5 ~ 10 years, which mean that there are many difficulties in cross-breeding of *P. tomentosa*. To overcome the difficulties of abortive pollen and little flower and further study the function of *PtAP*3, sense and antisense expression vectors are constructed, and some transgenic tobacco transformants have been obtained by *Agrobacterium*; further molecular identification and morphologic observations are under way. At same time, the transgenic work of *P. tomentosa* is being carried out. In future the transformants from *P. tomentosa* may reduce abortive pollen thus promoting pollen living ability, which will also provide a new route for cross breeding of *P. tomentosa*.

Besides *P. tomentosa*, a native greenbelt species, plays a key role in beautifying the environment in north China, although pollen and catkin production contamination is another problem. Especially male clones with good reproductive ability release more pollen, and is one of the relatively serious sources of allergy, which increases the density of dust breathed in to bodies. In northern cities, the catkins from female clones in every spring have become an important factor affecting the environment and have attracted wide attention from society. As the part of anti-sense sequence (with intron sequence) of *PtAP*3 gene is ligated into pBI121 and 35S promoters, *Ptap*3 is constructed to transform *P. tomentosa* clones to suppress its expression. Introduced intron in anti-sense construct will help cosuppress expression of *PtAP*3 gene, thus bringing about disturbing the floral development of *P. tomentosa* along with poplars and willows to improve the environment.

In this research, *PtAP*3 homologue gene was isolated and anti-sense expression vector was constructed and transformed into *P. tomentosa* and tobacco. At present, *LEAFY* and *AGAMOUS* involved in flowering genes are being isolated, so that the rule of spatio-temporal expression will be analyzed as a whole. This will provide a molecular mechanism to explain flower development of *P. tomentosa*, and also contribute on long breeding cycle and pollen contamination. This study has provided transgenic safety of *P. tomentosa* by genetic engineering and affords us useful references for genetic improvement of other trees in the *Salix* section.

毛白杨 *PtLFY* 在花芽发育中的表达模式与花芽形态分化*

摘　要　本研究以毛白杨花芽为材料，采用 RT-PCR 技术分离克隆了毛白杨 *PtLFY* cDNA 序列，测序结果表明该序列全长 1314bp，包含一个开放阅读框，编码 377 个氨基酸。Alignment 分析显示该基因与拟南芥等物种 *LFY/FLO* 同源基因所编码的氨基酸相似性达到 68%~75%。蛋白结构预测分析表明，在 PtLFY 蛋白 N-端和 C-端具有 2 个高度保守的区域，其中 PtLFY-C 端由 7 个 α-helix 组成 Helix-turn-helix 结构。采用 Real-time qRT-PCR 技术检测了 *PtLFY* 在雌雄花芽发育过程中的表达模式，结果显示从 9 月 13 日到翌年 1 月 25 日，*PtLFY* 在毛白杨雌雄花芽中持续稳定表达，到 2 月 25 日表达量少许下调，但该基因在雄花芽中的相对表达量明显高于雌花芽。解剖分析结果表明，雄花芽形态分化进程明显早于雌花芽，这种差异可能与 *PtLFY* 在雌雄花芽发育过程的差异表达存在密切联系。本研究对于阐明 *PtLFY* 在毛白杨雌雄花芽发育和开花中的分子作用机制具有重要的理论意义，为进一步开展毛白杨开花调控研究奠定了基础。

关键词　毛白杨，*PtLFY*，表达模式，花芽，形态分化，qRT-PCR

毛白杨(*Populus tomentosa*)是我国特有的白杨派树种，雌雄异株，但偶见雌雄同株现象(张志毅等，1992)。在开花发育方面，董源(1982，1984)就进行了毛白杨的胚胎学观察研究；朱大保(1990)对毛白杨的有性生殖能力进行了研究；张志毅等(1992，2000)先后对毛白杨开花结实特性和三倍体毛白杨的有性生殖能力进行了研究。杨树作为木本模式树种，与拟南芥(*Arabidopsis thaliana*)有许多不同之处，杨树具有多年生、童期长达 7~10 年、生命周期长达 100~200 年等特点(Braatne et al. , 1996)。另外杨树在生命中首次经历开花事件后，一年一次的季节性开花现象伴随杨树的整个生殖阶段(Yuceer et al. , 2003)。这些特点决定了杨树具有自身的花发育规律。杨树的花结构也不同于拟南芥等模式植物具有的四轮花器官结构，仅由苞片和雄蕊或雌蕊两轮结构组成(Sheppard et al. , 2000)，这种结构差异的分子机制尚不清楚。近年来，有关花分生组织特征基因和花器官特征基因在拟南芥、金鱼草(*Antirrhinum majus*)等许多植物中相继被分离。这些基因的发现，使得人们对植物的成花和花器官的发生与发育机理的认识进入到分子水平。作为花分生组织特征基因，*LEAFY* (*LFY*) 控制拟南芥花序梢分生组织向花分生组织的转变(Weigel et al. , 1992; Weigel and Nilsson, 1998; Peña et al. , 2001)。*LFY* 的表达水平被认为是开花转变的关键因子。尽管 Rottmann 等(2000)从毛果杨中分离克隆了 *PTLF* 基因，并在杨树和拟南芥中进行了转化研究，但杨树 *LFY* 同源基因在花芽发育过程的表达模式尚不清楚，关于该基因的表

* 本文原载《林业科学》，2010，46(2)：32-38，与安新民、王冬梅、王泽亮、王静澄、曹冠琳和薄文浩合作发表。

达与杨树花芽分化关系的研究还未见报道。为此开展毛白杨 *LFY* 同源基因 cDNA 的克隆及其表达规律的研究，并对毛白杨雌雄花芽的形态分化规律进行解析，分析毛白杨 *LFY* 同源基因的表达与花芽分化的内在关系，对于揭示毛白杨开花的分子调控机制具有重要的理论意义，为进一步通过基因工程手段实现缩短育种周期和抑制杨树花粉与飞絮污染的研究奠定基础。

1 材料与方法

1.1 材料

供试毛白杨(*Populus tomentosa*) 雌雄成年植株定植于北京林业大学苗圃，分别采集不同发育阶段的雌雄花芽，一部分固定在 FAA 固定液中于 4℃保存，用于花芽形态解剖观察分析，另一部分液氮速冻贮于-70℃冰箱中，用于 RNA 提取。

表 1 用于毛白杨 *PtLFY* 基因克隆和 RT-PCR 分析的引物

编号	引物序列	退火温度	说明
P1U	5′-TCGGCACGAGGGCAGATAGATATG-3′	60℃	用于 *PtLFY* cDNA 的分离
P1D	5′-TTAGAGCCTTTAAGCATATCATGAT-3′		
P2U	5′-GTTTCTCTGAGGAGCCAGTACAGC-3′	55℃	用于 *PtLFY* RT-PCR 分析
P2D	5′-CTTAGTGGGGCATTTTTCTCCCCT-3′		
P3U	5′-CTGCCCGTTGCTCTGATGATTCA-3′	55℃	18S rRNA 引物用作 RT-PCR 内参
P3D	5′-CCTTGGATGTGGTAGCCGTTTCT-3′		

1.2 方法

1.2.1 *PtLFY* cDNA 的克隆和测序

雌雄花芽总 RNA 的提取采用改良的 CTAB 法(Chang et al., 1993)，逆转录与 cDNA 第一链的合成按照 SMART IV cDNA Library Construction Kit (BD Clontech)说明书进行。50μL PCR 反应体系包括 1μLcDNA 第一链，各 0.2μmol·L^{-1} 上游引物 P1U 和下游引物 P1D(表 1)，1×PCR 反应缓冲液，2mmol·L^{-1} $MgCl_2$，200μmol·L^{-1} dNTPs 和 2 单位 Taq DNA 聚合酶。PCR 循环条件为 94℃预变性 3 min，最后 72℃延伸 7 min，94℃变性 30s，60℃退火 30s，72℃ 延伸 40s，共进行 35 轮循环。DNA 序列委托上海生工生物工程技术服务有限公司测定。DNA 序列分析采用 Seq Aid Ⅱ和 Bioedit 软件包进行。

1.2.2 序列同源性分析与蛋白结构预测

LFY 同源基因氨基酸序列来自 GenBank (http://www.ncbi.nlm.nih.gov/entrez/query.fcgi)，Alignment 分析采用 CLUSTALX 1.81 进行，PtLFY 蛋白二级结构预测采用 ANTHEPROT 2000 V6.0 软件进行，保守序列和功能域分析采用 ExPASy 软件 (http://cn.expasy.org/) (Gasteiger et al., 2003)，PtLFY 蛋白三级结构预测分析采用 3D-JIGSAW 2.0 (http://bmm.cancerre searchuk.org/~3djigsaw/)，进一步采用 RasMol V 2.7.2.1 进行优化。

1.2.3 Real-time qRT-PCR 分析

按照 Chang et al.（1993）的方法，分别提取毛白杨不同发育时期雌雄花芽的总 RNA，用 RQ1 DNase I（Promega）对总 RNA 进行处理以去除 DNA 污染。RNA 浓度的测定采用 SPEKOL 1300spectrophotometer（Jena）。根据 SuperScript™ III Platinum® Two-Step qRT-PCR Kit With SYBR® Green（Invitrogen）说明，分别取预处理后不同样品总 RNA 1μg，进行 cDNA 第一链合成，然后将第一链 cDNA 稀释 5 倍作为 PCR 的反应模板，进行 qPCR 反应。20μL qPCR 反应体系包括 1μL 稀释 5 倍的 cDNA 第一链，各 0.2$\mu mol \cdot L^{-1}$ 上游引物 P2U 和下游引物 P2D（表 1），10μL 2×SYBR Green PCR Master Mix（Invitrogen），反应在 OPTICON2（MJ research）中进行。qRT-PCR 热循环条件为 50℃、94℃各 2min，最后 72℃延伸 7 min，94℃变性 20s，55℃退火 20s，72℃ 延伸 20s，共进行 35 轮循环。以 18S rRNA 作为内参基因，采用上游引物 P3U 和下游引物 P3D（表 1）进行 PCR 反应，反应体系和热循环条件同上。以上实验均为 3 次重复。

1.2.4 花芽解剖结构分析

定期采集毛白杨雌雄花芽样品，按照李正理（1996）和林加涵（2000）的方法，对样品进行如下处理：固定—脱水—透明—浸蜡—包埋—切片—贴片—脱蜡—复水—番红、固绿染色—封藏，切片厚度为 10μm，用加拿大树胶封片。在 OlympusAx70 光学显微镜下观察、拍照记录。

2 结果与分析

2.1 毛白杨 *PtLFY* 基因的分离克隆

为了从毛白杨中分离 *PtLFY* 同源基因，首先根据美洲黑杨 *LFY* 同源基因（*PTLF*）设计合成了一对 PCR 引物（P1U，P1D），利用 RT-PCR 技术从毛白杨雄花芽 mRNA 中分离 *PtLFY*。PCR 结果如图 1 所示，扩增出长度大约为 1.3 kb 条带。将该片段克隆并测序。结果表明该基因序列长度为 1314bp，包含一个完整的阅读框（ORF），编码 377 氨基酸（图 2）。

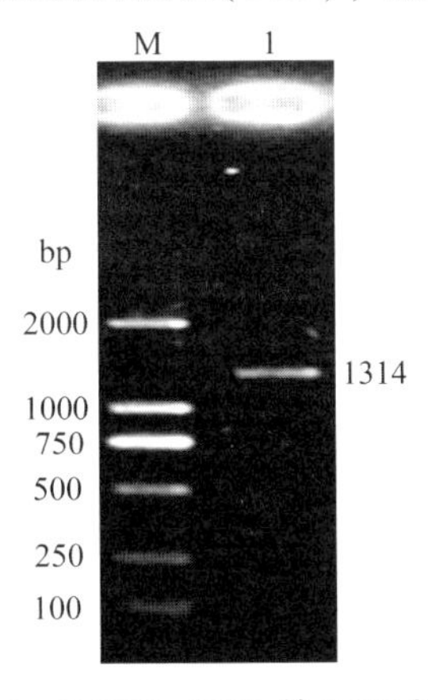

图 1 *PtLFY* cDNA 的 PCR 扩增

M. DL-2000 marker；1. PCR 产物 PCR product

```
                                                                                       T     1
TCGGCACGAGGGCAGATAGATATGGATCCGGAGGCTTTCACGGCGAGTTTGTTCAAATGGGACACGAGAGCAATGGTGCCACATCCTAAC    91
                         M  D  P  E  A  F  T  A  S  L  F  K  W  D  T  R  A  M  V  P  H  P  N    23
CGTCTGCTTGAAATGGTGCCCCCGCCTCAGCAGCCACCGGCTGCGGCGTTTGCTGTAAGGCCAAGGGAGCTATGTGGGCTAGAGGAGTTG   181
R  L  L  E  M  V  P  P  P  Q  Q  P  P  A  A  A  F  A  V  R  P  R  E  L  C  G  L  E  E  L    53
TTTCAAGCTTATGGTATTAGGTACTACACGGCAGCAAAAATAGCTGAACTCGGGTTCACAGTGAACACCCTTTTGGACATGAAAGATGAG   271
F  Q  A  Y  G  I  R  Y  Y  T  A  A  K  I  A  E  L  G  F  T  V  N  T  L  L  D  M  K  D  E    83
GAGCTTGATGAAATGATGAATAGTTTGTCTCAGATCTTCAGGTGGGATCTTCTTGTTGGTGAGAGGTATGGTATTAAAGCTGCTGTTAGA   361
E  L  D  E  M  M  N  S  L  S  Q  I  F  R  W  D  L  L  V  G  E  R  Y  G  I  K  A  A  V  R   113
GCTGAAAGAAGAAGGCTTGATGAGGAGGATCCTAGGCGTAGGCAATTGCTCTCTGGTGATAATAATACAAATACTCTTGATGCTCTCTCC   451
A  E  R  R  R  L  D  E  E  D  P  R  R  R  Q  L  L  S  G  D  N  N  T  N  T  L  D  A  L  S   143
CAAGAAGGTTTCTCTGAGGAGCCAGTACAGCAAGACAAGGAGGCAGCAGGGAGCGGTGGAAGAGGGACATGGGAGGCAGTGGCAGCGGGG   541
Q  E  G  F  S  E  E  P  V  Q  Q  D  K  E  A  A  G  S  G  G  R  G  T  W  E  A  V  A  A  G   173
GAGAGGAAGAAACAGCCAGGGCGGAAGAAAGGCCAAAGAAAGGTGGTGGACCTTGATGGAGATGATGAACATGGCGGTGCTATCTGTGAG   631
E  R  K  K  Q  P  G  R  K  K  G  Q  R  K  V  V  D  L  D  G  D  D  E  H  G  G  A  I  C  E   203
AGACAGCGGGAGCACCCATTCATTGTAACAGAGCCTGGTGAAGTGGCACGTGGCAAAAAGAACGGTCTTGATTACCTCTTCCATTTATAT   721
R  Q  R  E  H  P  F  I  V  T  E  P  G  E  V  A  R  G  K  K  N  G  L  D  Y  L  F  H  L  Y   233
GAACAGTGTCGTGATTTCTTGATCCAAGTCCAAAGCATTGCCAAGGAGAGAGGAGAAAAATGCCCCACCAAGGTGACAAATCAGGTGTTT   811
E  Q  C  R  D  F  L  I  Q  V  Q  S  I  A  K  E  R  G  E  K  C  P  T  K  V  T  N  Q  V  F   263
AGGTATGCCAAGAAGGCAGGAGCAAGCTACATCAACAAGCCCAAAATGAGACACTACGTGCACTGCTATGCTTTACATTGCCTCGATGGG   901
R  Y  A  K  K  A  G  A  S  Y  I  N  K  P  K  M  R  H  Y  V  H  C  Y  A  L  H  C  L  D  G   293
GACGCATCTAATGCACTTAGGAGAGCGTTCAAGGAGAGAGGAGAGAATGTTGGGGCATGGAGACAGGCTTGTTACAAGCCCCTTGTAGCC   991
D  A  S  N  A  L  R  R  A  F  K  E  R  G  E  N  V  G  A  W  R  Q  A  C  Y  K  P  L  V  A   323
ATCGCCTCTCGCCAAGGCTGGGACATAGACTCCATTTTCAATGCTCATCCTCGGCTTGCCATTTGGTATGTTCCGACCAAGCTCCGTCAA  1081
I  A  S  R  Q  G  W  D  I  D  S  I  F  N  A  H  P  R  L  A  I  W  Y  V  P  T  K  L  R  Q   353
CTTTGTTATGCAGAGCGCAATAGTGCCACCTCTTCAAGCTCTGTCTCTGGTACTGGAGCTCACCTGCCGTTTTGAGTTCTTAATTATGCC  1171
L  C  Y  A  E  R  N  S  A  T  S  S  S  S  V  S  G  T  G  A  H  L  P  F  <                   377
AAGATAAATACTCCTATCTCTCTATAAAATTGTCAAAATGTATGTTGTAGCAGGTCAGGACAAAGTATTGGTTGATGGAGGATGGTTCAT  1261
TAAATCTCACATCCTTGAGTATTTATATATCATGATATGCTTAAAGGCTCTAA                                      1314
```

图 2　毛白杨 *PtLFY* cDNA 及推测的氨基酸序列

2.2　序列同源性分析与蛋白结构预测

在 GenBank 进行 Blast 检索，结果表明毛白杨 *PtLFY* 编码的氨基酸与毛果杨 *PTLF*（U93196）、金鱼草 *FLO*（M55525）、烟草 *NFL*1（U15798）和 *NLF*2（U15799）、苹果 *ALF*1（AB056158）和 *ALF*2(AB056159)、矮牵牛 *ALF*（AF030171）、桉树 *ELF*1（AF034806)所编码的氨基酸同源性分别为 99%、75%、70%和 70%、68%和 69%、68%、68%。通过与其他物种的同源基因比较发现，在这些同源基因的 N-端与 C-端各具有一个高度同源保守区，分别在氨基酸 37~162 和 203~360 之间，这是 *LFY* 及其同源基因的一个显著特征。

为分析预测 *PtLFY* 蛋白的分子功能，联合使用 ANTHEPROT 2000V6.0、Protein Analysis System（ExPASy）（SIB）（http：//cn. expasy. org/）（Gasteiger et al.，2003）和 3D-JIGSAW 2.0 对 PtLFY 可能的二级、三级结构进行了模拟构建。结果表明 PtLFY-C 端保守区由 2 个 β-sheet 和 7 个 α-helix 结构组成，其空间结构如图 3C 所示，与 HTH（Helix-Turn-Helix）结构极其相似，其中 α2-helix（图 3A）、α3-helix（图 3B)在 PtLFY 蛋白与目标 DNA 结合的特异序列识别中非常重要，参与了 PtLFY 与 DNA 大沟和小沟的识别。拟南芥 LFY 蛋白 α2-helix 中 Asn（N)291 和 α3-helix 中 Lys（K）307 参与了 DNA 大沟的碱基特异识别，Arg（R)237 介导了 DNA 小沟的识别(Hamès et al.，2008)。PtLFY-C 蛋白中相应结构 α2-helix 中 Asn（N)258 与 α3-helix 中 Lys（K)277 以及 Arg（R）206，可能发挥类似的功能。

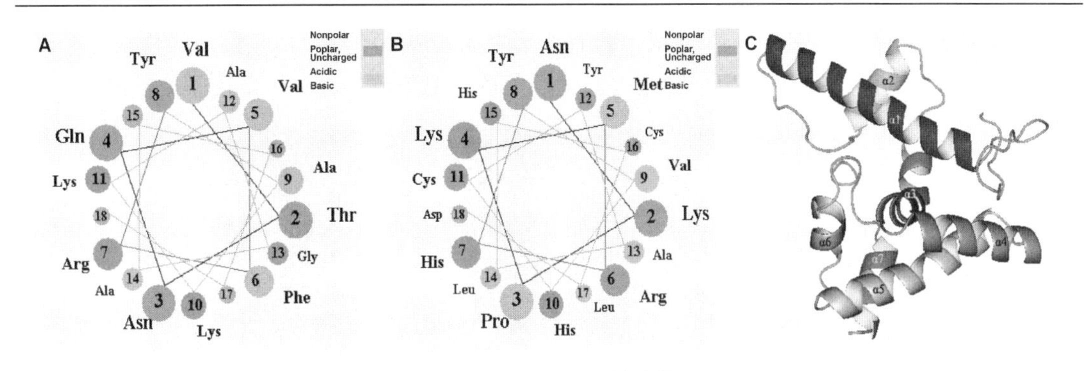

图 3 *PtLFY*-C 端的结构

A. α2 螺旋空间结构；B. α3 螺旋空间结构；C. *PtLFY*-C 端的三级结构

2.3 *PtLFY* 在花芽发育过程中的表达模式

为了探明 *PtLFY* 在毛白杨雌雄花芽发育过程中的表达规律，我们分别收集了 9 月 13 日、11 月 24 日、12 月 10 日、12 月 24 日、1 月 25 日及 2 月 25 日六个不同发育时期的花芽，分别提取其总 RNA，进一步定量均一化，以 18S rRNA 为内参基因，进行 Real-time qRT-PCR 分析，分析结果如图 5 所示。自当年 9 月 13 日至翌年 1 月 25 日，*PtLFY* 在雄花芽中持续稳定高丰度表达，到 2 月 25 日时，该基因的表达量有轻微下调(图 4A)，而在同期的雌花芽中，*PtLFY* 表达与在雄花芽中具有相似的趋势，但其相对表达量远低于在雄花芽中的表达量(图 4B)。

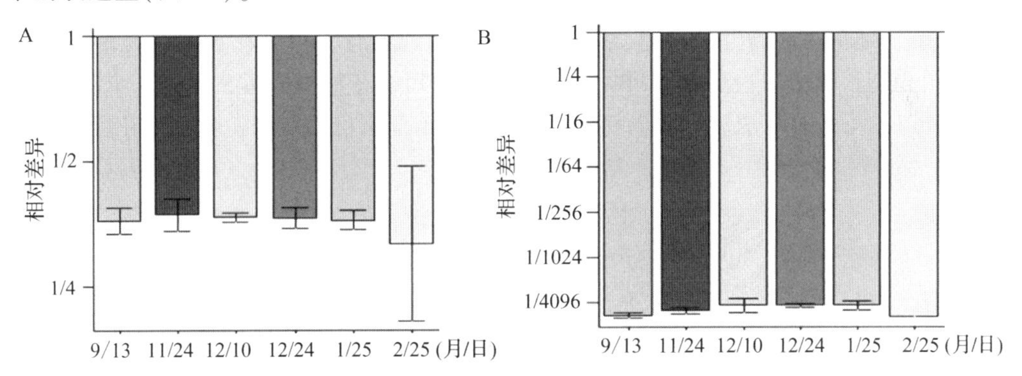

图 4 *PtLFY* 基因在毛白杨雌雄花芽发育过程中的表达模式

A. 雄花芽发育过程中 *PtLFY* 基因的实时 qRT-PCR 分析；B. 雌花芽发育过程中 *PtLFY* 基因的实时 qRT-PCR 分析

2.4 花芽解剖结构分析

为了解毛白杨花芽形态分化的基本规律，我们分别对不同发育时期的雌雄花芽进行了显微观察分析，毛白杨雌花芽的形态分化如图 5 所示。在北京地区，毛白杨雌花芽形态分化始于 7 月底 8 月初，此时花序原基开始形成（图 5A）；此后发育进程加快，小花原基分化，到 8 月 29 日，子房雏形初步形成，侧膜胎座出现，花序基本成形（图 5B）；9 月 28 日，部分子房从外形看已经发育完整（图 5C）。之后经过 3 个多月的发育，子房内部结构不断充实完善（图 5D, E），到翌年 1 月 25 日，含有 2 个心皮的单室子房和柱头形成，柱头二分裂，花柱中空，基生胎座，胚珠倒生于胎座上，珠柄发达，单珠被，通常珠心较厚（图 5F）。

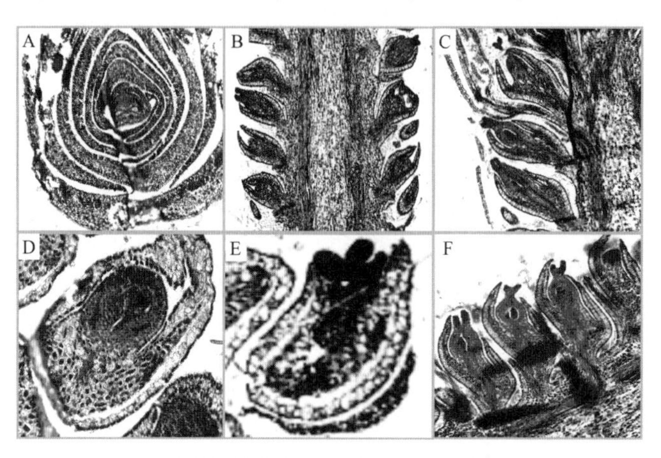

图 5　毛白杨雌花芽不同发育阶段的解剖分析

A. 花芽纵切：花原基出现(8 月 4 日，×100)；B. 花芽纵切：子房雏形（8 月 29 日，×100)；C. 花芽纵切：子房形成(9 月 28 日，×100)；D. 子房纵切(11 月 24 日，×100)；E. 子房纵切（12 月 24 日，×100)；F. 子房纵切：可见二分叉的柱头和倒生的胚珠(1 月 25 日，×100)

毛白杨雄花芽的形态分化如图 6 所示。雄花器官的发育并不集中在花芽的顶端，在环状的花原基边缘不断膨大的突起的组织环，中间的细胞分化慢，其结果是一个凹陷的分生组织区在花的发端出现（图 6A）。经过大约一个半月的发育，圆盘状的分生组织伸出，雄蕊原基出现，雄花序形成(图 6B)。之后经过约半月的发育，每个花盘内花药形状已经非常明显(图 6C)，花药进一步发育(图 6D)，到 10 月旬，花粉囊成形，可见 4 个花粉囊室。到 12 月中旬，花粉囊完全成形，横切面呈蝴蝶形，4 个花粉囊室清晰可见，花粉渐趋成熟（图 6F）。

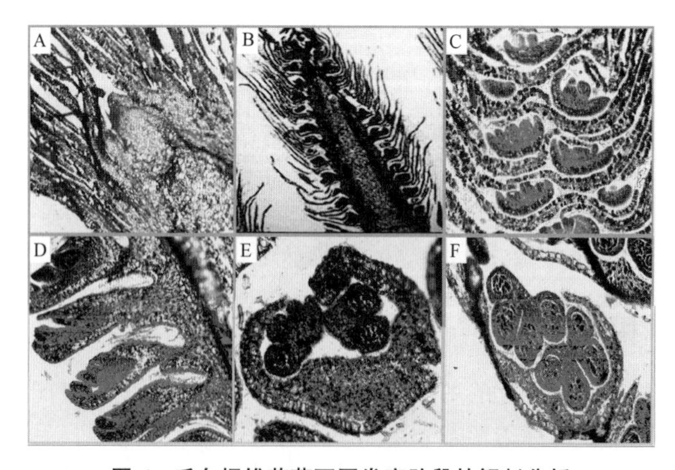

图 6　毛白杨雄花芽不同发育阶段的解剖分析

A. 花芽纵切：花的发端(6 月 4 日，×100)；B. 花序纵切（7 月 20 日，×100)；C. 花序局部纵切：雄蕊原基(8 月 4 日，×100)；D. 局部花序纵切：发育的花药（9 月 13 日，×100)；E. 花粉囊横切(10 月 17 日，×100)；F. 花粉囊横切：花粉囊成形，花粉渐成熟（12 月 24 日，×100)

3 讨论

本研究中从毛白杨花芽 mRNA 中分离克隆了 *PtLFY*，并与之前分离的基因组 DNA 序列进行了比较，证实该基因确由 3 个外显子和 2 个内含子组成，编码 377 个氨基酸(An et al.，2005)，它与拟南芥 *LFY*、桉树 *ELF*1、矮牵牛 *ALF*、苹果 *ALF*1 和 *ALF*2、烟草 *NFL*1 和 *NLF*2、金鱼草 *FLO* 及毛果杨 *PTLF* 具有 58%～99%的同源性。通过比较发现不同物种 *LFY* 同源基因的 N-端 和 C-端各有一个非常保守的区域。*LFY* 基因 C-端保守区的晶体结构已经得到解析，由 7 个 α-helix 构成，其中 α2 和 α3 螺旋在与靶标 DNA 结合识别过程中至关重要，从分子结构水平说明了转录因子 *LFY* 在开花调控中的作用机制(*Hamès* et al.，2008)。经过系列生物信息学软件预测分析表明，*PtLFY* 基因 C-端氨基酸具有类似的 Helix-turn-helix 结构(图 3)，这种序列上的高度同源性和结构上的高度相似性说明 *PtLFY* 在杨树开花发育中具有相似的功能。与一些具有双拷贝 *LFY* 同源基因的苹果、桉树、银杏、烟草等双子叶植物不同，毛白杨基因组 DNA 中只有单拷贝的 *PtLFY* 基因(An et al.，2005)。由此可见，*PtLFY* 在控制毛白杨开花转变过程中的作用无可替代。

Rottmann et al.(2000)研究表明，*PTLF* 不仅在发育中的花序中强烈表达，而且在叶原基、幼叶(在靠近花序的顶端营养芽中最明显)、幼小实生苗中也有表达。在其他木本植物中，*LFY* 同源基因的表达也不总是在生殖发育过程中表达，在桉树和葡萄的叶原基中也有表达(Southerton et al.，1998；Dornelas et al.，2004；Carmona et al.，2002)。在猕猴桃、葡萄和苹果中观察到了 *LFY* 同源基因季节性的表达现象(Walton et al.，2001；Carmona et al.，2002；Wada et al.，2002)。此外，通过 *PTLF* 启动子驱动 GUS 基因的研究发现，GUS 在杨树的枝条中观察到强烈的表达（Wei et al.，2006)。本研究，着重研究了 *PtLFY* 在毛白杨雌雄花芽发育过程的表达模式，从 9 月 13 日至翌年 2 月 25 日，无论雌雄花芽均检测到 *PtLFY* 基因持续稳定的转录本，而且在雄花芽中 *PtLFY* 基因的相对表达量远远高出同期雌花芽(图 4)。该结果为阐明 *PtLFY* 在花发育分子机制的中作用提供了新的证据。

研究发现毛白杨雌雄花芽形态分化进程差异较大，雄花芽的形态分化明显早于雌花芽。雄花的原基的出现始于 6 月初(图 6A)，雌花原基的出现则晚 2 个月(图 5A)；雄花序雏形形成 7 月中旬(图 6B)，而雌花序形成则在 8 月底(图 5B)。之后雌雄花蕊进入快速发育时期，到 12 月 24 日时雄蕊发育基本完善，花粉囊横切面呈蝴蝶形，雄蕊发育花粉逐渐成熟(图 6F)。1 个月后，子房发育较为完善，可见二分叉的柱头和倒生的胚珠(图 5F)。一方面，实时 qRT-PCR 分析结果为上述雌雄花芽发育进程的差异提供了分子证据，*PtLFY* 在毛白杨雌雄花芽发育过程中的表达模式与雌雄花芽形态分化进程相耦合，*PtLFY* 基因持续稳定高水平的表达促进杨树雌雄花芽的形态分化，说明了 *PtLFY* 基因在杨树花发育中的重要作用。毛白杨雌雄花芽形态分化进程的差异解释了雄花开花早于雌花的原因；另一方面，毛白杨雌雄花芽形态分化进程的差异为解释自然界雄花开放早于雌花现象提供了证据。

本研究不仅对于阐明 *PtLFY* 在毛白杨雌雄花发育中的分子机制具有重要的理论意义，而且对于进一步开展毛白杨开花调控、缩短育种周期、控制花粉飞絮污染等方面具有潜在的应用价值。

Expression Profiling of *PtLFY* in Floral Buds Development Associated with the Floral Bud Morphological Differentiation in *Populus tomentosa*

Abstract　A full-length cDNA denominated *PtLFY* was obtained from *Populus tomentosa* using RT-PCR technique. *PtLFY* is 1314bp and contains an open reading frame encoding a 377 amino acid polypeptide. Sequence analysis by blast showed that PtLFY shares 68%～75% homology in amino acid sequence with *Arobidopsis LFY* and other *LFY/FLO* homologues. It was found that PtLFY protein contains 2 conserved domains at N-terminal and C-terminal by alignment analysis. PtLFY-C region adopts a seven-α helix fold structure that binds target promoter DNA elements. The expression patterns of *PtLFY* in developmental floral buds were examined using real-time quantitative RT-PCR. It showed that the continuous and stable transcripts of *PtLFY* were detected in both male and female floral buds of *P. tomentosa* from September 13th to January 25th. , and that of slight down regulation was detected on February 25th. . However, the relative expression of *PtLFY* in male floral buds was obviously much higher than in female floral buds. The anatomical results of both male and female floral buds showed that initiation and differentiation of male floral buds is obviously earlier than female floral buds. It revealed that there is close relationship between expression of *PtLFY* and morphological differentiation of floral buds in *P. tomentosa*. This study will contribute to understanding the molecular mechanism of *PtLFY* in development of both male and female floral buds and flowering in *P. tomentosa*, also will benefit regulation of flowering.

Keywords　*Populus tomentosa*, *PtLFY*, expression profiling, floral buds, morphological differentiation, qRT-PCR

毛白杨 *PtSEP3-1* 基因启动子的克隆分析及其表达载体构建*

摘　要　*SEP*（*SE*、*LATA*）类基因属于花器官发育 ABCDE 模型中的 E 类基因，拟南芥中的研究表明该类基因可能具有控制花器官形态发育以及激活其他类型基因的功能，是一类花发育过程中的关键基因。因此，研究杨树 SEP 类基因启动子表达特性对于杨树的开花调控研究具有重要意义。本文根据毛白杨 *SEP3* 基因和毛果杨基因组序列设计引物，通过 PCR 获得了 *PtSEP3-1* 基因上游 2000bp 的序列。序列分析结果表明该序列具有启动子的基本元件 TATA-box 和 CAAT-box，还包含大量光响应元件 ACE、Box I 和 Box 4 等，此外还有脱落酸响应元件 ABRE，赤霉素响应元件 GARE-motif 以及胁迫响应元件 HSE、TC-rich repeats 等。进一步构建了一个以 *PtSEP3-1* 启动子驱动 *GUS* 基因的植物表达载体 *pPtSEP3-1* protest，为该启动子的功能鉴定奠定了基础。

关键词　毛白杨，*SEP3* 基因，启动子，表达载体

毛白杨（*Populus tomentosa* Carr.）为我国特有的白杨派树种，具有速生优质的特点，是我国北方主要的行道树种和用材树种。但是在开花季节雄株散粉，雌株产生白色飞絮，严重污染城市环境的同时也给毛白杨的育种和推广工作带来了困难。为了通过基因工程的手段调控毛白杨的开花和花器官的发育，就有必要对毛白杨花发育过程中的关键基因进行研究，而启动子的选择又直接关系到基因工程调控的特异性及调控效率等方面。启动子是位于结构基因 5′端上游的一段 DNA 序列，能与 RNA 聚合酶及转录因子特异结合，控制基因转录起始时间和表达的程度，是基因转录最主要的一种调节方式。启动子中包含多种重要的顺式作用元件，其类型直接影响着基因的表达量及表达位点（张春晓等，2004）。因此，启动子的分离与分析不仅是研究基因表达调控的重要环节，也是构建基因表达载体实现基因表达调控的关键所在。目前在植物表达载体中使用的大多为组成型的强启动子。单子叶植物转基因主要使用来自玉米的 Ubiquitin 启动子和来自水稻的 Actin 启动子；双子叶植物一般使用花椰菜叶病毒 CaMV 35S 启动子或胭脂碱氨酸合成酶 Nos 启动子（侯丙凯等，2001）。组成型启动子可以使外源基因持续恒定地表达，但是外源基因的组成型表达不仅造成资源的浪费，重复使用同一种启动子驱动两个或两个以上的外源基因还可能引起基因沉默或共抑制现象（Elmayan and Vaucheret，1996），此外还会引发转基因植物安全性问题，因此目前对特异性启动子的研究和应用越来越受到重视（Potenza et al.，2004）。Coen 和 Meyerowitz（1991）在对模式植物拟南芥、矮牵牛和金鱼草突变体及其基因功能分析的基础

* 本文原载《基因组学与应用生物学》，2010，29（2）：239-244，与王静澄、李昊、崔东清、刘军梅和安新民合作发表。

上，提出了花器官发育的 ABC 模型。经过逐步完善，ABC 模型已发展为 ABCDE 模型（Theissen, 2001）。目前，这一模型作为解释花器官发育的基本分子机制，已得到广泛认可。SEP 类基因属于这一模型中的 E 类（Pelaz et al. , 2000；Theissen, 2001），对 *PtSEP*3 超表达的毛白杨转基因植株的研究显示它控制心皮发生，也影响了叶的发生和形态发育过程（樊金会等, 2007）。目前关于 SEP 类基因启动子的研究还较少，而关于毛白杨 *PtSEP*3-1 基因启动子的克隆分析研究尚未见报道。

前期的芯片杂交分析表明（数据尚未发表），*PtSEP*3-1 基因在杨树花芽中高丰度表达，暗示了其可能与杨树花器官发生过程密切相关。本研究根据杨树基因组序列并结合杨树 *SEP*3 基因序列设计特异引物，通过 PCR 克隆获得了毛白杨 *SEP*3 基因启动子，运用生物信息软件对该启动子的功能元件进行了分析，进一步构建了植物表达载体 *PtSEP*3-1 pro-test，为 *PtSEP*3-1 基因启动子的功能鉴定与基因的表达调控研究奠定了基础。

1　结果与分析

1.1　毛白杨 *PtSEP*3-1 基因上游序列的获得

通过 PCR 得到大小约为 2000bp 的条带（图 1A），与预期片段大小相近。电泳分离回收后克隆至 PCR2. 1-TOPO 载体，重组子通过 *Eco*R I 酶切能够释放出与 PCR 产物一样大小的 DNA 片段，说明该基因片段已经克隆入 PCR 2. 1-TOPO 载体（图 1B）。测序结果表明，克隆所得片段 3′ 端近 100bp 序列与 *SEP*3 基因 cDNA5′端序列一致，由此说明该片段来源于 *SEP*3 基因上游的启动子区域。

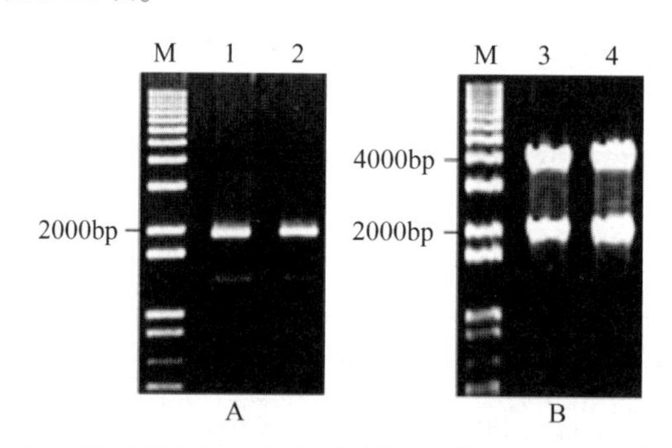

图 1　毛白杨 *PtSEP*3-1 基因启动子的克隆及 *Eco*R I 酶切鉴定

注：M：1 kb DNA ladder (Invitrogen)；A：PCR 扩增；B：重组子 *Eco*R I 酶切鉴定；1，2：PCR 扩增产物；3，4：酶切产物

1.2　毛白杨 *PtSEP*3-1 基因启动子序列分析

用 TSSP-TCM 软件（Shahmuradov et al. , 2005）对 *PtSEP*3-1 基因起始密码子上游序列进行分析，预测出转录起始点为 A，位于起始密码子上游 650bp 处，符合转录起始点为 A 的一般规律（Joshi, 1987）。

用 PLACE (Higo et al. , 1999) 和 Plant CARE (Lescot et al. , 2002) 软件分析此序列，预测出该序列具有启动子的基本转录元件：TATA box 位于转录起始点（TSS）–22bp 处，

CAAT-box 分别位于-220bp、-494bp、-1301bp 和-1305bp 处。此外，该序列还含有：7 个 Box 4(位于-52bp，-277 bp，-281bp，-285bp，-961bp，-1204bp 和-1287 bp)，3 个 Skn-1-motif（位于-404bp，-1077bp 和-1278bp)，2 个 GARE-motif（位于-688bp 和-905bp)，2 个 HSE（位于-184bp 和-347 bp)，1 个 I-box（位于-420bp)，1 个 ACE（位于-537 bp)，1 个 3-AF3 binding site（位于-921bp)，1 个 CAT-box（位于-1105bp)，1 个 MYB 结合元件 MBS（位于-1259 bp)，还有 6 个双相连作用元件 BoxI/CAAT-box（位于-118bp)，A-box/CCGTCC-box（位于-430bp)，CAAT-box/TC-rich repeats（位于-453 bp)，ACE/G-Box（位于-537 bp)，GT1-motif/MBSI（位于-605bp）和 ABRE/G-Box（位于-797 bp)。其中 ACE、BoxI、G-Box、GT1-motif 以及 I-box 等多个元件为光响应元件，Box 4 为与光响应有关的 DNA 保守域，CAT-box 和 CCGTCC-box 为分生组织表达调控元件和分生组织特异性活化调控元件，ABRE 为脱落酸响应元件(Zheng et al.，2007)，GARE- motif 为赤霉素响应元件，MBSI 与类黄酮合成基因调节有关，此外还含有胁迫诱导元件，HSE 为高温胁迫响应元件，TC-rich repeats 为抗病和逆境胁迫响应元件。

1.3 植物表达载体的构建

将含有 *PtSEP*3-1 启动子重组质粒(引入了 *Sac* I 和 *Kpn* I 酶切位点）和经过改造的 pCAMBIA 1305 质粒载体(切掉了驱动 *GUS* 基因的 35S 启动子)分别进行 *Sac* I 和 *Kpn* I 双酶切鉴定，回收纯化目的片段，通过 T4-DNA 连接酶连接，经酶切验证连接正确，获得含 *PtSEP*3-1 promoter:: *GUS* 结构的植物表达载体，命名为 *pPtSEP*3-1 protest（图 2)。

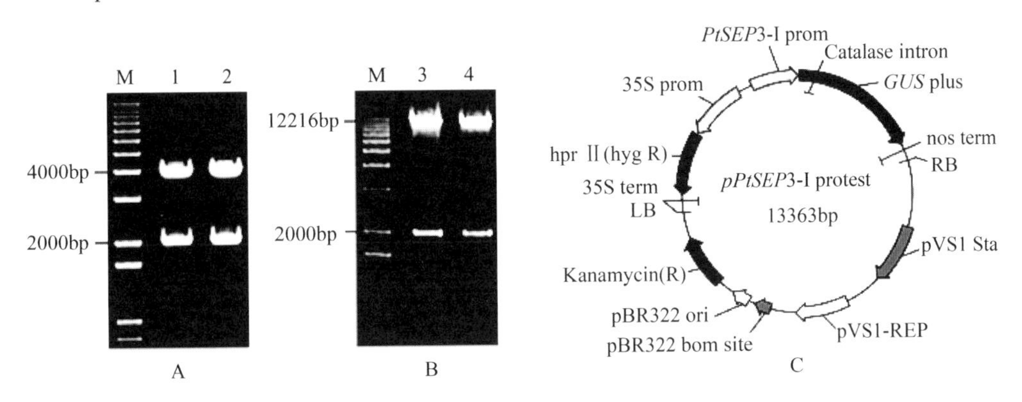

图 2 毛白杨 *PtSEP*3-1 promoter:: *GUS* 植物表达载体构建

注：M：1kb DNA ladder（Invitrogen)；A：*PtSEP*3-1 启动子重组质粒 *Sac* I/*Kpn*I 双酶切；B：重组表达载体 *pPtSEP*3-1 protest *Sac* I/*Kpn*I 双酶切；C：*pPtSEP*3-1 protest 植物表达载体结构；1，2：*tSEP*3-1 启动子阳性克隆质粒 DNA 双酶切；3，4：*pPtSEP*3-1 protest 双酶切

2 讨论

杨属植物基因组相对较小，组培体系成熟，生长周期相对其他木本植物较短，因此被作为木本植物研究中的模式树种(Jansson and Douglas，2007)。而毛果杨全基因组序列测序

的完成(Tuskan et al. , 2006)，为分子生物学与生物信息学相结合的研究方法提供了基础，大大促进了杨树功能基因及其启动子的分离和鉴定工作。

ABCDE 模型认为拟南芥雌蕊发育是由 C 类和 E 类基因共同决定的(Mizukami and Ma, 1992)，有研究表明，属于 E 类的 *SEP* 基因可以激活属于 C 类的 *AG* 基因的表达(Castillejo et a1. , 2005)，对某些 B 类基因的表达可能也有一定影响(Rainer and Günter, 2009)，毛白杨中也可能存在类似的机制。本文克隆的毛白杨 *PtSEP*3-1 基因启动子 TSS 以上包含 1309 bp，软件分析显示其中包含启动子基本转录元件，其中 TATAbox 位于 TSS 上游-22bp 处，最近的 CAAT-box 位于 TSS 的-220bp 处，已分离的植物启动子大多与此类似。在该启动子中发现了赤霉素响应元件 GARE-motif，脱落酸响应元件 ABRE，分生组织特异性活化和分生组织表达调控元件 CAT-box 及 CCGTCC-box，以及大量的光响应元件，这与 *SEP*3 基因受成花诱导表达(Cseke et al. , 2005)是一致的。

目前，本研究已构建了一个 *PtSEP*3-1 promoter::*GUS* 结构的植物表达载体 *pPtSEP*3-1 protest。为了进一步研究该启动子的功能，鉴定其能否驱动下游基因在转基因植物中高效表达，哪个表达区段的作用更明显，还需将本研究所构建的 *pPtSEP*3-1 protest 载体转化到毛白杨等植物中进行稳定表达，相关工作目前正在进行中。

3 实验材料与方法

3.1 毛白杨基因组总 DNA 的分离

使用天根公司 DNA 提取试剂盒，以毛白杨叶片为材料提取总 DNA。

3.2 引物设计与 PCR 扩增

根据 NCBI 中(http://www.ncbi.nlm.nih.gov/nuccore/90903288)登录的毛白杨 *PtSEP*3 基因序列结合毛果杨序列设计引物，并由生工生物工程(上海)有限公司合成。其中上游引物：5′-AGAGCTCAGTTCAATCAATACTTGACTAC-3′；下游引物：5′-TGGTACCATCTTCTCCCTCTCTCTCTC-3′。20μL PCR 反应体系包括：100ng 毛白杨基因组 DNA、10μL 2× GC Buffer、各 0.4μL 上下游引物、1.6μL 2.5mmol/L dNTPs 以及 0.5unit LA *Taq* (TaKaRa)。PCR 反应条件为：预变性 94℃ 5min；变性 94℃ 1min，退火 57℃ 0.5min，延伸 72℃ 2.5min，35 个循环；72℃再延伸 10min 后中止反应。

3.3 目的片段克隆及测序分析

PCR 产物利用 1.2% (g/v)琼脂糖凝胶电泳检测，使用试剂盒(天根公司)回收、纯化目的片段，克隆目的片段于 PCR 2.1-TOPO 载体上(Invitrogen)，转化大肠杆菌 Topl0 感受态细胞，经菌落 PCR 筛选获得阳性克隆。质粒 DNA 的提取、纯化、酶切和转化等操作参照相应试剂盒的说明书进行(质粒提取试剂盒购于天根公司，限制性内切酶购于 Promega 公司)。抗生素及菌落 PCR 筛选获得阳性克隆，送由北京金唯智公司测序。

3.4 启动子序列分析

以 TSSP-TCM、PLACE 和 PlantCARE 软件对所得序列进行分析，预测转录起始点及功能元件。

3.5 植物表达载体构建

以 3.3 中获得的阳性克隆菌株并提取质粒，经 *Sac* I 和 *Kpn* I 双酶切后，与同样双酶切的经过改造的植物表达载体 pCAMBIA 1305 连接，连接产物转化大肠杆菌感受态细胞，经抗性筛选和菌落 PCR 检测获得阳性克隆，提取阳性重组质粒 DNA，进行酶切鉴定，获得重组植物表达载体。

Cloning and Construction of Expression Vectors of *PtSEP*3-1 Promoter from *Populus tomentosa*

Abstract *SEP* (*SEPALLATA*)-class genes confer E-function in ABCDE model that depicts its roles in floral organ development. Previous studies on Arabidopsis have shown that, genes of this class, key component during floral development, might feature the function of determing floral organogenesis and activating other genes. Therefore analyse expression characteristics of the poplar SEP-class genes promoter are significant for its regulation of flowering. Here, a 2000bp long sequence that locates at upstream region of *PtSEP*3-1 was isolated by using PCR with primers that designed from genomic sequences of *P. trichocarpa* and *SEP*3 of *P. tomentosa*. Sequence analysis revealed this sequence contains basic promoter elements, TATA-box and CAAT-box. Light response *cis*-acting elements, ACE, Box I, Box 4, ABA response element ABRE, GA response elements GARE-motif and stress-induced cis-acting elements including HSE, TC-rich repeats were also represented. Moreover, *PtSEP*3-1 protest vector carrying *PtSEP*3-1 promoter:: *GUS* was constructed, laying a foundation for functional identification of the *PtSEP*3-1 promoter.

Keywords *Populus tomentosa*, *SEP*3 gene, promoter, expression vector

Isolation of a *LEAFY* Homolog from *Populus tomentosa*: Expression of *PtLFY* in *P. tomentosa* Floral Buds and *PtLFY*-IR-mediated Gene Silencing in Tobacco (*Nicotiana tabacum*) *

Abstract To understand the genetic and molecular mechanisms underlying floral development in *P. tomentosa*, we isolated *PtLFY*, a *LEAFY* homolog, from a *P. tomentosa* floral bud cDNA library. DNA gel blot analysis showed that *PtLFY* is present as a single copy in the genomes of both male and female individuals of *P. tomentosa*. The genomic copy is composed of three exons and two introns. Relative expression levels of *PtLFY* in tissues of *P. tomentosa* were estimated by RT-PCR, our results revealed that *PtLFY* mRNA is highly abundant in roots and both male and female floral buds. A low level of gene expression was detected in stems and vegetative buds, and no *PtLFY*-specific transcripts were detected in leaves. *PtLFY* expression patterns were analyzed during the development of both male and female floral buds in *P. tomentosa* via real-time quantitative RT-PCR. Continuous, stable and high-level expression of *PtLFY*-specific mRNA was detected in both male and female floral buds from September 13th to February 25th, but the level of *PtLFY* transcripts detected in male floral buds was considerably higher than in female floral buds. Our results also showed an inverted repeat *PtLFY* fragment (*PtLFY*-IR) effectively blocked flowering of transgenic tobacco plants, and that this effect appeared to be due to post-transcriptional silencing of the endogenous tobacco *LFY* homologs *NFL*1 and *NFL*2.

Key words flowering, gene silencing, *LEAFY* homolog, *Populus tomentosa*, qRT-PCR

Introduction

Chinese white poplar (*Populus tomentosa* Carr.) is a tree species native to a large area of northern China. It is an important species in forest production and forest reclamation projects along the Yellow River. In addition, *Populus* (poplar) species are an important model system for molecular genetic studies of woody plants.

Two commonly used model plant genera, *Populus* and *Arabidopsis*, are both eudicots (Soltis et al., 1999; Wikström et al., 2001) and are characterized by a monopodial shoot system (Bradley et al., 1997; Reinhardt and Kuhlemeier, 2002; Yuceer et al., 2003). However, the two genera differ in many ways (Boes and Strauss, 1994). *Arabidopsis* is an annual herbaceous plant

* 本文原载《植物细胞报告》(*Plant Cell Reports*), 2011, 30(1): 89—100, 与安新民、王冬梅、王泽亮、李博、薄文浩和曹冠琳合作发表。

and completes its life cycle in two months, with only a short juvenile phase followed by the production of flowers and seeds in the reproductive phase (Somerville and Koornneef, 2002). In addition, *Arabidopsis* plants do not undergo specific phases of seasonal vegetative growth and floral development in the reproductive phase (Hsu et al., 2006). Moreover, the *Arabidopsis* flower conforms to the general angiosperm pattern in consisting of four types of floral organ arranged in a series of concentric whorls. From the outermost whorl inwards, the flower consists of sepals in whorl 1, petals in whorl 2, stamens in whorl 3, and carpels in whorl 4(Weigel and Meyerowitz, 1994; Yanofsky, 1995; Ng and Yanofsky, 2001; Lohmann and Weigel, 2002).

In contrast, poplars are perennial trees with a lifespan of about 100~200 years and a long juvenile phase (Braatne et al., 1996). In general, poplar seedlings begin flowering after at least 7~10 years and thereafter annual flowering occurs during the reproductive phase. After completion of flowering and fruiting, shoots initiate early vegetative buds (vegetative zone I), floral buds (floral zone), and late vegetative buds (vegetative zone II) in a sequential manner, indicating that the shoot goes through repeated phase-change cycles between vegetative and reproductive growth (Yuceer et al., 2003). Such recurrent developmental transitions between vegetative and reproductive growth are absent in *Arabidopsis* (Boss et al., 2004). Flower development in poplars also differs markedly from that of *Arabidopsis* in that the male and female flowers are borne on separate trees from axillary inflorescences. Instead of four concentric whorls of organs, poplar flowers have only two whorls, comprising a reduced perianth cup surrounding either the stamens or carpels (Boes and Strauss, 1994; Sheppard, 1997; Rottmann et al., 2000).

Both physiological changes and expression of floral genes in the shoot apex are involved in the floral transition. The switch from vegetative to reproductive growth requires activation of genes involved in flower differentiation (Böhlenius, 2007). Several floral meristem identity genes have been isolated from *Arabidopsis* and other model plant species. One of these genes, *LFY* (*LEAFY*) plays a crucial role in the transition from vegetative to reproductive development in *Arabidopsis* (Weigel et al., 1992; Weigel and Nilsson, 1995). Another floral meristem identity gene, *AP*1 (*APETALA*1), is required for sepal and petal development and is also involved in controlling the transition mentioned above (Mandel et al., 1992). Loss-of-function mutations in *LFY* lead to plants in which shoots replace most flowers (Weigel et al., 1992). *LFY* expression is first detectable in leaf primordia and reaches maximal levels in young floral meristems (Blazquez et al., 1997; Blazquez et al., 1998). Constitutive expression of either *LFY* or *AP*1 results in flowering of transgenic *Arabidopsis* plants in vitro in just 10 days (Mandel and Yanofsky, 1995; Weigel and Nilsson, 1995). When the *LFY* gene from *Arabidopsis* is expressed constitutively in hybrid aspen (*P. tremula*×*P. tremuloides*), plants flowered in vitro within 7 months (Weigel and Nilsson, 1995). Overexpression of *MdMADS*5, an *APETALA*1-like gene of apple, causes early flowering in transgenic *Arabidopsis* (Kotoda et al., 2002). Expression of *Arabidopsis LFY* and *AP*1 genes in *Citrus* induces flowering within the first year. This shortening of the juvenile period is stable and occurs in both zygotic and nucellar-derived seedlings (Peña et al., 2001). However, the influence of flowering genes is more complex in poplar. Overexpression of *PTLF*, a *LFY* homolog from *P. trichocarpa*, in several poplar hybrids induced precocious flowering in a tiny proportion of transgenic lines, although overexpression of *PTLF* in *Arabidopsis* accelerated flowering (Rottmann et al., 2000).

In the present study, a *LFY* gene homolog, *PtLFY*, was isolated from a *P. tomentosa* floral bud cDNA library. The tissue-specific expression pattern of *PtLFY* was determined by RT-PCR, while the time course of *PtLFY* expression in developing floral buds of both male and female plants was quantified by qRT-PCR. Furthermore, the potential function of the *PtLFY* gene was investigated using an inverted repeat structure of *PtLFY* (*PtLFY*-IR) to induce PTGS in a heterologous plant species, tobacco.

Materials and methods

Plant materials

Samples of vegetative and reproductive tissues were collected during 2004 and 2005 from both female (5082) and male (LM50) trees of *P. tomentosa* growing in the Beijing Forestry University nursery. Young leaves were used for extraction of genomic DNA. Floral buds of *P. tomentosa* were collected at different developmental stages, from floral bud initiation to maturity, between September 13th 2004 and February 25th 2005 for the RNA extraction assay. After each collection, the samples were frozen rapidly in liquid nitrogen and stored at −80℃ until use. Besides, root, stem and leaf samples were taken from one-month-old tissue-cultured plantlets of *P. tomentosa*, followed by immediate RNA extraction.

Sterile tobacco (*Nicotiana tabacum*) leaf discs were used for *Agrobacterium*-mediated transformations. Young leaves of transgenic tobacco plants were used for extraction of genomic DNA and RNA.

cDNA library construction and cloning of *PtLFY*

To construct a cDNA library, total RNA was extracted from a mixed sample of floral buds at different developmental stages using a CTAB-based method as described previously (Chang et al., 1993). The mRNA corresponding to the respective total RNA was then isolated and purified with the PolyATtract® mRNA Isolation System (Promega). Subsequently, the cDNA library was constructed with the mRNA as template using the SMART™ cDNA Library Construction Kit (Clontech).

Primers used for screening of *PtLFY* and DNA probe labeling were designed from the nucleotide sequence of *PTLF* (GenBank accession no. U93196), a *LFY* homolog from *Populus balsamifera*, and were synthesized by Sangon Co. (Table 1). *PtLFY* was obtained by PCR screening of the *P. tomentosa* female floral bud cDNA library. The 20μL PCR reaction mixture was composed of 1×PCR buffer (10mM Tris-HCl, pH 8.3, 50mM KCl, 2.5mM $MgCl_2$), 1μL SM buffer

(0.01% gelatin, 50mM Tris-HCl, pH 8.0, 0.1M NaCl, 8mM $MgSO_4$) containing recombinant bacteriophages, 0.4μL of each 10μM forward primer (P1F) and reverse primer (P1R), and 1 unit Taq DNA polymerase. Thermal cycling was performed at 94℃ for 5min, then 94℃ for 30s, 60℃ for 20s, 72℃ for 30s min for 30 cycles, and 72℃ for 1min and finally kept at 4℃, using a GeneAmp® PCR System 970(ABI). The positive clones of the Lambda phage clones screened by PCR were sequenced directly with an ABI 377 DNA sequencer (Applied Biosystems).

Protein structure and phylogenetic relationship analysis

The amino acid sequences of *LFY* homologous genes were retrieved from GenBank (http://www.ncbi.nlm.nih.gov/entrez/query.fcgi) and aligned using Clustal X 1.81. The secondary structure of the PtLFY protein was predicted using ANTHEPROT 2000 6.0, and the conserved protein domain sequences were analyzed with the Proteomics Server of the Expert Protein Analysis System (ExPASy) of the Swiss Institute of Bioinformatics (http://cn.expasy.org; Gasteiger et al., 2003). The tertiary structure of the PtLFY protein was predicted using 3D-JIGSAW 2. (http://bmm.cancerresearchuk.org/~3djigsaw/), and viewed with RasMol 2.7.2.1. Genetic distance matrices were obtained from the alignments and neighbor-joining trees constructed with bootstrap sampling of 1000 replications using MEGA 4.1(Tamura et al., 2007).

DNA gel blot analysis

A 1076bp fragment amplified by the P2F and P2R primers (Table 1) was used as the PCR probe for DNA gel blot analysis. Fifteen micrograms of genomic DNA from both male and female *P. tomentosa* were digested with the restriction enzymes *Bam*H I, *Hin*d III and *Eco*R I, and electrophoresed on a 0.8% agarose gel. After blotting onto a positively charged nylon membrane by capillary transfer using 20×SSC, the blots were hybridized with a DIG-labeled DNA probe at 58℃, and washed at high stringency (2×SSC, 0.1% SDS, at 15-25℃ for 2×5min, then 0.5×SSC, 0.1% SDS at 65-68℃ for 2×15min). Immunological detection was performed with the DIG DNA Labeling and Detection Kit (Roche) according to the manufacturer's protocol.

Table 1 Primers used for screening the cDNA library, RT-PCR and qRT-PCR, DNA probes, construction of the expression vector, PCR amplification, DNA gel blot and RT-PCR analysis of transformants

Primers	Oligonucleotide
For screening cDNA library, RT-PCR and qRT-PCR	
P1F forward	5′-ATGGATCCGGAGGCTTTCACGGC-3′
P1R reverse	5′-AGCGGAGAAAAATGCCCCACTAAG-3′
For DNA blots analysis	
P2F forward	5′-GAAGTGGCACGTGGCAAAAAGAA-3′
P2R reverse	5′-ACGGAGTTTGGTGGGCACATACCA-3′

（续）

Primers	Oligonucleotide
For inverted repeat sequence construction	
P2aR reverse	5′-TTCTAGACGGAGTTTGGTGGGCACA-3′ (introduced *Xba* I)
P2bR reverse	5′-GGATCCACGGAGTTTGGTGGGCACA-3′ (introduced *Bam*H I)
For identification of transgenic tobacco	
P3F forward	5′-GGATTGATGTGATATCTCCACTG-3′ (from CaMV35S promoter)
P3R reverse	5′-CCACAGTTTTCGCGATCCAGACT-3′ (from sequence of gus)
For RT-PCR analysis of transgenic tobacco	
P4F forward	5′-GTGCAGCAGCAAGAGAGAGAAGC-3′ (Tobacco *NFL1* & *NFL2*)
P4R reverse	5′-CTCCGTCACGATAAAAGGATGCTC-3′ (Tobacco *NFL1* & *NFL2*)
P5F forward	5′- CTGCTGGAATTCACGAAACA-3′ (Tobacco *ACTIN*)
P5R reverse	5′-GCCACCACCTTGATCTTCAT-3′ (Tobacco *ACTIN*)

RT-PCR and qRT-PCR analysis

Total RNA was extracted from root, stem, leaf, vegetative bud and floral bud respectively according to the method described previously (Chang et al., 1993). Five micrograms of total RNA was pretreated with RQ1 DNase I (Promega) to remove genomic DNA contaminants. The concentration of total RNA was measured using a SPEKOL 1300 spectrophotometer (Jena). First-strand cDNA was synthesized using 1.0μg DNase-treated total RNA, Superscript III (Invitrogen), and oligo$(dT)_{20}$ in a total volume of 20μL. The first-strand cDNA was diluted 1∶10 with ddH_2O, and 2μL of the diluted cDNA was used as a template for RT-PCR and qRT-PCR analysis. RT-PCR system were performed in a total volume of 20μL, with 2μL of 10×PCR buffer, 1.6μL of 2.5mM dNTPs and each 0.4μL of 10μM P1F and P1R (Table 1), and 1 unit Taq DNA polymerase in the reaction. PCR conditions were the same as used previously for the library screening. qRT-PCR reactions were performed in a total volume of 20μL, with 0.4μL of 10μM P1F and P1R (Table 1), and 10μL 2×SYBR® Green PCR Master Mix (Invitrogen), using a DNA Engine Opticon 2 system (MJ Research). The qRT-PCR program included a preliminary step of 2min at 50℃, 94℃ respectively, followed by 35 cycles of 94℃ for 15s, 56℃ for 3s, and 72℃ for 3s, with final extension at 72℃ for 7 min. No-template controls for each primer pair were included in each run. According to a previous study, the poplar *ACTIN* gene (GenBank accession: AY261523.1) is a stably-expressed internal control (Zhang et al., 2008; Zheng et al., 2009), therefore it was employed as an internal reference gene to normalize small differences in template amounts with the forward primer 5′-CTCCATCATGAA ATGCGATG-3′ and reverse primer 5′-TTGGGGCTAGTGCTGAGATT-3′. At least three different RNA isolations and cDNA syntheses were used as replicates for the qRT-PCR.

Construction and transformation of *PtLFY-IR* structure in tobacco

To examine the biological function of *PtLFY*, the PCR primers P2F, P2aR and P2bR (Table 1), designed according to the sequence of the highly conserved region between the tobacco *LFY* homologs *NFL*1 (GenBank accession no. U16172), *NFL*2(U15799) and *PtLFY*, were used to amplify *PtLFY* fragments from genomic DNA. A common *Nco* I restriction enzyme site was found in each *PtLFY* fragment that amplified using primers P2F and P2aR, and P2F and P2bR. The resulting fragments were digested with *Xba* I and *Nco* I, and *Bam*H I and *Nco* I, respectively. In addition, pBI121 plasmid DNA was digested with *Xba* I and *Bam*H I. The corresponding fragments were purified using the QIAquick™ Gel Extraction Kit (QIAGEN) and ligated with T4 DNA ligase (Promega) at 16℃ overnight, then transformed into competent cells of *E. coli* TG1. The fragment containing the inverted repeat (IR) sequence of the *PtLFY* fragment was verified by digestion with *Bam*H I and *Xba* I, and by PCR amplification using primers P3F and P3R (Table 1). The *PtLFY*-IR structure was assembled into the binary vector pBI121, which gave plasmid pBI121-*PtLFY*-IR. Subsequently, pBI121-*PtLFY*-IR was introduced into the *Agrobacterium tumefaciens* strain GV3101 according to a direct and efficient liquid nitrogen freezing-thawing method reported previously (Tzfira et al., 1997). *Agrobacterium tumefaciens* GV3101 manipulation and *N. tabacum* leaf disc transformation were performed as described previously (Li et al., 1992).

Identification of transformants by PCR and DNA gel blots analysis

Genomic DNA was extracted from young leaves of transformants and used as the template for PCR identification. The PCR reaction conditions were similar to those used previously with the exception that primers P3F and P3R were used, and thermal cycling was performed at 94℃ for 5min, then at 94℃ for 4s, 58℃ for 4s, 72℃ for 40s for 30 cycles and 72℃ for 10min. The PCR product amplified with primers P3F and P3R was then used as DNA probe against transgenic tobacco DNA. DNA gel blotting and immunological detection was performed as described above.

Detection of *NFL*1 & *NFL*2 gene silencing in transgenic tobacco plants

Total RNA was extracted from the whole plantlet of transgenic tobacco using SV Total RNA Isolation System (Promega). First-strand cDNA was synthesized using Reverse Transcription System (Promega), and diluted 1∶10 with ddH_2O. Semi-quantitative RT-PCR was performed in 20μL reactions containing 0.4μL of 10μM P4F and P4R primers which were designed according to the conserved sequence of *NFL* homologs (Table 1), 1μL of cDNA template, 2μL of 10×PCR buffer, 1.6μL of 2.5mM dNTPs and 1 unit Taq DNA polymerase in reaction system. Thermal cycling was performed at 94℃ for 3 min, then at 94℃ for 20s, 60℃ for 20s, 72℃ for 20s for 30 cycles and 72℃ for 7 min. The tobacco *ACTIN* gene (GenBank accession: U60491) was selected as an endogenous reference gene to normalize small differences in template amounts because of its relatively expression levels. Sequences of primers P5F and P5R are given table 1.

Phenotype analysis of transgenic tobacco plants

The verified transgenic tobacco plants were propagated and synchronized using vegetative stem cuttings containing an axillary bud from primary transformants. Transplanting of the rooted transgenic tobacco plantlets into an artificial soil mix (humus : perlite : vermiculite = 3:1:1) in individual pots, the plantlets were covered with clear plastic and the pots placed in a shaded greenhouse. After hardening for two to three weeks and once the plantlets had produced four to five leaves (June 15th), the plants were moved to the outside of the greenhouse and covered with clear plastic to maintain humidity for two weeks. Subsequently, the plastic was removed for a period each day to harden the plants. Plant height and number of leaves of each T_0 transgenic line were recorded on August 11th, August 21st, September 1st, September 11th, September 21st, and October 2nd. In addition, the date was recorded when flower primordia were first visible.

Results

Cloning and DNA blot analysis of *PtLFY* from *P. tomentosa*

Three positive phage plaques were isolated by PCR screening. The longest of the *PtLFY* cDNA fragments was 1314bp encoding 377 amino acids. Comparison of the cDNA sequence against our previously submitted *PtLFY* genomic sequence (GenBank accession: AY211519) from *P. tomentosa* showed that the gene is composed of three exons and two introns (Fig. 1A). Using Seqaid II and the SPL (search for potential splice sites) tool available at http://dot.imgen.bcm.tmc.edu:9331/gene-finder/gf.htm, the donor and acceptor sites of the first intron (596bp) were found to be at 458 and 1054bp. The donor and acceptor sites of the second intron (669 bp) were at 1388 and 2057 bp. The positions of splice sites were the same as those of *LFY* homologs of other plants (Frohlich and Parker, 2000), but the length of the introns differed greatly. For example, the intron of *Arabidopsis LFY* is 910bp long in the position of the second intron. The splice junctions in *PtLFY* followed the "GT……AG" rule, the splice junctions and restriction endonuclease sites are shown in Fig. 1A.

To determine the number of *PtLFY* copies in the *P. tomentosa* genome, hybridization of genomic DNA from both male (LM50) and female (5082) clones were performed. The DNA blots revealed the same pattern of hybridization in both male and female clones. A single band was visualized in three different digestions in both male and female clones. This indicated that there was only one gene copy in the genomic DNA, and no additional bands were observed at a lower hybridization temperature and with reduced washing stringency (Fig. 1B).

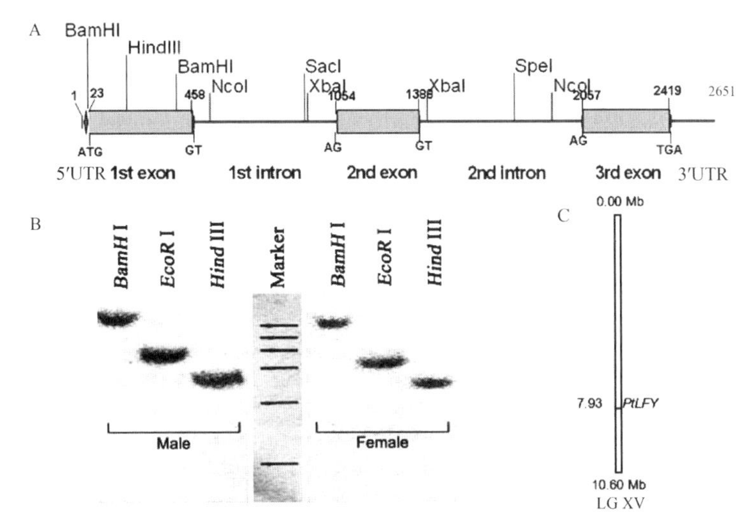

Fig. 1 The structure and restriction endonuclease map (A), DNA gel blot analysis (B) and physical chromosomal localization (C) of *PtLFY* from *P. tomentosa*

Prediction of protein structure and phylogenetic relationships of *PtLFY*

A comparison of the amino acid sequences of *PtLFY* and other *FLO/LFY* homologs *LFY* (Weigel et al., 1992), *FLO* (Coen et al., 1990), *VFL* (Carmona et al., 2002), *SdF* (Fernando et al., 2003), and *AFL1* (Haettasch et al., 2006) showed the presence of two highly conserved regions, and the C-terminal conserved region of *PtLFY* protein is mainly composed of seven α-helix and two β-sheet structures (Fig. 2).

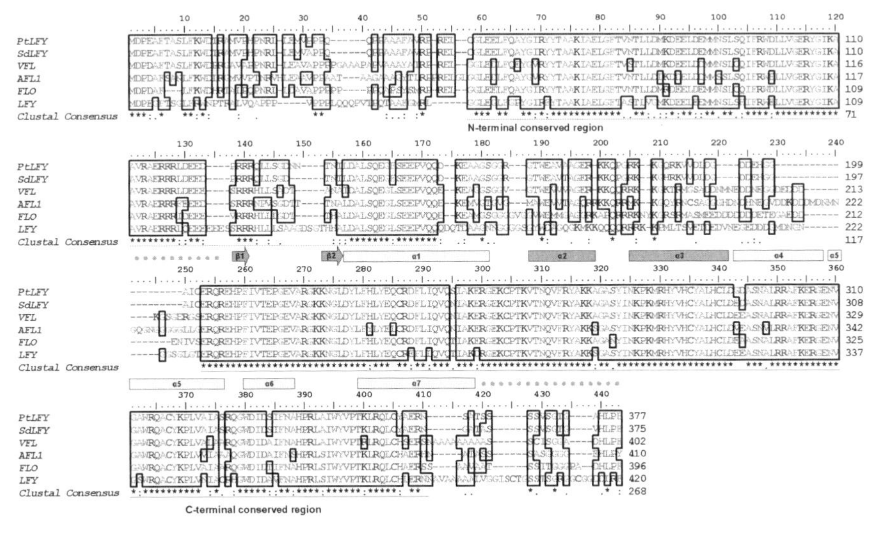

Fig. 2 Sequence alignments of *PtLFY* with *SdLFY* (GenBank accession no. AY230817), *VFL* (AF450278), *AFL1* (AB162028), *FLO* (M55525) and *LFY* (NM_ 125579)

The red line indicates the conserved region. Asterisks and dots indicate totally conserved and conservatively replaced amino acids, respectively. The arrows indicate the predicted β-sheets, and rectangles indicate positions of the predicted the α-helices.

The phylogenetic relationships of the *FLO/LFY* homologs are represented in Fig. 3. The earliest divergence occurred between the angiosperms and gymnosperms, and another important event was the divergence of monocotyledons and dicotyledons. The duplicated homologs are indicated to have arisen from relatively recent events (e. g. as in maize, tobacco, pear, loquat and apple), with the exception of a duplication that predated species diversification within Maloideae. No duplication of *PtLFY* has occurred in *P. tomentosa*.

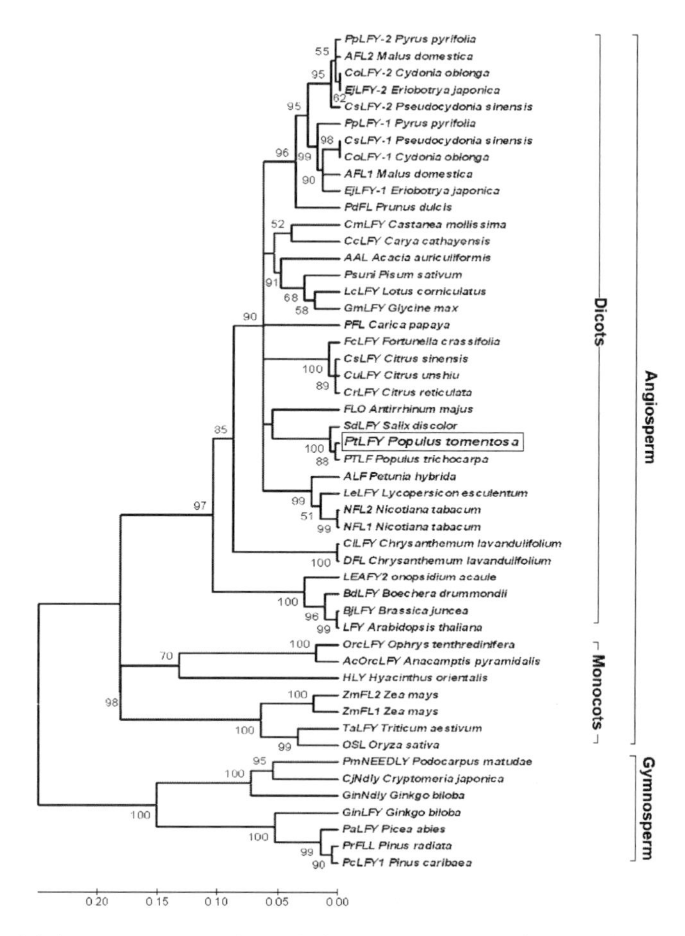

Fig. 3 Neighbor-joining tree representing relationships of *PtLFY* (framed) with gymnosperm and other angiosperm *LFY/FLO* homologs

Bootstrap support values (%) from 1000 replications are indicated when over 50%. GenBank accession number for each sequence: *AAL* (AY229891), *AcOrcLFY* (AB088457), *AFL*1 (AB162028), *AFL*2 (AB056159), *ALF* (AF030171), *BdLFY* (AY734564), *BjLFY* (DQ471932), *ClLFY* (AY672542), *CjNdly* (AB074568), *CmLFY* (DQ989225), *CoLFY*-1 (AB162031), *CoLFY*-2 (AB162037), *CrLFY* (DQ995349), *CsLFY* (AY338976), *CsLFY*-1 (AB162032), *CsLFY*-2 (AB162038), *CuLFY* (DQ995347), *DFL* (AY559245) *EjLFY*-1 (AB162033), *EjLFY*-2 (AB162039), *FcLFY* (DQ497003), *FLO* (M55525), *GinLFY* (AF108228), *GinNdly* (AF105111), *GmLFY* (DQ448809), *HLY* (AY520841), *LcLFY* (AY770393), *LEAFY*2 (AF184589), *LeLFY* (AF197934), *LFY* (NM_ 125579), *NFL*1 (U16172), *NFL*2 (U15799), *OrcLFY* (AB088454), *OSL* (AF065992), *PaLFY* (AY701763), *PcLFY*1 (AY640316), *PdFL* (AY947465), *PFL* (DQ054794), *PmNEEDLY* (AY957473), *PpLFY*-1 (AB162029), *PpLFY*-2 (AB162035), *PrFLL* (AF109149), *Psuni* (AF010190), *PTLF* (U93196), *SdLFY* (AY230817), *TaLFY* (AB231889), *ZmFL*1 (AY179883), *ZmFL*2 (AY789046).

Expression pattern of *PtLFY* in *P. tomentosa*

The differential expression of *PtLFY* in various tissues of *P. tomentosa* was detected by RT-PCR, and the result indicated that *PtLFY*-specific mRNA was relatively abundant in seedling roots and both male and female floral buds. *PtLFY*-specific transcripts were only faintly detected in seedling shoots and vegetative buds and appeared to be absent from leaves (Fig. 4A). Furthermore, the relative difference of *PtLFY* expression in female and male floral buds indicated by qRT-PCR analysis is shown in Fig. 4B and 4C. The *PtLFY* expression patterns during both male and female floral bud development were similar. In both male and female floral buds *PtLFY* expression was continuous and stable from September 13th to February 25th. However, the expression of *PtLFY* in male floral buds was 25-fold higher than that in female buds.

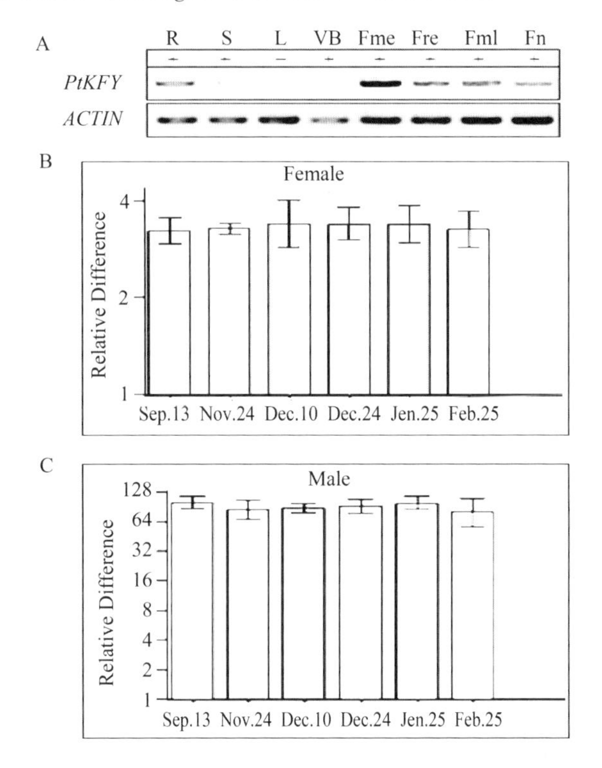

Fig. 4 The expression patterns of *PtLFY* in *P. tomentosa*

(A) Expression of *PtLFY* in different tissues of *P. tomentosa*. R, S and L (root, stem and leaf from one-month-old tissue-cultured plantlets, respectively), VB (vegetative bud), Fme (floral bud-male-early stage, on July 5th), Ffe (floral bud-female-early stage, on July 5th), Fml (floral bud-male-late stage, on March 10th in next year), Ffl (floral bud-female-late stage, on March 10th in next year), + detectable, −undetectable. The expression patterns of *PtLFY* during floral bud development in *P. tomentosa* derived from real time quantitative RT-PCR analysis in female floral buds (B) and male floral buds (C).

Potential function of *PtLFY* revealed by post-transcriptional gene silencing

To investigate the function of *PtLFY* gene *in planta*, a plasmid containing a *PtLFY* inverted repeat (IR) sequence that released from *PtLFY* fragment (from 668 to 1076bp) was constructed (Fig. 5A) and used for genetic transformation of tobacco plants. Both PCR amplification and DNA gel blot analysis showed that *PtLFY*-IR structure had been successfully integrated into the tobacco genome (Fig. 5B). Notably, it was observed that the resulting transgenic tobacco plants presented a significantly low *NFL*1 and *NFL*2 transcript level (slight/faint) when compared to the wild-type control, suggesting that *PtLFY*-IR in tissues was responsible for silencing the endogenous genes (Fig. 5C).

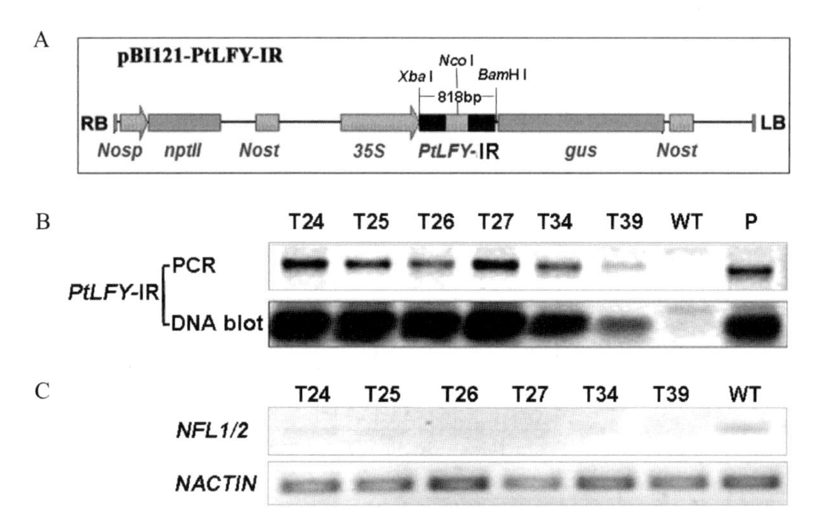

Fig. 5 Molecular identification of transgenic tobacco plants carrying the *PtLFY*-IR silencing construct

(A) Schematic representation of pBI121-*PtLFY*-IR for gene silencing of *NFL*1 and *NFL*2 in transgenic tobacco. RB (T-DNA right border), LB (T-DNA left border), *Nosp* (Nos promoter), *nptII* (Neomycin phosphotransferase II), *Nost* (Nos terminator), 35*S* (CaMV 35S promoter), *gus* (β-glucuronidase gene), *PtLFY*-IR (inverted repeat *PtLFY* fragment, two black boxes in *PtLFY*-IR respectively represent two 285bp-long inverted sequences from the 3rd exon of *PtLFY*, and the box intermediated represents a 248bp-long sequence from the 2nd intron of *PtLFY*). (B) Transgenic tobacco plants identified by PCR amplification and DNA gel blot analysis. T_{24}, T_{25}, T_{26}, T_{27}, T_{34} and T_{39} are six transgenic lines, WT is the wild type, P is positive control (the recombinant plasmid pBI121-PtLFY-IR) (C) The suppression of *NFL*1 and *NFL*2 expression in transgenic tobacco plantlets identified by semi-quantitative RT-PCR (using tobacco *ACTIN* as reference control). T_{24}, T_{25}, T_{26}, T_{27}, T_{34} and T_{39} are six transgenic lines, WT indicates wild type.

In order to observe the flowering characteristics of tobacco plants expressing the *PtLFY-IR* silencing construct, some transgenic tobacco lines and wild-type (WT) plants were transferred from the greenhouse to natural growing conditions for further analysis (Fig. 6). Only one transgenic line (T_{35}) flowered out of a total of 15 lines, whereas all 10 WT plants flowered (Table 2). Compared with WT plants, the transgenic tobacco plants produced fewer leaves (Fig. 6C) and were slightly shorter (Fig. 6D), but almost no variation in leaf shape or size was observed on the transgenic plants (Fig. 6A, 6B).

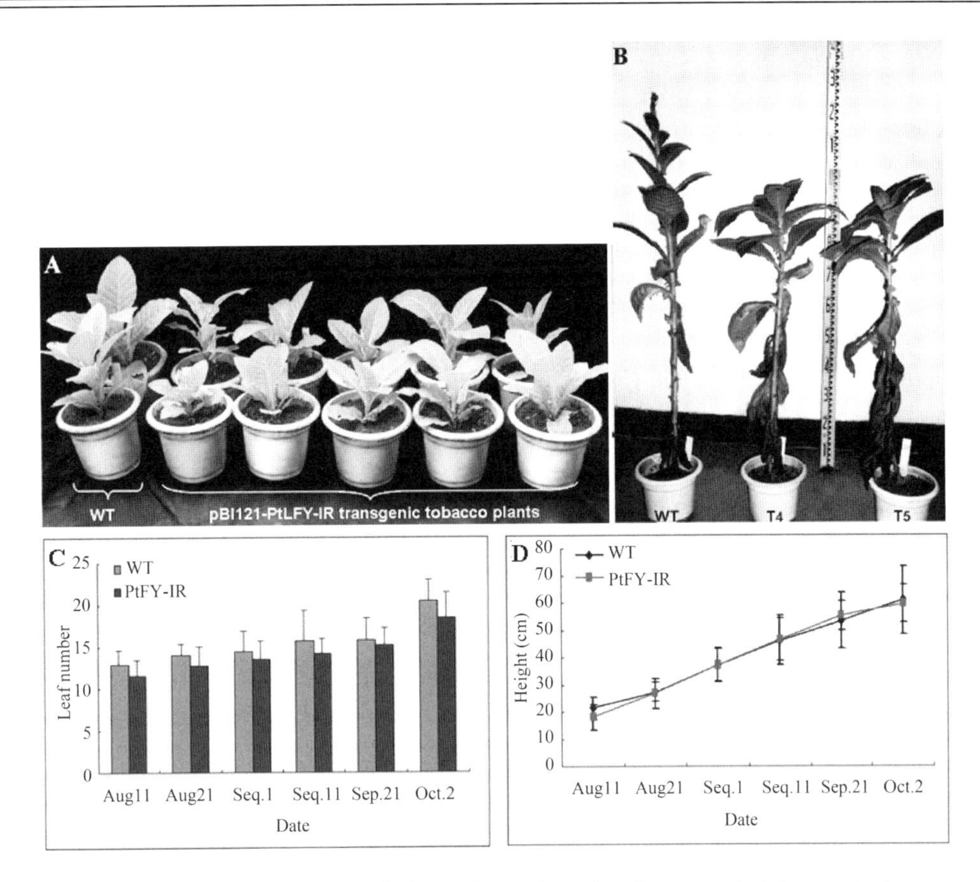

Fig. 6 Phenotypic characteristics and growth traits of transgenic tobacco plants carrying *PtLFY*-IR grown in natural conditions

(A) Transgenic plants compared with wild-type tobacco plants (60-day-old after synchronized culture of vegetative stem cuttings). (B) Comparison of height in transgenic and wild-type tobacco plants (120-day-old after synchronized culture of vegetative stem cuttings). T4 and T5 (corresponding to very the fourth and fifth plants from left to right in front row in Fig. 6A) represent two transgenic lines. (C) Change in leaf number during the development of transgenic tobacco plants and wild-type tobacco plants. (D) Height of transgenic and wild-type tobacco plants from August 11th to October 2nd (the ruler unit: cm). No significant differences between controls (wild type) and transgenic tobacco plants were observed ($P>0.05$) according to the ANOVA FISHER's LSD test.

Discussion

A number of *LFY* (*LEAFY*) homologs have been isolated from many plant species, including annual and herbaceous dicots such as *Arabidopsis thaliana LFY* (Weigel et al., 1992), *Antirrhinum majus FLO* (Coen et al., 1990), and *Petunia hybrida ALF* (Souer et al. 1998); monocots such as *Oryza sativa RFL* (Junko et al., 1998) and *Zea mays ZmFL*1 and *ZmFL*2 (Bomblies et al., 2003); the gymnosperm *Pinus NEEDLY* and *PrFLL* (Mellerowicz et al., 1998; Mouradov et al., 1998) as well as many other woody angiosperms including *Populus trichocarpa PTLF* (Rottmann et al. 2000). *LFY* is found in all land plants, which evolved during the past 400 million years (Maizel et al., 2005). These *LFY* homologs have two distinct characteristics: they are composed of three exons and two introns, and have two highly conserved regions (the N- and C-terminals) in diverse species (Maizel et al., 2005). The stability and conserved nature of the primary structure of the protein indicated that *LFY* homologs have a similar function even in different species. In *Arabidopsis*, *LFY* is a crucial integrator of endogenous and environmental signals, including hormonal cues, photoperiodic changes and exposure to cold temperature. Its expression is up-regulated in response to these signals (Chae et al., 2008) and, in turn, *LFY* acts to coordinate initial expression of floral homeotic genes (Weigel and Meyerowitz, 1993; Nilsson et al., 1998; Blazquez and Weigel, 2000). *LFY* is involved in regulating the transition from an inflorescence meristem to a floral meristem (Weigel et al., 1992), and it serves as a developmental switch for floral initiation in a diversity of plants (Weigel et al., 1995). Our tertiary structure predictions based on the inferred protein sequence imply that LFY adopts a novel seven-helix fold that binds to DNA as a cooperative dimer, forming base-specific contacts in both the major and minor grooves. Cooperativity is mediated by two basic residues and this may explain the effectiveness with which LFY triggers sharp developmental transitions (Hamès et al., 2008). As in the *Arabidopsis* LFY protein, the PtLFY C terminus is composed of a seven-helix fold (Fig. 2), indicating that *PtLFY* might play a role in triggering the floral transition in *P. tomentosa*.

The DNA gel blot analysis of restriction endonuclease-digested *P. tomentosa* genomic DNA showed that both the male and female *P. tomentosa* genome contains only one copy of a *LFY* homolog (Fig. 1B). Gymnosperm species have two homologs of *LFY* genes, whereas angiosperm species generally possess only one copy of the gene (Frohlich and Parker, 2000). The *Ginkgo biloba* tree has two homologous genes, *GinLFY* (GenBank accession no. AF108228) and *GinNdly* (AF105111). The *Pinus radiata* genome contains the homologs *NEEDLY* and *PRFLL* (Mellerowicz et al. 1998; Mouradov et al., 1998). *NEEDLY* is expressed during vegetative and reproductive development, while *PRFLL* is expressed preferentially in male cones during reproductive development. However, two *LFY* copies have been reported to occur in some angiosperm species such as tobacco (Ahearn et al., 2001), apple (Wada et al., 2002), maize (Bomblies et al., 2003), and pear (Esumi, 2004). In addition, the neighbor-joining tree presented in Figure 3 showed that the duplication event that gave rise to the apple *LFY/FLO* homolog occurred before the divergence of the Maloideae. No *PtLFY* duplication event has occurred during the evolution of *P. tomentosa*, implying that *PtLFY* gene might be indispensable and irreplaceable for floral bud initiation and floral organ development.

P. tomentosa is a dioecious tree, although the sporadic occurrence of bisexual (perfect) flowers has been noted in some *Populus* species (Boes and Strauss, 1994). Previous studies on the expression pattern of *PTLF*, a *LFY* homolog of *Populus trichocarpa*, showed the gene was expressed most strongly in developing inflorescences, also in leaf primordia, very young leaves, apical vegetative buds and seedlings (Rottmann et al., 2000). The pattern of expression does, in general, agree with that seen in *Arabidopsis* and *Antirrhinum* (Weigel et al., 1992; Coen et al., 1990; Blazquez et al., 1997). In our study, however, RT-PCR analysis showed a high level of *PtLFY* transcripts in roots and developing female and male floral buds, lower levels in stems and vegetative buds, and virtually no expression in leaves (Fig. 4A). A high and stable quantity of *PtLFY* transcripts accumulated in both male and female developing floral buds, and that the relative expression level of *PtLFY* in developing male floral buds is much higher than that in developing female floral buds (Fig. 4B&C). The differential expression of *PtLFY* in floral buds might have something to do with the observation that the developmental progression of male floral buds occurs prior to that of female floral buds (An et al., unpublished data). Alternatively, this difference might indicate that a much higher level of *PtLFY* expression could be required for male floral bud development in *P. tomotosa*. In poplar, there are more androecium primordia present in male floral buds than there are gynoecium primordia in female floral buds. Thus, a much higher and constant level of *PtLFY*-specific mRNA may be necessary to ensure stamen morphogenesis. A previous *in situ* hybridization study also showed that a *PTLF* antisense probe hybridized strongly to the floral meristems and developing flowers in both male and female poplars (Rottmann et al., 2000). Although *Hevea brasiliensis* also has male, female and bisexual flowers, in that species all floral meristems were shown to express *HbLFY* transcripts equally (Marcelo and Adriana, 2005). In contrast, the distinction between male and female inflorescences in maize apparently requires the differential expression of distinct maize *FLO/LFY* paralogs (Bomblies et al., 2003). In other woody species, the expression patterns of *LFY* homologs are not always related to reproductive development. *LFY* homologs are expressed in leaf primordia of *Eucalyptus* and grape (Southerton et al., 1998; Carmona et al., 2002; Dornelas et al., 2004). In transgenic poplar, the *PTLF* promoter was found to direct the highest level of GUS gene expression in shoots (Wei et al., 2006). In contrast, we found a relatively low level of *PtLFY* expression in stems, with expression the highest in both male and female floral buds and roots in *P. tomentosa* (Fig. 4A). In addition, seasonally-dependent expression of *LFY* homologs has been observed in other woody species such as kiwifruit, grape and apple (Walton et al., 2001; Carmona et al., 2002; Almada et al., 2009; Wada et al., 2002).

This study has contributed to a better understanding of the biological role of *PtLFY* during reproductive development, and also explored a new method of controlling flower initiation in *P. tomentosa*. The high expression level in developing male and female floral buds implies that *PtLFY* might play an important role in development of floral buds. To reveal the potential function of *PtLFY*, a common nucleotide sequence among *PtLFY*, *NFL1* and *NFL2* that encodes the C-terminal region, which is comprised of several conserved secondary domains such as α2, α3, α4, α5, α6 and α7 helices, was targeted for RNAi analysis (Fig. S1, S2). The helix-turn-helix (HTH, helices α2 and α3) motif is involved in sequence-specific contact between LFY and both the minor and major grooves of the target DNA sequence (Hamès et al., 2008). The *PtLFY*-IR silencing construct was designed based on conserved sequence of the target region (Fig. 5A), and transgenic tobacco lines were obtained that carried the *PtLFY*-IR structure (Fig. 6A). Semi-quantitative RT-PCR analysis showed that the expression of endogenous *NFL1* and *NFL2* was reduced by the presence of *PtLFY*-IR in the transgenic tobacco lines (Fig. 5C). Fourteen out of 15 transgenic tobacco lines possessed a non-flowering phenotype (Table 2), indicating that *PtLFY* has a similar function to *Arabidopsis LFY in planta*. This result suggested that the *PtLFY*-IR silencing construct reduced expression of the endogenous *NFL1* and *NFL2*, and thereby inhibited flowering in transgenic tobacco. Similarly, suppression of Sus activity caused by construct *Sus*-IR in the ovule epidermis led to a fiberless phenotype in transgenic cotton plants (Ruan et al., 2003). This approach to generate hpRNAs in plants is known as inverted repeat (IR) silencing triggers (Chuang and Meyerowitz, 2000; Wesley et al., 2001). The most potent variation is a hairpin in which the terminal loop is initially formed by a short intron. Although there is some variation, the success rate for gene silencing has been reported to exceed 90% in some publications (Chuang and Meyerowitz, 2000; Wesley et al., 2001; Kerschen et al., 2004). Therefore, hpRNAi has been widely adopted for effecting gene knock-down in many plant species (Ossowski et al., 2008). In another respect, a large number of non-coding RNAs (e. g. microRNA) have been shown to be conserved among diverse plant species, implying that an identical or similar RNAi sequence could function across different plant families (Sunkar and Jagadeeswaran, 2008). The *PtLFY*-IR-mediated gene silencing of *LFY* orthologs in tobacco presented in our study suggests a down-regulation of transcription that could be due to an RNAi mechanism. In previous study, clustered shoots on crown, growth rate change and not flowering were seen in some transgenic poplar lines with strong *PTLF* expression, early flowering by overexpression of *PTLF* were still observed in others (Rottmann et al., 2000). Consequently, we infer that the *PtLFY*-IR construct may be of general use to control flowering in poplars.

P. tomentosa is a versatile species important in wood production, ecological rehabilitation, pulping for paper, and biomass production. It also provides shade and aesthetic beauty in rural and urban areas. Given that the production of flowers might be energetically expensive and possibly have a negative impact on vegetative growth, along with the known allergenic properties of pollen, and other environmental issues such as catkins production, the regulation of flowering in *P. tomentosa* is an important issue that needs to be considered when planting this species. Our results provided evidence for the important role of *PtLFY* in flowering of *P. tomentosa*, but further investigation is needed to elucidate the molecular mechanisms underlying the function and regulation of ptLFY in *P. tomentosa*.

Table 2　Comparison of flowering traits in T_0 transgenic tobacco lines and wild-type tobacco plants grown in natural conditions

Transgenic lines	Number of plants	Date of flowering	Number of flowering plants	Proportion of flowering lines (%)
T1	3		0	
T3	3		0	
T4	4		0	
T5	3		0	
T7	3		0	
T9	4		0	
T22	3		0	
T23	3		0	
T24	4		0	
T25	3		0	
T26	3		0	
T27	4		0	
T34	3		0	
T35	3	Sept. 11	3	6.7
T39	3		0	
WT[a]	10	Aug. 26	10	100

[a] WT = wild-type tobacco plants.

Ectopic Expression of a Poplar *APETALA3*-like Gene in Tobacco Causes Early Flowering and Fast Growth*

Abstract A MADS-box gene, designated *PtAP3*, was isolated from a floral bud cDNA library derived from *Populus tomentosa*. Analysis by multiple alignments of both nucleotide and amino acid sequences, together with phylogenetic analysis, revealed that *PtAP3* is an ortholog of *Arabidopsis AP3*. Analysis of RNA extracts from vegetative and reproductive tissues of *P. tomentosa* by RT-PCR indicated that *PtAP3* is expressed in roots, stems, leaves and vegetative and floral buds. Notably, the expression of *PtAP3* was found to fluctuate during floral bud development between September and February and differences in expression were observed between male and female buds. In the former, a gradual down-regulation during this period, interrupted by a slight up-regulation in December, was followed by a sharper up-regulation on February. In developing female floral buds, expression was stable from September to November, sharply up-regulated in December, and then gradually down-regulated until February. The functional role of *PtAP3* was investigated in transgenic tobacco plants. Of 25 transformants, nine displayed an earlier flowering phenotype compared with the wild type plants. Furthermore, most of the transgenic tobacco had faster growth and more leaves than untransformed controls. The traits proved to be heritable between the T0 and T1 generations. Our results demonstrate a regulatory role of the *PtAP3* gene during plant flowering and growth and suggest that the gene may be an interesting target for genetic modification to induce early flowering in plants.

Key words *APETALA3*, *Populus tomentosa*, early flowering, fast growth, tobacco

1 Introduction

Current knowledge concerning the ABCDE gene model comes largely from extensive studies on the model plant *Arabidopsis*. Within the B class genes, *APETALA3* (*AP3*) and *PISTILLATA* (*PI*) are of interest because they have been showed to act as identity genes required for petal and stamen morphogenesis (Jack et al., 1992). Both *AP3* and *PI* gene products belong to the MADS-box super family of proteins, which are known as the MIKC-domains (TheiBen et al., 2000). The MADS-box domain contains approximately 60 aa and is highly conserved across the entire super family. It is essential for binding at conserved DNA sites containing the sequence CC (A/T)$_6$GG, known as CArG elements, and is considered to play a role in protein dimerization (Riech-

* 本文原载 *Biotechnology Letters*, 2011, 33(6): 1239-1247, 与安新民、叶梅霞、王冬梅、王泽亮、曹冠琳和郑会全合作发表。

mann et al., 1996). The gene products of *AP3* form heterodimers that recognize and bind to these sites (Hill et al., 1998).

The particular importance of understanding the flowering mechanism in poplar is due to the economic and ecological importance of the species for forest production, reclamation and biomass (Polle and Douglas, 2010). *Populus* differs from *Arabidopsis* in many aspects. It is a woody perennial tree, with a long juvenile phase and a long life span (Hsu et al., 2006). It flowers annually or seasonally during the reproductive developmental phase (Yuceer et al., 2003) and each seasonal flowering period is interrupted by a vegetative period. Furthermore, the structures of poplar flowers are quite distinct from those of *Arabidopsis* (Rottmann et al., 2000). At present, the functional role of *AP3* and/or its orthologs in the development of woody plant flowers is still unclear. In this study, we report the isolation of *PtAP3*, an *AP3*-like gene in *P. tomentosa*, and the analysis of its function both *in situ* and in heterogeneous transformed tobacco plants. Our data provides new insights into the molecular mechanism underlying the development of floral buds in *P. tomentosa* and, additionally, one possible approach to accelerating early flowering in plants via genetic modification.

2 Materials and methods

2.1 Plant materials

Vegetative and floral buds of *P. tomentosa* were rapidly frozen in liquid nitrogen after collection from the adult trees and stored at −80℃ until use. Additionally, root, stem and leaf samples were taken from 30-day-old tissue-cultured plantlets of *P. tomentosa*, followed by immediate RNA extraction. Sterilized tobacco (*Nicotiana tabacum*) leaf discs were used as transgenic acceptor materials. The young leaves of transgenic tobacco plants were used for extraction of DNA and RNA.

2.2 Cloning and characterization of *PtAP3*

The primers P1F and P1R (Table 1) used for the screening of *PtAP3* were designed according to *PTD* (accession AF057708), an *AP3* homolog from *P. trichocarpa*. *PtAP3* was obtained by PCR screening of the *P. tomentosa* male floral bud cDNA library. SM buffer containing recombinant bacteriophages was used as DNA template. Thermal cycling was performed at 94℃ for 5min, 94℃ for 30s, 60℃ for 20s, 72℃ for 30s, for 35 cycles, then at 72℃ for 5min. The positive recombinant bacteriophage clones were isolated by PCR.

The amino acid sequences of *AP3* homolog genes were retrieved from GenBank (http://www.ncbi.nlm.nih.gov/entrez/query.fcgi) and aligned with CLUSTALX1.81 and Bioedit software. Conserved domain sequences and functional domains were analyzed using the Expert Protein Analysis System (ExPASy) proteomics server of the Swiss Institute of Bioinformatics (SIB) (http://cn.expasy.org/). The tertiary structure of the PtAP3 protein was built using software 3D-

JIGSAW 2.0(http://bmm.cancerresearchuk.org/~3djigsaw/).

Table 1 Primers used for screening the cDNA library, labelling DNA probes, construction of the expression vector, PCR amplification, DNA gel blot analysis, RT-PCR analysis of poplar tissues and qRT-PCR analysis of transformants.

Primers	Oligonucleotide *	Features
P1F forward	5′-TTGGATCCATGGTCGTGGAAAGA-3′(*Bam*H I)	
P1R reverse	5′-AAGAGCTCTCAAG GAAGGCGAAGTT-3′(*Sac* I)	For screening cDNA library
P2F forward	5′-GAAGCAAAACTAGAGGATCTACAGG-3′	For RT-PCR analysis
P2R reverse	5′-TCAAGGAAGGCGAAGTTCAT-3′	
P3F forward	5′-CTCCATCATGAAATGCGATG-3′ (*P. tomentosa ACTIN*)	
P3R reverse	5′-TTGGGGCTAGTGCTGAGATT-3′ (*P. tomentosa ACTIN*)	
P4F reverse	5′- TTGGATCCATGGTCGTGGAAAGA-3′(*Bam*H I)	For PCR, DNA blot analysis
P4R reverse	5′-TAAGCTTGATGTGGAGGGGCTAAT-3′	For qRT-PCR analysis
P5F forward	5′- CTGCTGGAATTCACGAAACA-3′ (Tobacco *ACTIN*)	
P5R reverse	5′-GCCACCACCTTGATCTTCAT-3′ (Tobacco *ACTIN*)	

* The added restriction sites are underlined.

2.3 Phylogenetic relationship analysis

Genetic relationships were analyzed using a number of B-class MADS-box protein sequences of plant species retrieved from GenBank. Multiple sequence alignment was performed with ClustalW, genetic distance matrices were obtained from the alignments and a neighbor-joining tree was constructed with bootstrap sampling of 1000 replications using MEGA 4.1 (Tamura et al., 2007).

2.4 RT-PCR analysis

Total RNAs were extracted from vegetative and reproductive tissues of *P. tomentosa* according to the method described previously (Chang et al., 1993). Total RNA was pre-treated with RQ1 DNase I (Promega) to remove contaminating genomic DNA. The first-strand cDNA was synthesized using 1.0μg of treated total RNA, Superscript III (Invitrogen) and oligo d $(T)_{20}$, subsequently diluted 1∶10 with ddH_2O as template. P2F and P2R primers (Table 1) were used for PCR amplification. Conditions of thermal cycling were identical to those described above. No-template controls for each primer pair were included in each run. *P. tomentosa* actin was used as an internal reference to normalize small differences in template amounts with P3F and P3R primers (Table 1).

2.5 Construction of expression vectors and transformation of tobacco

Both the plasmid DNA containing *PtAP3* and the binary vector pBI121 DNA were digested with both *Bam*H I and *Sac* I. The corresponding fragments were recycled and purified using QIAquick™ Gel Extraction Kit (QIAGEN) and ligated with T4-DNA ligase (Promega) at 16℃ overnight. The ligation product was then transferred into competent cells of *E. coli* TG1. The pBI121-*PtAP3* was verified by PCR (using P4F and P4R primers) (Table 1) and digestion. Tobacco leaf disc transformation was performed using the *Agrobacterium*-mediated method.

3 Molecular identification of transgenic tobacco plants

3.1 PCR and DNA gel blot analysis

Genomic DNA was extracted from young leaves of transformed plants and used as the template for PCR identification. The PCR reaction system and conditions of thermal cycling were identical to those described in the previous sections. PCR products were separated by electrophoresis on a 0.8% agarose gel and blotted onto positively charged nylon membrane by capillary transfer in the presence of 20×SSC buffer. A 203 bp fragment used as DNA probe for gel blot analysis was amplified from genomic DNA of *P. tomentosa* using P4F and P4R primers (Table 1). The blots were hybridized with a DIG-labeled DNA probe at 58℃, and washed at high stringency. Immunological detection was performed with the DIG DNA Labeling and Detection Kit (Roche) according to the manufacturer's protocol.

3.2 Quantitative RT-PCR analysis

Total RNA was extracted from transgenic tobacco leaves using the SV Total RNA Isolation System (Promega). Reverse transcription was performed as described in previous sections. The first-strand cDNA was diluted 1∶10 with ddH_2O as template. P2F and P2R were used for qPCR analysis. The qPCR reaction was performed using PowerSYBR® Green PCR Master Mix (Invitrogen) on a DNA Engine Opticon 2 system (MJ Research). The program included a preliminary step of 2min at 50℃, predenaturation at 94℃ for 2min, followed by 35 cycles of 94℃ for 15s, 60℃ for 30s and 72℃ for 30s, with a final extension of 7 minutes at 72℃. No-template controls for each primer pair were included in each run. Tobacco *actin* was used as an internal reference to normalize small differences in template amounts with P5F and P5R primers (Table 1). Three replicates were employed for the qPCR analysis of each sample.

3.3 Phenotypic analysis of transgenic tobacco plants

Twenty five transgenic plantlets were transplanted into individual pots with artificial soil mix as growing support. The plants were moved outside the greenhouse when they had produced four to five leaves (June 15). Plant height and number of leaves of each T_0 transgenic line were recorded on August 11 and 21, September 1, 11 and 21, and October 2. The date when flower primordia were first visible was also recorded.

4 Results and discussion

A positive phage plaque was isolated from the *P. tomentosa* male floral bud cDNA library by PCR. Sequencing revealed the length of *PtAP*3 cDNA to be 717 bp, encoding 238 amino acids. The sequence alignments suggest that *PtAP*3 belongs to the MADS-box gene family. It contains M-, I-, K- and C-domains, which are typical characteristics of MIKC-type genes. The predicted tertiary structure of the MADS-box domain of PtAP3 consists of one helix and two sheet structures.

MADS-box genes play crucial roles in plant growth and development (Causier et al., 2002). The M-domain itself is the most highly conserved of the four major MADS-box protein domains and has been widely studied across the taxonomic kingdoms (Leseberg et al., 2006). Proteins incorporating the M-domain are involved in DNA binding and dimerization with other MADS-box proteins (Diaz-Riquelme et al., 2009). The protein PtAP3 reported here shares 82% amino acid similarity with PTD, an AP3 homolog from *P. trichocarpa*. The data described in the present report will add to existing knowledge in this field (Sheppard et al., 2000).

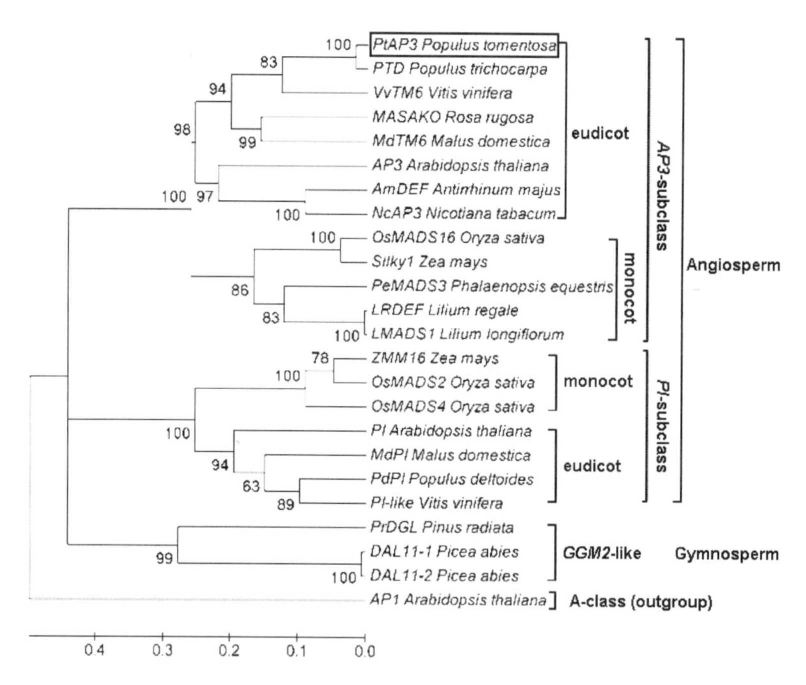

Fig. 1 Neighbor-joining tree representing relationships of *PtAP*3 with gymnosperm and other angiosperm MADS-box gene homologs

Bootstrap support values (%) from 1000 replications are indicated when over 50%. GenBank accession number for each sequence: *PtAP*3 (AY210488), *AP*1 (BT004951), *AP*3 (AY142590), *PI* (NM_ 122031), *VvTM*6 (BQ979341), *PI-like* (DQ059750), *MASAKO* (AB055966), *MdTM*6 (AB081093), *MdPI* (AJ291491), *Silky*1 (AF181479), *ZMM*16 (NM_ 001111666), *OsMADS*2 (L37526), *OsMADS*4 (L37527), *OsMADS*16 (AF077760), *PeMADS*3 (AY378150), *LMADS*1 (AF503913), *LRDEF* (AB071378), *PTD* (AF057708), *PdPI* (EU029172), *DAL*11-1 (AF158539), *DAL*11-2 (AF158540), *PrDGL* (AF120097), *AmDEF* (AB516402), *NcAP*3 (X96428).

To better understand the relationships between *PtAP3* and the B-class MADS-box genes of other species, a neighbor-joining tree was constructed from *PtAP3* together with 23 previously reported B-class MADS-box genes from *Arabidopsis*, gymnosperm, eudicot and monocot species. Similarly to the previous studies (Winter et al., 2002), the B-class MADS-box genes were classified into *AP3*, *PI* and *GGM2* subclasses (the *AP3* and *PI* subclasses associated with the angiosperm group and the *GGM2* subclass with the gymnosperm group). The analysis shows that *PtAP3* falls into the eudicot and monocot *AP3* subclass of B-class MADS-box proteins (Fig. 1). The result suggests that *PtAP3* shares similar functions with those of other genes in the same clade. In other words, *PtAP3* is likely to fulfil a B-function role during the development of floral organs in *P. tomentosa*.

The expression profiles of *PtAP3*, analyzed by RT-PCR, indicate that the gene is expressed in various tissues of *P. tomentosa* including roots, tender stems and leaves, vegetative buds and male and female floral buds. In floral buds, its relative expression levels fluctuated and differences were apparent between the two sexual forms (Fig. 2A, B and C). In developing male floral buds, a gradual down-regulation between September 13 and January 25 was interrupted by a slight up-regulation, demonstrated by the data point in December and expression was then sharply up-regulated on February 25(Fig. 2B). In contrast, transcript levels in developing female floral buds were stable from September 13 to November 24, sharply up-regulated in December and then gradually down-regulated until February 25th (Fig. 2C). These results imply that *PtAP3* is associated with sexual differentiation of *P. tomentosa* floral organs.

Twenty five independent T_0 generation transgenic tobacco plants and 10 wild-type were transferred to conditions exterior to the greenhouse for further study. The average height of all these transgenic plants was statistically higher than that of wild type during development (Fig. 4A) and the transgenic

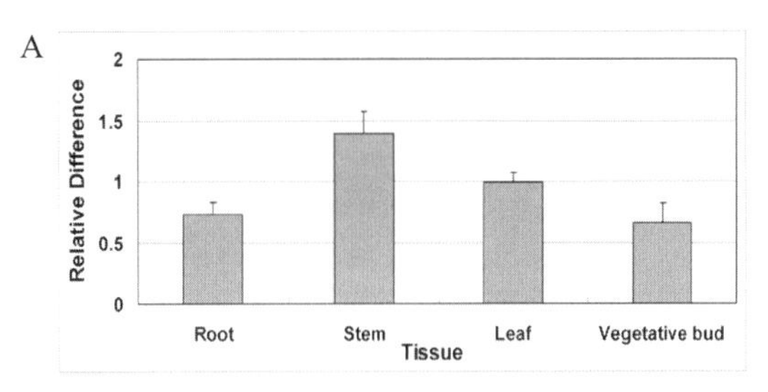

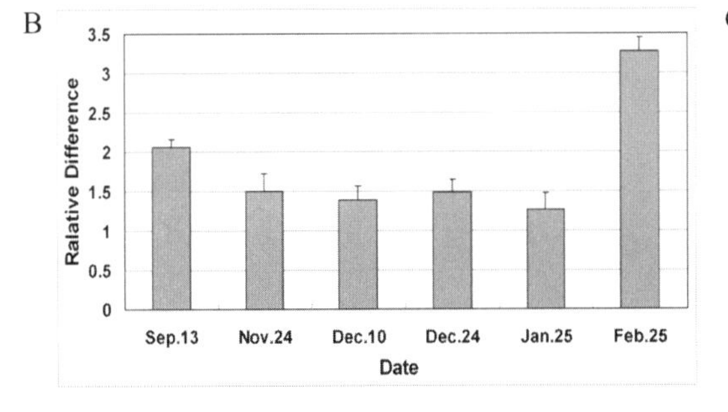

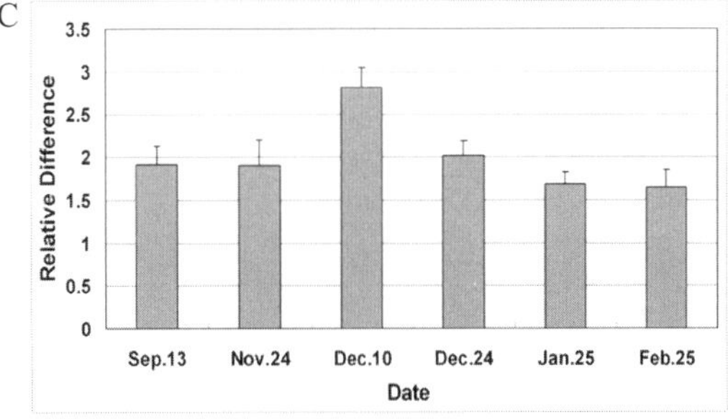

Fig. 2 Expression patterns of *PtAP3* in *P. tomentosa*

A. Expression in different tissues. Roots, stems and leaves were collected from one-month-old tissue- cultured plantlets. Vegetative buds were collected from adult trees. B & C. The expression patterns of *PtAP3* during floral bud development derived from real time RT-PCR analysis in male floral buds (B) and female floral buds (C).

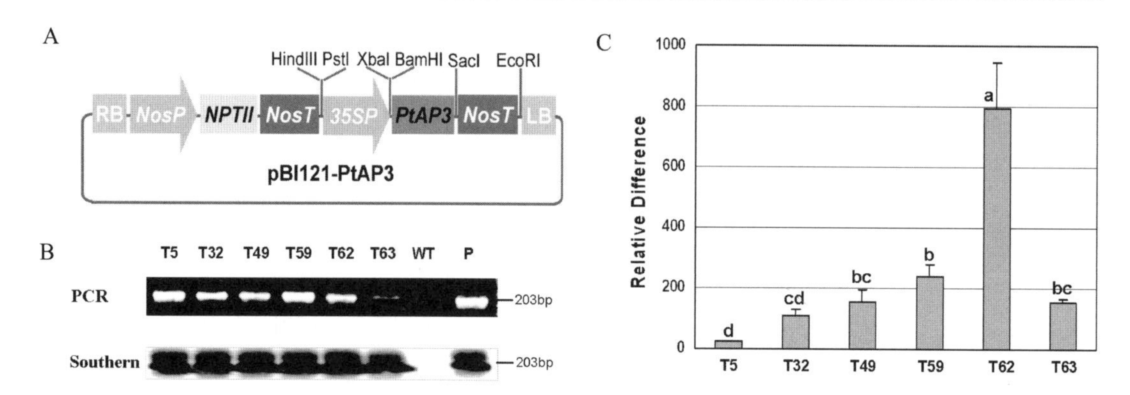

Fig. 3 Molecular identification of transgenic tobacco plants carrying the 35S:: *PtAP3* construct

A. Schematic representation of pBI121-PtAP3 for over-expression of *PtAP3* in transgenic tobacco. RB (T-DNA right border), LB (T-DNA left border), *Nosp* (Nos promoter), *npt*II (Neomycin phosphotransferase II), *Nost* (Nos terminator), 35S (CaMV 35S promoter), *gus* (b-glucuronidase gene), *PtAP3* (an *AP3*-like gene from *P. tomentosa*). B. Transgenic tobacco plants identified by PCR amplification and DNA gel blot analysis. T5, T32, T49, T59, T62 and T63 are six transgenic lines, WT is the wild type, P is positive control (the recombinant plasmid pBI121-*PtAP3*). C. The over-expression of *PtAP3* in transgenic tobacco plants identified by quantitative RT-PCR (using tobacco *ACTIN* as reference control). T5, T32, T49, T59, T62 and T63 are six transgenic lines. Significance was assessed using the ANOVA Fisher's LSD test ($*P<0.05$).

plants also had more leaves than the control plants (Fig. 4B). Nine flowered earlier than the wild-type plants by between 4 and 25days (Table 2; Fig. 4C). No obvious phenotypic variation of floral organs was observed in the transgenic tobacco compared with the wild type (Fig. 4D). Leaves, however, appeared to be narrower, longer and uneven (Fig. 4E). No difference was observed in fruit shape (Fig. 4Fa, Fb). The T_1 generation transgenic plants derived from T_0 seeds flowered earlier than wild type (Fig. 4G), indicating that the early flowering trait had been inherited in the T_1 generation.

Table 2 Comparison of flowering traits in T_0 transgenic tobacco lines and wild-type tobacco plants grown in natural conditions

Transgenic line	Date of flowering	Number of leaves at time of flowering	Final height (cm)	Note
1	Aug. 3	20. 0	110. 0	Early flowering
2	Aug. 1	16. 0	105. 0	Early flowering
3	Aug. 6	19. 0	145. 0	Early flowering
4	Aug. 10	19. 0	126. 0	Early flowering
5	Sep. 1	20. 0	82. 0	No early flowering
6	Sep. 3	15. 0	82. 0	No early flowering
7	Aug. 14	20. 0	112. 0	Early flowering
8	Aug. 16	23. 0	124. 0	Early flowering
9	Aug. 29	17. 0	86. 0	No early flowering
10	Sep. 12	16. 0	77. 0	No early flowering
11	Aug. 26	18. 0	85. 0	No early flowering

(Continured)

Transgenic line	Date of flowering	Number of leaves at time of flowering	Final height (cm)	Note
12	Sep. 3	16. 0	80. 0	No early flowering
13	Sep. 12	15. 0	72. 0	No early flowering
14	Aug. 29	19. 0	84. 0	No early flowering
15	Sep. 15	15. 0	68. 0	No early flowering
16	Sep. 16	15. 0	72. 0	No early flowering
17	Aug. 31	16. 0	82. 0	No early flowering
18	Sep. 16	17. 0	80. 0	No early flowering
19	Sep. 13	18. 0	82. 0	No early flowering
20	Aug. 30	16. 0	84. 0	No early flowering
21	Aug. 22	15. 0	92. 0	Early flowering
22	Sep. 11	18. 0	78. 0	No early flowering
23	Sep. 10	19. 0	72. 0	No early flowering
24	Aug. 18	18. 0	100. 0	Early flowering
25	Aug. 20	16. 0	100. 0	Early flowering
WT	Aug. 26	14. 5(average)	69. 2(average)	Control

In the present study, despite varying levels of expression of *PtAP*3, all the transgenic tobacco plants grew faster and produced more leaves, while 9 of the 25 plants flowered earlier. We speculate that the essential genetic factor for such phenotypic traits is the conserved MADS-box domain. The MADS-box domain of the *PtAP*3 protein bears very close homology to those of the *Arabidopsis* proteins *AtAP*3 and *AtAP*1 (Fig. S1) and it is expected that its function during flower development is similar to that of the *Arabidopsis* proteins. The observed effects of over expression of *PtAP*3 in transgenic tobacco plants suggest that one of the protein's major functions concerns the regulation of flowering time. It is possible that faster growth and more leaves also contribute to the transition to flowering.

Our study will contribute to the greater understanding of the genetic factors controlling flower development in poplar. The *PtAP*3 gene represents a valuable element to further this knowledge not only in poplar but possibly other woody species. Based on these findings, our work will be extended to the production of transgenic poplar expressing altered levels of the endogenous poplar gene to further investigate its involvement in flowering in this species. Control of flowering time is also an important factor in other domains of plant culture. For instance, speed to flowering has an important effect on production costs in floriculture, as it determines the productivity of a defined area of greenhouse space (Chandler and Brugliera, 2010). Although some control of flowering can be achieved by manipulation of photoperiod and/or use of plant growth regulators, ectopic expression of *PtAP*3 might be one approach using genetic modification to achieve early flowering induction in these plants.

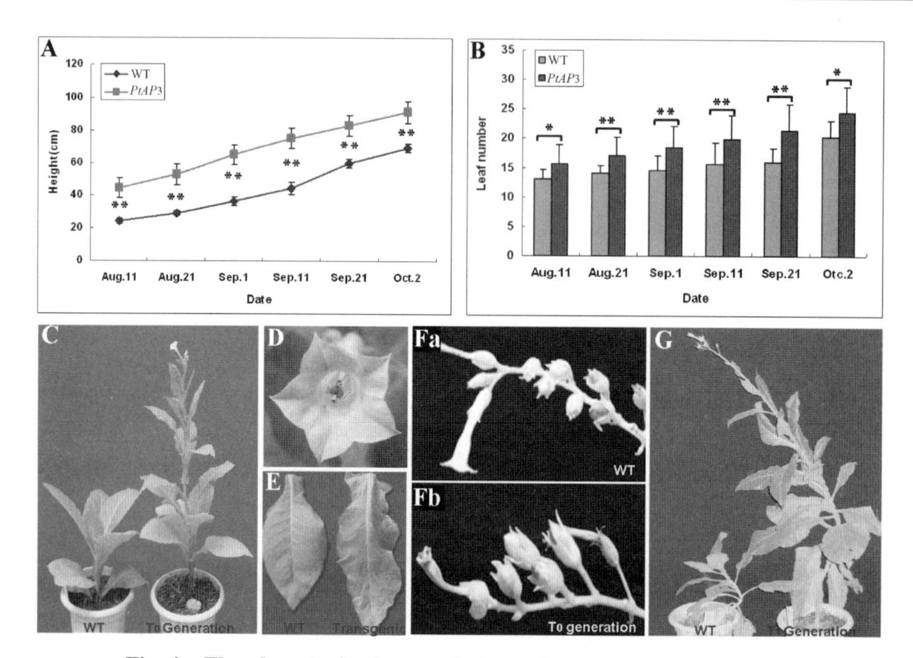

Fig. 4 The phenotypic characteristics and growth traits of tobacco plants carrying pBI121-*PtAP3* grown in natural conditions

A. The height of transgenic and wild type tobacco plants from August 11th to October 2nd. Significance was assessed using the ANOVA Fisher's LSD test (** P<0.01) B. Change in leaf number during the development of transgenic tobacco plants and wild type. Significance was assessed using the ANOVA Fisher's LSD test (* P<0.05, ** P<0.01). C. Early flowering phenotype produced in T_0 generation of transgenic tobacco. D. A flower of transgenic tobacco (T_0 generation). E. The altered leaf shape of a T_0 generation transgenic tobacco plant. F. Fa and Fb were fruits from wild type and transgenic tobacco plants (T_0 generation), respectively. G. The early flowering trait was inherited in the T_1 generation of transgenic tobacco.

外源突变基因*AGM3*过表达对烟草开花的抑制和花器官发育的影响*

摘　要　为了解*AGM3*基因在异源植物中对开花及花器官发育的影响，采用农杆菌介导法将dominant negative mutation(DNM)结构基因35S-*AGM3*-E9导入烟草(*Nicotiana tabacum*)，经PCR和Southern检测获得了一批阳性转化植株。荧光定量分析结果显示，*AGM3*在各个转基因株系中均有表达，且不同株系间表达量差异极显著。调查结果分析表明：与野生型植株相比，46.7%的转基因株系未成花，33.3%的转基因株系开花时间平均推迟29.7 d，20%的转基因株系开花时期与野生型相同。对花器官形态的观察发现，转基因烟草出现花萼数目减少、花冠产生深裂、花瓣和雄蕊形状改变、雄蕊数目增多或减少等变异。这些研究结果表明35S-*AGM3*-E9有效地抑制了花的发育，为利用转基因手段获得不育材料的研究奠定了基础。

关键词　*AGM3*，抑制开花，花器官发育，烟草

Coen和Meyerowitz以拟南芥(*Arabidopsis thaliana*)和金鱼草(*Antirrhinum majus*)等植物为研究对象，将花器官特征基因分为A、B、C三类。拟南芥*AGAMOUS*(*AG*)是最早鉴定出的C类基因(Coen and Meyerowitz，1991)。花发育早期*AG*在3、4轮中特异表达(Yanofsky et al.，1990；Drews and Meyerowitz，1991)，它与*APETALA3*(*AP3*)、*PISTILLAT*(*PI*)、*SEPALLATA*(*SEP*)联合控制雄蕊发育，与*SEP*共同控制心皮的发育(Honma and Goto，2001；Theissen，2001)。*ag*突变体的雄蕊和雌蕊变异为花萼和花瓣并形成重瓣花(Bowman et al.，1991a)；转反义*AG*的拟南芥(Mizukami and Ma，1995)、矮牵牛(*Petunia hybrida*)(van der Krol et al.，1993)、烟草(Kempin et al.，1993；Mande et al.，1992)和番茄(*Lycopersicon esculentum*)(Pnueli et al.，1994)等均发生花器官的变异。

数十种植物的*AG*基因已被分离(高志红等，2008)，对*AG*同源基因的研究是当今分子生物学的热点，Kato和Hibino克隆了大桉(*Eucalyptus grandis*)*AG*同源基因*EgAGL*，并发现*EgAGL1*和*EgAGL2*在花芽强烈表达(Kato and Hibino，2009)。Cartolano等发现*AG*在内轮花器官的表达增强，会导致*AG*的异位表达，使外轮花器官变异(Cartolano et al.，2009)。Narumi等将拟南芥的嵌合*AG*阻遏子转化蓝猪耳(*Torenia fournieri*)，导致了花形态的变异(Narumi et al.，2008)。目前大多数*AG*同源基因的研究集中在功能验证、解析作用机理以及探讨进化关系等方面，通过抑制AG蛋白从而控制植物开花的研究鲜见报道，育种中能够直接借鉴的成果不多。

与常见的转录水平的基因沉默不同，dominant negative mutation(DNM)是一种在蛋白水平上使基因失去作用的新手段，它利用突变的蛋白与内源蛋白产生竞争从而抑制其功能

* 本稿原载《中国生物工程杂志》，2011，31(6)：49-57. 与曹冠琳、安新民、龙萃和薄文浩合作发表。

(Sheppard, 1994)。Mizukami 等曾报道删除了 C 末端的 AG 蛋白会使拟南芥花表型类似于 *ag* 突变体，删除 K 区和 C 区的 AG 蛋白会导致雄蕊和心皮数目增多，说明被删剪的 AG 抑制了内源 AG 的作用(Mizukami et al., 1996)。本研究采用的 *AGM*3 基因是拟南芥 *AG* 的突变结构，其氨基酸序列与 AG 以及烟草 NAG1(AG 同源)相比，在 MADs Domain 和 C Domain 末端发生突变(图 1)。MADs Domain 的主要作用是与靶 DNA 结合，C Domain 的主要作用是保证 AG 控制花器官的功能正常发挥(Mizukami et al., 1996)。作为一种转录因子，AG 蛋白结合靶 DNA 之前必须与其他蛋白形成聚合体，AGM3 的作用原理是它的过量表达会与 AG 争夺其他蛋白并充分结合，但 AGM3 的两处突变导致它形成的蛋白聚合体不能结合在 DNA 上，也不能行使功能，从而产生对 AG 的竞争抑制。

为了在较短时间内验证 *AGM*3 对异源植物开花的抑制，本研究将 35S-*AGM*3-E9 转化模式植物烟草，并考察了 *AGM*3 在烟草中的表达以及其对烟草开花的抑制作用。Kempin 等克隆的烟草 *AG* 同源基因 *NAG*1 依次在花分生组织、在雄蕊和心皮表达，异位表达 *NAG*1 可使萼片向雄蕊转变，使花瓣向心皮转变。反义表达 *NAG*1 使雄蕊不同程度地向花瓣的转变，而心皮的变异不明显(Kempin et al., 1993)。NAG1 和 AG 蛋白有 73%的同源性，在 MADs Domain 的同源性达到 100%(图 1)，MADs Domain 是个序列特异性 DNA 结合位点，预示着 NAG1 和 AG 识别相同的核苷酸序列。

```
                 N Domain
AG     HFLQLLQISYFPENHFPKKNKTFPFVLLPPTAITAYQSELGGDSSPLRKSGRGKIEIKRI    60
AGM3   ...............................GSMAYQSELGGDSSPLRKSGRGKIEIKRI    29
NAG1   .................................MDFQSDLTREISPQRKLGRGKIEIKRI    27

                 MADs Domain
AG     ENTTNRQVTFCKRRNGLLKKAYELSVLCDAEVALIVFSSRGRLYEYSNNSVKGTIERYKK   120
AGM3   ENTTNRQVTFCKRRNGLLEEAYELSVLCDAEVALIVFSSRGRLYEYSNNSVKGTIERYKK    89
NAG1   ENTTNRQVTFCKRRNGLLKKAYELSVLCDAEVALIVFSSRGRLYEYANNSVKATIERYKK    87

                 I Domain
AG     AISDNSNTGSVAEINAQYYQQESAKLRQQIISIQNSNRQLMGETIGSMSPKELRNLEGRL   180
AGM3   AISDNSNTGSVAEINAQYYQQESAKLRQQIISIQNSNRQLMGETIGSMSPKELRNLEGRL   149
NAG1   ACSDSSNTGSISEANAQYYQQEASKLRAQIGNLQNQNRNMLGESLAALSLRDLKNLEQKI   147

                 K Domain
AG     ERSITRIRSKKNELLFSEIDYMQKREVDLHNDNQILRAKIAENERNNPS.....ISLMPG   235
AGM3   ERSITRIRSKKNELLFSEIDYMQKREVDLHNDNQILRAKIAENERNNPS.....ISLMPG   204
NAG1   EKGISKIRSKKNELLFAEIEYMQKREIDLHNNNQYLRAKIAETERAQQQQQQQMNLMPG    207

                 C Domain
AG     GSNYEQLMPPPQTQSQPFDSRNYFQVAALQPNNHHYSSAGRQDQTALQLV   285
AGM3   GSNYEQLMPPPQTQSQPFDSRNYFQSRHCNLTITITHPPVAKTKPLSSCK   254
NAG1   SSSYE.LVPPP....HQFDTRNYLQVNGLQTNNHYT....RQDQPSLQLV   248
```

图 1　AG、AGM3 和 NAG1 氨基酸序列比对

AG: Amino acid of AGAMOUS in *Arabidopsis thaliana*; AGM3: Dominant negative mutation of AG; NAG1: AG homologue of *Nicotiana tabacum*. The MADs Domain and C Domain of AG are indicated by double-ended arrows above the sequences; The black frame indicates variation sites of AGM3; Amino acids conserved in two of the three sequences at least are shaded gray; The alignment was generated using the DNAMAN program.

1 材料与方法

1.1 植物材料、质粒与菌种

遗传转化受体植物材料为野生型烟草 W38(*Nicotiana tabacum* cv. W38)的组织培养苗。含 DNM 结构基因 35S-*AGM*3-E9 的重组质粒引进于俄勒冈州立大学(Oregon State University),质粒以 pART27 载体为基本骨架,其 T 区结构如图 2 所示,*AGM*3 是拟南芥 *AG* 基因的突变结构,其氨基酸序列及突变区域如图 1 所示。遗传转化所用菌种为根癌农杆菌(*Agrobacterium tumefaciens*) GV3101。植物材料及农杆菌均由本实验室保存。

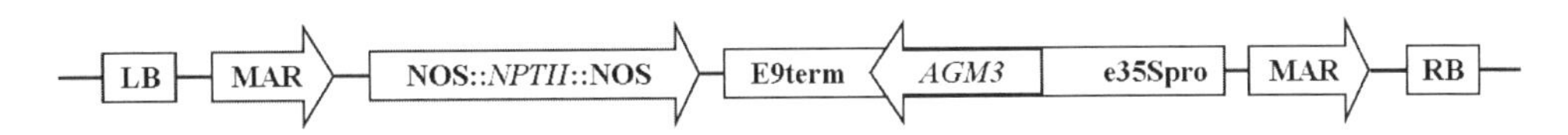

图 2 含 35S-*AGM*3-E9 结构的载体 T 区图示

MAR: Matrix attachment region; *NPTII*: Neomycin phosphotransferase II gene; E9term: Terminator from *rbc*S of pea (*Pisum sativum*); e35Spro: Enhanced CaMV 35S promoter; LB: Left border of T-DNA; RB: Right border of T-DNA.

1.2 DNM 结构 35S-*AGM*3-E9 在烟草中的转化

烟草的转化采用农杆菌介导的叶盘转化法。将含 35S-*AGM*3-E9 的重组质粒转化农杆菌 GV3101 感受态并培养于含抗生素的 YEB 平板上,对长出的单菌落进行 PCR 检测和酶切验证。将验证过的农杆菌摇至 OD_{600} = 0.4~0.6 并立即用于烟草的侵染。烟草的转化参考了本实验室建立的方法(林元震,2006),将剪成 1~2cm^2 的烟草叶片在分化培养基(MS + 6-BA 1.0mg·L^{-1}+ NAA 0.1mg·L^{-1})上预培养 48h 后,浸于菌液中侵染 5min 并暗培养 2~4d,最后转接到筛选分化培养基(MS + 6-BA 1.0mg·L^{-1}+ NAA 0.1mg·L^{-1}+ Kan 50mg·L^{-1}+ Cef 250mg·L^{-1})中培养至不定芽分化。将不定芽转接到生根培养基(1/2MS + IBA 0.3mg·L^{-1}+ Kan 50mg·L^{-1}+ Cef 250mg·L^{-1})(曹冠琳等,2010),将生根的抗性植株继代扩繁以用于后续检测。

1.3 转基因烟草的分子检测

转基因烟草叶片基因组 DNA 的提取采用 CTAB 法,阳性对照为质粒 DNA,阴性对照为未经转化的烟草基因组 DNA。*AGM*3 基因的引物为:Forward primer 5′-AAGGATC CATGGCGTACCAATCGG-3′和 Reverse primer 5′-CTGAATTCTTACACTAACTGGAG AGCGC-3′。PCR 反应条件为:94℃变性 30s,56℃退火 30s,72℃延伸 1min,共进行 30 个循环。PCR 产物用 1%的琼脂糖电泳检测。

采用 DIG High Prime DNA Labeling and Detection Starter Kit Ⅰ(Roche)试剂盒对 PCR 检测呈阳性的植株进行 PCR-Southern 和 Dot-Southern blot 检测。以重组质粒上扩增的 *AGM*3 (774bp)片段为模板,在 37℃下用 DIG-High Prime 标记 20h 制备探针。在 PCR-Southern 检

测中，将转基因烟草的 PCR 产物电泳，然后用 20×SSC 转移到 Positive - charged Nylon 膜上；在 Dot-Southern blot 检测中，将基因组 DNA 直接点于尼龙膜上。尼龙膜经紫外固定后加入探针，55℃下杂交 4~10h。在 20~25℃下用 2×SSC/0.1% SDS 洗膜(2×5min)，然后在 68℃下用 0.5×SSC/0.1% SDS 洗膜(2×15min)(曹冠琳等，2010)，最后进行免疫检测。

1.4　转基因植株的 qRT-PCR 分析

采用 SV Total RNA Isolation System(Promega)试剂盒提取转基因烟草总 RNA。取 1μg 总 RNA 为模板，用 A3500 Reverse Transcription System(Promega)试剂盒合成 cDNA。采用 SYBR Premix Ex Taq(Takara)进行 qPCR，以 *ACTIN* 为内参基因进行 *AGM3* 相对表达量的检测。*AGM3* 基因引物为 Forward primer 5′-TCAGGAACTTGGAAGGCAGATTAGA-3′和 Reverse primer 5′-TGAGATTGCGTTTGAGGTGGTG-3′，内参 *ACTIN* 引物为 Forward primer 5′-AGACTGCAAAGAGTAGCTCTTCTGTTGA-3′ 和 Reverse primer 5′-CATGAT GGAATTGTAAGTTGTTTCGTG-3′。qPCR 在 Opticon™(MJ research™)检测仪中进行，反应条件为 95℃变性 5s，62℃退火 20s，72℃延伸 15s，进行 40 个循环，每个反应重复 3 次。

1.5　转基因烟草的表型分析

将经过分子检测的转基因烟草以及野生型对照的组培苗移栽并进行生长情况和开花状况的观察。移栽地点为北京林业大学苗圃，移栽所用花盆高 21cm，直径 20cm，培养基质为草炭土、蛭石、珍珠岩以质量比 1∶1∶1 混合，移栽时尽量保证组培苗的生长状态一致。移栽后对烟草进行株高等营养生长状态的跟踪测量和记录。待植株进入生殖生长，对转基因烟草的开花时间，花表型特征进行观察和记录。数据分析在 SPSS 13.0 中进行。

2　结果与分析

2.1　转基因烟草的分子鉴定

为筛选转基因植株，对分化出的 Kan 抗性苗进行 PCR 检测。检测结果如图 3A 所示，部分抗性植株扩增出约为 774bp 的特异性条带，与阳性对照的 PCR 产物相比长度一致，而野生型的 DNA 未扩增出相应条带，初步推测这些抗性植株为阳性。

为证实 PCR 产物为外源 *AGM3*，用 DIG 标记的 *AGM3* 探针对 PCR 产物进行 PCR-Southern 检测，在高严谨度的杂交和洗膜条件下，获得了较为理想的信号。结果如图 3B 所示，1~7 泳道中被检测的转化植株均获得较强的 Southern 信号，且长度与阳性对照一致，而野生型对照没有杂交信号。为进一步证实外源基因已整合到烟草基因组中，用基因组 DNA 与 DIG 标记的 *AGM3* 探针进行 Dot-Southern blot 杂交，结果如图 3C 所示，1~7 号转基因植株以及阳性对照均有较强的显色反应，而野生型对照呈阴性。经 PCR、PCR-Southern 和 Dot-Southern blot 验证，共证实转 *AGM3* 基因的烟草 15 株。

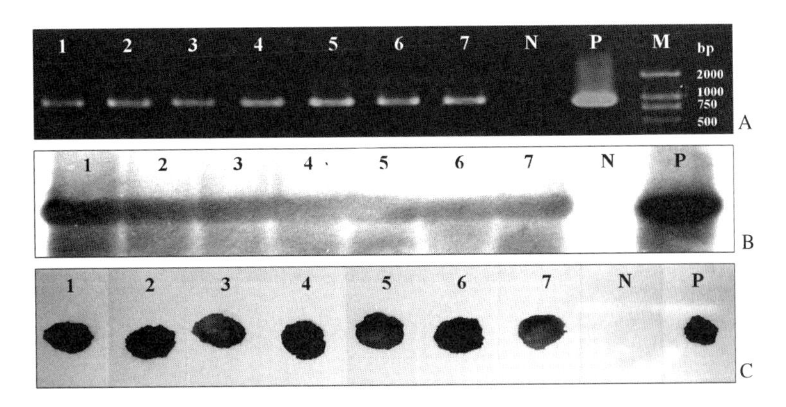

图 3　转 DNM 结构 35S-*AGM3*-E9 烟草的 PCR 和 Southern 检测

A：Transgenic tobacco plants verified by PCR；B：PCR-Southern analysis of the transgenic tobacco plants；C：Dot-Southern blot analysis of the transgenic tobacco plants；1-7：Transgenic plants；N：Non-transformed plant as negative control；P：Positive control (plasmids)；M：DL2000 Marker.

2.2　*AGM3* 基因在转基因烟草中的表达

为检测 *AGM3* 在转基因烟草中的表达，进行了实时荧光定量分析。提取转基因植株总 RNA，反转录为 cDNA 后，采用 SYBR Green 法，以 *ACTIN* 为内参进行相对表达量的检测。

结果如图 4 所示，*AGM3* 在转基因组株系 AGM3-01、AGM3-05 和 AGM3-10 中的相对表达量明显高于其他株系，在 AGM3-07、AGM3-13、AGM3-15 中的相对表达量低于其他株系。AGM3-05 的相对表达量最高，达内参的 37.53 倍，而 AGM3-13 的相对表达量最低，只有内参的 3.02 倍。方差分析显示这些转基因株系间 *AGM3* 相对表达量差异呈极显著($P<0.01$)。

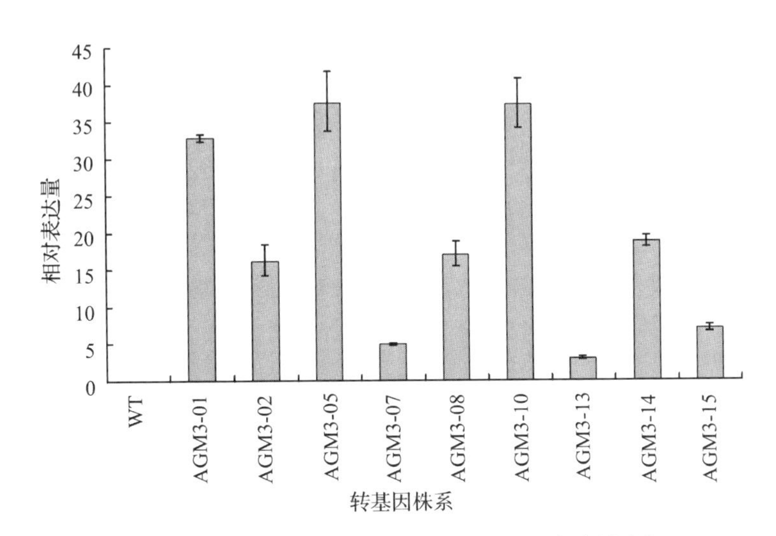

图 4　转 35S-*AGM3*-E9 烟草外源基因相对表达量分析

2.3　转基因烟草的生殖与生长特性

2.3.1　*AGM3* 对烟草开花时间的影响

为考察 *AGM3* 的对烟草花发育的影响，将转基因烟草和野生型对照的开花时间进行比较。野生型植株均产生花器官(图 5A)，开花时间为移栽后的 124.7±1.5d，而转基因株系中只有 53.3%最终开花(图 5B)，其余 46.7%的株系在其长周期内均未开花(图 5C)。按开花受抑制的程度不同，可将转基因株系分为以下三种情况：第一，与野生型开花时间一致。3 个(20%)株系在移栽后的 124～128d 开花(图 5D)，平均开花时间为 126.3±2.1d，与野生型相比差异不显著(P=0.326)；第二，开花时间明显晚于野生型。5 个(33.3%)转基因株系在 142～165d 之间开花，平均时间为 154.4±9.6d(图 5E)，与野生型相比差异极显著(P=0.002)，平均推迟 29.7 d；第三，未产生花器官，这样的株系有 7 株(46.7%)(图 5F)。开花时间的改变说明 *AGM3* 对烟草开花有明显的抑制作用，可导致部分转基因植株开花延迟，甚至使一些转基因植株不产生花器官，从而失去育性。

图 5　转 35S-*AGM3*-E9 烟草的表型观察

A: Terminal inflorescence of wild type tobacco; B: Terminal inflorescence of transgenic tobacco; C: Top of the main stem of non-flowering transgenic tobacco; D: Transgenic line of *AGM3*-13 flowered at almost the same time with wild type control; E: Flowering time of transgenic line *AGM3*-02 was delayed obviously; F: Transgenic line of *AGM3*-01 did not flower.

2.3.2　转基因植株花器官形态的变化

为考察 *AGM3* 对花器官形态的影响，对 8 个开花的转基因株系花器官形态进行了观察，发现有 3 株转基因烟草的花器官出现变异(图 6)。野生型烟草为总状聚伞花序，花萼

5 中裂，花冠漏斗状 5 浅裂，辐射对称，5 枚雄蕊冠生，花药纵裂(Mande et al., 1992；曹冠琳等, 2010)(图 6A, B)。转基因烟草花器官变异主要存在如下几种情况：第一，花萼裂数减少（图 6C)；第二，花冠出现一或两个深裂，且花瓣形状不规则。部分深裂处着生 1~3 枚多余的小型花瓣，使花瓣总数增多(图 6D, E)。第三，雄蕊数目少于(图 6E)或多于野生型，为 4~7 枚，而且有 1~2 枚花药着生于花冠裂缝处，此处花药没有花丝或花丝整体沿裂缝附着(图 6F, G)。变异的花药可开裂散粉(图 6F)，或仅有花粉囊的部分组织(图 6G)。部分花冠深裂处着生丝状组织，不能分辨其为变异的花丝或是花瓣(图 6H)。转基因烟草未观察到心皮形态的改变(图 6F, G)。

图 6 转 35S-*AGM3*-E9 烟草与野生型花器官形态的比较

A: Side view of wild type flower; B: Front view of wild-type flower; C-H: Floral variation of transgenic tobacco (Arrows indicate the positions of mutations); C: Decrease of sepals number; D: A deep cleft on corolla and morphological changes of petals; E: Two deep clefts on corolla and numerical change of stamens; F: Anthers growing on both sides of the cleft and anther dehiscence; G: Partial tissue of anther growing on one side of the cleft; H: Filamentous tissue growing on the cleft of the corolla.

2.3.3 转基因烟草营养生长的变化

对移栽后 125d 内转基因烟草的营养性生长进行观察，未发现营养器官形态的变异，但是部分转基因烟草高生长受影响，按影响程度不可分为三类：第一，6 个(33.3%)转基因株系平均高度 170.59±9.06cm，与野生型(平均高度 180.37±10.13cm)相比差异不显著(P=0.184)；第二，3 个(16.7%)株系平均高度 122.29±19.63cm，与野生型相比差异显著(P=0.010)，平均低于野生型 58.08cm；第三，6 个(33.3%)株系均高 52.11±12.57cm，与野生型相比差异极显著($P \ll 0.01$)，低于野生型 128.27cm。图 7 记录了移栽后 55~125d 内野生型、AGM3-01、AGM3-02 和 AGM3-13 高生长动态变化。从曲线图中可以看出，与野生型高度类似的株系 AGM3-02 和 AGM3-13 在 3 月 11 日至 3 月 25 日开始迅速增长，在 4

月 8 日至 4 月 23 日开始长势渐缓，与野生型的高度变化类似，说明 AGM3-02 和 AGM3-13 高生长未受明显影响。AGM3-01 一直处于缓慢生长的状态，说明 AGM3-01 在移栽后的高生长始终受到明显抑制。

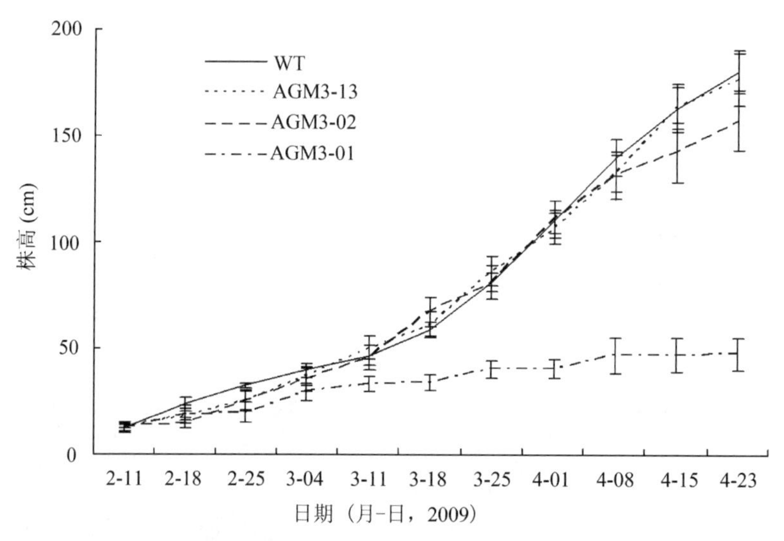

图 7　转 35S-*AGM3*-E9 烟草高生长动态变化

WT: Wild type tobacco as control; AGM3-01, AGM3-02, AGM3-13: Transgenic lines.

3　讨论

不同物种中 *AG* 同源基因的作用具有保守型。郭余龙等将陆地棉(*Gossypium hirsutum*)的 *AG* 同源基因 *GhMADS3* 在烟草中异位表达产生花器官变异(郭余龙等，2007)。Tani 等通过对桃(*Prunus persica*) *AGAMOUS*-like 以及其他几个 MADs-box 基因的研究推测这些基因在多年生木本植物和模式植物中的作用一致(Tani et al.，2009)。多种植物中 *AG* 同源基因异位或反义表达产生的变异与模式植物情况非常类似(Benedito et al.，2004；Kitahara et al.，2004；Kyozuka et al.，2002；Wang et al.，2006；Fan et al.，2007)，这些成果为 *AGM3* 转化异源植物烟草提供依据。

除控制花器官的发育外，*AG* 基因也调控花分生组织发育，表达反义 *AG* RNA 的拟南芥，其花分生组织活动的终止性受到影响(Mizukami and Ma，1995)，*AG* 基因的这种作用为本实验中转基因烟草的变异提供依据。很多研究发现过量表达 *AG* 同源基因的可导致早花(Fan et al.，2007；Lemmetyinen et al.，2004；Yu et al.，2002；Yoo，2006)，但是鲜有报道指出单独抑制 *AG* 基因可以推迟或完全抑制开花(曹冠琳等，2010)。与传统的基因删除手段不同，本研究所采用的 DNM 的方法具有迅速、稳定、适用范围广等优点，研究结果证实了 *AGM3* 转基因烟草同野生型相比，出现了不开花、开花推迟或花形态变异的现象，而转空载体 pART27 的植株的分子检测和表型观察结果均与野生型一致，可以排除载体骨架的影响。这些结果都说明 *AGM3* 基因在抑制烟草花发育上起到了明显的效果。

Bowman 等发现表达反义 *AG* 基因使雄蕊着生花瓣表皮细胞，或使心皮外表皮有萼片特征的长细胞(Bowman et al. , 1991b)。Kempin 等也发现反义表达 *NAG*1 的烟草雄蕊向花瓣转变，而心皮无明显变异，Kempin 认为这是由于反义基因的表达量不足以抑制 *NAG*1 的活性(Kempin et al. , 1993)。本研究中转基因烟草的心皮亦无形态改变，而且雄蕊仅形态变异，并未完全转变为花瓣，可能是因为 NAG1 的作用没有被完全抑制，依然具有控制内两轮花器官发育的功能。Fan 等将风信子(*Hyacinthus orientalis*)的 *AGL*6 在拟南芥中异位表达导致了雄蕊数目减少(Fan et al. , 2007)，Lemmetyinen 等发现垂枝桦(*Betula pendula*)中正反义表达 *AG* 同源基因均导致雄蕊缺失(Lemmetyinen et al. , 2004)，本研究也观察到了雄蕊数目的减少，其中的机理还有待进一步考察。

本研究还观察到萼片数目和花瓣形态的变异，这在 Kempin 的研究中没有出现，可能是因为 DNM 的方法作用于蛋白，比基因特异性沉默的抑制范围更广，甚至可能影响一个基因家族(Lagna and Hemmati, 1998)。一些基因与 *AG* 基因有类似的结构，例如研究发现 D 类基因和 C 类基因的结构类似，都属于 *AG* 类基因(de Folter et al. , 2005)，在烟草花器官中也克隆到其他 *AG* 类基因(谢灿等，1999)，这些类似基因的活性可能受到 DNM 影响从而导致相应的变异。

此外 *AG* 与 *EMBRYONIC*、*APETALA*1 处于 1 个负调模式中，*LEAFY* 基因正调节 *AG* 的表达，*APETALA*2 和 *AG* 在花器官中互相拮抗，*TERPROOF* 和 *SEEDSTICK* 可能是 *AG* 的下游基因(高志红等，2008)，*LEUNIG* 在 1、2 轮花器官中负调节 *AG* 的表达(Liu Zhongchi and Meyerowitz, 1995)。WUSCHEL 是维持植物干细胞数量稳定的信号分子，AG 蛋白通过抑制 *WUSCHEL* 基因从而促进花的形成，而 bZIP 转录因子 PERIANTHIA 正是通过直接调控 *AG* 的表达使花干细胞作用终止(Das P et al. , 2009)。鉴于 *AG* 在调控网络中的重要作用，揭示 *AGM*3 对这些相关基因的影响将是进一步研究的重点。

本研究对利用基因工程获得不育系具有重要意义，在育种中解决转基因安全隐患，减少生殖生长对营养组织的消耗，抑制杨柳科植物飞絮、花粉污染等方面都具有一定的应用价值。但是应该注意到部分转基因烟草出现了高生长受抑制的现象，这对于某些生产需求来说是不利的。一般情况下 *AG* 同源基因基因只在花中特异表达，而本研究中 *AGM*3 在转基因烟草中组成型表达，很可能对根、茎、叶等营养器官造成不利影响。已有研究者将 *AG* 增强子与 mini35S 启动子融合构成花组织特异性启动子，它驱动的 *Diphtheria toxin A* 和 *Barnase* 使转基因拟南芥的雄蕊和心皮消失，并且没有影响营养性生长(Liu Zongrang, and Liu Zhongchi, 2008)。因而将花器官特异性启动子和 DNM 相结合，才有可能最有效地避免营养性状的变异。此外，可将转基因的手段和传统育种方法相结合，对转基因株系进行开花和生长性状的观察和筛选，更有效地选育出符合不同生产需求的株系。

Over-expression of *AGM*3, a Foreign Mutant Gene of *AGAMOUS*, Represses Flowering and Alters Floral Organ Development in Transgenic Tobacco

Abstract In order to investigate floral inhibition through over-expression of *AGM*3 in heterologous plants, a dominant negative mutation construct gene, 35S-*AGM*3-E9 was transformed into *Nicotiana* *tabacum* via the *Agrobacterium*-mediated method. The results of PCR and Southern analysis showed that the gene of *AGM3* was integrated into tobacco genome. Furthermore, the data of real-time qRT-PCR analysis showed that *AGM3* was expressed in all the transgenic lines and the transcripts of *AGM3* were significantly different among these transgenic lines. The results of the investigation indicated that 46.7% of the transgenic lines did not flower during their life cycles, and flowering time of the 33.3% lines was delayed 29.7 days averagely, while the rest 20% of transgenic lines flowered at the same time with the wild type control. In addition, over-expression of *AGM3* in transgenic tobacco plants caused decreased number of sepals, deep clefts on corolla, morphological changes of petals and stamens, and numerical change of stamens. The facts suggested that over-expression of *AGM3* effectively suppress floral development, which laid a foundation for genetically engineered sterility.

Keywords *AGM3*, dominant negative mutation, represses flowering, floral organ development, tobacco

第四部分

杨树抗逆生理生化与分子生物学

The Changes of G6PDHase, ATPase and Protein During Low Temperature Induced Freezing Tolerance of *Populus suaveollens* *

Abstract The effect of cold acclimation on the freezing tolerance of *P. suaveollens* seedlings and the changes of G6PDHase, ATPase, and protein in branches of *P. suaveollens* seedlings during cold acclimation were studied. In addition, the seedlings are pretreated with protein synthesis inhibitor (cycloheximide) before cold acclimation in order to examine whether cold acclimation-induced proteins play a role in the development of freezing tolerance. The results show that freezing tolerance of *P. suaveollens* seedlings could be obviously induced by cold acclimation, this cold acclimation will require experiencing two stages in order to acquire higher freezing tolerance. Although the first stage of cold acclimation, involving a temperature of −10℃, has few effects on the increase in freezing tolerance, it might provide the basis for acclimation at 20℃ of the second stage and the acquisition of freezing tolerance. Cold acclimation distinctly increased not only the activities of G6PDHase and ATPase, and the content of protein in branches of seedlings but also the freezing tolerance of seedlings. After 2 days of deacclimation, the freezing tolerance of seedlings decrease to the level of nonacclimtion, but the activities of G6PDHase and ATPase, and the content of protein in branches of seedlings still maintained a little higher than those in b ranches of nonacclimated seedlings. Protein synthesis and the acquisition of freezing tolerance are inhibited by cycloheximide. Further analysis found that low temperature-induced changes in the activities of G6PDHase and ATPase, and the content of protein directly correlated with the development of freezing tolerance.

Keywords *Populus suaveollens*, cold acclimation, freezing tolerance, survival rates, protein

1 Introduction

Freezing temperature constitutes one of the most important environmental constrains limiting the productivity and distribution of plants. Many studies have shown that higher plants exhibit an increase in freezing tolerance when exposed to low temperature. This process, termed cold acclimation, involves a series of physiological and metabolic changes, including alterations in the concentration of carbohydrates, soluble sugars, proteins and free amino acids, changes in isozyme activities as well as in membrane composition and it has been suggested that these changes occuring during cold acclimation have more or less relation to the increase in freezing tolerance (Guy,

* 本文原载《中国林业研究》(*Forestry Studies in China*), 2000, 4(2): 17-26, 与林善枝合作发表。

1990; Hughes et al. , 1990; Thomashow, 1990) . Until recently, the mechanism of cold acclimation for increasing freezing tolerance has not been investigated clearly. Although various biochemical responses of plants to low temperature have been widely documented and reviewed, little is known about the relation among the changes of energy, the contents of protein and freezing tolerance, especially in woody plants.

In order to further investigate the molecular mechanism of cold acclimation and freezing tolerance, the changes of the activities of G6PDHase and ATPase, the content of protein, and the freezing tolerance of *Populus suaveollens* seedlings during cold acclimation at −10℃ and −20℃ are examined, and the possible relations between the change of protein content and the increase of freezing tolerance at energymetabolic angle are studied. These may be useful in making scientific researches on the freezing tolerance of woody plants.

2 Metarials and methods

2.1 Plant material and experimental conditions

Poplar (*P. suaveollens*) plants were established from dormant branch cuttings collected from Heilongjiang Province in China on March 2000. The cuttings were rooted and grown in the greenhouse in the pots containing a 2∶1(v/v) mixture of soil and sand. After the seedlings were grown for 5 weeks in the greenhouse, they were placed outdoor for 3 weeks. Part of eightweek old seedlings that were maintained at 20℃ with an 8h photoperiod for 14 days was referred to as nonacclimated seedlings, the others were further divided into two groups. One group of seedlings held at −10℃ with an 8h photoperiod for 6 days were referred to as preacclimated seedlings, part of the preacclimated seedlings further transferred to −20℃ with an 8h photoperiod for 6 days were referred to as cold-acclimated seedlings. The other group of seedlings first pretreated with cycloheximide (CH) at 20℃ for 6 days were transferred to the conditions of cold acclimation and these seedlings were referred to as inhibitor-pretreated seedlings. Finally, part of cold-acclimated seedlings transferred to the conditions of nonacclimation for 2 days were referred to as deacclimated seedlings. Branches were randomly harvested from nonacclimated, preacclimated, cold-acclimated, and inhibitor-pretreated seedlings, the bark tissue from the collected branches were removed and immediately frozen on dry ice until used for protein, ATPase and G6PDHase extraction.

2.2 Evaluation of freezing tolerance and survival ratio of seedlings

Survival ratio was examined from nonacclimated, preacclimated, cold-acclimated, and inhibitor-pretreated seedlings. Some of the abovementioned seedlings were cooled gradually to −15℃, −20℃, −23℃, −25℃, −27℃, −30℃, −32℃, and −35℃, respectively, in a programmed freezer . The chamber was cooled at a constant rate of 0. 5℃ · min^{-1} and then held for 12h at each temperature. These seedlings were transferred to 20℃ and then thawed for 2 days. The standard of seedlings survival rates depends upon whether these seedlings regrow normally and the degree of freezing injury of bark tissue and buds at −35℃ for 12h. Freezing tolerance was determined on the basis of the temperature at which 100% of seedlings survived and the survival rates of seedlings at −35℃ for 12h.

2.3 Protein extraction and measurement

Protein from bark of branches was obtained in borate buffer (50mM sodium brote, 50mM ascorbic acid, 1mM phenylmethylsulfonyl fluoride, 1% - mercaptoethanol, 0.5% sodium dodecyl sulfate, pH9.0) at 4℃ as described earlier (Coleman et al., 1991, GU Ruisheng et al., 1999). The homogeneous liquid was extracted at 4℃ for 2h, and then centrifuged at 4℃ with 12 000g for 30min. The resulting supernatant was collected. 2.5 volumes of acetone pretreated in − 20℃ was added to the supernatant and held at −20℃ for 1h. The precipitate was collected by centrifugation at 12000g for 20min, and then dried at −20℃ for 20min. The protein pellet was resuspended in Tris-HCl buffer[12.5% (v/v) 0.5M Tris-HCl, pH6.8, 5%(v/v) β-mercaptoethanol, 10% (v/v) glycerol, 72.5% (v/v) deionized water] and the supernatant was collected at 4℃ with 12000g for 10min. The resulting supernatant was used as the souble protein and the protein concentrations were assayed by using the Bradford (1979) dye-binding technique with bovine serum albumin (BSA) as a standard.

2.4 G6PDHase extraction and analysis

G6PDHase was extracted from bark of branches and analyzed using the technique described by Li Jinshu (1985).

2.5 ATPase extraction and analysis

ATPase was extracted from bark of branches and analyzed using the method described by Aoyan Guangcha (1985).

3 Results

3.1 Effects of low temperature on the survival rates and freezing tolerance of seedlings

From Table 1, the preacclimated and cold acclimated seedlings had an increase in the survival rate and freezing tolerance compared to nonacclimated seedlings and then decreased to the level of nonacclimated seedlings after 2 days of deacclimation. The survival rates of seedlings at −35℃ for 12h increased from 0% in nonacclimated seedlings to 30% in preacclimated seedlings, and to 100% in cold-acclimated seedlings. The temperature at which 100% of seedlings survived are −15℃ in nonacclimated and deacclimated seedlings, − 20℃ in preacclimated seedlings and −35℃ in cold-acclimated seedlings.

In order to demonstrate in detail the effect of low temperature on the development of freezing tolerance of seedlings, we determined the kinetic changes of freezing tolerance at −20℃ for 1~6 days and at 20℃ for 2 days (Figure 1) . At 1 day after transferring seedlings from −10℃ to −20℃, freezing tolerance of seedlings started to increase distinctly from −20℃ to −23℃ and continued to gradually increase during a period of 2 to 5 days to reach a maximum level (−35℃) on the 6th day. After 2 days of deacclimation, the acquired freezing tolerance of seedlings during acclimation of −20℃ decreased rapidly to the level of nonacclimated seedlings (−15℃) .

Table 1 The changes of the survival rates and freezing tolerance of seedlings under different conditions

Treatment	The survival rates of seedings at −35℃ for 12h(%)	The temperature at which 100(%) of seedlings survived(℃)
Nonacclimated	0	−15
Preacclimated	30	−20
Cold-acclmated	100	−35
Deacclimated	0	−15

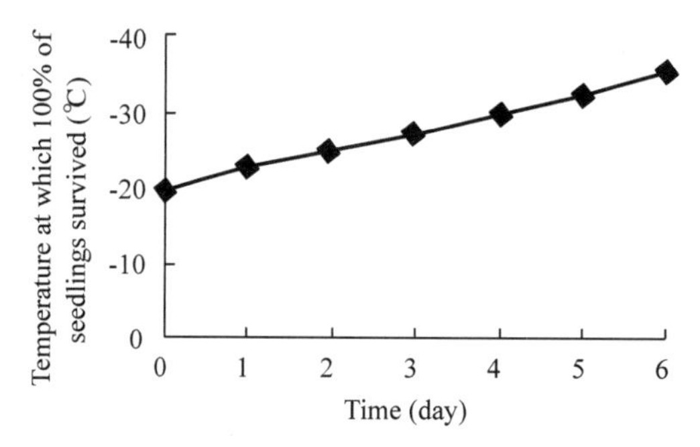

Fig. 1 The effect of cold acclimation of −35℃ on the freezing tolerance of seedlings

(Preacclimated conditions were used as zero time point)

3. 2 Effects of low temperature on the protein content in branches of seedlings

Table 2 shows that the protein content increased from 0. 451mg · g^{-1} FW in branches of nonacclimated seedlings to 0. 554mg · g^{-1} FW in branches of preacclimated seedlings, and to 0. 937mg · g^{-1} FW in branches of cold-acclimated seedlings. That is to say, the protein content has increased by 30% in branches of preacclimated seedlings and by 120% in branches of cold-acclimated seedlings as compared with that of branches of nonacclimated seedlings. After 2 days of deacclimation at 20℃, the protein content decreased from 0. 937mg · g^{-1} FW in branches of cold-acclimated seedlings to 0. 466mg · g^{-1} FW in branches of deacclimated seedlings, but the level of protein was still higher than that of nonacclimated seedlings.

Table 2 The changes of the content of protein and the activities of G6PDHase and ATPase in branches under different conditions

Treatment	Protein content (mg · g^{-1}) FW	G6PDHase activity (n(enzyme unit) · g^{-1} FW)	ATPase activity (Umpi · g^{-1} apoenzyme · h^{-1})
Nonacclimated	0. 426	47. 2	33. 7
Preacclimated	0. 553	67. 4	44. 8
Cold-acclmated	0. 937	134. 9	78. 1
Deacclimated	0. 466	58. 3	39. 6

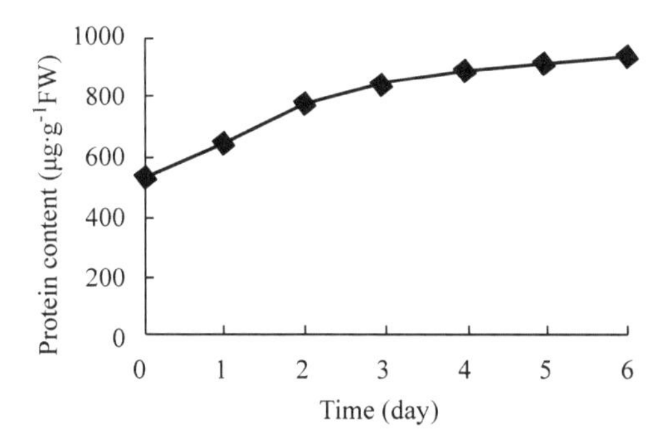

Fig. 2 The effect of cold acclimation of −20℃ on protein content of seedlings

(Preacclimated conditions were used as zero time point)

To examine the changes of protein in branches during cold acclimation, we determined the kinetics of protein content in branches during cold-acclimation at −20℃ (Figure 2) . The protein content in branches increased from 0. 554mg · g^{-1} FW to 0. 653mg · g^{-1}FW as the 1 day holding temperature decreased from −10℃ to −20℃ , and further increased gradually during a period of 2 to 5 days to reach the highest amount (0. 937mg · g^{-1} FW) on the 6th day.

3. 3 Effects of low temperature on the activities of G6PDHase and ATPase in branches of seedlings

The branches of seedlings exhibited a significant increase in the activities of G6PDHase and ATPase after 6 days of preacclimation or cold acclimation compared to nonacclimated ones, respectively, preacclimated branches showed a relative lesser. 2 days of deacclimation at 20℃ resulted in a rapid decrease in the activities of G6PDHase and ATPase. Comparatively speaking, the decrease in G6PDHase activity was more marked, but they still maintained a higher content than those of nonacclimated branches (Table 2) .

The activities of G6PDHase and ATPase in branches were daily on the increase during cold acclimation at −20℃ and reached the maximum levels of about 134. 9n · g^{-1}FW and 78. 1Umpi · g^{-1} apoenzyme · h^{-1} on the 6th day, respectively. Comparing the G6PDHase with ATPase, there was obvious enhancement in the G6PDHase activity during cold acclimation. For example, the G6PDHase activity had been enhanced by 110% in branches of cold-acclimated seedlings, whereas, the ATPase activity in branches of the cold-acclimated seedlings had been enhanced only by 70%(Table 2 and Figure 2) .

3.4 Effect of cycloheximide (CH) pretreatment on the protein content and freezing tolerance

P. suaveollens seedlings pretreated with 20mg · kg^{-1} cyclohex imide (CH) during nonacclimation for 6 days were transferred to −10℃ for 6 days and further transferred to −20℃ for 6 days, the changes of the protein content and freezing tolerance of seedlings were showed in Table 3 . After CH pretreatment, the protein content and freezing tolerance decreased to the levels of nonacclimation, in other words, the temperature at which 100% seedlings survived decreased from −35℃ to −10℃ , at the same time, the content of protein in branches had decreased about 55%. The results indicated that the protein synthesis and the acquisition of freezing tolerance could be inhibited by CH.

Table 3　The effect of cyclocheximide (CH) pretreatment on the protein content and freezing tolerance

Treatment	Protein content (mg · g^{-1}FW)	The temperature at which 100% of seedlings survived(℃)
Nonacclimated	0.426	−15
Cold-acclmated	0.937	−35
CH-pretreated	0.426	−15

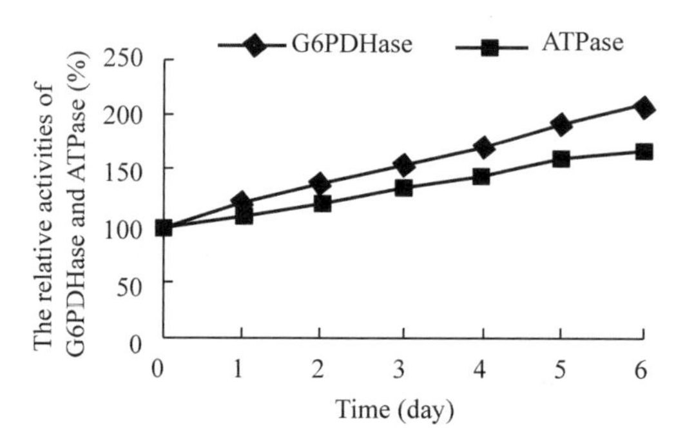

Figure 3　The effect of cold acclimation of −20℃ on the protein content of seedlings

(Preacclimated content and freezing tolerance)

4 Dicussion

4.1 Relationship between cold-acclimation and freezing tolerance

Many plants have an ability to cold-acclimate and to develop freezing tolerance, as exposure to low temperature is a major trigger for the induction of freezing tolerance (Levitt, 1980). The acclimation response is a complex phenomenon and freezing tolerance in plants is considered to be a quantitative genetic trait, mainly influenced by the genotype of the plant and the acclimating temperature used (Hughes et al., 1990; Dunn et al., 1994). Although it has been reported that cold-acclimation induced an increase in freezing tolerance in many plant species (Levitt, 1980; Guy, 1990, Thomashow, 1990), not all plants are able to acclimate for freezing tolerance and not all temperatures are suitable for the induction of freezing tolerance. For example, a low temperature of 6℃ to 2℃ that can cold-acclimate genetically competent plants (e. g., cereals) will damage a chill-sensitive species (e. g., maize) (Hughes et al., 1996). Studies by Shen Zhenyan et al. (1983) had shown that in order to acquire higher freezing tolerance, the tomoto seedlings should be first transferred to a cooler with a temperature of 5~10℃ for 5 days and then further transferred to −2℃ for 5 days.

Our results clearly demonstrated that exposure of the nonacclimated seedlings to low temperature of −10℃ or exposure of the preacclimated seedlings to −20℃ resulted in an increase in the survival rates and freezing tolerance of seedlings, but the seedlings of acclimation at −10℃ showed a relative lower (Table 1). Moreover, freezing tolerance of seedlings had been enhanced gradually during cold-acclimation at −20℃ and reached the maximum level on the 6th day (Figure 1). In addition, our previous studies found that the seedlings started to exhibit damage when they were directly exposed to −20℃ for 4h; if the seedlings that have been preacclimated at −10℃ for 6 days were transferred to −20℃ for 6 days, they have acquired the highest freezing tolerance (−35℃). These facts indicated that the induction and the developmental process of freezing tolerance in *P. suaveollens* seedlings require a certain temperature and duration. The acquisition of the highest freezing tolerance induced by adaptive cold acclimation will require two stage experiences. Although the first stage of cold acclimation, involving a temperature of −10℃ with 6 days, had few effect on the increase of freezing tolerance, it might provide the basis for acclimation at −20℃ of the second stage and the acquisition of freezing tolerance. A shift to lower temperature (from −10℃ to −20℃) had much effect on the acquisition of freezing tolerance. Therefore, to choose positive temperature which can effectively acclimate but will not damage plant is significant for the development of freezing tolerance.

4.2 Relationship between low temperature-induced changes in the content of protein and freezing tolerance

It has been reported that an increase of protein content in many plants induced by cold acclimation was directly correlated with the freezing tolerance, the increase of protein content during cold acclimation was accompanied by the enhancement of freezing tolerance (Guy, 1987; 1990, Kazuoka et al., 1992; Graham et al., 1982; Pan Jie et al., 1994; Lin Shanzhi, 1997). Moreover, the increase of protein content during cold acclimation may play a specific role in conferring freezing tolerance in plants (Guy et al., 1989, 1990; Perras et al., 1989; Arora et al., 1994). But fewer investigator pointed out that the amount of soluble protein has not essentially increased during cold acclimation; the increase of freezing tolerance may not be due to the increase in soluble protein (Levitt, 1980; Uemura et al., 1996). I t was known from our experiments that the amount of protein in branches gradually increased during clod acclimation at −20℃ and decreased after 2 days of deacclimation at 20℃, but still maintained a relative higher compared to nonacclimated branches (Figure 2). The branches of cold-acclimatied seedlings contained the highest amount of protein (0.937mg · g^{-1} FW) and was the most freezing tolerance(−35℃); the branches of nonacclimated seedlings was the least freezing tolerance (−15℃) and contained the lower amount of protein (0.462mg · g^{-1}FW) (Table 1, Table 2). Further analysis found that an increase in protein content during clod acclimation at−20℃ was obviously paralleled by the acquisition of freezing tolerance (Figure 1, Figure 2). For example, the rise in protein amount to a maximum on day 6 concided with the highest freezing tolerance on the 6th day. These results showed that the protein was accumulated in response to low temperature. An increase in protein content during cold acclimation may associate with the establishment of freezing tolerance, and may play a specific role in the cold-acclimation process and in the regrowth of the cold-acclimated seedlings after deacclimation (Neven et al. 1993).

To examine whether the increase in protein during cold acclimation at −20℃ was correlated with the acquisition of freezing tolerance, we determined the changes of protein content and freezing tolerance in the cycloheximide (CH)-pretreated seedlings during clod acclimation at −20℃ (Table 3). After CH-pretreatment, the temperature at which 100% seedlings survived decreased from −35℃ to −10℃. At the same time, the content of protein in branches decreased by 55%. The fact that the decrease of freezing tolerance was accompanied by the decrease of protein content indicates that the decrease of freezing tolerance may due to the inhibition of protein synthesis resulting in a decrease of protein content. An increase in the content of cold acclimation-induced protein was directly correlated with the acquisition of freezing tolerance.

It is, therefore, of interest to considering how low-temperature-induced proteins might contribute to the establishment of freezing tolerance in plants. On the basis of recent studies, the characteristics of the role of proteins can be summarized as follows: the direct affection on extracellular ice formation; the indirect alteration in the percentage of water frozen at a given temperature by restricting the rate of growth of extracellular ice crystals; the depression in the freezing point of the body fluid and the stabilization in cell membranes against either rupture or loss of its semipermeablity(Duman et al., 1991; Kazuoka et al., 1992; Boot he et al., 1995; Hughes et al., 1996).

4.3 Role of NADpH2 and ATP in freezing tolerance

ATPase in the cell membranes is thought as a functional protein and plays a critical role in solute transport, energy metabolism and so on (Mito et al. 1996). Early studies of changes in ATPase with cold acclimation reported increase in the activity of ATPase that paralleled the acquisition of freezing tolerance and a decline in the activity of ATPase as freezing tolerance was lost

(Jian Lingchen et al. , 1983, Zhao et al. , 1993; Salzman et al . , 1993; Lin Shanzhi, 1997) . Studies by Dai Jinpin et al (1991) has shown that the plasma membrane ATPase in cucumber seedlings has an acquisition of freezing tolerance during cold aaclimation at 5℃ . Recent reports have implicated cold liability of endomembrane ATPase as a possible cause of cold-induced injury in chilling sensitive plants, that is to say, ATPase in cell membranes may be the primary sites of freezing injury(Steponkus, 1984; Yoshida et al. , 1989; Jian Lingchen, 1992; Boothe et al. , 1995; Lin Shanzhi, 1997) . We observed that an increase in ATPase activity induced by cold acclimation was paralleled not only by the increase in protein content but also by the acquisition of freezing tolerance (Figure 1, Figure 2 and Figure 3) . It appears that ATPase may play an important role in the induction of freezing tolerance, presumable at least in part by energizing the synthesis of protein associated with freezing tolerance.

Sagisaka (1985) first reported that the G6PDHase activity in poplar twigs exhibited a marked increase in fall and winter followed by a gradual decrease in spring, and suggested that an increase in G6PDHase activity was required for the tolerant to freezing. As reported previously (Jian Lingchen, 1990; Liu Hongx ian et al. , 1991) and confirmed in this paper (Figure 3), the activity of G6PDHase in branches gradually increased during cold acclimation at −20℃ and decreased after 2 days of deacclimation at 20℃ . Moreover, this increasing trend was paralleled by both the increase of the protein content and the acquisition of freezing tolerance (Figure 1, Figure 2 and Figure 3) . The analysis of the changes of the G6PDHase activity and the ATPase activity found that the increasing trend of the G6PDHase activity and the ATPase activity was very similar (Figure 3) . These results indicated that increase in the G6PDHase activity and the ATPase activity induced by cold acclimationan was directly correlated with both the increase of the protein content and the acquisition of freezing tolerance. It is because that G6PDHase is a key regulatory enzyme of pentose phosphate cycle, cold acclimation-induced increase in G6PDHase activity accomplanied by the increase of ATPase activity will accelerate pentose phosphate cycle, by which many intermediates for use in protein synthesis and NADpH2 are produced, NADpH2 can be used as a specific electron donor for many biosynthesis reaction and also be oxidated to ATP by bio-oxidation-systems, ATP can be hydrolyzed to ADP or AMP by ATPase and then generate many energy for biosynthesis. Therefore, an alteration in metabolic pathway as the result of an increase in G6PDHase activity induced by cold acclimation was closely associated with the energy dissipation as the result of an increase in ATPase activity. The higher NADPH2 levels and increasing ATP supply are essential for the synthesis of protein and the induction of freezing tolerance.

5 Conclusion

This study indicates that freezing tolerance of *P. suaveollens* seedlings can be effectively induced by cold acclimation that will require experiencing two stages in order to acquire the highest freezing tolerance. Although the first stage of cold acclimation, involving a temperature of −10℃ , has few effects on the increase of freezing tolerance, it may provide the basis for acclimation at−20℃ of the second stage and the acquisition of freezing tolerance. The fact that the increase in activities of G6PDHase and ATPase as a result of adaptive cold acclimation parallel not only to the increase of protein content but also to the enhancement of freezing tolerance shows that there is a close correlation among them. An marked increase of G6PDHase and ATPase induced by adaptive cold acclimation may be a form of energy generation, and may be a adaptive reaction to low temperature environment. Moreover, they may provide the energy for the synthesis of protein associated with the development of freezing tolerance, and for the activation in enzymes correlated with the membrane stability. So this may be just the cause of cold acclimation-induced increase in freezing tolerance of *P. suaveollens* seedlings.

To further elucidate the mechanism of the acquisition of freezing tolerance induced by cold acclimation in *P. suaveollens* seedlings, it is necessary to isolate cold acclimation-induced specific protein that may be involved in the development of freezing tolerance, to determine the nucleotide sequence of this protein and to identify the function of this protein from its deduced amino acid sequences.

转 Bt 基因杨树对美国白蛾幼虫中肠解毒酶及乙酰胆碱酯酶的影响*

摘　要　以转 Bt 基因欧洲黑杨叶片饲喂 4~5 龄美国白蛾幼虫，对其中肠几种重要解毒酶及乙酰胆碱酯酶的活性进行了测定，结果表明，中肠多功能氧化酶受到强烈抑制，饲喂 4h 后，多功能氧化酶的活性比对照降低 50.89%，48h 后，该酶活性仅为对照的 1/6。酯酶及羧酸酯酶的活性也受到明显抑制，饲喂 48h 后，上述两种酶的活性分别比对照降低 73.58%和 56.55%。谷胱甘肽 S-转移酶变化比较复杂，饲喂 24h 后其活性出现峰值；48h 后与对照持平。中肠乙酰胆碱酯酶活性无明显变化。转基因叶片对幼虫中肠解毒酶活性的抑制可能是其毒杀害虫的作用机制之一。

关键词　转 Bt 基因杨树，美国白蛾，解毒酶

通过基因工程手段获得转基因抗虫植物是森林害虫防治的一条新途径。1991 年，Mc Cown 首次报道将 Bt(*Bacillus thuringiensis*)毒蛋白基因导入银白杨×大齿杨获得转基因抗虫植株，此后国内研究者相继将 Bt 基因转入欧洲黑杨、美洲黑杨及欧美杨等获得一批转 Bt 基因抗虫树种(田颖川等，1993；陈颖等，1995；王学聘等，1997)。在抗虫性方面也有相关报道(Karl et al.，1995)，转 *CryIAc* 基因的欧洲黑杨对舞毒蛾和杨尺蠖的毒杀死亡率可达 80%~90%(田颖川等，1993)，而且在大田条件下能显著降低虫口密度和林地土壤中的虫口密度(胡建军等，1999)。

农药毒理学研究表明，害虫摄入杀虫剂以后，其体内酶。特别是与解毒相关的酶的活性会发生相应变化。昆虫体内解毒酶一般指酯酶、羧酸酯酶(CaE)、多功能氧化酶(MFO)和谷胱甘肽 S-转移酶(GST)等。乙酰胆碱酯酶(AChE)通常被认为是杀虫剂作用的靶标。上述几种酶的活性直接影响外源有毒物质在昆虫体内的作用。害虫摄入转基因抗虫植物后，其中肠解毒酶和乙酰胆碱酯酶的变化关系到转基因树种的使用寿命、害虫的抗性发展以及害虫综合防治措施的制定，但有关这方面的研究至今尚未见报道。

1　材料与方法

1.1　材料

(1)植物材料：转 Bt 基因欧洲黑杨(*populus nigra* L.)叶片由中国林业科学研究院韩一凡研究组提供。

---

* 本文原载《东北林业大学学报》，2001，29(3)：28-30，与丁双阳、李怀业和李学锋合作发表。

(2)供试昆虫：供试美国白蛾(*Hyphantria cunea* Drury)幼虫采自辽宁省宽甸县县城街道行道树，室内饲养1d以上，试验时选取发育正常的4~5龄幼虫、饥饿4h后饲喂转基因叶片，对照饲喂当地杨树(*populus canadensis* Moench)叶片。饲喂后定时解剖测定幼虫中肠酶活性。

1.2　测定方法

(1)羧酸酯酶和酯酶的活性测定：参照Asperen(1962)及陈巧云等(1978)的方法，取不同饲喂之间的美国白蛾4~5龄幼虫10头，用0.04mol/L pH 7.0的磷酸缓冲液(PBS)将其中肠冰浴匀浆，于4℃下10000r/min离心10min，取上清液作为酶源。取适量稀释后的酶液1.0mL，加入0.04mol/L pH7.0的PBS 0.5mL，再加入0.003mol/L的α-乙酸萘酯底物(含10^{-6}mol/L毒扁豆碱、测酯酶时不含毒扁豆碱)5.0mL，混匀后于37℃恒温水浴中反应30min，然后加入1.0mL显色液(1%固蓝盐B溶液∶5%十二烷基磺酸钠溶液=2∶5)，摇匀后静置30min，于600nm处比色，以α-萘酚作标准曲线，以加酶量为横坐标，吸光度(OD)值为纵坐标做标准曲线，计算每毫升酶液生成的α-萘酚，测定酶源蛋白质含量(g/L)，计算出羧酸酯酶和酯酶的活力。

(2)多功能氧化酶：参照Hansan和Hodgon(1971)的方法。酶源制备，用0.2mol/L pH7.8的磷酸缓冲液冰浴匀浆，其余同前。取适量稀释后的酶液1.0mL，加入含有2.5mg NADPH和10μL 0.1M NTAN的缓冲液2mL(预保温5min)，于34℃水浴振荡30min，然后用1.0mL 1mol/L的HCl终止反应。用5mL乙醚萃取，取乙醚层，再用3mL 0.5N NaOH萃取。测定NaOH溶液层410nm的吸光值，不加NADPH的为对照。

(3)谷胱甘肽S-转移酶：参考Oppenoorth(1979)的方法。酶源制备，用0.06mol/L pH7.0的磷酸缓冲液(含2mmol/L EDTA)冰浴匀浆，其余同前。取适量稀释后的酶液3mL，加入0.2mol/L的CDNB 20μL和0.4mol/L的GSH 100μL，25℃下保温20min，于340nm下测定吸光度值。

(4)乙酰胆碱酯酶：采用Gorun等(1978)改进的Ellman方法。酶源制备，用0.1mol/L pH7.6的磷酸缓冲液冰浴匀浆，其余同前。取适量稀释后的酶液1.0mL，加入0.004mol/L的ATCh 1.0mL，25℃下保温15min，于412nm下测定吸光度值。

(5)酶源蛋白质含量：采用考马斯亮蓝G-250方法测定，本文中酶活性均以每克蛋白表示。

2　结果与分析

2.1　中肠酯酶和羧酸酯酶活性变化

选取发育正常一致的转Bt基因欧洲黑杨叶片饲喂美国白蛾4~5龄幼虫，并于饲喂后不同时间测定其中肠酯酶和羧酸酯酶活性。从测定结果(见表1)可以看出，转Bt基因杨树叶片对美国白蛾幼虫中肠酯酶和羧酸酯酶活性有明显抑制作用。饲喂时间越长，酶活力下降越多。饲喂48h后。酯酶活力比对照降低73.58%，羧酸酯酶活力下降了56.55%。

2.2 中肠多功能氧化酶活性变化

多功能氧化酶是昆虫体内重要的解毒酶之一，从转 Bt 基因杨树叶片对美国白蛾幼虫中肠多功能氧化酶活性的影响(见表1)可知，转 Bt 基因杨树叶片对美国白蛾幼虫中肠多功能氧化酶活性有强烈抑制作用。饲喂后仅 4h，多功能氧化酶的活性就比对照降低了 50.89%，48h 后仅为对照的 1/6。试验中也观察到，饲喂 48h 后幼虫的行动明显比对照迟缓，可能与解毒酶活性受到抑制有关。

表1 转 Bt 基因杨树叶片对美国白蛾幼虫中肠酯酶和羧酸酯酶活性的影响

时间(h)	酯酶活性($mmol \cdot g^{-1} \cdot min^{-1}$)	羧酸酯酶活性($mmol \cdot g^{-1} \cdot min^{-1}$)
4	3.21±0.15	1.61±0.17
12	2.53±0.24	1.64±0.11
24	1.96±0.16	1.49±0.20
36	1.50±0.19	1.09±0.03
48	0.79±0.02	0.73±0.01
CK	2.99±0.31	1.68±0.28

注：表中数据为 3 次试验统计结果(平均值±标准差)。

2.3 谷胱甘肽 S-转移酶活性变化

由转 Bt 基因杨树叶片对美国白蛾幼虫中肠谷胱甘肽 S-转移酶活性的影响(见表2)可知，在前 2 个取样点，即饲喂后的 4h 和 12h，谷胱甘肽 S-转移酶活性表现为受到抑制、其酶活性分别比对照降低 20.40% 和 42.24%，此后酶活性大幅度提高，24h 时达到对照的 1.3 倍，以后又降低，48h 时与对照基本持平。谷胱甘肽 S-转移酶这种变化的生理机制有待进一步研究。

表2 转 Bt 基因杨树叶片对美国白蛾幼虫中 3 种酶活性的影响

时间(h)	多功能氧化酶活性($mol \cdot g^{-1} \cdot min^{-1}$)	谷胱甘肽 S-转移酶活性($mmol \cdot g^{-1} \cdot min^{-1}$)	乙酰胆碱酯酶活性($mmol \cdot g^{-1} \cdot h^{-1}$)
4	5.24±0.41	0.66±0.03	1.13±0.29
12	4.79±0.07	0，45±0.03	1.01±0.23
24	3.95±0.12	0.07±0.07	1.25±0.14
36	2.65±0.09	0.48±0.01	1.14±0.34
48	1.78±0.07	0.82±0.06	1.02±0.27
CK	10.67±0.21	0.82±0.09	1.16±0.14

注：表中数据为 3 次试验统计结果(平均值±标准差)。

2.4　中肠乙酰胆碱酶活性变化

用转基因杨树叶片饲喂幼虫后，由其中肠乙酰胆碱酶活性变化(见表2)可以看出，Bt毒素对该酶无明显影响。不同取样时间乙酰胆碱酶的酶活性在数值上虽然比对照略有升高或降低，但均没有显著差异。

3　讨论

酯酶、羧酸酯酶和多功能氧化酶等是昆虫体内重要的解毒酶，对分解外源毒物、维持正常生理代谢起重要作用。Justin 等(1989)曾报道，用 Bt 处理棉铃虫(*Helicoverpa armigera* Hubner)和斜纹贪夜蛾(*Spodoptera litura* Fabricius)幼虫，发现它们对硫丹、久效磷、氰戊菊酯及氯氰菊酯的敏感性增加了，并推测可能是由于 Bt 对棉铃虫、斜纹夜蛾解毒酶系统发生作用而造成的。本研究结果显示，转 Bt 基因抗虫杨树叶片对美国白蛾幼虫中肠酯酶、羧酸醋酶和多功能氧化酶有明显抑制作用，而且随饲喂时间的延长抑制作用加强。通过抑制解毒酶活性来干扰昆虫正常的生理代谢，这可能是转 Bt 基因杨树毒杀害虫的重要机制之一。

乙酰胆碱酯酶是神经突触部位清除乙酰胆碱、维护神经正常传导的重要酶类，其活性是衡量昆虫神经生理活性的主要指标。由于许多杀虫剂以乙酰胆碱酯酶为靶标、因而该酶与昆虫的解毒作用密切相关。已有报道认为在离体条件下 Bt δ-内毒素可能作用于突触前膜，阻断神经传导(Cooksey, 1969; Singh and Gill, 1985; Cheung et al. , 1985)。从本研究的结果看，饲喂转 Bt 基因杨树叶片后，美国白蛾幼虫中肠乙酰胆碱酯酶活性与对照相比差异不显著，其原因有待进一步研究。

Effects of Bt Transgenic Poplar on Detoxification Enzyme and AChE in America White Moth Larvae

Abstract　The acticities of several detoxification enzymes and acetylcholinesterase (AChE) of America white moth fourth-fifth instar larvae feeding leaves of Bt transgenic poplar (*Populus nigra* L.) were investigated. The results showed that the acticity of mix-function oxidase in the midgut of the larvae was decreased by 50. 89% after 4 hours and 48hours later it was only 1/6 of control. The activities of esterase and carboxylesterase (CarE) were also decresed by 73. 58% and 56. 55%. The activity of AChE did not change obviously. The acticity of glutathione s-transferase (GST) increased remarkably after 24 hours feeded with transgenic poplar. Thus the detoxicification enzyme system in the midgut was inhibited and further toxic reaction of Bt protein could take place. This may be one of the important mechanisms of Bt transgenic poplar.

Keywords　transgenic poplar, america white moth, detoxification enzymes

转 *CpT* I 基因杨树对美国白蛾幼虫中肠解毒酶及乙酰胆碱酯酶的影响*

摘　要　以转 *CpT* I 基因毛白杨回交杂种((*Populus tomentosa* ×*Populus bolleana*) ×*Populus tomentosa*)叶片饲喂 4~5 龄美国白蛾(*Hyphantria cunea* Drury)幼虫，对其中肠解毒酶和乙酰胆碱酯酶活性进行了测定。结果表明，酯酶、羧酸酯酶的活力受到明显抑制，且随时间的延长抑制强度增加，饲喂48h 后，上述两种酶的活力分别比对照降低 76.42%和 73.91%；多功能氧化酶在饲喂的前 12h，表现为受到抑制，最大抑制率为 35.78%，24h 后酶活力反而高于对照；谷胱甘肽 S-转移酶和乙酰胆碱酯酶活力受到抑制，而且两者表现极为相似，均在饲喂后 4h 抑制作用最强，酶活力分别比对照降低 51.40%和 40.57% ，但在整个试验过程中，均没有表现出随时间加强的趋势。

关键词　转 *CpT* I 基因杨树，美国白蛾，酯酶，CarE，MFO，AChE，GST

通过基因工程手段获得转基因抗虫植物是森林害虫防治的一条新途径。在多种转 Bt 基因植物获得成功的同时，转其他基因植物也相继问世，*CpT* I 基因便是其中之一。豇豆胰蛋白酶抑制剂(*CpT* I)是丝氨酸蛋白酶抑制剂类 Bowman-Birk 型双头抑制剂，具有广泛的抗虫谱，对鳞翅目、鞘翅目及双翅目的害虫均有毒杀作用。Hilder 分离并克隆了 *Cp2T* I 的 cDNA，并率先获得能稳定遗传的抗虫烟草植株(Hilder et al.，1987)。国内刘春明等克隆了 *CpT* I 基因并转化成功，获得抗虫转基因烟草(刘春明等，1992)。另外，转 *CpT* I 基因棉花也表现出较好的抗虫能力(李燕娥，1998)。北京林业大学森林资源与环境学院林木育种研究室与中科院遗传所朱祯研究室合作获得了转 *CpT* I 基因毛白杨，迄今为止，有关转 *CpT* I 基因植物的抗虫性，特别是对害虫生理生化方面的影响尚未见报道。

1　材料与方法

植物材料：转 CpTI 基因毛白杨回交杂种((*Populus tomentosa* ×*P. bolleana*) ×*P. tomentosa*)系北京林业大学森林资源与环境学院林木育种研究室所有。

研究方法：供试美国白蛾(*Hyphantria cunea* Drury) 幼虫采自辽宁省丹东市宽甸县县城街道行道树，室内饲养 1d 以上，试验时选取发育正常的 4~5 龄幼虫，饥饿 4h 后饲喂转基因叶片，对照饲喂未转基因毛白杨叶片。于饲喂后 4、12、24、36、48h 定期解剖，取幼虫中肠测定酶活性。

测定方法：

* 本文原载《北京林业大学学报》，2001，29(5)：100-102，与丁双阳、李怀业、李学锋和高恒合作发表。

(1)羧酸酯酶(CarE)和酯酶的活性测定

参照 Asperen 及陈巧云等的方法(Asperen V K.，1962；陈巧云等，1978)，取不同饲喂时间的美国白蛾4~5龄幼虫10头，用0.04mol/L pH7.0的磷酸缓冲液，冰浴匀浆，于4℃下10000r/min离心10min，取上清液作为酶源。取适量稀释酶液，加入0.04mol/L pH7.0的磷酸缓冲液0.5mL，再加入0.003mol/L的α-乙酸萘酯底物(含10^{-6}mol/L毒扁豆碱，测酯酶时不含毒扁豆碱)5.0mL，混匀后于37℃恒温水浴中反应30min，然后加入1.0mL显色液(质量分数为1%的固蓝盐B溶液:质量分数为5%的十二烷基磺酸钠溶液=2:5)，摇匀后静置30min，于600nm处比色。以α-萘酚作标准曲线，以加酶量为横坐标，吸光度值为纵坐标做标准曲线，计算每毫升酶液生成的α-萘酚，测定酶源蛋白质含量，计算出羧酸酯酶(CarE)和酯酶的活力。

(2)多功能氧化酶(MFO)

参照 Hansan 和 Hodgson 的方法(Hansan et al.，1971)，酶源制备，用0.2mol/L pH7.8的磷酸缓冲液冰浴匀浆，其余同上。取适量稀释后的不同体积酶液，加入含有2.5mg NADPH 和10μL 0.1 M NTAN 的磷酸缓冲液2mL(预保温5min)，于34℃水浴振荡30min，然后用1.0mL 1mol/L的HCl终止反应，用5mL乙醚萃取，取乙醚层，再用3mL 0.5N NaOH萃取，测定NaOH溶液层410nm的吸光度值，不加NADPH的为对照。

(3)谷胱甘肽S-转移酶(GST)

参考 Oppenoorth 的方法(Oppenoorth，1979)，酶源制备，用0.06mol/L pH7.0的磷酸缓冲液(含2mmol/L EDTA)冰浴匀浆，其余同上。取适量稀释后的酶液3mL，0.2mol/L的CDNB20μL和0.4mol/L的GSH 100μL，25℃下保温20min，于340nm下测定吸光度值。

(4)乙酰胆碱酯酶(AChE)

采用 Gorun 等改进的 Ellman 方法(Gorun et al.，1978)，酶源制备，用0.1mol/L pH7.6的磷酸缓冲液冰浴匀浆，其余同上。取适量稀释后的酶液1mL，加0.004mol/L的ATCh 1.0mL，25℃保温15min，加DTNB3.0mL显色，于412nm下测定其吸光度值。

(5)酶源蛋白质含量

采用考马斯亮蓝G-250染色法测定，以牛血清白蛋白(BSA)为标准蛋白。

2 结果与分析

2.1 中肠酯酶和羧酸酯酶活性变化

选取发育正常一致的转 *CpT* I 基因毛白杨叶片饲喂美国白蛾4~5龄幼虫，并于饲喂后4、12、24、36、48h测定其中肠酯酶和羧酸酯酶活性。从测定结果(见表1)可以看出，转CpTI基因杨树叶片对美国白蛾幼虫中肠酯酶和羧酸酯酶活性有明显抑制作用，而且饲喂时间越长，酶活力下降越多。饲喂48h后，酯酶活力仅为对照的23.58%，羧酸酯酶活力也大幅度下降，饲喂48h后酶活力是对照的26.09%。

表 1　转 *CpT* I 基因杨树叶片对美国白蛾幼虫中肠酯酶和羧酸酯酶活性的影响

时间(h)	酯酶活性($mmol \cdot g^{-1} \cdot min^{-1}$)	羧酸酯酶活性($mmol \cdot g^{-1} \cdot min^{-1}$)
4	3.83±0.24	2.93±0.15
12	2.71±0.16	1.93±0.16
24	2.36±0.23	1.79±0.37
36	1.68±0.31	1.23±0.11
48	0.75±0.02	0.78±0.16
CK	3.18±0.14	2.99±0.24

注：表中数据为 3 次试验统计结果(平均值±标准差)。

2.2　中肠多功能氧化酶活性变化

转 *CpT* I 基因杨树叶片对美国白蛾幼虫中肠多功能氧化酶(MFO)活性的影响见表 2，从开始饲喂至 12h 这段时间内，MFO 活性呈现受抑制状态，12h 时抑制作用最强，酶活性比对照降低 35.78%，此后酶活性提高，饲喂后 24、36、48h 的酶活性测定结果均高于对照，48h 时，处理是对照的 1.32 倍，这是否与 MFO 的可诱导性有关尚待进一步研究。

表 2　转 *CpT* I 基因杨树叶片对美国白蛾幼虫中肠 3 种酶活性的影响

时间(h)	多功能氧化酶活性 ($umol \cdot g^{-1} \cdot min^{-1}$)	谷胱甘肽 S-转移酶活性 ($umol \cdot g^{-1} \cdot min^{-1}$)	乙酰胆碱酯酶活性 ($umol \cdot g^{-1} \cdot h^{-1}$)
4	2.89±0.50	0.42±0.06	0.73±0.11
12	2.09±0.29	0.52±0.10	1.13±0.16
24	4.00±0.25	0.45±0.03	0.85±0.10
36	3.68±0.10	0.76±0.06	0.93±0.09
48	4.31±0.36	0.55±0.03	0.90±0.09
CK	3.26±0.39	0.85±0.04	1.22±0.11

注：表中数据为 3 次试验统计结果(平均值±标准差)。

2.3　谷胱甘肽 S-转移酶活性变化

转 *CpT* I 基因杨树叶片对美国白蛾幼虫中肠谷胱甘肽 S-转移酶(GST)活性的影响见表 2，在整个试验过程中，GST 活性均低于对照，表明在此期间 GST 是受到抑制的，但与酯酶和羧酸酯酶的活性变化明显不同的是 GST 活性受抑制的程度并没有随着时间的推移而加强，就实验测得的数据来看，饲喂后 4h，GST 受抑制的程度最大，其酶活力比对照降低 51.40%，以后呈现波动状态，抑制幅度在 11.71%~48.25%。

2.4　中肠乙酰胆碱酯酶活性变化

用转 *CpT* I 基因杨树叶片饲喂美国白蛾幼虫后，其中肠乙酰胆碱酯酶(AChE)活性变化见表 2，从表中数据可以看出，*CpT* I 对该酶表现抑制作用，其中饲喂后 4h 抑制作用最大，为 40.57%，其余各取样点抑制幅度在 8.19%~29.89%，但抑制作用没有表现出随时间增大的趋势。

3 讨论

美国白蛾，又名秋幕毛虫或秋幕蛾，是重要的国际性检疫害虫，主要危害果树、行道树和观赏树木等阔叶树及农作物。据调查，在美国受害的阔叶树就达 100 多种，日本被害植物有 317 种，我国辽宁省美国白蛾发生区被害植物 94 种，杨树是受害较重的树种之一。目前，疫区上没有彻底治理美国白蛾的方法，本研究发现转 *CpT* I 毛白杨对美国白蛾幼虫中肠解毒酶系有抑制作用，可望为今后的害虫治理工作提供理论依据。

以往的研究结果认为，豇豆胰蛋白酶抑制剂(*CpT* I) 通过与害虫消化酶分子活性中心结合，形成酶-抑制剂复合物，阻断或减弱消化酶的蛋白水解作用；同时，酶-抑制剂复合物能刺激消化酶过量分泌，通过神经系统的反馈，使昆虫产生厌食反应，最终造成昆虫的非正常发育或死亡。昆虫摄入含有 *CpT* I 基因的毛白杨叶片后，体内解毒酶系是否受到影响和影响程度如何至今尚未见报道。本研究结果表明，美国白蛾幼虫摄入转 *CpT* I 毛白杨叶片后，体内解毒酶系活性受到影响，影响程度因酶而异。酯酶和羧酸酯酶活性受到明显抑制，而且随时间的延长抑制程度加强，由此推测 *CpT* I 在影响害虫中肠细胞结构(另文发表)、阻碍细胞能量代谢和酶蛋白合成的同时，可能直接作用于酯酶和羧酸酯酶，致使酶活性降低。谷胱甘肽-S 转移酶表现为全程抑制，但随时间无明显变化；多功能氧化酶表现为前期受抑制，后期受刺激而活性提高；乙酰胆碱酯酶作为清除神经突触部位的乙酰胆碱、维护神经正常传导的重要酶类，被认为是有机磷和氨基甲酸酯等杀虫剂作用的靶标，在本研究中乙酰胆碱酯酶也表现为全程受到抑制，但与时间也无明显相关性。这 3 种酶的变化是 *CpT* I 直接作用的结果，还是由于昆虫中毒后生命力下降、生理代谢功能紊乱所致，还需要进一步的实验证实。

Effects of *CpT* I Transgenic Poplar on Detoxicification Enzyme and AChE of Fall Webworm

Abstract The activities of several detoxicification enzymes and AChE of fourth ~ fifth instar larvae of forest insect pest *Hyphantria cunea* (Drury) feeded with leaves of *CpT* I trangenic poplar [(*Populus tomentosa* ×*P. bolleana*) ×*P. tomentosa*] were investigated. The results indicated that the activities of esterase and CarE were inhibited by 76. 42% and 73.91% after 48 hours feeded. The activity of mix-function oxidase was inhibited in the first 12 hours after treatment, but from then on the activity of FMO increased and at the end of treatment it was 1. 3 fold compared with CK. The activitiesof GST and AChE were also inhibited obviously. The inhibition was the strongest after 4 hours feeded. The activities ofthe two kinds of enzyme decreased 51. 40% and 40. 57%, but they didn't change by the time during the whole treatment .

Key words transgenic poplar, *Hyphantria cunea*, esterase, CarE, MFO, AChE, GST

在低温诱导毛白杨抗冻性中 CaM 含量和 G6PDHase 及 ATPase 活性的变化*

摘 要 该文对毛白杨幼苗在低温锻炼中钙调蛋白(CaM)、6-磷酸葡萄糖脱氢酶(G6PDHase)、腺苷三磷酸酶(ATPase)、幼苗存活率及抗冻性的动态变化过程进行了测定。结果表明，为了获得较高的抗冻性，生长期毛白杨幼苗的低温锻炼可分 3 阶段进行，其中第 3 阶段的-3℃锻炼对毛白杨幼苗抗冻性发育最有效；低温锻炼提高了幼苗叶片和枝条中 CaM 含量和 G6PDHase 及 ATPase 活性，同时也提高了幼苗的存活率和抗冻性，但毛白杨幼苗叶片的 CaM 含量和 G6PDHase 及 ATPase 活性提高程度较枝条明显。进一步分析发现，在-3℃低温锻炼期间，上述各项指标变化与幼苗抗冻性的提高存在着明显的相关性，CaM 可能参与 G6PDHase 和 ATPase 活性的调节。

关键词 毛白杨，低温锻炼，钙调蛋白，存活率，抗冻性

自 1830 年 Gopper 开始低温抗性研究至今已有 170 多年历史，期间许多学者对低温锻炼中碳水化合物、脂类、氨基酸、膜保护酶及可溶性蛋白等物质代谢变化与植物抗寒冻性的关系进行了广泛研究(王毅等，1994；王荣富等，1989；刘祖祺等，1990)，但关于植物体内能量的消长变化与抗寒冻性之间的关系报道较少，且仅限于草本植物。低温锻炼可提高植物的抗寒冻性已被许多实验所证实(简令成，1992；沈征言等，1983；潘杰等，1994；刘鸿先等，1991；Guy et al.，1987)，但其机理至今尚未清楚。1976 年，钙调蛋白(CaM)被发现以来，CaM 在应答和调节植物感受外界逆境方面的作用已日益引起人们的重视。三倍体毛白杨具有速生优质等优良特性，但抗冻性相对较弱，其分布和栽培范围不超过北纬 41°。本文以三倍体毛白杨无性系为试材，研究了它们在低温锻炼中 CaM，G6PDHase，ATPase 及幼苗存活率和抗冻性的动态变化，以探讨 CaM 与 G6PDHase 及 ATPase 和植物抗冻性间的可能关系；并从代谢途径改变及能量变化的角度来探讨低温锻炼提高毛白杨抗冻性的可能机理；为毛白杨抗冻性改良与栽培提供科学依据。

1 材料与方法

1.1 材料

选用北京林业大学毛白杨研究所苗圃半年生、高约 40~50cm、生长健壮、长势一致的毛白杨 BT17 无性系扦插盆栽苗，在 2000 年 7~8 月进行试验。

---

* 本文原载《北京林业大学学报》，2001，23(5)：4-9，与林善枝、李雪平合作发表。

1.2 材料处理

将供试毛白杨幼苗分两组，每组 10 株，其中一组置于 25~30℃条件下作为对照，每天光照 6h；另一组再分成两小部分(光照条件同上)，其中一部分移入-1℃锻炼 1~7d，并将其中锻炼 5d 的部分幼苗移入-2℃锻炼 1~7d；另一部分直接移入-2℃锻炼 1~7d；将上述两部分幼苗移入-3℃锻炼 5d。最后，将经-3℃锻炼 5d 后的部分幼苗移至 25~30℃条件下进行脱锻炼 2d。

1.3 幼苗存活率及抗冻性测定

参照潘杰等(1994)的方法，并略有修改。将上述各处理后的幼苗置于低温生化培养箱中，以 0.15℃/min 的速率降至-12℃，停留 12h 后转到 25~30℃条件下，2d 后测定幼苗的存活率，膨胀度的维持与是否生长以及褐变程度是测定存活率的标准。以-12℃处理 12h 后的幼苗存活率和 100%幼苗存活温度的高低来表示抗冻性的大小。

1.4 G6PDHase 的提取与活性测定

按欧阳光察(1985) 的方法，以每分钟 OD_{340} 增加 0.01 为一个酶活性单位(U)，酶活性以 U/mg 鲜重表示。

1.5 线粒体 Na^{+}-K^{+} ATPase 的提取与活性测定

按李锦树(1985) 的方法，以 UPmg 鲜重作为 ATPase 活性单位。

1.6 CaM 的提取与含量测定

按赵升皓等(1988) 的方法，以 μg/g 鲜重为单位。

1.7 相关性分析

按林德光(1993) 的方法。

2 实验结果

2.1 不同低温锻炼对毛白杨幼苗存活率及抗冻性的影响

从表 1 中可以看出，未锻炼(对照) 毛白杨幼苗在-12℃处理 12h 后的存活率为 0，100%幼苗存活的温度为-3℃，处理 2 和处理 4 的幼苗存活率分别为 20%和 30%，抗冻性由对照的-3℃分别提高到-4℃和-5℃，其后即使再延长时间，幼苗的存活率和抗冻性还维持不变(如处理 3 和 5)。在处理 6 条件下，幼苗的存活率也仅为 45%，抗冻性仅比对照提高了 3℃(-3℃→-6℃)；而在处理 7 条件下，幼苗的存活率和抗冻性已开始明显提高，它们分别比对照提高了 55%和 4℃；不过只有处理 8 的条件才更有利于幼苗抗冻性完全发育，此时的存活率可达 100%，抗冻性可比对照提高了 8℃(-3℃→-12℃)；脱锻炼 2d 后，幼苗的存活率和抗冻性又降至未锻炼的水平(处理 9)。上述结果表明，毛白杨幼苗的低温锻炼可分 3 个阶段进行，其中第 3 阶段(-3℃) 低温锻炼对提高幼苗抗冻性的作用最明显。深入分析发现，在-3℃锻炼期间，随着锻炼天数的延长，毛白杨幼苗的存活率和抗冻性得到不断提高，并于第 5d 达最大值(100，-12℃)，这表明幼苗抗冻性的发育和提高是在低温锻炼过程中逐渐获得(图 1，2)。

表 1　不同低温锻炼对毛白杨幼苗存活率、抗冻性、CaM 含量以及 G6PDHase 和 ATPase 活性的影响

处　理		-12℃处理12h 幼苗存活率(%)	幼苗 100%存活的温度(℃)	CaM 含量(μg · g⁻¹)		G6PDHase(U · mg⁻¹)		ATPase 活性(U · mg⁻¹)	
				枝条	叶片	枝条	叶片	枝条	叶片
1	未锻炼(对照)	0	-3	52. 7	41. 6	39. 5	31. 8	30. 4	24. 8
2	-1℃锻炼 5d	20	-4	68. 5	58. 2	49. 4	41. 4	35. 3	29. 8
3	-1℃锻炼 7d	20	-4	68. 9	58. 6	49. 6	43. 0	35. 5	30. 0
4	-2℃锻炼 5d	30	-5	76. 4	64. 5	55. 3	46. 1	38. 0	32. 3
5	-2℃锻炼 7d	30	-5	76. 8	64. 7	55. 8	46. 4	38. 5	32. 4
6	-1℃锻炼 5d+(-2)℃锻炼 5d	45	-6	84. 3	70. 7	59. 3	50. 9	42. 6	37. 2
7	-1℃锻炼 5d+(-2)℃锻炼 5d+(-3)℃锻炼 1d	55	-7	110. 7	104. 0	78. 6	74. 5	47. 2	44. 7
8	-1℃℃锻炼 5d+ -2℃锻炼 5d+(-3)℃锻炼 5d	100	-12	163. 4	149. 8	104. 7	96. 7	66. 9	64. 5
9	脱锻炼	0	-3	52. 5	46. 7	39. 2	37. 1	29. 9	27. 4

注：CaM 含量，G6PDHase，ATPase 活性均为鲜重。

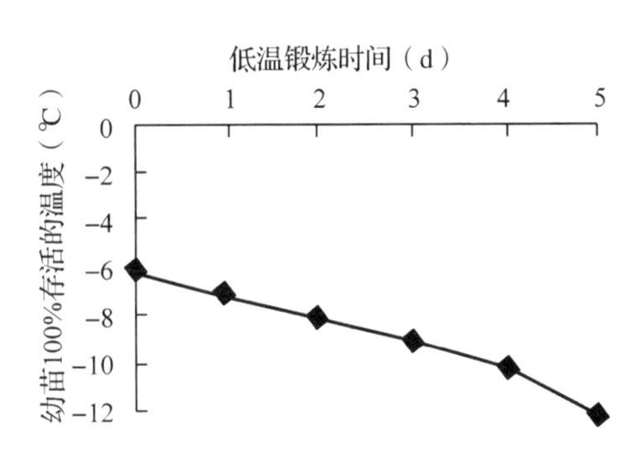

图 1　-3℃低温锻炼对毛白杨幼苗抗冻性的影响

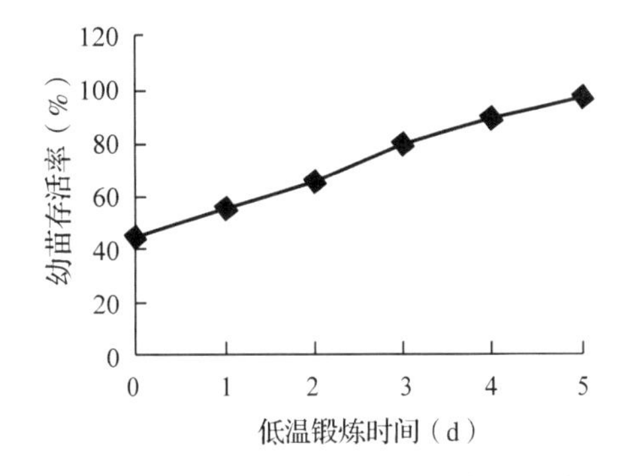

图 2　-3℃低温锻炼对毛白杨幼苗存活率的影响

2. 2　不同低温锻炼对钙调蛋白(CaM) 含量的影响

同未锻炼相比，毛白杨幼苗在各种低温锻炼期间，其叶片和枝条的 CaM 含量均有所增加(表 1)。处理 2 和处理 4 幼苗的 CaM 含量增加程度相对较小，其中处理 2 叶片和枝条的 CaM 含量只分别比对照增加了 40%和 30%，处理 4 也仅分别比对照增加了 55%和 45%，其后即使再延长锻炼时间，幼苗叶片和枝条的 CaM 含量几乎不变(如处理 3 和 5)。在处理 6 条件下，幼苗叶片和枝条的 CaM 含量增加程度虽大于处理 2 和处理 4，但仅分别比对照增加了 70%和 60%。而在处理 7 条件下(即依次经过-1℃和-2℃各锻炼 5d 后转入-3℃锻炼的第 1 天)，幼苗叶片和枝条的 CaM 含量增加较为明显，分别比对照增加了 150% 和 110%。不过只有处理 8 的条件才更有利于幼苗 CaM 含量的增加，其叶片和枝条的 CaM 含量分别比对照增加了近 260% 和 210%。上述结果表明，不同的低温锻炼对幼苗 CaM 含量的效应是不同的，先后经过-1、-2 和-3℃3 个阶段的低温锻炼后对提高毛白杨幼苗 CaM 含量的效果最好，其中第 3 阶段-3℃低温锻炼对 CaM 含量增加的效应最大。而且在不同低温锻炼中叶片的 CaM 含量增加程度均较枝条明显。进一步分析发现，在-3℃锻炼期间，随着锻炼天数的增加，幼苗叶片和枝条 CaM 含量也不断增加，并于第 5 天达到含量最大值(图 3)。脱锻炼 2d 后，幼苗枝条 CaM 含量虽有所下降但仍维持在未锻炼水平，而叶片 CaM 含量却明显高于未锻炼水平(表 1)。

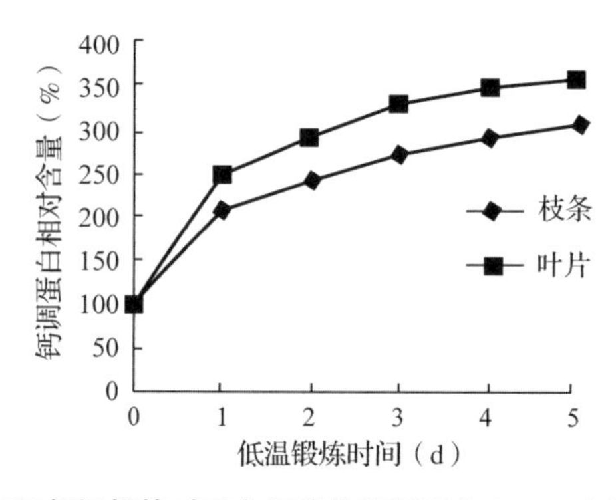

图 3 -3℃低温锻炼对毛白杨幼苗钙调蛋白(CaM)含量的影响

2.3 不同低温锻炼对毛白杨幼苗 G6PDHase 和 ATPase 活性的影响

与未锻炼(对照) 相比，毛白杨幼苗在各种低温锻炼期间，叶片和枝条的 G6PDHas 和 ATPase 活性均有所提高(表 1)，说明各种低温锻炼均能促进幼苗体内 PPP 途径和 ATP 水解作用的加强。但不同的低温锻炼对幼苗叶片和枝条的 G6PDHase 和 ATPase 活性的效应是不同的，从它们的活性提高程度来看，处理 2<处理 4<处理 6<处理 7<处理 8，其中处理 2、处理 4 和处理 6 的幼苗叶片和枝条的 G6PDHase 和 ATPase 活性只分别比对照提高了

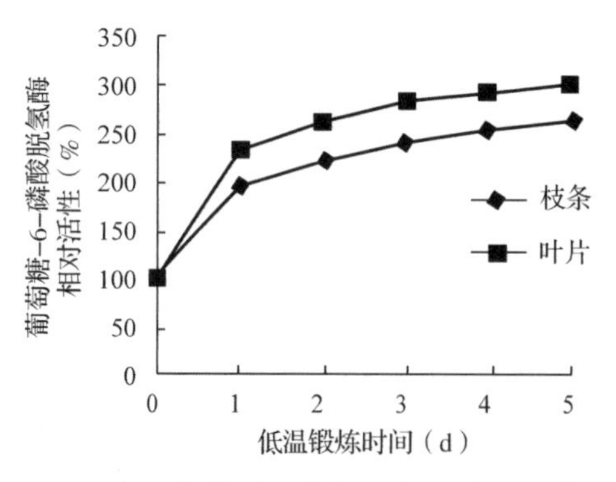

图 4 -3℃低温锻炼对毛白杨幼苗 6-磷酸葡萄糖脱氢酶(G6PDHase)活性的影响

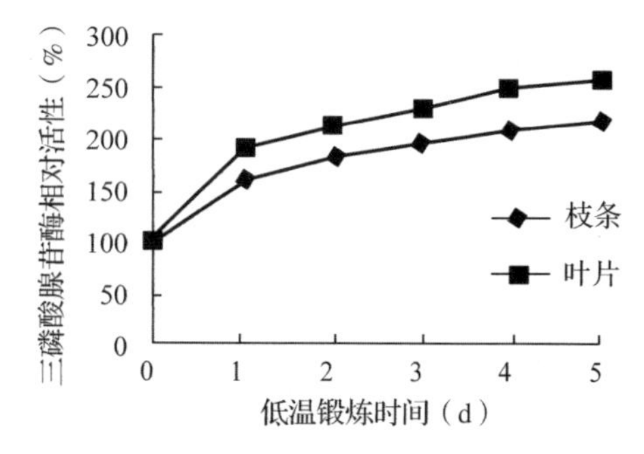

图 5 -3℃低温锻炼对毛白杨幼苗腺苷三磷酸酶(ATPase)活性的影响

30%和25%与20%和16%，45%和40%与30%和25%，60%和50%与50%和40%，而处理8却分别比对照提高了204%和165%与160%和120%，其提高程度分别相当于处理2的6.8倍和6.6倍与8.0倍和7.5倍、处理4的4.5倍和4.1倍与5.3倍和4.8倍、处理6的3.4倍和3.3倍与3.2倍和3.0倍；在处理7条件下(即依次经过-8℃和-2℃各锻炼5d后转入-3℃锻炼的第1天)，幼苗叶片和枝条的G6PDHase活性提高虽较为明显(136%和99%与80%和55%)，但其提高程度也仅分别为处理8的2P3和3P5与1P2和5P11。另外在处理2和处理4后再延长锻炼时间，幼苗叶片和枝条的G6PDHase和ATPase活性几乎不变(如处理3和5)。由此可见，处理8(即依次经过-1℃，-2℃和-3℃ 3个阶段的低温锻炼)对提高幼苗叶片和枝条的G6PDHase和ATPase活性的效果最好，其中第3阶段-3℃低温锻炼对幼苗叶片和枝条的G6PDHase和ATPase活性提高的效应最大；而且对于相同低温锻炼而言，幼苗叶片的G6PDHase和AT2Pase活性提高程度均较枝条明显，叶片和枝条的G6PDHase活性提高程度均比各自的ATPase明显。进一步分析发现，随着-3℃低温锻炼时间的延长，幼苗叶片和枝条的G6PDHase和ATPase活性均逐渐提高，并于第5天达到活性最大值(图4，5)，但叶片和枝条的G6PDHase活性提高程度及活性最大值均大于各自的ATPase；脱锻炼2d后，枝条G6PDHase和ATPase活性虽有下降但仍接近对照水平，而叶片G6PDHase和ATPase活性却明显高于对照(表1)。反映了相同低温锻炼对同一植株的G6PDHase和ATPase活性效应是不同的，以及同一植株不同组织在低温锻炼或脱锻炼中代谢途径的改变与能量的变化也存在一定的差异。

3 讨论

植物抗寒性是植物长期适应低温环境而形成的一种潜在的遗传特性，植物抗寒冻基因只是一种诱发性基因，只有在特定条件(主要是低温和短日照)的作用下，才能启动抗寒冻基因的表达从而发展为抗寒冻力(刘鸿先等，1991；Guy et al.，1987；欧阳光察，1985；李锦树，1985；赵升皓，1988；林德光，1993；简令成，1987)。沈征言等(沈征言等，1983)的试验表明番茄必须经过5~10℃和0~2℃的两步锻炼后才能使幼苗抗冷性得到完全发育。本试验发现，为了获得较高的抗冻性，毛白杨幼苗的低温锻炼可分3阶段进行，其中第3阶段-3℃低温锻炼对幼苗抗冻性发育最有效，而第1阶段-1℃锻炼和第2个阶段-2℃锻炼虽无法促进幼苗抗冻性完全发育，但它们能使幼苗在一定程度上增强对低温的适应性，为第3阶段-3℃低温锻炼的进行提供基础和可能性，这说明毛白杨幼苗抗冻性的诱导及发育需要一定的程序、温度及锻炼时间，不适宜的温度即使再延长锻炼天数也无法诱导抗冻性完全发育；抗冻性的发育是一个逐步获得的过程，而且低温诱导幼苗抗冻性的提高也有一定限度(表1，图1)。

3.1 CaM水平变化与G6PDHase及ATPase活性的关系

CaM是一种分布较广、具有多种功能的调节蛋白，在Ca^{2+}信使系统信息传递过程中起着重要作用。许多研究表明，CaM能调节许多依赖于Ca^{2+} CaM酶类的活性，其水平变化可能是植物对外界刺激反应的重要机制之一(孙大业等，1991)。近年来研究发现，ATPase

活性受 Ca^{2+}-CaM 依赖的磷酸化作用的控制，Ca^{2+}-CaM 系统参与 ATPase 诱导产生；外源 CaM 对 G6PDHase 和 ATPase 有激活作用，并且可被 CaM 专一性抑制剂绿丙嗪抑制，这说明 G6PDHase 和 ATPase 可能是 Ca^{2+}-CaM 的靶酶，CaM 可能参与 G6PDHase 和 ATPase 活性的调节(孙大业等，1991；李德红等，1998；孙大业等，1998；李美如登，1996；唐军等，1995；林善枝，1997)。我们试验发现，低温锻炼所引起的幼苗 CaM 含量与 G6PDHase 及 ATPase 活性的动态变化非常相似，即呈现出逐渐增加的趋势(图 3 至图 5)。从提高程度来看，幼苗叶片和枝条的 CaM 相对含量提高程度均大于各自相对应的 G6PDHase 和 ATPase；从出现含量(活性) 峰值时间进程来看，CaM 含量最大值出现时间与 G6PDHase 和 ATPase 活性最大值是同步的；通过相关性分析发现，在低温锻炼中幼苗的 CaM 水平变化与 G6PDHase 和 ATPase 活性提高成极显著正相关(表 2)。由此可认为，CaM 可能参与 G6PDHase 和 ATPase 活性的调节，且与幼苗抗冻性发育密切相关，但其机理有待于进一步深入研究。

3.2 ATPase 及 ATP 和抗冻性的关系

一些报道指出，低温锻炼能引起 ATPase 活性发生变化，且与植物抗寒冻性的提高密切相关(林善林，1997；Mito，1996；Steponkus，1984；简令成，1981；Zhao，1993；Salzman，1993；戴金平，1991)。Levitt(1980)认为 ATPase 稳定、不易受低温影响的植物较抗冷。我们的实验结果指出，在低温锻炼中，伴随幼苗 ATPase 活性提高的同时也提高了幼苗存活率和抗冻性(图 1，2，4)，说明低温锻炼可能促进了 ATPase 或激活了 ATPase，加强了幼苗体内 ATP 水解作用，从而为与抗冻性形成和幼苗存活率提高有关的 RNA 和蛋白质(酶) 等生物合成提供能量。比较分析叶片和枝条中 ATPase 活性变化发现，幼苗枝条的 ATPase 活性在低温锻炼和脱锻炼中变化不大，而叶片 ATPase 活性提高程度大于枝条，表明了低温锻炼能使枝条中 ATPase 获得耐低温特性，保持了相对稳定的 ATPase 活性反应，反映出同一植株不同组织对低温胁迫的敏感性和能量代谢上的差异，这也是幼苗叶片在低温下比枝条易发生冻害的主要原因之一。

表 2 低温锻炼中 CaM 水平变化与其可能调节酶类活性变化的相关性

种类	依赖于 Ca^{2-}-CaM 调节的酶类	
	G6PDHase	ATPase
枝条	0.863**	0.832**
叶片	0.76**	0.686**

注：**表示与 CaM 极显著相关(相关系数 $r > P_{0.01} = 0.1684$)。

3.3 G6PDHase 及 $NADPH_2$ 和抗冻性的关系

许多研究指出，在低温锻炼中，G6P 代谢转变成磷酸戊糖代谢，即提高了 PPP 途径，其产生的某些中间产物(RU5P 和 E4P 等) 是重要的生物合成原料，所产生的 $NADPH_2$ 是许多生物合成反应的专一性电子供体，以及它可通过细胞色素系统或转氢酶系统重新氧化产生能量(ATP) 供给生物体利用。目前已在多种植物的低温锻炼中发现了 G6PDHase 活性

和 $NADPH_2$ 水平的提高，这对低温锻炼的进行和植物抗寒冻性的提高是非常必要的(刘鸿先等，1991；简令成，1990；Sagisaka，1985)。本试验发现，在低温锻炼中，毛白杨幼苗叶片和枝条的 G6PDHase 活性呈现出相似于 ATPase 活性的逐渐提高趋势(图 4，5)，说明低温锻炼所引起的体内代谢途径改变与能量变化有关。G6PDHase 是 PPP 途径的关键性调控酶，低温锻炼所引起的 G6PDHase 活性提高可能是通过 PPP 途径加强来产生更多的 $NADPH_2$，而 $NADPH_2$ 又可通过生物氧化系统产生大量 ATP 来满足 ATPase 活性提高的需要，这种高含量的 ATP 能使植物进入一个特殊的充能状态发育高水平的抗冻性，以确保植株及时利用 ATP 合成与抗冻性发育有关的 RNA、蛋白质和脂类等物质，从而增强了植株对低温的适应性，最终导致幼苗存活率和抗冻性的提高。这表明低温锻炼能引起的幼苗存活率和抗冻性的提高与体内代谢途径和能量代谢的适应性变化密切相关。由此可认为毛白杨幼苗在低温下可能主要是通过 PPP 途径来产生高水平的 $NADPH_2$，才能满足 ATPase 活性提高及水解作用加强的需要，释放出更多能量供给生物合成；$NADPH_2$ 既是生物合成的供氢体，又是新的能量形式，它与 ATP 间存在着密切的关系，两者都是抗冻性发展的推动力。低温锻炼提高毛白杨幼苗抗冻性的一个表现是能较快而又明显地提高了 CaM 含量，进一步调节激活了可能依赖于 Ca^{2+}-CaM 的酶类(G6PDHase 和 ATPase) 活性，而这些酶活性变化的综合结果是使它们能在低温下行使其功能如提高了促进幼苗体内 PPP 途径与加强了 ATP 水解作用，从而为抗冻性的形成与提高提供基础。

Changes of Content of CaM, and Activities of G6PDHase and ATPase during Low-temperature-induced Freezing Tolerance of *Populus tomentosa* Seedlings

Abstract The changes of calmodulin (CaM), glucose-6-phosphate dehydrogenase (G6PDHase), adenosin triphosphatase (ATPase), survival rates and freezing tolerance in *Populus tomentosa* seedlings under cold acclimation were studied. The results showed that the cold acclimation of *Populus tomentosa* seedlings needed to experience three stages in order to acquire the highest freezing tolerance, and the third stage of cold acclimation at −3℃ had greater effect on the development of freezing tolerance. Cold acclimation increased the content of CaM, the activities of G6PDHase and ATPase in leaves and branches, the survival rates and freezing tolerance of seedlings. But the enhancing degree of CaM, G6PDHase and ATPase in leaves was larger than that in branches. Further analysis found that the changes of the above-mentioned indexes were closely related to the freezing tolerance of seedlings; CaM possibly regulated the activities of G6PDHase and ATPase.

Keywords *Populus tomentosa*, cold acclimation, calmodulin, survival rates, freezing tolerance

Role of $CaCl_2$ in Cold Acclimation Induced Freezing Resistance of *Populus tomentosa* Cuttings*

Abstract *Populus tomentosa* cuttings were treated with 1mmol · L^{-1}, 5mmol · L^{-1}, 10mmol · L^{-1} or 15mmol · L^{-1} of $CaCl_2$ for 1-7d, respectively, for studying the effects of different concentrations of $CaCl_2$ on freezing resistance. Results indicated that 10mmol · L^{-1} of $CaCl_2$ has greater effect than other concentrations on the enhancement of freezing resistance, and the optimum time of pretreatment was 5d. In addition, cuttings used for cold acclimation at −3℃ were pretreated with or without 10mmol · L^{-1} of $CaCl_2$, 3mmol · L^{-1} of Ca^{2+} chelator EGTA, 0.05mmol · L^{-1} of CaM antagonist CPZ or 0.1mmol · L^{-1} of Ca^{2+} channel inhibitor $LaCl_3$. The changes in CaM and freezing resistance of all cuttings were investigated. The results showed that cold acclimation at−3℃ increased CaM content and decreased theminimum temperature for 100% survival. The $CaCl_2$ pretreatment enhanced the effect of cold acclimation and obviously increased CaM content and decreased theminimum temperature for 100% survival, but this effect was strongly inhibited by the EGTA, CPZ or $LaCl_3$. It is concluded that the effect of $CaCl_2$ on freezing resistance is associated with its concentration and time of pretreatment, Ca^{2+}-CaM may be involved in the induction of freezing resistance of the cuttings.

Keywords *Populus tomentosa* cuttings, cold acclimation, $CaCl_2$, CaM, freezing resistance

1 Introduction

Plant membranes are the primary targets of injury arising from low temperature (Hallgren and Oquist, 1990; Odlum and Blake, 1996). It has been suggested that calcium may be as the stabilizer of cell membrane, which exercise the protective function on the cell wall and microsomal membranes (Legge et al., 1982; Minorsky, 1985; Rickauer and Tanner, 1986; Liang and Wang, 2001). Li et al. (1996) have observed that the effect of $CaCl_2$ on the survival rates of rice seedlings could be associated with the increase of cellular total soluble Ca^{2+} content caused by $CaCl_2$ treatment. Extraneous $CaCl_2$ pretreatment resulted in an increase of membrane stability and freezing resistance of rice seedlings under low temperature (Duan et al., 1999; Liang and Wang, 2001). Since the hypoyhesis that calcium may be as primary messenger transducer of low temperature in plants has first been proposed by Minorsky (1985), much attention has been focused on the studies of the role of calcium calmodulin in the response and adaptation of plants to the envi-

* 本文原载《中国林业研究》(*Forestry Studies in China*), 2002, 4(3): 38-42, 与林善枝合作发表。

ronment (Gong et al., 1990; Poovaiah et al., 1993; Bush, 1995; Li et al., 1997). However, little is known about the effect of the extraneous $CaCl_2$ pretreatment on freezing resistance in woody plants. In particular, the mechanism of protective function of calcium on cell membrane in woody plants is poorly understood. In this study, freezing resistance of *P. tomentosa* cuttings treated with different concentrations of $CaCl_2$ was investigated to explore the effect of the extraneous $CaCl_2$ on freezing resistance in woody plants. In addition, the changes of CaM content and freezing resistance during cold acclimation combined with some effectors (e. g. $CaCl_2$, EGTA, $LaCl_3$ and CPZ) pretreatment were studied to determine whether the increase of CaM content was related to the enhancement of freezing resistance.

2 Materials and methods

2.1 Plant materials

Populus tomentosa plants were established from dormant branch cuttings collected from Heilongjiang Province in northeastern China. The cuttings were rooted and grown in greenhouse in pots containing a 2:1(v/v) mixture of soil and sand. After raised for 5 weeks in greenhouse, they were placed outdoors for 3 weeks.

2.2 Selection of $CaCl_2$ concentration suitable for the increase of freezing resistance

In order to study the effects of different concentrations of $CaCl_2$ on freezing resistance, *P. tomentosa* cuttings were treated with 1mmol · L^{-1}, 5mmol · L^{-1}, 10mmol · L^{-1} and 15mmol · L^{-1} of $CaCl_2$ for 1~7d at 25℃ with an 8h photoperiod and a light intensity of 30μmol · m^{-2} · s^{-1}, respectively.

2.3 Pretreatment and cold acclimation

Eight week old cuttings were divided into three groups. The first group of 20 cuttings placed at 25℃ for 10d was referred to as non acclimated (Control or NA) cuttings. The second group of 30 cuttings held at −3℃ for 5d was referred to as cold acclimated (CA) cuttings. To determine the role of $CaCl_2$ in the induction of freezing tolerance, the third group of 60 cuttings was further divided into three sub groups, each group of 20 cuttings, watered their roots daily with 10mmol · L^{-1} $CaCl_2$, 3mmol · L^{-1} EGTA, 0.05mmol · L^{-1} CPZ and 0.1mmol · L^{-1} $LaCl_3$ at 25℃ for 5d, respectively, and then these cuttings were transferred to −3℃ for 5d and were referred to as cold acclimation combined with effector pretreatment (E-CA) cuttings. All handlings were done in an 8h photoperiod and a light intensity of 30μmol · m^{-2} · s^{-1}.

2.4 Freezing resistance evaluation

The cuttings with $CaCl_2$-pretreatment, NA, CA and CA-E used for examining freezing resistance were directly transferred to a second chamber set at 0℃. After equilibration at 0℃ for 12h, the chamber temperatures were lowered stepwise to −1℃, −3℃, −5℃, −7℃, −10℃, −12℃, −14℃ and−16℃, respectively, at a constant rate of 1.0℃ $min^{-1}$and then held for 12h at each temperature. At the same time, the cuttings were directly transferred from each treated temperature to 25℃ for 2d. Finally, the survival rates of cuttings were determined. Survival rates were scored whether these cuttings regrowing normally and the extent of freezing injury of bark tissue leaves and buds after treatment at each temperature for 12 h. Freezing resistance was evaluated according to the minimum temperature at which 100% of cuttings survived.

2.5 CaM extraction and analysis

Leaves and branches were ground in Tris-HCl buffer (Tris-HCl (pH 8.0) 50mmol · L^{-1}, EGTA 1mmol · L^{-1}, phenylmethylsulfony l fluoride 0.5mmol · L^{-1}, NaCl 0.15mmol · L^{-1}, $NaHSO_3$ 20mmol · L^{-1}) at 4℃. The homogeneous liquid was extracted at 90℃ for 2min, and then centrifuged at 4℃ with10000g for 30min. The resulting supernatant was used as the soluble CaM and its concentration was assayed using the technique described by Zhao et al. (1988).

All the determinations were performed at least in triplicates and the data presented were the average values.

3 Results

3.1 Effect of $CaCl_2$ on freezing resistance

As shown in Figure 1, the lowest temperature for 100% survival of cuttings treated with $CaCl_2$ of 1mmol · L^{-1} and 5mmol · $L^{-1}$for 7d had decreased by 1℃ and 2℃, respectively, compared with the control. One day after treating with $CaCl_2$ of 10mmol · L^{-1}, the lowest temperature for 100% survival started to decrease from−3℃ to −5℃ and gradually reached the maximum level (−8℃) on the 5th day, and then increased to −7℃ on the 7th day. However, when the cuttings

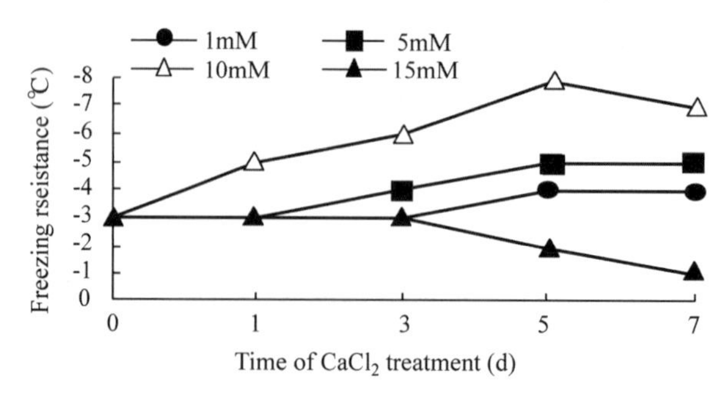

Fig. 1 Effects of $CaCl_2$ with different concentrations on freezing resistance of *P. tomentosa* cuttings

were treated with $CaCl_2$ of 15mmol · L^{-1} for 7d, the lowest temperature for 100% survival decreased by 2℃ compared with the control.

3.2 Effects of $CaCl_2$ and cold acclimation on freezing resistance

After 5d of cold acclimation at −3℃, the lowest temperature for 100% survival decreased from −3℃ in non acclimated cuttings (NA) to −10℃ in cold acclimated cuttings (CA), and to −14℃ in the cuttings of cold acclimation combined with 10mmol · L^{-1} $CaCl_2$ pretreatment (CaClCA) (Fig. 1). However, after the pretreatments with EGTA, $LaCl_3$ or CPZ, respectively, the lowest temperature for 100% survival obviously increased to the level of the control (Fig. 2).

The time pattern of change in freezing tolerance at−3℃ during 1~5d was studied (Fig. 3). One day after transferring cuttings to −3℃, the lowest temperature for 100% survival started to decrease from −3℃ to −5℃ and gradually reached aminimum level (−10) on the 5th day.

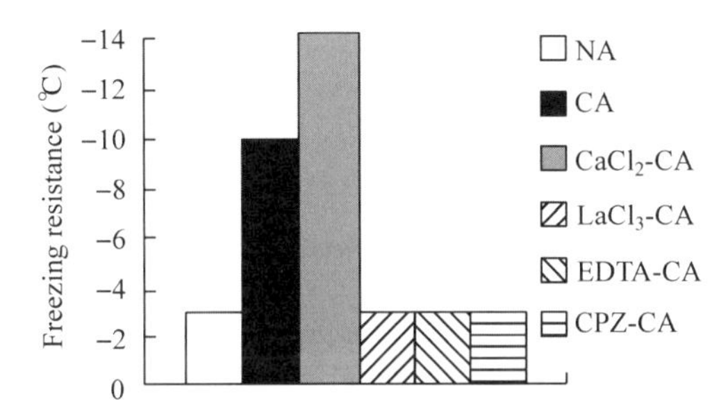

Fig. 2 Change in freezing resistance of *P. tomentosa* cuttings

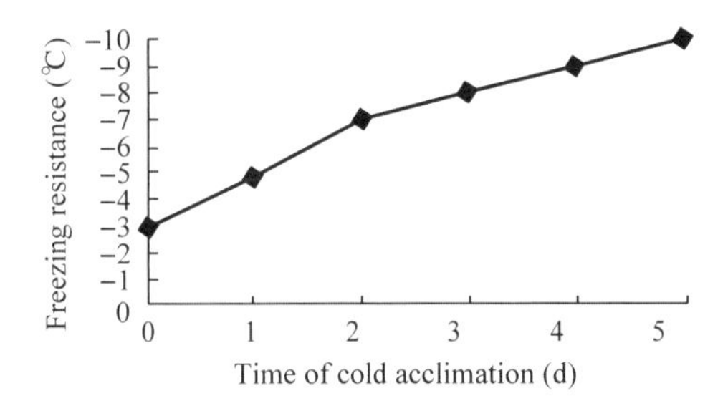

Fig. 3 Change in freezing resistance of *P. tomentosa* cuttings during cold acclimation at −3℃

In addition, to further examine the effect of 10mmol · L^{-1} $CaCl_2$ on the enhancement of freezing resistance of cuttings, the change in freezing resistance of cuttings pretreated with 10mmol · L^{-1} $CaCl_2$ for 1~5d during cold acclimation at −3℃ was determined. From Table 1, the lowest temperature for 100% survival decreased with the increase of time for $CaCl_2$ pretreatment and reached aminimum level of −14℃ on the 5th day, and decreased by 4℃ compared with CA cuttings.

3.3 Effects of $CaCl_2$ and cold acclimation on CaM content

Compared with the controls, cuttings of both cold acclimation combined with or without $CaCl_2$ pre treatment increased the content of CaM. However, cuttings of $CaCl_2$-CA had 26.1% and 29.9% higher CaM in leaves and branches than that of CA. A low CaM content was observed in the cuttings pretreated with EGTA, $LaCl_3$ or CPZ (Fig. 4). In addition, it is found that CaM content increased with the time increasing of cold acclimation at -3℃ in a period of 5d (Fig. 5).

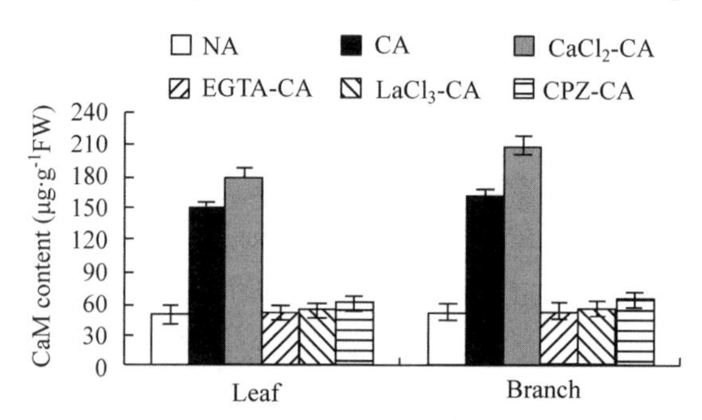

Fig. 4 Effects of Cold, $CaCl_2$, EGTA, $LaCl_3$ or CPZ on CaM contents in the leaves and branches of *P. tomentosa* cuttings.

Table 1 Changes of CaM content and freezing resistance of *P. tomentosa* cuttings pretreated with 10mmol · L^{-1} $CaCl_2$ for 1~5d during cold acclimation at -3℃

Treatment	Freezing resistanee(℃)	CaM content(μg · g^{-1}FW)	
		Leaf	Branch
CA(Control)	-3.0	41.6	52.7
CA(5d)	-10.0	149.8	163.4
$CaCl_2$-CA(1d)+CA(4d)	-10.0	153.4	169.3
$CaCl_2$-CA(2d)+CA(3d)	-11.0	160.1	177.5
$CaCl_2$-CA(3d)+CA(2d)	-12.0	166.4	184.3
$CaCl_2$-CA(4d)+CA(1d)	-13.0	172.4	189.3
$CaCl_2$-CA(5d)	-14.0	178.8	194.9

To study the possible correlations among the changes of CaM content, the accumulation of Ca^{2+} and the increase of freezing resistance during cold acclimation, we determined the changes of CaM content and freezing resistance in the cuttings pretreated with 10mmol · L^{-1} of $CaCl_2$ for 1~5d during cold acclimation at -3℃. The result showed that CaM content and freezing resistance increased with increasing exposure to $CaCl_2$ pretreatment up to 5d (Table 1), indicating that an increase of CaM content and freezing resistance during cold acclimation associated with the accumulation of Ca^{2+}.

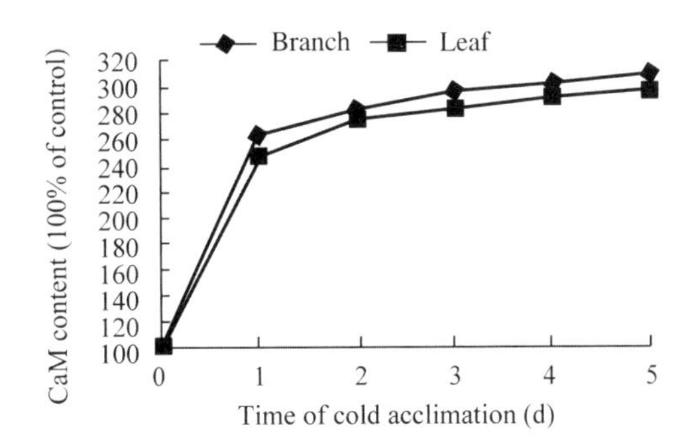

Fig. 5 Change of CaM content in leaves and branches of *P. tomentosa* cuttings during cold acclimation at −3℃.

4 Discussion

A hypothesis of chilling in plants has first been proposed by Minorsky (1985) that calcium may be as primary messenger transducer of low temperature, this hypothesis was supported by Price et al. (1994), Monroy et al. (1993, 1995), Knight et al. (1996). The observations that $CaCl_2$-induced elevation of survival rates of r ice seedlings with the increase of cellular total soluble Ca^{2+} content caused by $CaCl_2$ treatment indicated that Ca^{2+} maybe involved in the increase of freezing tolerance (Li et al., 1996). Zeng et al. (1999) reported that the enhancement of chilling resistance of rice seedlings induced by cold acclimation was inhibited by Ca^{2+} chelator EGTA and CaM antagonist CPZ, suggesting that Ca-CaM messenger system was involved in chilling resistance formation. Moreover, many studies have revealed that cold acclimation induced an accumulation in Ca^{2+} and CaM, which likely contributed significantly to the development of cold tolerance (Bush, 1995; Li et al., 1997; Li et al., 1996; Lin et al., 2001; 2002). In present study, it is found that $CaCl_2$ pretreatment had enhanced the effect of cold acclimation on the increase of CaM content and freezing resistance of *P. tomentosa* cuttings, but its effect was strongly inhibited by Ca^{2+} chelator EGTA, Ca^{2+} channel inhibitor $LaCl_3$ or CaM antagonist CPZ, respectively. Similar findings were reported in some plants (Knight et al., 1991; Monroy et al., 1993; Li et al., 1996). Moreover, it appeared that the increasing process of CaM content was in accordance with the increase of freezing resistance during cold acclimation at−3℃ (Fig. 3, 5). Thus, our results show that the enhancing effect of $CaCl_2$ pretreatment on the increase of freezing resistance prior to freezing acclimation likely associates with Ca^{2+}-induced increase of CaM. It is, therefore, possible that the increase of CaM content may be associated with extraneous Ca^{2+}, Ca^{2+} CaM may be involved in the induction of freezing tolerance.

In addition, it is found that the treatment of 10mmol · L^{-1} $CaCl_2$ has more effect than that of the other concentrations on the decrease of theminimum temperature for 100% survival, and during the treatment of 10mmol · L^{-1} $CaCl_2$ for 1 ~ 7d, theminimum temperature for 100% survival reaches aminimum level (−8) on the 5th day (Fig. 1), indicating that the $CaCl_2$ effect on freezing resistance of *P. tomentosa* cuttings is correlated with its concentration and duration of treatment. The concentration optimum of $CaCl_2$ treatment is 10mmol · L^{-1}, and the time optimum of $CaCl_2$ treatment is 5d.

Response of Antioxidant Defense System in *Populus tomentosa* Cuttings Subjected to Salt Stress *

Abstract To explore the antioxidant defense system of *Populus tomentosa* cuttings subjected to salt stress, the cutt ings were treated with three different concentrations (85mmol · L^{-1}, 170mmol · L^{-1}, 260mmol · L^{-1}) of NaCl for 3~54h, and the changes in superoxide dismutase (SOD) and peroxidase (POD) activities, and malondialdehyde (MDA) content in the leaves of cuttings were investigated. The results showed that SOD and POD activities significantly decreased during 3h at the beginning of all NaCl treatment, and then slightly decreased during a period of 9~54h, but a lower decrease in SOD activity than in POD activity was observed in all treated cuttings. MDA content distinctly increased and reached a maximum value during 15h after the beginning of salt stress, and then decreased, but its level was higher than the control. The decreased SOD and POD activities and the increased MDA content affected by the treatment of 260mmol · L^{-1} NaCl were more obvious than that affected by the treatment of 85mmol · L^{-1} or 170mmol · L^{-1} NaCl. It is possible that the decrease in SOD and POD activities under salt stress resulted in the enhancement of membrane lipid peroxidation, which seems to be the key factor that leads to the increase of MDA content and the occurrence of oxidative damage to cuttings.

Keywords *Populus tomentosa*, antioxidant defense system, NaCl, SOD, POD, MDA

1 Introduction

Soil sanity constitutes one of the most important environmental factors that affects the normal growth and the productivity of plants. In general, plants have evolved several cellular defense mechanisms to overcome or alleviate the oxidative damage caused by active oxygen species (AOS) during periods of normal growth and moderate stress. These mechanisms include scavenging the AOS by endogenous antioxidants such as superoxide dismutase (SOD), peroxidase (POD), catalase (CAT) and so on. However, when plants are subjected to severe environmental stresses, the production of AOS can exceed the capacity of the antioxidant system to neutralize them, which leads to the alteration on the balance between the production of AOS and their detoxification by the antioxidative system (Hernandez et al., 1993; 1995; Gomez et al., 1999). The elevated levels of AOS can enhance membrane lipid peroxidation, which can cause oxidative damage to cellular membrane (Halliwell and Gutteridge, 1989; Chen, 1991; Prasad et al., 1994; Prasad, 1996;

* 本文原载《中国林业研究》(*Forestry Studies in China*), 2002, 4(2): 16-20, 与林善枝、林元震和张谦合作发表。

Hauptmann and Cadenzas, 1997; Knag and Saltveit, 2001). In recent years there are many reports about the variation of the antioxidant defense system in crops under salt stress (Hernandez et al., 1993 and 1995; Gossett et al., 1994; Shalata and Tal, 1998; Comba et al., 1998; Meneguzzo et al., 1999; Taha et al., 2000; Shalata et al., 2001). But up to now very little is known about salt stressinduced oxidative damage in poplars, especially in *Populus tomentosa*.

P. tomentosa which has fast growth rates, high yields and short rotation times, is a special native species in China, and widely used as an important afforestation and landscape tree species in the north of China. But the limitation of inherited salt resistance of *P. tomentosa* has made them grow very difficult on salinity soil.

In this study, the changes of SOD and POD activities, and MDA content in *P. tomentosa*, cuttings treated with three different concentrations of NaCl were investigated to explore the response of antioxidant defense system to salt stress and the mechanism of salt injury.

2 Materials and methods

2.1 Plant materials and pretreatment

Populus tomentosa plants were established from dormant branch cuttings collected from Heilongjiang Province in northeastern China. The cuttings were rooted and grown in the greenhouse in the pots containing a 2:1(v/v) mixture of soil and sand.

After 8-week old cuttings were removed from the pots, the roots were carefully and gently washed with distilled water several times. The first part of 15 cuttings used for the control were raised in distilled water in three glass cups, 5 cuttings per cup, and the distilled water was replaced every day. The second part of 45 cuttings used for NaCl treatment were grown in glass cups containing 85mmol · L^{-1}, 170mmol · L^{-1}, or 260mmol · L^{-1} of NaCl solution for 3~54h, 5 cuttings per cup, 15 cuttings pertreatment. The NaCl solution was replaced every day. All handlings were done in an 8h photoperiod and a light intensity of 30μmol · m^{-2} s^{-1}. Leaves were randomly collected from the control and all NaCl-treated cuttings at different times, and immediately frozen on dry ice until used for analysis of the content of MDA and the activities of SOD and POD. Each experiment was repeated three times and consisted of three replicates.

2.2 Preparation of extract

Extracts for determination of SOD, POD and MDA were prepared from 1g of leaves and homogenized under 4℃ in 5mL of extraction buffer containing 62.5mM potassium-phosphate buffer (pH7.8), 1% (w/v) PVPP. The homogenates were centrifuged at 15000g for 20min and the supernatant fraction was used for the assays.

2.3 Assay of enzyme and protein determination

SOD was determined by monitoring the inhibition of photochemical reduction of nitroblue tetrazolium(NBT) as described previously (Liu et al. , 1985). One unit (U) of SOD was defined as the amount of enzyme that produced a 50% inhibition of NBT photochemical reduction. POD activity was determined by measur ing the increase in absorption at 470nm according to Liu et al. (1985). The reaction was carried out at 25℃ for 20min in a 3mL reaction mixture containing 0. 02mL tissue extract, 0. 1M potassium-phosphate buffer (pH 7. 0), 20μL of guaiacol. The reaction was started by the addition of 20μL H_2O_2. One unit (U) of POD was defined as the amount of enzyme that increased a 0. 01 of A470 permin under the assay condition. Protein concentration was assayed by using the methods of Bradford (1976) with bovine serum albumin (BSA) as a standard.

2. 4 Measurement of MDA

MDA was measured by the thiobarbituric reaction according to Liu et al. (1985).

3 Results

3. 1 Change in the activities of SOD and POD

As shown in Fig. 1 and 2, there was a similar trend at three NaCl concentrations with a decrease in SOD and POD activities of cuttings, but SOD and POD activities affected by salt stress were different in cuttings treated with three NaCl concentrations. Compared with the control, after salt stress for 54h, a decrease in SOD and POD activities of 38. 8% and 59. 5% was found in cuttings treated with 85mmol · L^{-1} NaCl, respectively, and a 42. 2% and 63. 1% of SOD and POD activities decreased were observed in cuttings treated with 170mmol · L^{-1} NaCl. Whereas, 260mmol · L^{-1} NaCl treatment caused a significant decrease in SOD and POD activities (51. 3% and 73. 0%, respectively). The results showed that all treated cuttings exhibited a lower decrease in SOD activity than in POD activity.

In addition, SOD and POD activities significantly decreased during 3 h at the beginning of all NaCl treatment and then slightly decreased during a period of 9~54h.

3. 2 Change in the content of MDA

With the increase of salt stress, MDA content distinctly increased and reached a maximum value during 15h after the beginning of salt stress, and then decreased, but its level was higher than the control. Compared with the control, MDA content had increased by 24. 9% in cuttings treated with 85mmol · L^{-1} NaCl, by 35. 2% with 170mmol · L^{-1} NaCl, and by 79. 3% with 260mmol · L^{-1} NaCl after salt stress for 54h. The results showed that effects of salt stress on MDA content were different among three NaCl concentrations, a high content of MDA was present in cuttings treated with 260mmol · L^{-1} NaCl.

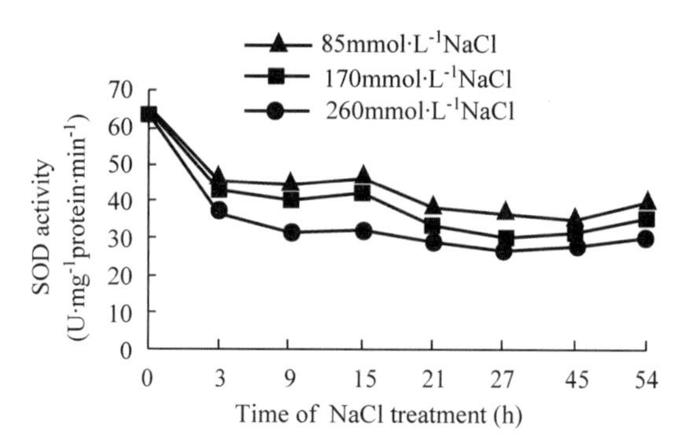

Fig. 1　Changes of SOD activity in leaves of cuttings treated with three NaCl concentrations

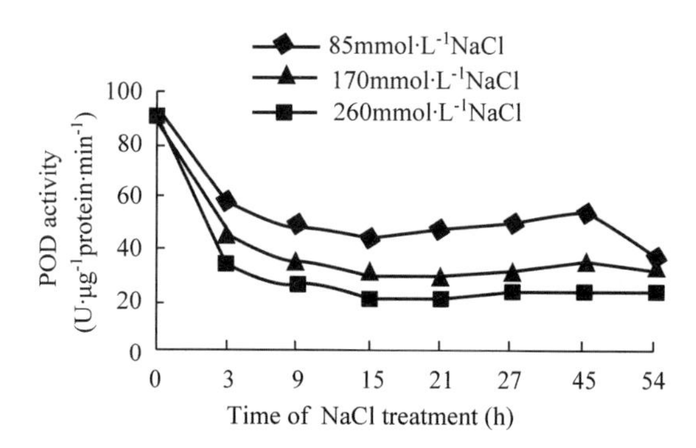

Fig. 2　Changes of POD activity in leaves of cuttings treated with three NaCl concentrations

4　Discussion

The alleviation of oxidative damage and increased resistance to salt stress is often correlated with a more efficient antioxidative system (Hernandez et al., 1993; 1995; Gossett et al., 1994; Shalata and Tal, 1998; Comba et al., 1998; Meneguzzo et al., 1999; Taha et al., 2000; Shalata et al., 2001). In the experiments, when the NaCl-induced oxidative stress was analyzed, a lower decrease in SOD activity than in POD activity was observed in all treated cuttings (Fig. 1 and 2), indicating that different kinds of antioxidant defense enzymes make different response to salt-stress. It was possible that the response of SOD to NaCl-stress may be more sensitive than SOD, namely, a lower decrease in SOD induced-by NaCl-stress may be regarded as a response to the generated AOS during salt stress due to rapidly decreased POD activity. Another possibility is that the AOS might not only be detoxified by SOD, but some subsequent steps might be followed. The product of superoxide scavenged by SOD is H_2O_2, which may be eliminated by conversion to H_2O by POD (Bowler et al., 1992; Asada 1994; Meneguzzo et al., 1999; Mittova et al., 2000; Shalata et al., 2001). Thus, it was suggested that oxidative damage of cuttings resulted from NaCl-stress be mainly linked to the significant decrease of POD activity.

In this study, an obvious increase in MDA content was detected in *P. tomentosa* cuttings subjected to 260mmol · L^{-1} compared with the two lower concentrations (85mmol · L^{-1} and 170mmol · L^{-1}) (Fig. 3). MDA, which is membrane lipid peroxidation products, is a diagnostic indicator for lipid peroxidation and oxidative damage resulting in the loss of membrane integrity (Draper and Hadley, 1990; Janero, 1990; Smirnoff, 1993). Thus, the effect of NaCl treatment of the higher concentrations on membrane lipid peroxidation was more obvious than those of the two lower concentrations (85mmol · L^{-1} and 170mmol · L^{-1}). The increase of MDA content in *P. tomentosa* cuttings under NaCl stress may be attributed to the enhancement of lipid peroxidation, as was the case for other plants (Gossett et al., 1994; Shalata and Tal, 1998; Shalata et al., 2001).

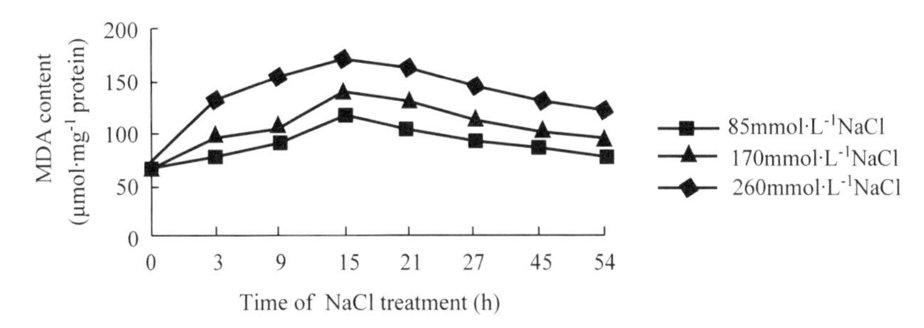

Fig. 3 Changes of MDA content in leaves of cuttings treated with three NaCl concentrations

The MDA content had increased with the decrease of the SOD and POD activities under salt stress (Fig. 1, 2 and 3). With the decrease in the activities of SOD and POD, the ability of the detoxification of AOS decreased and the membrane lipid peroxidation enhanced. Thereby, the decrease in the SOD and POD activities under salt stress seems to be the key factor that leads to the increase of MDA content and the occurrence of oxidative damage of cuttings. In addition, MDA content rapidly increased and reached a maximum value during 15h after the beginning of salt stress and then decreased (Fig. 3). On the other hand, SOD and POD activities significantly decreased during 3h at the beginning of NaCl treatment, and then slightly decreased during a period of 9~54h (Fig. 1 and 2). The results showed that during the first period of salt stress, the significant decrease in SOD and POD activities could cause the great production of AOS. As a result of the elevation of AOS levels, the enzymatic antioxidant defense system was out of balance, severely metabolic function was disrupted, which lead to the degradation of a variety of biological important substances such as membrane lipids, amino acids, proteins and carbohydrates (Janero 1990; Halliwell and Gutteridge 1989; Hauptmann and Cadenzas 1997). Although SOD and POD activities slightly decreased during 9~54h of salt stress, their levels were insufficient to neutralize the elevated levels of AOS and to protect cuttings from oxidative damage.

It could be concluded that the increased synthesis and accumulation of AOS might be a contributing factor in the development of salt stress damage, the development of salt-dependent oxidative stress in cellular membrane could be the results of the enhanced membrane lipid peroxidation caused by the decreased activities of SOD and POD, which might just be one cause of oxidative damage to *P. tomentosa* cuttings under salt stress.

Cold Acclimation Induced Changes in Total Soluble Protein, RNA, DNA, RNase and Freezing Resistance in *Populus tomentosa* Cuttings*

Abstract The changes in the contents of total soluble protein and RNA, the activity of RNase in leaves and branches of *Populus tomentosa* cuttings at various periods (viz: cold acclimation, deacclimation, chilling stress and the recovery after chilling stress), and the survival rate and the freezing resistance of cuttings during cold acclimation at −3℃ were investigated. Results showed that cold acclimation not only increased the contents of total soluble protein and RNA, the survival rates and the freezing resistance of cuttings, decreased the activity of RNase, but also reduced the declining degree of total soluble protein and RNA contents, and the increasing level of RNase caused by chilling stress as compared with the controls. In addition, cold acclimation augmented the increase in the level of total soluble protein and RNA, and facilitated the decrease of RNase during the recovery periods. Further analysis found that the DNA content of all treatments kept relative stability at various periods. The changes in total soluble protein, RNA and RNase were closely related to the freezing resistance of cuttings. It appears that the increase of RNA content caused by cold acclimation induced decrease of RNase activity may be involved in the accumulation of total soluble protein and the induction of freezing resistance of cuttings.

Keywords *Populus tomentosa*, cold acclimation, freezing resistance, total soluble protein, nucleic acid

1 Introduction

Many plants exhibit an increase in freezing tolerance after exposure to low temperature, a process known as cold acclimation. Over the past several years, much attention has been focused on the studies of physiological and metabolic changes caused by cold acclimation, including alterations in lipid composition, changes in enzyme activities, and increase in the concentrations of proline, soluble protein and soluble sugar (Guy, 1990; Thomashow, 1990; Antikainen et al., 1994; Fowler et al., 1981). In recent years, physiological and molecular studies have showed that certain cold acclimation induced the synthesis or accumulation of specific proteins as result of altered gene expression in many herbaceous and cereal plant species (Guy et al., 1987; Shrhan et al., 1987; Guy, 1990; Marentes et al., 1993; Sieg et al., 1996; Arora et al., 1996; Griffith

* 本文原载《中国林业研究》, 2002, 4(2): 9-15, 与林善枝合作发表。

et al. , 1997; Antikainen et al. , 1997; Bravo et al. , 1999). However, little is known about the changes in nucleic acid and ribonuclease, especially about the possible correlations between the changes of total soluble protein, RNA, DNA, RNase and the increase of freezing resistance induced by cold acclimation, especially in woody plants.

In this paper, the correlations mentioned above were investigated in detail to explore the molecular mechanism of cold acclimation and freezing tolerance in woody plants.

2 Materials and methods

2.1 Plant materials and experimental conditions

Populus tomentosa plants were established from dormant branch cuttings collected from Hebei Province in northern China in March 2000. The cuttings were rooted and grown in greenhouse in pots containing a 2∶1(v/v) mixture of soil and sand. After the cuttings were raised for 5 weeks in green house, they were placed outdoors for 3 weeks. Eight-week old cuttings were divided into two groups, 60 cuttings per group. The first group of cuttings placed at 25℃ with an 8h photoperiod and a light intensity of 30umol · m^{-2} · s^{-1} for 14d was referred to as non-acclimated cuttings. The second group of cuttings held at −3℃ with an 8h photoperiod and a light intensity of 30umol · m^{-2} · s^{-1} for 5d, was referred to as cold acclimated cuttings. Some cold acclimated cuttings were transferred to 25℃ with an 8h photo-period . Finally, all these above mentioned cuttings were transferred to −14℃ for 3d, and then were held at 25℃ within 8h photoperiod and a light intensity of 30umol · m^{-2} · s^{-1} for 2d and were referred to as deacclimated cuttings. Branches and leaves were randomly harvested from cuttings at various periods, and used for analyses of total soluble protein, RNA and DNA contents, and RNase activity.

2.2 Evaluation of survival rate and freezing resistance of cuttings

All of the above treated cuttings used for examining freezing resistance were cooled stepwise to 0℃, −1℃, −2℃, −3℃, −4℃, −5℃, −6℃, −7℃, −8℃, −9℃ and−10℃, respectively, in a programmed freezer at a constant rate of 0. 5℃ · min^{-1} and then held for 12h at each temperature . These cuttings were transferred to 25℃ and then thawed for 2d. Survival rates of cuttings were scored upon whether these cuttings grew normally and the degree of freezing injury of bark tissue, leaves and buds after treatment at −10℃ for 12h . Freezing resistance was determined by theminimum temperature for 100% survival.

2.3 Total soluble protein extraction and analysis

Total soluble protein was extracted according to Coleman et al. (1991) and Gu et al. (1999) . The content of total soluble protein was assayed by the method of Bradford (1976) using bovine serum albumin (BSA) as a standard.

2.4 DNA and RNA extraction and measurement

DNA and RNA were extracted from leaves and branches, and measured using the procedure described by Zhu et al. (1993).

2.5 RNase preparation and determination

RNase was extracted and determined using the procedure described by Zhu et al. (1993). There action of enzyme activity was carried out at 30℃ in a 1.0mL reaction mixture containing 10mmol · L^{-1} KCl, 1mmol · L^{-1} $MgSO_4$, 500mmol · L^{-1} saccharose, 2mg · L^{-1} yeast RNA. The reaction was started by the addition of 1.0mL enzyme preparations after at least 10min of pre-incubation, and then stopped by adding 0.5mL of 10% (w/v) ice cool solution of $HClO_4$ and $(NH_4)_6MO_7O_{24}$. One unit (U) of enzyme activity was defined as the amount of the phosphate (Pi) released from the substrate permin.

All the determinations were performed at least in triplicates and the data presented were the average values.

3 Results

3.1 Changes in survival rate and freezing resistance of cuttings

From Table 1, the cold acclimated cuttings had an increase in the survival rate and freezing resistance compared with the control (nonacclimated) cuttings. After 5d of cold acclimation at −3℃, theminimum temperature for 100% survival cuttings decreased from −3℃ in nonacclimated cuttings (control) to −10℃ in cold acclimated cuttings. The survival rate of cuttings at −10℃ for 12h increased from 0% in the control cuttings to 100% in the cold acclimated cuttings. However, after 2d of deacclimation at 25℃, the increase of survival rate and freezing resistance of cuttings induced by cold acclimation decreased rapidly to the level of the control cuttings.

In order to demonstrate in detail the effect of cold acclimation on the development of freezing resistance of cuttings, the time pattern of freezing resistance at −3℃ during 1～5d was studied (Figure 1). One day after transferring cuttings to −3℃, the minimum temperature for 100% survival cuttings started to decrease from −3℃ to −5℃ and gradually reached a maximum level (−10℃) on the 5th day.

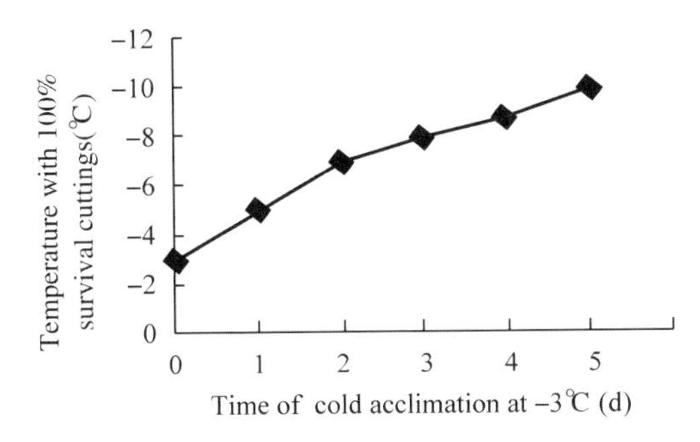

Fig. 1 Effect of cold acclimation at -3℃ on freezing resistance of *P. tomentosa* cuttings

Table 1 The effects of cold acclimation and deacclimation on the survival rates and freezing resistance of *Populus tomentosa* cuttings

Treatment	Survial rates of cuttings at -10℃ for 12h	Temperature with 100% survival cuttings(℃)
Control	0	-3
Cold acclimation	100	-10
Deacclimation	0	-3

3.2 Change in the content of RNA

As shown in Figure 2, RNA content increased with the time of cold acclimation in a period of 5d. After 5d of cold acclimation at -3℃, the content of RNA in leaves and branches of acclimated cuttings had increased by 29.2% and 40.1%, respectively, compared with the controls. Whereas, after 2d of deacclimation at 25℃ the decrease of RNA content in leaves and branches of acclimated cuttings occurred, but its content was still higher than that of the controls (Figure 3).

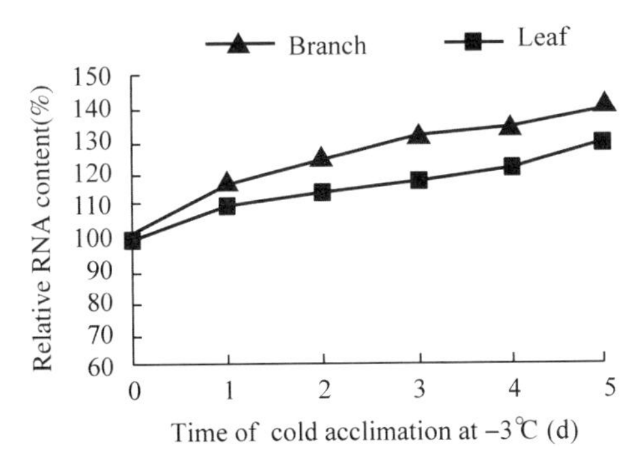

Fig. 2 Effect of cold acclimation on RNA content in leaves and branches of *P. tomentosa* cuttings

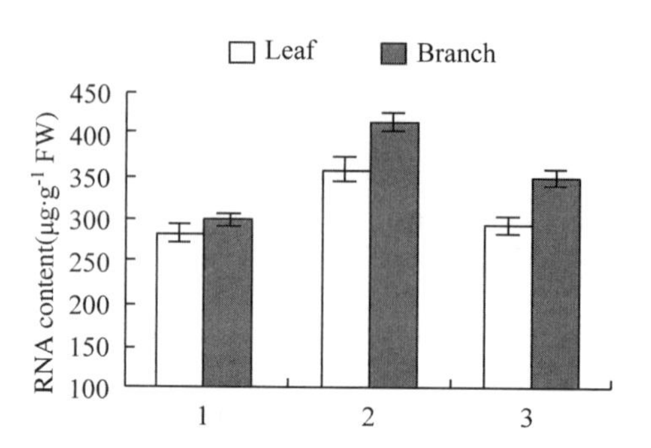

Fig. 3　Changes of RNA content in leaves and branches of *P. tomentosa* cuttings during cold acclimation and deacclimation

1. Conlrol; 2. Cold acclimation; 3. Deaclcimation

During chilling stress at−14℃, the RNA content of leaves and branches has decreased by 40. 4% and 34. 8% in the control cuttings, by 31. 1% and 24. 9% in the cold acclimated cuttings, respectively. On the 3rd day of recovery at 25℃, the increasing level of RNA content in leaves and branches of cold acclimated cuttings was much greater than those of the controls (Figure 4) .

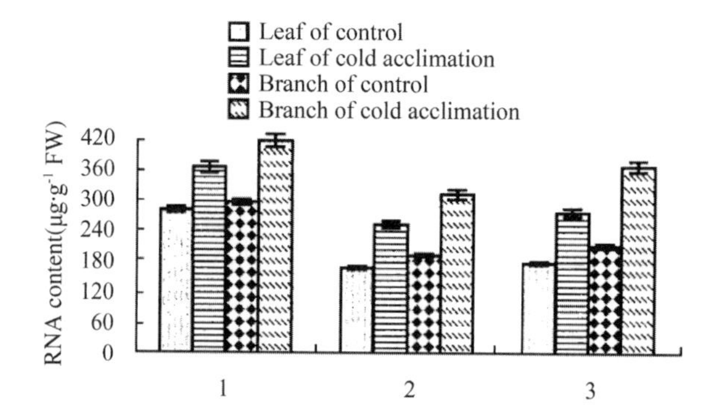

Fig. 4　Changes of RNA content in leaves and branches of *P. tomentosa* cuttings

1. Before chilling stress; 2. After chilling stress; 3. Recovery

3. 3　Change in the content of DNA

No matter when the cuttings were at cold acclimation or at deacclimation, the content of DNA in leaves and branches was closed to the level of the controls (Figure 5). Furthermore, a small decrease of its content caused by chilling stress and a less increase during the recovery period were observed (Figure 6). That is to say, the contents of DNA in leaves and branches of all treated cuttings were kept stable whether the cuttings were at cold acclimation, deacclimation, chilling stress, or recovery compared with RNA content.

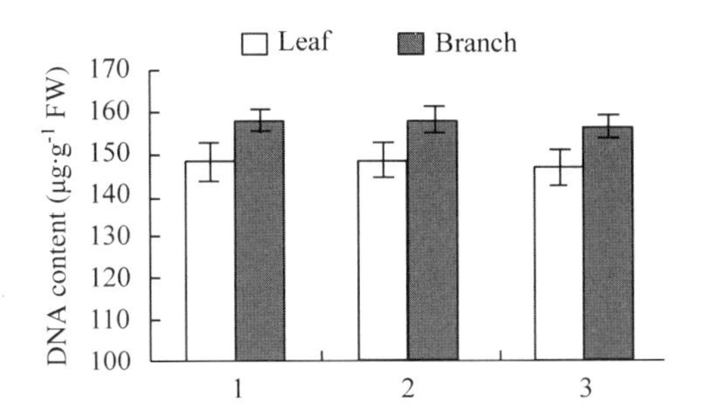

Fig. 5 Changes of RNA content in leaves and branches of *P. tomentosa* cuttings during cold acclimation and deacclimation

1. Control; 2. Cold acclimation; 3. Deacclimation

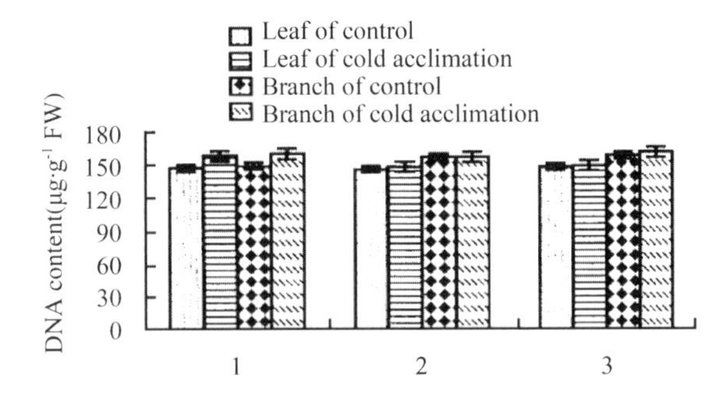

Fig. 6 Changes of DNA content in leaves and branches of *P. tomentosa* cuttings

1. Before chilling stress; 2. After chilling stress; 3. Recovery

3.4 Change in the activity of RNase

There was an obvious decrease in RNase activity of cold acclimated cuttings, its activity decreased with the time of cold acclimation at a period of 5d (Figure 7). After 5d of cold acclimation, the RNase activity in leaves and branches had exhibited a 30.0% and 42.1% decrease, re-

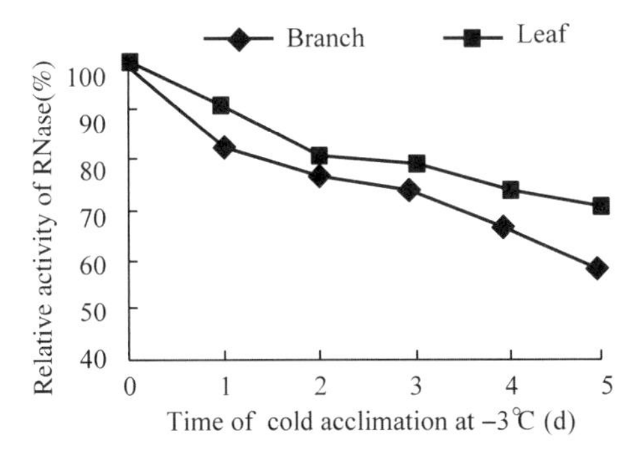

Fig. 7 Effect of cold acclimation on RNase activity in leaves and branches of *P. tomentosa* cuttings

spectively, compared with the Controls. Furthermore, the level of RNase increased when cuttings were deacclimated for 2d (Figure 8) . However, the RNase activity in leaves and branches had increased by 38. 9% and 31. 9% in the control cuttings, by 27. 5% and 26. 3% in cold acclimated cuttings, respectively during chilling stress at−14℃ followed by a decrease on the 3rd day of recovery at 25℃. The decreasing level of RNase activity in leaves and branches of the cold acclimated cuttings was much greater than those of controls (Figure 9).

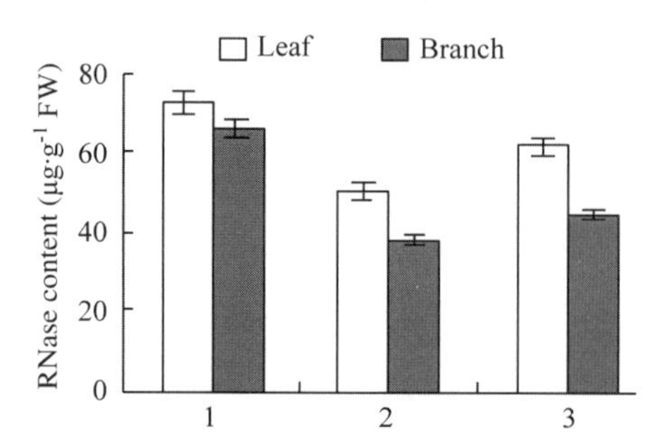

Fig. 8 Changes of RNase activity in leaves and branches of *P. tomentosa* cuttings during cold acclimation and deacclimation.

1. Control; 2. Cold acclimation; 3. Deacclimation

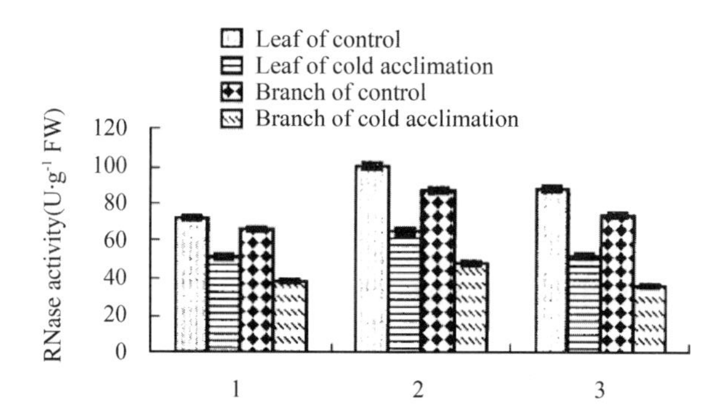

Fig. 9 Changes of RNase activity in leaves and branches of *P. tomentosa* cuttings

1. Before chilling stress; 2. After chilling stress; 3. Recovery

3. 5 Change in the content of total soluble protein

Compared with the controls, the content of RNA in leaves and branches was gradually increased during cold acclimation and reached to the maximum level of 617. 5ug · g^{-1}FW in leaves and 690. 4ug · g^{-1}FW in branches on the 5th day, which were about 146. 5% and 156. 4% of the control level, respectively (Figure 10). However, when cold acclimated cuttings were transferred to 25℃ for 2d, a 32. 3% and 21. 5% less total soluble protein content in leaves and branches were observed compared with their cold acclimated cuttings. After 3d of chilling stress at −14℃, the total soluble protein content of leaves and branches has decreased by 60. 1% and 44. 4% in the control cuttings, by 45. 1% and 40. 0% in cold acclimated cuttings, respectively. In addition, the content of total soluble protein in leaves and branches of cold acclimated cuttings had increased by 12. 0% and 15. 0% during the recovery period, their increasing level were much greater than those of the control cuttings (Figures 11 and 12).

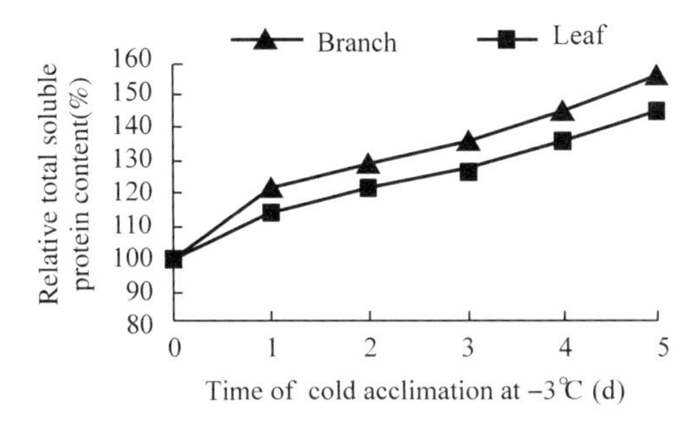

Fig. 10 Effect of cold acclimation on soluble protein content in leaves and branches of *P. tomentosa* cuttings

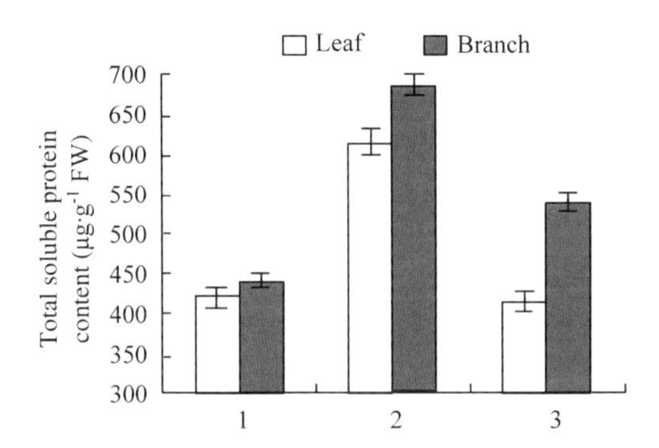

Fig. 11 Changes of DNA content in leaves and branches of *P. tomentosa* cuttings during cold acclimation and deacclimation

1. Control; 2. Cold acclimation; 3. Deacclimation

To examine whether the increase in total soluble protein during cold acclimation at −3℃ was correlated with the acquisition of freezing resistance, we determined the changes in total soluble protein content and freezing resistance of cuttings pretreated with 20μg/L cycloheximide (CH) during cold acclimation at −3℃. After cold acclimation combined with CH-pretreatment (CH-cold) for 5d, the survival temperature for 100% survival cuttings decreased from −10℃ in cold to −3℃ in CH-cold. In accordance, the total soluble protein content in leaves and branches of CH-cold had decreased by 31. 7% and 36. 7%, respectively, compared with that of cold acclimated cuttings (Table 2).

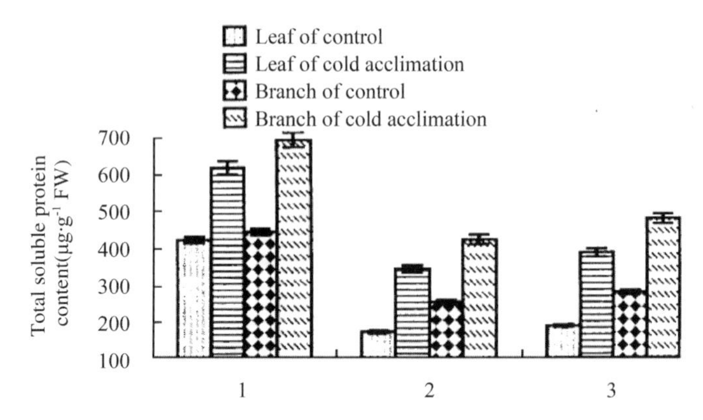

Fig. 12 Changes of total soluble protein content in leaves and branches of *P. tomentosa* cuttings

1. Before chilling stress; 2. After chilling stress; 3. Recovery

Table 2 Effect of CH on the content of total soluble protein and freezing resistance in *P. tomentosa* cuttings during cold acclimation at −3℃

Treatment	Temperature with 100% survival cuttings(℃)	Total soluble protein content(ug/gFW)	
		Leaf	Branch
Control	−3.0	421.4	441.5
5d of cold	−10.0	617.5	690.4
5d of CH-cold	−3.0	421.4	441.5

4 Discussion

It has been reported that the process of cold acclimation not only resulted in an increase in RNA content (Gusta et al., 1972; Sarhan et al., 1975; Durham et al., 1991; Zuo et al., 1999), but also induced the change in translatable messenger RNA (Guy et al. 1985; Meza-Basso et al., 1986; John-son-Flanagan et al., 1987; Mohapatra et al., 1987; Hahn et al. 1989; Danyluk et al., 1990; Kazuoka et al., 1992). In present study, cold acclimation increased the content of RNA in *P. tomentosa* cuttings, reduced the declining degree of RNA content caused by chilling stress, and augmented the increase in the level of RNA during the recovery period of chilling stress compared with control seedlings(Figures 3 and 4). Whereas little change in DNA content was observed at the above periods (Figures 5 and 6), suggesting that the change of environmental temperature had no effect on the DNA content, as reported for other plants (Chen et al., 1980; Gilmour et al., 1988; Zuo et al., 1999). In addition, the trend of change in the RNA content is contrary to that of RNase activity during cold acclimation (Figures 2 and 7). That is to say, the RNA content increased with the decrease of RNase activity and a decline in the content of RNA as RNase activity increase. The decreased RNase activity declined the degradation of RNA during cold acclimation. Thus the accumulation of RNA may be the result of the decrease or inhibition of RNase activity caused by cold acclimation (Gusta et al. , 1972; Brown et al. , 1973; Sasaki et al. , 1976; Sarhan et al. , 1985). We also observed that the cold acclimated cuttings elevated total soluble protein content, and its increasing trend was paralleled to the accumulation of RNA (Figures 2 and 10). Further analysis found that there was a high correlation coefficient of 0. 80 between total soluble protein and RNA . Results showed that the accumulation of RNA might be correlated to the increase in total soluble protein content during cold acclimation. Similar findings were reported in *Larixprinapia rupprechtii* (Zuo et al. , 1999). Although DNA is main genetic material and its level has not been affected easily by temporary low temperature, cold acclimation induced elevation at transcriptional level of DNA results in an increase in total RNA content, especially in translatable messenger RNA, which facilitates the synthesis of protein.

Recently, some studies revealed that specific protein synthesis caused by gene expression seemed to be an important mechanism involved in the induction of freezing tolerance during cold acclimation (Guy, 1990; Liu et al. , 1991; Griffith et al. , 1992; Marentes et al. , 1993; Pan et al. , 1994; Houde et al. , 1995; Antikainen et al. , 1996; Bravo et al. , 1999) . In our experiments, the levels of total soluble protein correlated with the degree of freezing tolerance during cold acclimation and deacclimation, the increase in the total soluble protein content accompanied by the acquisition of freezing tolerance during cold acclimation (Figures 1 and 10) . The fact that cold acclimation induced the increase of total soluble protein and the acquisition of freezing tolerance was strongly inhibited by cycloheximide, a well known powerful inhibitor of protein synthesis, indicated that the decrease of freezing tolerance might be the result of the inhibition of protein synthesis by cycloheximide. The increase in the content of total soluble protein was closely associated with the enhancement of freezing tolerance during cold acclimation and may play a critical role in the process of cold acclimation. On the basis of recent studies in many plants, the characteristics of the role of antifreeze protein can be summarized as follows: the direct affection on extracellular ice crystal formation; the indirect alteration in the percentage of water frozen at a given temperature by restricting the rate of growth of extracellular ice crystals; the depression in the freezing point of the body fluid and the stabilization in cell membranes against either rupture or loss of its semipermeablity (Kazuoka et al. , 1992; Hughes et al. , 1996; Anti-kainen et al. , 1997; Griffith et al. , 1997; Bravo et al. , 1999).

毛白杨幼苗低温锻炼过程中 Ca^{2+} 的作用及细胞 Ca^{2+}-ATP 酶活性的变化*

摘　要　观察了低温锻炼以及随后的常温生长中毛白杨幼苗质膜及线粒体 Ca^{2+}-ATP 酶活性、CaM 含量和抗冻性的变化。结果表明，单纯低温锻炼在一定程度上提高了毛白杨幼苗质膜及线粒体 Ca^{2+}-ATP 酶活性、CaM 含量和抗冻性，减小了低温胁迫所引起的质膜及线粒体 Ca^{2+}-ATP 酶活性和 CaM 含量的下降程度，促进了胁迫后恢复过程中质膜及线粒体 Ca^{2+}-ATP 酶活性和 CaM 水平的迅速回升。在低温锻炼的同时，用 $CaCl_2$ 处理能加强低温锻炼的效果，但这种效应可被 EGTA、LaCl3 和 CPZ 处理所抑制。

关键词　毛白杨，低温锻炼，$CaCl_2$，Ca^{2+}-ATPase，钙调蛋白，抗冻性

自从 1976 年钙调蛋白(CaM) 被发现以来，Ca^{2+} 与其重要受体蛋白 CaM 构成的钙信使系统在植物感受逆境信号中所起的作用已日益引起人们的重视(Poovaiah 和 Reddy，1993；郝鲁宁和余叔文，1992；龚明等，1990；Bush，1995；李卫等，1997；曾韶西和李美如，1999)。低温是一种重要的逆境，低温锻炼对提高植物抗低温逆境能力具有十分重要的作用(简令成，1992；潘杰等，1994；刘鸿先等，1991；Guy 等，1987；曾韶西和李美如，1999)。低温锻炼所引起的细胞质内 Ca^{2+} 浓度和 CaM 含量的增加与植物抗冻性的提高有关(孙大业等，1998；李卫等，1997)，而 Ca^{2+} 与 CaM 结合后所形成的 Ca^{2+}-CaM 复合物可能参与了 Ca^{2+}-ATP 酶活性的调节(Fukumoto 和 Venis，1986；Rai-Caldogon et al.，1992；Wimmers et al.，1992；Tawfik 和 Palta，1993；林善枝等，2001)。但到目前为止，还未见有关低温锻炼的同时用 $CaCl_2$ 处理所引起的 CaM 水平变化及其与 Ca^{2+}-ATP 酶活性的关系的报告。本文以新品系毛白杨幼苗为试材，欲通过研究在低温锻炼的同时用 $CaCl_2$、钙离子螯合剂 EGTA、钙离子通道阻断剂 $LaCl_3$、钙调素拮抗剂 CPZ 处理对幼苗 CaM 含量、质膜及线粒体 Ca^{2+}-ATP 酶活性及抗冻性的影响，并研究这些指标在低温胁迫下及随后的回常温恢复生长中的变化，从中探讨 CaM 水平变化及其与 Ca^{2+}-ATP 酶活性的关系，Ca^{2+} 和 CaM 及 Ca^{2+}-ATP 酶在低温锻炼中对提高毛白杨幼苗抗冻性的作用机理，以及幼苗发生冻害的可能机制，从而为毛白杨抗冻性改良与解决生产实际问题提供科学依据。

* 本文原载《植物生理与分子生物学学报》，2002，28(6)：449-456，与林善枝合作发表。

1 材料与方法

1.1 材料

供试材料取自北京林业大学苗圃，为当年生、生长健壮、长势一致、高 40~50cm 盆栽土培的毛白杨(*Populus tomentosa*) 无性系 BT17 幼苗。将供试幼苗分 3 组，第 1 组幼苗置于25~30℃下，并用4 种效应剂：$CaCl_2$ 10mmol/L、钙离子螯合剂 EGTA 3mmol/L、钙离子通道阻断剂 $LaCl_3$ 100μmol/L 和钙调素拮抗剂 CPZ 50μmol/L 分别处理 5d，每种处理 40 盆幼苗，每盆 1 株，每天每盆浇 50ml，然后移入-3℃进行锻炼 5d，这些幼苗称为效应剂预处理的低温锻炼幼苗；第 2 组幼苗共 40 盆，每盆 1 株，先置于 25~30℃下生长 5d，每天每盆浇蒸馏水 50ml，然后再移入-3℃进行锻炼 5d，这些幼苗称为低温锻炼的幼苗；第 3 组幼苗共 40 盆，每盆 1 株，置于 25~30℃下生长，前 5d 每天每盆浇蒸馏水 50ml，这些幼苗作为对照。最后从上述 3 组处理中分别取出 5 盆幼苗置于-14℃低温生化培养箱内进行低温胁迫处理，3d 后取出置于 25~30℃下恢复生长 3d。以上各种处理每天光照 6h，光照强度为 30μmol/(m^2 · s)。在单纯低温锻炼后、结合效应剂处理的低温锻炼后、低温胁迫后及恢复第 3 天分别取幼苗的叶片和枝条进行各项指标的测定。本实验在 2000 年 7~8 月进行，材料处理及各项指标测定均重复 3 次，取其平均值。

1.2 幼苗半致死温度(LT_{50}) 测定

幼苗存活率测定参照潘杰等(1994) 的方法，并略有修改。对照以及结合 $LaCl_3$、EGTA 或 CPZ 处理的低温锻炼幼苗均分别用-3.0、-5.0、-7.0、-9.0、-11.0℃处理；单纯低温锻炼幼苗分别用-10、-12、-14、-16、-18℃处理；结合 $CaCl_2$ 处理的低温锻炼幼苗分别用-14、-16、-18、-20、-22℃处理。每个温度各处理 12h，然后转到 25~30℃下，2h 后测定幼苗的存活率，膨胀度的维持与是否生长以及冻害程度是测定存活率的标准。用指数函数对所得到的不同温度下的幼苗存活率进行拟合，求出幼苗存活率为 50%的温度作为 LT_{50}。以 LT_{50} 的高低来表示抗冻性的相对大小。

1.3 质膜制备和 Ca^{2+}-ATP 酶活性测定

质膜制备按焦新之等(1988) 的方法。质膜制剂纯度的鉴定参照 O' Neill 和 Spanswick (1984) 的方法，以 Na_3VO_4 作为质膜 ATPase 专一抑制剂。质膜 Ca^{2+}-ATPase 活性测定参照 Lin 和 Morales (1977) 方法，0.5ml 反应液中 ATP 3mmol/L、$CaCl_2$ 3mmol/L、$(NH_4)_6MO_7O_{24}$ 1mmol/L、Tris2Mes 50mmol/L (pH 6.5)、0.01%Triton X-100、20μg 膜蛋白，用 ATP 启动反应，37℃下反应 30min，然后用 0.5mL 10%的冰冷 TCA 中止酶反应。

1.4 线粒体制备和 Ca^{2+}-ATP 酶活性测定

线粒体制备按欧阳光察(1985) 的方法。线粒体制剂纯度的鉴定参照 Dupont 等(1982) 的方法，以 NaN_3 作为线粒体 ATPase 专一抑制剂。线粒体 Ca^{2+}-ATPase 活性测定参照曾韶西和李美如(1999)方法，并略有修改。0.5ml 反应液中 ATP 5mmol/L、$CaCl_2$ 4mmol/L、$(NH_4)_6MO_7O_{24}$2mmol/L、Tris-Mes 50mmol/L（pH6.0）、0.01% Triton X-100、20μg 膜蛋白，用 ATP 启动反应，37℃下反应 30min，然后用 0.5ml 10%的冰冷 TCA 中止酶反应。

1.5 钙调素(CaM)提取与含量测定

分别称取 1.0g 幼苗叶片和枝条，加入 1.5ml 提取液(Tris-HCl 50mmol/L（pH 8.0）、EGTA 1mmol/L、PMSF 0.5mmol/L、NaCl 0.15mol/L、$NaHSO_3$ 20mmol/L)，于冰浴中充分研磨，匀浆液在 90℃水浴中处理 3min，冷却后在 7000g 下离心 30min，上清液即用于 CaM 的测定。CaM 的含量测定按赵升皓等(1988)的方法。

1.6 蛋白质含量测定

蛋白质含量测定按 Bradford（1976)的方法。

2 结果

2.1 低温锻炼及各种效应剂处理对毛白杨幼苗抗冻性的影响

低温锻炼 5d 后，毛白杨幼苗 50%存活的温度（LT_{50}）由未锻炼幼苗的-6.2℃下降到-14.3℃；而结合 $CaCl_2$ 处理的低温锻炼后幼苗的 LT_{50} 由未锻炼幼苗的-6.2℃下降到-18.1℃，与单纯的低温锻炼相比，LT50 下降了 3.8℃(表 1)。说明在低温锻炼的同时用 $CaCl_2$ 处理可加强低温锻炼提高幼苗抗冻性的作用。而结合 EGTA、$LaCl_3$、CPZ 处理的低温锻炼幼苗，其 LT50 接近未锻炼对照水平。这表明这些通过螯合 Ca^{2+}、阻断其通道或与 CaM 拮抗来 Ca^{2+}阻挡发生作用的物质能够消除 Ca^{2+}加强低温锻炼诱导的幼苗抗冻性形成的作用，从而证明 Ca^{2+}和 CaM 参与毛白杨幼苗抗冻性的低温诱导过程。

表 1 低温锻炼及结合 $CaCl_2$、EGTA、$LaCl_3$、CPZ 处理的低温锻炼对毛白杨幼苗半致死温度的影响

Treatments	LT_{50} of cuttings(℃)
Control	-6.2
Chilling hardening 5d	-14.3
Chilling hardening 5d	-18.1
Chilling hardening 5d	-6.7
Chilling hardening 5d	-6.9
Chilling hardening 5d+CPZ pretreatment 5d+freezing acclination 5d	-6.1

2.2 低温锻炼及各种效应剂处理对毛白杨幼苗 CaM 含量的影响

经或未经 $CaCl_2$ 处理的低温锻炼幼苗叶片和枝条的 CaM 含量均高于未锻炼幼苗水平，但结合 $CaCl_2$ 处理的低温锻炼幼苗其 CaM 含量有较大幅度增加，与单纯低温锻炼幼苗相比，叶片和枝条 CaM 含量分别增加了 39.15 和 48.74μg/g FW。而结合 EGTA、$LaCl_3$ 和 CPZ 处理的低温锻炼仅略高于未锻炼水平(图 1)。说明 Ca^{2+} 对 CaM 的积累有促进效应。-14.0℃低温胁迫 3d 后，未锻炼幼苗以及有无 $CaCl_2$ 处理的低温锻炼幼苗其叶片和枝条 CaM 含量均有所下降。与胁迫前相比，结合 $CaCl_2$ 处理的低温锻炼幼苗其叶片和枝条的 CaM 含量分别下降了 43.13 和 38.32μg/g FW，而单纯低温锻炼则分别下降了 50.85 和 47.27μg/g FW，但两者下降程度都小于未锻炼幼苗；恢复生长 3d 后，结合 $CaCl_2$ 处理的低温锻炼幼苗 CaM 含量回升程度大于单纯低温锻炼幼苗，尤以枝条回升更为显著(图 2)。说明在低温锻炼的同时用 $CaCl_2$ 处理可加强低温锻炼提高幼苗 CaM 含量，减少低温胁迫引起的 CaM 含量的下降程度，且能使 CaM 含量在恢复期迅速回升，这也许正是 Ca^{2+} 作用的结果。

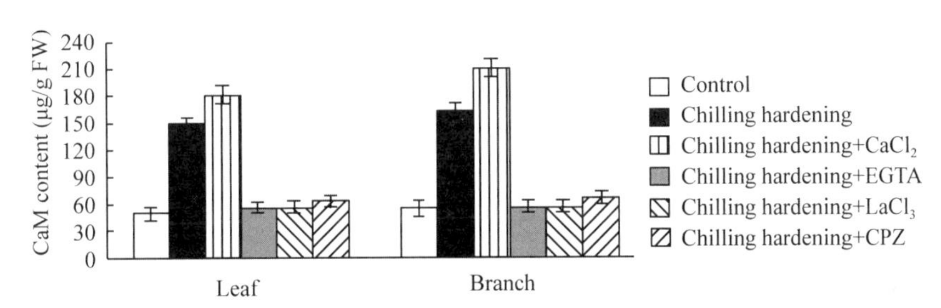

图 1 低温锻炼以及结合 $CaCl_2$、EGTA、$LaCl_3$、CPZ 处理的低温锻炼对毛白杨幼苗叶片和枝条 CaM 含量的影响

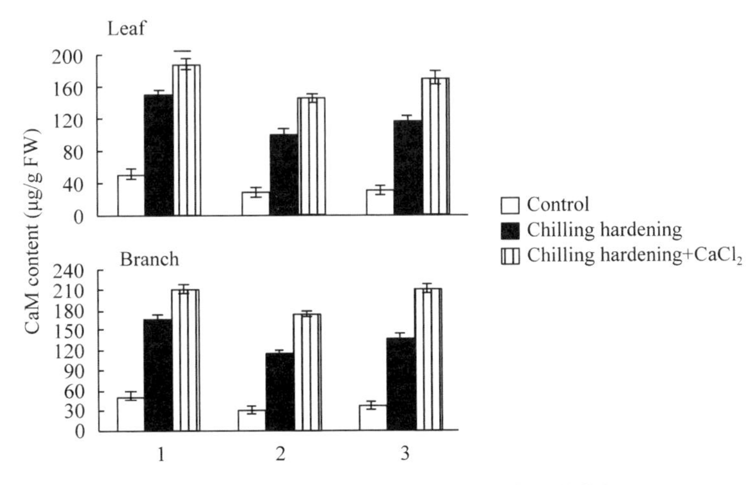

图 2 毛白杨幼苗叶片和枝条 CaM 含量的变化

2.3　质膜 Ca^{2+}-ATP 酶活性的变化

质膜 Ca^{2+}-ATP 酶是 Ca^{2+}泵的主要组分，是控制胞质内游离 Ca^{2+}浓度的主要调节者，它能有效地将因逆境刺激所升高的过量胞质 Ca^{2+}及时运到胞外而使胞质 Ca^{2+}迅速回落到较低水平，对膜的稳定性起着重要作用。与单纯低温锻炼相比，在低温锻炼的同时用 $CaCl_2$ 处理可使叶片和枝条质膜 Ca^{2+}-ATP 酶活性分别提高 24.12 和 37.25μmol/(Pi mg · protein h)，说明 $CaCl_2$ 处理对低温锻炼提高或激活质膜 Ca^{2+}-ATP 酶活性的作用有明显的促进效应。但这种效应可被 EGTA、$LaCl_3$ 或 CPZ 所抑制，体现在幼苗质膜 Ca^{2+}-ATP 酶活性明显降低（图 3）。表明 Ca^{2+}和 CaM 可能参与了质膜 Ca^{2+}-ATP 酶活性的调节。在 -14.0℃低温胁迫下，各种处理幼苗叶片和枝条质膜 Ca^{2+}-ATP 酶活性均有所降低，其中枝条 Ca^{2+}-ATP 酶活性下降相对较小。$CaCl_2$ 处理的低温锻炼幼苗，无论是叶片还是枝条，质膜 Ca^{2+}-ATP 酶活性下降程度都小于单纯低温锻炼幼苗，但两者下降程度均小于未锻炼幼苗。回常温恢复 3d 后，各种处理幼苗质膜 Ca^{2+}-ATP 酶活性均有所提高，尤以结合 $CaCl_2$ 处理的低温锻炼的幼苗枝条提高程度更为明显（图 4）。说明 $CaCl_2$ 处理能使幼苗质膜 Ca^{2+}-ATP 酶保持相对较高的活性水平，甚至在能伤害幼苗的低温下也有激活幼苗质膜 Ca^{2+}-ATP 酶活性的作

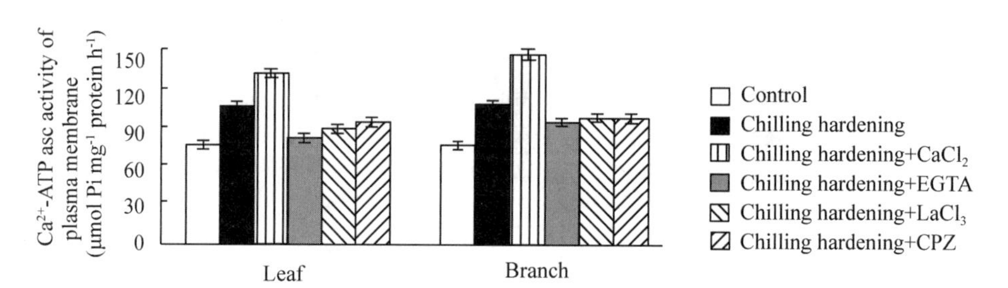

图 3　低温锻炼以及结合 $CaCl_2$、EGTA、$LaCl_3$、CPZ 处理的低温锻炼对毛白杨幼苗叶片和枝条质膜 Ca^{2+}-ATP 酶活性的影响

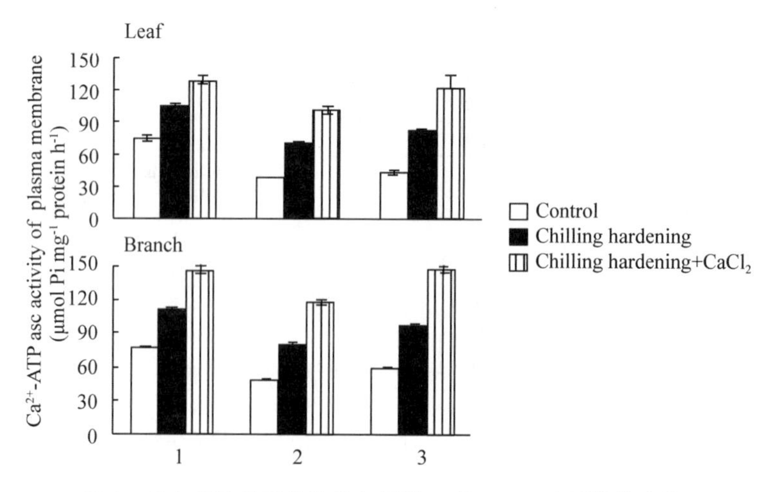

图 4　毛白杨幼苗叶片和枝条质膜 Ca^{2+}-ATP 酶活性的变化

1. 低温胁迫前；2. 低温胁迫后；3. 恢复后

用，这也许是钙信使系统对胞质 Ca^{2+} 浓度升高的一种反馈控制反应。反映了相同浓度的 Ca^{2+} 以及相同条件的低温锻炼对同一植株枝条的质膜 Ca^{2+}-ATP 酶的作用效果较叶片明显，以致使枝条无论在低温锻炼中还是在低温胁迫下或恢复生长期均能保持相对较高的质膜 Ca^{2+}-ATP 酶活性和较低水平的胞质 Ca^{2+}。

2.4 线粒体 Ca^{2+}-ATP 酶活性的变化

植物线粒体 Ca^{2+}-ATP 酶具有依赖于 Ca^{2+} 而水解 ATP 的活性，它在调节细胞和线粒体 Ca^{2+} 水平上起着十分重要的作用。图 5 表明，在低温锻炼的同时用 $CaCl_2$ 处理可明显提高幼苗线粒体 Ca^{2+}-ATP 酶活性，与单纯低温锻炼相比，其叶片和枝条的线粒体 Ca^{2+}-ATP 酶活性分别提高了 15.48 和 21.56μmol Pimg^{-1} protein h^{-1}，比未锻炼幼苗的分别提高了 28.45 和 39.71μmol Pimg^{-1} protein h^{-1}；而在低温锻炼的同时加 EGTA、$LaCl_3$ 或 CPZ 处理则明显抑制了低温锻炼对幼苗线粒体 Ca^{2+}-ATP 酶活性的提高。说明 Ca^{2+} 和 CaM 可能参与线粒体 Ca^{2+}-ATP 酶活性的调节。

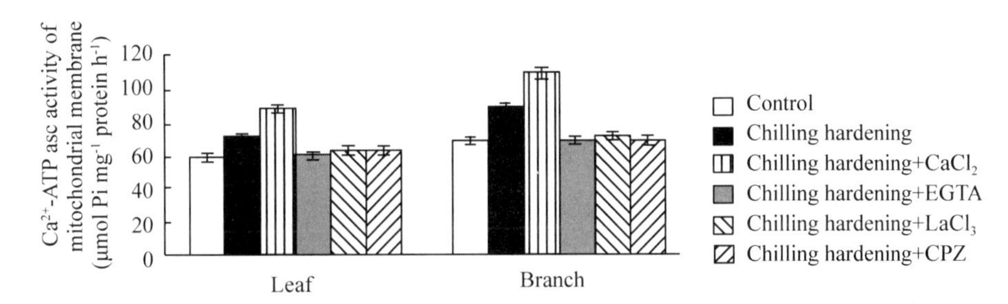

图 5 低温锻炼以及结合 $CaCl_2$、EGTA、$LaCl_3$、CPZ 处理的低温锻炼对毛白杨幼苗叶片和枝条线粒体膜 Ca^{2+}-ATP 酶活性的影响

-14.0℃低温胁迫能不同程度引起各种处理幼苗叶片和枝条线粒体 Ca^{2+}-ATP 酶活性降低，其中枝条 Ca^{2+}-ATP 酶活性下降相对较小，$CaCl_2$ 处理的低温锻炼幼苗，无论是叶片还是枝条，线粒体 Ca^{2+}-ATP 酶活性下降程度都小于单纯低温锻炼幼苗，但两者下降程度均小于未锻炼幼苗。回常温恢复 3d 后，各种处理幼苗线粒体 Ca^{2+}-ATPase 活性均有所提高，尤以结合 $CaCl_2$ 处理的低温锻炼的幼苗枝条提高程度更为明显(图 6)。说明不论在低温锻炼中还是在低温胁迫下或恢复生长期，在低温锻炼的同时用 $CaCl_2$ 处理均能使幼苗线粒体 Ca^{2+}-ATP 酶保持相对较高的活性水平。反映出枝条无论在低温锻炼中还是在低温胁迫下或恢复生长期均能保持相对稳定的线粒体 Ca^{2+}-ATP 酶活性反应，以提供较多的能量用于抵御外界低温的侵袭，这正是枝条比叶片较抗冻的主要原因之一。

3 讨论

CaM 是一种分布较广、具有多种功能的调节蛋白，可作为 Ca^{2+} 的重要受体蛋白，在 Ca^{2+} 信使系统信息传递过程中起着重要作用，Ca^{2+} 与 CaM 结合后使 CaM 活化，然后激活并调节一些依赖 Ca^{2+}-CaM 酶类(靶酶)的活性，推动相应的生理生化过程(Hepler 和 Wayme，

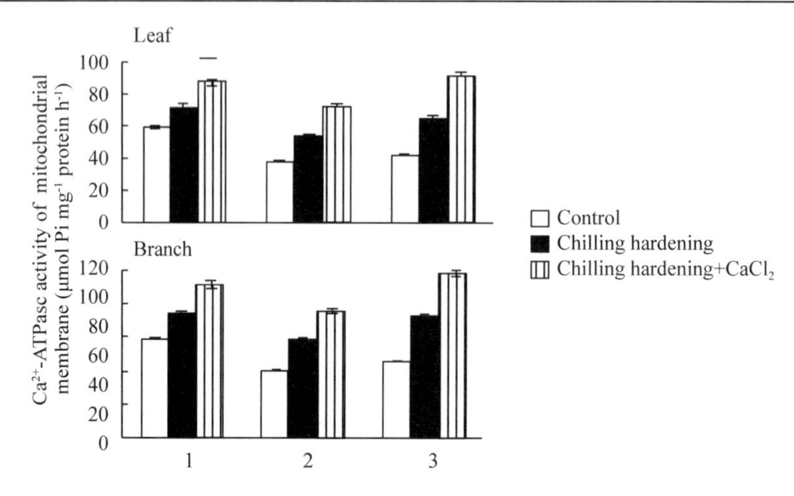

图 6　毛白杨幼苗叶片和枝条线粒体膜 Ca^{2+}-ATP 酶活性的变化

1. 低温胁迫前；2. 低温胁迫后；3. 恢复后

1985；Rickauer 和 Tanner，1986；孙大业等，1998；李美如等，1996；曾韶西和李美如，1999；Bonza et al.，2000；Chung et al.，2000）。许多研究表明，在低温锻炼过程中有 CaM 积累，它可能是植物适应低温胁迫时的一个普遍反应（李卫等，1997；Erlandson 和 Jensen，1989）。Monroy 等（1993）及 Knight 等（1991）在低温驯化的同时用钙离子螯合剂 EGTA、钙离子通道阻断剂 La^{3+}、质膜拉伸通道抑制剂 Gd^{+}和钙调素拮抗剂 W7、CPZ 及 TFP 处理，结果发现这些效应剂强烈地抑制了低温诱导的抗冻力形成；曾韶西和李美如（1999）对冷和盐处理水稻幼苗在处理前用钙离子螯合剂 EGTA 和钙调素拮抗剂 CPZ 进行预处理，观察到明显地抑制由冷和盐处理诱导的提高抗寒力的作用。这些研究为说明 Ca^{2+}-CaM 信使系统在植物低温驯化过程中起着重要的调节作用提供了直接的证据。李美如等（1996）报告了外源 $CaCl_2$ 处理可加强冷锻炼的作用效果，且明显地提高水稻幼苗结合态钙的水平和抗冷性。

本文结果发现在低温锻炼的同时用 $CaCl_2$ 处理，幼苗 CaM 含量和抗冻性的提高程度明显大于单一的低温锻炼，而 EGTA、CPZ 或 $LaCl_3$ 处理则强烈地抑制了低温锻炼对幼苗抗冻性和 CaM 含量的提高作用（图 1，表 1）。表明钙离子处理能促进 CaM 的合成，而钙离子螯合剂或钙离子通道阻断剂的处理则抑制了 CaM 的合成，在毛白杨幼苗抗冻性的低温诱导过程中也伴随着 CaM 的积累，而且这种积累是依赖 Ca^{2+}的。

近年来研究发现，Ca^{2+}在植物抗逆境过程中具有重要作用，Minorsky（1985）曾推测 Ca^{2+}可能起低温信息传导物的作用。启动以 Ca^{2+}和 CaM 为核心的信使系统的一个中心环节是胞内 Ca^{2+}浓度的改变（Poovaiah et al.，1987；郝鲁宁和余叔文，1992），维持胞内 Ca^{2+}的低稳态水平则是 Ca^{2+}发挥其信使作用的前提条件，而控制胞质内游离 Ca^{2+}浓度涨落主要由定位在质膜和细胞器外膜的 Ca^{2+}-ATPase 来执行（龚明等，1990；曾韶西和李美茹，1999）。从本实验结果可以看出，在低温锻炼的同时用 $CaCl_2$ 处理不仅能明显提高幼苗质膜及线粒体 Ca^{2+}-ATP 酶的活性（图 3，5），而且也伴随着 CaM 含量和抗冻性的提高（图 1，表 1），但这种 Ca^{2+}的加强低温锻炼诱导的幼苗抗冻性形成和质膜及线粒体 Ca^{2+}-ATP 酶活性提高的作用效应则可被钙离子螯合剂 EGTA、钙离子通道阻断剂 $LaCl_3$ 或 CaM 拮抗剂 CPZ 处理所消除（图 1、3、5，表 1）。这表明质膜和线粒体 Ca^{2+}-ATP 酶可能是 Ca^{2+}-CaM 的作用靶酶，Ca^{2+} 与 CaM 结合形成有活性 Ca^{2+}-CaM 复合物后进一步激活并调节质膜和线粒体 Ca^{2+}-ATP 酶，而这种调节的结果是加快质膜 Ca^{2+} 泵运转，增强 ATP 水解作用，促进细胞生理代谢发生适应性变化，从而使得与抗冻性形成有关的 RNA、蛋白质和脂类等物质合成加强，最终导致幼苗抗冻性的提高。据此可以认为，$CaCl_2$ 处理加强低温锻炼诱导的幼苗抗冻性形成与质膜和线粒体 Ca^{2+}-ATP 酶活性提高密切相关，Ca^{2+}-CaM 信号系统参与幼苗质膜和线粒体 Ca^{2+}-ATP 酶活性的调节与抗冻性的诱导过程。

低温胁迫虽能引起经或未经 $CaCl_2$ 处理的低温锻炼幼苗质膜和线粒体 Ca^{2+}-ATP 酶活性的降低，但结合 $CaCl_2$ 处理的低温锻炼幼苗降低程度不如单纯低温锻炼幼苗明显（图 4、6）。这表明结合 $CaCl_2$ 处理的低温锻炼能较好地维持或激活幼苗质膜和线粒体 Ca^{2+}-ATP 酶活性，使幼苗能更有效地把因低温胁迫刺激所升高的过量胞质 Ca^{2+} 及时地运到胞外或运入胞内钙库储藏起来，从而使胞质 Ca^{2+} 浓度迅速降低到一种静息态（即所谓的低稳态）水平，最终导致细胞生理代谢和信息传递功能的正常进行。与结合 $CaCl_2$ 处理的低温锻炼幼苗相比，单纯低温锻炼幼苗，尤其是未锻炼幼苗由于在低温胁迫下，质膜和线粒体 Ca^{2+}-ATP 酶活性显著下降，致使被升高的过量胞质 Ca^{2+} 不能及时外运，造成细胞一定的伤害。说明质膜和线粒体 Ca^{2+}-ATP 酶活性的提高是钙信使系统对胞质 Ca^{2+} 浓度升高的一种反馈控制行为，而结合 $CaCl_2$ 处理的低温锻炼能明显提高幼苗抗冻性可能与其具有较强维持 Ca^{2+} 低稳态的能力有关。这正是在低温锻炼的同时用 $CaCl_2$ 处理能更有效地提高毛白杨幼苗抗冻性的主要原因之一。

Calcium Effect and Changes of Ca^{2+}-ATPase Activities During Chilling Hardening in *Populus tomentosa* Cuttings

Abstract *Populus tomentosa* cuttings were established from dormant branches collected from Heilongjiang Province, China. To explore the role of calcium calmodulin messenger system in the transduction of low temperature signal in woody plants, *P. tomentosa* cuttings used for chilling hardening at −3℃ were pretreated with $CaCl_2$(10mmol/L), Ca^{2+} chelator EGTA (3mmol/L), Ca^{2+} channel inhibitor $LaCl_3$(100μmol/L) or CaM antagonist CPZ (50μmol/L). The changes in calmodulin (CaM) content, Ca^{2+}-ATPase activities of mitochondrial and plasma membrane, and LT50 of cuttings were investigated to elucidate the physiological mechanisms by which trees adapt to chilling. The results showed that the content CaM, the Ca^{2+}-ATPase activities of mitochondrial and plasma membrane as well as chilling resistance of cuttings were increased by chilling hardening at−3℃ (Fig. 1, 3 and 5; Table 1). Treatment with $CaCl_2$at the time of chilling hardening enhanced the effect of chilling hardening and markedly increased the content of CaM, the Ca^{2+}-ATPase activities of mitochondrial and plasma membrane and the chilling resistance, but this enhancement was abolished by Ca^{2+} chelator EGTA, Ca^{2+} channel inhibitor $LaCl_3$ or CaM antagonist CPZ (Figs. 1, 3 and 5; Table 1), indicating that the calcium-calmodulin messenger system was involved in the course of chilling resistance formation. The leaves and branches of cuttings pretreated with $CaCl_2$ had increased by 39. 15 and 48. 74μg/g FW of CaM content, by 24. 12 and 37. 25μmol Pi mg^{-1} protein h^{-1} of Ca^{2+}-ATPase activities of plasma membrane, and by 15. 48 and 21. 56μmol Pi mg^{-1} protein h^{-1} of Ca^{2+}-ATPase activities of mitochondrial membrane, respectively, as compared with chilling hardening (Figs. 2, 4 and 6). In addition, LT50 of cuttings was lowered from −14. 3℃ in chilling hardening cuttings to −18. 1℃ in the cuttings pretreated with $CaCl_2$ (Table 1). The addition of $CaCl_2$ at the same time of chilling hardening reduced the declining degree of CaM content, Ca^{2+}-ATPase activities of mitochondrial and plasma membrane caused by chilling stress at −14℃, and enhanced the increase level of CaM content and of Ca^{2+}-ATPase activity in the recovery periods (Figs. 2, 4 and 6). Furthermore, the change in CaM content was closely correlated to the Ca^{2+}-ATPase activities of mitochondrial and plasma membrane, and correlated to the chilling resistance of cuttings after both chilling hardening with or without $CaCl_2$ pretreatment. It is suggested that the enhancement of chilling resistance of cuttings induced by chilling hardening be related to the effective activation of Ca^{2+}-ATPase activities of mitochondrial and plasma membrane. Ca^{2+} calmodulin is involved in the regulation of the Ca^{2+}-ATPase activities of mitochondrial and plasma membrane, and the induction of chilling resistance of cuttings.

Keywords *Populus tomentosa*, chilling hardening, $CaCl_2$, Ca^{2+}-ATPase, calmodulin, chilling resistance

Identification of Expression of *CpT* I Gene in Transgenic Poplars at Protein Level*

Abstract The contents of total soluble protein and cowpea trypsin inhibitor (*CpT* I) in the browse and metaphylla of transgenic hybrid triploid poplars [(*Populus tomentosa*×*P*. *bolleana*)×*P*. *tomentosa*] transformed with *CpT* I gene were determined in order to study the products of *CpT* I gene expression at protein level. The results indicated that the amount of total soluble protein was greater in transgenic poplars than in non-transgenic poplars, but was more in the metaphylla than in browse. The expression of *CpTI* gene resulted in an obvious increase in *CpT* I content, whereas *CpT* I was not detected in non-transgenic poplars. It was found that there were high amount of total soluble protein and *CpT* I in 3 clones of TG07, TG04 and TG71 compared with other transgenic clones. In addition, the analysis of protein by SDS-PAGE showed that a specific protein band of about 11.3 kD corresponding to the 80 amino acids encoded by the *CpT* I gene was observed in transgenic poplars on the gel of protein, which was not detected in non-transgenic poplars.

Keywords *Populus tomentosa*, transformat ion, *CpT* I , *CpT* I gene, total soluble protein

1 Introduction

Trees are susceptible to many insect pests, and the breeding for resistance is a difficult task in longlived trees. Genetic engineering provides the opportunity to transfer new specific traits of interest into valuable genotypes. In trees, resistance to insect pests has been achieved by employing two types of gene products: protease inhibitors, and insecticidal crystalline proteins (from *Bacillus thur ingiensis*). Proteinase inhibitor proteins are found throughout all life forms and comprise of one of the most abundant classes of proteins in the world. It has been proved that proteinase inhibitors have serious negative effects on the growth and development of a great variety of insects by forming enzyme-inhibitor complex to reduce the proteolytic activities and nutrition ingestion. Since their inhibitory properties were revealed, a lot of proteinase inhibitor genes were cloned to serve the genetic engineering of insect-resistance to trees (Pearce et al., 1983; Hilder et al., 1987; Lee et al., 1986; Cleveland et al., 1987; Williamson et al., 1987; Fox 1986; Thornburg et al., 1987; Hammond et al., 1987; Walling et al., 1986; Joudrier et al., 1987; Sanchez-Serrano et al., 1987; Graham et al., 1986).

CpT I is one kind of serine proteinase inhibitors existing in legume plants. The first demon-

* 本文原载《中国林业研究》(*Forestry Studies in China*), 2002, 4(2): 33-37, 与林元震、张谦和林善枝合作发表。

stration of the effectiveness of it to enhance insect resistance was revealed by the *CpT* Ⅰ gene transformed tobacco in 1987 (Hilder et al., 1987). From then on, many proteinase inhibitor gene transformed plants were produced, and their insect resistance was confirmed by insect trials (Liu et al., 1993; Xu et al., 1996; Zhao et al., 1995; Confalonieri et al., 1998; Tian et al., 2000; Li et al., 2000). Although many successful transformations were tested by molecular method, such as southern blotting, northern blotting, PCR (polymerase chain reaction) and western blotting etc., few researches were launched to examine the expression of the foreign genes using its hydrolytic property and the total soluble protein by SDS-PAGE, and little was known about whether there are spatial differences in the expression of foreign genes and native genes in the host genome.

P. tomentosa is a special native species in China, it has characteristics of fast growth, high yield and short rotation, and is widely used as an important afforestation and landscape species in the north of China. In 1998, some (*Populus tomentosa*×*P. bolleana*)×*P. tomentosa* clones were transformed with *CpTI* gene (Hao et al., 2000), in order to examine whether the foreign genes in the transgenic plants are expressed actively and stably at protein level. In addition, the spatial difference in gene expression and the effects of incorporation of foreign genes on gene expression were discussed.

2 Materials and methods

2.1 Plant materials

Hybrid poplar clones of [(*Populus tomentosa*×*P. bolleana*)×*P. tomentosa*] were transformed by the Agrobacterium tumenf aciens with the T-DNA of plasmid pBinΩSCK harboring the *CpTI* gene and kanamycin resistance gene NPT from Tn5 constructed, both of which were under the control of constitutive CaMV 35S promoter and the T-nos terminator at 3′ end. The existence of SKTI signal peptide sequence in the 5′ region and KDEL coding sequence at the 3′ end determined high expressing of the integrated gene. The plants used for the determination of the contents of total soluble protein and *CpT* Ⅰ were established from dormant branches of transformed clones with the kanamycin resistence feature and positive reaction to PCR and PCR-southern blotting. The cuttings were rooted and grown in greenhouse in the pots containing a 2∶1(v/v) mixture of soil and sand. The leaves of both transgenic and nontransgenic plants were collected and immediately frozen on dry ice until being used for the analysis of total soluble protein and *CpT* Ⅰ.

2.2 Extraction and analysis of total soluble protein

Total soluble protein from 0.5g of leaves from transgenic poplars (TG04, TG07, TG08, TG16, TG71) and the control was extracted in 3mL of Tris-HCl buffer containing 62.5mM Tris-HCl (pH 6.8), 1mM phenylmethylsulfonyl fluoride (PMSF), 5% (v/v) mercaptoethanol, 0.5% (w/v) sodium dodecyl sulfate (SDS), 5% (v/v) glycerol with the method of Guy et al.,

(1992) with a few modifications. The homogenate was extracted at 4℃ for 2h, and then centrifuged at 4℃ with 12000g for 30min. 2. 5 unit volumes of acetone pretreated at −20℃ was added to the resulting supernatant and held at −20℃ for 1h. The precipitate was collected by centrifugation at 12000g for 20min, and then dried at −20℃ for 20min. The protein pellet was resuspended in 0. 3mL of Tris−HCl buffer containing 0. 5M Tris−HCl (pH 6. 8), 5% (v/v) mercaptoethanol, 10% (v/v) glycerol and the supernatant was collected at 4℃ with 12000g for 10min. The resulting supernatant was used as the total soluble protein. The content of total soluble protein was assayed using the method of Bradford (1976) with bovine serum albumin (BSA) as a standard. The total soluble protein was analyzed with SDS-PAGE on 18% linear polyacrylamide gradient gel and coomassie blue staining described by Laemmli (1970).

2. 3 Extraction and determination of *CpT* I

The inhibitory assay of trypsin proteinase inhibitor with casein as the substrate of trypsin was performed as described previously (Hao et al. , 2000; Xu et al. , 1996; Johhston et al. , 1995). The bovine trypsin inhibitors of 0, 0. 4, 0. 8, 1. 2, 1. 6, 2. 0 and 2. 4μg were mixed with the trypsin, respectively, and incubated at 27℃ for 30min, followed by the addition of 100μg substrate of trypsin. The casein with the hydrolysis was incubated for 1h at 30℃. Data derived from the OD_{280} were used to draw the standard inhibitory curve. The same procedure for construction of the standard inhibitory curve was used in the inhibitory reaction of the plant *CpTI* to trypsin except that the bovine inhibitor was replaced with 50μg extracts from the metaphylla and browse of each line. The quantification of plant *CpT* I was calculated according to the data of OD280 got from the standard inhibitory curve.

3 Results

3. 1 Change in total soluble protein content

According to Table 1, only 0. 218mg·g^{-1} FWand 0. 437mg·g^{-1}FW of total soluble protein were detected in browse and metaphylla of non−transgenic plants, respectively. Whereas, the average amounts of total soluble protein in browse and metaphylla of transgenic plants were 0. 414mg·g^{-1} FW and 0. 574mg·g^{-1} FW respectively, showing an increase of 89. 91% and 31. 35% respectively, compared with non−transgenic plants. Among the 5 transgenic clones, the content of total soluble protein exceeding the average amounts was detected in browse of TG04, TG16 and TG17, and in metaphylla of TG04, TG08, TG16 and TG17. The lowest amounts of total soluble protein in browse and metaphylla were 0. 352mg·g^{-1} FW and 0. 465mg·g^{-1} FW, respectively, and were found in TG08, which were more than 0. 134mg·g^{-1}FW and 0. 03mg·g^{-1}FW of total soluble protein in browse and metaphylla respectively, compared with non-transgenic plants. The results showed that the total soluble protein content in browse and metaphylla were higher in five transgenic clones than in non-transgenic plants.

Table 1 Changes in the contents of total soluble protein and *CpT* I

Clone	Total soluble protein content ($mg \cdot g^{-1}FW$)		CpTI content ($\mu g \cdot g^{-1}FW$)	
	Browse	Metaphylla	Browse	Metaphylla
TG04	0.451	0.585	13.87	14.06
TG07	0.360	0.608	16.74	16.00
TC08	0.352	0.465	14.65	14.65
TG16	0.438	0.611	12.59	11.46
TG71	0.469	0.601	16.78	16.55
CK	0.218	0.437	0	0

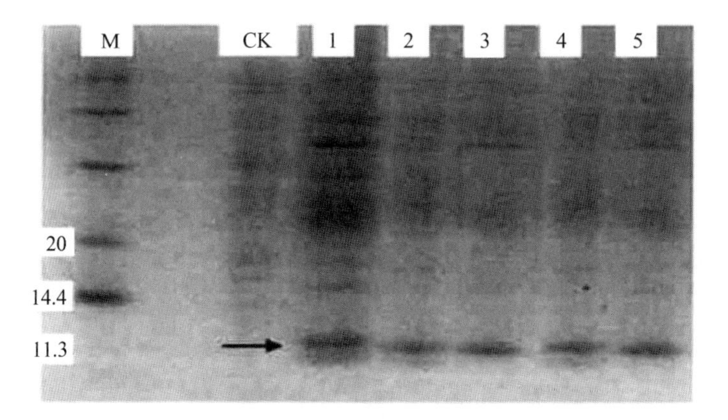

Fig. 1 Analysis of protein in leaves by SDS-PAGE

M is marker protein, CK is the protein of the control, lines of 1 ~ 5 are the protein from 5 transgenic clones

3.2 Assay of total soluble protein by SDS-PAGE

As shown in Plate 1, one clear specific band of protein with 11.3 kD was only observed in 5 clones of transgenic plants, but not in control plants. In addition, the other proteins exhibited an increase in staining intensity in all transgenic clones compared with the control.

3.3 Change in *CpT* I content

As shown in Table 1, most transgenic clones have more than 14.0$\mu g \cdot g^{-1}$FW of *CpT* I in browse and metaphylla, accounting for more than 3% of the total protein. Evenmore than 16μg of *CpT* I was detected in two kinds of leaves from clone TG07 and TG71. These results indicated that there was large amount of *CpT* I in both metapylla and browse of each clone, but the amount of *CpT* I in browse and metaphylla was very close, it seemed that there was no significant difference in the *CpT* I content in the two kinds of foliage from the same transgenic clones. However, no *CpT* I was detected in the control plant, indicating the absence of the *CpT* I gene. Moreover, there was more than or close to 14.0μg of *CpT* Ⅰ in 1g of both fresh browse and metaphylla from 3 outstanding clones TG04, TG07 and TG71, further confirmed their advantage in insect-resistance.

4 Discussion

Due to the limitation of present gene transformation, many researches have reported that the incorporation of foreign gene into the host genome might result in the occurrence of recombination, non-sitespecific integration and multiple duplications of foreign genes, and further lead to an alteration on the expression of some native genes in the host genome and the foreign genes (Zambryski et al., 1982; Bishop et al., 1989; Vaden et al., 1990; Artelt et al., 1991; Offringa et al., 1993; Ohba et al., 1995; Risseuw et al., 1995; Risseuw et al., 1997; de Neve et al., 1997; de Buck et al., 1999).

In the study, the increase of total soluble protein was found in the leaves of transgenic plants compared with the control, and was much more than that of *CpT* Ⅰ induced by the expression of *CpT* Ⅰ gene in the leaves of all transgenic clones (Table1), indicating that the expression of *CpT* Ⅰ gene may result in expression of some native genes in the poplar genome. The finding that nearly all of proteins, with few exceptions, had an increase in staining intensity caused by the expression of *CpT* Ⅰ gene (Fig. 1), which further supports the above conclusion.

In comparison to the control, the specific protein band of about 11.3 kD corresponding to the 80 encoded amino acids was detected in all tested transgenic clones, confirming at protein level that *CpT* Ⅰ gene was inserted into poplar genome and was expressed stably in the transgenic poplars.

A different *CpT* Ⅰ content was observed in all tested transgenic clones. But, as for the same clone, no significant difference of *CpT* Ⅰ content was found in browse and metaphylla. The expressing effect of *CpT* Ⅰ gene in all tested transgenic clones showed significant differences, but it had no temporal and spatial differences.

Test of Insect-Resistance of Transgenic Poplar with *CpT* I Gene*

Abstract Both non-transgenic hybrid triploid poplars [(*Populus tomentosa*×*P. bolleana*)× *P. tomentosa*] and transgenic ones expressing cowpea trypsin inhibitor were cut at the base of the stem to produce auxoblasts, and used as source of leaves for insect feeding trials performed on 3 major insect species of poplar, the forest tent caterpillar (*Malacosoma disstria* L.), gypsy moth (*Lymantria dispar* L.) and willow moth (*Stilpnotia candida* Staudinger). The height and basal diameter of trees were measured by the end of that year (2000). The results indicated that the growth elements of transgenic poplars were not interfered by the incorporat ion of the *CpT* I gene. Intriguingly, the height and basal diameter of the clone TG04 were much greater than that of the control. The transgenic foliage consumed by insects induced the increase of larval mortality, and decrease of larval wet weight gain, faecal output, pupal weight and egg deposition. Among them 3 transgenic clones, TG04, TG07 and TG71 received special attention for their outstanding insect resistance compared with other transgenic clones, which showed that the *CpT* I gene in them was expressed more actively and stably than in others.
Keywords cowpea trypsin inhibitor, *Populus tomentosa*, *Lepidoptera*, insect resistance

1 Introduction

Cowpea trypsin inhibitor (*CpT* I), is a small polypeptide belonging to the bowman-birk type of double-headed serine protease inhibitors (Hilder et al., 1989; Ryan 1989; Laskawiski et al., 1980; Liu et al., 1993). They could bind tightly to active site of digestive enzymes localized in the midgut of insect to form an enzyme-inhibitor complex through which hydrolyzing activities of the digestive enzymes of insects are inhibited (Broadway et al., 1986), and the minimal counter-evolution for insects eating foliage containing *CpT* I was determined by the fact that the active site of digestive enzyme is comparably conserved (Ryan 1989; Hilder et al., 1989). High efficacy of *CpT* I and other trypsin proteinase inhibitors in affecting growth and development of insects has been observed (Xu et al., 1996; Johnston et al., 1993; Johnston et al., 1995; Jongsma et al., 1996; Zhao et al., 1995; Liu et al., 1992; Boulter et al., 1993). *CpTI* gene was firstly transferred to tobacco in 1987, insect trials indicated that the transgenic tobacco was highly resistant to *Heliothis virescens* (Hilder et al., 1987). Since then, a lot of proteinase inhibitor transformed

* 本文原载《中国林业研究》(*Forestry Studies in China*), 2002, 4(2): 27-32, 与张谦、林善枝和林元震合作发表。

plants were produced, and their insect resistance abilities were confirmed by insect trials (Liu et al., 1993; Xu et al., 1996; Zhao et al., 1995; Confalonieri et al., 1998; Tian et al., 2000; Li et al., 2000).

In 1998, many genetically engineered plantlets of [(*Populus tomentosa*×*P. bolleana*)×*P. tomentosa*] clones were produced by the transformation with *CpT* I gene (Hao et al., 2000). In order to screen out clones with fine growth traits and high insect-resistance for the popularization and application of transgenic poplars, the insect trials performed on 3 major poplar insect species with five coefficients, larval mortality, faecal output, wet weight gain, pupal weight, and egg deposition were experimented in this research and the results obtained from the trials were discussed.

2 Materials and methods

2.1 Plant materials

Hybrid poplar clone [(*P. tomentosa*×*P. bolleana*)×*P. tomentosa*] was transformed in 1999 by the *Agrobacterium tumefaciens* with the T-DNA of plasmid pBinΩSCK harboring the *CpT* I gene and kanamycin resistance gene NPT from Tn5 constructed in Zhuzheng Laboratory in Institute of Genetics, the Chinese Academy of Science (CAS), both of which were under the control of constitutive CaMV 35S promoter and the T-nos terminator in 3′ end. The existence of SKTI singal peptide sequence in the 5′ region and KDEL coding sequence in the 3′ end determined the high expression of the integrated gene.

In the spring of 2000, both transgenic and nontransgenic plants were cut into 10cm from the base of the stems to produce auxoblasts for propagating the transgenic plants. The foliage derived from the same position of auxoblasts were excised at the base of the petiole, sealed in a plastic sandwich bag, and then taken to the laboratory of Beijing Forestry University.

2.2 Insect materials

The forest tent caterpillar, *Malacosoma disstria*, were collected from only one group in *Malus spectabilis* located in Beijing Forestry University which promised all larvae were in the same early 3rd-instar and reared with the foliage of *Malus spectabilis* for two days before the insect trials starting. The gypsy moth, *Lymantria dispar* L. were reared from the egg bands provided by the Forest Protection Laboratory of the University to early 3rd-instar prior to the trials, the larvae were fed with the foliage of *Diospyros kaki* L. The first generation of 5th-instar willow moth larvae, *Stilpnotia candida* Staudinger, were gathered from the *Populus canadensis* Moench (*P. deltoids* × *P. nigra*) in front of the Forest Protection Station on campus of Beijing Forestry University and transferred to the glass containers after an intervals without food provision for 12h to avoid the gut of diet before weighing the insects. One pair of pupae fed with foliage from the same transformant were selected for laying eggs from which the second generation of larvae were raised. The egg bands held in a glass container were conveyed to our greenhouse at 27℃ and a photoperiod of 16h/8h

(light/dark) .

2.3 The measures of the growth and development of transgenic clones

The height and the basal diameter of auxoblasts were measured by the end of 2000.

2.4 Insect bioassay

Four non-choice feeding assays were conducted from April 18, May 18, June 9 and June 15 using early 3rd-instar forest tent caterpillar, the first generation of later 5th-instar willow moth, early 3rd-instar gypsy moth and the second generation of early 3rd-instar willow moth, respectively.

For each insect bioassay, a group of insects held in glass vial (7.5cm diameter, 12.5cm height) sealed the mouth with net cover (three replicate containers per treatment) were allowed to feed on the leaves of each line (the control plant included). To maintain the humidity of air and leaf, one wet cotton ball on which was a piece of filtration paper was placed on the bottom of each container. The feeding foliage was replaced with new one every other day. To evaluate the toxicity of each line to insect pests, the larval mortality of trial performed on each larva species was recorded and corrected as described previously (Tian et al., 1993), the mean wet weight gain was calculated as the difference between the initial mean wet weight and the final mean wet weight prior to pupation. To measure the food intake (Johnston et al., 1993), the faeces of forest tent caterpillar and willow moth from April 30 to May 5, July 13 to July 20, respectively, were collected and oven-dried at 60℃ for 72h before weighting them. By the end of the trial performed on willow moth, the male and female pupal weight were measured respectively, the egg deposition was counted to estimate their reproductivity.

3 Results

3.1 Tree investigation

As shown in Table 1, the height and basal diameter of the control were 156cm and 241cm respectively, and those of transgenic poplars were more than or very close to them, which indicated

Table 1 The measurement of the basal diameter and the height of the trees

Clone	Basal diameter(mm)	Height(cm)
TG04	195±11	259±12
TG07	150±7	245±9
TG08	159±13	256±5
TG16	148±4	247±8
TG71	154±7	236±10
CK	156±9	241±6

that the basal diameter and height of transgenic clones were not significantly less than that of non-transgenic ones. Even more, the clone TG04 whose basal diameter and height reached 195mm and 259cm respectively, were much greater than the control in these two growth elements.

3.2 Insect bioassay

3.2.1 Fatal effectiveness

As shown in Table 2, more than 50% of forest tent caterpillar and gypsy moth were killed by transgenic foliage, and over 80% of the larvae of forest tent caterpillar consuming TG04 and TG07 were caused to death (Table 1), which indicated that the insects consuming transgenic foliage had greater mortality than those eating non-transgenic leaves.

Table 2 Larval mortality of three insect species

Clone	Forest tent caterpillar(%)	Gypsy moth (%)	Willow moth (%)
TG04	80.19	53.85	44.52
TG07	85.39	76.92	81.13
TC08	52.51	65.39	26.41
TG16	67.49	53.85	32.07
TG71	62.50	52.87	43.20
CK	0	0	0

The insect bioassay performed on gypsy moth within 5d caused significantly higher mortality to the larvae given transgenic foliage than to those given non-transgenic one (Table 2), the corrected larval mortality of gypsy moth fed transgenic foliage was over 52%, which indicated that gypsy moth was most susceptible to the transgenic plants. Intriguingly, two days later, the whole assay was forced to end as the mortality of larvae provided with transgenic plants was all over 86% without manifest divergence. Less fatal rate as the control plants caused to the feeding larvae, it also reached 76.67%.

Although the first generation of later 5th-instar willow moths exhibited no significant difference in larval mortality (data not shown), the second generation of early 3rd-instar larvae showed manifest discrepancy on fatal rate. The larval mortality of willow moth eating foliage from TG04, TG07 and TG71 was all over 43%. Moreover, as much as 81.13% of larvae of willow moth were killed by the end of the trial (Table 2).

3 out of 5 transgenic clones, TG04, TG07, TG71 caused more larvae to death than other 2 clones, and the TG07 was screened out as the best one as it could give rise to the death of over 75% of 3 kinds of larvae.

3.2.2 Foliage consumption

In the early 2 days, no significant differences in foliage intake of forest tent caterpillar and willow moth were observed, but with elongation of the time, the difference occurred (data not shown). During the trial performed on willow moth, this fact was further confirmed by the observation of dramatic divergence in foliage consumptions (Fig. 1).

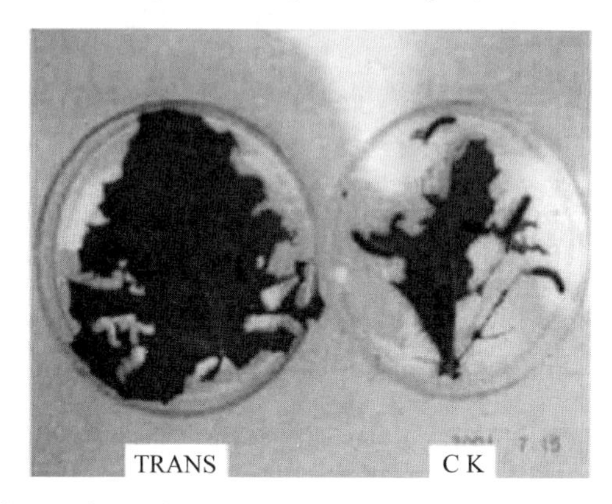

Fig. 1 Comparison of the feeding effect performed on willow moth

TRANS: foliage from transgenic poplar, CK: foliage from nontransgenic poplar

3.2.3 Faecal output

An average of 755mg of feces per larva was excreted by the larvae of forest tent caterpillar consuming foliage from the control within 8d, 1025mg by those of willow moth (Fig. 2). While, less than 320mg of feces was excreted by the larvae of forest tent caterpillar eating foliage from transgenic clones, 720mg by those of willow moth, which showed that the larvae consuming transgenic foliage made less feces than those eating non-transgenic foliage. Compared with other transgenic clones, only 36mg, 67mg and 189mg of feces were excreted by the larvae of forest tent caterpillar consuming the foliage from TG04, TG07 and TG71 respectively, 461mg, 254mg and 342mg by those of willow moth correspondingly (Fig. 2). These results indicated that the TG07, TG04 and TG71 were more effective in reducing the output of feces excreted by the larvae eating the foliage from them. Because the development stage of willow moth selected for the collection of feces was later than that for forest tent caterpillar, the larvae of willow moth in the same 8d excreted more feces than that of forest tent caterpillar.

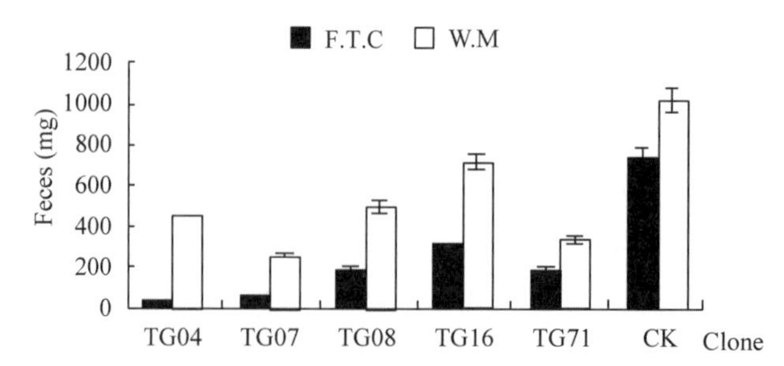

Fig. 2 The 8d period of mean faecal weight of forest tent caterpillar (F. T. C) and willow moth (W. M.)

3.2.4 Wet weight gain

By the end of the trials, the mean wet weight of the larvae consuming foliage from the control was increased by 141mg and that of willow moth by 723mg. However, the mean wet weight gain of the larvae of forest tent caterpillar and willow moth consuming foliage from transgenic clone were less than 80mg and 260mg respectively, from these results we have gotten the fact that the mean wet weight gain of larvae consuming transgenic foliage was much less than that of larvae eating non-transgenic foliage. Among 5 transgenic clones, TG07 and TG04 resulted in the increase of larvae wet weight gain for the forest tent caterpillar by less than 40mg, so did TG07 and TG71 for that of the willow moth, and the efficacy of these 3 clones in reducing the larval wet weight gain of insects was revealed (Table 3).

Table 3 The mean wet weight gain of larvae mg

Clone	Forest lent caterpillar	Willow moth
TG04	27±6	163±29
TG07	37±4	23±4
TG08	76±11	118±14
TG16	58±7	256±11
TC71	66±10	40±8
CK	141±14	723±42

3.2.5 Pupal weight gain and egg deposition

The pupae of willow moth fed with transgenic foliage, whether the pupae are male or not, were almost double weight of that of insects fed with nontransgenic foliage (Table 4), which were further demonstrated by significant differences in the physical size of pupae in Fig. 3.

This fact also occurred to the number of eggs deposited by the matured willow moths (Table 4). It was revealed that the mature willow moths eating the foliage from transgenic poplars deposited an average of only 68 eggs, while those consuming non-transgenic poplars produced 132 eggs.

Table 4 Investigation of pupal weight and egg deposition

	CK		TRANS.	
	Female	Male	Female	Male
Pupal weight(mg)	378±15	217±21	190±13	129±25
No. of eggs	132±24		68±11	

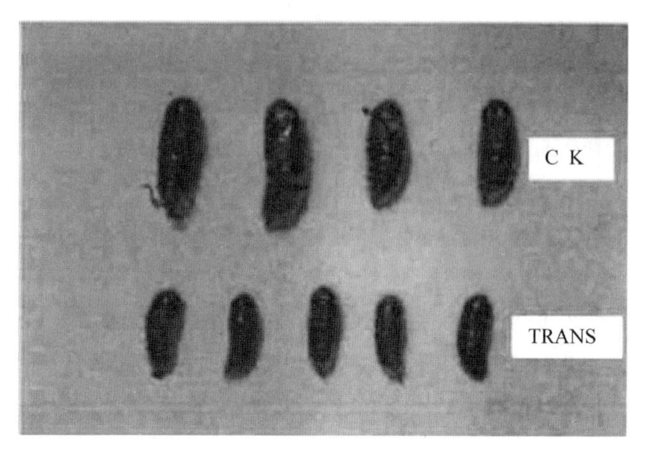

Fig. 3　The comparison of the pupae of willow moth

CK is the pupae from the larvae eating leaves of the control,
TRANS. is the pupae from the larvae eating leaves of transgenic clones

4　Discussion

To investigate whether the growth and development of transgenic clones following dormancy was interfered by the incorporation of a foreign gene into them, the height and basal diameter of transgenic auxoblasts were measured. It is found that no transgenic clones are significantly less than the control plant in both basal diameter and height, which indicates that the growth elements of the genetically engineered plants are not interfered by the incorporation of the foreign gene into them. However, the improvement in the basal diameter and height of clone TG04 seems that the incorporation of foreign gene can enhance its growth.

Trypsin is the major digestive enzyme localized in the midgut of *Lepidopteran*, when its inhibitor, *CpT* Ⅰ, is ingested into the digestive channel, their proteinase activities can be affected by the formation of the enzyme-inhibitor complex in which the active site of the enzyme lies. The accumulated undigested food can lead to the occurrence of more digestive enzyme signaled by the feed-back mechanism and the occurrence of being sick of food (Ryan et al., 1989), the prolonged ingestion of inhibitor without sufficient nutrient supply finally exerts seriously negative effects on growth, development and reproductivity. These predictions are identified by the increase of fatal effectiveness, and the decrease of foliage intake (data not shown), wet weight gain, faecal output, pupal weight and egg deposition of insects fed with transgenic foliage in our experiment. Similar results were observed in transgenic tobacco, rice and cotton (Hilder et al., 1987; Xu et al., 1995; Zhao et al., 1995).

Compared with other transgenic clones, 3 transgenic lines TG04, TG07 and TG71 shows greater insect-resistance by the fact that they significantly induce the increase of the larval mortality, and the decrease of the faecal output, pupal weight and egg deposition, which indicates that the *CpT* Ⅰ gene in them is expressed more actively and stably than that in other clones.

The forest tent caterpillar and willow moth show no significant difference in foliage ingestion during the period of 1 ~ 2 d of trials, but with the elongation of trials, the difference in food intake among larvae eating the foliage from different clones occurres, which seems to be in consistent with the deduction that the effect of proteinase inhibition occurres after the ingestion of transgenic foliage. Further analysis finds that the insect trials performed on the forest tent caterpillar and willow moth are lasted to the pupation, while that performed on the gypsy moth is forced to end after 7 d due to the high and unanimous mortality of larvae eating foliage from any clone (the control included), which indicates that there is significant difference in fatal effectiveness of transgenic foliage to different kinds of insects. In addition, the high mortality occurred on gypsy moth consuming the foliage from transgenic poplars after 7 d of the trial may result from the inheritance of the high insect-resistance of *P. tomentosa* from which (*P. tomentosa* ×*P. bolleana*)×*P. tomentosa* or iginated.

Another thing mentioned is the selection of larval development stage for the insect-feeding bioassays, many related researches chose the 1st-instar larvae as the experimental subject, but the 1st-instar larvae is very susceptible to the change of environment and food, which determined the need to choose more adaptive older larvae. Hence, in our research the 3rd-instar larvae are selected as trial insects through which more reliable results can be observed.

The improvement of insect-resistance of the selected transgenic lines was proved by this research and theminimal counter-evolution for insects eating foliage with *CpT* Ⅰ was determined by its special insect-resistant mechanism. But there is no clone whose fatal effectiveness to any target insects is over than 80%, not to mention 90% as reports of transgenic poplars with Bt gene, which indicates that some measures should be taken to accomplish the increase of expression level of *CpTI* gene or combination of other anti-insect gene with *CpT* Ⅰ gene. Besides, whether these features would be applied to bearing trees, and whether the long-term intake of poisonous foliage with immediate mortality would induce the insect to transfer their spectra of digestive enzyme by site-directed mutagenesis remain uncertain.

To resolve these problems and further improve the insect-resistance and stability of transgenic poplars, the field evaluation of these transformants and incorporating binary insecticidal genes with different anti-insect mechanism are our research orientation (Karl et al., 1995; Zhao et al., 1995; Li et al., 2000; Tian et al., 2000), and now, the establishment of the test-stand for the transformed poplars and the transformation of binary genes into them are underway.

The Role of Calcium and Calmodulin in Freezing-Induced Freezing Resistance of *Populus tomentosa* Cuttings*

Abstract To explore the role of calcium-calmodulin messenger system in the transduction of low temperature signal in woody plants, *Populus tomentosa* cuttings after being treated with $CaCl_2$(10mmol/L), Ca^{2+} chelator EGTA (3mmol/L), Ca^{2+} channel inhibitor $LaCl_3$(100mmol/L) or CaM antagonist CPZ (50mmol/L) were used for freezing acclimation at −3℃. The changes in the calmodulin (CaM) and malonaldehyde (MDA) contents, the activities of superoxide dismutase (SOD), peroxidase (POD) and Ca^{2+}-dependent adenosine triphosphatase (Ca^{2+}-ATPase) of mitochondrial membrane as well as freezing resistance (expressed as LT_{50}) of cuttings were investigated to elucidate the physiological mechanisms by which trees adapt to freezing. The results showed that freezing acclimation increased the CaM content, the activities of SOD, POD and Ca^{2+}-ATPase of mitochondrial membrane as well as freezing resistance of cuttings, and decreased the MDA content as compared with control cuttings. Treatment with $CaCl_2$ at the time of freezing acclimation enhanced the effect of freezing acclimation on the above-mentioned indexes, but this enhancement was abolished by Ca^{2+}chelator EGTA, Ca^{2+}channel inhibitor $LaCl_3$ or CaM antagonist CPZ, indicating that the calcium-calmodulin messenger system was involved in the course of freezing resistance development. The presence of $CaCl_2$ at the same time of freezing acclimation also reduced the degree of decline in CaM content, and in SOD, POD and Ca^{2+}-ATPase activities caused by freezing stress at −14℃, and enhanced the level of increase in CaM content, and in SOD, POD and Ca^{2+}-ATPase activity in the recovery periods at 25℃. The change in CaM content was found to be closely correlated to the levels of SOD, POD and Ca^{2+}-ATPase, and to the degree of freezing resistance of cuttings during freezing acclimation either with or without $CaCl_2$ treatment. It was suggested that the increase of CaM content induced by $CaCl_2$treatment promote the formation of Ca^{2+}-CaM complexes, which effectively activates the activities of SOD, POD and mitochondrial Ca^{2+}-ATPase and then further result in the adaptive changes associated with the development and enhancement of freezing resistance. Thus, It could be concluded that Ca^{2+}-calmodulin may be involved in the regulation of the increase in SOD, POD and Ca^{2+}-ATPase activities, and the induction of freezing resistance of cuttings.

Keywords *Populus tomentosa*, cuttings, freezing acclimation, calmodulin, $CaCl_2$, freezing resistance

Freezing temperature is one of the most important environmental factors limiting the

---

* 本文原载 *Journal of Plant Physiology and Molecular Biology*, 2004, 30(1): 59-68, 与林善枝、林元震、张谦和郭晥合作发表。

productivity and distribution of plants. Many plants increase in freezing resistance in response to low temperature pretreatment, a process known as freezing acclimation. Over the past several years, much attention has been focused on the studies of physiological and metabolic changes during freezing acclimation, including alterations in lipid composition and soluble sugar concentration, changes in enzyme activities, etc. Although these studies have been more or less concerned with the increase in freezing resistance, and various biochemical responses of plants to low temperature have been widely documented and reviewed, theme chanism of freezing acclimation for increasing freezing resistance has not been fully understood (Lin and Zhang, 2000).

Most plants have evolved several cellular defense systems to keep the balance between the production of active oxygen species (AOS) and their detoxification by the anti-oxidative system during periods of normal growth, but temperature stress causes elevated levels of AOS, which can damage cellular membrane, and severely disrupt metabolic function (Prasad et al., 1994; Anderson et al., 1995; Prasad, 1996; Li et al., 2000). Many studies showed that freezing acclimation caused an increase in activity of cell antioxidant enzymes such as SOD, POD and CAT, which resulted in the enhancement of freezing resistance (Hodges et al., 1997; Pinhero et al., 1997; Sala, 1998; Shen et al., 1999; Kang and Saltveit, 2001; Zhang et al., 2002). It has been reported that the increase in the activity of Ca^{2+}-ATPase of plasmolemma, tonoplast and chloroplast membrane ran parallel to the acquisition of freezing resistance during freezing acclimation (Zeng and Li, 1999; Lin and Zhang, 2002b), but reports about Ca^{2+}-ATPase of mitochondrial membrane are limited in number. Recently, it was found that calcium-calmodulin played an important role as a second messenger in growth, development, and adaptation of plants to the environment (Hao and Yu, 1992; Poovaiah and Reddy, 1993; Bush, 1995; Li et al., 1997; Lin and Zhang, 2002a, b). However, little is known about the correlation between the change in CaM content, the increase in activity of SOD, POD and Ca^{2+}-ATPase of mitochondrial membrane, and the degree of increase in freezing resistance induced by freezing acclimation, especially in woody plants.

In this paper, the correlation mentioned above was investigated in detail using *Populus tomentosa* cuttings to elucidate the physiological mechanism of freezing resistance induced by freezing acclimation, and to explore the role of calcium-calmodulin messenger system in the transduction of freezing temperature signal in woody plants.

1 Materials and Methods

1.1 Plant material

Populus tomentosa cuttings were obtained from Hebei province, China. The rooted cuttings were grown in pots containing a 2:1(v/v) mixture of soil and sand. After the cuttings were grown for 5 weeks in a greenhouse, they were placed outdoors for 3 weeks. These cuttings were used as experiment materials.

1.2 Freezing acclimation, Freezing stress and recovery

Eight-week-old cuttings were divided into three groups. The first group of 100 cuttings placed at 25℃ was referred to as non-acclimated (NA or control) cuttings. The second group of 100 cuttings directly held at −3℃ for 5d were referred to as freezing−acclimated (FA) cuttings. To determine the role of CaM in the induction of freezing resistance, the third group of 400 cuttings were further divided into four subgroups, 100 cuttings per subgroup, theirs roots watered with $CaCl_2$ 10mmol/L, EGTA 3mmol/L, $LaCl_3$100mmol/L or CPZ 50mmol/L at 25℃ for 5d, 200mL per pot per day, and then directly transferred to −3℃ for 5d for freezing acclimation combined with effector pretreatment (FA−E). Finally, the NA, FA and FA−E cuttings were placed at −14℃ for 3d, and then were allowed to recover at 25℃ for 2d. Cuttings of all treatments were grown under an 8−h photoperiod and a light intensity of 30μmol/(m^2 · s).

1.3 Evaluations of survival rates and LT_{50} of cuttings

All tested cuttings used for the evaluation of survival rates and LT_{50} were treated with different temperatures. Temperatures −3℃, −5℃, −7℃, −9℃ and −11℃ were used for the cuttings of NA, FA−$LaCl_3$, FA−CPZ and FA−EGTA cuttings, −14℃, −16℃, −18℃, −20℃ and −22℃ used for FA and FA−$CaCl_2$ cuttings. Cuttings were held at each temperature for 12h. Then all tested cuttings were directly transferred from every treated temperature to 25℃ for 2d. Survival cuttings were defined as an 80% of survival leaves and buds of cuttings after treatment at every temperature for 12h, and then the amounts of them were calculated. The percentage of survival cuttings is taken as survival rates.

The data of survival rates of cuttings subjected to treatment with each temperature were used to fit exponential function by using the logarithm of survival rate as a liner function of the time of cold treatment, and then the temperature for 50% survival of cuttings (LT_{50}) was calculated. LT_{50} is taken as a measure of the freezing resistance.

1.4 CaM extraction and analysis

One gram of leaves were ground at 4℃ in 1.5mL of extraction buffer containing Tris−HCl 50mmol/L (pH 8.0), EGTA 1mmol/L, phenyl methylsulfonyl fluoride (PMSF) 0.5mmol/L, NaCl 0.15mmol/L, 1% (*w/v*) PVPP, $NaHSO_3$ 20mmol/L. After being kept at 90℃ for 2min, the mixture was centrifuged at 4℃ with 10000×*g* for 30min. The supernatant was analyzed for the soluble CaM and its concentration was assayed by using the ELISA technique described by Zhao et al. (1988). The antibody used for ELISA was Promega product.

1.5 Mitochondria preparation and Ca^{2+}−ATPase assays

Mitochondria were isolated from leaves as described by Ouyang (1985), and its purity was tested by the method of Dupont et al. (1982). The activity of Ca^{2+}−ATPase was measured according to Zeng and Li (1999). The enzyme activity was assayed at 37℃ for 30min in a 0.5mL reaction mixture containing ATP 5mmol/L, $CaCl_2$ 4mmol/L, $(NH_4)_6MO_7O_{24}$ 2mmol/L, Tris-Mes (pH6.5) 50mmol/L, 0.01%Triton X-100, and 20mg membrane protein. The reaction was started by the addition of 0.02mL of Ca^{2+}-ATPase solution, and then stopped by adding 0.5mL of 10%(*w/v*) TCA. One unit (U) of enzyme activity was defined as the amount of the inorganic phosphate (Pi) released from the substrate permin. Protein was assayed by the method of Bradford (1976) using bovine serum albumin (BSA) as a standard.

1.6 SOD, POD and MDA extraction and determination

Extracts for determination of SOD, POD and MDA were prepared from 1g of leaves and homogenized under 4℃ in 5mL of extraction buffer containing 62.5mmol/L potassium-phosphate buffer (pH 7.8), 1% (*w/v*) PVPP. The homogenates were centrifuged at 15000×*g* for 20min and the supernatant was used for the assays. SOD was determined by monitoring the inhibition of photochemical reduction of nitro blue tetrazolium (NBT) as described previously (Liu and Zhang, 1994). One unit (U) of SOD was defined as the amount of enzyme that produced a 50% inhibition of photochemical reduction of NBT. POD activity was determined by measuring the increase in absorption at 470 nm according to Liu and Zhang (1994). The reaction was carried out at 25℃ for 20min in a 3mL reaction mixture containing 0.02mL tissue extract, potassium-phosphate buffer 0.1mmol/L (pH 7.0), and guaiacol 20mmol/L. The reaction was started by the addition of 20μL of H_2O_2. One unit (U) of POD was defined as the amount of enzyme that caused a 0.01 increase of A470 permin under the assay condition. MDA was measured by the thiobarbituric acid reaction according to Draper and Hadley (1990).

All the determinations were performed at leastin triplicates and the average values were presented.

2 Results

2.1 Change in freezing resistance

After 5d of freezing acclimation at −3℃, the LT_{50} of cuttings decreased from −6.2℃ for unacclimated (control) cuttings to −14.3℃ for freezing acclimated (FA) ones. Treatment with $CaCl_2$ at the same time of freezing acclimation enhanced the effect of freezing acclimation, the LT_{50} of cuttings was lowered from −14.3℃ to −18.1℃. However, after freezing acclimation with EGTA, $LaCl_3$ or CPZ treatment, the LT_{50} of cuttings increased and closed to the level of control cuttings (Table 1).

In order to demonstrate in detail the effect of freezing acclimation on freezing resistance of cuttings, the time course of freezing acclimation at −3℃ in a period of 1 ~ 5d was studied (Fig. 1). One day after transferring cuttings to −3℃, LT_{50} of cuttings started to decrease from −6.2℃ to −7.1℃ and gradually decreased to −14.3℃ on the 5th day, showing the induction and development of freezing resistance of cuttings during freezing acclimation.

Table 1 Effects of freezing acclimation aloneandin combination with $CaCl_2$, EGTA, $LaCl_3$, or CPZ treatment on LT_{50} of *P. tomentosa* cuttings

Treatment	LT_{50} of cuttings (℃)
Control	-6.2
Freezing acclimation (FA) 5d	-14.3
FA 5d + $CaCl_2$ treatment 5d	-18.1
FA 5d + EGTA treatment 5d	-6.7
FA 5d + $LaCl_3$ treatment 5d	-6.9
FA 5d + CPZ treatment 5d	-6.4

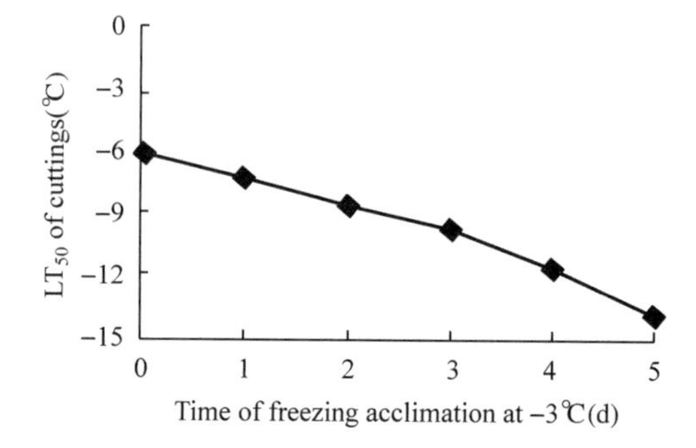

Fig. 1 Change in LT_{50} of cuttings with time of freezing acclimation

2.2 Change in CaM content

After 5d of freezing acclimation at -3℃, the CaM content in leaves of FA cuttings increased about 1.0-fold compared with control cuttings. The presence of $CaCl_2$ at the time of freezing acclimation enhanced the effect of freezing acclimation, and CaM content in FA-$CaCl_2$-treated cuttings was about 1.2-fold of FA ones. Much lower CaM content was found in the EGTA-, $LaCl_3$-or CPZ-treated cuttings than in the FA-$CaCl_2$-treated cuttings (Table 2). In addition, CaM content gradually increased with time during freezing acclimation at -3℃ (Fig. 2).

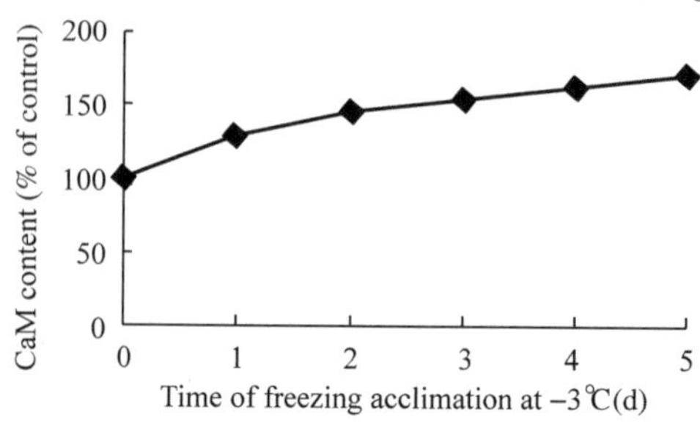

Fig. 2 Effect of freezing acclimaton on CaM content of *P. tomentosa* cutting

Table 2 The changes in CaM and MDA contents, and in activities of Ca^{2+}-ATPase, SOD and POD in *P. tomentosa* cuttings under different conditions

Treatments	Ca^{2+} ATPase [U/(g·h)]	SOD (U/g FW)	POD (U/g FW)	CaM (mg/g FW)	MDA (mmol/g FW)
Nonacclimation (control or NA)	24.9±0.8	650.0±1.5	140.0±1.0	50.0±1.1	160.5±0.9
Freezing acclimation (FA) 5d	34.0±1.0	760.2±0.4	178.5±0.8	140.8±1.5	97.2±1.0
FA 5d+$CaCl_2$ pretreatment 5d	41.1±0.5	855.0±1.0	224.1±1.5	178.6±1.4	80.0±1.3
FA 5d+EGTA pretreatment 5d	26.2±1.1	670.0±0.9	147.3±1.0	51.2±1.0	159.0±0.9
FA 5d+$LaCl_3$ pretreatment 5d	27.8±1.2	678.1±1.0	149.0±1.8	54.0±0.7	158.3±0.7
FA 5d+CPZ pretreatment 5d	27.6±1.0	660.2±1.2	142.4±1.3	51.4±0.9	158.0±1.1

During freezing stress at −14℃, there was adecrease in CaM content of all treated cuttings followed by an increase on the 2nd day of recovery at 25℃. But a smaller decline in CaM content caused by freezing stress, and a greater increase in CaM content induced by 2d of recovery at 25℃ was observed in the FA$CaCl_2$ treated cuttings than in the control cuttings, the degree of change in CaM content of FA cuttings lied between the above two kings of cuttings (Fig. 3).

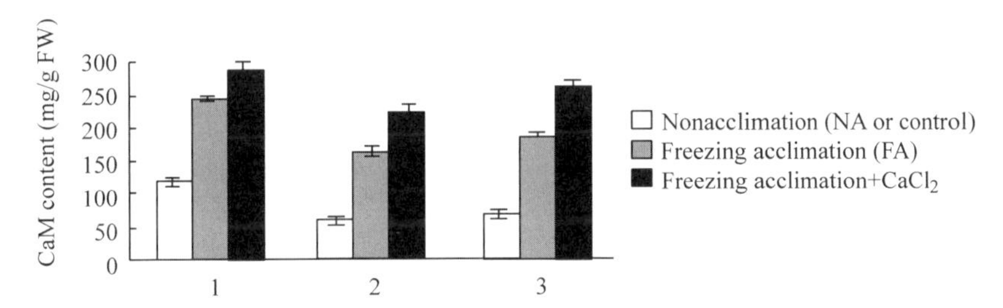

Fig. 3 Changes in CaM content of *P. tomentosa* cuttings

1. Before freezing stress; 2. After freezing stress; 3. After recovery from freezing stress

2.3 Change in Ca^{2+}-ATPase activity of mitochondrial membrane

There was a gradual increase in the activity of mitochondrial membrane Ca^{2+}-ATPase of FA cuttings during freezing acclimation at −3℃ (Fig. 4). After 5d of freezing acclimation, the activity of mitochondrial membrane Ca^{2+}-ATPase was about 1.4-fold of control. The presence of $CaCl_2$ enhanced the effect of freezing acclimation, and the activity of Ca^{2+}-ATPase in FA-$CaCl_2$-treated cuttings increased about 1.7-and 1.2-fold, respectively, compared with control and FA cuttings, whereas, the Ca^{2+}-ATPase activity similar to the level of control was observed in the EGTA-, $LaCl_3$-or CPZ-treated cuttings (Table 2).

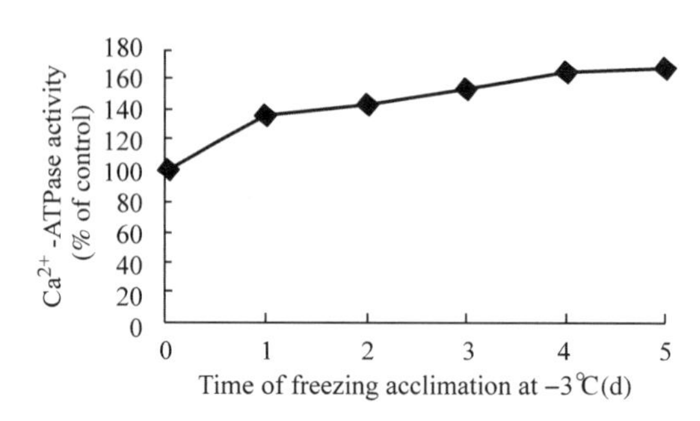

Fig. 4　Effect of freezing acclimation on mitochondrial membrane Ca^{2+}-ATPase activity of *P. tomentosa* cutting

After 3 d of freezing stress at −14℃, a low degree of decrease in activity of mitochondrial membrane Ca^{2+}-ATPase was found in FA-$CaCl_2$cuttings than in FA cuttings, especially than in control ones. Two days of recovery at 25℃ resulted in an increase in mitochondrial membrane Ca^{2+}-ATPase activity, a greater increase in Ca^{2+}-ATPase activity was observed in FA$CaCl_2$ cuttings than in FA ones, the extents of increase in the above two kinds of cuttings were much greater than in control ones (Fig. 5).

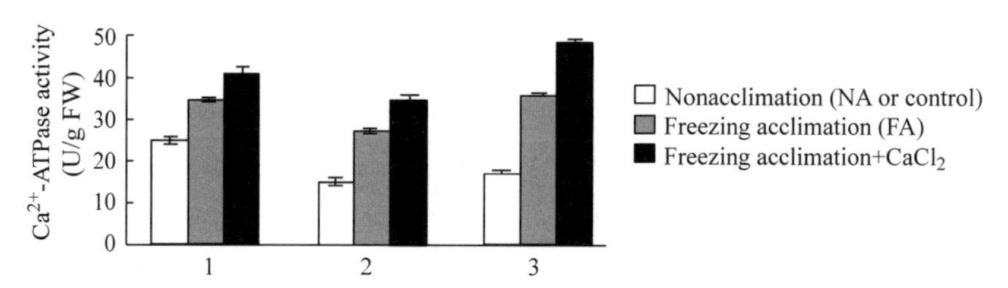

Fig. 5　Change in mitochondrial membrane Ca^{2+}-ATPase activity of *P. tomentosa* cuttings

1. Before freezing stress; 2. After freezing stress; 3. After recovery from freezing stress

2.4　Changes in activity of SOD and POD

The activities of SOD and POD in FA cuttings gradually increased with time during freezing acclimation at −3℃ (Fig. 6). After 5d of freezing acclimation, the activities of SOD and POD in FA cuttings were about 1.2-and 1.3-fold of control cuttings, respectively. The presence of $CaCl_2$ during freezing acclimation enhanced the effect of freezing acclimation on the increase in SOD and POD activities. However, lower SOD and POD activities were observed in cuttings treated with EGTA, $LaCl_3$ or CPZ (Table 2).

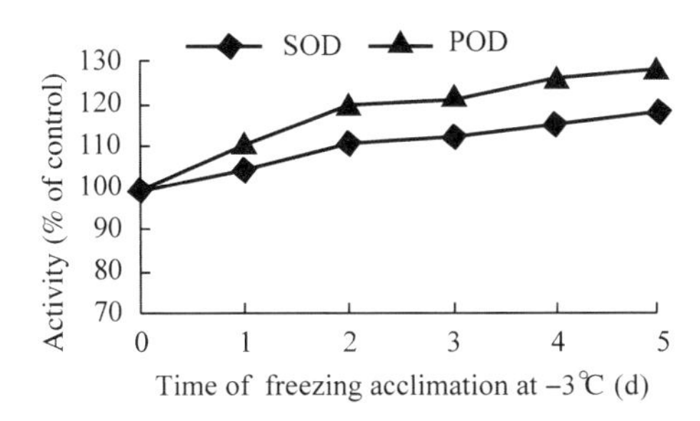

Fig. 6 Effect of freezing acclimation on SOD and POD activity of *P. tomentosa* cutting

Three days of freezing stress at −14℃ resulted in a decrease in SOD and POD activities of all treated cuttings, but the degree of decline in FA−$CaCl_2$cuttings was lower than that in FA cuttings. The increase in SOD and POD activity caused by the recovery from freezing stress were much greater in FA−$CaCl_2$ cuttings than in FA cuttings, especially than in control ones (Figs. 7, 8).

2.5 Change in MDA content

The MDA content of FA cuttings gradually decreased with time during freezing acclimation at −3℃ (Fig. 9). After 5d of freezing acclimation, the MDA content of FA cuttings was about three-fifths of control ones. Lower MDA content was observed in FA$CaCl_2$ cuttings than in FA cuttings. In contrast, MDA contents of EGTA−, $LaCl_3$−or CPZ−treated cuttings were much higher than those in FA cuttings treated or untreated with $CaCl_2$(Table 2). During freezing stress at −14℃, the increase in MDA content was found in all treated cuttings, the increase of MDA content in FA cuttings was greater than in FA−$CaCl_2$ cuttings, the highest level was observed in NA cuttings. But the 2d recovery from freezing stress resulted in a decrease in MDA content, the lower amount of MDA was found in FA−$CaCl_2$ cuttings (Fig. 10).

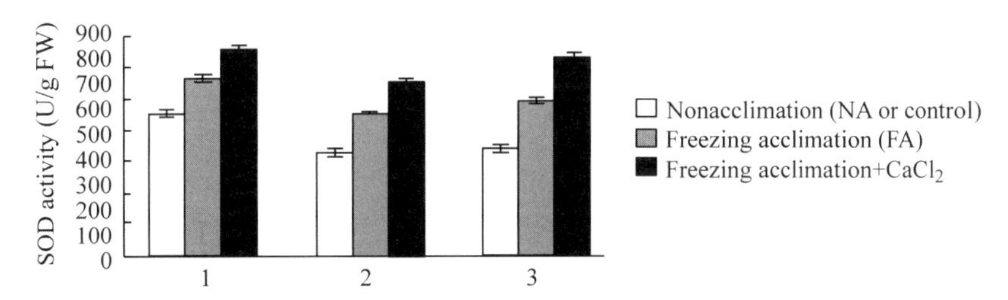

Fig. 7 Change in SOD activity of *P. tomentosa* cuttings

1. Before freezing stress; 2. After freezing stress; 3. After recovery from freezing stress

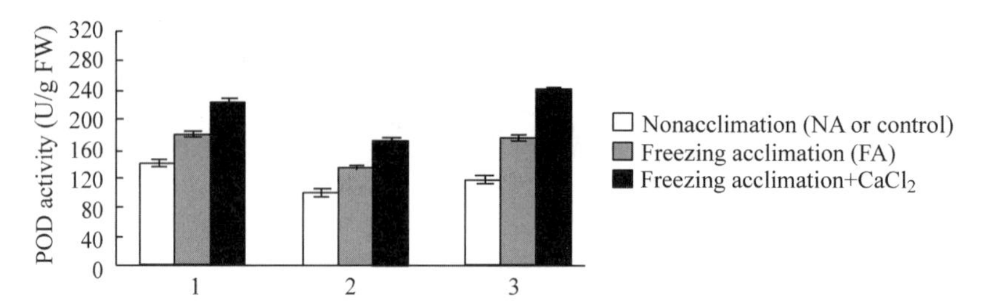

Fig. 8　Change in POD activity of *P. tomentosa* cuttings

1. Before freezing stress; 2. After freezing stress ; 3. After recovery from freezing stress

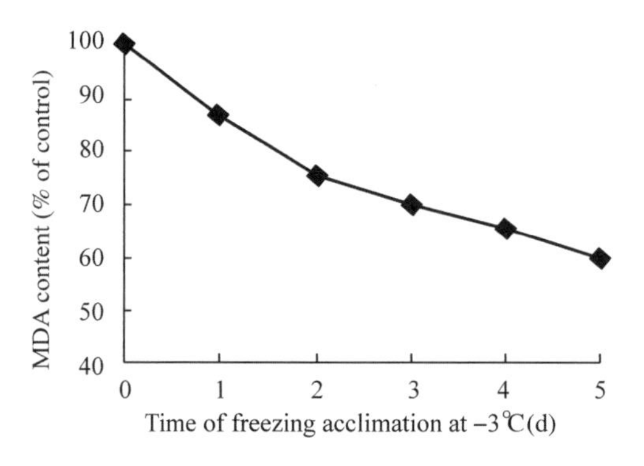

Fig. 9　Effect of freezing acclimation on MDA content of *P. tomentosa* cuttings

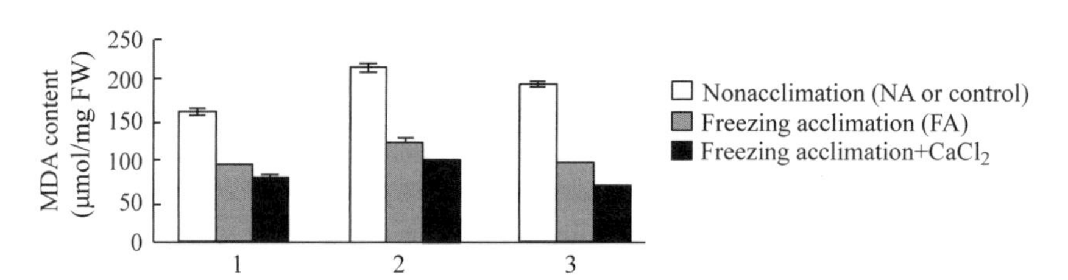

Fig. 10　Change in MDA content of *P. tomentosa* cuttings

1. Before freezing stress; 2. After freezing stress ; 3. After recovery from freezing stress

3　Discussion

Minorsky (1985) first hypothesized that calcium may function as a primary signal transducer of low temperature stress in plants, which was supported by the results of Price et al. (1994), Monroy et al. (1993, 1995) and Knight et al. (1996). Many studies have revealed that freezing acclimation induces an accumulation of Ca^{2+} and CaM, which may contribute significantly to the development of freezing resistance (Bush, 1995; Li et al., 1996; Li et al., 1997; Lin et al., 2001; Lin and Zhang, 2002a, b). However, knowledge of how woody plants sense and transduce low temperature signal is still very limited. In the present study, it was found that treatment with $CaCl_2$ at the time of freezing acclimation enhanced the effect of freezing acclimation, and markedly increased CaM content and lowered the LT_{50} of *P. tomentosa* cuttings, and this enhancement was abolished by treatment with Ca^{2+} chelator EGTA, Ca^{2+} channel inhibitor $LaCl_3$ or CaM antagonist CPZ, indicating that the enhancing effect of $CaCl_2$ treatment on the increase of freezing resistance prior to freezing acclimation may be associated with Ca^{2+}-induced increase in CaM content. Similar findings had been reported for other plant (Monroy et al., 1993; Li et al., 1996; Zeng and Li, 1999; Lin et al., 2001). It is, therefore, possible that the accumulation of CaM may be associated with the increase of cellular total soluble Ca^{2+} content during freezing acclimation, Ca^{2+}-CaM messenger system may be involved in the course of freezing resistance development (Zeng and Li, 1999; Lin and Zhang, 2002b). The study by Zeng and Li (1999) showed that the increase in Ca^{2+}-ATPase activity of plasmolemma and tonoplast membrane in roots and leaf chloroplasts in rice seedlings induced by freezing acclimation was strongly inhibited by EGTA or CPZ. Recently, it was revealed that Ca^{2+}-ATPase activity was regulated by Ca^{2+}-CaM-dependent phosphorylation (Harper et al., 1998; Bonza et al., 2000; Chung et al., 2000; Lin et al., 2001; Lin and Zhang, 2002c). Li et al. (1996) observed that $CaCl_2$ treatment increased the activities of SOD, POD and CAT in rice seedlings, reduced the degree of decline in SOD, POD and CAT activities caused by chilling stress. The increase in SOD, POD and CAT activities induced by $CaCl_2$ treatment could be inhibited by CPZ in rice seedlings (Liang and Wang, 2001) and tobacco seedlings (Zhang et al., 2002). In this study, it appeared that the increasing trend of CaM content in *P. tomentosa* cuttings was paralleled to the increase in SOD, POD or mitochondrial Ca^{2+}-ATPase activities during freezing acclimation (Figs. 2, 4 and 6). The effect of freezing acclimation on the increase in CaM content, and in the activities of SOD, POD and mitochondrial Ca^{2+}-ATPase were markedly enhanced by $CaCl_2$ treatment, but this enhanced effect of $CaCl_2$ was abolished by EGTA, $LaCl_3$ or CPZ (Table 2). Moreover, treatment with $CaCl_2$ at the same time of freezing acclimation reduced the degree of decline in the activities of SOD, POD and mitochondrial Ca^{2+}-ATPase as well as in CaM content caused by freezing stress at -14℃, and enhanced the increase level in the above-mentioned indices in the recovery periods (Figs. 3, 5, 7 and 8). Theses results showed that the change in CaM content was closely correlated to the activities of mitochondrial v-ATPase, SOD or POD after freezing acclimation both with and without $CaCl_2$ pretreatment. It was suggested that SOD, POD and mitochondrial Ca^{2+}-ATPase be Ca^{2+}-CaM-dependent target enzymes, the effective activation of SOD, POD and mitochondrial Ca^{2+}-ATPase be related to the elevation of CaM content.

Low temperature stress resulted in an increase in cytoplasmic Ca^{2+} concentration, which may disturb physiological metabolism system, and caused cell death (Zeng and Li, 1999). It has been known that the maintenance of a low level of cytoplasmic calcium is required for the start of the function of the calcium messenger system. There is evidence that the ability to maintain calcium homeostasis may be related to the effective activation of Ca^{2+}-ATPase located in plasmalemma and cellular organelle membrane (Hao and Yu, 1992; Poovaiah et al., 1993). In the present study, the presence of $CaCl_2$ at the same time of freezing acclimation enhanced the effect of freezing acclimation on the increase in mitochondrial Ca^{2+}-ATPase activity and freezing resistance (Tables 1 and 2). Although the decrease in mitochondrial Ca^{2+} ATPase activity caused by freezing stress at −14℃ was found in FA and FA-$CaCl_2$ cuttings, the degree of decline in Ca^{2+}-ATPase activity of FA-$CaCl_2$ cuttings was smaller than that of FA ones (Fig. 5). The results showed that the enhancement of freezing resistance in FA-$CaCl_2$ cuttings might be related to the effective activation of mitochondrial Ca^{2+}-ATPase, because activated Ca^{2+}-ATPase could bring back rapidly the raised cytoplasmic Ca^{2+} concentration from freezing stress to the state of calcium homeostasis, leading to the maintenance of normal functions of the calcium messenger system and physiological metabolism (Zeng and Li, 1999). Therefore, it was suggested that there be stronger ability of maintaining calcium homeostasis in FA-$CaCl_2$ cuttings. In addition, we observed that MDA content decreased with an increase in SOD and POD activities during freezing acclimation (Figs. 6 and 9). Thus, the increase in SOD and POD activities seems to be the key factor leading to the reduction of membrane lipid peroxidation and the decrease in MDA content during freezing acclimation (Fan et al., 1995, Hodges et al. 1997, Lin et al., 2001), which further result in the elevation of membrane stability and freezing resistance of cuttings during freezing acclimation.

It was found that the rise of mitochondrial Ca^{2+}-ATPase activity during freezing acclimation ran parallel to increase in SOD and POD activities, which was closely correlated to the decrease in LT_{50} of cuttings (Figs. 1, 2 and 6). Moreover, the decrease in SOD and POD activities caused by freezing stress and the increase in these activities in the recovery periods was in accordance with those of mitochondrial Ca^{2+}-ATPase activity (Figs. 5, 7 and 8). It seems that the changes in SOD and POD activities were closely associated with mitochondrial Ca^{2+}-ATPase. In combination with our recent work that the increase in Ca^{2+}-ATPase activity of plasmalemma and mitochondrial membrane in FA cuttings during freezing acclimation accelerate the hydrolysis of ATP and the release of energy (Lin and Zhang, 2002c), it can be concluded that mitochondrial Ca^{2+}-ATPase plays an important role in the induction of freezing resistance in *P. tomentosa* cuttings, presumably at least in part by providing energy for the biosynthesis of substances and the activation of enzymes associated with the enhancement of freezing resistance.

Summing up, treatment with $CaCl_2$ at the same time of freezing acclimation enhanced the effect of freezing acclimation and markedly increased the activities of SOD, POD and mitochondrial Ca^{2+}-ATPase, CaM content as well as freezing resistance of *P. tomentosa* cuttings, and this enhancement was abolished by EGTA, $LaCl_3$ or CPZ. It was suggested that the increase in CaM content induced by $CaCl_2$ treatment promotes the formation of Ca^{2+}-CaM complexes, which effectively activates the activities of SOD, POD and mitochondrial Ca^{2+}-ATPase and then further result in the adaptive changes associated with the development and enhancement of freezing resistance. Thus, it can be concluded that calcium-calmodulin messenger system is involved in the regulation of SOD, POD and mitochondrial Ca^{2+}-ATPase, and the induction of freezing resistance of cuttings.

Role of Glucose-6-Phosphate Dehydrogenase in Freezing-induced Freezing Resistance of *Populus suaveolens**

Abstract To explore the role of glucose-6-phosphate dehydrogenase (G6PDH, EC 1.1.1.49) in the enhancement of freezing resistance induced by freezing acclimation, G6PDH was purified from the leaves of 8-week-old *Populus suaveolens* cuttings. The G6PDH activity in the absence or the presence of reduced dithiothreitol (DTT_{red}) were determined, and the changes in superoxide dismutase (SOD), peroxides (POD) and cytosolic G6PDH activities, malondialdehyde (MDA) content as well as freezing resistance (expressed as LT_{50}) of *P. suaveolens* cuttings during freezing acclimation at −20℃ were investigated. The results showed that the purified G6PDH was probably located in the cytosol of *P. suaveolens*. Freezing acclimation increased the activities of SOD, POD and cytosolic G6PDH, and decreased the MDA content and LT_{50} of cuttings, while 2 d of de-acclimation at 25℃ resulted in a decrease in SOD, POD and cytosolic G6PDH activities, and caused an increase in MDA content and LT_{50}. The change in cytosolic G6PDH activity was found to be closely correlated to the levels of SOD, POD and MDA, and to the degree of freezing resistance of cuttings during freezing acclimation. It is suggested that the enhancement of freezing resistance of cuttings induced by freezing acclimation is related to the distinct increase in cytosolic G6PDH activity, which may be involved in the activation of SOD and POD, and the induction of freezing resistance of cuttings.

Keywords *Populus suaveolens*, freezing acclimation, freezing resistance, LT_{50}, G6PDH

G6PDH is the main regulated enzyme that catalyzes the first irreversible reaction of the pentose phosphate pathway (PPP). Its main physiological function is to provide NADPH for reductive biosyntheses (Copeland and Turner, 1987; Graeve et al., 1994; von Schaewen et al., 1995; Dennis et al., 1997). It was found that the G6PDH activity in poplar twigs exhibited a marked increase in fall and winter followed by a gradual decrease in spring (Sagisaka, 1972). Sagisaka (1985) reported that inactivation of G6PDH resulted in injuries of poplar twigs during frozen storage. An increase in G6PDH activity induced by low temperature was observed in alfalfa (Krasnuk et al., 1976), ryegrass (Bredemeijer and Esselink, 1995), soybean (van Heerden et al., 2003), banana (Lin et al., 2001), and *Populus tomentosa* (Lin and Zhang, 2001). But the actual role of G6PDH in the enhancement of freezing resistance induced by freezing in plants has not been studied yet. Little is known about the correlation between the elevation of G6PDH activity,

* 本文原载《植物生理与分子生物学学报》, 2005, 31(1): 34-40, 与刘文凤、林元震、张谦和朱保庆合作发表。

the increase in SOD and POD activities, and the degree of increase in freezing resistance induced by freezing acclimation, especially in woody plants.

Populus suaveolens, a freezing-resistant arbor plants, can survive under a temperature of approximately −43.5℃ in winter in Daxinganling, Northeast of China, and therefore is a good material to study the mechanism of freezing resistance of woody plants (Lin, 2001). In the present study, the changes in the SOD, POD and G6PDH activities, MDA content as well as LT_{50}, and the correlation mentioned above were investigated in detail using *P. suaveolens* cuttings to explore the physiological role of G6PDH in the enhancement of freezing resistance induced by freezing acclimation.

1 Materials and Methods

1.1 Plant material

Populus suaveolens cuttings were obtained from Heilongjiang province, China. The rooted cuttings were grown in pots containing a 2∶1 (*V/V*) mixture of soil and sand. After the cuttings were grown for 5 weeks in a greenhouse, they were placed outdoors for 3 weeks, and them used as experiment materials.

1.2 Freezing acclimation and de-acclimation

Eight-week-old cuttings were divided into three groups, 500 branches of 100 cuttings in 25 pots per group. The first group placed at 25℃ was referred to as non-acclimated (NA or control) cuttings. The second group held at −20℃ for 6d was referred to as freezing-acclimated (FA) cuttings, and the FA cuttings exposed to 25℃ for 2 d were referred to as de-acclimated (DA) cuttings. Cuttings of all treatments were grown under an 8−h photoperiod and a light intensity of 30μmol/(m^2·s).

1.3 Measurement of survival rates and LT_{50} of cuttings

All tested cuttings used to measure the survival rates and LT_{50} were treated with different temperatures. Temperatures−20, −23, −26, −29, −32 and−35℃ were used for NA and DA cuttings, −35, −38, −41, −44, −47, −50 and−53℃ used for FA cuttings. Cuttings were held at each temperature for 12h. Then all treated cuttings were directly transferred from the temperature of each treatment to 25℃ for 2d. Surviving cuttings were defined as an 80% of surviving leaves and buds after treatment each temperature for 12h, and then the survival rates were calculated. The data of survival rates of cuttings subjected to treatment with each temperature were used to fit exponential function by using the logarithm of survival rate as a linear function of the time of cold treatment, and then the temperature for 50% survival of cuttings (LT_{50}) was calculated. LT_{50} is taken as a measure of freezing resistance.

1.4 G6PDH extraction and determination

G6PDH was extracted according to the procedure of Graeve et al. (1994) with some modifications, and its purity was tested by the method of Debnam and Emes (1999). Two grams of leaves from *P. suaveolens* cuttings acclimated at −20℃ for 6d were ground under liquid nitrogen and suspended in 1.5mL of ice-cold extraction buffer containing Tris-HCl 0.15mol/L (pH 8.0), NADP0.1mmol/L, PMSF 1mmol/L, β-mercaptoethanol 3mmol/L, 6-aminocaproic acid 1mmol/L, benzamidine 2mmol/L and 2% (*W/V*) insoluble PVP. The homogenates were centrifuged at 4℃ at 16000×*g* for 20min, and then solid ammonium sulfate was added to the supernatant to 40% of saturation. The solution was stirred for 30min and precipitated proteins were removed by centrifugation for 30min at 4℃ at 15000×*g*. The resulting supernatant was collected, to which solid ammonium sulfate was added to 70% saturation, and centrifuged as above. The insoluble material was dissolved in 2mL of icecold buffer composed of Tris-HCl 0.03mol/L (pH 8.0), NADP 0.1mmol/L, PMSF 1mmol/L and β-mercaptoethanol 5mmol/L, 6-aminocaproic acid 1mmol/L and benzamidine 2mmol/L, centrifuged at 4℃ at 14000×*g* for 10min and the supernatant was used for the analysis of G6PDH. The activity of enzyme was measured in a butter containing Tris-HCl 0.15mol/L (pH 8.0), NADP 0.1mmol/L and G6P 3mmol/L. The reaction was started by the addition of 30mL enzyme preparation after 10min of pre-incubation at 30℃. One unit (U) of enzyme activity was defined as the amount of enzyme that increased 0.01 of A_{340} permin under the assay condition.

Whether the G6PDH was in cytosol or in plastid is distinguished by measuring G6PDH activity in the absence or the presence of reduced dithiothreitol (DTT_{red}) under the standard reaction condition according to Johnson (1972).

Protein content was measured by the method of Bradford (1976) using bovine serum albumin as astandard.

1.5 SOD, POD and MDA extraction and determination

SOD, POD and MDA were extracted from 1g of leaves as described by Lin et al. (2002). SOD was determined by monitoring the inhibition of photochemical reduction of nitroblue tetrazolium (NBT) as described previously (Liu and Zhang, 1994). One unit (U) of SOD was defined as the amount of enzyme that produced a 50% inhibition of photochemical reduction of NBT. POD activity was determined by measuring the increase in absorption at 470 nm according to Liu and Zhang (1994). One unit (U) of POD was defined as the amount of enzyme that caused a 0.01 increase in A_{470} permin under the assay condition. MDA was measured by the thiobarbituric acid reaction according to Draper and Hadley (1990).

All the determinations were performed at least in triplicates and the average values were presented.

2 Results

2.1 Effect of DTT_{red} on G6PDH activity

After pre-incubation with DTT_{red} 70mmol/L for 100min, G6PDH activity was less than 5% lower than the maximal activity of control (Fig. 1), indicating that pre-incubation with DTT_{red} had no effect on the G6PDH activity and the enzyme was in cytosol of *P. suaveolens*.

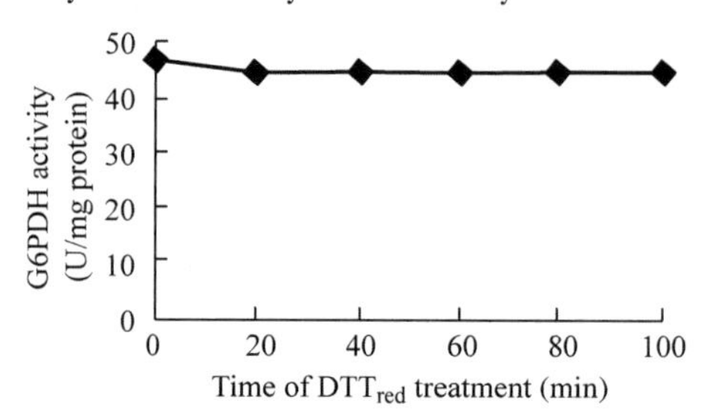

Fig. 1　Effect of DTT_{red} on G6PDH activity

2.2 Change in freezing resistance

During freezing acclimation at −20℃, there was a gradual decrease in LT_{50} of cuttings (Fig. 2). After 6d of freezing acclimation, the LT_{50} of cuttings lowered more than 16℃ compared with control cuttings. But after 2d of de-acclimation at 25℃, the LT_{50} of cuttings returned to about the control level (Table 1).

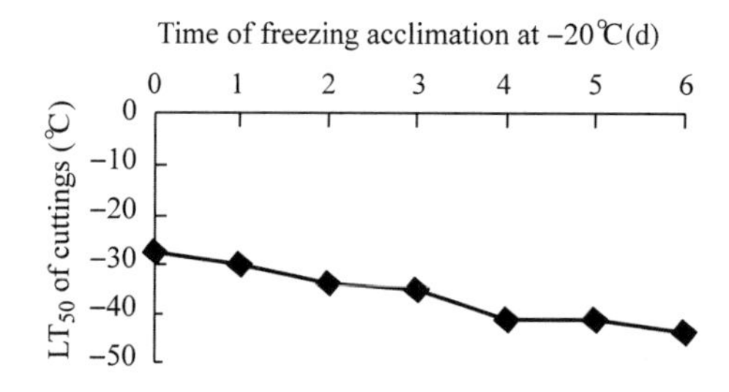

Fig. 2　Change in LT_{50} of cuttings with time of freezing acclimation at −20℃

Table 1　Changes in LT_{50}, MDA content, and the activities of SOD, POD and G6PDH in *P. suaveolens* cuttings

Treatments	LT_{50}(℃)	G6PDH (U/mg protein)	SOD (U/mg protein)	POD (U/mg protein)	MDA (mmol/g FW)
Control	−27.1	47.2±1.2	80.6±2.3	22.3±1.7	22.1±1.0
FA	−43.5	143.9±2.3	113.9±1.8	35.9±1.5	10.2±0.5
DA	−26.6	40.2±4.9	81.1±2.5	22.5±1.8	21.4±0.8

2.3 Changes in the activities of G6PDH, SOD and POD

The activities of SOD, POD and cytosolic G6PDH in cuttings gradually increased with time of freezing acclimation in a period of 6d (Fig. 3), but a larger increase was observed in cytosolic G6PDH than in SOD activity, especially than in POD activity (Fig. 3). Two days of de-acclimation at 25℃ resulted in a decrease in SOD, POD and cytosolic G6PDH activities (Table 1). In addition, the activity of cytosolic G6PDH increased with the decrease in LT_{50} of cuttings (Fig. 4).

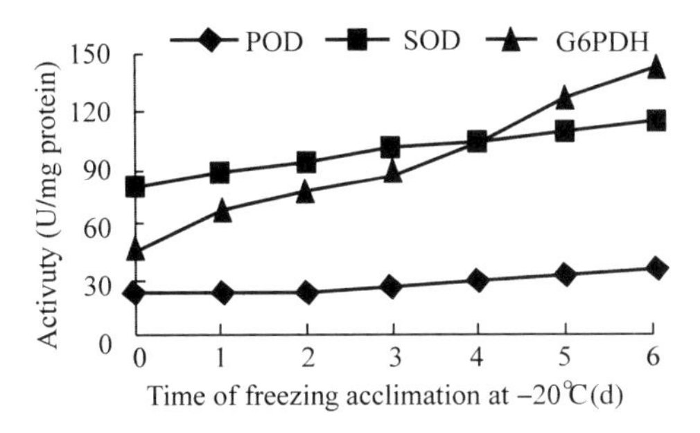

Fig. 3 Effect of freezing acclimation at −20℃ on the activities of G6PDH, SOD and POD in cuttings of *P. suaveolens*

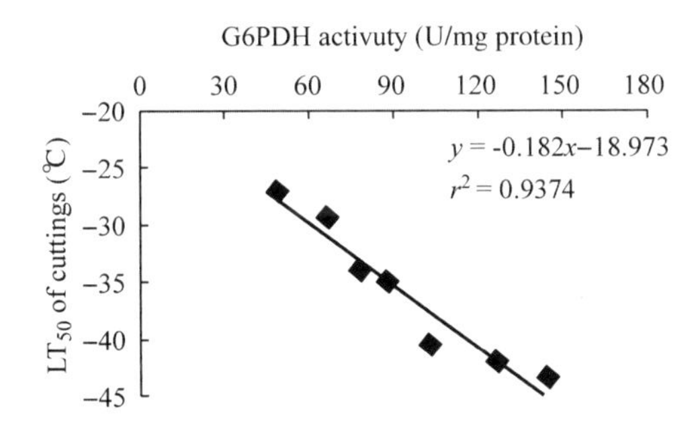

Fig. 4 Relationship between G6PDH activity and LT_{50} in *P. suaveolens* cuttings during freezing acclimation at −20℃

2.4 Change in MDA content

Contrary to the increasing trend of G6PDH, SOD and POD activities, a gradual decrease was observed in MDA content during freezing acclimation at −20℃ (Figs. 3 and 5), moreover, there was a very significant negative correlation between the SOD and POD activities and MDA content correlation (Fig. 6). Two days of deacclimation at 25℃ resulted in a significant increase in MDA content (Table 1).

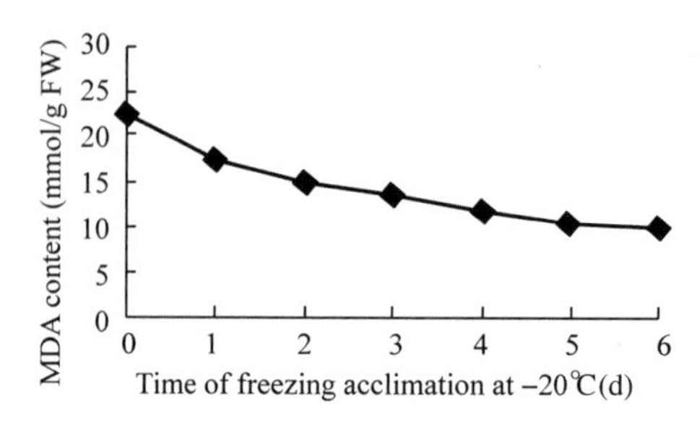

Fig. 5 Effect of freezing acclimation at −20℃ on MDA content in *P. suaveolens* cuttings

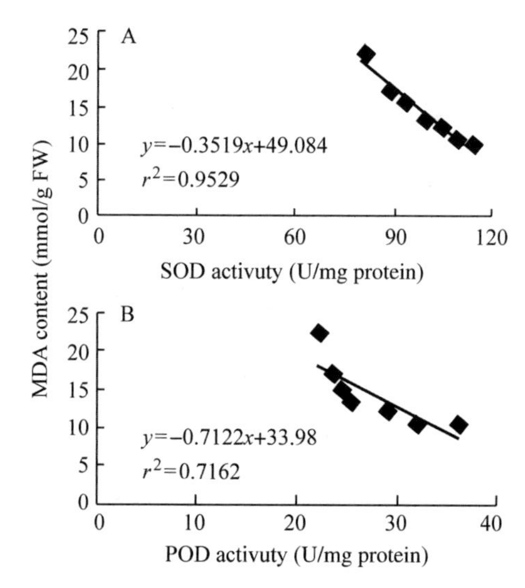

Fig. 6 Relationship between MDA content and the SOD (A) and POD (B) activities in *P. suaveolens* cuttings during freezing acclimation at −20℃

3 Discussion

In plants, G6PDH activity is present in both cytosol and plastid (Dennis and Miernyk, 1982). One can distinguish the G6PDH in cytosol or in plastid is routinely by measuring G6PDH activity in the absence or the presence of reduced dithiothreitol (DTT_{red}) (Johnson, 1972; Wendt et al., 2000). In this work, theG6PDH activity of *P. suaveolens* cuttings was not inactivated by pre-incubation with DTT_{red}(Fig. 1), as was the case for the cytosolic G6PDH of pea, spinach and potato (Fickenscher and Scheibe, 1986; Graeve et al., 1994; von Schaewen et al., 1995; Schnarrenberger et al., 1995; Wendt et al., 2000), indicating that the purified G6PDH from *P. suaveolens* was probably the cytosolic isoform.

Although low temperature induced increase in G6PDH activity was observed in some herbaceous plants, up to now reports about the situation in woody plants are very limited. We have found a larger increase in G6PDH activity and freezing resistance observed in freezing-resistant cuttings of *P. suaveolens* than in freezing-sensitive ones of *P. tomentosa* during freezing acclimation (Lin and Zhang, 2003). In addition, the increase in G6PDH activity under low temperature was most probably of cytosolic origin (Lin and Zhang 2003; van Heerden et al., 2003). In this study, we observed that freezing acclimation induced increase in cytosolic G6PDH activity in *P. suaveolens* cuttings was followed by an obvious decrease after 2d of deacclimation at 25℃ (Table 1), indicating that the increase in cytosolic G6PDH activity during freezing acclimation is an active response to low temperature. Similar findings had been reported in ryegrass (Bredemeijer and Esselink, 1995), soybean (van Heerden et al., 2003), banana (Lin et al., 2001), and *P. tomentosa* (Lin and Zhang, 2001).

Many studies have revealed that freezing acclimation resulted in an increase in the activities of cell antioxidant enzymes, which is related to the decrease in MDA content and the enhancement of freezing resistance (Bridger et al., 1994; Fan and Guo, 1995; Saruyama and Tanida, 1995; Hodges et al., 1997; Shen et al., 1999; Kang and Saltveit, 2001; Lin et al., 2001, 2004a). In this study, we observed that MDA contentdecreased with an increase in the activities of SOD and POD during freezing acclimation, and there was a significant negative correlation between MDA content and the activities of SOD and POD ($r=-0.976^{**}$ and -0.846^{**}) (Figs. 3, 5 and 6), indicating that the increase in SOD and POD activities seems to be the key factor leading to the less product of membrane lipid peroxidation during freezing acclimation. Thus, the results in this paper enlightened us that an increase in SOD and POD activities caused by freezing acclimation could result in a decrease in membrane lipid peroxidation and an elevation of cell membrane stability of cuttings.

Many studies have showed that freezing acclimation induces an increase in G6PDH activity, which may contribute significantly to the enhancement of freezing resistance (Krasnuk et al., 1976; Bredemeijer and Esselink, 1995; van Heerden et al., 2003). However, knowledge of the physiological role of G6PDH in the enhancement of freezing resistance is still very limited. Our recent work indicated that the G6PDH activity increased with an increase in SOD and POD activities during freezing acclimation, which was closely correlated to the decrease in MDA content and LT_{50} in *P. tomentosa* and banana (Lin and Zhang, 2001, 2003; Lin et al., 2001). In this study, the rise of cytosolic G6PDH activity during freezing acclimation ran parallel to an increase in SOD and POD activities and to a decrease in LT_{50} of *P. suaveolens* cuttings (Figs. 2, 3 and 4), and there was a very significant positive correlation between G6PDH and POD and SOD (0.997^{**} and 0.980^{**}) respectively, while a significant negative correlation between G6PDH and LT_{50} ($r=-0.968^{**}$) was shown in cuttings. The results showed that an increase in POD and SOD activities and a decrease in LT_{50} induced by freezing acclimation might be associated with an increase in cytosolic G6PDH activity of cuttings. G6PDH is the main regulated enzyme catalyzing the first irreversible reaction of the pentose phosphate pathway (PPP). In general, the main physiological function of G6PDH is to generate NADPH and some important intermediate metabolite for a variety of anabolic metabolisms and detoxifation reactions (Copeland and Turner, 1987; Graeve et al., 1994; Dennis et al., 1997; van Heerden et al., 2003; Lin and Zhang, 2003). It has been reported that the response of plants to low temperature resulted in the increase in the capacity of PPP and the activity of its key regulatory enzyme G6PDH, which are required for the induction of freezing resistance (Sadakane and Hatano, 1982; Bredemeijer and Esselink, 1995; van Heerden et al., 2003). Thus, it is concluded that freezing acclimation could result in an increase in the cytosolic G6PDH activity and the capacity of PPP to generate NADPH in *P. suaveolens* cuttings. Many studies have revealed that in most plants subjected to environmental stresses such as low temperature and drought, the decreased activity of G6PDH and the less production of NADPH could result in a decline in the activities of some NADPH-dependent antioxidant enzymes (ascorbate peroxidase, dehydroascorbate reductase and glutathione reductase), which would result in severe damage of cell membrane structure (Kocsy et al., 2001; Kevers et al., 2004; Lin et al., 2004b). The study by Sagisaka (1985) showed that inactivation of G6PDH resulted in injuries of poplar twigs under low temperature. Thus, it is suggested that cytosolic G6PDH may play an important role in the enhancement of freezing resistance in *P. suaveolens* cuttings, presumably at least in part by providing NADPH as a reductant for the activation of some NADPH-dependent antioxidant enzymes associated with the enhancement of cell membrane stability.

Summing up, the significant increase in cytosolic G6PDH activity of cuttings caused by freezing acclimation first enhance the capacity of PPP to generate NADPH, which effectively activates the activities of SOD and POD and promotes many important enzymatic reactions, and then further result in the elevation of membrane stability and freezing resistance of cuttings. Thus, it can be concluded that the distinct increase in cytosolic G6PDH activity induced by freezing acclimation may be involved in the activation of SOD and POD, and the induction of freezing resistance of cuttings.

Resistance of Transgenic Hybrid Triploids in *Populus tomentosa* Carr. Against 3 Species of Lepidopterans Following Two Winter Dormancies Conferred by High Level Expression of Cowpea Trypsin Inhibitor Gene*

Abstract Hybrid triploid poplars [(*P. tomentosa* ×*P. bolleana*)×*P. tomentosa*] genetically engineered with cowpea trypsin inhibitor (*CpT* I) gene have been out-planted in field for two years. They were used to detect their efficacy against 3 species of poplar defoliators: forest tent capterpillar, *Malacosoma disstria* Hübner, gypsy moth, *Lymantria dispar* Linnaeus and willow moth, *Stilpnotia candida* Staudinger by using detached leaves and for the purpose of identifying the *CpT* I gene at the molecular level. Foliage of transgenic poplars elicited an increase in larval mortality rate and a decrease in foliage consumption, wet weight gains, faeces excretion, deposited pupae number and pupae weight, thus indicating its effectiveness in affecting the growth, development and fecundity of larvae rather than only directly killing them. PCR and southern blotting analyses confirmed the stable incorporation of *CpT* I gene while proteinase inhibitory assays disclosed its high level expression in the two-field-season of transgenic trees. Efficacious insect resistance and higher content of *CpT* I in foliage were found in transgenic clone TG04, TG07, TG08 and TG71, demonstrating a correspondence between the insect resistance level and the *CpT* I content in the foliage of transgenic poplar.

Keywords *Populus tomentosa* Carr., *CpT* I gene, transgenic poplar, *Lepidoptera*, insect resistance, proteinase inhibitory assay, PCR, Southern blotting.

Introduction

Chinese White Poplar (*Populus tomentosa* Carr.) is one of the indigenous tree species of white poplar (Section *Leuce*) in China. It is found in the country's middle eastern region, and covers about one million km^2(approximately 1/9 of the total area of China). It is characterized by rapid growth, high yield and short rotation and is commonly known for the superior timber that it yields, and is widely used as an important afforestation and landscape species in the north of China (Zhu and Zhang, 1997). Moreover, the newly bred hybrid triploid poplar [(*P. tomentosa*× *P. bolleana*)×*P. tomentosa*] (Zhang et al., 1992; 1997b) has demonstrated great improvement in photosynthetic rate (Li et al., 2000a) and wood quality (Pu et al., 2002; Xing et al., 2002) in

* 本文原载 *Silvae Genetica*, 2005, 54(3): 108-116, 与张谦、林善枝和林元震合作发表。

addition to the traits in diploid *P. tomentosa* Carr. , and is a good candidate for genetic engineering as a result of the establishment of an improved regeneration system(Hao et al. , 1999a; Lu et al. , 2001) and its high sterility rate (Zhang et al. , 2000), thus promising a dramatically decreased possibility of transgene dispersal from engineered poplars to wild types through pollen hybridization and also ensuring high biosafety of engineered poplars to environment.

However, *P. tomentosa* Carr. , like other poplar species, is vulnerable to many insects. It is the host to more than 200 species of insects, most notably to 30 kinds of Lepidoptera and Coleoptera (including underground insects, leaf or twig damaging insects and stem borers), which pose the most serious threat (Xu, 1988). Over the past few decades, various methods have been utilized to reduce this insect problem. The traditional spray application of insecticidal proteins and chemical toxins is effective in that it kills the insects instantly, but it has devastating long-term effects for the environment. Traditional breeding for insect resistance is a difficult task in trees with long rotations and limited resistant hereditary resources (Xu, 1988; Whitman et al. , 1996; Zhang, 1997a). But genetic engineering provides an opportunity to transfer new specific traits of interest into valuable genotypes (Campbell et al. , 2003). To date, scientists world-wide have been able to genetically engineer poplars resistant to insects by relying mainly upon genes encoding insecticidal proteins of *Bacillus thuringiensis* δ-endotoxin (Mccown et al. , 1991; Wu et al. , 1991; Tian et al. , 1993, 2000; Pannetier et al. , 1997; Chen et al. , 1995) and those encoding proteinase inhibitor (Klopfenstein et al. , 1991; Leple et al. , 1995; Urwin et al. , 1995; Heuchelin et al. , 1997; Confalonieri et al. , 1998; Hao et al. , 1999b; Li et al. , 2000b; Delle Donne et al. , 2001; Yang et al. , 2003).

Cowpea trypsin inhibitor (*CpT* Ⅰ) is a small polypeptide belonging to the Bowman-Birk type of double-headed serine proteinase inhibitors (Hammond et al. , 1984; Hilder et al. , 1989; Ryan, 1989, 1990; Liu et al. , 1993), and has high efficacy in affecting growth and development of insects including *Lacanobia oleracea* (Bell et al. , 2001) and a wide range of other Coleoptera, *Diabrotica* species, *Anthronomus grandis*, Lepidoptera and Orhoptera (Hilder et al. , 1990). Moreover, it is less likely to induce resistance of insects due to the fact that it binds competitively to the comparatively conserved active site of digestive enzymes, and is highly sensitive to pepsin, thus promising it safety to mammals (Ryan et al. , 1989; Hilder et al. , 1989). Therefore, it is considered to be an ideal candidate for yielding transgenic plants resistant to predators (Ghoshal et al. , 2001). This was firstly demonstrated by Hilder et al. (1987) when cowpea trypsin inhibitor (*CpT* Ⅰ) gene was transferred from *Vigna unguiculata* to tobacco. The transgenic tobaccos conferred resistance to a wide range of insect pests, such as Lepidoptera *Heliothis* and *Spodoptera*. It was followed by the genetically engineered rice (XU et al. , 1996), tobacco (Ghoshal et al. , 2001) and hybrid triploid poplar [(*P. tomentosa*×*P. bolleana*)×*P. tomentosa*] (Hao et al. , 1999b), etc.

In the hope of providing a new poplar variety to meet the pressing afforestation need in China, the hybrid triploid poplar [(*P. tomentosa*×*P. bolleana*)×*P. tomentosa*] was further improved in resistance against insect attack by genetically engineering it with *CpT* Ⅰ gene at our university, and integration and expression of the foreign gene in the host genome had been characterized by our preliminary studies performed on regenerated shoots and plantlets (Hao et al., 1999b). They were then planted in the field for more than two field seasons without insect feeding bioassays or further studies at molecular level. Two-field-season of plantation in field increased our interest in figuring out the efficacy of transgenic poplars to defoliators and our curiosity about the integration status and the expression level of *CpT* Ⅰ gene in host genome.

In this study, we detail the use of insect feeding bioassays to evaluate the efficacy of two-field-season of genetically engineered poplars against 3 species of poplar defoliators and of PCR, Southern blotting and proteinase inhibitory assays to identify *CpT* Ⅰ gene and its encoded protein at the molecular level. Increased insect resistance of transgenic poplars was identified, stable incorporation and active expression of *CpT* Ⅰ gene in engineered poplars were confirmed, efficacious insect resistance and higher content of *CpT* Ⅰ were detected in 4 transgenic clones revealing a correspondence between the insect resistance level and the *CpT* Ⅰ content in the foliage of transgenic poplars.

Materials and methods

Plant materials

Transgenic hybrid triploid poplars [(*P. tomentosa*×*P. bolleana*)×*P. tomentosa*] were produced by *Agrobacterium tumefaciens*-mediated transformation as described by Hao et al. (1999b). The plasmid T-DNA (pBin Ω SCK) harbored the *CpT* Ⅰ gene and kanamycin resistance gene *NPT*Ⅱ from Tn5, both of which were under the control of a constitutive CaMV 35S promoter in 5′ end and a T-nos terminator in 3′ end. A SKTI signal peptide and a KDEL coding sequence were fused in the 5′ region and 3′ end of *CpT* Ⅰ gene, respectively, thus forming a Signal- *Cp*TI-KDEL construct. Between the promoter and the modified *CpT* Ⅰ gene construct is a 68bp long Omega element from a Tobacco Mosaic Virus (TMV) gene encoding a 126kD protein, which is located in the untranslation region of upstream transcription sequence, and has the potential to enhance the expression of downstream gene.

Both transformed and untransformed (control) poplars were propagated from micro-cultured shoots and 3~10 individual seedlings for each clone (control included) and were then transplanted in Houbajia field, in a northern suburb of Beijing in fall of 2000.

Insect materials

The early 3rd-instar forest tent caterpillar, *Malacosoma disstria* Hübner, were collected from branches of *Malus micromalus* Makino at our university and reared with the foliage of *Malus micromalus* Makino for 2d. The gypsy moth, *Lymantria dispar* (L.) was reared from the egg bands provided by Forest Protection Laboratory at our university until early 3rd-instar, at which point the larvae were fed with the foliage of *Diospyros kaki* (L.). The first generation willow moth, *stilpnotia candida* staudinger were gathered in their later 5th-instar from the *Populus Canadensis* Moench (*P. deltoides* ×*P. nigra*) in front of Forest Protection Station at our campus and were transferred to the greenhouse for feeding assay. 3 egg bands produced by the first generation willow moth were collected for the preparation of the second- generation larvae and the larvae were reared until 3rd-instar. All larvae were reared in a controlled environmental chamber at 27℃ and a photoperiod of 16 :8(L :D) h.

Insect bioassay and data analysis

Four non-choice laboratory feeding assays were launched on 18 April, 18 May, 9 June and 28 June using early 3rd-instar forest tent caterpillar, later 5th-instar willow moth, early 3rd-instar gypsy moth and early 3rd instar willow moth, respectively. The leaves with the same size were excised at the base of the petiole from each tree, sealed in plastic sandwich bags, and taken to the laboratory. The same poplars were used for all 4 assays.

For each insect bioassay, a group of 10 to 20 defoliators held in a glass vial (7.5cm diameter, 12.5cm height) were allowed to feed on the leaves of each clone (3 replicate treatments per clone). To maintain the humidity of air and leaves, a piece of wet filtration paper was placed on the bottom of each container. The feeding foliage was replaced with a new one every other day. Larvae were removed from original rearing leaves the day before the launching of insect bioassays to void the mid-gut of insects before weighing them.

The mortality rates of larvae feeding on the foliage of each clone was recorded and analyzed to evaluate the toxicity of each transgenic clone to defoliators.

All original defoliators in each vial and the final survival larvae were grouply weighed, respectively, to calculate the mean wet weight gains of insects. Faeces excreted in a period of 8d were collected and dried overnight in an oven at 60℃ and the mean dry weight gains were used as a measure of food intake (Johnston et al., 1993). By the end of the trial performed on the second-generation of willow moth larvae, male and female pupae weight were measured respectively, the egg numbers deposited were counted to estimate their fecundity.

All of the data from insect bioassays were analyzed using one-way analysis of variance (Anova) and subsequent Fisher's multiple comparisons if there are significant differences as described by Rosner (2004).

Assay of cowpea trypsin inhibitor

Leaf protein extraction was performed as described previously (Perlak et al., 1991). Assay of trypsin proteinase inhibitor with the casein as the substrate of trypsin was performed as also described previously (Hao et al., 1999; Xu et al., 1996). 0, 0.4, 0.8, 1.2, 1.6, 2.0 and 2.4μg of cowpea trypsin inhibitors were mixed with trypsin respectively, and incubated at 27℃ for 30minutes, followed by the addition of 100μg substrate of trypsin. The casein with the hydrolysis was incubated at 30℃ for 1h. Data derived from the OD280 were used to draw the standard inhibitory curve. The same procedure for construction of standard inhibitory curve was used in the inhibitory reaction of plant *CpT* Ⅰ to trypsin, except that the cowpea trypsin inhibitor was replaced with 50μg extracts from the young leaves and old leaves of each clone. The quantification of plant *CpT* Ⅰ was calculated according to the data of OD_{280} achieved from the standard inhibitory curve.

PCR and Southern analysis

DNA was extracted from poplar leaves as described by Rogers and Bendich (1985). PCR analysis was performed with gene specific primer combination 5′-ATGAAGAGCACCATCTTCTTTGCTC-3′ and 5′-CTTACTCATCATCTTCATCCC TGG-3′ designed according to sequence of the *CpT* Ⅰ gene cloned by Liu et al. (1993). PCR amplification was performed on an ABI thermocycler (9700) and in PCR buffer (10mM Tris pH 8.0, 0.01 percent gelatin, 0.1 percent Triton X-100, 80mM KCl, 3.5mM $MgCl_2$) containing 25pM of each primer, 250μM of each dNTP (Pharmacia Ultra pure), 20~100 ng of target DNA, 5 units of Taq DNA polymerase (Promega). Reaction volumes were made up to 50μL. Amplification of *CpT* Ⅰ genes was achieved by an initial denaturation step at 94℃ for 5min followed by addition of the polymerase and 34 cycles of 94℃, 30s; 55℃, 60s; 72℃, 60s, with a final extension step of 10min at 72℃. Amplified products were separated on 1.5% agarose gel in 1×TAE buffer. Fragments were viewed under UV light after gel staining with ethidium bromide and photographed using an Olympus Digital Camera. DNA fragment sizes were determined by comparison to [lambda] *Hin*d III digested and Pharmacia 100bp ladder size markers.

As for Southern blotting analysis, 10μL (1μg/μL) plant genomic DNA was subject to overnight digestion by 1μl *Hin*d III (5U/μl) at 37℃, electro phoresed through 1.0% (w/v) agarose, and blotted onto hybond-N^+ membrane (Amersham) using standard protocol (Sambrook et al., 1989). The DNA was cross-linked to membrane using HL-2000 HybriLinker (UVP) prior to pre-hybridization for 1h at 42℃ in DIG Easy Hyb solution provided in DIG High Prime DNA Labeling System and Detection Starter Kit I (Roche). DNA probe of *CpT* Ⅰ gene was prepared by artificially synthesizing a 100bp fragment of *CpT* Ⅰ gene specific sequence due to the unavailability of positive *Agrobacterium tumefaciens* containing the gene of interest and then labeling it through mixing it with the DIG-High Prime contained in the Kit overnight at 37℃, and hybridization performed at 42℃ for 16h. The membrane was rinsed twice in ample 2×SSC, 0.1% SDS at room temperature under constant agitation for 5min, and washed 2×15min in 0.5×SSC, 0.1% SDS (pre-warmed to wash temperature) at 65℃ under constant agitation. It was followed by a 5min rinse in Washing buffer, 30min successive incubation in both Blocking solution and Antibody solution, 2×15min washing in Washing solution and 5min equilibration in Detection buffer. Hybridization bands were revealed by incubation without agitation in freshly prepared color substrate solution containing NBT/BCIP.

Results

Fatal effectiveness

To determine the fatal effectiveness of transgenic poplars to Lepidoptera, 3 species of defoliators, 12 transgenic poplar clones and one untransformed control were used to conduct insect bioassays. The mortality rates of forest tent caterpillar, gypsy moth and willow moth were recorded after 20d, 5d and 38d of incubation, respectively. Analyses revealed significant interaction between genotype and larval mortality rates of three species of insects ($P=3.2E-10$, $4.4E-4$ and $9.8E-9$, for forest tent caterpillar, gypsy moth and willow moth, respectively). All larvae feeding on foliage from transgenic poplars demonstrated greater mortality rates than the untransformed control. After checking the average mortality rates of 3 species of larvae feeding on foliage from the same poplar clone, significant differences in fatal toxicity to larvae were found among all poplar clones ($P=0.016$). The mortality rates of larvae feeding on foliage of distinct poplar clone ranged from only 12.04% (Control) to as high as 83.33% (TG07). The larvae exposed to foliage from clone TG04, TG07, TG16 and TG71 exhibited larval mortality rates higher than 57% (Table 1).

In April, the foliage of 8 transgenic clones caused a larval mortality rate of more than 50% in forest tent caterpillars after 20d feeding assays. Among them, 4 transgenic clones: TG04, TG07, TG20 and TG53 were fatal to over 80% insects and TG53 was found fatal to even 95.56% insect pests (Table 1), while, the larvae given the foliage of clone TG10 disappeared only 22.22%, demonstrating its weakness in defending against predators (Table 1).

In May, 3rd-instar gypsy moth was found to be more susceptible to transgenic poplars than were both forest tent caterpillar and willow moth. More than 60% larvae consuming foliage of transgenic clone TG04, TG07, TG08, TG16 and TG71 died within 5d (Table 1), which was significantly higher than that for untransformed control (only 13.33%) ($P=4.4E-4$). The highest mortality rate (up to 80%) was found on larvae fed with foliage from transgenic clone TG07, the lowest (40%) for clone TG53 except the control. Interestingly, on 7th day, the whole assay was forced to end seeing as the mortality rates of all larvae given transgenic plants was over 86% without significant difference and that for control also reached 76.67%.

Contrary to 3rd-instar gypsy moth, the first generation later 5th-instar willow moths appeared to be highly resistant to transgenic poplars and finished their developmental cycle without any abnormality (data not shown), and the second-generation defoliators feeding on foliage of transgenic poplars kept the insect bioassay pupating without high larval mortality rates. Nevertheless, significant differences in larval mortality rates were found in the second-generation larvae consuming foliage from distinct poplar clones ($P=9.8E-9$). The mortality rates of larvae eating foliage from clone TG04, TG07 and TG71 was all over 50%, and even up to 83.33% larval mortality rate were observed for clone TG71 (Table 1).

Table 1 Average larval mortality (%) with standard errors of different insect species after feeding on field-collected foliage of control and genetically engineered hybrid poplars

Transgenic clone	*M. disstria* (Hübner)	*L. dispar* (Linnaeus)	*S. candida* (Staudinger)	Average
TG02	46. 67±10. 00^{b}	46. 67±15. 28^{a}	26. 67±10. 41^{bc}	40. 00±2. 89^{abc}
TG03	68. 89±10. 18^{cde}	50. 00±10. 00^{ab}	25. 00±5. 00^{ab}	47. 96±22. 02^{bc}
TG04	82. 22±7. 698^{ef}	60. 00±10. 00^{abc}	51. 67±7. 64^{ef}	64. 63±15. 80^{cd}
TG07	86. 67±13. 33^{ef}	80. 00±20. 00^{c}	83. 33±2. 89	83. 33±3. 33^{d}
TG08	52. 63±7. 75^{bc}	70. 00±17. 32^{bc}	35. 00±8. 66^{bcd}	52. 54±17. 50^{bc}
TG10	22. 22±10. 18^{a}	40. 00±10. 00^{a}	35. 00±10. 00^{bcd}	32. 41±9. 17^{ab}
TG16	71. 11±7. 95^{de}	60. 00±17. 32^{abc}	40. 00±10. 00^{ade}	57. 04±15. 77^{bcd}
TG20	86. 67±13. 33^{ef}	40. 00±10. 00^{a}	23. 33±7. 64^{ab}	50. 00±32. 83^{bc}
TG34	48. 89±7. 76^{b}	53. 33±11. 55^{ab}	41. 67±10. 41^{de}	47. 96±5. 89^{bc}
TG53	95. 56±7. 70^{f}	40. 00±0^{a}	30. 00±10. 00^{bcd}	55. 19±35. 32^{bcd}
TG71	62. 50±5. 35^{bcd}	60. 00±17. 32^{abc}	50. 00±8. 66^{ef}	57. 50±6. 61^{bcd}
TG80	48. 89±16. 78^{b}	53. 33±5. 77^{ab}	63. 33±10. 41^{f}	55. 19±7. 40^{bcd}
Control	11. 11±3 85^{a}	13. 33±11. 55	11. 67±2. 89^{a}	12. 04±1. 16^{a}

Average represents the average larval mortality of 3 defoliators; within each column, means with the same letter are not significantly different (P=0. 05); ANOVA FISHER's LSD test

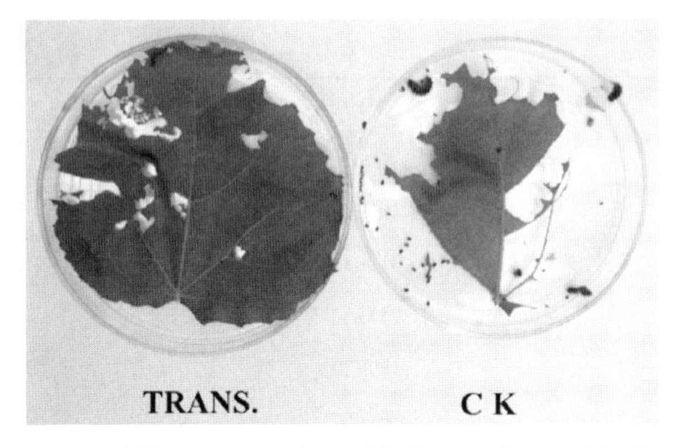

Fig. 1 Comparison of foliage consumption of 3rd-instar *S. candida* Staudinger larvae after feeding for 30h. TRANS. represents foliage of transgenic clone TG07, CK is foliage of untransformed control.

Based on the results disclosed above, 7 transgenic clones with comparatively higher efficacy against 3 species of defoliators were screened out of all transgenic poplars for the subsequent further analysis, and 3 clones TG04, TG07 and TG71 were considered as the ideal transgenic poplars in afforestation for their being fatal toxicity to more than 50% larvae, particularly the best clone TG07.

Foliage consumption and faeces excretion

To investigate the effects of transgenic poplar on food intake and faeces excretion of larvae, the foliage consumption was observed first. As shown in Figure 1, 3rd-instar willow moth, like other two species of defoliators, consumed less field-collected foliage containing *CpT* I gene, and were much smaller in body size than those larvae fed with untransformed control foliage. Later, the faeces excreted by both forest tent caterpillar and willow moth feeding on foliage of poplars for a period of 8d were collected and measured. Larvae feeding on transgenic poplars excreted significant fewer faeces ($P=2.3E-19$ and $P=4.7E-11$, for forest tent caterpillar and willow moth, respectively) (Table 2). An average of 755mg of faeces were totally excreted by individual larva of forest tent caterpillar consuming foliage from control within 8d, and 1025mg by that of willow moth (Table 2), while, less than 320mg of faeces were collected from individual larva of forest tent caterpillar and 720mg from willow moth exposed to foliage from transgenic clones during the same period. Transgenic clone TG04, TG07 and TG71 elicited only 36mg, 67mg and 189mg of faeces per larva of forest tent caterpillar, respectively, and 461mg, 254mg and 342mg per larva of willow moth correspondingly (Table 2) indicating their greater effectiveness in reducing the excretion of faeces than control and other transgenic clones.

Table 2 Average excreted faeces weight (mg) with standard errors of 2 species of larvae after feeding on field collected foliage of control and transgenic hybrid poplars for a period of 8d

Transgenic clone	*M. disstria* (Hübner)	*S. candida* (Staudinger)
TG04	$36±7^{a}$	$461±49^{b}$
TG07	$67±11^{a}$	$254±36^{a}$
TG08	$191±18^{b}$	$504±42^{b}$
TG16	$311±15^{d}$	$719±78^{d}$
TG53	$254±38^{c}$	$519±42^{bc}$
TG71	$189±22^{a}$	$342±51^{a}$
TG80	$289±18^{cd}$	$628±64^{cd}$
Control	755±58	1025±140

Within each column, means with the same letter are not significantly different ($P=0.05$); ANOVA FISHER's LSD test.

Wet weight gain

Along with the analysis of mean faeces weight, the mean wet weight gains of larvae were calculated to indirectly evaluate the effects of transgenic poplars on metabolism, growth and development of defoliators. Larvae feeding on foliage from transgenic poplars had significantly lower mean wet weight gains than did those exposed to control foliage (P= 1. 4E−18 and 1. 3E−20, for forest tent caterpillar and willow moth, respectively) (Table 3). The mean wet weight of forest tent caterpillar consuming foliage from control was increased by 141mg after 20d feeding, and 723mg for willow moth after 28d rearing. However, only less than 84mg and 360mg of mean wet weight gains were revealed in forest tent caterpillar and willow moth consuming foliage from transgenic clone, respectively. These results indicated that the transgenic foliage induced the reduction of mean wet weight gains of larvae. Further analysis found that TG04 and TG07 resulted in the increase in larval wet weight gains of forest tent caterpillar by less than 40mg, as did TG07 and TG71for that of the willow moth, and the efficacy of these 3 clones in reducing the larval wet weight gains was further confirmed (Table 3).

Table 3 Average wet weight gains (mg) with standard errors of 2 species of larvae after feeding on field-collected foliage of control and genetically engineered hybrid poplars

Transgenic clone	*M. disstria* (Hübner)	*S. candida* (Staudinger)
TG04	27±6[a]	163±29
TG07	37±4[a]	23±4[a]
TG08	76±11[c]	118±14
TG16	58±7[b]	256±11[b]
TG53	62±4[b]	228±9[b]
TG71	66±10[b]	40±8[a]
TG80	83±6[c]	359±16
Control	141±14	723±42

Within each column, means with the same letter are not significantly different (P=0. 05); ANOVA FISHER's LSD test.

Pupae weight and deposited egg number

On the 29th day of the bioassay performed on willow moth, pupae were firstly discovered in vials given control foliage. 5d later they were discovered in vials containing foliage of transgenic poplars. On 38th day of the inoculation, all pupae in vials containing transgenic foliage and those given control foliage were collected and grouply weighed respectively to further investigate effects of transgenic poplars on the growth and development of larvae by calculating their mean weight. Two types of willow moth pupae given foliage from transgenic poplars had almost half weight of pupae developed from the larvae exposed to control foliage (P=1. 8E−4 and 0. 02, for female and male pupae, respectively) (Table 4), and were much smaller in physical size in comparison to control pupae (Figure 2).

Table 4 Average pupae weight (mg) and deposited egg number with standard errors of *S. candida* Staudinger after feeding on field-collected foliage

	Control		Transgenic	
	Female	Male	Female	Male
Pupae weight/(mg)	378±15	217±21	190±13	129±25
Egg number	132±24		68±11	

Control represents *S. candida* Staudinger feeding on foliage of untransformed poplars, Transgenic is *S. candida* Staudinger feeding on foliage of transgenic poplars; Female and male represent corresponding gender of *S. candida* Staudinger.

After the above analysis, the pupae developed from larvae given foliage of transgenic poplars and those from control foliage were grouply incubated in 2 distinct containers in greenhouse. It is profitable for their independent hatching and copulating within distinct group. The fecundity of adult willow moths was evaluated by counting their deposited egg number. The number of deposited eggs of adult willow moth developed from larvae exposed to control foliage (132 eggs) was almost double of that for transgenic poplars (68 eggs) ($P = 0.03$) (Table 4). These results demonstrated that foliage of transgenic poplars has evident effectiveness in retarding physical development of insects and affecting their fecundity as well.

Figure 2 Comparison of pupae size of pupated *S. candida* Staudinger after feeding on field-collected foliage. TRANS. Represents pupae developed from larvae after feeding on foliage of transgenic poplars, CK is pupae developed from larvae after feeding on untransformed control foliage.

PCR and Southern blotting analysis

Insect bioassay was followed by molecular identification at DNA level to determine the integration status of *CpT* I gene in the host genome. Genomic DNA extracted from 6 transgenic poplar clones (TG04, TG07, TG08, TG16, TG53 and TG71) and one control poplar were used to perform PCR analysis with *CpT* I gene specific primer combination. As shown in Figure 3, a clear and identical DNA fragment of about 415bp specific to *CpT* I gene was observed in all transgenic lanes, while no corresponding DNA fragment was shown in non-transgenic lane. Identical results were obtained in 3 repeated experiments.

Following PCR analysis, Southern blotting experiment was performed on *Hin*d III digested genomic DNA from 4 transgenic poplar clones (TG04, TG07, TG16 and TG71) and one control poplar. It was demonstrated that there were clear and strong hybridization signals ranging from 1.5kb to 4.0kb in size on lanes for all transgenic poplar clones. However, no corresponding hybridization signal in the lane for the untransformed poplar clone was observed indicating the absence *CpT* I gene in the genome of control poplar (Figure 4). The position of hybridization signals on the film differed from every other transgenic clone, which demonstrated the random incorporation loci of each copy of *CpT* I gene in the genome of different poplar clone. Interestingly, the main hybridization signal observed on lane for TG16 was much weaker than those for other 3 transgenic clones and its position was the lowest on the film. In addition, a very dim hybridization signal was detected on the film in the lane for TG16 and its place was very close to the electrophoresis start line which could be contributed to incomplete digestion of plant genomic DNA. The above results confirmed the presence and stable incorporation of *CpT* I gene in the host genome of two-field-season trees.

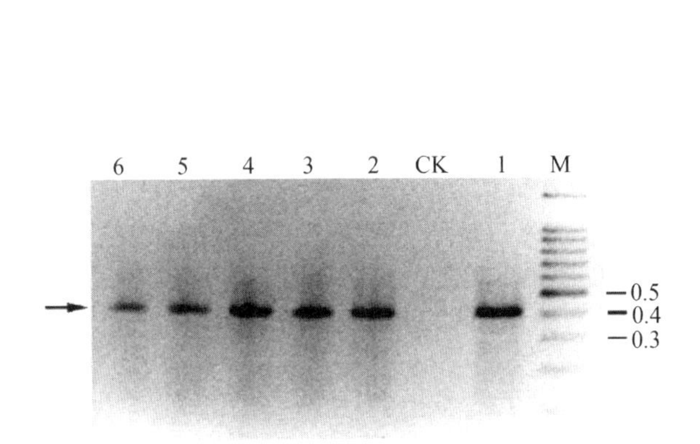

Figure 3 PCR analysis of plant genomic DNA using *CpT* I gene specific primer combination. Mis 2-kb ladder DNA Marker, CK is genomic DNA of untransformed control poplar; 1~6 represents genomic DNA of transgenic clone TG04, TG07, TG08, TG16, TG53 and TG71, respectively, arrow (left) indicates the specific bands corresponding to *CpT* I gene, numbers (right) indicate the DNA size.

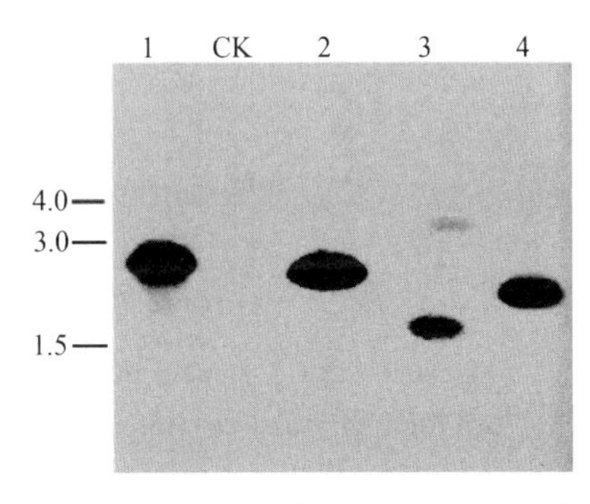

Figure 4 Southern blot analysis of *CpT* I gene with plant genomic DNA. 10μg of *Hin*d III restricted genomic DNA was probed with a 100bp artificially synthesized and DIG High Prime labeled *CpT* I gene specific DNA sequence. CK is genomic DNA of untransformed control poplar; 1 ~ 4 represents genomic DNA of transgenic clone TG04, TG07, TG16 and TG71, respectively; the numbers (left) represent 4-kb Marker.

Determination of *CpT* I content

After insect bioassay and molecular identification at DNA level, the *CpT* I content in both young and old leaves of poplars with comparatively high insect resistance were determined by proteinase inhibitory assays to investigate the role of *CpT* I played in insect resistance and the expression feature of this gene in host genome. It was found that *CpT* I content in leaves from 7 tested transgenic poplar clones ranged from 9. 56μg · g^{-1} FW to 16. 74μg · g^{-1} FW (Figure 5) and *CpT* I content in 2 types of leaves was very close. Most transgenic clones have more than 14. 0μg · g^{-1} FW of *CpT* I in both young leaves and old leaves. More than 16μg · g^{-1} FW of *CpT* I was detected in 2 kinds of leaves from clone TG07 and TG71. These results demonstrated that there was high amount of *CpT* I in both young and old leaves of each transgenic clone highly resistant to defoliators, and seemed that there was no significant difference in the expression level of *CpT* I gene in 2 types of foliage from the same transgenic clone. However, no *CpT* I was detected in control plant, indicating the absence of the *CpT* I gene. Moreover, *CpT* I content detected in 2 types of leaves from clone TG04, TG07, TG08 and TG71 was much more or a little less than 14. 0μg · g^{-1} FW, thus accounting for their high insect resistance.

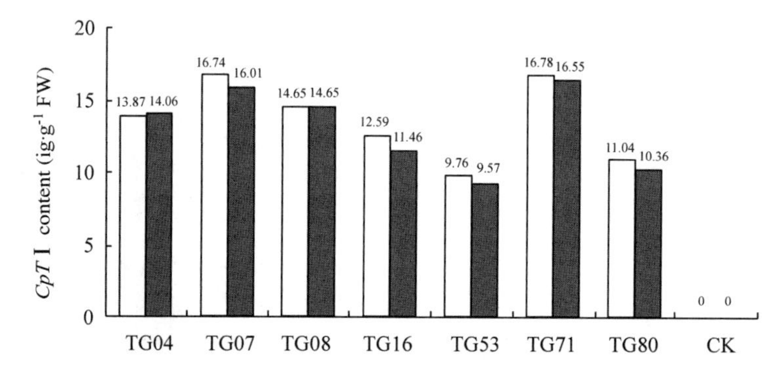

Figure 5 *CpT* I content in young leaves and old leaves (μg · g^{-1}FW) detected by proteinase inhibitory assays. Hatched and black bars represent young leaves and old leaves, respectively, numbers on top of each bar indicate corresponding *CpT* I content.

Discussion

The role of cowpea trypsin inhibitor (*CpT* I) as defensive compounds against insect pests of the orders Lepidoptera, Coleoptera and Orthoptera is well established by artificial diet bioassays (Broadway et al., 1986; Johnson et al., 1993; Bell et al., 2001) and insect feeding trials on genetically engineered plants (Hild et al., 1987; Boulter et al., 1989; Ghoshal et al., 2001). Here, we have proved that following 2 winter dormancies transgenic poplars conferred resistance to 3 Lepidopterans by highly and constitutively expressing *CpT* I protein in the foliage. Although the transgenic foliage did not elicit mortality rates of willow moth after feeding trials as high as that exposed to foliage expressing *Bacillus thuringiensis* δ-endotoxin (Mccown et al., 1991; Wu et al., 1991; Tian et al., 1993; Chen et al., 1995), it was found to be fatal to gypsy moth and comparatively deleterious and toxic to forest tent caterpillar, and resulted in huge decrease in foliage consumption, wet weight gains, faeces excretion, deposited pupae number and pupae weight of willow moth, indicating *CpT* I being efficacious in retarding growth and physical development of insects and impairing their fecundity by affecting inner metabolism rather than directly killing them completely. This insect resistance phenotype was proved particularly evident on 4 transgenic clones: TG04, TG07, TG08 and TG71.

Within the mid gut of insect, *CpT* I binds competitively to the binding site of a target enzyme to form an enzyme-inhibitor (EI) complex and subsequently renders the enzyme incapable of binding to and cleaving to peptide bonds of proteins (Broadway et al., 1992). However, an ingestion of proteinase inhibitors does not eliminate proteolytic digestion in the midgut of insect, but leads to the accumulation of undigested food. It is followed by hyper production of proteolytic enzymes as a result of feedback regulation, which, in turn, induce reduced availability of essential amino acids for protein synthesis, finally resulting in retarded growth and development (Lawrence et al., 2002). However, processing these procedures in the gut of larvae eating foliage of transgenic poplars is completely dependent on the presence of high level *CpT* I protein expressed in the foliage. Being under control of constitutive CaMV 35S promoter and the modification with addition of an Omega element, a SKTI signal peptide and a KDEL coding sequence have the potential to enhance the expression level of *CpT* I gene dramatically. This prediction was confirmed by our proteinase inhibitory assays performed on the protein extracts of leaves from 7 transgenic clones, particularly 4 clones with higher resistance against predators.

Moreover, the detection of high level *CpT* I in leaves and the identification of *CpT* I gene using PCR and Southern blotting analysis further confirmed that the stable integration and expression status of foreign *CpT* I gene in the hosts have not been changed by the two field-season of growing in the field.

In screening for fine transgenic clones, 3 transgenic clones: TG04, TG07 and TG71 were picked out immediately due to their outstanding performances in all indexes used for evaluating insect resistance. TG16, TG80 or TG53, rather than TG08, should be selected as the fourth prominent transgenic clone in terms of fatal toxicity to tested larvae (Table 1), while further analysis found that TG08 was superior to all of them with regard to the subsequent evaluation indexes including faeces excretion (Table 2), mean weight gains (Table 3) and *CpT* I content (Figure 5), which thereby determined its selection as the final fine transgenic clone.

With regard to developmental stage of tested larvae, many insect feeding bioassays performed on transgenic plants expressing *Bt* δ-endotoxin usually choose the 1st-instar larvae as testing insects (Tian et al., 1993; Chen et al., 1995), and consequently revealed extremely high larval mortality rates. Yang et al. (2003) conducted feeding trials using various in star larvae, and found gradual decrease in larval mortality rates as the larvae developed. Our studies were performed on 3rd-instar larvae, and disclosed middle level larval mortality rates. It seems that developmental stage plays an important role in the response of larvae to the feeding, and that lower-in star larvae is much more susceptible to the change of environment and food than higher-in star larvae, particularly 1st-instar larvae due to their weakness. Therefore, the best way to acquire convincing results of insect feeding assays would be to perform such test using larvae at different developmental stage. Besides the lab tests under controlled condition, revealing the performances of transgenic poplars in the field is much more important because of their final usage in afforestation, and has been under way. Their fine performances in lab tests ever led to an expectation that the growth status of transgenic poplars should be better than untransformed poplars as a result of reduction in damage to leaves, twigs, roots or stems caused by insect pests. However, according to our current field investigation, no significant difference in growth status and damage degree between transgenic and untransformed poplars has been discovered (data not shown). One possible reason is that both transgenic and untransformed poplars in our small area of test field separated from other forest stands are not ideal hosts to local insect pests, or cannot provide a fitting environment for their survival. Moreover, this result from field investigation also identified that the growth status of transgenic poplars has not been interfered by the incorporation of foreign *CpT* I gene. Nevertheless, what their success rate in insect resistance and growth status would be once they were actually used in afforestation remains uncertain, and requires further investigation, because the change in environment, the management and the interference of surrounding creatures (particularly human beings etc) have the potential to make what has been observed in lab tests under controlled conditions inapplicable to actual field condition.

Assessment of Rhizospheric Microorganisms of Transgenic *Populus tomentosa* with Cowpea Trypsin Inhibitor (*CpT* Ⅰ) Gene*

Abstract To have a preliminary insight into biosafety of genetically transformed hybrid triploid poplars (*Populus tomentosa* ×*P. bolleana*) ×*P. tomentosa* with the cowpea trypsin inhibitor (*CpT* Ⅰ) gene, two layers of rhizospheric soil (from 0 to 20cm deep and from 20 to 40cm deep, respectively) were collected for microorganism culture, counting assay and PCR analysis to assess the potential impact of transgenic poplars on non-target microorganism population and transgene dispersal. When the same soil layer of suspension stock solution was diluted at both 1∶1000 and 1∶10000 rates, there were no significant differences in bacterium colony numbers between the inoculation plates of both transgenic and non-transgenic poplars. The uniform results were revealed for both soil layer suspension solutions of identical poplars at both dilution rates except for non-transgenic poplars at 1∶10000 dilution rates from the same type of soil. No significant variation in morphology of both Gram-positive and Gram-negative bacteria was observed under the microscope. The potential transgene dispersal from root exudates or fallen leaves to non-target microbes was repudiated by PCR analysis, in which no *CpT* Ⅰ gene specific DNA band was amplified for 15 sites of transgenic rhizospheric soil samples. It can be concluded that transgenic poplar with the *CpT* Ⅰ gene has no severe impact on rhizospheric microorganisms and is tentatively safe to surrounding soil micro-ecosystem.

Keywords transgenic poplar, *CpT* Ⅰ gene, rhizospheric microorganisms, ecological risk, soil profile

1 Introduction

Genetic engineering provides an opportunity to transfer new specific traits of interest (for example, those for insect pest resistance) into valuable genotypes within a short period of time, and greatly reduces the cost and environmental risk by promoting crop yield and decreasing the use of chemical insecticides (Peferoen, 1997; Baute et al., 2002; Bourguet et al., 2002). It is also more effective in controlling insect pests due to its high specificity to target organisms. Hence, genetic engineering developed rapidly during the past decades.

However, since the publication by Losey et al., (1999), in which *Bt* gene inserted maize cultivation proved to have an impact on non-target lepidopterans, especially the larvae of the monarch butterfly (*Danaus plexipus*), increasing attention has been paid to biosafety of genetically en-

* 本文原载《中国林业研究》, 2005, 7(3): 28-34, 与张谦、林善枝、林元震和杨乐合作发表。

gineered organisms (GMO) and the debate about its potential risks and benefits continues to the present (Losey et al., 1999; Hails, 2000). Along with the benefits mentioned above, GMOs with insect-resistant genes, such as *Bt* gene, also involve many potential risks, including the evolution of resistant insect pests against them (Mallet and Porter 1992, Tabashnik 1994), their impacts on non-target organisms (Pilcher et al., 1997; Yu et al., 1997; Hilbeck et al., 1998; Losey et al., 1999; Stanley-Horn et al., 2001; Bourguet et al., 2002) and transgene dispersal from GMO to its related wild types through hybridization with pollen or recombination (Hoffmann et al., 1994; Kling, 1996; Hails, 2000; Matus-Cádiz et al., 2004; Gustafson et al., 2005). Therefore, public concern that the use of transgenic plants may pose a threat to non-target organisms (Hodgson, 1999; Wraight et al., 2000) with the potential of ecological risks is on the rise.

In addition to these aboveground effects, underground impacts of GMOs have also been recognized as a result of recent methodological advances in soil microbic ecology (Bruinsma et al., 2003; de Vries et al., 2004). It has focused on the identification of GMOs-driven effects on the microbial communities and processes in soil that are essential to key terrestrial ecosystem functions. The foreign gene dispersal through root exudates was for the first time identified from transgenic *Brassica napus*, *B. nigra*, *Datura innoxia* and *Vicia narbonensis* by co-culture, in which a *hph* gene was transferred to mycelial material of *Aspergillus niger* (Hoffmann et al., 1994).

The release of root exudates from transgenic plants into rhizospheric soil generally results in a variation in the rhizospheric microbial population. The study on transgenic tobacco with *PI* gene revealed an increase in nematode population and a dramatic decrease in *Collembola* population in rhizospheric soil of transgenic plants (Oger et al., 1997). It was followed by many related studies with similar results (Qian et al., 1995; Glandorf et al., 1997; Griffiths et al., 2000; Cowgill et al., 2002; Turrini et al., 2004; Baumgarte and Tebbe, 2005).

Although the potential ecological risk of *Bt* gene transferred organisms has been extensively investigated, less data regarding the potential impact of *CpT* I gene incorporated organisms on ecosystem have been revealed (Donegan et al., 1999; Cotter et al., 2004). With the emergence and spread of increasing *CpT* I gene GMOs, it is urgent to carry out related studies to figure out the possible roles of *CpT* I gene transferred organisms in the environment, by which suitable measures to reduce or avoid the risks of a *CpT* I gene inserted GMOs can be proposed.

In Beijing Forestry University, cowpea tyrpsin inhibitor (*CpT* I) gene transferred hybrid triploid poplars (*Populus tomentosa* ×*P. bolleana*) ×*P. tomentosa* have been obtained (Hao et al., 1999). The confirmation of foreign gene incorporation in the host genome and its expression (Lin et al., 2002; Zhang et al., 2004) and the screening for transgenic poplar clones with high target insect resistance have been accomplished (Zhang et al., 2002). However, its possible impact on surrounding organisms remains unclear. Since the triploid hosts are highly sterile (Zhang et al., 2000), the possibility of transgene dispersal from genetically engineered poplars to wild type relatives by hybridization through pollen or recombination is extremely low. However, the root exudates and fallen leaves containing expression products of specific *CpT* I gene and marker *NPT*

Ⅱ gene have potential impacts on non-target organisms in soil. It finally focused our attention on identifying transformants-driven potential impacts on rhizospheric microorganisms through microorganism culture, counting assay and PCR analysis to assess the potential impact of transgenic poplars on non-target microorganisms and transgene dispersal.

2 Materials and methods

2.1 Soil sample collection

Prior to soil sample collection, soil profiles at the rhizosphere of both transgenic and non-transgenic poplars were constructed by digging vertical holes with a scoop. These were divided into two soil horizons, namely an upper layer and a lower layer (0 to 20cm and 20 to 40cm deep beneath the soil surface). The soil samples were collected from the upper and lower layers randomly (three replications per trial).

2.2 Counting assay of rhizospheric microorganism

Dilution plate count was adopted as the method for microorganism counting assay and statistical analysis, and was performed as described by Luo (1990) and Chen (1990). The suspension stock solution of microorganisms was prepared by suspending 1g of rhizospheric soil collected from each site with 10mL of distilled H_2O and subsequently incubated for 30min under constant agitation (about $70r \cdot min^{-1}$) at room temperature. It was followed by the preparation of a dilution series of microorganism suspension solution of 10^{-1}, 10^{-2}, 10^{-3} and 10^{-4} by adding 0.9, 9.9, 99.9 and 999.9mL of distilled H_2O into 0.1mL of suspension stock solution, respectively.

After the preparation of a suspension solution, 0.1mL of diluted suspension solution from each site was transferred to 15mL of potato dextrose agar (PDA) media in a disk 9cm in diameter and inoculated in a chamber for 24h at 28℃ (three replications per treatment). Colony forming units (CFU) in each disk were recorded for statistical analysis.

2.3 Microscopic observation of rhizospheric microorganisms

Following counting assay, Gram staining was performed on rhizospheric microorganisms isolated randomly from individual plaque for recording Gram staining features and the morphology of bacteria was observed under a microscope as described by Liu et al. (2003a).

2.4 Plant genomic DNA isolation

Half a gram of leaves was collected on transgenic poplar lines TG07 and taken to the lab in dry bags. DNA was isolated as described by Rogers and Bendich (1985) and purified with Wizard PCR Preps DNA Purification System Kit (Promega).

2.5 PCR analysis

2.5.1 Primer synthesis

Primers were designed according to the sequence of CaMV 35S promoter and Nos terminator flanking *CpT* I gene (5′-GGATTGATGTGATATCTCCACTGAC-3′, 5′-CTTTATTGCCAAATGTTTGAACGAT-3′), and synthesized on an Applied Biosystems 392 DNA synthesis instrument. Following cleavage from solid support and deprotection, no further purification was necessary and oligonucleotides were stored in 80% ammonia at 20℃.

2.5.2 PCR condition

Reactions were performed in a PCR buffer (10mmol · L^{-1}Tris pH 8.0, 0.01% gelatin, 0.1% Triton X-100, 80mmol · L^{-1} KCl, 3.5mmol · L^{-1} $MgCl_2$) containing 25pmol · L^{-1} for primer, 250μmol · L^{-1} for dNTP (Pharmacia ultra pure), 20 ~ 100 ng of target DNA, 5 units of Taq DNA polymerase. Reaction volumes were made up to 50μL. Amplification of *CpT* I genes was achieved by an initial denaturation at 94℃ for 5min followed by an addition of the polymerase and 30 cycles of 94℃, 3s, 60℃, 6s, 72℃, 6s, with a final extension for 10min at 72℃ using a Hybaid Omnigene Thermal Cycler with simulated tube control (calibration factor 150).

2.5.3 PCR analysis

Aliquots (3 ~ 10μL) of the PCR products were fractionated through 1% agarose for 40min at 7 V · cm^{-1} in Tris-Borate (TBE) buffer and stained with ethidium bromide as described by Hamill et al. (1991). DNA gels were photographed using an Olympus digital camera. DNA fragment sizes were determined by comparison to λ Hind III digest and Pharmacia 10bp ladder size markers.

3 Results

3.1 Statistical analysis of rhizospheric microorganisms

After 30min of incubation with gentle and constant agitation at room temperature, the dilution series of soil suspension solution were transferred to potato dextrose agar (PDA) media for inoculation at 28℃. Twenty-four h later, many bacteria colonies were observed across entire plates and significant differences in colony number on the plates inoculated with soil solutions at various dilution rates. They were numerous and impossible to count manually when diluted at 1:10 or 1:100 rate (data not shown). In contrast, there were no bacteria colonies or only a few on the plates with soil solution at 1:100000 dilution rate (data not shown). In order to acquire the statistical results of bacteria colonies closer to the real value, only the colonies observed on plates with soil solution at 1:1000 and 1:10000 dilution rates were recorded and statistically analyzed. For the soil solution at 1:1000 dilution rate, the bacteria colonies observed on plates inoculated with the same layer soil solution of both transgenic and non-transgenic poplars were found close to each other (*P* values for upper layer and lower layer were 0.5831 and 0.9417, respectively). Compared with the bacteria colonies for lower layer soil solution, more colonies were found on plates for upper layer rhizospheric soil solution prepared from both transgenic and non-transgenic poplars (Table 1). However, statistical analysis indicated that there were no significant differences between the two layers of soil solution for both types of poplars (P values for transgenic and non-transgenic poplars were 0. 3505 and 0. 2381, respectively).

Table 1 Analysis of rhizosphere microorganisms by dilution plate count g^{-1}

Poplar type	Soil source	Colonies at 2 dilution rates	
		1:1000	1:10000
Transgenic poplar	0~20cm layer	160. 33±7. 02	19. 33±13. 65
	20~40cm layer	135. 00±40. 95	25. 33±7. 37
Non-transgenic poplar	0~20cm layer	176. 33±45. 94	27. 33±2. 52
	20~40cm layer	137. 00±17. 52	19. 00±4. 36

For soil solution at 1∶10000 dilution rates, more complicated results were identified. With regard to the upper layer soil solution, more colonies were found on plates for rhizosphere bacteria of non-transgenic poplars. Opposite results were observed for lower layer soil solutions. Nevertheless, no marked differences were revealed between the two types of poplars by statistical analysis (P values for upper layer and lower layer were 0. 3746 and 0. 2694, respectively). Similar to soil solution at 1∶1000 dilution rates, no significant differences in bacterium colony number between both soil layer dilution solutions of transgenic poplars were demonstrated (P value was 0. 5396). However, the unique significant differences were found for two layers of soil solution of non-transgenic poplars (P value was 0. 0456). Based on these findings, it can be concluded that the bacterial or microbic populations were not disturbed by the growth of transgenic poplars inoculated with the *CpT* I gene.

3. 2 Microscopic observation of microorganisms

The morphology of rhizospheric microorganisms was also investigated following the population counting assay. A bacterium colony in a transgenic plate plus a corresponding colony in non-transgenic plates with identical external configuration was regarded as a bacterium combination. Given this criterion, 8 combinations of bacteria colonies were isolated from both transgenic and non-transgenic plates for Gram staining. It was determined that bacteria within a combination (from transgenic and non-transgenic plates, respectively) have identical Gram staining features and only two out of eight colonies were Gram-positive bacteria. Observation under the microscope revealed that there were enormous blue rod-shaped bacteria in Gram-positive reaction solution (Fig. 1), and red rod-shaped bacteria in the Gram-negative reaction solution (Fig. 2). Although variation in bacterium size were observed due to the differences at developmental stages, no significant difference in morphology between bacteria of transgenic and non-transgenic plates was observed. These results demonstrated that the morphology of soil microbes was not affected by the root exudates or fallen leaves of transgenic poplars.

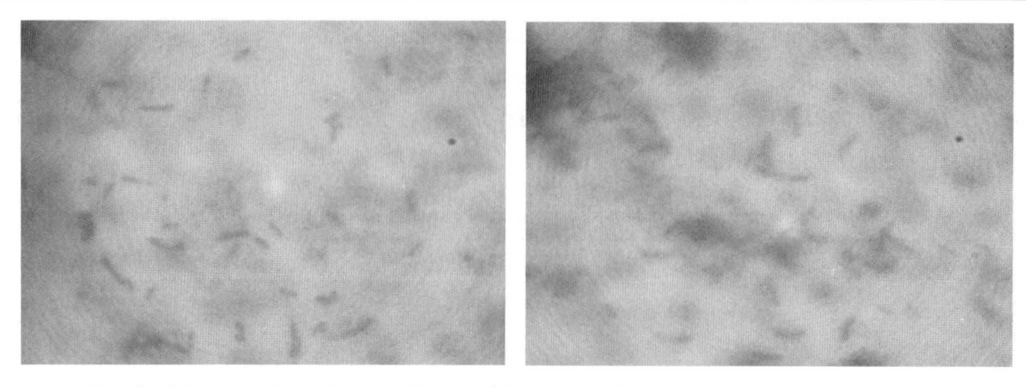

Fig. 1 Microscopic analysis of Gram-positive bacteria morphology by Gram staining

Upper: bacteria grown from rhizospheric soil of transgenicpoplar;

Lower: bacteria grown from rhizospheric soil of non-transgenic poplar.

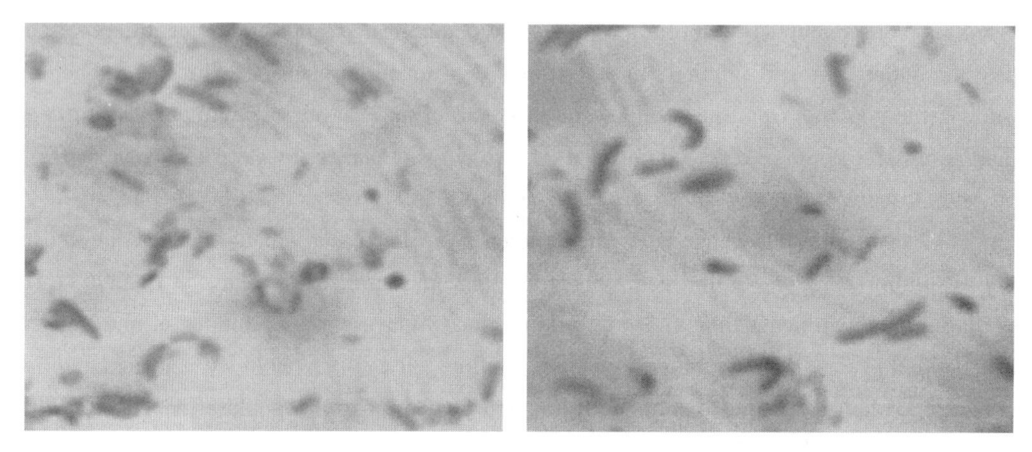

Fig. 2 Microscopic analysis of Gram-negative bacteria morphology by Gram staining

Upper: bacteria grown from rhizospheric soil of transgenic poplar;

Lower: bacteria grown from rhizospheric soil of non-transgenic poplar.

3.3 *CpT* I gene analysis of rhizospheric microorganism by PCR

Genes escaping from roots of transgenic poplars into rhizospheric soil or even into rhizospheric microorganisms through root exudates was detected by PCR analysis using specific primers flanking the *CpT* I gene. The upper primer was at the end of a CaMV 35S promoter and the down primer at the beginning of a Nos termintor. The gel electrophoresis of amplified products showed that a clear and sharp DNA band specifically corresponding to the *CpT* I gene and its flanking sequence (about 450 bp) were observed on gel lane for plant genomic DNA of transgenic clone TG07. However, no corresponding DNA band was found on the lanes for suspension stock solution of rhizospheric soil collected from one non-transgenic poplar and 5 transgenic poplars (Fig. 3). The same results were repeated in three independent trials. Based on the above results, it was easy to infer that the *CpT* I gene had not been transferred to rhizospheric microorganisms of these three non-

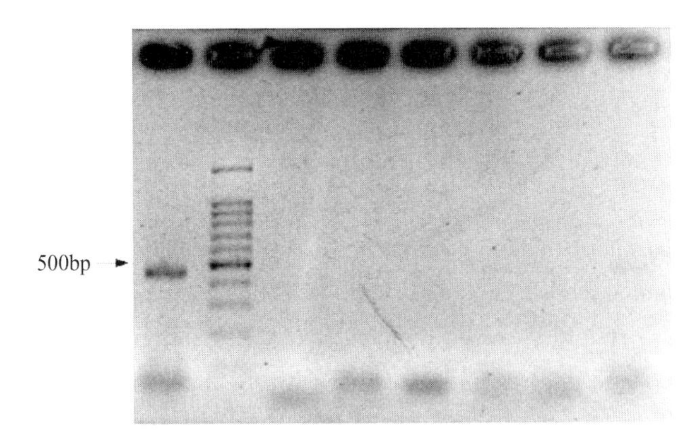

Fig. 3 Detection of rhizosphere soil solution sample of transgenic poplars by PCR analysis

transgenic poplars and 15 transgenic poplars.

4 Discussion

The exposure of *Bt* maize having indirect impacts on non-target lepidopterans (Losey et al., 1999) enkindled the public concern about the risk of GMOs, and launched vast related studies to reveal their potential effects and role in the environment as well as functional mechanisms (Kohli et al., 2003). It resulted in the discovery of several unpredicted phenomena, including the evolution of resistant insects, impacts on non-target organisms and transgene escape from GMOs to related wild type relatives (Pilcher et al., 1997; Yu et al., 1997; Hilbeck et al., 1998; Losey et al., 1999; Stanley-Horn et al., 2001; Bourguet et al., 2002; Matus-Cádiz et al., 2004; Gustafson et al., 2005).

However, the majority of above discoveries were derived from the *Bt* gene-transformed plants. Whether it would apply to proteinase inhibitor (PI) gene-transformed plants, especially *CpT* I gene, activated the decision to explore the potential risks of a two-field season of *CpT* I gene transferred poplars. Since the triploid hosts are highly sterile (Zhang et al., 2000), the possibility of transgene dispersal from genetically engineered poplars to wild type relatives by hybridization through pollen or recombination is extremely low, which makes it unnecessary to study transgene dispersal through pollen hybridization or recombination.

As for expressed products of foreign genes, especially the Bt gene, recent studies have revealed its retention in the soil through root exudates and fallen leaves, the transfer via food chains and the potential to cause negative effects on soil, organisms and nutrient cycles in water (James, 1997; Wang et al., 2002; de Vries et al., 2004). The study on the behavior of Bt toxin in soil derived from root exudates identified the delay of biological degradation through its binding to active agents on soil surface (Saxena et al., 1999; Saxena and Stotzky, 2000). Although the explicit conclusions about the impact of Bt toxin infiltrating into the soil in natural environments still remained unclear, it apparently confirmed the spatial transfer of transgenic products through root exudates, fallen leaves and natural transformation (Wang et al., 2002; de Vries et al., 2004). Based on the above two reasons, our study finally shifted to focus on rhizospheric microorganisms of transgenic poplars.

Assessment of the impact of GMOs on rhizospheric microbic population in soil has become a hot topic (Qian and Ma, 1995). It has been identified that plant root exudates provides nutrients for the survival of rhizospheric microorganisms by excreting various types of compounds into the soil. Carbohydrates, amino acids, fatty acids and nucleotides acids are four types of main ingredients, and organic acids, steroids, growth hormones, flavones and enzymes are also involved (Xie et al., 2003). Rhizospheric microorganisms can sensitively apperceive variations in plant root exudates mentioned above by excreting their own exudates and consequently interacting with those of plant roots and exactly indicate the impact of plant root exudates on the environment (Liu et al., 2003b). This indication method is dependent in host instance of on the amount variation in rhizospheric microorganisms (Zhu et al., 2003). It provides a scientific foundation for detecting the impact of transgenic plants on rhizospheric microorganisms using population variation variations as criteria.

In addition to proteins of interest expressed by the *CpT* Ⅰ gene, there is another aggressive protein, expressed by the marker *NPT* Ⅱ gene, which confers the host resistance against kanamycin. The potential impact by the pervasion of *NTP* Ⅱ protein across rhizospheric soil of transgenic poplars on rhizosphere microorganisms is another highlight. Recent studies have confirmed their persistence in soil for up to 137 d after burying the transgenic leaves in soil (Widmer et al., 1997). Therefore, it is essential to determine the population and morphology of rhizospheric microorganisms in the first place, and consequently assess the potential impact of transgenic poplars on soil ecosystems. In our research, no significant differences are demonstrated in population and morphology of rhizospheric microorganisms between transgenic and non-transgenic poplars by statistical analysis. Moreover, at a dilution rate of 1∶1000, very close numbers of bacteria colonies in both types of plates were observed for both layers of soil samples. Similar results are achieved in Bt gene transferred potatos and corn (Donegan et al., 1995; 1996; Zangerl et al., 2001). A preliminarily conclusion is that the transgenic poplars have no marked impact on the rhizospheric microorganisms and are tentatively safe to soil and nearby micro-ecosystems. This result is further confirmed by our PCR analysis with a *CpT* Ⅰ gene specific primer combination, in which no clear and sharp DNA band corresponding to a *CpT* Ⅰ gene is amplified from 15 soil samples.

At the same transgenic or non-transgenic poplar site, no significant difference in populations of rhizospheric microorganisms is demonstrated for both layers of soil samples. The same results take place for the soil samples at both dilution rates except for the non-transgenic site at the dilution rate of 1∶10000. The variation in bacterium colony numbers resulting from the soil sample at the dilution rate of 1∶1000 is less than that for dilution rate at 1∶10000, which seems that a dilution rate of 1∶1000 is optimal among five treatments.

Construction and Characterization of cDNA Library from Water-Stressed Plantlets Regenerated in vitro of *Populus hopeiensis* *

Abstract In order to isolate and clone water-stress-responsive genes, total RNA was extracted from water-stressed plantlets regenerated in vitro of *Populus hopeiensis* using a QIAGEN RNeasy Plant Mini Kit. CDNA, synthesized by LD-PCR with the SMART cDNA Library Construction Kit, was in vitro packaged into a phage λTriplEx2 vector. The resulting primary library and amplified library have a titer of 1.68×10^6 and 1.69×10^9 pfu · mL^{-1} respectively. The combination ratio reached 98.8% and the average size of inserts was about 80bp. In addition, the percentage of inserted fragments (>400bp) was approximately 90%. The results indicate that a cDNA library has been successfully constructed.

Keywords cDNA library, *Populus hopeiensis*, water-stressed plantlet, characterization

1 Introduction

Drought is a negative environmental condition which severely limits the growth and development of plants. By alteration of gene expression, plants respond and adapt to drought or water stress to survive at molecular, biochemical and cellular levels as well as at the physiological level. As the most important life-style of land ecosystem, forest trees have long been studied for their responses to water stress. Many proteins involved in the adaptation of trees to drought, such as Dehydrin and BspA, have been characterized by SDS-PAGE and 2D-PAGE from water-stressed *Populus tremula* and *Pinus pinaster* (Pelah et al., 1995, 1997; Costa et al., 1998; Wang et al., 2002). Through the cDNA-AFLP technique, differentially expressed mRNA transcripts have also been identified from dehydrated *Prunus amygdalus* and *Pinus pinaster*, and with the differential fragments, the corresponding full-length genes have been obtained by screening the cDNA library (Campalans et al., 2001; Dubos and Plomion, 2003; Dubos et al., 2003). However, no comprehensive mechanism of plants including forest tree adaptation to drought has been established up to now.

Populus hopeiensis is a drought-tolerant and chilling-tolerant tree species, mainly distributed in Northwest and North China (Xu, 1988). In order to isolate water-stress-responsive genes and study the mechanism of drought tolerance at a molecular level, the authors constructed a cDNA li-

* 本文原载 *Forestry Studies in China*, 2005, 7(3): 39-42, 与王泽亮、林善枝、林元震撼张谦合作发表。

brary from water-stressed plantlets regenerated in vitro of *P. hopeiensis*.

2 Materials and methods

2.1 Plant materials and water-stress treatment

Explants from hydroponically grown shoots of *P. hopeiensis* were sterilized and cultured on a half-strength MS medium supplemented with 0.5mg·L^{-1} BA and 0.1mg·L^{-1} NAA at 25±1℃ under a 16-h photoperiod using cool-white fluorescent light. For rooting, the regenerated shoots were subcultured on half-strength MS medium supplemented with 0.4mg·L^{-1} IBA. After 6~7 weeks of subculture, the rooted plantlets were wilted by continuous monitoring of weight loss on a balance at room temperature to 85% of their original fresh weight according to Pelah et al. (1997) with some modifications, then kept in closed plastic bags for an additional 3h.

2.2 RNA extraction and determination of yield and quality

The water-stressed plantlets were ground in liquid nitrogen, then a maximum of 100mg of powder was extracted for total RNA using the QIAGEN RNeasy Plant Mini Kit. Finally, the total RNA was diluted into 35μL of RNase-free water.

To analyze yield and quality, 15μL of the total RNA sample was taken up in a final volume of 3mL, then determined for the values of OD230, OD260 and OD280 by spectrophoto graphy and 5μL of the RNA sample was tested by agarose gel electrophoresis.

2.3 Construction and amplification of cDNA library

A cDNA library was constructed using the SMART cDNA Library Construction Kit (Clontech). Appropriate volumes of total RNA were reversely transcribed into a single-strand cDNA and then the double-strand cDNA was synthesized by LD-PCR. The resulting cDNA was digested with Proteinase K and *Sfi* Ⅰ, and afterwards tested for size fractionation by a CHROMA SPIN-400 Column. Each fraction was carefully collected and the profile was checked by agarosegel electrophoresis. The first three fractions containing usable cDNA were pooled, and ligated to the *Sfi* Ⅰ-digested λTriplEx2 vector according to three different ratios of cDNA and the vector. The cDNA ligation products were *in vitro* packaged into λphage using the Packagene Lambda Packaging System (Promega). The obtained primary library can be stored at 4℃ for 7 d, or in 7% DMSO (V/V) at-70℃ for up to one year.

To amplify the primary library, enough lysate and 500μL of overnight *E. coli* 'XL1-Blue' culture were combined and incubated at 37℃ for 15min. Four and a halfmL of melted top agar was added to the infected bacteria, and then the mixture was spread onto the surface of a 15mm LB/$MgSO_4$ agar plate. The plate was inverted and incubated at 37℃ for 6~12h until the plaques began to touch each other, then 12mL of 1×λ dilution buffer was added to the plate and the plate was stored at 4℃ overnight. After incubating the plate at room temperature for 1h on a shaker

($75r \cdot min^{-1}$), the lysate was collected and combined with 10mL of chloroform, then vortexed for 2min. Cell and agar debris were removed by a centrifugation at $7000r \cdot min^{-1}$ for 10min, and the supernatant (amplified library) can be stored at 4℃ for up to six months or in 7% DMSO (V/V) at−70℃ for up to one year.

2.4 Characterization of cDNA library

To obtain the titer of the primary library, 1μL of 20×diluted phage was combined with 200μL of overnight *E. coli* 'XL1-Blue' culture, and incubated at 37℃ for 15min. TwomL of melted top agar was added to the infected bacteria, then the mixture was spread onto the surface of a 9mm $LB/MgSO_4$ agar plate. The plate was inverted and incubated at 37℃ for 12 ~ 18 h until the plaques were visible. The plaques were counted and the titer of the primary library was calculated. For titering the amplified library, the dilution factor was 10^5 and 5μL of diluted phage was mixed with the bacteria culture. To determine the recombination ratio of the primary library, 40μL of $20mg \cdot mL^{-1}$ IPTG and 40μL of $20mg \cdot mL^{-1}$ X−gal were added into the melted top agar before plating the mixtures.

To obtain the length of the inserted fragments, 18 white plaques were randomly picked from five plates in order to determine the recombination ratio of the primary library. This mixture was dissolved in 20μL of 1×λ dilution buffer and then denatured at 94℃ for 10min. The inserts were amplified using a λTriplEx sequencing primer supplied with the Kit. PCR was performed in the final volume of 20μL, containing 9.9μL of ddH_2O, 2μL of 10×buffer, 1.6μL of $2.5mmol \cdot L^{-1}$ dNTP, 1μL of each $4\mu mol \cdot L^{-1}$ primer, 4μL of template and 0.5μL of $2.5\ U \cdot \mu L^{-1}$ *Taq* DNA polymerase. Amplifications were carried out at the following cycling parameters: preliminary denaturation (5min, 94℃), then 30 cycles of denaturation (40s, 94℃), annealing (40s, 57℃) and extension (2min, 72℃) and final extension (10min, 72℃). The resulting products were analyzed by 1.1% agarose gel electrophoresis.

3 Results and discussion

3.1 Extraction of total RNA

High-quality total RNA is crucial to ensure a well-representative cDNA library. Hence during the extraction of RNA, besides all the general precautions of handing RNA, all procedures must be carried out quickly, to prevent the plant powder from thawing before adding the homogenized lysate. In this study, the concentration of RNA reached $0.84\mu g \cdot \mu L^{-1}$ and about 30μg of total RNA was isolated from less than 100mg of plant material. By spectrophotography, the values of OD_{260}/OD_{280} and OD_{260}/OD_{230} were determined as 1.944 and 2.019 respectively, which indicated that the extracted total RNA was pure. The integrity of total RNA was checked by agarose gel electrophoresis, shown in Fig. 1. The two bands of 28s rRNA and 18s rRNA are present in the picture, and their brightness ratio is ca. 1.5 : 1, so the integrity of RNA also meets the requirement

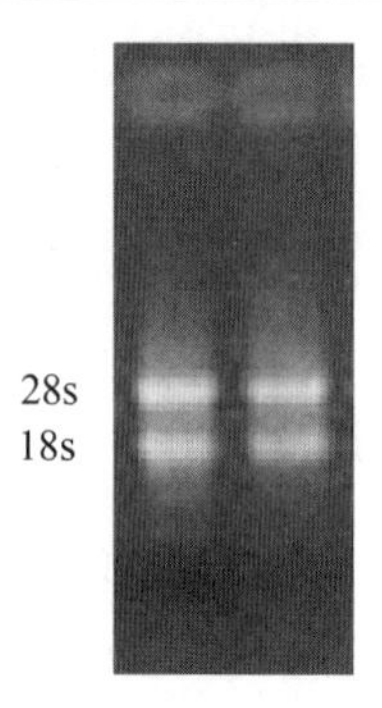

Fig. 1 1. 0% agarose gel electrophoresis of total RNA

of a cDNA library construction.

3. 2 Achievement of cDNA

FiveμL of a cDNA sample by reverse transcription and LD-PCR was run on 1. 1% agarose gel to check cDNA quality. The result is presented in Fig. 2. The cDNA appeared as an approximately 0. 1 ~ 5 kb smear on the gel, which showed that the representativeness of cDNA was good.

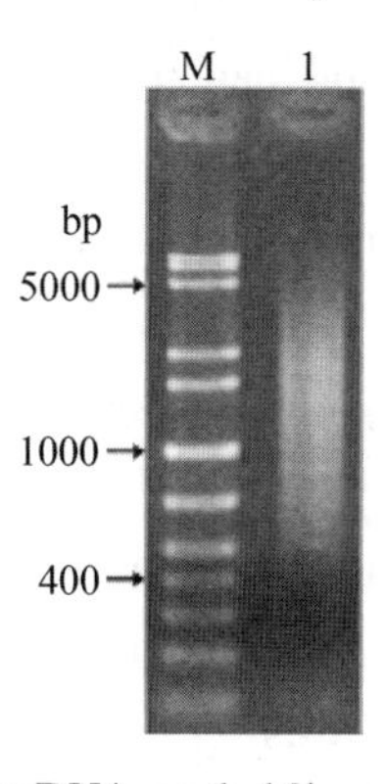

Fig. 2 dscDNA on 1. 1% agarose gel

Lane M: 1kb plus DNA ladder marker; Lane 1: dscDNA

cDNA containing small fragments would result in a library that has a preponderance of very small inserts or apparently nonrecombinant clones. To remove small cDNA fragments and residual primers, the digested cDNA with proteinase K and *Sfi* I was tested for size fractionation by passing through CHROMA SPIN-400 column, 16 fractions were collected, then the profile was tested on 1. 1% agarose gel to ensure which fraction contained the appropriate cDNA. As shown in Fig. 3, the flow-throughs of tubes numbered 6 to 10 contain cDNA. To guarantee the quality of the library, only the first three fractions were pooled and used to precipitate cDNA.

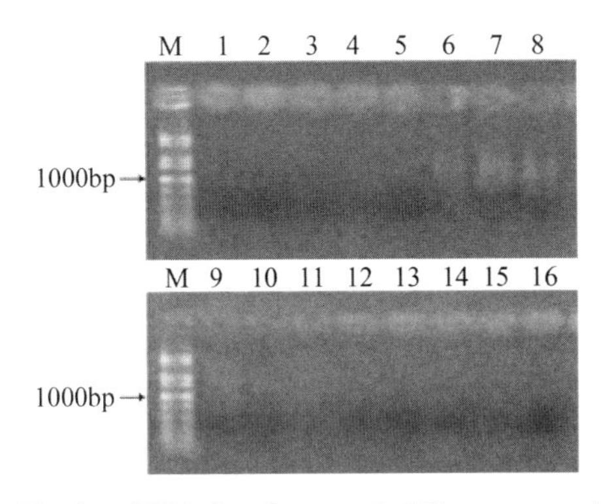

Fig. 3 cDNA fractions on 1.1% agarose gel

Lane M: 1kb plus DNA ladder marker; Lanes 1-16: cDNA fractions

3.3 Characterization of cDNA library

To obtain the optimal ratio of cDNA to vector, 0.5, 1.0 and 1.5μL of cDNA sample were respectively combined with 1.0μL of 0.5μg · μL^{-1} vector in the ligation reactions. The titers of the three resulting libraries was 1.68×10^6, 0.91×10^6 and 0.53×10^6 pfu · Ml^{-1}, so the optimal ratio of cDNA to vector was 0.5:1 in the present study, which also demonstrated that too much cDNA in the ligation reaction would reduce the titer of the resulting cDNA library. And the first primary library was used to produce the amplified library, which has a titer of 1.69×10^9 pfu · mL^{-1}.

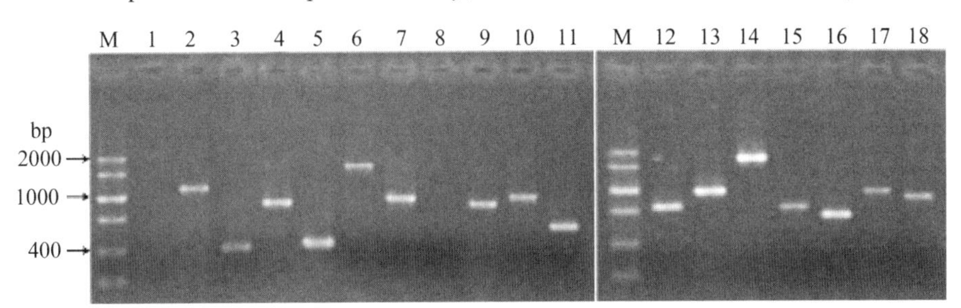

Fig. 4 Detection of insert size in the primary cDNA library of *Populus hopeiensis*

Lane M: DNA marker; Lanes 1-18:

PCR products of plaques randomly picked from blue/white screening plates

The insert profile of cDNA library was obtained by agarose gel electrophoresis shown in Fig. 4. After the removal of about 150 bp vector sequences, the average size of inserts was *ca.* 800bp, and the percentage of inserted fragments of more than 400bp was approximately 90%. There were three inserts whose average size was more than 1500bp.

All results indicate that the quality of the cDNA library is high, which lays a foundation for isolating water-stress-responsive genes and studying the mechanism of drought tolerance of *P. hopeiensis*.

具有光肩星天牛内切聚葡糖酶结合活性短肽的筛选[①]

摘　要　内切葡聚糖酶(endoglucanases)是光肩星天牛幼虫肠道的主要纤维素消化酶。本研究以光肩星天牛内切葡聚糖酶的同工酶 AgEG2 为靶分子，从随机多肽噬菌体展示库中筛选与 AgEG2 有亲和活性的短肽，通过 3 轮筛选，短肽序列 TPHRSPL 出现频率为 33.7%，而且展示该短肽的噬菌体均对 AgEG2 有很高的结合能力。进一步合成短肽 TPHRSPL，并对肠道纤维素酶提取液进行了 Western 分析，结果表明该短肽能特异结合内切葡聚糖酶的同工酶 AgEG1 和 AgEG2，而与粗酶液中其它蛋白组分均无结合特性。表明筛选获得的短肽 TPHRSPL 对光肩星天牛内切聚葡糖酶具有特异结合亲和性。该短肽为研究光肩星天牛纤维素酶的特性及开发天牛的生物防治制剂奠定了基础。

关键词　光肩星天牛，内切聚葡糖酶，随机多肽噬菌体展示库，短肽，亲和性

光肩星天牛(*Anoplophora glabripennis*)广泛分布于我国 24 个省(自治区，直辖市)危害杨(*Populus*)、柳(*Salix*)、榆(*Ulmus*)、槭(*Acer*)、槐(*Sophora*)、桑(*Morus*)等多种林木，是我国杨树最重要的蛀干害虫(张星耀和骆有庆，2003)。天牛幼虫钻蛀树干隐蔽危害，世代长且不整齐，天敌种类少，因而控制难度极大，常规的防治方法很难奏效。我国三北防护林生态工程，许多以杨树为主的林分由于光肩星天牛的猖獗危害，造成大量树木被迫砍伐，甚至整个林分被毁灭，给我国的杨树人工林带来严重威胁(骆有庆等，1999；2002)。

纤维素酶是天牛幼虫肠道的主要消化酶类，是研究天牛的生物学特性及开发抗天牛生物制剂的靶标之一。作者对光肩星天牛幼虫肠道内的纤维素酶体系进行研究表明，光肩星天牛幼虫肠道内具有完整的纤维素酶体系，其中内切-β-1, 4-葡聚糖酶活性最高，且具有广泛适宜的 pH 值和温度范围及较高的热稳定性等特性。通过酶活性化学显色反应，分离出内切-β-1, 4-葡聚糖酶的两种同工酶，分别命名为 AgEG1(26kD)和 AgEG2(39kD)(Chen et al., 2002)。借鉴蛋白酶抑制剂在害虫控制方面的成功经验，开发天牛纤维素酶的专性抑制剂或提高天牛杀虫剂的效果是一个重要策略。但目前对天牛纤维素酶的研究主要集中在纤维素酶的来源、组成及其特性等方面(李庆，1991；殷幼平等，1996；2000；蒋书楠等，1996)。仅吴明等(2003)报道了 Cu^{2+} 对松墨天牛(*Monochamus alternatus*)纤维素酶的抑制作用。

近年来，噬菌体展示技术(phage display)的迅速发展，已成为分子间相互作用研究的有力工具，也为筛选生物活性肽、蛋白质、受体以及开发新型药物建立了新的方法(Devlin et al., 1990；McGregor et al., 1996；Smith et al., 1997)。运用噬菌体展示技术筛选出天牛

---

① 本文原载《林业科学研究》，2006，19(3)：267-271，与陈敏、卢孟柱和王敏杰合作发表。

纤维素酶的结合短肽，可为天牛纤维素酶的生物学特性研究及其抑制剂、杀虫剂的开发提供基础。

本研究首次利用噬菌体展示技术进行了天牛纤维素酶结合短肽筛选的尝试。以 AgEG2 作为靶分子对随机七肽噬菌体库进行筛选，获得特异结合光肩星天牛内切葡聚糖酶的短肽序列，并对其结合特性进行了研究。本文还对该短肽用于天牛纤维素酶特性的研究及其抑制剂的开发前景进行了讨论。

1 材料和方法

1.1 噬菌体展示库

随机七肽噬菌体展示库 Ph. D. -7^{Tm}试剂盒购自 New England Biolabs（NEB）。该噬菌体展示库以 M13 噬菌体为载体，将编码随机 7 肽的 DNA 序列插入 M13 噬菌体的 *P*Ⅲ基因，外源随机 7 肽展示于外壳蛋白 *P*Ⅲ的 N 端。文库容量为 2×10^{12} Pfu（plaque forming unit）。

1.2 光肩星天牛幼虫

光肩星天牛幼虫采自天津地区的杨树被害木，室内采用柳树(*Salix* sp.) 和糖槭(*Acer saccharum* Marsh.)等枝条人工饲养，备用。挑选生长健康、3 龄以上的幼虫作为实验材料。

1.3 AgEG2 的分离和纯化

按照作者介绍的方法进行 AgEG2 的分离纯化(Chen et al.，2002)。从非变性聚丙烯酰胺凝胶的相应 AgEG2 位置切取胶块，加适量 0. 1mol·L^{-1} $NaHCO_3$(pH 8. 6) 溶液洗脱 3 次，每次 30min，最终酶液浓度稀释到 100μg·mL^{-1} 备用。回收酶液进行 SDS-PAGE(12%)分析检验其纯度，并用蛋白分子量标准估计分子量。同时制备筛选过程中的对照洗脱液：进行 AgEG2 分离纯化时，留出两个泳道不加样，电泳后切取与酶带相同大小的空白凝胶，用 0. 1mol·L^{-1} $NaHCO_3$ 洗脱，方法同酶液的制备。

1.4 噬菌体展示库的筛选

以 AgEG2 作为靶分子对随机七肽噬菌体展示库进行淘选，筛选程序参照试剂盒说明书进行，略有改动。将 150μL AgEG2 的 $NaHCO_3$ 溶液加入酶标板微孔中，4℃包被过夜，同时包被空白胶的 $NaHCO_3$ 洗脱液作为对照。倒掉包被液，微孔中加满封闭液(0. 1mol · L^{-1} NaH-CO_3，5mg·mL^{-1} BSA，0. 02% NaN_3)，4℃ 孵育至少 1h。采用酶液洗脱方法制备空白聚丙烯酰胺凝胶的 TBS（50mmol·L^{-1} Tris-HCl（pH 7. 5)，150mmol·L^{-1} NaCl）洗脱液。取该洗脱液 100μL 稀释 10μL 噬菌体库(约 2×10^{11} Pfu)，并在 4℃ 孵育 2h，以吸附原肽库中与聚丙烯酰胺凝胶洗脱液特异结合的噬菌体。倒掉酶标板中的封闭液，用 TBST(TBS+ 0. 1% Tween20)洗 6 次后，加入上述与聚丙烯酰胺凝胶洗脱液吸附过的噬菌体七肽库，4℃ 缓慢振荡 4h，用 TBST 洗 10 次，特异结合的噬菌体用 100μL 洗脱缓冲液(0. 2mol·L^{-1} Glycine-HCl，pH2. 2，1mg·mL^{-1} BSA)洗脱 3 次，每次 10min，洗脱液立即用 1mol·L^{-1} Tris-HCl 缓冲液(pH 9.1) 中和。取少许洗脱液(约 1μL)感染大肠杆菌(*Escherichia coli* (Migula) Castellani and Chelmers) ER2738 测定噬菌体效价，以噬菌斑形成单位(Pfu)表示，其余噬菌体通过感染宿主菌扩增后进入下一轮筛选。重复以上筛选程序 2 轮，从第二轮开始将 TBST 中的吐温浓度增加为 0.5%。测定每轮筛选获得的噬菌斑数(产出量)以计算噬菌体的回收比。噬菌体的回收比=噬菌体的产出量/噬菌体的投入量。噬菌体的扩增、纯化和定量按照试剂盒的使用手册进行。

1.5　多肽序列分析

随机挑取第 3 轮筛选获得的噬菌体克隆，按照试剂盒介绍的方法制备 DNA 测序模板。用 310 型 DNA 自动测序仪(美国 ABI)进行了 DNA 序列分析。推导展示短肽的氨基酸序列。

1.6　单克隆噬菌体与 AgEG2 结合活性测定

以上噬菌体单克隆分别进行扩增后，按照上述筛选方法测定与 AgEG2 的结合能力。10^{11}Pfu 新扩增的单克隆噬菌体加入包被 AgEG2 的酶标板中，按筛选程序进行结合和洗涤，洗涤条件同第三轮筛选。测定洗脱液中噬菌体效价(噬菌体的产出量)。每个噬菌体克隆设三个重复，以相同投入量的随机七肽噬菌体展示库作为对照。计算噬菌体的回收比。

1.7　合成短肽与 AgEG2 的结合分析

由赛百胜公司合成上述短肽 TPHRSPL(命名为 P2)，并在短肽 N 端标记生物素(biotin)以便于检测。短肽用去离子灭菌水溶解后于-20℃分装保存。按梁国栋报道的方法(梁国栋，2001)，以短肽 P2 作为探针与光肩星天牛幼虫的纤维素酶提取液进行 Western 分析，检测该短肽的结合特性。光肩星天牛幼虫粗酶液进行非变性 PAGE (12%)分离后，立即将酶带电转印到尼龙膜(Hybond-N+，Amershan pharmacia 产品)上，为保持酶的活性，缓冲液中不加甲醇，转印条件为：恒流 100mA，1h。转移结束后取出尼龙膜，用 TBS 漂洗 1~5min。用 10mL 以 TBS 10 倍稀释的封闭液 I (购自 Roche 公司)于室温封闭 1h，用 TBS 漂洗 2~5min 后，加入用 10mL 封闭液稀释的 P2 探针($1\mu g \cdot mL^{-1}$)，4%缓慢振荡 2h。用 TBS 洗膜并用碱性磷酸酶缓冲液($100mmol \cdot L^{-1}$ NaCl，$100mmol \cdot L^{-1}$ Tris，$5mmol \cdot L^{-1} MgCl_2$，pH 值 9.5)平衡后，将膜放入用封闭液按 1∶5000 稀释的链霉素标记的碱性磷酸酶溶液(SP-AP，Promega)，室温缓慢振荡 1h 后，用适量 BCIP/NBT (Roche)溶液于黑暗中进行显色反应，直到出现清晰的条带为止。用无菌水冲洗后进行拍照、保存。

2　结果

2.1　AgEG2 的分离和纯化

从非变性聚丙烯酰胺凝胶中洗脱回收的 AgEG2 酶液，以羧甲基纤维素作为底物进行酶活性测定，仍保持内切葡聚糖酶活性，且分离的酶液在 SDS-PAGE 分析中呈单一酶带，经蛋白分子量标准估计分子量约为 39kD (图 1)。以此分离纯化的酶液作为靶分子，进行噬菌体随机多肽库的筛选。

2.2 噬菌体展示库的淘选

从随机七肽噬菌体展示库中取 2×10^{11} Pfu 噬菌体(约包含 2.8×10^{9} 不同短肽的噬菌体克隆)进行筛选，并采取空白胶预吸附的方法去除噬菌体肽库中与聚丙烯酰胺结合的噬菌体。从表1可见，三轮筛选中噬菌体回收比不断增高，由第1轮的 6.5×10^{-6} 升至第三轮的 3.9×10^{-3}，升高了600倍，表明筛选的富集效果是显著的，有专一结合的噬菌体存在。而对照每轮筛选的噬菌体回收比保持在 $10^{-5}\sim10^{-6}$ 左右，说明没有噬菌体富集现象。

表1 每轮筛选的噬菌体的回收比

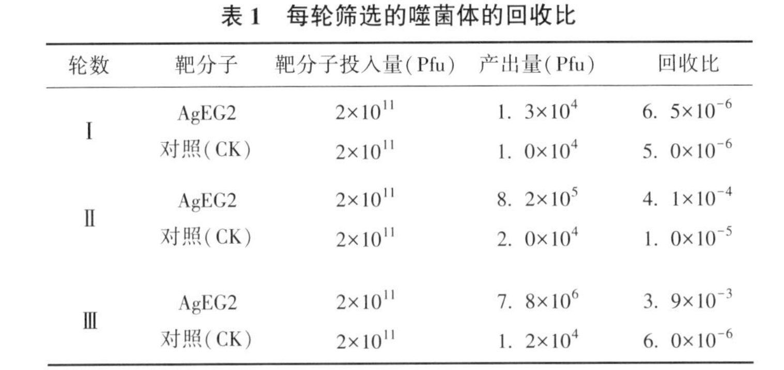

轮数	靶分子	靶分子投入量(Pfu)	产出量(Pfu)	回收比
Ⅰ	AgEG2	2×10^{11}	1.3×10^{4}	6.5×10^{-6}
	对照(CK)	2×10^{11}	1.0×10^{4}	5.0×10^{-6}
Ⅱ	AgEG2	2×10^{11}	8.2×10^{5}	4.1×10^{-4}
	对照(CK)	2×10^{11}	2.0×10^{4}	1.0×10^{-5}
Ⅲ	AgEG2	2×10^{11}	7.8×10^{6}	3.9×10^{-3}
	对照(CK)	2×10^{11}	1.2×10^{4}	6.0×10^{-6}

97KD
66KD
45KD
30KD
20.1KD
14.4KD
A M

图1 SDS—PAGE 分析纯化的 AgEG2

A：AgEG2，M：蛋白分子量标准，从上到下分别为97KD、66KD、45KD、30KD、20.1KD和14.4KD

2.3 结合短肽的序列分析

从第三轮筛选的洗脱液中随机挑取了11个噬菌体单克隆(b1~b11)，分别进行扩增后，对每个单克隆进行DNA序列测定。根据试剂盒使用说明书提供的外源序列插入位点，从噬菌体DNA序列中查找出随机七肽的基因序列，推导出七肽的氨基酸序列(表2)。b3、b6、b8和b9四个噬菌体克隆含有完全相同的短肽TPHRSPL序列，出现频率为36%，此外，b5也包含PHR同源序列。以上结果表明经过3轮筛选，特异结合的克隆得到了富集。从表2还可看出，b6克隆所插入的外源基因序列与b3、b8和b9不同，但它们表达相同的短肽序列，表明结合作用取决于短肽的氨基酸序列而不依赖于噬菌体本身。

表2 短肽的DNA序列和氨基酸序列

克隆	DNA序列	氨基酸序列
b1	5′-CATTTGCTTATTCCTCATCCT-3′	HLLIPHP
b2	5′-GCTTTGGCTCAGAAGGGTCTT-3′	ALAQKGL
b3	5′-ACTCCGCATCGTTCTCCTCTG-3′	TPHRSPL
b4	5′-CCGAGTCATCTTCATCTTTAT-3′	PSHLHLY
b5	5′-AGTTATGATCCTCCTCATCGT-3′	SYDPPHR

（续）

克隆	DNA 序列	氨基酸序列
b6	5′-ACTCCTCATCGGTCTCCTCTT-3′	TPHRSPL
b7	5′-ACTGCTAATACGCATAGGACT-3′	TANTHRT
b8	5′-ACTCCGCATCGTTCTCCTCTG-3′	TPHRSPL
b9	5′-ACTCCGCATCGTTCTCCTCTG-3′	TPHRSPL
b10	5′-TGGATGGCTTTTCAGAATACG-3′	WMAFQNT
b11	5′-CTTCATTTGCCGACTCCTGCG-3′	LHLPTPA

2.4　筛选出的噬菌体与 AgEG2 的亲和力分析

上述挑取的 11 个克隆(b1~b11)进行扩增后分别与靶分子进行结合筛选，检测富集噬菌体与靶分子的结合能力，以相同投入量的噬菌体肽库作为对照，结果如图 2 所示。b3、b6、b8 和 b9 克隆的回收比为 4×10^{-2}~6×10^{-2}，而其余克隆及对照的噬菌体回收比约 10^{-4}~10^{-6}，前者是后者的 400~40000 倍，表明富集的噬菌体克隆与 AgEG2 的亲和力显著高于其它克隆。

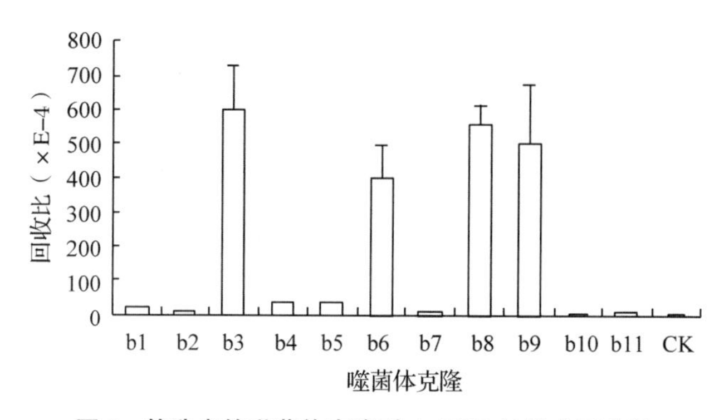

图 2　筛选出的噬菌体克隆对 AgEG2 的结合回收比

（b1-b11 为噬菌体克隆，CK 为未筛选的噬菌体肽库。结果为 3 次实验的平均值）

2.5　P2 与内切葡聚糖酶的结合作用

为进一步验证短肽 P2 的结合特性，合成了该短肽序列并进行了 Western 分析(图 3)。从图 3B 可以看出，短肽 P2 仅与尼龙膜上的 AgEG1 和 AgEG2 两条蛋白带有结合作用，而对纤维素酶提取液中的其它蛋白组分没有吸附作用，表明短肽对光肩星天牛的内切葡聚糖酶有特异结合作用。由图 3A 可见，光肩星天牛粗酶液中 AgEG1 的含量大于 AgEG2，而图 3B 中，短肽 P2 对 AgEG1 的杂交信号反而比 AgEG2 弱，表明该短肽与 AgEG2 的结合能力比对 AgEG1 的结合能力更强。

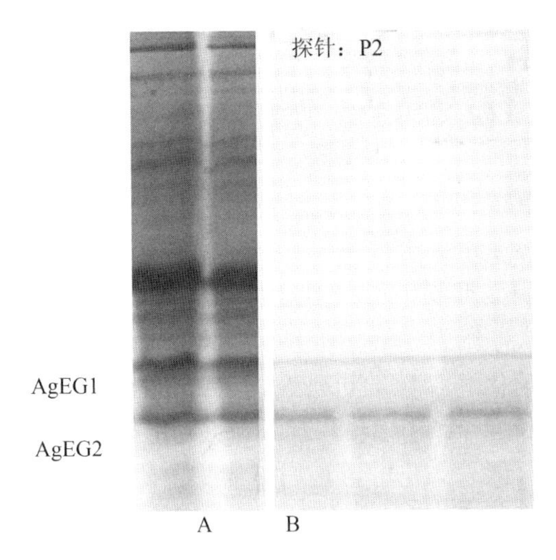

图 3　合成肽(P2)与内切葡聚糖酶的特异结合作用

(A：光肩星天牛幼虫纤维素酶粗酶液 native-PAGE；B：短肽 P2 作为探针与纤维素酶提取液印迹膜进行 Western 分析)

3　讨论

本研究利用噬菌体展示技术进行了筛选天牛纤维素酶特异结合活性短肽的尝试。由于对天牛内切葡聚糖酶的结构特性缺乏了解，因此研究选用了库容量较大的随机七肽噬菌体库。纤维素酶是由多种组分构成的复合酶系(Bayar, 1985)，组成复杂，因此纯化困难，通常的蛋白纯化方法易导致酶分子的天然构象发生改变而失去活性。本研究以非变性聚丙烯酰胺凝胶中洗脱回收的内切葡聚糖酶作为靶分子，既简便易行，又能很好地保持其原来的构型和酶活性。但酶液中残留的大量聚丙烯酰胺凝胶短链分子会吸附肽库中与其特异结合的噬菌体，造成了前期筛选所获的富集噬菌体，经检测均对聚丙烯酰胺特异结合，而与酶无结合作用。为避免酶液中短链聚丙烯酰胺分子影响筛选结果，对筛选程序进行了改进，将噬菌体展示原库以及每轮筛选的噬菌体首先与空白聚丙烯酰胺凝胶的洗脱液吸附，以去除噬菌体展示库中与聚丙烯酰胺分子特异结合的噬菌体，然后再与靶分子进行吸附，确保所筛选出的结合序列是酶分子的特异结合序列。每轮筛选的噬菌体回收比、最后一轮筛选的单克隆噬菌体结合力分析均表明，改进的筛选方法能成功去除噬菌体库中与聚丙烯酰胺分子特异结合的噬菌体克隆，所筛选出的富集噬菌体是通过与酶分子的特异结合而获得，这证明以上改进是有效的，与靶分子特异结合的噬菌体得到了有效富集。

为证明所获短肽 TPHRSPL 可特异结合光肩星天牛内切葡聚糖酶，合成了短肽 P2。通过 Western 分析，发现 P2 不但结合 AgEG2，而且和 AgEG1 也有亲和结合作用，而对粗酶液中的其它酶和蛋白却没有结合作用。这证明了该短肽本身对内切葡聚糖酶有较强的结合作用，同时也表明 AgEG1 和 AgEG2 为同工酶甚至为同位酶，它们可能会有较高的氨基酸序列同源性及类似的结构。许多研究表明，同一物种的内切葡聚糖酶同工酶往往有高度的序列同源性，例如，马铃薯线虫 *Globodera rostochiensis* (Wollenweber)的两个内切葡聚糖酶同工酶 GrEG1 和 GrEG2 的相对分子质量分别为 49.7kD 和 42kD，而其 cDNA 在 5′端有 95%的同源性(Smant et al., 1998)。同样，黄胸散白蚁 *Reticulitermes speratus* (Kolbe)的两个内切葡聚糖酶同工酶 RsEG1 和 RsEG2 的氨基酸序列有 98%的同源性(Tokuda et al., 1999)。

初步测定了 P2 与 AgEG1 和 AgEG2 结合后对其活性的影响，结果酶活性没有明显的改变。原因可能是短肽与酶分子的结合位点不是其活性作用的关键部位，不至于影响其催化功能。虽然对酶活性没有明显抑制作用，但这些与 AgEG1 和 AgEG2 特异结合的短肽同样具研究和运用前景：①短肽可作为标记，研究天牛内切葡聚糖酶的诱导效应及在肠道的分布特性，从而揭示天牛的消化和食性的机理，为开发天牛消化酶抑制剂奠定生物学基础；②短肽可作为"向导"，与能抑制纤维素酶活性的物质相连，使之更有效地找到靶部位，增强对酶的抑制作用，或结合能杀天牛的毒蛋白，避免由于天牛食性强、食物滞留时间短而造成毒蛋白在肠道中发挥作用不充分；③短肽与其它大分子物质相连，与纤维素酶分子结合后形成物理障碍，影响纤维素酶的催化活性。因此，特异结合短肽的获得为开发杀天牛生物制剂或增加其杀虫效果奠定了基础。

Selecting and Identification of Binding Peptides of the Endoglucanases from *Anoplophora glabripennis*

Abstract　Endoglucanases are the main cellulolytic enzymes in the gut of *Anoplophora glabripennis*. In this study, random peptide phage display technology was employed to screen peptides that bound the AgEG2, a member of endoglucanase isozymes. Phage clones displaying peptide TPHRSPL accounted for 33.7% of the selected phage population after three rounds of screening and showed higher phage recovery than the other clones in the binding assay. Peptide TPHRSPL was chemically synthesized and tested for its binding activity to AgEG2. The synthetic peptide exhibited high binding specificity for AgEG1 and AgEG2. This indicated that peptide TPHRSPL had the affinity to the endoglucanase of *A. glabripenni*, which could be used to study the biological role of the enzyme in the gut, and had the potential to be developed into biological control agents of *A. glabripenni*.

Keywords　*Anoplophora glabripennis*, endoglucanases, random peptides phage display library, binding peptides, affinity

Prokaryotic Expression Analysis of an NBS-type *PtDRG*01 Gene Isolated from *Populus tomentosa* Carr. *

Abstract In order to investigate the protein features of an NBS gene (*PtDRG*01, *EF*157840) isolated from *Populus tomentosa* Carr., the full-length open reading frame was fused into a prokaryotic expression vector pGEX-KG. PCR analysis and double endonuclease digestion showed that the recombinant vector was successfully constructed and transferred into an expression host *E. coli* strain XA_{90}. It was indicated by SDS-PAGE analysis that IPTG treatment successfully induced the expression of a fusion protein of about 79 kD, which was consistent with the predicted value. In addition, the prokaryotic expression system was also optimized. The result suggests that 1mmol/L IPTG treatment for 4h at 37℃ was most effective, and the product was predominately soluble and not extra-cellular secreting. Moreover, the fusion protein was purified with an affinity chromatography column using Glutathione Sepharose 4B. This work will lay a foundation for further studies on biological functions of the *PtDRG*01 gene.

Keywords *Populus tomentosa* Carr., NBS gene, prokaryotic expression, protein purification

1 Introduction

Populus tomentosa has a number of good characteristics, such as rapid growth, excellent resistance to diseases, and superior wood quality; therefore, it is highly valued and considered as the best among the native poplar species in China (Zhu and Zhang, 1997). It is widely cultivated in north China. However, a number of diseases seriously affected its popularization, with the expansion of cultivation areas (Yuan, 1998). *Melampsora magnusiana* causes one of the most serious diseases as *Melampsora magnusiana* Wanger, among most pathogens. It infects leaves and buds of young plants and makes trees form tiny yellow spots, and thus reduces the efficiency of photosynthesis (Yu et al., 2004), and the affected leaves drop early. A severe infection can cause large necrosis and badly affect growth, with the result that the infected leaves and buds die, the dry weight is reduced by 29%~32%, the volume is reduced by 31%~42% and the increment growth is reduced by 65% (Pei et al., 2003). Further, it may result in serious decline of resistance to the autumn frost, *Xanthomonas populi*, *Valsa sordida* Nit. and so on (Dowkiw et al., 2003).

* 本文原载《中国林业动态》(*Frontiers of Forestry in China*), 2009, 49(2): 216-222, 与李琰、张谦、饶星、李海霞、刘婷婷和安新民合作发表。

So far, studies on the disease resistance of *P. tomentosa* remains so weak that the study on screening of disease-resistant genetic resources, selecting resistant varieties, mapping and cloning of resistance gene are still a blank field, in sharp contrast to the urgent needs of disease control in the production of *P. tomentosa* (Zhang et al., 2005). This is far behind the study on the disease resistance of Tacamahaca and Aigeiros (Stirling et al., 2002; Zhu et al., 2002; Dowkiw et al., 2004; Lescot et al., 2004; Yin et al., 2004). Therefore, researches in this area are of great significance.

Using the hybrid clones of *Populus* with high disease resistance screened by inoculation of *Melampsora magnusiana* Wanger in vivo as the material in previous researches, we used NBS conservative domain on an R disease-resistant gene to design degenerate primers, and obtained an entire gene named *PtDRG*01. *PtDRG*01 is 2324 bp, encodes 678 amino acids and has a molecular weight of 79kD. For further testing of its expression characteristics and verification of its functions, we constructed the prokaryotic expression vector of PtDRG01, established an efficient expression system of the gene in Escherichia coli and detected the specific protein with target length by optimizing the induced conditions and analysis of the expression characteristics. After the fusion protein was purified with the affinity chromatography column glutathione sepharose 4B, we obtained the purified target protein, which thus provides material for further study on the functions of proteins. This work lays the foundation for further studies on biological functions of the *PtDRG*01 gene and revelation of their transmission mechanism of the disease resistance signal.

2　Materials and methods

2.1　Materials

An NBS-type *PtDRG*01 gene isolated from *P. tomentosa* was cloned in our lab. The *PtDRG*01 gene was provided with the accession number EF157840 in GenBank (Zhang, 2007). Expression vector pGEX-KG was purchased from GE healthcare, the *E. coli* strain XA90 was preserved by our lab.

2.2　Methods

2.2.1　Construction and identification of the prokaryotic expression vector of *PtDRG*01

Gene specific primers were separately designed based on the homing sequence and the terminator sequence of the coding region on *PtDRG*01, and the nucleotide sequence of the restriction sites of *Nco*I and *Xba*I were added at both ends of the primers. The nucleotide sequences of specific primers of the *PtDRG*01 gene is ProU (*Nco*I): 5′-CATGCCATGGGTA TGCAGAAAGAAAAACGCAAGCAA-3′ and ProD (*Xho*I): 5′-CCGCTCGAGTCAGCCC AAGATAAAATCAGATGG-3′. The primers were synthesized by Shanghai Sangon Biological Engineering Technology And Service Co, Ltd. PCR products were digested by restriction enzymes *Nco*I and *Xho*I (Promega) and coupled-reacted with the expression vector PGEX-KG digested by the same enzymes. Finally, the recombinant vector was transferred into an expression host *E. coli* strain XA90 and cultured in a LB liquid medium with ampicillin (Amp, 100mg/L). The positive clones were screened by gene-specific PCR amplification.

2. 2. 2 *PtDRG*01 induced gene expression and SDS-PAGE analysis

A single positive bacterial colony XA_{90} was inoculated in LB medium containing 100mg/L Amp, shaken cultured at 180r/min, 37℃ until OD_{550} reached 0. 6. Then IPTG was added to a final concentration of 1mmol/L. The culture was incubated at 37℃ for 3 ~ 4h. Its final concentration was 1mmol/L by adding IPTG, and then it was cultured for 3 ~ 4h (Li et al., 2004). The expressed product was analyzed with SDS-PAGE on 10% linear polyacrylamide gradient gel (Low Molecular Weight Markers Proteins were provided by TaKaRa).

Referring to the methods in our lab (Lin et al., 2005a), 5% stacking gel with 10% separation gel 1mL of bacteria solution was centrifuged at 13000r/min for 5min and the supernatant was discarded and the bacteria were harvested. The pellet was resuspended in 2mL of ddH_2O and ultrasonically disrupted at 30% for 6s at the interval of 3s, for a total of 20min. Then cell lysate was centrifuged at 13000r/min, and 25μL of supernatant was commixed with an equal volume of a 2× SDS gel-loading buffer containing 1mol/L Tris-HCl (pH 6. 8), 4. 0mL of 10% SDS, 2. 5mL β-mercaptoethanol, 2. 0mL glycerol, 1. 0mL of 0. 1% bromophenol blue and 9. 5mL of ddH_2O. Thirty-fiveμL of the expressed product was analyzed by SDSPAGE, commixed with an equal volume of a 2×SDS gelloading buffer and analyzed by SDS-PAGE. The electrophoresis lasted for 4h, 80V constant voltage in the stacking gel and 100V constant voltage in the separation gel. After acrylamide gel was stripped, the gel was stained overnight with coomassie brilliant blue R250, and placed in a destaining solution; the destaining solution was replaced until the protein band was clear. The results were analyzed by a gel imaging system.

2. 2. 3 The most effective condition for induced protein expression

The activated bacteria solution was added using 1mmol/L IPTG when OD_{550} reached 0. 6, shaken cultured at 200r/min, 37℃. OnemL of the sample was ultrasonic disrupted at different times (1 ~ 5h) and different temperatures (25, 30, 37℃) respectively. The expressed product was analyzed with SDS-PAGE. The percentage of target protein accounting for the total bacterial protein was analyzed by a gel imaging system.

2. 2. 4 Secretory analysis of the target protein of *PtDRG*01

4 mL of the bacteria solution in the most effective condition was centrifuged at high speed for 15min at 4℃. The pellet and the supernatant were harvested respectively. The pellet was resuspended with a 2×SDS gel-loading buffer, boiled and centrifuged, and analyzed by SDS-PAGE. The supernatant was filtrated by a 0. 25μm filter membrane, placed on ice, added into PMSF and Dnase, and precipitated with $(NH_4)_2SO_4$ (68g/100mL). After the solution dissolved fully, it was centrifuged at 40000r/min for 20min. The supernatant was discarded and the pellet dissolved in 1mL of ddH_2O, 20μL of which was commixed with an equal volume of a 2×SDS gel-loading buffer and analyzed by SDS-PAGE.

2.2.5 Dissolubility analysis of the target protein of *PtDRG*01 gene

4 mL of the bacteria solution in the most effective condition was centrifuged at high speed for 15min at 4℃ and the supernatant was discarded. The pellet was resuspended in protein buffer (pH = 8), lysozyme-cleaved for 20min, and centrifuged at 10000r/min for 15min at 4℃. The pellet and the supernatant were harvested, respectively, commixed with an equal volume of a 2× SDS gelloading buffer and analyzed by SDS-PAGE.

2.2.6 Purification of the target protein of *PtDRG*01with the affinity chromatography column

A single positive bacterial colony XA90 was inoculated into LB medium containing 100mg/L Amp, and shaken cultured at 180r/min, 37℃ until OD_{550} reached 0.6. Then IPTG was added to a final concentration of 1mmol/L. The culture was incubated for 3~4h. Its final concentration was 1mmol/L by adding IPTG, and then it was cultured for 3~4h (Lin et al., 2005b). TwomL of impure protein samples, extracted from the cell lysate from ultrasonic disruption, was added into the GE healthcare with glutathione sepharose 4B, stood for 30min and fully integrated. The impure protein was eluted by 30mL of PBS buffer containing 140mmol/L NaCl, 2.7mmol/L KCl, 10mmol/L Na_2HPO_4, 1.8mmol/L KH_2PO_4; GST fusion protein was eluted with elution buffer (0.61g of Tris and 0.31g of glutathione reductase dissolved in 60mL of dH_2O, pH was 8.0 by adding HCl 5mol/L, and the final volume was 100mL by adding dH_2O); the samples of the eluting peak were collected by an automatic collector and concentrated after dialysis with 50% glycerol. The result was analyzed with SDS-PAGE.

3 Results and analysis

3.1 Results of construction of prokaryotic expression vector of *PtDRG*01 and identification of enzyme digestion

In this study, two gene-specific primers were added with the restriction sites for *Nco*I and *Xho*I respectively. The expression vector and PCR amplification products were digested by two incision enzymes (Faivre et al., 2006), and then the two enzyme products were linked to get a recombinant plasmid. The recombinant plasmid was transferred into an expression host E. *coli* strain XA_{90}, and was inoculated overnight at 37℃ in LB medium containing 100mg/L Amp. A single bacterial colony was selected to extract the plasmid DNA and verify the linkage effect. From Fig. 1, we can see that we obtained two specific fragments by digesting the recombinant plasmid with the incision enzymes of *Nco*I and *Xho*I. According to the control Marker, we can see that the length of the target gene was about 2000bp, indicating that the gene had been successfully integrated into the expression vector and the recombinant plasmid was correct.

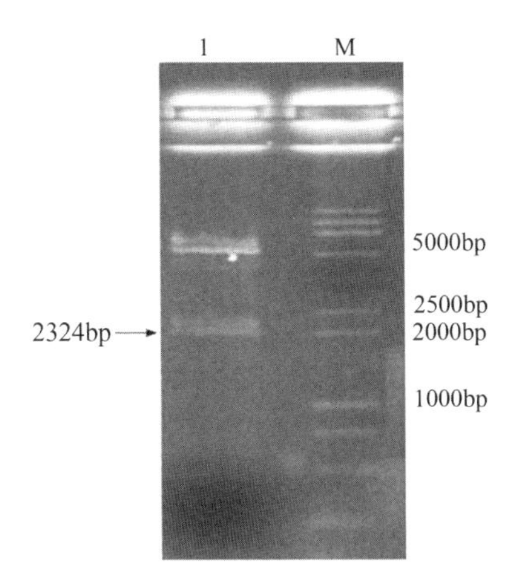

Fig. 1 Identification of recombinant *pGEX-KG-Pt*01

M: DNA marker; 1: production of *Nco* I/*Xho* I digestion

3.2 Screening for the optimum conditions of induced expression

3.2.1 Screening for the optimum time of IPTG induction

When OD_{600} of the cell density of the recombinant strain, negative control, and positive control reached 0.6, 1mmol/L of IPTG was added to induce the expression of exogenous protein, and the result was analyzed with SDS-PAGE after 1~5h. It showed that *E. coli* with the recombinant plasmid was induced respectively using IPTG; the protein band with a molecular weight of about 80kD markedly increased, and the expression amount of the protein increased with induction time, the maximum appearing at 4h, followed by a downward trend (Fig. 2). Therefore, the specific proteins with a target length could be detected in the range of 1~5h in IPTG induced time, showing that the *PtDRG*01 gene had a complete encoding box, and the maximum expression amount of the protein appeared at 4h. With the analysis software of the gel imaging system, we contrasted the

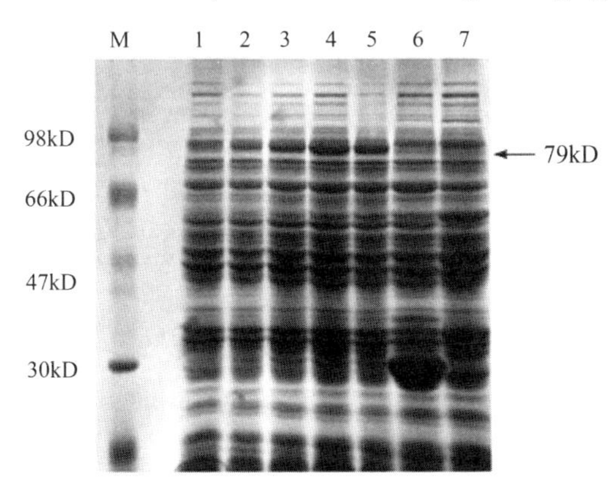

Fig. 2 Expression of recombinant in different induced time.

M: protein marker; 1-5: expression of *pGEX-KG-Pt*01 by IPTG induction after (1~5h);

6: expression of XA_{90} cell with pGEX; 7: expression of XA_{90}.

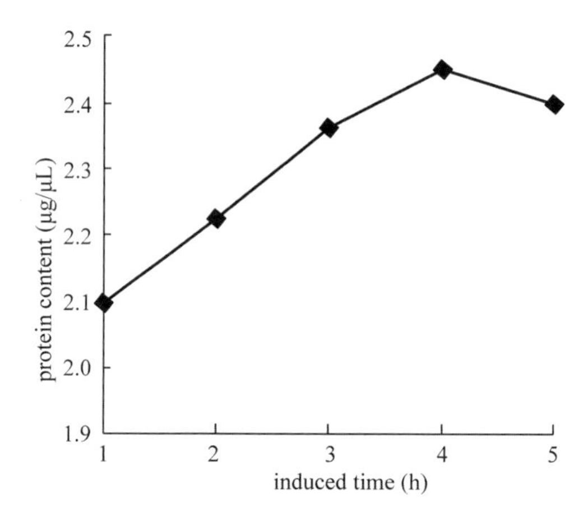

Fig. 3 Change in the expression of the recombinant in different induced time

results with the standard protein and got the curve of the expression (Fig. 3).

3.2.2 Screening for the optimum temperature of IPTG induction

The induced temperature is another important factor for the target protein expression. When OD_{600} of the cell density of the recombinant strain, negative control and positive control reached 0.6, 1mmol/L of IPTG was added to induce the expression of exogenous protein at 25, 30, and 37℃, respectively, and the result was analyzed with SDSPAGE after 4h. The results are presented in Fig. 4, showing that the expression amount of the protein increased with the increase in induced temperature, and the maximum appeared at 37℃. The 37℃ temperature was optimum for prokaryotic expression of the *PtDRG*01 gene. The result analyzed by the analysis software of the gel imaging system is shown in Fig. 5, which is consistent with the results above.

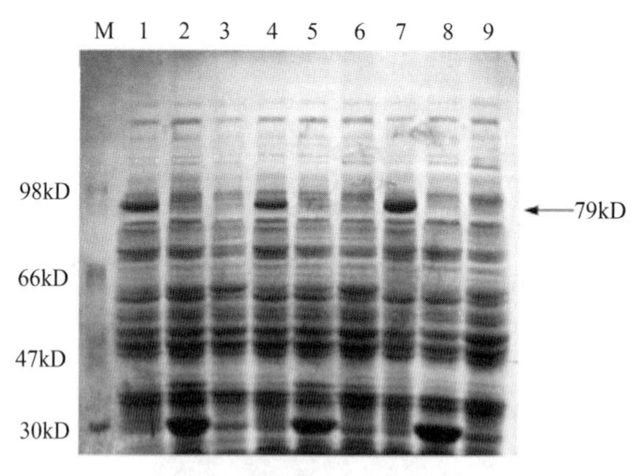

Fig. 4 Expression of recombinant at different induced temperatures

M: protein marker; 1, 4, 7: *pGEX-KG-Pt*01 recombinant protein at 25, 30, 37℃; 2, 5, 8: expression of XA90 cell with pGEX; 3, 6, 9: expression of XA90.

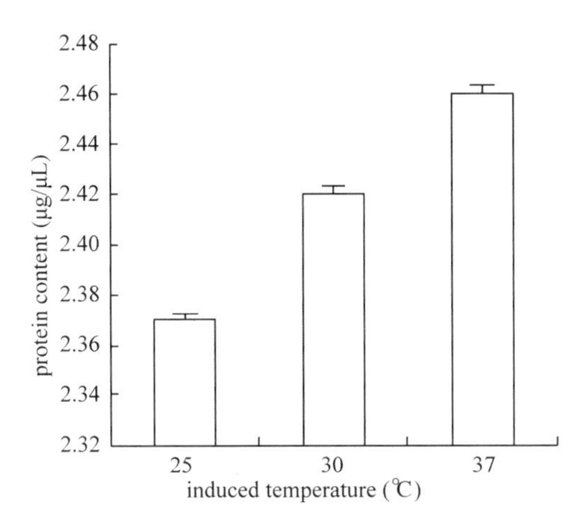

Fig. 5 Change in the expression of the recombinant at different induced temperatures

3.3 Extra-cellular secretion

If there is a signal sequence in the upper stream of the target gene, the fusion protein can be secreted out of the cell and we could directly collect a large number of medium to extract the fusion protein (Lin et al., 2005b). The extra-cellular medium of the recombinant strains and the intra-cellular proteins were collected respectively. The result was analyzed with SDS-PAGE. The expressed fusion protein was found in the intracellular extract but not in the medium of the recombinant strain, positive control or negative control (Fig. 6), indicating that the resulting protein was intra-cellular secreting.

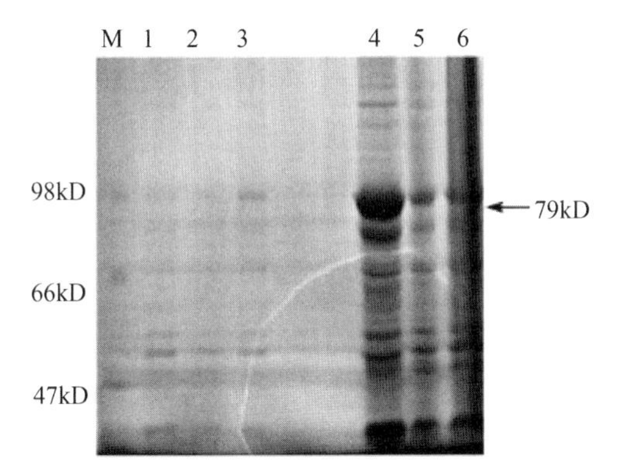

Fig. 6 Secreting analysis of fusion protein

M: protein marker; 1, 2, 3: the culture medium of pGEX-KG-Pt01, *pGEX*, XA_{90} cell;
4, 5, 6: the extractant of pGEX-KG-Pt01, *pGEX*, XA_{90} cell.

3.4 Potential of solution

As the GST protein expressed by the vector pGEX was a water-soluble protein (Zhang et al., 2005), we could speculate that the fusion protein was also soluble. The recombinant and control cells were broken by lysozyme, and the supernatants and the pellet were analyzed by SDSPAGE. The expressed fusion protein was found in the supernatants but not in the pellet (Fig. 7), indicating that the fusion protein was not in the inclusion bodies but was expressed in a soluble way.

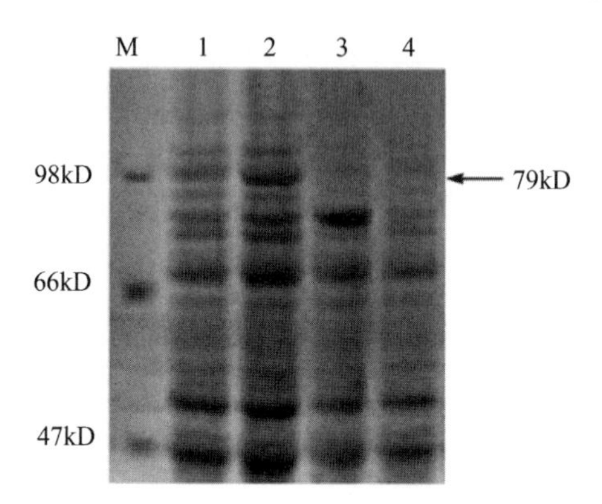

Fig. 7 Dissolving analysis of fusion protein

M: protein marker; 1, 2: the filtrate of XA_{90} cell and *pGEX-KG-Pt*01 cell; 3, 4: the deposition of XA_{90} cell and *pGEX-KG-Pt*01 cell.

3.5 Purification and Identification

The recombinant strains induced by IPTG were collected and broken ultrasonically in an ice bath. The fusion protein was extracted with ddH_2O and purified with the affinity chromatography column using glutathione sepharose 4B, concentrated after dialysis with 50% glycerol. The result was analyzed with SDS-PAGE (Fig. 8). The impure protein was eluted with elution buffer. The fusion protein of *PtDRG*01 gene was purified and successfully induced the expression of a fusion protein of about 79kD.

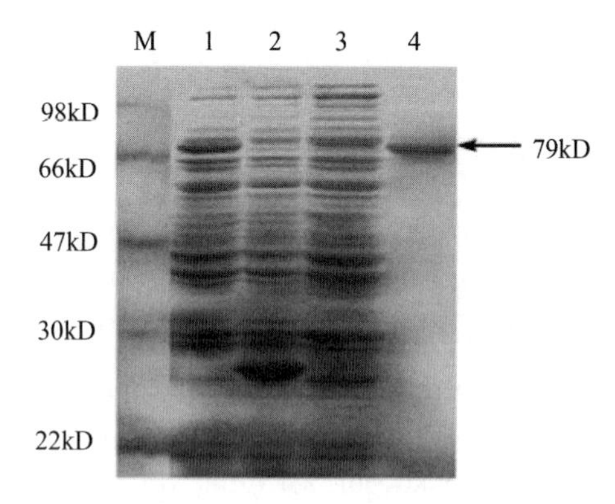

Fig. 8 Purification of recombinant protein

M: protein marker; 1, 4: *PtDRG*01 protein before and after purified; 2: pGEX protein; 3: XA_{90} protein.

4 Discussion

The *PtDRG*01 gene isolated from *P. tomentosa* had high homology with the NCBI gene, NBS-type genes of a large number of plants (including *P. trichocarpa*, *P. deltoides*, *P. balsamifera* subsp *trichocarpa*, as well as *P. tremula* and other poplars) logged in EST database, and the RGA gene (Zhang et al., 2001). Moreover, the *PtDRG*01 protein had high homology with many well-known plant disease resistance proteins, in particular with the disease-resistant protein N of the tobacco mosaic virus (Liu et al., 2003); their amino acid sequences contained the complete structures of TIR and NBS. We found that the *PtDRG*01 gene had ATP-binding sites in the NBS domain, using a large number of bioinformatics analysis; therefore, the *PtDRG*01 gene might be associated with the ATP enzyme reaction and affect ATP activity inside the cell, which plays a key role in the transmission of the disease resistance signal. We studied the prokaryotic expression to verify this.

At present, prokaryotic expression systems for different genes are quite different. The buckwheat trypsin inhibitor gene (BTI) was cloned into the expression vector pQE231, expressed in *E. coli* M15 host cells, and the recombinant BTI induced by IPTG, with the result showing that most of the recombinant protein found was soluble (Li et al., 2007). The *GlgC* gene was cloned to prokaryotic vector pET-28a-c (+) and then recombinant vector pET-glgC was transformed into the host cells of *E. coli* BL21, and the specific protein was produced by the host (Jia et al., 2006). The B7.2(*lgV* + *C*) gene was cloned to prokaryotic vector pGEX-4T-3, and the results indicated that the fusion protein reached the maximum level after 3mmol/L IPTG treatment for 5h (Yan et al., 2002). The prokaryotic vector pGEX–KG used in this study was an *E. coli* vector for expressing the fusion protein of glutathione S-transferase (GST), mainly due to the following: ① pGEX–KG contains an inducible strong promoter tac, which is a mixed promoter with lac and trp. When the target gene is linked under the downstream of a strong promoter, the foreign gene *PtDRG*01 would have a highly efficient expression in *E. coli*. ② When the highlevel expression of foreign genes inhibit the growth of the host cells, the repressor would protect the host so that the host bacteria could accumulate to a considerable number, and the expressed protein accumulates greatly in a short time and reduces the degradation of expression products by instantaneous derepression. IPTG combined with the repressor lac to induce the expression of the downstream (Sivaguru et al., 2003). PGEX–KG contained *lac*I gene, resulting in a large number of lac repressors and ensuring that the transcription of the downstream occurred only by IPTG induction. ③ The product of the expression vector was conducive to the separation and purification; the fusion protein was purified with the affinity chromatography column glutathione sepharose 4B in non-denaturing conditions, to maintain the integrity of the protein. ④ The coding region of the vector contained the foreign gene *PtDRG*01 and an encoding gene of the vector, the fusion protein, contained a glutathione transferase of about 20kD encoded by the vector. Thrombin restriction sites existed between the GST gene and the target gene, and the product could be digested by thrombin (Fan et al., 2007), so as to remove the GST part in the fusion protein and generate the *PtDRG*01 protein.

One mmol/L IPTG treatment for 4h at 37℃ was most effective for the expression, and the main reasons may be related to the host bacteria: *E. coli* is a kind of bacteria with a body temperature, and its optimum growth temperature was about 37℃; the growth can be divided into four phases: lag phase, exponential phase, stationary phase and decline phase. At first, the strains were activated overnight for about 12~16h in our study, and then cultured in fresh

culture medium for 2h. The strains were exactly in the stationary phase after 1mmol/L IPTG treatment for 4h, when the copy of the foreign gene recombinant plasmid increased a great deal and the expression of a fusion protein increased.

In the bioinformatics analysis of the PtDRG01 protein, we found that its isoelectric point was 8.165 and the hydrophilic amino acids accounted for the majority, indicating the protein was hydrophilic and alkalineresistant. In addition, a large number of α-helix, β-sheet, curls and flexible zones existed in the secondary structure of the protein. Some foreign researches have shown that many exogenous proteins are expressed in *E. coli*, and the proteins cannot spontaneously fold to generate a certain spatial structure with a specific function, but exist in cells with inclusion bodies, a kind of insoluble precipitation, which is not conducive to the biological activity of foreign proteins (Elena et al., 2006). In this study, we extracted proteins from the pellet and the cell lysate respectively, and the result was analyzed with SDS-PAGE. The desired protein was only in the supernatant of the cell lysate, proving that the recombinant protein did not form inclusion bodies when expressed in cells, and was soluble.

This work lays the foundation for further study on biological functions of the *PtDRG*01 gene and the method of its realization.

After the purification of fusion proteins with the affinity chromatography column, we focused on the exact function of the protein and the impact and mode of action of the ATP activity, and revealed the significance of the function for the disease-resistant signal transmission pathway. On the other hand, we will take the *PtDRG*01 fusion protein as the antigen to obtain specific antibodies, and prepare it for western blotting. It can also be used for molecular testing of the *PtDRG*01 gene transforming into other plants, in order to identify the function of the *PtDRG*01 gene in disease resistance and the signaling pathway.

Identification and Characterization of CBF/DREB1-related genes in *Populus hopeiensis*[*]

Abstract The dehydration-responsive element-binding factor (DREB) is a plant-specific family of transcription factors and plays an important role in plant's response and adaptation to abiotic stress. In the present work, two highly similar CBF/DREB1-like genes, designated as *PhCBF4a* and *PhCBF4b*, were identified from *P. hopeiensis*. These two genes contain all conserved domains known to exist in other *CBF/DREB1* genes. And in the AP2domain, there is only one different amino acid residue between *PhCBF4a* and *PhCBF4b*, Alanine or Valine, the nonpolar amino acid, suggesting that *PhCBF4a* and *PhCBF4b* may have similar DNA binding ability. Their expression is induced by water-loss treatment, and their expression patterns are similar. Moreover, with genomic DNA as template, the presence of the same bands in PCR products as those in expression pattern analysis indicated that *PhCBF4a* and *PhCBF4b* exist in genome of *P. hopeiensis*. Their detailed functions were discussed and would need further study.

Key words CBF, DREB, abiotic stress, *Populus hopeiensis*

1 Introduction

Plants are exposed to environmental stresses such as drought, high salt, and low temperature, which cause adverse effects on the growth of plants and the productivity of crops. Being sessile, they have adapted to respond to these stresses at the molecular and cellular levels as well as at the physiological and biochemical levels, thus enabling them to survive. A number of genes are induced by these stresses, which products function not only in stress tolerance but also in the regulation of gene expression and signal transduction in stress responses. Deciphering the mechanisms by which plants perceive environmental signal and its transmission to cellular machinery to activate adaptive responses is of critical importance for the development of rational breeding and transgenic strategies leading to ameliorate stress tolerance in crops and trees (Agarwal et al., 2006; Yamaguchi-Shinozaki and Shinozaki, 2006; Shinozaki and Yamaguchi-Shinozaki, 2007).

During the past few years, substantial progress has been made toward understanding how environmental stresses regulate gene expression of plants. Particularly, an important plant-specific

---

* 本文原载《林业研究》(*Forestry Studies in China*), 2008, 10(3): 143-148, 与王泽亮、安新民、李博、任媛媛、江锡兵和薄文浩合作发表。

transcription factor, named AP2/EREBP, has been identified, which were classified into five groups-DREB subfamily, ERF subfamily, AP2 subfamily, RAV subfamily and others in *Arabidopsis* (Sakuma et al. , 2002). DREB1/CBF-like transcription factors belonged to DREB subfamily, and were thought to be involved in ABA-independent signal transduction. DREB1/CBF specifically bound to the DRE/CRT (5′-TACCGACAT-3′) *cis*-acting elements and activated the transcription of downstream genes driven by the DRE/CRT sequence (Yamaguchi-Shinozaki and Shinozaki, 2006). In *Arabidopsis*, expression of *DREB1/CBF* genes was induced by cold stress but not by drought and high-salinity stress (Liu et al. , 1998). As a result, DREB1/CBF factors were thought to function in cold-responsive gene expression pathway. But later, some researchers found that *CBF4/DREB1D* (Haake et al. , 2002), *DDF1/DREB1F* and *DDF2/DREB1E* (Magome et al. , 2004), which also were members of DREB1/CBF subfamily, could be induced by drought stress and ABA treatment or induced by high-salinity stress in *Arabidopsis*. All these suggested the complexity of stress-responsive gene expression regulation in plant.

In addition to *Arabidopsis*, the identification and characterization of homologous *DREB* genes are currently on going in various plants, including wheat, rice, maize, soybean, ryegrass and grape (Shen et al. , 2003; Dubouzet, et al. , 2003; Xiong and Fei 2006; Xiao et al. , 2006 and 2008). All these proteins showed significant sequence similarity in the conserved DNA binding domain (AP2/ERF) found in the EREBP/AP2 proteins. The ERF/AP2 domain, a region of about 60 amino acids, consisted of a three-stranded anti-parallel beta-sheet and an amphipathic alpha-helix packed approximately parallel to the beta-sheet (Allen et al. , 1998). It was shown that the beta-sheet played a critical role in binding of ERF/AP2 to *cis*-acting elements, in a particular way different from the beta-sheets of other known DNA-binding domains in the number and arrangement of beta-strands (Allen et al. , 1998).

As the most important life-style of land ecosystem, forest trees have tremendous economic and ecological value, as well as unique biological properties of basic scientific interest (Sterky et al. , 2004; Bradshaw et al. , 2000). The genus *Populus* is consisted of 40 species distributed widely in diverse habitats throughout the northern hemisphere. Being rapid growth, prolific sexual reproduction, facile transgenesis and cloning, small genome, and present available genome sequences, it has been adopted as a model system for forest tree biology (Bradshaw et al. , 2000; Gail, 2002; Amy et al. , 2004). *P. hopeiensis* Hu et Chow is a drought and chilling-tolerant poplar species, which is mainly distributed in northwest of China and north China (Xu, 1988). In the present study, we identified two *DREB1/CBF*-relate-d genes from water-stressed plantlets of *P. hopeiensis*, designated as *PhCBF4a* and *PhCBF4b*, which were highly similar in nucleotide and amino acid sequence and may play a important role in dehydration-regulated gene expression in *P. hopeiensis*.

2 Materials and methods

2.1 Plant materials and water-stress treatment

Explants from hydroponically grown shoots of *P. hopeiensis* were sterilized and cultured on half-strength MS medium supplemented with 0.5mg · L^{-1} BA and 0.1mg · L^{-1} NAA at 25±1℃ under a 14h photoperiod using cool-white fluorescent light. For rooting, the shoots regenerated were subcultured on half-strength MS medium supplemented with 0.4mg · L^{-1} IBA.

After 6~7 weeks of subculturing, the rooted plantlets were wilted by continuous monitoring of weight loss on a balance at room temperature to 90%, 80%, 70%, and 50%, of their original fresh weight respectively according to Pelah et al. (1997) with some modifications, then kept in closed plastic bags for an additional 3h.

2.2 RNA, genomic DNA isolation and first strand cDNA synthesis

The water-stressed plantlets were ground in liquid nitrogen, then a maximum of 30mg powder was extracted for total RNA using SV Total RNA Isolation System(Promega, Madiso-n, USA). At last, total RNA was eluted into 50uL of RNase-free water. RNA concentration and quality were measured with a spectrophotometer. For the extraction of genomic DNA, the Genomic DNA Extraction Kit (Tiangen, Beijing, China) was employed.

1ug total RNA was used to synthesis first strand cDNA with Reverse Transcription System (Promega, Madison, USA) according to the manual.

2.3 Isolation of CBF gene from *P. hopeiensis*

The partial sequence of CBF gene was firstly amplified with the above first strand cDNA as template using the primers CBF4F (5′-TCCAACAGGAGAACTCAAGA) and CBF4R (5′-GGCAATAACATCCCTTCTGC) which were designed according to *CBF*4 gene from *Arabidopsis thaliana* and the poplar genome sequences (http://genome.jgi-psf.org/Poptr1/Poptr1.home.html). The reaction was performed using the following conditions: preliminar -y denaturation (5min, 94℃), then 30 cycles of denaturation (30s, 94℃), annealing (30s, 60℃) and extension (1min, 72℃), and final extension (5min, 72℃), and then stored at 4℃.

The 5′ and 3′ ends of CBF gene were obtained using the BD SMART RACE cDNA Amplification Kit (Clontech, Palo Alto, CA) following the manufacturer's instructions. The gene-specific primers for RACE amplification were as follows: *CBF*4-5: 5′-CTCAACGCCA -AAGCAGCAACATCA, *CBF*4-5N: 5′-CCCTAAACTTCTTTCTTCCTG CCCTC for 5′ end and *CBF*4-3: 5′-TGTTGCTGCTTTGGCGTTGAGGG, *CBF*4-3N: 5′-TGATTCTGCCTGGA -GGTTGCCTA for 3′end.

All the amplified fragments were tested by agarose gel electrophoresis, purified with QIAquick Gel Extraction Kit (Qiagen, Germany), and then cloned into pMD19-T plasmid vector (Tarkara, Dalian, Cnina) and sequenced.

2.4 Bioinfomatic analysis

General sequence analyses were performed with software DNAMAN5.2.2(Lynnon Biosoft). Multiple sequence alignments were conducted with ClustalX1.83 using default parameters. The phylogenic tree was analyzed and viewed by Phylip3.67 and Mega3.0 respectively. The deduced proteins were analyzed for structural motifs using the PROSITE server (Http://www.expasy.org/prosite) and for secondary structure prediction using the Predict Protein server (Http://www.predict protein.org/newwebsite/submit.php). And the web tool at http://cti.itc.virginia.edu/~cmg/Demo/wheel/wheelapp.html) was used for prediction of the amphipathic alpha-helix domain.

2.5 RT-PCR expression analysis

The PCR reactions were performed using the gene specific primers with the Takara Ex *Taq* (Takara, Dalian, China), in the final volume of 20uL, containing 13.25uL ddH_2O, 2.0uL of 10 ×buffer, 2.0uL of 2.0mM dNTP, 0.8uL of each 10uM primer, 1uL of template, 0.15uL of 5U · uL^{-1} *Taq* DNA polymerase. Amplifications were carried out at the following cycling parameters: preliminary denaturation (5min, 90℃), then 30 cycles of denaturation (20s, 94℃), annealing (20s, 58℃) and extention (40s, 72℃), and final extension (7min, 72℃), then stored at 4℃ until used. The same primers were also used to amply the responding bands from genomic DNA in *P. hoipeiensis*.

3 Results

3.1 Isolation and sequence analysis of *CBF*4 genes in *P. hopeiensis*

A 685bp cDNA fragment was isolated from rooted plantlets of 30% water loss in *P. hopeiensis* (Fig. 1A). To obtain thc full length gene, 5′ and 3′ RACE (Rapid Amplification of cDNA Ends) were employed to extend to both ends of the putative CBF genes (Fig. 1B). After sequencing and assembling, it' s interesting that two highly similar cDNA sequences with polyA tail were obtained, which were 1061bp and 1041bp respectively. Compared with the longer cDNA sequence, the shorter had 15bp and 5bp fragments deleted, in 3′ encoding region and 3′ UTR respectively. Both full length sequences contain a complete ORF, of 248 and 243 amino acids respectively. Genomic PCR using primers designed from the 5′ and 3′ UTR and the subsequent sequencing of the PCR product revealed that the genes have no introns (Fig. 5 C), the same as the matching gene in *P. trichocarpa* (Data not shown). Sequence alignment showed that these two genes are similar to *CBF* genes reported in other species (Data not shown). Moreover, phylogenic analysis with four *Arabidopsis* CBF genes revealed that the isolated two genes clustered with AtCBF4 firstly, and may be orthologs of *AtCBF*4 (Fig. 2). So we designated the two genes as *PhCBF*4*a* and *PhCBF*4*b*.

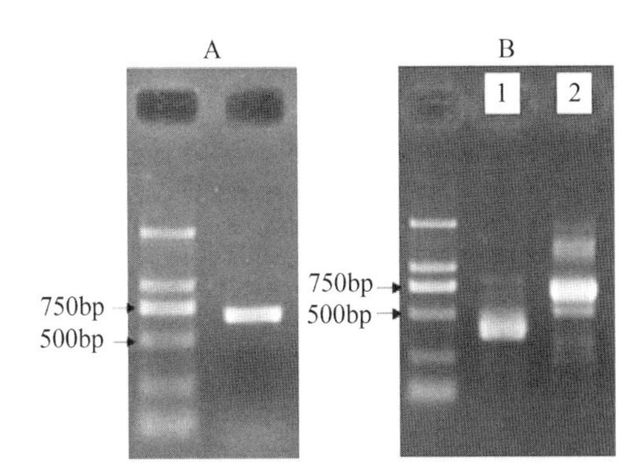

Fig. 1 Isolation of *CBF*4 gene in *P. hopeiensis*

A: The partial fragment. B: The 5′ RACE (1) and 3′ RACE (2) fragments.

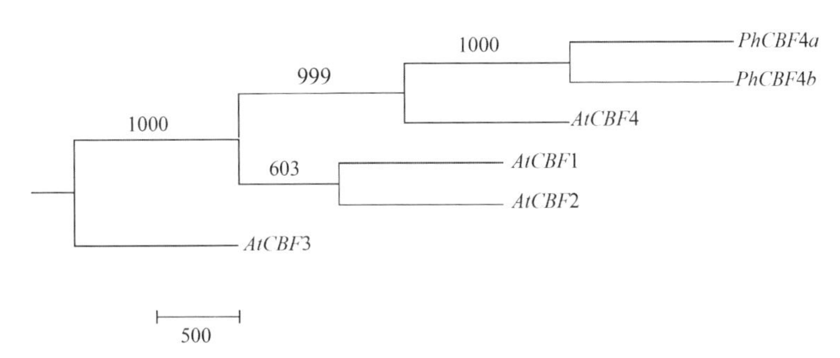

Fig. 2 Phylogenic tree calculated using *PhCBF4a*, *PhCBF4b* and four *Arabidopsis CBFs*

Bootstap values are shown on branches

*PhCBF*4*a* and *PhCBF*4*b* share 95% overall amino acid sequence identity with each other. In addition to the *indel* fragment, there are also ten different amino acids across the two proteins. Moreover, the *indel* fragment doesn't change the encoding region of it's downstream sequences. Each protein contains a conserved AP2 DNA binding domain specifically characterized by the YRG and RAYD elements (Okamuro et al., 1997), and a nuclear location signal (Dingwall and Laskey, 1991) close to the N-terminus, which had been also found in other CBF proteins. In the AP2 domain, a valine in position 14 and a glutamic acid in position 19, that were found in the DREB subfamily of AP2/EREBP family and may play important roles in DNA binding specificity, are also conserved (Sakuma et al., 2002). In addition, the CBF signature sequences, PKK/RPAGRxKFxETRHP and DSAWR (Jaglo et al., 2001), which bracket the AP2 DNA binding domain and are conserved in CBF-like proteins across many species, are also present in the two proteins.

3.2 Secondary structure prediction of *PhCBF*4*a* and *PhCBF*4*b*

With the online tools, the predicted secondary structures of the two proteins fit completely the expected AP2 domain, showing a three-stranded beta-sheet and an amphipathic alphahelix. Almost all residues known to be probably involved in the interact ions with the *cis*-acting element

(Allen et al., 1998), namely R82, R85, R87, R97, R104, G83, K91, E95, W106 and T109, are present, except W89 and Y120, which are replaced by a serine (S89) and a histidine (H120) residue respectively. A schematic diagram of the putative alpha-helix of *PhCBF4a* and *PhCBF4b* shows that the hydrophobic portion, facing the beta-sheet, is nearly totally conserved, except the Y120, the probable unique residue of the helix involved in DNA contacts (Allen et al., 1998).

3.3 Expression pattern analysis

The expression patterns of *PhCBF4a* and *PhCBF4b* in response to water stress were analyzed by semi-quantitative RT-PCR. The results show that the expression of these two genes is induced by water-stress treatment (Fig. 3A). And their expression patterns are similar, which are accumulated to the highest level when the plantlets lost 30% fresh weight. But the transcript of PhCBF4a may be more stable, showing a high level when water loss of plantlets reached 50%. Moreover, from genomic DNA of *P. hopeiensis*, two similar bands were also amplified.

4 Discussion

After the discovery of *DREB/CBF* genes in *Arabidopsis* and their role in abiotic stress response (Stockinger et al., 1997; Liu et al., 1998), many homologous genes have been identified in other plants. Here we report the identification and characterization of two similar *CBF*4 genes in *P. hopeiensis*. Putative amino acid sequence analysis of these two proteins definitely assigned them to the CBF/DREB1 subfamily. They contain the nuclear location signal close to the N-terminus, AP2 DNA binding domain characterized by YRG and RAYD elements, and two the CBF signature sequences bracketing the AP2 domain. In addition, phylogenic analysis with four *Arabidopsis* CBFs indicates that these two proteins are probable orthologs of AtCBF4. As a result, the isolated two genes were designated as *PhCBF4a* and *PhCBF4b*.

PhCBF4a and *PhCBF4b* are nearly identical in amino acid sequences, except the *indel* fragment in 3′ encoding region and other ten amino acid residues difference. Especially, in the AP2 domain, there is only one different amino acid, Alanine or Valine, the nonpolar amino acid, suggesting that *PhCBF4a* and *PhCBF4b* may have similar binding ability with *cis*-element. Compared with previous reports (Allen et al., 1998; Latini et al., 2006), there are two different residues, i.e., S89 and H120, in AP2 domain of PhCBF4a/b. In *Arabidopsis*, Rice and *Triticum aestivum*, different residues combination was present in the corresponding positions of AP2/EREBP proteins (Sakuma et al., 2002; Dubouzet, et al., 2003; Shen et al., 2003). Moreover, overexpression of AP2/EREBP genes could improve the tolerance of transgenic plants to drought, high-salt and cold, suggesting that S89 and H120 play the same roles as W89 and Y120, although that Y120, which is uncharged amino acid, or H120, which is charged, was the unique residue of the predicted alpha-helix involved in DNA contacts (Latini et al., 2006), and that this amino

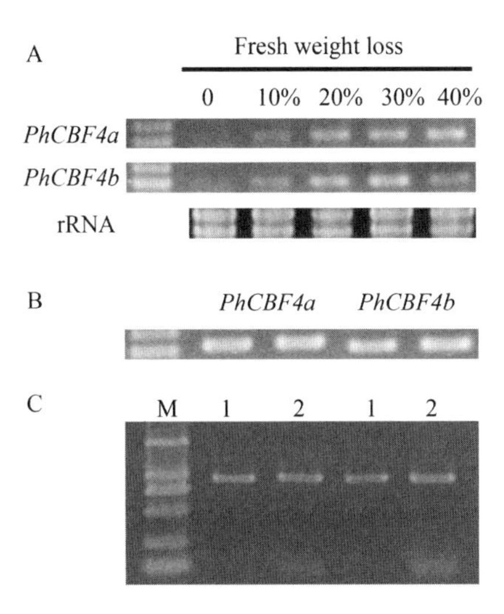

Fig. 3 PCR analyses of *PhCBF4a* and *PhCBF4b*

A: Expression patterns of *PhCBF4a* and *PhCBF4b* under water-stress treatment. B: The same two bands as those in A were amplified from genomic DNA. C: The fragments were obtained with the primers designed from the 5′and 3′UTR. The marker was DL2000 and the templates used were 1st cDNA(1) and genomic DNA(2).

acid residue just locates in the hydrophobic portion of the helix.

In our experiments so far, expression patterns of *PhCBF4a* and *PhCBF4b* in response to water stress are similar, although the transcript of *PhCBF4a* may be more stable. From cDNA and genomic DNA, both the two bands could be obtained by PCR, indicating that the presence of *PhCBF4a* and *PhCBF4b* isn't due to the regulation at the transcriptional or post-transcriptional level, but that these two genes exist in genome of *P. hopeiensis* indeed.

Why do two high similar *PhCBF4* genes exist in genome of *P. hopeiensis*? In *Arabidops is*, *CBF2/DREB1C* was thought to function as a negative regulator of *CBF1/DREB1B* and *CBF3/DREB1A* expression (Novillo et al., 2004), whereas Gilmour et al. (2004) thought that the three *AtCBFs* had redundant functional activities. In 2006, Zhao et al. (2006) isolated two groups DREB-like genes in *Brassica napus*, and argued that the trans-active group I genes were expressed at the early stage of cold stress to open the DRE-mediated signaling pathway, whereas the trans-inactive group II genes were expressed at the later stage and close the signal pathway. But in *P. hopeiensis*, whether this is also the case will need further study.

Successful *Agrobacterium*-mediated Transformation of *Populus tomentosa* with Apple SPDS Gene*

Abstract The problem of salinized soils has become one of the most serious constraints to agricultural and forest productivity. With the purpose of enhancing salt stress tolerance of *Populus tomentosa*, we transformed this tree species with spermidine synthase (SPDS) genes derived from an apple by an *Agrobacterium*-mediated method. Four transgenic clones were confirmed by PCR and Southern blot analysis. As well, the expression of introduced SPDS genes was analyzed by real-time quantitative PCR.

Keywords salt tolerance, spermidine synthase gene, transformation, *Populus tomentosa*

1 Introduction

The problem of salty soils represents one of the most significant constraints to agricultural and forest productivity both in China and elsewhere. The issue has also seriously affected the ecosystem. It has become an important topic in the research of the future development in agriculture and forestry. The issue is how to alleviate the salt stress and make use of the large amounts of salty soils and salty water resources. Almost all of our traditional methods have some flaws. For example, methods to reconstruct salty soils using suitable irrigation and drainage, aspersions with fresh water and chemical amelioration are hard to popularize for reasons such as high expenditures and little effect (Lin, 2004). Another traditional method, domesticating wild plants, it is also difficult to realize because of a very limited choice of plant species (Zhai et al., 1989). Along with the development of biotechnologies, gene transfer approaches have been employed to improve the stress tolerance of plants. These approaches efficiently speed up the selection and cultivation progress of new species and have opened a new window for research in improving salt tolerance of plants by genetic engineering.

Polyamines (PAs) are a kind of secondary production in the metabolism of an organism. They play an import role in regulating plant growth (Dai, 1988), controlling morphogenesis and enhancing plant stress adaptation, both biotic and abiotic (Bouchereau et al., 1999; Pang et al., 2007). In plants, the most common PAs are diamine putrescines (Put), triamine spermidines

* 本文原载《中国林业研究》(*Forestry Studies in China*), 2008, 10(3): 153-157, 与刘婷婷、庞晓明和龙萃合作发表。

(Spd) and tetramine spermines (Spm), which have a tight relationship with plant response to salt stress (Krishnamurthy and Bhagwat, 1989; Kaur-Sawhney et al., 2003). In plants, the first step of PA biosynthesis is the formation of Put. There are normally two pathways for Put biosynthesis, i. e., an ornithine and an arginine pathway. The key enzymes involved in polyamine biosynthesis in plants include ornithine decarboxylase (ODC, EC 4. 1. 1. 17), arginine decarboxylase (ADC, EC 4. 1. 1. 19), S-adenosyl methionine decarboxylase (SAMDC, EC 4. 1. 1. 50), spermidine synthase (SPDS, EC 2. 5. 1. 16) and spermine synthase (SPMS, EC 2. 5. 1. 22). Put is formed directly from L-ornithine and L-arginine by ODC and ADC, respectively. SAMDC decarboxylates S-adenosyl methionine (SAM) and the decarboxylated SAM provides Put and Spd with one and two aminopropyl groups to form Spd and Spm in a reaction catalyzed by SPDS and SPMS, respectively.

During the past few years, genes encoding these polyamine biosynthetic enzymes have been cloned and characterized from different plant sources. Several reports have shown the enhanced salinity stress tolerance in transgenic plants that overexpress ADC (Roy and Wu, 2001), ODC (Kumria and Rajam, 2002) or SAMDC (Roy and Wu, 2002; Waie and Rajam, 2003) transgenes. In addition, we found recently that in several plant species such as tobacco (Wi et al., 2006), *Arabidopsis thaliana* (Kasukabe et al., 2004), potato (Kasukabe et al., 2006) and European pear (Wen et al., 2008) overexpressing *SPDS* transgenes resulted in high tolerance to multiple environmental stresses, including chilling and freezing temperatures, salinity, drought, etc, compared with their wild type counterparts.

However, such attempts are still limited and there is no information available on the stress responses of transgenic *Populus tomentosa* overexpressing the *SPDS* transgene. *P. tomentosa* is the peculiar breed in *Leuce* Duby in China. It is so popular and widely distributed that it has become the primary category of afforestation in the northern areas of the county due to its excellent qualities. However, because of its low salt-tolerance, its popularization and utilization has been seriously constrained. Genes for *SPDS* could play an important part in the strategies to improve stress tolerance of crop plants by using gene transfer technology. Thus, in our present study, *Populus tomentosa* was transformed with the apple *Mdspds* gene, with the purpose of enhancing salt tolerance of the transgenic plant.

2 Materials and methods

2.1 Materials

2.1.1 Agrobacterium strains and plasmids

The construct of pBI121 with an apple *Mdspds* gene insert was provided by Dr. Moriguchi (National Institute of Fruit Research, Japan). The *Agrobacterium tumefaciens* strain LBA4404 was employed in this study.

2. 1. 2 Plant material and culture media

Axenic plants of *Populus tomentosa*, maintained under in vitro growth conditions at 26±1℃ and 16h photoperiod in the laboratory, were used for our present study.

The differentiation medium consisted of an MS medium supplemented with 0. 1mg/L NAA, 1. 0mg/L 6-BA, 30g/L sucrose and adjusted to pH 5. 8. The rooting medium was 1/2MS medium supplemented with 0. 4mg/L IBA and 20g/L sucrose.

2. 2 Methods

2. 2. 1 Sensitivity experiment of kanamycin (Km)

Leaf explants were cut to the costates with a scalpel and multiplied on the differentiation medium with different concentrations of Km (0, 10, 20, 30, 40 and 50mg/L). Every treatment was repeated 30 times from 30 clones. After about 50 d, the status of shoot differentiation was observed and recorded.

At the same time, the grown shoots (1~2cm in length) were cultivated on the rooting medium with the same concentrations of Km (i. e. , 0, 10, 20, 30, 40, 50mg/L). Every treatment was repeated 30 times from 30 clones. After 50d, the rooting situation was recorded.

2. 2. 2 Agrobacterium culture and preparation for engineering germ liquid

Agrobacterium tumefaciens was grown on a YEB medium supplemented with 50mg/L Km and 50mg/L rifampycin (Rif) for two days at 28℃ in the dark. Then, the bacterial colony was suspended overnight in the liquid YEB medium supplemented with the antibiotics mentioned above and shaken at 180~200r/min. The overnight culture product (1mL) was added to a fresh liquid YEB medium (50mL) and suspended again under the same conditions. After 5~6h, the bacterial culture was grown to an A600 of 0. 2~0. 4 and used for infection.

2. 2. 3 Transformation and regeneration of transgenic Plants

SPDS genes were transformed according to the Agrobacterium-mediated leaf-disk method. Sterile leaves, in roughly the same shape and color, were dipped in the engineering germ liquid for 15~20min. The excrescent liquids were discarded and the leaf disks were inoculated into a co-cultivation medium (the same as the differentiation medium), cultured for 2~3d in darkness at 25℃. When the leaf disk fringes bulged, they then cultured on the selective medium (differentiation medium with 30mg/L Km and 250mg/L ceftomine (Cef)). When buds were generated, both leaves and buds were transferred onto the differentiation media with 50mg/L Km, 250mg/L Cef and selected for a second time. The well-grown shoots (2~3cm in length) were excised carefully and transferred onto a rooting medium supplemented with 50mg/L Km and 250mg/L Cef.

2. 2. 4 PCR and Southern blot analysis

Genomic DNA was isolated from leaf explants using CTAB. The primer sequences used were: 5′-GATGTGATATCTCCACTGACGTAAG-3′ and 5′-ATCTGCTTGTTGG ATGCTACTG-3′. The PCR program included denaturation at 94℃ for 5min followed by 35 cycles of denaturation at 94℃ for 30s, annealing at 58℃ for 30s and synthesis at 72℃ for 1min and finally 1 cycle of 10min at

72℃. The products were separated by electrophoresis on a 1% (w/v) agarose gel, stained by ethidium bromide and visualized under ultraviolet light.

For Southern blot analysis, the *SPDS* gene probe was prepared using a DIG-11-dUTP labeled purified PCR production. Then total DNA (10μg) was respectively restricted with *Hind*III and *Bam*HI. The digested DNA was separated by electrophoresis on a 0.8% (w/v) agarose gel and transferred to a positively charged nylon membrane. After prehybridization at 42℃ for 30min, hybridization was carried out for 16~18h at 42℃. The membrane was washed and then an immunity test was carried out, following the manual of DIG DNA Labeling and Detection Kit (Roche).

2.2.5 Real-time quantitative PCR analysis

Total RNA was extracted by using the SV Total RNA Isolation System (Promega, USA) and was reverse transcripted by using a Reverse Transcription System (Promega, USA) following the instruction of the kit. The synthesized cDNA was stored at −20℃ for further use.

Measurement of the expression quantity of the *SPDS* gene adopted a relatively quantitative method, which fixed the quantity of the target gene (*SPDS*) and the reference gene (*ACTIN*) simultaneously and separately. After that, the relative quantities of the target gene to the reference gene were evaluated. PCR amplification was carried out in a total volume of 20μL, which included 0.5μL SYBR Green 1 on the Opticon 2(M J RESEARCH INCORPORATED). Every reaction was repeated three times. The specialty of this amplification production is the reference to a melting curve. After the experiment, the data were analyzed with the Option Monitor Analysis Software.

3 Results and discussion

3.1 Determination of optimal concentration of Km

Given the fact that the *SPDS* gene was bonded with NPT II, which is kanamycin-resistant, and given that the exogenous gene had been transferred into the *P. tomentosa* genome, the shoots would have grown onto the medium with an optimal concentration of Km. However various plants adapt to different concentrations of Km, so it is necessary to acknowledge the critical concentration of Km on differentiation and rooting.

As shown in Table 1, the concentration of Km clearly affected leaf differentiation and the rooting of adventitious shoots. Even the lowest concentration of Km could restrain leaf differentiation. On the control medium without Km supplement, the regeneration ratio reached 93.33%, while on the medium with 10mg/L Km, the explants regeneration ratio fell to 36.67%. With an increase in the Km concentration, the regeneration ratio decreased sharply. When the Km concentration reached 30mg/L, leaf differentiation no longer occurred, nor were calli formed. Furthermore, the color of the leaves turned yellow. The results demonstrate that 30mg/L Km was the optimal concentration to obtain transformed cells and tissues efficiently.

Nevertheless, the rooting of the adventitious shoots of *P. tomentosa* behaved more sensitively to Km concentration than did leaf differentiation. On the control medium without kanamycin, sterile shoots grew healthily and the roots appeared very stocky on all shoots. On the other hand, the addition of 10mg/L Km resulted in very few plants rooting; the roots turned out to be very short and thin, without any side roots. When the Km concentration increased to 20mg/L, the adventitious shoots did not root or grow at all and some leaves even turned white. Therefore, the present study demonstrated that 10mg/L Km was the optimal concentration for rooting of the strain of *P. tomentosa* used by us.

3.2 Generation of transgenic plants

A highly efficient regeneration system is the foundation of a successful genetic transformation. It can be seen from our experiment that the 93.33% differentiation ratio and 100% rooting ratio of *P. tomentosa* indicate that we have developed a mature and efficient system. Before the transformation experiments, the optimal concentration of kanamycin was determined for selecting transformed plants. The putative transgenic buds were regenerated on a co-cultivation medium with 30mg/L Km about 30d later. They were subcultured one time per week. Roots developed after 40~50d, which was about 15~20d later, compared with the rooting time of untransformed control explants. A total of fifteen Km-resistant *in vitro* plantlets with roots were obtained.

In order to develop a highly efficient genetic transformation system, this study improved the traditional transferring procedure. The details are as follows:

1) With the experience of forerunners as reference, we adopted a liquid bacterium (A_{600} = 0.2~0.4) for infection to make the agrobacterium at the logarithm growth period with the strongest activity (Hao et al., 1999). We ensured that the liquid bacterium could fully infect the cut cells.

2) We strictly controlled the co-cultivate time. If the infecting time is too short, the agrobaterium will not have enough time to infect sufficiently the cut cells, which would decrease the transfer ratio. In contrast, if the infection time is too long, the agrobacterium could overgrow, which would lead to the death of leaves. We considered the optimum co-cultivate timing as the time when the edge of leaves began to present an obviously macroscopic bacterium culture. Zhou et al. (1997) also considered the same method.

3) Our research adopted the grads selection method. Since the high concentration kanamycin plays a certain restraining role to explant differentiation, we first selected antibiotics from low concentrations, in order to prohibit some transfer genetic plants from being killed, if the concentration were too high. After that, we strictly selected them through 50mg/L Km, eliminated artificial positive adventitious shoots. This method both enhanced the efficiency of leaf differentiation and also improved the transfer frequency.

Table 1 Effects of different Km concentrations on differentiation of adventitious shoots and formation of adventitious roots

Km (mg/L)	Regeneration ratio (%)	Regeneration state	Rooting ratio (%)	Rooting state
0	93. 33(28/30)	Green leaf, a great deal of shoots	100. 00(30/30)	Grown well
10	36. 67(11/30)	Dark green leaf, adventitious bud appeared in part of the shoots	16. 67(5/30)	Grow nornmlly, rooting initiated in part of the shoots
20	6. 67(2/30)	Light green leaf, calli formed at cut surface, bud scarcely appeared	0	No growth, no rooting
30	0	Light yellow leaf, not differentiating	0	With light yellow leaves, no rooting

3. 3 Molecular characterization

Total DNA was isolated from fifteen Km-resistant plants and un-transgenic plants. Plasmid pBI121 with the Mdspds gene insert was used as a positive control and the Un-transgenic explants as negative control. After PCR amplification, four plants out of 15 were positive.

As shown in Fig. 1, a clear band of 1200bp was revealed in four transgenic strains and the positive control plasmid; there was no amplification in the negative control, demonstrating that Mdspds was integrated into the genome of *P. tomentosa*.

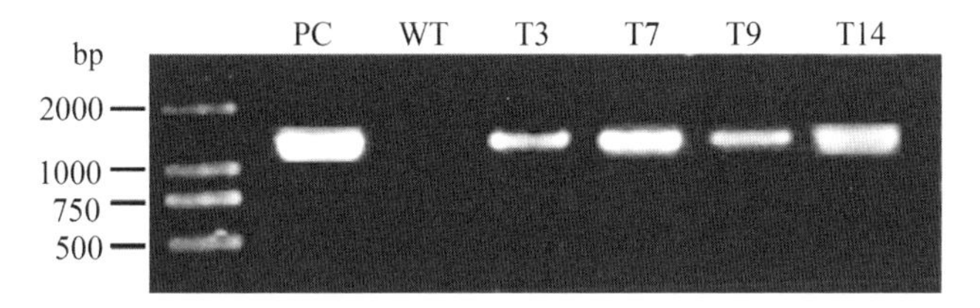

Fig. 1 PCR amplification of four positive plants detected by electrophoresis.

Lane 1: marker DL2000; Lane 2: positive control (plasmid);
Lane 3: negative control (wild type); Lanes 4~7: transform plants.

To test further whether these PCR-positive plants were true transgenic clones, the plants were also analyzed by Southern blotting analysis. Total DNA of these positive clones was digested overnight with the restriction enzymes *Bam*H I and *Hin*d III at 37℃ and was used in subsequent Southern hybridization experiments. The agarose gel electrophoresis revealed a well-distributed smear for all the digested samples, indicating that the genome DNA was fully digested by enzymes (Fig. 2). The Southern blotting result of clone T3 is shown in Fig. 3. No bands were detected in the non-transformed control plants, whereas one band was observed in the transgenic plants T3 digested with *Bam*H I and *Hin*d III, respectively, indicating that only one copy of *SPDS* gene has been integrated in the poplar genome.

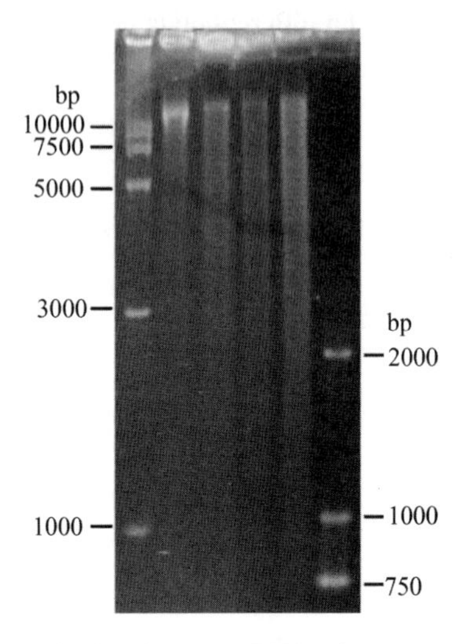

Fig. 2　Genome DNA digested with restriction enzyme

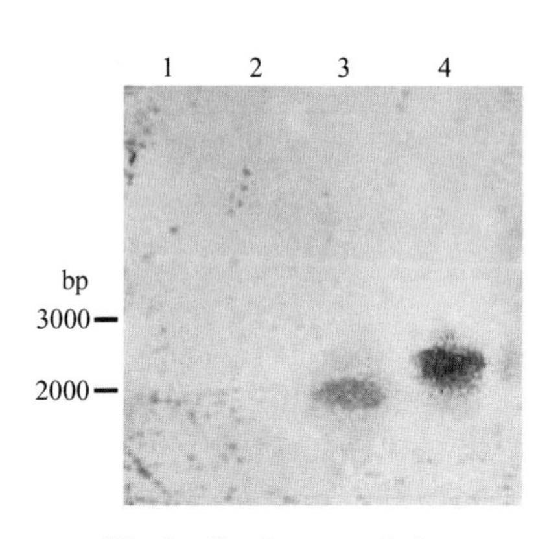

Fig. 3　Southern analysis.

Lanes 1～2: control (un-transformed plants); Lanes 3～4: transform plants T3.

3.4　Real-time quantitative PCR

To reveal further the expression levels of the exogenous gene Mdspds in transgenic clones, realtime quantitative PCR was adopted to conduct a relatively quantitative analysis (Fig. 4). Reverse PCR (RT-PCR) is a highly special and sensitive method to detect genetic expression. This method is able to distinguish precisely the difference between gene expressions, through comparing the PCR production in the exponential growth period. The housekeeping gene ACTIN was used as a reference gene, with relatively invariable expression levels. The target gene and reference gene were simultaneously detected in real time and then the relative quantities of the target gene were evaluated against the reference gene. The melting curves obtained had single apices, indicating that there was no noise fluorescence. Its quantity was fixed precisely in our research.

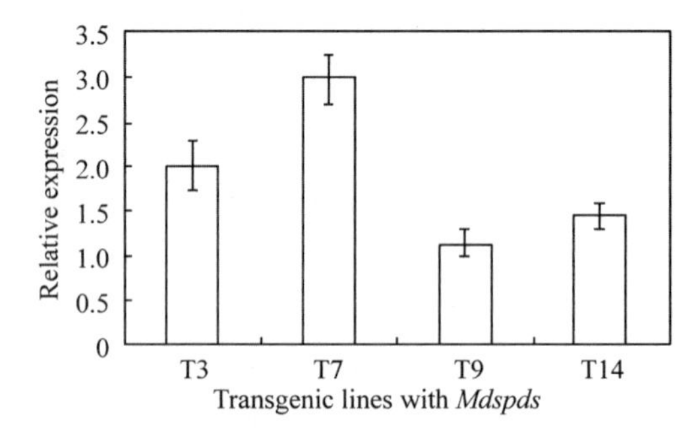

Fig. 4　Relative expression of *Mdspds* gene in transgenic *P. tomentosa*

We experimented with a series of molecular tests on transgenic plants and *Mdspds* gene expression quantity detection, which established a stable foundation for future anti-salt examination.

Functional Analysis of 5′ Untranslated Region of a TIR-NBS-encoding Gene from Triploid White Poplar*

Abstract Genome-wide analyses have identified a set of TIR-NBS-encoding genes in plants. However, the molecular mechanism underlying the expression of these genes is still unknown. In this study, we presented a TIR-NBS-encoding gene, *PtDrl*02, that displayed a low level of tissue specific expression in a triploid white poplar [(*Populus tomentosa*×*P. bolleana*)×*P. tomentosa*], and analyzed the effects of the 5′ untranslated region (UTR) on gene expression. The 5′ UTR sequence repressed the reporter activity of β-glucuronidase (GUS) gene under *PtDrl*02 promoter by 113.5-fold with a staining ratio of 2.97% in the transgenic tobacco plants. Quantitative RT-PCR assays revealed that the 5′ UTR sequence decreased the transcript level of the *GUS* reporter gene by 13.3-fold, implying a regulatory role of 5′ UTR in transcription and/or mRNA destabilization. The comparison of GUS activity with the transcript abundance indicated that the 5′ UTR sequence decreased the translation efficiency of target gene by 88.3%. Additionally, the analysis of the transgenic P-985/UTRΔ/GUS plants showed that both the exon1 sequence and the leading intron within the 5′ UTR region were responsible for the regulation of gene expression. Our results suggested a negative effect of the 5′ UTR of *PtDrl*02 gene on gene expression.

Keywords TIR, NBS, 5′ UTR, regulation, intron, poplar

Introduction

Plants have developed an effective innate immune system to protect them from invading pathogens. One of the most important components of the immune system is the disease resistance (*R*) proteins which can specifically recognize the pathogen-derived elicitor, and subsequently trigger the defense responses, including localized cell death (the hypersensitive response) and systemic acquired resistance (Chisholm et al., 2006; Kohler et al., 2008). These R proteins, in the majority of the cases, are produced by the NBS-LRR class of *R* genes that comprise one of the largest gene families in plant genomes (Meyers et al., 2003; Zhou et al., 2004; Ameline-Torregrosa et al., 2008; Kohler et al., 2008; Yang et al., 2008). The NBS-LRR class of genes can be further divided into two subgroup including TIR-NBS-LRR and non-TIR-NBS-LRR based on their N-terminal region whether or not homology to the Toll/interleukin-1 receptor (TIR) domain (Mey-

* 本文原载 *Molecular Genetics and Genomics*, 2009, 282(4): 381-394, 与郑会全、林善枝、张谦和雷杨合作发表。

ers et al. , 1999). In addition to the contribution of the NBS-LRR genes, numerous of studies also addressed that a certain amount of NBS-LRR-related genes/proteins may coordinately play a pivotal role with the NBS-LRR *R* genes/proteins in the plant defense system. Of which the TIR-NBS class of genes is the most impressive in that they may encoded a set of adapter molecules interacting with TIR-NBS-LRR *R* proteins as well as downstream signaling components during plant defense response (Meyers et al. , 2002; Tan et al. , 2007; Burch-Smith and Dinesh-Kumar, 2007).

Genome-wide analysis have identified 21 TIR-NBS-encoding genes in *Arabidopsis thaliana* (Meyers et al. , 2002; Meyers et al. , 2003), 13 in *Populus trichocarpa* (Tuskan et al. , 2006; Kohler et al. , 2008), 14 in *Vitis vinifera* (Yang et al. , 2008), and 28 in *Medicago truncatula* (Ameline-Torregrosa et al. , 2008), suggesting that the TIR-NBS-encoding genes are prevalent in the plant kingdom. Intriguingly, it was observed that these TIR-NBS-encoding genes, in most of the case, were expressed at low levels or even absent in unchallenged plants (Meyers et al. , 2002; Tan et al. , 2007; Kohler et al. , 2008), which strikingly resemble to that of the majority of the cloned plant *R* genes and the NBS-LRR-encoding members in plant genome (Grant et al. , 1995; Dixon et al. , 1996; Salmeron et al. , 1996; Hammond-Kosack and Jones, 1997; Parker et al. , 1997; Milligan et al. , 1998; Mes et al. , 2000; Shen et al. , 2002; McHale et al. , 2006; Tan et al. , 2007; Kohler et al. , 2008). However, the molecular mechanism underlying the low expression pattern of TIR-NBS-encoding genes is still unknown.

The 5′ untranslated region (5′ UTR) is now recognized as a pivotal intramolecular module that can directly impact the gene expression in eukaryotes. The regulatory effect can be positive or negative, which mainly depends on the sequence features of the 5′ UTR such as the presence of regulatory *cis*-elements, intron sequence, upstream open reading frame (uORF) or uATG, and the secondary structures. Commonly, the secondary structures of the 5′ UTR have been characterized as down-regulators of gene expression. They are capable of inhibiting gene transcription process (Curie and McCormick, 1997), accelerating mRNA degradation rate (Muhlrad et al. , 1995; Stefanovic et al. , 2000; Cannons and Cannon, 2002), or reducing translation efficiency (Kozak, 1989; Vega Laso et al. , 1993; Muhlrad et al. , 1995; Hess and Duncan, 1996; Wood et al. , 1996; McCarthy, 1998; Hua et al. , 2001; Myers et al. , 2004; Brenet et al. , 2006; Bunimov et al. , 2007).

Chinese white poplar (*P. tomentosa* Carr.) is a very important indigenous tree species in China and is widely used for forestation and landscape enhancement in northern China (Zhu and Zhang, 1997). The triploid white poplar [(*P. tomentosa*×*P. bolleana*)×*P. tomentosa*] was developed from the diploid white poplar by our laboratory in the 1990s (Zhang et al. , 1992 and 1997). It has better growth performance, improved disease resistance, higher photosynthesis rate and wood quality (Zhang et al. , 2005 and 2008) when compared with the diploid poplar. In addition, this poplar has 57 chromosomes (Zhu and Zhang, 1997), providing more genetic resources for gene isolation (Zhang et al. , 2008). In previous study, we have isolated a total of 59

NBS-encoding resistance gene analogs (RGAs) from this triploid white poplar (Zhang et al., 2008). Herein, we obtained a full length cDNA sequence of a RGA (DQ324288) that putatively encoded a TIR-NBS protein, designated as *PtDrl*02. The *PtDrl*02 gene displayed low level of expression when compared to the endogenous *ACTIN* gene, and showed tissue-specific expression pattern. Sequence analysis indicated that the *PtDrl*02 gene contained a relative long 5′ UTR sequence predicted to have complex structures at both the DNA and RNA level. To characterize the potential effect of the *PtDrl*02 5′ UTR on gene expression, we conducted transient and stable expression assays in this study, and demonstrated that the 5′ UTR played a negative role in modulating gene expression at both the transcription level and the translation level.

Materials and methods

Plant materials and growth conditions

A triploid white poplar clone 'L9' [(*Populus tomentosa* × *P. bolleana*)×*P. tomentosa*] (Zhang et al., 1992 and 1997; Zhang et al., 2008) was used as source material for this study. The poplars were propagated by cutting and raised in the pots in a nursery at Beijing Forestry University under natural daylight condition, and the 18-month-old plants with identical growth status were selected for sample collection. The second to fourth apical leaves (AL), mature leaves (ML) on the middle position of the poplar plantlets and petioles (P) as well as green young stems (YS), barks (B) and roots (R) were collected for either gene isolation or gene expression analysis.

Tissue-cultured tobacco (*Nicotiana tabacum* cv. W38) plants were raised on Murashige-Skoog (MS) medium (Murashige and Skoog, 1962) supplemented with 30g/L sucrose, 5.5g/L agar, 0.1mg/L NAA and adjusted to pH 5.8, and maintained in a growth chamber with a 16/8h light/dark photoperiod at 23℃. Leaf samples and green young stems as well as roots were dissected from the tissue-cultured tobaccos. These detached samples were used for *Agrobacterium*-mediated transient expression assay. The fully developed leaves harvested from the tissue-cultured tobacco were used for genetic transformation.

Total RNA extraction and first-strand cDNA synthesis

Total RNA was extracted from the poplar or tobacco sample using the SV Total RNA Isolation System (Promega, Madison, WI, USA) and treated with an RNAse-free DNAse I to eliminate the residual genomic DNA according to the manufacturer's instructions (Promega, Madison, WI, USA). The total RNA was then evaluated following agarose gel electrophoresis and spectrophotometrical analysis. About 300 ng of total RNA was used to generate the first strand cDNA in a total volume of 20μL. The reaction was performed using a SuperScript™ Ⅲ Platinum Two-Step qRT-PCR Kit with SYBR Green (Invitrogen Corporation, Carlsbad, CA, USA) following the manufacturer's protocol.

Isolation of *PtDrl*02 full-length cDNA and its genomic clone

Depending on DQ324288 the full-length cDNA sequence was obtained by rapid amplification of cDNA end (RACE) methods with a poplar leaf RNA as templates. Both 5′ and 3′ RACE were conducted using a BD SMART™ RACE cDNA Amplification Kit (Clontech, Palo Alto, CA, USA), according to the manufacturer's instructions. The gene-specific primers employed were as follows: 5′ RACE primary primer (GSP-1), 5′-TTCAGCTGTTCCGGATGAGCCACAT-3′, and nest primer (NGSP-1), 5′-GACGAA TTCGCTCTTTGATCAG-3′; 3′ RACE primary primer (GSP-2), 5′-GGATGCCAGGAATAG GAAAGACGACA-3′, and nest primer (NGSP-2), 5′-CATCCGGAACAGCTGAATGCATT G-3′. The RACE-PCR products were cloned into the pGEM-T easy vector (Promega, Madison, WI, USA) and sequenced. The identified 5′ and 3′ end sequences of DQ324288 were then assembled *in silico* using DNAMAN (Lynnon Biosoft, USA). Based on that, the full-length cDNA sequence was finally cloned from a poplar leaf cDNA pool and verified by sequencing.

To isolate a full length *PtDrl*02 genomic clone, poplar genomic DNA was extracted from the mature leaves using a Plant Genomic DNA Kit (TIANGEN, Beijing, P. R. China) and subjected to PCR amplification procedures. The genomic sequence corresponding to the cDNA was amplified using primer pairs of 5′-ACGCGGGGACCCCATTTCTC-3′ and 5′-AATAAAGGGCAAAATAATG-3′. The 5′ flanking region of *PtDrl*02 gene was obtained as previously described (Zheng et al., 2007).

Semi-quantitative RT-PCR

Semi-quantitative RT-PCR was carried out in a 50μL reaction system containing 3μL of cDNA, 1μL forward primer (10μmol/L), 1μL reverse primer (10μmol/L), 5μL 10×*LA Taq* polymerase buffer (Mg^{2+} Plus), 0.5μL *LA Taq* DNA polymerase (5U/μL; TaKaRa, Dalian, Liaoning, P. R. China), 1μL dNTP (10mmol/L), and 38.5μL double distilled water. The thermal cycling conditions were: (a) predenaturing at 94℃ for 5min; (b) denaturing at 94℃ for 30s, primer annealing at 56℃ for 30s and extension at 72℃ for 40s, repeat for 35 cycles; (c) final extension at 72℃ for 10min. The Poplar *ACTIN* gene (Accession: AY261523.1) was selected as the endogenous control gene. The following primers were used: *PtDrl*02 forward primer, 5′-CCCATTTCTCCGATTCAATAACTT-3′, and reverse primer, 5′-CTTTTCTCCACTCCTTCACCAACT-3′; *ACTIN* forward primer, 5′-CTCCATCATGAA ATGCGATG-3′, and reverse primer, 5′-TTGGGGCTAGTGCTGAGATT-3′. Gene-specific PCR products were observed on 1% agarose gel stained with ethidium bromide using the Gene Genius Imaging System and quantified with the GeneTool software (Gene Company Limited). Each PCR assay for the same tissue sample was carried out for three biological replicates and each replicates two technological repeats in separate experiments.

Real-time quantitative RT-PCR

Real-time quantitative RT-PCR was performed using the SuperScript™ Ⅲ Platinum® Two-Step qRT-PCR Kit with SYBR® Green (Invitrogen Corporation, Carlsbad, CA, USA) in an Opticon2 thermocycler (MJ Research, Bio-Rad) following the manufacturer's instructions. Three microlitre of cDNA and 1μL of 10μM primer solution (forward and reverse, respectively) were used for each real-time PCR in the 50μL reaction systems. The thermal cycling conditions were the same for all primers: initial incubation 50℃ for 2min, predenaturing at 94℃ for 5min, followed by 40 cycles of 94℃ for 30s, 58℃ for 30s, 72℃ for 30s and plate read step, then final extension at 72℃ for 10min and product melting curve 70~95℃. The generated melting curve was employed as a significant parameter to check the specificity of the amplified fragment. To quantify expression level of target gene (*GUS*), the tobacco *ACTIN* gene(Accession: U60491)was selected as endogenous control gene because of the relative stability of its signal among replicates and between genotypes. Real time primers were as follows: *GUS* forward primer, 5′-ATCCGGTCAGTG GCAGTGAAGG-3′, and reverse primer, 5′-CAGCGTAAGGGTAATGCGAG-3′; *ACTIN* forward primer, 5′-CTGCTGGAATTCACGAAACA-3′, and reverse primer, 5′-GCCACC ACCTTGATCTTCAT-3′. All reactions were carried out in triplicate, and the generated real-time data were analyzed using the Opticon Monitor Analysis Software 3. 1 tool.

Construction of GUS-reporter vectors

For construction of P-985/UTR/GUS, P-985/UTRΔ/GUS and P-985/GUS, nucleotide sequences of P-985/UTR, P-985/UTRΔ and P-985 were PCR-amplified from the obtained *PtDrl*02 genomic clone using a common forward primer containing the *Xho* I restriction site (underlined), 5′-CCGCTCGAGTTAATACCTTCCCTTTACGCAACCA-3′ (Pf1), and reverse primers containing the *Bam*H I restriction sites (underlined), 5′-CGGGATCCGCAGA ATTTAAAGAACAATCAATTAG-3′ (Pr03), 5′-CGGGATCCCTGTATATCAAAGAGTTA GCTATG-3′ (Pr02), 5′-CGGGATCCAAATCGAATCAGAAGTTCATGTG-3′ (Pr01), respectively. The amplified product was then digested with *Xho* I and *Bam*H I (Promega, Madison, WI, USA) and purified with TIANquick Midi Purification Kit(TIANGEN, Beijing, P. R. China)and then fused to the *β*-glucuronidase (GUS) reporter gene of the modified pBI121 vector (Clontech, USA) harboring an additional *Xho* I site immediately downstream of the *Hin*d III site, which was previously digested with *Xho* I and *Bam*H I to release the 35S promoter. The resulting recombined vectors were then verified by PCR detection, restriction fragment analysis and sequencing. For biological transformation, the P-985/UTR/GUS, P-985/UTRΔ/GUS and P-985/GUS constructs were finally introduced into *Agrobacterium tumefaciens* strain EHA105 via the freezing-thaw method (An, 1987).

Agrobacterium-mediated transient assay

A positive colony of *Agrobacterium* carrying the P-985/UTR/GUS construct was inoculated into 60mL of fresh YEB medium (5g beef extract, 1g yeast extract, 5g peptone, 5g sucrose, 4g $MgSO_4 \cdot 7H_2O$, 15g agar/H_2O/L) supplemented with rifampicin (50mg/L) and kanamicin (50mg/L), at 28℃ for 24h. *Agrobacterium* cells were collected by centrifugation for 15min at 4000r/min, resuspended in MS medium and adjusted the OD_{600} to 0.8 for tissue transformation. Leaf segments, stem pieces (8 ~ 15mm in length) with partial petiole, and roots (8 ~ 15mm in length) were dissected from the tobacco plants and immediately dipped into the *Agrobacterium*/MS suspension, incubated for 10min at 28℃, and then blotted dry with sterile filter papers and cultured in petri-dishes containing co-cultivation medium (MS medium) at 23℃ under 16/8h light/dark photoperiod for 48h. For histochemical analysis of GUS activity, the transformed samples were rinsed for three times with the sterile H_2O supplemented with Carbenicillin (500mg/L) to remove the excess *Agrobacterium* and then subjected to the X-Gluc solution as described by Jefferson et al., (1987). After overnight incubation at 37℃, stained samples were bleached with 70% (v/v) ethanol and observed directly or with OLYMPUS BX51 microscope. Tissue samples transformed with the binary vector pCAMBIA-1305.2 (www.cambia.org) by the same method were used as positive control, and the untransformed samples were used as negative control. Because both of the P-985/UTR/GUS construct and the pCAMBIA-1305.2 plasmid contain a leading intron sequence, the false positive staining of GUS activity is avoid although the expression may be uneven.

Plant transformation and synchronization

For genetic transformation of tobacco, a leaf disc transformation method (Horsch et al., 1985) was employed. The putative transgenic plantlets resistant to kanamycin were further confirmed by PCR as well as preliminary GUS staining (see the *Agrobacterium*-mediated transient assay except for the rinse step). The verified transgenic tobaccos were then propagated and synchronized using vegetative stem cutting containing axillary bud from primary transformants in the MS medium. One-month-old *in vitro* grown plants were used for further experiments.

Fluorometric GUS assay

Fluorometric GUS assay was performed as described by Jefferson et al. (1987). The entire stem tissue of tobacco was ground in liquid nitrogen and homogenized in freshly prepared GUS extraction buffer (50mM NaH_2PO_4, pH 7.0, 10mM EDTA, 0.1% Triton X-100, 0.1% (w/v) sodium laurylsarcosine, 10mM β-mercaptoethanol). After centrifuging for 10min at 12000rpm at 4℃, GUS activity of the supernatant was determined using 4-methylumbelliferyl glucuronide (4-MUG) as substrate. Fluorescence of GUS-catalysed hydrolysis reaction product, 4-methylumbelliferone (4-MU), was measured with the TECAN GENios system. Protein concentration in the supernatant was assessed by the Bradford (1976) method, using bovine serum albumin (BSA) as a standard. GUS activity was normalized to protein concentration of each supernatant extract and calculated as pmol of 4-MU per milligram of soluble protein perminute.

Computer-assisted analysis

Sequence alignment analyses were carried out with the BLAST programs in NCBI (http: //www. ncbi. nlm. nih. gov/) (Altschul et al. , 1990) and the search tool in Pfam (http: //pfam. sanger. ac. uk/) (Finn et al. , 2006); The CLUSTAL X program (Thompson et al. , 1997) was used for multiple alignments of amino acid sequences. The online programs, including PLACE (http: //www. dna. affrc. go. jp/PLACE/signalscan. html) (Higo et al. , 1999), PlantCARE (http: //bioinformatics. psb. ugent. be/webtools/plantcare/html/) (Lescot et al. , 2002), NSITE-PL and ScanWM-P (Softberry, http: //linux1. softberry. com/berry. phtml) as well as UTRScan (http: //www. ba. itb. cnr. it/BIG/UTRScan/) (Pesole and Liuni, 1999), were employed to predict the *cis*-elements located in either the promoter region or 5′ UTR of the gene. The Mfold RNA/DNA folding program available through Rensselaer bioinformatics web server (http: //www. bioinfo. rpi. edu/applications/mfold/) was used to predict 5′ UTR secondary structures of the gene (Zuker, 2003).

The *PtDrl*02-related genes were predicted following a BLAST search of draft genomic sequence of *P. trichocarpa* at the JGI *Populus* Genome database (http: //genome. jgi-psf. org/Poptr1/Poptr1. home. html). The *PtDrl*02 homology sequences available in GenBank database (NCBI) were used as the query sequences, and the resulted highly hit genomic regions (approximately 10-kb sequences) were respectively subjected to gene prediction FGENESH program (Softberry) to identify the putative genes. The full length genes starting from a transcriptional start site (TSS) while ending with a translational stop codon were selected for further investigation. The BLASTX program in NCBI (Altschul et al. , 1990) and the Pfam search tool (Finn et al. , 2006) were used to identify whether the putative gene encodes a TIR-NBS protein similar to PtDrl02.

Results

Characterization of *PtDrl*02 gene

Based on the sequence of DQ324288, we obtained a full-length cDNA sequence from cDNA pool of the unchallenged poplar leaves, and designated as *Populus tomentosa* disease resistance-like (*PtDrl*02) gene. The cDNA sequence was predicted to have an open reading frame (ORF) (1929-bp) that encoded a protein of 642 amino acids with a molecular weight of approximately 73 kDa (Fig. 1). Blast analysis revealed that both the *PtDrl*02 cDNA sequence and its putative amino acid sequence were very similar to the TIR-NBS-like sequences in GenBank (NCBI) (Supplemental Table 1 and 2), and the *PtDrl*02 has two complete protein domains: TIR and NBS (Fig. 1). The detailed motifs of the present TIR and NBS domain of *PtDrl*02 were further identified by multiple alignment analysis comparing to six of the represented *Populus* TIR-NBS proteins. These results suggested that the *PtDrl*02 gene was a member of the *Populus* TIR-NBS-encoding gene family.

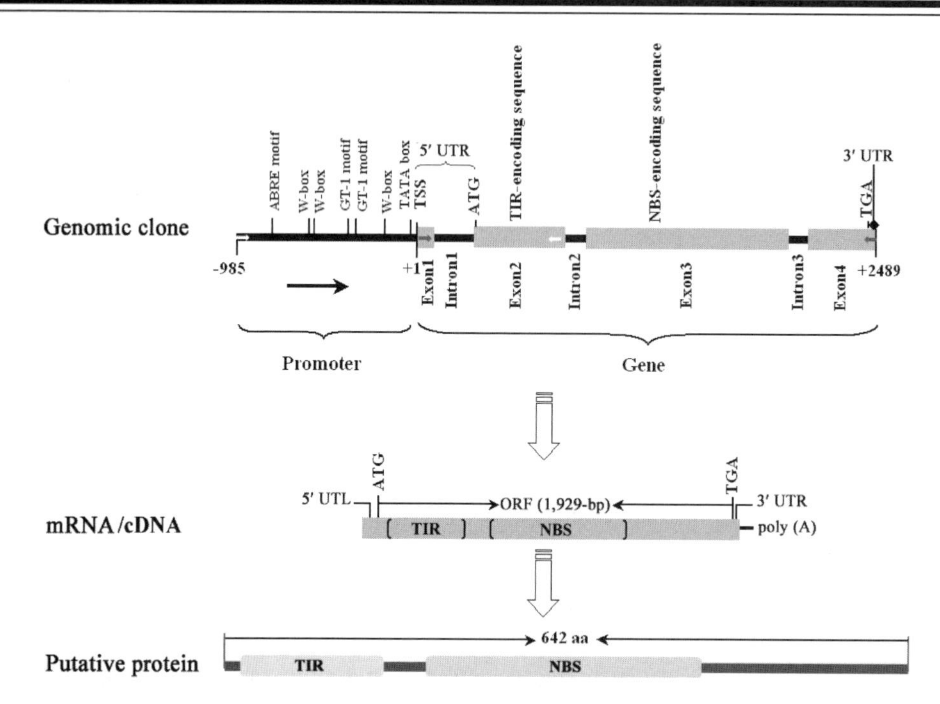

Fig. 1　Schematic representation of *PtDrl*02 gene and its deduced protein.

The genomic, mRNA and predicted protein components are shown in detail. TSS, transcriptional start site; 5′ UTR, 5′ untranslated region; 3′ UTR, 3′ untranslated region; 5′ UTL, 5′ untranslated leader; ORF, open reading frame; TIR, Toll/interleukin-1 receptor domain; NBS, nucleotide binding site domain. The locations of the primer pairs used to amplify the genomic fragment and promoter sequence of *PtDrl*02 gene are marked with red and white arrow pairs, respectively. The gray lines within the putative protein structure of PtDrl02 represent the non-TIR and NBS regions.

The basal expression patterns of *PtDrl*02 gene were first examined in 18-month-old poplar plantlets using a sensitive semi-quantitative RT-PCR method (see materials and methods). As shown in Fig. 2A, the *PtDrl*02 gene was expressed in apical leaves (AL), mature leaves (ML), petioles (P), and the green young stems (YS), while no transcript was detected in the barks (B) and roots (R). Notably, transcript abundance of *PtDrl*02 gene in these tissues was significantly lower than that of the reference *ACTIN* gene(Fig. 2B).

To study the mechanism underlying the low expression pattern of *PtDrl*02 gene, we isolated and analyzed the *PtDrl*02 genomic fragment (Fig. 1). The *PtDrl*02 gene consisted of four exons that were separated by three typical plant introns (AT rich, > 60%; donor/acceptor boundary, GT/AG). The promoter, which is a major module of the *PtDrl*02 gene, was 986-bp long, and predicted to have a canonical TATA box (TGTCTATATA) and a cluster of defense-related *cis*-elements. No any negative and/or tissue-specific elements were found within the *PtDrl*02 promoter as well as the following 5′ untranslated region (309-bp) (5′ UTR) that comprised the exon1 sequence (the 5′ untranslated leader of mRNA) and the leading intron (Intron1). However, it was observed that the 5′ UTR DNA sequence contained multiple stem-loop structures, and that the 5′ UTR RNA sequence could form a stable secondary structure with folding free energy (ΔG) of -45.10kcal mol^{-1}(Fig. 3). Similar secondary features were also found in the 5′ UTR sequence of another five poplar TIR-NBS-encoding genes (Fig. 4). These finding implied that the *PtDrl*02 5′ UTR sequence might have a regulatory role in gene expression.

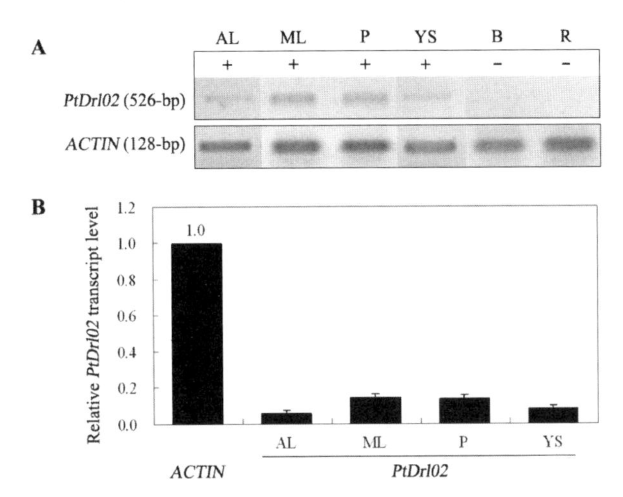

Fig. 2 Semi-quantitative RT-PCR analysis of the *PtDrl*02 gene in triploid white poplar.

A. RT-PCR analysis of the *PtDrl*02 gene expression in various tissues. AL, the second to fourth apical leaves; ML, mature leaves on the middle position of the poplar plantlets; P, petioles; YS, green young stems; B, barks; R, roots. +, detectable; −, undetectable. The poplar *ACTIN* (Accession: AY261523. 1) gene is selected as the endogenous control. Size of the PCR product is shown. B. Relative transcript level of the *PtDrl*02 gene. Expression level of *PtDrl*02 gene in panel A is quantified by referring to the *ACTIN* control gene that is arbitrarily set to 1. 0 for standardization. The expression data calculated by GeneTool software was corrected with the ratio 4. 109 because the size of the PCR product of *PtDrl*02 gene was rather different from that of the *ACTIN* gene. Each RT-PCR assay for the same tissue sample was carried out for three biological replicates and each for two technological repeats in separate experiments. The mean and standard deviation of the relative *PtDrl*02 transcript level in respective tissue are present.

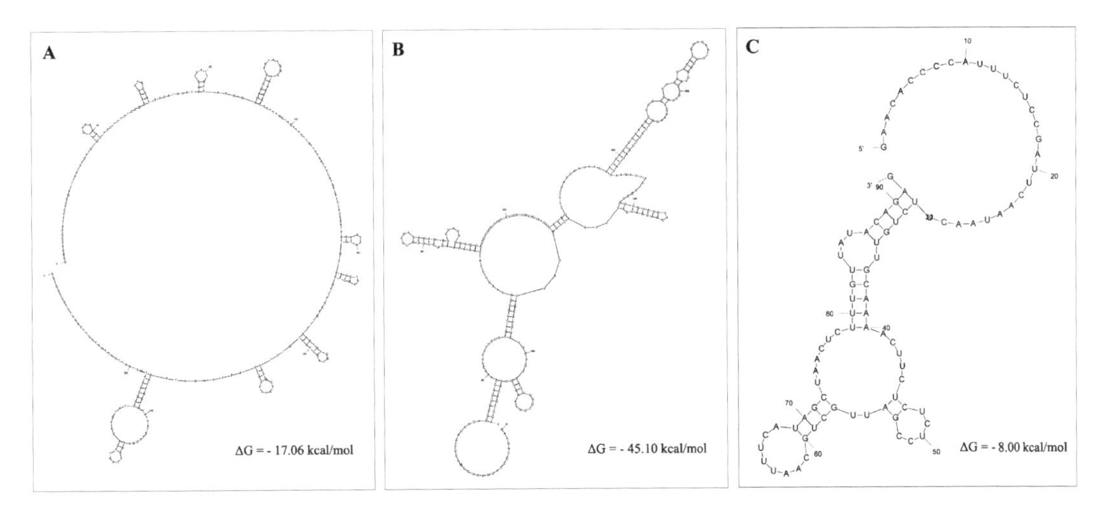

Fig. 3 Predicted secondary structure of the *PtDrl*02 5′ UTR.

A. The secondary structure of *PtDrl*02 5′ UTR DNA sequence; B. The secondary structure of *PtDrl*02 5′ UTR RNA sequence; C. RNA secondary structure of the exon1 sequence from *PtDrl*02 5′ UTR; The secondary structures were predicted using the Zuker Mfold algorithm (Zuker, 2003), and the free energy (ΔG) of respective structure is shown.

A

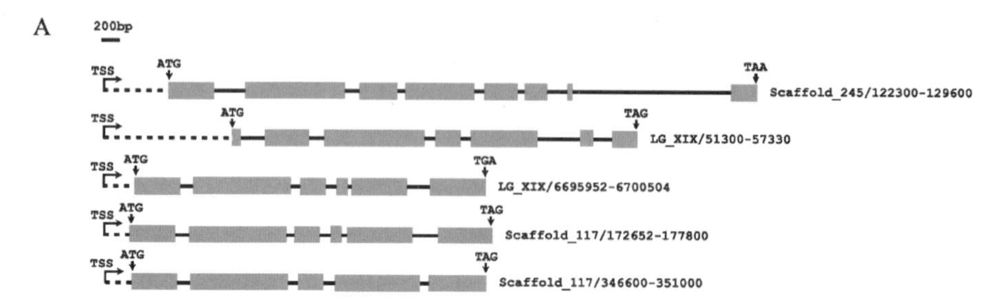

B

Gene	Scaffold_245/122300-129600		LG_XIX/51300-57330		LG_XIX/6695952-6700504		Scaffold_117/172652-177800		Scaffold_117/346600-351000	
	DNA	RNA	DNA	RNA	DNA	RNA	DNA	RNA	DNA	RNA
Folding structure of 5′ UTR										
Free energy (kcal/mol)	- 66.34	- 154.70	- 87.42	- 279.10	- 25.83	- 62.60	- 16.56	- 46.70	- 20.78	- 46.70
AT/AU content (%)	62.2		70.7		65.7		66.6		65.2	

Fig. 4 The 5′ UTR of five *PtDrl*02-related TIR-NBS-encoding genes in *Populus*.

A. Schematic representation of the genes, including Scaffold_ 245/122300-129600, LG_ XIX/51300-57330, LG_ XIX/6695952-6700504, Scaffold_ 117/172652-177800 and Scaffold_ 117/346600-351000. These genes were *in silico* predicted from the available *Populus* (*P. trichocarpa*) genome sequences in JGI database (Tuskan et al., 2006), which were identified to encode the TIR-NBS proteins homology to PtDrl02. The names of the genes represent their chromosome locations. The translational start codon (ATG) and stop codon (TAA/TAG/TGA) are shown. The coding exons and internal introns are indicated with the orange rectangles and black lines, respectively. The imaginary underlines represent the 5′ untranslated region (5′ UTR) of the genes, ranging from the transcriptional start site (TSS) to the translational start codon. Among the five *PtDrl*02-related TIR-NBS-encoding genes, Scaffold_ 117/346600-351000 is most homologous to the *PtDrl*02 gene with 91% nucleotide identities and 85% encoded protein homology, respectively.

B. Predicted DNA and RNA secondary structures of the 5′ UTRs. The secondary structures were predicted using the Zuker Mfold algorithm (Zuker 2003), and the free energy (ΔG) of each structure is shown. AT/AU content of respective 5′ UTR sequence is present.

Transient expression of chimeric *GUS* gene in tobacco tissues

To determine the effect of 5′ UTR sequence of *PtDrl*02 gene on gene expression, chimeric vector P-985/UTR/GUS (Fig. 5) that contains the full 5′ UTR sequence, the β-glucuronidase (GUS) gene, and the promoter sequence was preliminarily subjected to the *Agrobacterium*-mediated transient assay. As shown in Fig. 6, the GUS activity from P-985/UTR/GUS construct can be detected in the transformed tobacco leaves, stems and petioles, but not in the roots. This expression pattern in tissues was in agreement with that of the expression profile of *PtDrl*02 gene in triploid white poplar, suggesting that the *PtDrl*02 5′ UTR sequence, together with its own promoter, was responsible for the expression of *PtDrl*02 gene.

A

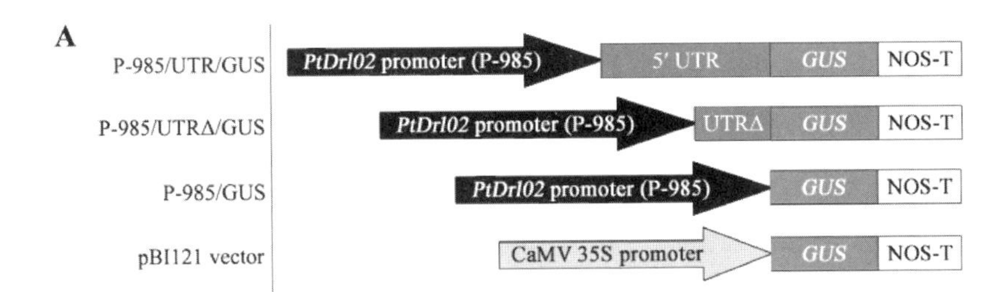

B

Transformant	Kan⁺(n)	PCR⁺(n)	GUS⁺(n)	GUS staining	G/P (%)
P-985/UTR	136	101	3	faint	2.97
P-985/UTRΔ	75	63	60	medium	95.24
P-985	70	61	59	high	96.72
CaMV 35S	65	60	60	strong	100

C

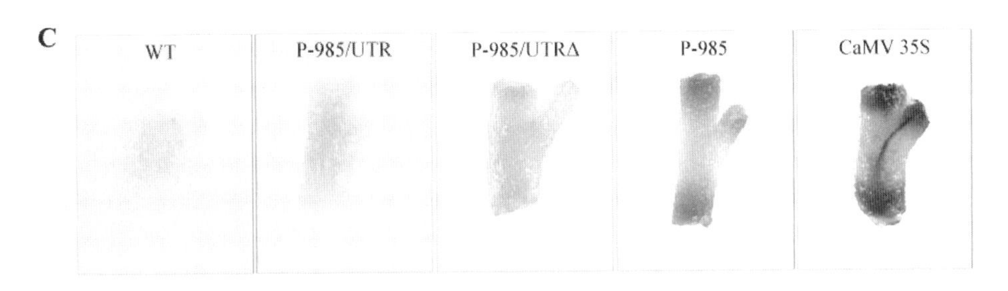

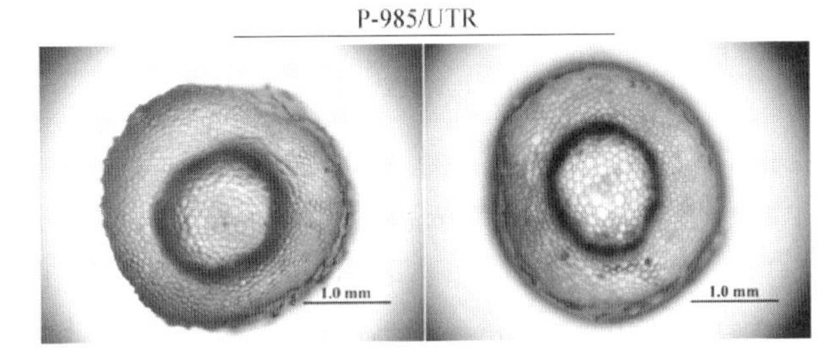

Fig. 5 Plant transformation.

A. Schematic diagrams of different GUS reporter constructs prepared for plant transformation. The different *PtDrl*02 regulatory sequences (P-985/UTR, P-985/UTRΔ and P-985) were inserted into the modified pBI121 vectors (see Materials and methods), replacing of the CaMV 35S promoter, to generate the P-985/UTR/GUS, P-985/UTRΔ/GUS and P-985/GUS construct, respectively. The pBI121 (CaMV 35S/GUS) vector was used as positive control.

B. Transformation of different GUS reporter constructs. Kan$^+$, kanamycin resistance; PCR$^+$, PCR detection with positive result; GUS$^+$, histochemical GUS staining with positive result; n, the total number of transformants examined; G/P, staining ratio relative to the PCR result in transgenic tobacco plants.

C. GUS staining of the stem tissues of transgenic tobacco plants. The transverse sections of the stem tissues from P-985/UTR plants are shown. The untransfomed tobacco plants (wild-type, WT) were used as negative control.

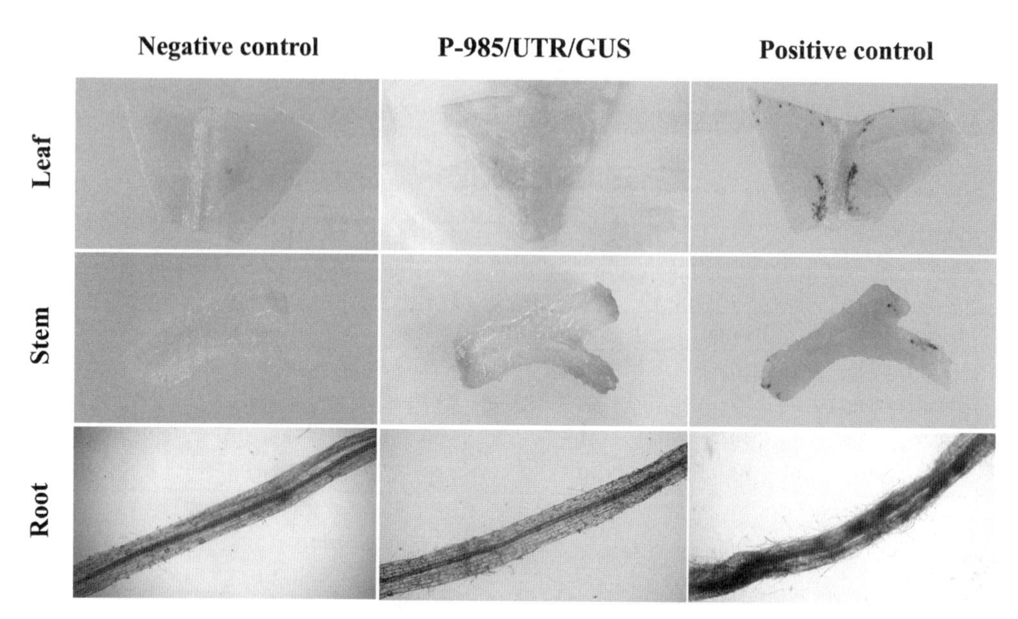

Fig. 6 Transient *GUS* expression in the dissected tobacco tissues.

The dissected tobacco tissues including the leaf segments and stem pieces with partial petiole as well as the roots were transformed with the P-985/UTR/GUS construct or the pCAMBIA-1305. 2 plasmid serving as positive control by using the *Agrobacterium*-mediated method. The untransformed tobacco samples were used as negative control. Transient *GUS* expression in respective tissue was determined by histochemical staining. The represent result, from four independent experiments, is shown.

The role of the *PtDrl*02 5′ UTR in regulating gene expression

The P-985/UTR/GUS construct was then stably transferred into tobacco plants, and a total of one hundred and one transgenic lines (P-985/UTR) were obtained. However, only three transgenic lines shown detectable (slight/faint) GUS activities and the expression was exclusively localized in the stems (Fig. 5). To determine whether the low level *GUS* gene expression was caused by the 5′ UTR sequence, we further constructed a P-985/GUS vector lacking the *PtDrl*02 5′ UTR sequence, and obtained a total of 59 P-985/GUS transgenic lines (P-985). These transgenic tobaccos exhibited an increase in GUS activities compared to the P-985/UTR plants (Fig. 5) and the expression of *GUS* gene was mainly localized in the leaf veins, petioles, stem tissues, and some root tips (data not shown) with the G/P ratio of 96.72%. The tissue distribution pattern of the GUS activities in P-985 plants was compatible with the results from RT-PCR analysis of native *PtDrl*02 gene, the transient GUS activities from the P-985/UTR/GUS construct, and the stable *GUS* expression (obscure in stems) in the P-985/UTR plants. These results indicated that the *PtDrl*02 5′ UTR is not responsible for the regulation of gene expression pattern in tissues, but plays a role in repressing the expression of target genes.

To elucidate the effect of the leading intron and exon1 sequence of *PtDrl*02 5′ UTR on gene expression, we also constructed a P-985/UTRΔ/GUS vector that just maintained the exon1 sequence, and generated P-985/UTRΔ plants via genetic transformation (Fig. 5). Most of the transgenic lines displayed an identical tissue expression pattern to P-985 plants. However, the GUS activities were weaker than that of the P-985 plants, but significantly stronger than that of P-985/UTR plants (Fig. 5).

Further quantitative analysis of GUS activities using fluorometric method revealed that the 5′ UTR sequence of *PtDrl*02 gene indeed reduced the GUS activities by up to 113.5-fold, and that the exon1 sequence caused about 5.5-fold reduction in GUS expression (Fig. 7). In contrast to exon1, the leading intron had more effect on gene expression, and resulted in an approximately 20.6-fold decrease in GUS activities.

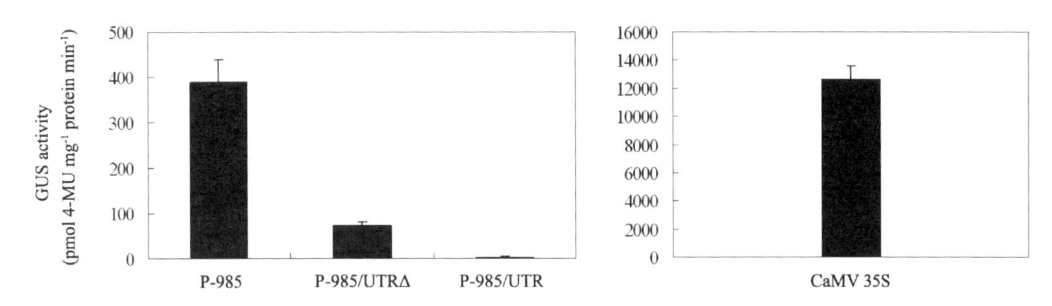

Fig. 7 GUS reporter activities in the stem tissues of different transgenic tobacco plants (P-985, P-985/UTRΔ and P-985/UTR). GUS activity from the CaMV 35S transformants is served as a comparison. Data are mean and standard deviation of ten transgenic lines.

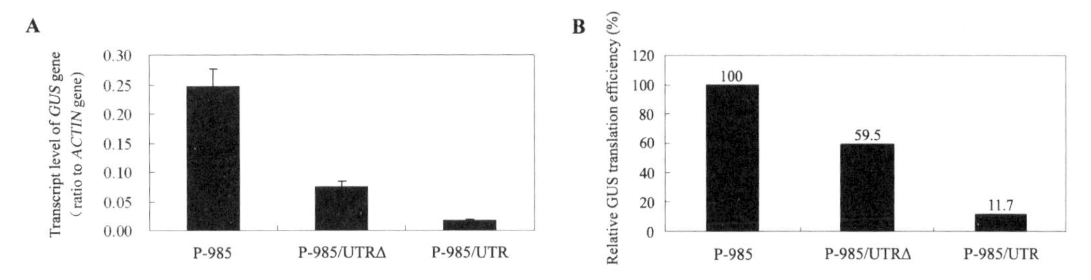

Fig. 8 Effect of the *PtDrl*02 5′ UTR and its derivative (UTRΔ) on mRNA accumulation and translation efficiency of *GUS* reporter gene expression in stem tissues from transgenic tobacco plants.

A. Quantitative analysis of the *GUS* transcript level relative to the *ACTIN* gene by Real time qRT-PCR assay. Data are mean and standard deviation of ten transgenic lines.

B. Effect of the *PtDrl*02 5′ UTR sequence on the translation efficiency of *GUS* reporter gene. The translation efficiency was evaluated by calculating the *GUS* enzymatic activity relative to the *GUS* mRNA level, and the value from the P-985 was arbitrarily set to 100 for standardization. Relative translation efficiency of the *GUS* reporter gene from the P-985/UTRΔ and P-985/UTR are shown.

The *PtDrl*02 5′ UTR represses the transcript level and translation efficiency of the *GUS* reporter gene

To investigate whether the *PtDrl*02 5′ UTR sequence repressed the gene expression at the transcription level, we measured the mRNA abundance of *GUS* gene in stem tissues of transgenic plants (P-985, P-985/UTRΔ and P-985/UTR) using semi-quantitative RT-PCR assay. P-985 plants displayed the strongest expression, while the P-985/UTR plants exhibited the lowest transcript abundance (data not shown). To have a more accurate estimate of the reduction at the mRNA level, we carried out qRT-PCR to quantify the *GUS* mRNA accumulation in transgenic plants. The *PtDrl*02 5′ UTR, the exon1 and the leading intron led to a decrease in transcript abundance of *GUS* reporter gene by 13. 3-fold, 3. 3-fold and 4-fold, respectively (Fig. 8A).

The translation efficiency of the *GUS* reporter gene in transgenic tobacco plants was measured by calculating the relative GUS activities to corresponding mRNA level. As shown in Fig. 8 B, the *PtDrl*02 5′ UTR, the exon1, and the leading intron caused decrease in translation efficiency of *GUS* reporter gene by 88. 3%, 40. 5% and 47. 8%, respectively.

Discussion

The *PtDrl*02 gene from triploid white poplar belongs to the TIR-NBS-encoding gene family, and its expression pattern resembles to that of most of the identified plant TIR-NBS-encoding/NBS-LRR-encoding/*R* genes, also displaying a low level of expression with tissue-specificity in unchallenged plants. Since the *R* gene expression can be negatively regulated by multiple factors/elements at various levels, leading to a low *R* gene activity (Li et al., 2007; Yi and Richards, 2007), we therefore hypothesized that the native *PtDrl*02 gene expression might also be tightly control by a set of repressors in poplars. Here, one of which, the 5′ UTR sequence, was under characterization. The results obtained in this study showed that the *PtDrl*02 5′ UTR sequence could play as an intramolecular module in regulating gene expression. Although it was not responsible for the spatial expression pattern of the regulated gene, it did confer a negative effect on gene expression at both the transcription and translation level in the unchallenged plants.

In addition to the contribution of promoter sequence on the gene transcription, the regulatory effect from the 5′ UTR sequence was also pivotal (Hultmark et al., 1986; Bolle et al., 1994; Curie and McCormick, 1997; Hua et al., 2001; Samadder et al., 2008). Curie and McCormick (1997) have showed that the *LAT*59 5′ UTR sequence could inhibit gene expression at transcription level and the affected inhibitor was further mapped to a putative stem-loop region that might significantly impede the transcription elongation process. In our study, the *PtDrl*02 5′ UTR sequence (at the DNA level) similarly has the DNA structure with stem-loop, and this 5′ UTR sequence has been determined to down-regulated the transcript level of the reporter gene (*GUS*). It was, therefore, conceivable that the present *PtDrl*02 5′ UTR sequence might also inhibit the transcription elongation process of downstream gene, which finally leaded to a low transcript level of *PtDrl*02 gene and *GUS* reporter gene in triploid white poplar and transgenic P-985/UTR plants respectively. Considered that the transcriptional expression of the genes is commonly determined by the dynamic relative rate of transcription and degradation, and the fact that there have been several studies highlighting the transcript half-life regulation from the structured 5′ UTR (Stefanovic et al., 2000; Hua et al., 2001; Cannons and Cannon, 2002), the possibility that the *PtDrl*02 5′ UTR sequence functions by reducing the transcript half-life of the genes still remains because the *PtDrl*02 5′ UTR also has a complex secondary structure at the RNA level.

Here, we also attended the regulatory effect of the leading intron on the gene transcription. Many studies have identified that introns within the 5′ UTRs can enhance the gene expression (Norris et al., 1993; Curie et al., 1993; Rose and Beliak, 2000; Rose, 2002; Morello et al., 2002; Curi et al., 2005; Kim et al., 2006; Chung et al., 2006; Samadder et al., 2008). The up-regulation is commonly designated as intron-mediated enhancement (IME) (Chung et al., 2006), and mostly attributed to the transcriptional events and the stabilization of RNAs, promoting the transcript abundance of target genes (Rose, 2004; Samadder et al., 2008). However, in our study, the removal of the *PtDrl*02 leading intron resulted in an increase in transcript abundance (Fig. 8A), consistently associated with the enhanced GUS activity in the identical transgenic plants. This effect strikingly disagreed with that of IME, implying that the leading intron of *Pt-Drl*02 gene functions using an alternative mechanism rather than IME. In fact, there have been several reports regarding the negative effect from the intron (Bentley and Groudine, 1986; Pan and Simpson, 1999; Taylor et al., 1997). It was proposed that the intron sequence can inhibit gene expression at the transcription level: examples include the *c-myc* gene, in which the first intron interacts with the HoxB14 protein and causes a transcription elongation block (Bentley and Groudine, 1986; Pan and Simpson, 1999), and the human histone H3. 3 gene (Taylor et al., 1997). Furthermore, Bousquet-Antonelli et al., (2000) revealed a nuclear pathway that "rapidly degrades unspliced pre-mRNAs in yeast", which mainly involved the exosome complex, and indeed the exosome-like complex has been reported in plants (Chekanova et al., 2002; Samadder et al., 2008). Thus, the involvement of *PtDrl*02 leading intron in transcription inhibition as well as pre-mRNA degradation is possible. Considering the conformation of the 5′ UTR, it was found that removal of the leading intron indeed dramatically disrupted the *PtDrl*02 5′ UTR secondary structure and this change produced a great but incomplete mitigation of inhibition on gene expression, and in fact the remaining inhibitory effect was attributed to the 5′ UTR exon1 which also made a great contribution to the secondary structure of the *PtDrl*02 5′ UTR. We thereby proposed that the leading intron, in addition to the exon1 sequence, functions as a component of the 5′ UTR and acts in a structure-dependent manner to decrease the transcript abundance.

Comparison of GUS activities values with the respective transcript levels indicated that the *Pt-Drl*02 5′ UTR also greatly decreased the translation efficiency (88. 3%) (Fig. 8B). This observation showed that the level of GUS activity reduction was attributed to the drop at the transcription level as well as the translation efficiency. To our surprise, the *PtDrl*02 5′ UTR construct (P-985/UTR/GUS), which had the intron, had a more profound inhibition when compared to the intron-free one (P-985/UTRΔ/GUS) that would yield identical mRNA. This result suggested an involvement of intron sequence in translational control, albeit it was an unexpected effect. Similar observations indeed have been made in animal and plant systems (Le Hir et al., 2003; Rose, 2004; Curi et al., 2005; Samadder et al., 2008), but all of them directly pointed to the IME. Thus, how the *PtDrl*02 5′ UTR intron mediated the translation repression is still an open question. The *PtDrl*02 5′ UTR exon1, which acted as the leader sequence of mature mRNA, did function by contributing 40.5% inhibition on translation efficiency. In general, the 5′ leader of mRNA can inhibit translation by its secondary structure, as well as upstream open reading frame (uORF) (Kozak, 1991; Vega Laso et al., 1993; McCarthy, 1998; Morris and Geballe, 2000; Meijer and Thomas 2002). The present *PtDrl*02 5′ UTR exon1 did not contain any uATG or uORF, but had one secondary structure at 3′ end of the sequence (Fig. 3C), immediately upstream of the translational start codon (ATG). It is proposed that structures with predicted free energies greater that-30 kcal/mol is sufficient for inhibiting translation *in vitro* (Zuker, 2000), and the structure with stability of-10 kcal mol^{-1} can reduce translation efficiency by approximately 50% (McCarthy, 1998). It is also proposed that the positioning of a given secondary structure close to the start codon could increase its effectiveness when compared to the inhibition observed when it is located close to the 5′ end of the leader (Vega Laso et al., 1993). The folding energy of *PtDrl*02 exon1 sequence is relatively high ($\Delta G = -8.00$kcal/mol). However, given its position, the inhibitory effect might be strengthened, finally resulting in a suppression of 40.5% on translation. Additionally, we cannot rule out the possibility that the interaction of unknown *PtDrl*02 mRNA binding protein with the exon1 structure could better stabilize the secondary structure such that the free energy is lower than−8.00 kcal/mol and to a certain extent, block scanning ribosomes to reach the start codon, reducing the translation efficiency.

Rahman et al. (2001) demonstrated one negative 5′ UTR regulatory segment binding to unknown factors in the human *Pax*5 *exon*1*A* gene by using the eletrophoretic mobility shift assay (EMSA), while Lin et al. (2004) firstly identified a novel *cis*-acting silencer in intron1 sequence that was crucial in the negative-regulation of *myf*-5 expression in zebrafish. Apart from the secondary structure, the possibility of presence of repressor elements within the *PtDrl*02 5′ UTR still remains albeit a search through database (UTResource, Pesole and Liuni, 1999) with the 5′ UTR sequence did not reveal any known functional elements. Thus, further studies to demonstrate whether the *PtDrl*02 5′ UTR contains any repressor elements will be conducted.

In this study, the *PtDrl*02 5′ UTR as well as its derivative (exon1) were placed under the control of its native promoter. The *PtDrl*02 promoter was found to be tissue-specific, and had been identified to contain two positive regulatory regions (-985/-669 and-669/-467) and one negative segment (-467/-244) (Zheng et al., unpublished). It would be interesting to see whether the UTR function depends on the interactions with its regulated promoter sequence. In Hua et al. (2001) study, it was suggested that the *At-P5R* 5′ UTR acted in an *At-P5R* promoter-dependent manner, and similar behavior also was found on *SAMDC* leading sequence (Hu et al., 2005). However, in some case, such as the *Ac* transposon gene, it was reported the negative effect of the *Ac* 5′ UTL (UTR) not only drives the expression of its own promoter, but also the 35S and NOS promoter (Scortecci et al., 1999). Further experiments focusing on the regulatory effect of 5′ UTR on heterogeneous promoters (e. g. CaMV 35S) will be performed through which the regulatory manner of *PtDrl*02 5′ UTR can be clarified.

Data from animal systems have demonstrated the involvement of the 5′ UTR in controlling gene expression typically for those of growth factors, transcription factors, and proto-oncogenes (van der Velden and Thomas, 1999), whose expression should be strongly and finely controlled, suggesting that 5′ UTR sometimes structured in a way to prevent harmful overproduction of regulatory proteins (Ringnér and Krogh, 2005). Indeed, some diseases are resulted from the mutations in 5′ UTR (Kozak, 2002; Pickering and Willis, 2005). In plants, several *R* genes are shown to have introns and/or uORF-containing 5′ UTR (Botella et al., 1998; Halterman et al., 2001; Tan et al., 2007; Schmidt et al., 2007). Our present work here also revealed another five *Populus* TIR-NBS-encoding genes (putative *R* genes), in addition to the *PtDrl*02, which also contain relatively long, AT-rich (> 60%) and structured 5′ UTR (Fig. 4). The presence of these 5′ UTR features may be indicative of expression regulation of these genes. However, mechanisms involved in 5′ UTR mediated control are still not well understood. More detailed studies should be carried out.

Functional Identification and Regulation of the *Ptdrl*02 Gene Promoter from Triploid White Poplar*

Abstract The *PtDrl*02 gene belongs to the TIR-NBS gene family in triploid white poplar (*Populus tomentosa*×*P. bolleana*)×*P. tomentosa*. Its expression pattern displays tissue-specificity, and the transcript level can be induced by wounding, methyl jasmonate (MeJA), and salicylic acid (SA). To understand the regulatory mechanism controlling *PtDrl*02 gene expression, we functionally characterized the *PtDrl*02 promoter region. Using the β-glucuronidase (GUS) as a reporter, we found that the *PtDrl*02 promoter directed gene expression mainly in the aerial parts of the plants and was confined to the cortex tissues of leaf veins, petioles, stems, and stem piths, showing a typical tissue-specific expression pattern. Deletion analysis revealed two positive regulatory regions (−985 to −669 and −669 to −467) responsible for the basal activity of the *PtDrl*02 promoter. Impressively, the sequence from −669 to −467 was shown to contain *cis*-element(s) responding to wounding and MeJA, while the promoter region between −244 and 0 could individually display wounding-responsiveness, and the fragment from −467 to −244 was required for SA-and NaCl-inducible expression of the *PtDrl*02 promoter. Additionally, it was found that the −985 to −669 sequence was the ABA-responding promoter fragment. These results suggested that the *PtDrl*02 promoter was modulated by multiple *cis*-regulatory elements in distinct and complex patterns to regulate *PtDrl*02 gene expression. Our study also suggested that the *PtDrl*02 gene 5′ untranslated region, as well as a *Populus* WRKY transcription factor, PtWRKY1, was involved in the regulation of *PtDrl*02 promoter activities.

Keywords *Populus*, TIR-NBS, *cis*-element, untranslated region, transcription factor

Introduction

The Toll/interleukin-1 receptor (TIR) domain is a conserved protein module that occurs in plants, animals, and insects (Jebanathirajah et al., 2002; Burch-Smith and Dinesh-Kumar, 2007). The TIR domain has been implicated in host defense, and is commonly thought to play a pivotal role during resistance response. In plants, there are at least three types of genes encoding TIR domain proteins (Meyers et al., 2002). The TIR-nucleotide binding site-leucine rich repeat (TIR-NBS-LRR) group is the most well known, possibly due to their disease resistance functions. The better-characterized members include tobacco *N* (Whitham et al., 1994), flax *L6* (Lawrence

* 本文原载《植物细胞报道》(*Plant Cell Reports*), 2010, 29(5): 449-460, 与郑会全、林善枝、张谦、雷杨和侯璐合作发表。

et al. , 1995) and *M* (Anderson et al. , 1997), *Arabidopsis RPP5* (Parker et al. , 1997) and *RPS4* (Gassmann et al. , 1999), and *Arabidopsis TAO1* (Eitas et al. , 2008). Two other TIR domain protein genes, TIR-NBS and TIR-X, might encode a set of adaptor molecules interacting with the TIR-NBS-LRR resistance proteins (*R* proteins), as well as downstream signaling components during defense response (Meyers et al. , 2002; Jebanathirajah et al. , 2002; Burch-Smith and Dinesh-Kumar, 2007; Tan et al. , 2007; Zheng et al. , 2009). A large number of TIR-encoding genes have been identified, including those represented in *Arabidopsis* (Meyers et al. , 2003), *Populus* (Kohler et al. , 2008), and grapevine (Yang et al. , 2008). Although the biological roles of most of these TIR-encoding genes have not been determined, they are generally thought to function as immune protein genes required for plant defense responses.

RNA profiling analysis revealed that the majority of the TIR-encoding genes were expressed at low levels and with tissue specificity in unchallenged plants, and that the expression level could be further enhanced under some specific conditions (Meyers et al. , 2002; Zipfel et al. , 2004; Tan et al. , 2007; Kohler et al. , 2008). These findings supported the fact that the TIR-encoding genes are tightly regulated, presumably for optimal plant defense and plant growth. Yi and Richards (2007) reported that expression of the *Arabidopsis RPP5* locus TIR-NBS-LRR *R* genes could be coordinately regulated by positive (transcriptional activation) and negative (RNA silencing) mechanisms. In particular, the *SNC1* gene was further demonstrated to be modulated at the transcriptional level by three regulatory modes: repression by the chromosomal structure, feedback amplification from *SNC1* on its own promoter sequences, and repression by *BON1* (Li et al. , 2007). Several other regulatory aspects, including alternative splicing (Schmidt et al. , 2007), 5′ untranslated region (UTR) modulation (Zheng et al. , 2009), protein conformational changes (Takken et al. , 2006), and subcellular localization, as well as a protein-protein association step (Burch-Smith et al. , 2007), were also reported. However, current knowledge concerning the molecular mechanism underlying the TIR-encoding gene expression regulation in plants is still limited, especially for that of the perennial woody plant species, *Populus*.

In a previous study, we identified a TIR-encoding gene, *PtDrl02*, which belongs to the TIR-NBS subfamily in the triploid white poplar (*Populus tomentosa*×*P. bolleana*)×*P. tomentosa*, and revealed that the *PtDrl02* 5′ UTR sequence conferred a negative effect on gene expression in unchallenged plants (Zheng et al. , 2009). In this continued study on *PtDrl02* gene function and regulation, we present the inducible expression patterns of the *PtDrl02* gene in response to several defense-related stimuli (wounding, methyl jasmonate (MeJA), and salicylic acid (SA)), and investigate the role of the *PtDrl02* promoter in gene expression regulation. We report the tissue-specific properties of the *PtDrl02* promoter and identify a set of regulatory regions responsible for its basal and inducible activities in response to different defense-related stimuli, as well as ABA and high salinity. In addition, a *Populus* WRKY transcription factor, PtWRKY1, is suggested to be involved in the regulation of *PtDrl02* promoter activity. A complex regulatory effect from the 5′ UTR sequence on the *PtDrl02* promoter activity under different stimulus-inducible conditions is also reported.

Materials and methods

Plant materials and growth conditions

A triploid white poplar clone 'L9' [(*Populus tomentosa* × *P. bolleana*)×*P. tomentosa*] (Zhang et al., 1992, 1997; Zhang et al., 2008) was used as the source material for this study. Poplar plants were propagated by cutting and raised in pots within a controlled environment chamber (photoperiod: 16/8h light/dark, minimum illumination: 0.2 mM/(s · m^2), day temperature: 20~30℃) at Beijing Forestry University. Four-month-old plants with identical growth status were then subjected to gene expression analysis.

Tissue-culture tobacco (*Nicotiana tabacum* cv. W38) plants were raised on Murashige-Skoog (MS) medium (Murashige and Skoog, 1962) supplemented with 30g/L sucrose, 5.5g/L agar, 0.1mg/L NAA and adjusted to pH 5.8. The plants were maintained in a growth chamber with a 16/8h light/dark photoperiod at 23℃. The fully developed tobacco leaves were then used for genetic transformation experiments.

Plant treatment

To test the effects of the different defense-related stimuli on *PtDrl*02 gene expression, the aerial parts between the second and fourth leaves of the poplar plantlets were sprayed with 200μM methyl jasmonate (MeJA) or 5mM salicylic acid (SA) solutions. The MeJA-treated poplar parts were then covered with a vinyl bag. For wounding treatment, the parallel poplar parts were mechanically wounded with pliers. Untreated poplar plantlets and plantlets treated with distilled water at identical parts to the MeJA and SA treatments were used as controls. The poplar samples were then harvested at time points 0, 6, 12, 24, and 48h post-treatment.

The *in vitro* transgenic tobacco plants (see below) were treated with 200μM MeJA, 5mM SA, or 100μM abscisic acid (ABA) and then incubated for 6, 6 and 24h, respectively. Mechanical wounding treatment was performed by pricking the stem with needles and incubating was continued for 12h. For high salinity treatment, the tobacco plants were removed from the in-vitro medium, and their roots were soaked in 300mM NaCl and incubated for 24h. Untreated tobacco plants and plants treated with distilled water were used as controls.

Total RNA extraction and Real-time quantitative RT-PCR

Total RNA was extracted from the poplar or tobacco sample using the SV Total RNA Isolation System (Promega, Madison, WI, USA) and treated with RNAse-free DNAse I to eliminate residual genomic DNA, according to the manufacturer's instructions (Promega, Madison, WI, USA). The total RNA was then evaluated by agarose gel electrophoresis and spectrophotometrical analysis.

Real-time quantitative RT-PCR (qRT-PCR) was performed using the SuperScript™ III Platinum Two-Step qRT-PCR Kit with SYBR Green (Invitrogen, Carlsbad, CA, USA) with approximately 300ng RNA as a template in an Opticon2 thermocycler (MJ Research, Bio-Rad) following the manufacturer's instructions. To quantify the expression level of target gene (*PtDrl02/GUS*), the Poplar/tobacco *ACTIN* gene (GenBank Accession No. AY261523. 1/U60491) was used as an endogenous control gene. Real-time PCR primers used were as follows: *PtDrl02* forward primer, 5′-ACCCCATTTCTCCGATTCA-3′, and reverse primer, 5′-CTGCATCATCTAAGGCAGCA-3′; *GUS* forward primer, 5′-ATCCGGTCAGTGGCAG TGAAGG-3′, and reverse primer, 5′-CAGCG-TAAGGGTAATGCGAG-3′; poplar *ACTIN* forward primer, 5′-CTCCATCATGAAATGCGATG-3′, and reverse primer, 5′-TTGGGGC TAGTGCTGAGATT-3′; tobacco *ACTIN* forward primer, 5′-CT-GCTGGAATTCACGAAA CA-3′, and reverse primer, 5′-GCCACCACCTTGATCTTCAT-3′. All reactions were carried out in triplicate, and the generated real-time data were analyzed using the Opticon Monitor Analysis Software 3. 1 tool.

Promoter-GUS chimeric vector construction and genetic transformation

A series of 5′ progressive deletions of the *PtDrl02* (previously designated as *PtDRG02*) gene promoter were generated by PCR using a *PtTIRp01* clone (Zheng et al., 2007) as a template with forward primers containing *Xho*I restriction sites (underlined): 5′-CCGCTCGAGTTAATACCTTC-CCTTTACGCAACCA-3′ (Pf1), 5′-CCGCTCGAGCTT TTGGATGAATTGTTAACG GAGG-3′ (Pf2), 5′-CCGCTCGAGTTCGTAGGCGAGCA ATAAAGTCCCA-3′ (Pf3), 5′-CC GCTCGAGTC-CAAAAACTCTCATACTCCACCAC-3′ (Pf4), and a common reverse primer containing a *Bam* H I restriction site (underlined) 5′-CGGGATCCAAATCGAATCAGAAGT TCATGTG-3′ (Pr01). Two promoter derivatives extending to the 5′ UTR sequence of the *PtDrl02* gene were created by PCR amplification as described previously (Zheng et al., 2009). Each of the PCR-amplified fragments was digested with *Xho* I and *Bam*H I (Promega, Madison, WI, USA) and purified with TIANquick Midi Purification Kit (TIANGEN, Beijing, P. R. China). They were then fused to the β-glucuronidase (GUS) reporter gene of the modified pBI121 vector (Clontech, USA) harboring an *Xho* I site immediately downstream of the *Hin*d III site, which was previously digested with *Xho*I and *BamH*I to release the 35S promoter. The resulting vectors, confirmed by DNA sequencing, were named as P-985/GUS, P-669/GUS, P-467/GUS, and P-244/GUS, P-985/UTR/GUS and P-985/UTRΔ/GUS.

For genetic transformation, the chimeric vectors were first introduced into *Agrobacterium tumefaciens* strain EHA105 via the freezing-thaw method (An, 1987), and then transferred into tobacco by the leaf disc transformation method as described by Horsch et al. (1985). The putative transgenic plantlets were confirmed by PCR, as well as preliminary GUS staining. The verified transgenic tobaccos were then propagated and synchronized (using vegetative stem cutting containing an axillary bud) from primary transformants in MS medium. One-month-old in-vitro-grown plantlets were used for subsequent experiments.

Histochemical and fluorometric GUS assay

For histochemical staining of GUS, fresh tissue samples were dissected from tobacco plants and immediately subjected to the X-Gluc solution (Jefferson et al., 1987). After overnight incubation at 37℃, stained samples were bleached with 70% (v/v) ethanol and observed with OLYMPUS BX51 and SZX12 microscopes.

A fluorometric GUS assay was performed as described by Jefferson et al. (1987). The tobacco stem tissues were ground in liquid nitrogen and homogenized in freshly prepared GUS extraction buffer (50mM NaH_2PO_4, pH 7.0, 10mM EDTA, 0.1% Triton X-100, 0.1% (w/v) sodium laurylsarcosine, 10mM β-mercaptoethanol). After centrifuging for 10min at 12000rpm at 4℃, the GUS activity of the supernatant was determined using 4-methylumbelliferyl glucuronide (4-MUG) as a substrate. The fluorescence of the GUS-catalyzed hydrolysis reaction product, 4-methylumbelliferone (4-MU), was measured with the TECAN GENios system. Protein concentration in the supernatant was assessed by the Bradford (1976) method, using bovine serum albumin (BSA) as a standard. GUS activity was normalized to the protein concentration of each supernatant extract and calculated as pmol of 4-MU per milligram of soluble protein perminute.

Agrobacterium-mediated effector-reporter transient expression assay

To construct the effector plasmid 35S/*PtWRKY*1, the coding region of *PtWRKY*1 (GenBank Accession No. GQ377421) gene, followed by a nopaline synthase terminator, was inserted into the plant expression vector pBI121, which had been previously digested with *BamH*I and EcoRI to release the *GUS* reporter gene and its terminator. For a negative control, empty effector plasmid 35S/Em was constructed by the replacement of *GUS* gene with a native sequence 5′-TCTAGAGGATCCAATTGCTACCGAGCTC-3′ in pBI121. The four 5′ progressive promoter deletion-GUS chimeric vectors described previously were used as reporter plasmids. The two different *Agrobacterium* EHA105 cultures (OD_{600} = 0.8) carrying effector or reporter plasmid were mixed at a 1∶1 ratio, and co-infiltrated into tobacco stems (about 0.5cm) using a vacuum chamber. The infiltrated samples were then blotted dry with sterile filter papers and maintained in petri-dishes containing co-cultivation medium (MS medium) at 23℃ under 16/8h light/dark photoperiod for 48h. The stem samples were then collected and subjected to gene expression analysis assay.

Results

*PtDrl*02 gene expression in response to defense-related stimuli

Expression of the *PtDrl*02 gene has been demonstrated at a detectable basal level in leaves, petioles, and young stems (Zheng et al., 2009). To establish whether or not the basal level could be enhanced by induction of defense-related stimuli, qRT-PCR was carried out using total RNA extracted from the poplar aerial parts between the second and fourth leaves at different time points after the treatments of wounding, MeJA, and SA (Fig. 1). The *PtDrl*02 transcript level could be substantially induced by wounding with a maximal level at 12h. Treatment with MeJA or SA also significantly increased the transcription of *PtDrl*02 gene, especially at 6h.

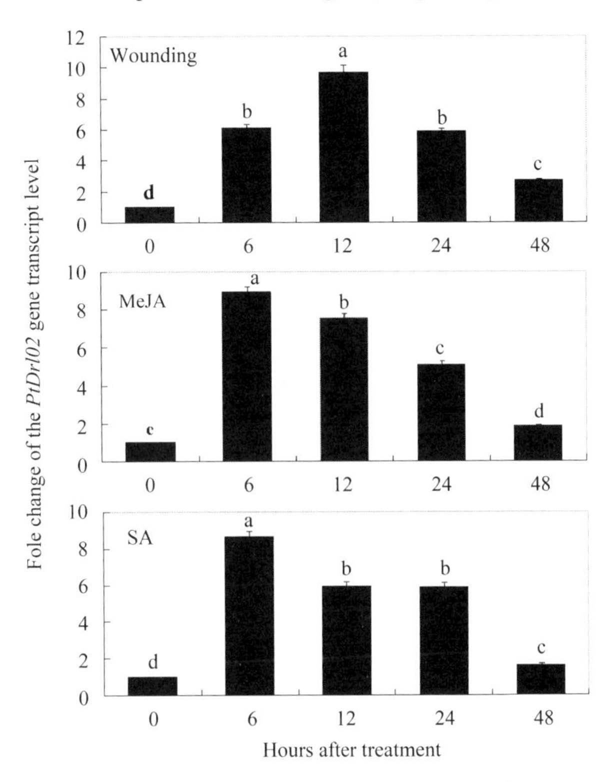

Fig. 1 Time course of the *PtDrl*02 transcript level in the aerial parts (between the second and fourth leaves region) of the triploid white poplar challenged with wounding, methyl jasmonate (MeJA), or salicylic acid (SA).

Significant differences among controls (untreated poplar plantlets and plantlets treated with distilled water at different time points) and the 0 h-treated poplar plantlets with regard to *PtDrl*02 transcript levels were not observed ($P > 0.05$) according to the ANOVA Fisher's LSD test. Thus, changes in the transcription of *PtDrl*02 were directly present multiples relative to 0h-treated data as 1.0. The standard deviations of fold change are also shown, and the letters above the columns indicate statistically significant differences ($P<0.05$) according to the ANOVA FISHER's LSD test.

GUS reporter gene expression from *PtDrl*02 promoter

To address the regulatory mechanism controlling the expression of *PtDrl*02 gene, the 986-bp full-length *PtDrl*02 promoter was fused to the *GUS* reporter gene in a plant expression vector (see materials and methods) and transferred into tobacco plants. The transgenic tobacco lines (n = 59), designated as P-985, were then subjected to histochemical GUS staining, which clearly revealed the tissue specificity of *PtDrl*02 promoter. *GUS* gene expression occurred mainly in the aerial parts of the plants, but was strictly confined to the cortex tissues of leaf veins, petioles, stems,

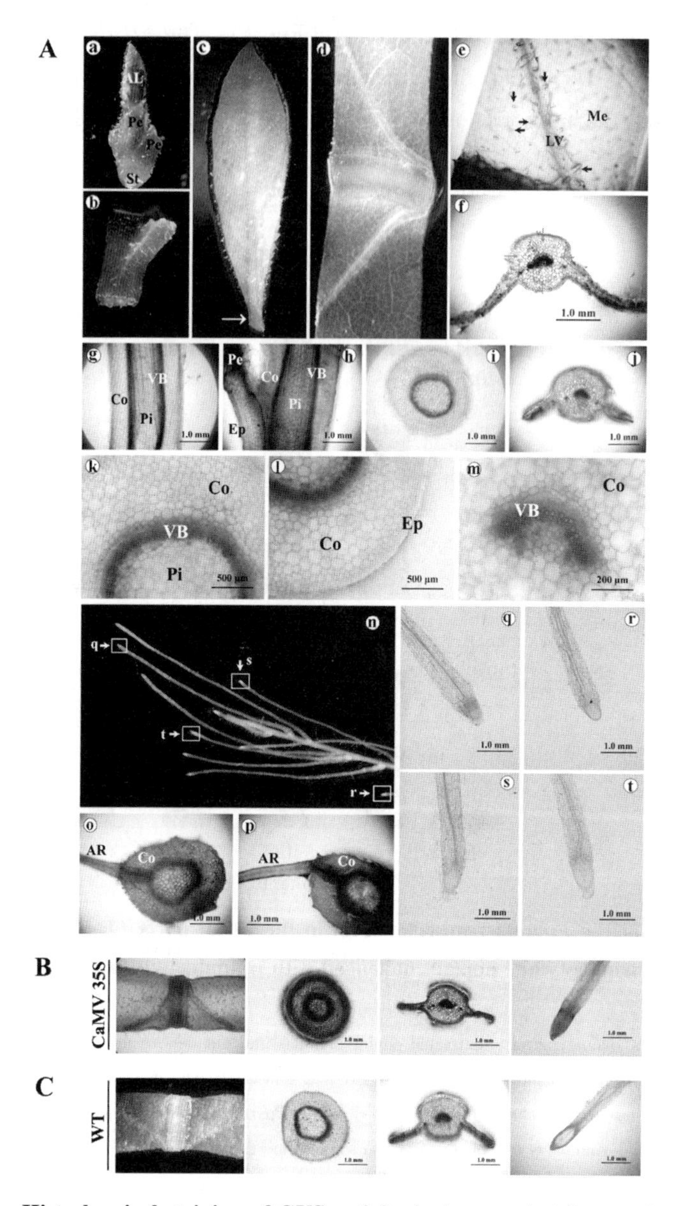

Fig. 2　Histochemical staining of GUS activity in transgenic tobacco plants.

A. In situ localization of GUS reporter in P-985 transgenic tobacco plants.

(a) apical shoot; (b) mature stem with partial petiole tissue; (c) immature leaf; (d-f) mature leaf sections; (g-i, k, l) mature stem sections; (j, m) transverse sections of petioles; (n, q-t) roots; (o, p) adventitious roots from stem tissues; Abbreviation: apical leaf (*AL*), petiole (*Pe*), stem (*St*), mesophyll (*Me*), leaf vein (*LV*), cortex (*Co*), pith (*Pi*), vascular buddle (*VB*), epidermis (*Ep*), adventitious roots (*AR*). The GUS staining region in (c) and (f) are indicated with a white arrow and red arrows, respectively. The dispersed leaf trichomes in (e) are marked with black arrows. The roots represented in (q-t) correspond to that in (n).

B. GUS staining from the CaMV 35S (pBI121 vector) transformants serving as positive controls.

C. GUS staining from the wild-type (*WT*) tobacco plants serving as negative controls.

and the stem piths. *GUS* was not expressed in the mesophyll cells, trichomes, epidermis, and vascular bundles (Fig. 2A). Intriguingly, the *GUS* gene expression in leaves seemed to behave in a development-associated pattern, because the mature leaves (Fig. 2A. d) displayed a stronger GUS activity than the immature ones (Fig. 2A. a and c). Histochemical staining also revealed some unexpected GUS expression in the transgenic tobacco plants, such as in the meristematic zone of some developing roots, and some adventitious roots (Fig. 2A. n and p). Collectively, the *PtDrl*02 promoter displayed a tissue-specific expression pattern, which differed from that of the cauliflower mosaic virus 35S (CaMV 35S) promoter that served as a positive control, directing constitutive *GUS* reporter gene expression in the transgenic tobacco plants (Fig. 2B).

To determine the regulatory regions responsible for the activity of the *PtDrl*02 promoter, a series of 5′ promoter deletion-GUS constructs were also produced, transferred into the tobacco plants, and then subjected to the histochemical GUS staining. The results indicated that promoter deletion from-985(relative to the transcription start site taken as +1) to -669, -467, and-244 (Fig. 3) did not significantly change the tissue-specificity of GUS reporter gene expression, but did reduce the overall GUS activities in the transgenic tobacco plants (data not shown). The fluorometric GUS assay clearly demonstrated that the deletion from -985 to-669 caused a moderate reduction (about 0. 4-fold) in GUS activity in transgenic tobacco stems, the predominant tissue that expressed the GUS gene. Additionally, it was found that further deletion from -669 to-467 caused an 8. 8-fold decrease in GUS activity in the stem tissues of transgenic plants. Intriguingly, the P-244plants did not exhibit a significant change in GUS activity compared to P-467 plants. These data indicated that the *PtDrl*02 promoter had two positive regulatory regions (-985 to -669, and-669 to-467) that are responsible for its basal activity.

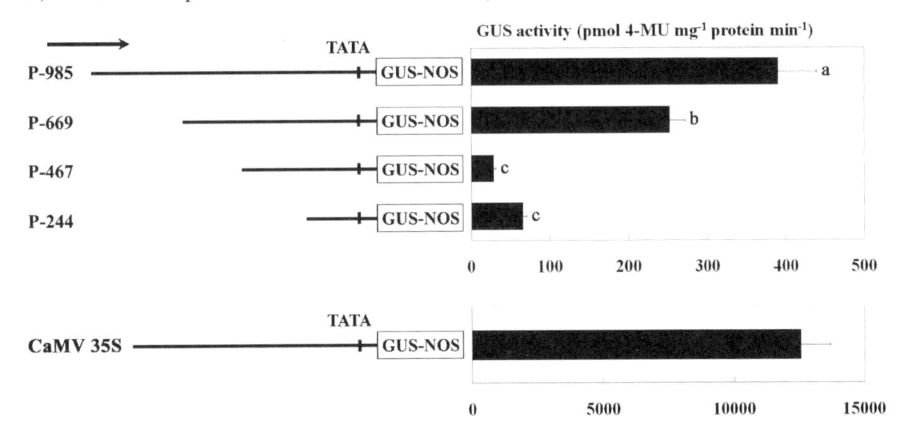

Fig. 3 5′ deletion analysis of the *PtDrl*02 promoter.

GUS activity in stem tissues of transgenic tobacco plants carrying respective promoter-GUS construct (indicated on the left) is shown. GUS activity from the CaMV 35S (pBI121 vector) transformants is provided as a comparison. A black arrow indicates the direction of transcription. Data are means and standard deviations of ten transgenic lines. The different letters above the columns indicate statistically significant differences ($P< 0.05$) according to the Fisher's least significant difference (LSD) test.

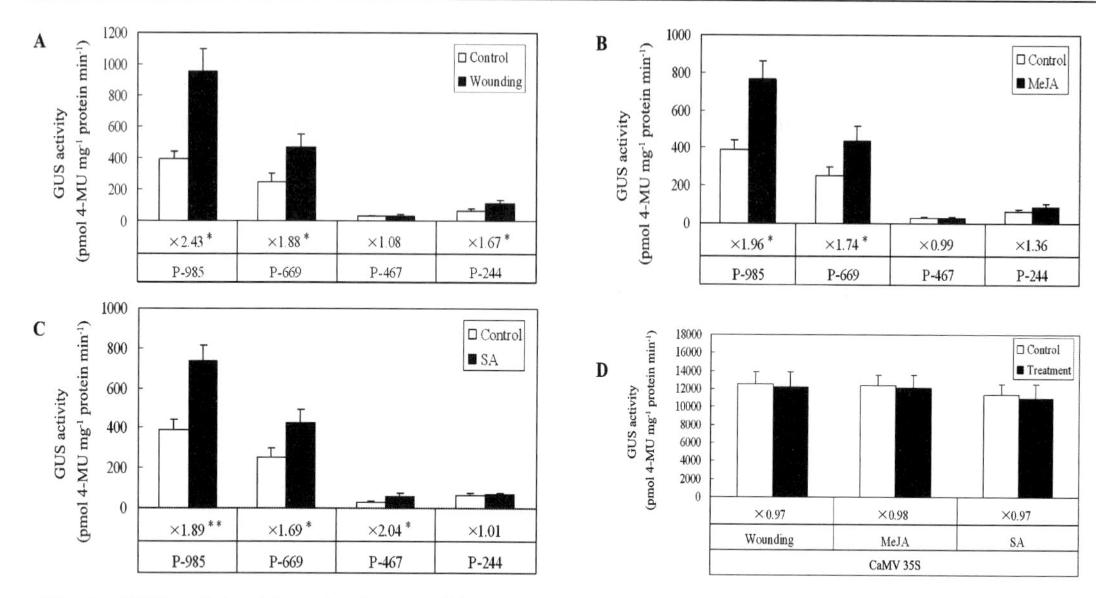

Fig. 4 GUS activity driven by the *PtDrl*02 promoters in stem tissues of transgenic tobacco plants in response to wounding, methyl jasmonate (MeJA), or salicylic acid (SA).

GUS activity from the CaMV 35S (pBI121 vector) transformants served as a comparison. Data are means and standard deviations of ten transgenic lines. The numbers below the bars indicate the fold changes of GUS activity. Significance of the changes produced after each treatment was assessed using Student's *t-tests* (* $P< 0.05$, ** $P< 0.01$).

Activation of the *PtDrl*02 promoter by defense-related stimuli

The expression of the *PtDrl*02 gene had been shown to respond to defense-related stimuli (wounding, MeJA, and SA); therefore, our next step was to test whether application of these stimuli could trigger the expression from the *PtDrl*02 promoter (Fig. 4A-C). Induction of GUS reporter activities was observed in the stem tissues of P-985 transgenic plants upon treatment by wounding, MeJA and SA. It was also found that the P-669 transgenic plants had similar inducible characteristics to P-985 plants, despite a lower inducible ratio. In contrast, the wounding-and MeJA-inducible expression pattern of *GUS* reporter gene seemed to disappear in the P-467 plants, which only responded to SA treatment. Intriguingly, it was observed that the P-244 plant exhibited a considerable ratio of enhancement of GUS activity in response to wounding, but not MeJA and SA. These results differed from that of the CaMV 35S promoter, which showed no significant inducible expression (Fig. 4 D). Collectively, the present results indicated that the sequences from-669 to-467 contained *cis*-element (s) responding to wounding and MeJA, the fragment from-244 to 0 harbored wounding-responsive element (s), and the region between-467 and-244 was required for SA-responsive activity of the *PtDrl*02 promoter.

Activation of the *PtDrl*02 promoter by abscisic acid (ABA) and high salinity (NaCl)

In addition to the challenge of the defense-related stimuli, the transgenic tobacco plants were also subjected to the ABA-and NaCl-inducible expression assays. As shown in Fig. 5, application of ABA activated GUS activity in the stem tissues of P-985 plants, while no significant ratio change was observed in the P-669, P-467, and P-244 plants. Treatment with NaCl produced an enhanced GUS activity in P-985, P-669, and P-467 transgenic tobacco stems, but not in P-244 plants. Thus, it was concluded that the sequences from-985 to-669 was the promoter region that responded to ABA, and that the sequences between-467 and-244 was required for NaCl-inducible activity of the *PtDrl*02 promoter.

Potential *cis*-elements responsible for the inducible activities of *PtDrl*02 promoter

In a previous study (Zheng et al., 2007), we identified a cluster of putative *cis*-elements within the *PtDrl*02 promoter. Integrating this information with the results of the present study

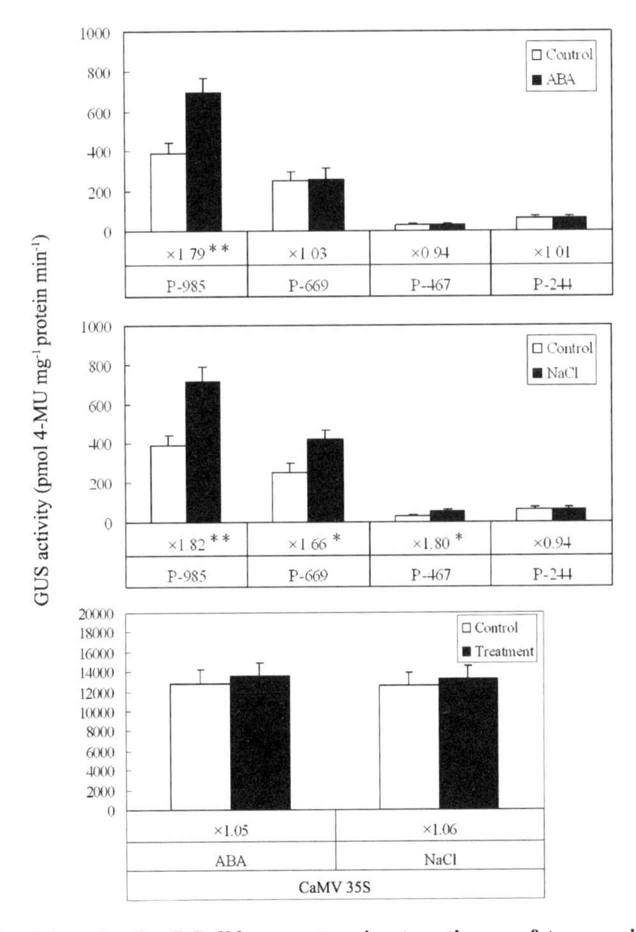

Fig. 5 GUS activity driven by the *PtDrl*02 promoters in stem tissues of transgenic tobacco plants in response to abscisic acid (ABA) or high salinity (NaCl).

GUS activity from the CaMV 35S (pBI121 vector) transformants served as a comparison. Data are means and standard deviations of ten transgenic lines. The numbers below the bars indicate the fold changes of GUS activity. Significance of the changes produced after each treatment was assessed using Student's *t-tests* (* $P < 0.05$, ** $P < 0.01$).

revealed a strong correlation between the inducibility of the *PtDrl*02 promoter and its internal *cis*-elements (Table 1). For example, the W-box motifs, known as defense-responsive elements, might be the regulatory motifs of the *PtDrl*02 promoter responsible for the response to the defense-related stimuli of MeJA and/or wounding. These W-box motifs occurred in the same orientation (+) and consistently contained a TGAC core sequence, as well as an additional 5′ T nucleotide, but they seemed to have a divergent context sequence (Supplemental Fig. 1).

Table 1　Sequence motifs within the *PtDrl*02 functional promoter regions.

Regulatory region	Corresponding indncer	Putative *cis*-element *	Element number
−985 to−669	ABA	ABRE motif(ACGTG)	1
−669 tp−167	Wounding, MeJA	W-box(TTGACT/TTGACA)	2
−467 to−244	SA, NaCI	GT−1 motif(GAAAAA)	2
−244 to 0	Wounding	W-box(TTGACA)	1

* The putative *cis*−elements of *PtDrl*02 promoter were identified based on online programs including PLACE (http://www.dna.affrc.go.jp/PLACE/signalscan.html) (Higo et al., 1999), PlantCARE (http://bioinformatics.psb.ugent.be/webtools/plantcare/html/) (Lescot et al., 2002), NSITE − PL and ScanWM-P (Softberry, http://linux1.softberry.com/berry.phtml), as described by Zheng et al. (2007) previously.

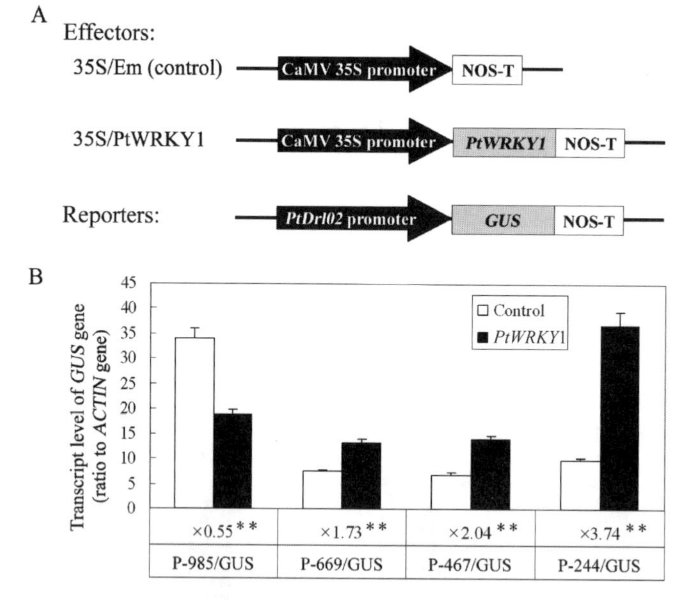

Fig. 6　*Trans*-regulation of the *PtDrl*02 promoter activity by the transcription factor *PtWRKY*1 gene in tobacco stems.

A. Schematic diagram of the effector and reporter constructs used in the *Agrobacterium*-mediated transient expression assays. The effector construct contains a CaMV 35S promoter fused to either the *PtWRKY*1-NOS-T (35S/PtWRKY1) or NOS-T gene (35S/Em; Control). The reporter constructs contained the 5′ deleted *PtDrl*02 promoter fragments fused to the *GUS* reporter gene (P-985/GUS, P-669/GUS, P-467/GUS, and P-244/GUS).

B. *GUS* transcription from different reporter constructs regulated by PtWRKY1 in an *Agrobacterium*-mediated transient expression system. The transcript level of *GUS* was determined by qRT-PCR using the *ACTIN* gene (GenBank Accession No. U60491) as an endogenous reference. Data are means and standard deviations obtained from four independent biological experiments. The numbers below the bars indicate the fold changes of *GUS* transcript level, and the significance of the changes was assessed using Student's *t-tests* (** $P< 0.01$).

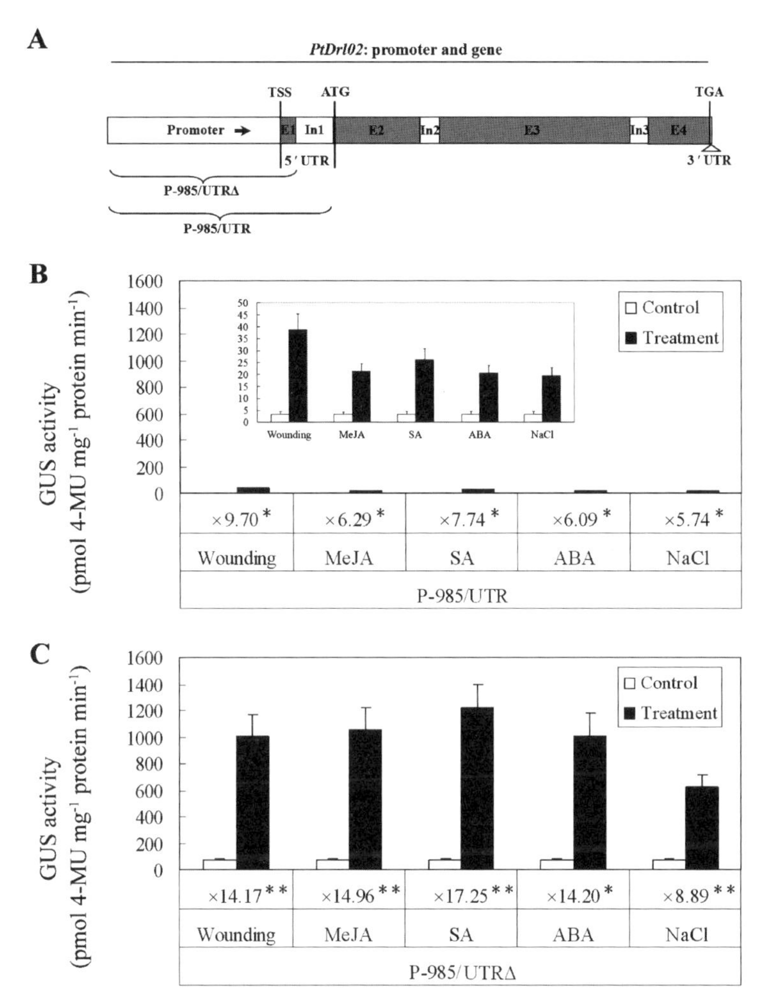

Fig. 7 (A) Architecture of P-985/UTR and P-985/UTRΔ from the *PtDrl*02 promoter-gene fragment (Zheng et al., 2009).

Abbreviation: transcription start site, *TSS*; the *ATG* translation start codon, *ATG*; the *TGA* translation stop codon, *TGA*; 5′ untranslated region, 5′ *UTR*; 3′ untranslated region, 3′ *UTR*; exon 1 to exon 4, E1 to E4; intron1 to intron 3, In1 to In3. (B, C) GUS activity in stem tissues of P-985/UTR and P-985/UTRΔ transgenic tobacco plants in response to wounding, methyl jasmonate (MeJA), salicylic acid (SA), abscisic acid (ABA), or high salinity (NaCl). Diagram with a low range of scale of the y-axis regarding the inducible activity of GUS in P-985/UTR transgenic tobacco stems is also shown in (B). Data are means and standard deviations of ten transgenic lines. The numbers below the bars indicate the fold changes of GUS activity. Significance of the changes produced after each treatment was assessed using Student's *t-tests* (* $P< 0.05$, ** $P< 0.01$).

The *PtDrl*02 promoter activity is regulated by the PtWRKY1 transcription factor

In a recent study, a WRKY transcription factor gene, *PtWRKY*1 (GenBank Accession No. GQ377421), was found to display a similar expression pattern to that of the *PtDrl*02 gene in triploid white poplar (Zheng et al., unpublished). The *PtWRKY*1 gene belongs to the group I WRKY gene family because it encodes a protein with two typical WRKY domains (Supplemental Fig. 2).

We investigated whether the PtWRKY1 factor had a regulatory effect on *PtDrl*02 promoter activity. We used an *Agrobacterium*-mediated effector-reporter transient expression assay, and the regulatory effect was assessed directly by the *GUS* reporter transcript level (quantified by qRT-PCR). As shown in Fig. 6, overexpression of the *PtWRKY*1 gene causes an approximately two-fold reduction in *GUS* transcription from the P-985/GUS construct, indicating a negative effect on the full-length promoter activity. However, this negative effect seemed to be completely abolished by the deletion of the promoter fragment from-985 to-669, because expression of the *PtWRKY*1 gene increased the *GUS* transcript level from the P-669/GUS truncated construct by 1.73-fold. Additionally, it was observed that another two truncated promoter constructs, P-467/GUS and P-244/GUS, also displayed upregulated *GUS* transcript levels in the presence of increased PtWRKY1 factor, but they had more profoundly increased ratios than in P-669/GUS construct.

The *PtDrl*02 promoter activity is affected by its 5′ UTR sequence

Previously, we demonstrated that the *PtDrl*02 5′ UTR sequence conferred a negative effect on the *PtDrl*02 promoter-directed *GUS* gene expression in unchallenged plants (Zheng et al., 2009). To further investigate the regulatory effect of the 5′ UTR on *PtDrl*02 promoter activity, the promoter-UTR transgenic tobacco plants, including P-985/UTR and P-985/UTRΔ, were subjected to the stimulus-induction assays. Similarly to the P-985 plants, both the P-985/UTR and P-985/UTRΔ plants displayed clear inducible GUS activity in stem tissues when exposed to the stimuli of wounding, MeJA, SA, ABA, and NaCl (Fig. 7). However, the repressive effect from the full-length 5′ UTR on PtDrl02 promoter activities was retained under these conditions, despite an augmented inducible ratio of GUS activities, as indicated in Fig. 7 B (versus Fig. 4 and Fig. 5). Deletion of the leading intron of the 5′ UTR largely alleviated the negative effect, and in the cases of wounding, MeJA, SA and ABA treatment, the P-985/UTRΔ plants had more absolute GUS activity than P-985 plants (Fig. 7 C versus Fig. 4 and Fig. 5).

Discussion

The *PtDrl*02 gene encodes a typical TIR-NBS protein in triploid white poplar, and showed a low level of basal expression and tissue-specificity in the unchallenged plants (Zheng et al., 2009), similar to most of the TIR-encoding genes investigated thus far (Tan et al., 2007; Kohler et al., 2008). The *PtDrl*02 gene also exhibited an inducible expression in response to different defense-related stimuli (wounding, MeJA, and SA) (Fig. 1), analogous to some plant TIR-encoding genes (Marathe et al., 2004; Zipfel et al., 2004; Tan et al., 2007). To further understand the mechanism by which the *PtDrl*02 gene expression is regulated, we focused on the functional role of the *PtDrl*02 promoter and demonstrated that the *PtDrl*02 promoter was essential for controlling the *PtDrl*02 gene expression both in tissues and in response to wounding, MeJA, and SA, as well as ABA and high salinity in *planta*. This finding increased our understanding of the regulation of *PtDrl*02 gene expression and provided new insights into the molecular mechanisms controlling TIR-encoding gene expression in *Populus*.

Histochemical analysis of GUS activity from the *PtDrl*02 promoter in transgenic plants allowed the definition of the more detailed expression pattern of *PtDrl*02 gene. In the mature leaves, GUS reporter gene expression was limited to the cortex tissue of the leaf vein system (Fig. 2A d-f), while similar cell-type specific expression was also found in petioles (Fig. 2 A j-m). In the stems, GUS activity was strongly evident in both the cortex and pith tissues (Fig. 2A g, h, i, k, l) during all growing stages. Specific expression of *PtDrl*02 gene in certain tissues, such as cortex and pith, might allow a pre-deposition of constitutive ability of the plants in response to certain biotic and/or abiotic stresses. Meanwhile, possible negative effects from PtDrl02 protein might also be avoided in tissues such as mesophyll and vascular bundles that are highly required for the normal processes of plants.

The deletion analysis of the *PtDrl*02 promoter provided evidence for the existence of multiple *cis*-regulatory elements, which were required for the stimulus-triggered *PtDrl*02 gene expression in triploid white poplar. Both the promoter fragments from-669 to-467 and-244 to 0 were shown to be wounding-responsive, while the-669 to-467 region also proved to be MeJA-responsive. Interestingly, these two regions had been shown to comprise novel W-box element(s) (Zheng et al., 2007; Table 1). The W-box is the binding site for a family of transcription factors termed WRKY proteins (Rushton et al., 1996). Accompanying studies of WRKY proteins, the functions of elements of this type have also been well defined. Most of them were shown to activate gene expression in a pathogen-, wounding-or plant hormone (e.g. JA and SA)-dependent manner (Rouster et al., 1997; Rushton and Somssich 1998; Eulgem et al., 1999; Rushton et al., 2002; Nishiuchi et al., 2004; Laloi et al., 2004; Rocher et al., 2005; Sobajima et al., 2007; Hiroyuki and Terauchi, 2008; Hwang et al., 2008). Thus, the identified W-boxes element(s) within the-669 to-467 and-244 to 0 regions of *PtDrl*02 promoter might be functional and necessary for the MeJA- and/or wounding-responsiveness of the promoter. Additionally, regarding the number (three) of W-box elements determined, we could speculate that they act synergistically, similar to the case reported by Eulgem et al. (1999). Ciolkowski et al. (2008) recently demonstrated that the W-box neighboring sequence had a regulatory effect on W-box activity. The *PtDrl*02 promoter W-box elements appeared to have distinct context sequences, but subtle resemblances still remained (Supplemental Fig. 1), including a conserved T nucleotide at positions 4-bp upstream and 2-bp downstream of W-box elements, and an identical ATTTA sequence located upstream of the first (−584 to −560) and third (−170 to −146) W-boxes. Further study is required to understand whether these characteristics are relevant to the W-box activities. In this report, we also found that the removal of the sequences from −467 to −244 completely abolished the SA-and NaCl-inducibility of the *PtDrl*02 promoter, indicating that this promoter region contained SA-responsive and NaCl-responsive elements. The two GT-1 motifs (GAAAAA) in this promoter region might be candidates, because the same GT-1 sequence was previously reported to be responsible for the pathogen-and

NaCl-induced activity of the promoter (Park et al. , 2004). In addition, we revealed that the−985 to−669 promoter region, which harbored an ABRE motif, accounted for the ABA-stimulated *PtDrl*02 promoter activity. This result implied that *PtDrl*02 gene expression might be directly modulated by ABA. Apart from the aforementioned inducible elements, the basal transcriptional modules included two activators (located in fragments from−985 to−669 and−669 to−467), which, to a certain extent, functioned to influence the inducible response of the promoter (Fig. 4 and Fig. 5). This result suggests the existence of cross-talk between the basal and inducible elements within the *PtDrl*02 promoter region. Furthermore, unknown negative element(s) within the −467 to −244 fragment might also be involved in the regulation of the *PtDrl*02 promoter activity upon wounding.

Using the *Agrobacterium*-mediated transient expression approach, we were able to demonstrate that the PtWRKY1 transcription factor was involved in the regulation of the *PtDrl*02 promoter activity. Full-length promoter expression was suppressed by the PtWRKY1 factor, while expression of *PtWRKY*1 gene significantly upregulated the activity of the truncated promoter regions (−669 to 0, −467 to 0, and−244 to 0) of *PtDrl*02 (Fig. 6). This result indicated that the PtWRKY1 factor has dual functional activity in regulating the *PtDrl*02 promoter expression. Indeed, dual functionality has been reported for several plant WRKY factors, including AtWRKY6(Robatzek and Somssich, 2002) and AtWRKY41(Higashi et al. , 2008), but the detailed molecular mechanisms remain unknown. One possible clue is that the promoter architecture, as well as the associated factors, might participate in determining the role of the WRKY transcription factors in transcriptional output, as suggested by Ciolkowski et al. (2008).

This study also showed that the *PtDrl*02 promoter activity was quantitatively reduced by its own 5′ UTR sequence under the stimulus-challenged conditions. The existence of this UTR had been previously demonstrated to inhibit gene expression at both the transcription level and the translation level in unchallenged plants, possibly due to its intrinsic secondary structure (Zheng et al. , 2009). Thus, it is possible that the secondary structure of the 5′ UTR was still retained under stimulus-challenged conditions, which led to a sustainable negative effect on *PtDrl*02 promoter expression. As expected, deletion of the leading intron within the *PtDrl*02 5′ UTR (P-985/UTRΔ) extensively alleviated the repressive effect of the full-length 5′ UTR (Zheng et al. , 2009; Fig. 7 C). However, to our surprise, the extent of the alleviation seemed to be greater than anticipated, especially for the cases challenged by wounding, MeJA, SA, and ABA (Fig. 7 C versus Fig. 4 and Fig. 5). One possible explanation is that the residual UTR exon1 sequence in P-985/UTRΔ functions with a positive effect, distinct from its previously defined role of negatively regulating gene expression under non-stress conditions (Zheng et al. , 2009). This result should be further investigated by additional studies.

*MdSPDS*1基因导入毛白杨的遗传转化体系优化研究*

摘　要　为了建立农杆菌介导的高效毛白杨遗传转化体系，并大量获得转*MdSPDS*1基因的转基因毛白杨，本文对*MdSPDS*1基因导入毛白杨的遗传转化体系中关键转化因子进行了优化，获得的优化转化体系如下：叶片预培养3d后，用$OD_{600}=0.4$的菌液侵染7min，再共培养3d。卡那霉素抗性芽的二次筛选中，卡那霉素的筛选压分别为20mg/L和50mg/L。实验获得毛白杨卡那霉素抗性植株共43株，经PCR检测，其中9株呈阳性，占抗性植株总数的20.9%，优化转化系统的阳性植株转化率达到9.38%。本研究为下一步鉴定基因功能和转基因植株耐盐性的分析奠定了基础。

关键词　*MdSPDS*1，遗传转化，毛白杨，农杆菌

毛白杨(*Populus tomentosa* Carr.)是我国特有的白杨派乡土树种，具有分布广、速生、材质优良及抗逆性强等特性(Zhu and Zhang，1997)，目前在生态防护林和工业用材林建设中发挥着极为重要的作用。然而，在一些较为极端的环境条件下尤其是在盐渍化程度较高的立地条件下，现有的毛白杨品种其生长性状往往不尽人意。利用基因工程技术将外源抗性基因转化到毛白杨中是目前实现毛白杨抗逆遗传改良的一条重要途径(林善枝等，2000)。为实现外源基因的稳定转化，目前在杨树上应用的遗传转化方法主要有农杆菌介导法(*Agrobacterium tumefaciens*-mediated genetic transformation methods)、PEG法、电激法和基因枪法(郝贵霞等，2000)。其中，农杆菌介导的遗传转化法是目前应用最广泛且结果较为理想、技术较为成熟的一种基因转化方法。与其他遗传转化方法相比，农杆菌介导法具有方法简单、成本低、转化效率高、导入DNA片段大等优点(王关林和方宏筠，2002)。然而，对杨树进行农杆菌介导的外源基因转化，通常存在转化频率低、基因型对转化条件影响大等问题，如何克服影响转化效率的障碍因子，建立高效的杨树遗传转化体系，尤其对一些具有重要经济性状的杨树品种而言具有重要的意义。

近年来，多胺(polyamines，Pas)合成途径中的相关基因在各种植物中被分离及克隆，其中*SPDS*1基因（亚精胺合成酶基因）已成功导入烟草、拟南芥、甘薯、西洋梨等植物中，研究表明，转基因植物的耐盐性得到了显著提高(Pang et al.，2007；Wi et al.，2006；Kasukabe et al.，2004；Kasukabe et al.，2006；Wen et al.，2008)。因此，将此类基因导入毛白杨有可能显著提高毛白杨的耐盐性。如果能提高农杆菌转化频率，得到大量的转基因植株，就能为下一步鉴定基因功能和转基因毛白杨耐盐性的分析奠定基础。

本研究通过对影响农杆菌介导毛白杨遗传转化频率的几个主要因素的研究，筛选适合

* 本文原载《北京林业大学学报》，2010，32(5)：21-26，与龙萃、庞晓明、曹冠琳和刘颖合作发表。

毛白杨遗传转化的因子组合，建立将 *MdSPDS*1 基因(apple spermidine synthase genes1，苹果亚精胺合成酶基因）导入毛白杨的遗传转化优化体系，为提高该树种的基因转化效率，加快毛白杨耐盐基因工程育种提供依据。

1　材料与方法

1.1　实验材料及培养基

材料取自山东省国营冠县苗圃全国毛白杨基因库，选择生长旺盛、无病虫害的毛白杨优良无性系 TC152 作为试验材料。经本实验室前期筛选，该无性系再生能力显著高于其他无性系(姚娜等，2007)，适合进行遗传转化研究。

携带有从苹果中克隆得到的亚精胺合成酶基因(*MdSPDS*1)表达载体的质粒是由 Dr. T. Moriguchi (National Institute of fruit research，Japan)惠赠，根癌农杆菌菌株 LBA4404 由本实验室保存。

试验所用的培养基成分详见表 1，其中基本培养基为 MS 培养基，碳源为 30g/L 蔗糖，5g/L 琼脂固化；组培生根培养采用 1/2MS(大量元素减半)为基本培养基，20g/L 蔗糖，5. 5g/L 琼脂固化。无菌苗培养光照时间为 16h/d，光照强度 1500~2000Lux(暗培养除外)。光培养和暗培养的培养温度均为 25±1℃。

表 1　试验所用的培养基成分

成分	预培养基	继代培养基	共培养基	筛选培养基	二次筛选培养基	生根培养基
6-BA(mg/mL)	2. 0	0. 3	2. 0	2. 0	2. 0	/
IBA(mg/mL)	0. 1	0. 1	0. 1	0. 1	0. 1	0. 4
Kan(mg/mL)	/	/	/	20	50	50
Cef(mg/mL)	/	/	/	250	250	150
蔗糖(g/L)	30	30	30	30	30	20
基本培养基	MS	MS	MS	MS	MS	1/2MS

1.2　实验方法

1.2.1　转化受体系统中抗生素和抑菌剂浓度的筛选

(1)取生长健壮、大小形状基本一致的毛白杨无菌组培苗叶片，剪 3~4 个切口深达主脉，接种于含不同浓度卡那霉素 (Kanamycin，Kan)的分化培养基上，分别设 Kan 终浓度为：0、10、20、30、40、50mg/L，每个浓度处理 30 片叶，每 10d 换 1 次培养基。培养 60 天后，根据不定芽分化及生长情况，筛选叶片 Kan 临界浓度。取生长健壮，高约 1~2cm 的不定芽接种于含不同浓度 Kan 的生根培养基中，每个浓度处理 30 个不定芽。60 天后，观察其生根情况，筛选适宜不定芽生根诱导的 Kan 临界浓度。

(2)实验取材和方法同(1)，在分化培养基和生根培养基中分别添加不同浓度头孢霉素 (Cefotaxime，Cef)(0、100、150、200、250、300、400mg/L)，观察不同浓度的头孢霉素对叶片不定芽诱导的影响，并统计培养 60d 后组培苗的生根率，确定不定芽生根 Cef 最佳筛选浓度。

1.2.2 叶盘法遗传转化的步骤

选取无菌苗上充分展开的、深绿色的、大小形状基本一致的叶片，垂直中脉平行横剪 3~4 个切口至中脉，背面向下接种在分化培养基上进行预培养 2d。预培养后的叶片浸泡在菌液中侵染 10min 后取出，用无菌滤纸吸干多余的菌液，在分化培养基上 28℃暗培养 2d。待叶片边缘出现肉眼可见的白色菌落后，转移到附加 Kan 20mg/L 和 Cef 200mg/L 的分化培养基上光照培养，进行抗性芽的筛选。当产生不定芽后，将叶片和不定芽一起移至 50mg/L Kan 的培养基中进行二次筛选。待抗性不定芽长到 2~3cm 时，切下并插入含有 Kan 50mg/L 和 Cef 150mg/L 的生根培养基上进行生根，两周左右可以长出不定根。在转化因子筛选试验中，除试验因子外，其他因子处理均按此方法进行，每处理 30 个叶片。

1.2.3 转化因子试验设计

(1)预培养时间对转化的影响。转化前，分别设置 0(对照)、1、2、3 和 4d 进行预培养，培养基为不定芽诱导培养基。

(2)菌液浓度的选择。农杆菌活化培养后，用液体 YEB 培养基稀释至 OD_{600} 为 0.2、0.4、0.6、0.8，比较不同浓度对转化率的影响。

(3)浸染时间对转化的影响。浸染时间分别设置为 5、7、10 和 15min 4 个处理。

(4)共培养时间对转化的影响。将侵染后的叶片置于预培养基上，放置于暗培养环境下，分别共培养 1、2、3、4、5d，然后转入筛选培养基中，60d 后进行统计。以转化后不定芽筛选培养基中 60d 后保留的 Kan 抗性芽百分率作为评价转化频率的指标。

转化率(%)=[(60d 后存活的抗性芽)/(60d 后存活的抗性芽+死亡的不定芽)]×100%

1.2.4 转基因植株的 PCR 扩增检测

采用 CTAB 法提取植物材料总 DNA。以质粒 DNA 作为阳性对照，以未转化的毛白杨 DNA 作为阴性对照，对抗性苗进行 PCR 检测，所用引物为：5′-GATGTGATATCTCCACT-GACGTAAG-3′, 5′-ATCTGCTTGTTGGATGCTACTG-3′；PCR 反应条件为：94℃预变性 5min，94℃变性 30s，58℃退火 30s，72℃延伸 1min，30 个循环，再 72℃延伸 10min。

2 结果与分析

2.1 转 *MdSPDS1* 基因毛白杨遗传转化体系的优化

2.1.1 卡那霉素(Kan)临界浓度选择

为确定毛白杨对 Kan 的本底抗性，开展了叶片诱导分化再生、芽诱导生根的临界浓度试验。由表 2 可知，较低浓度的卡那霉素即可抑制叶片再生不定芽。在不含 Kan 的对照培养基上，96.7%的外植体再生大量丛生芽。但是仅添加了 10mg/L 的 Kan 后，外植体的分化率即降低至 36.7%，随着 Kan 浓度的增加，分化率迅速下降。当 Kan 浓度大于 20mg/L 时，叶片既不分化出芽也不产生愈伤组织。故 20mg/L Kan 即可作为选择转化植株的临界浓度。

卡那霉素对毛白杨的不定芽生根也有很大影响。仅10mg/L的Kan就使组培苗的生根率、根系及不定芽生长均受到严重抑制，60d时，生根率仅16.7%。当Kan达到20mg/L以后，大部分不定芽因不能生根而逐渐发黄死亡。因此，毛白杨不定芽生根的Kan临界耐受浓度确定为10mg/L。当Kan浓度更高(50mg/L)时，不定芽叶片基本全部白化死亡，无法正常生存。在本实验中，将50mg/L Kan作为区分转基因植株和非转基因植株的筛选浓度，用以去除Kan抗性芽中的假阳性不定芽。

表2　不同浓度的卡那霉素对叶片再生不定芽和不定芽生根的影响

Kan(mg/L)	叶片再生率(%)	叶片再生状况	生根率(%)	白化苗率(%)
0	96.7(29/30)	叶片颜色深绿，大量丛生芽发生，芽数多、强壮、浓绿	100(30/30)	0(0/30)
10	36.7(11/30)	叶片基本深绿，部分分化不定芽，芽较强壮、淡绿色	16.7(5/30)	53.3(16/30)
20	6.7(2/30)	叶片浅绿，切口膨大形成愈伤，极少出芽，芽黄绿色	0(0/30)	73.3(22/30)
30	0(0/30)	叶片浅绿，既不分化出芽，也不产生愈伤	0(0/30)	100(30/30)
40	0(0/30)	叶片变黄，接近死亡	0(0/30)	100(30/30)
50	0(0/30)	叶片发褐死亡	0(0/30)	100(30/30)

2.1.2　头孢霉素(Cef)浓度选择

不同浓度的Cef对叶片分化试验结果见表3，随着Cef浓度的提高，分化率逐渐下降，当达到300mg/L时，观察到大部分叶片的膨大，分化率仅为3.3%；当Cef的浓度达到400mg/L时，叶片分化基本停止。因此，在分化培养基中，250mg/L Cef是合适的抑菌浓度。

不同浓度的Cef对幼芽生根影响的试验表明，超过150mg/L后对幼芽的生根影响逐渐加大。此外，随着Cef浓度的提高，幼芽诱导出的根及须根数量由多变少。由此可见，选择浓度为150mg/L的Cef，抑制农杆菌生长的同时，对幼芽的生根不会产生明显的抑制作用。

表3　头孢霉素浓度对叶片再生不定芽和不定芽生根的影响

Cef(mg/L)	叶片再生率(%)	叶片再生状况	生根率(%)	菌落状况
0	93.3(28/30)	叶片深绿，大量丛生芽发生	100(30/30)	+
100	90(27/30)	对叶片分化影响不明显	73.3(22/30)	+
150	83.3(25/30)	叶片深绿，芽生长正常	43.3(13/30)	-
200	76.7(23/30)	叶片深绿	16.7(5/30)	-
250	63.3(19/30)	叶片浅绿，生长基本正常	13.3(4/30)	-
300	23.3(7/30)	叶色浅绿，分化明显减慢	3.3(1/30)	-
400	0(0/30)	分化基本停止，叶片只是膨大	0(0/30)	-

注："+"表示农杆菌生长；"-"表示农杆菌未生长。

2.1.3　预培养时间和共培养时间的选择

为了确定预培养时间，本试验设计了0~4d的时间梯度。从表4可以看出，随着预培养时间的延长，叶片的分化率呈现先升高后降低的趋势。当预培养时间为 3d 时，叶片的分化率最高。当预培养时间为 4d 时，叶片的分化率反而降低，可能是由于叶片伤口细胞再次由脱分化状态进入分化状态，分化组织已经形成，不利于外源 DNA 的进入和整合。同时随着预培养时间的延长，加大逃逸芽出现的概率，所以综合考虑，确定预培养时间为 3d。共培养时间对转化的影响见表 4，最佳的共培养为 3d，并确保在叶片与培养基接触处出现肉眼可见的菌落。

表 4　预培养时间和共培养时间对转化的影响

接种	预培养时间				共培养时间			
叶片数（个）	预培养时间(d)	60d 后存活的抗性芽数(个)	60d 后死亡的抗性芽数(个)	转化率（%）	共培养时间(d)	60d 后存活的抗性芽数(个)	60d 后死亡的抗性芽数(个)	转化率（%）
30	0	6	62	8.82	1	0	82	0
30	1	9	75	10.71	2	7	71	8.97
30	2	8	59	11.94	3	11	75	12.79
30	3	13	82	13.68	4	2	49	3.92
30	4	7	76	8.43	5	0	0	0

2.1.4　菌液浓度和侵染时间对转化率的影响

本试验通过统计不同的菌液浓度梯度下毛白杨叶片的转化效率(表 5)发现，农杆菌浓度过高，叶片产生过敏性反应，导致在培养中会逐渐变黄坏死，褪绿首先发生在最易产生不定芽的叶脉切口处，最后蔓延至整个叶片，完全不能产生不定芽。综合考虑叶片不定芽诱导数量及转化率，确定菌液 OD_{600} 值为 0.4 时效果最佳。通过 4 个不同的浸染时间下毛白杨叶片转化效率的调查，确定最佳的浸染时间为 7min。

表 5　农杆菌菌液浓度和浸染时间对转化的影响

接种	农杆菌菌液浓度				农杆菌菌液浸染时间			
叶片数（个）	菌液浓度（OD_{600}）	60d 后存活的抗性芽数(个)	60d 后死亡的抗性芽数(个)	转化率（%）	侵染时间（min）	60d 后存活的抗性芽数(个)	60d 后死亡的抗性芽数(个)	转化率（%）
30	0.2	1	79	1.25	5	5	79	5.95
30	0.4	10	56	15.15	7	12	68	15.00
30	0.6	3	63	4.54	10	2	23	8.00
30	0.8	0	8	0	15	0	0	0

2.2　转化植株的获得及 PCR 检测鉴定

总结上述优化农杆菌菌株转化试验，改良毛白杨转化程序具体步骤如下：叶片经过预培养 3d 后，用 OD_{600}=0.4 的菌液侵染 7min，再共培养 3d 后，移至含 Kan 20mg/L 和 Cef 250mg/L 的筛选培养基上培养，20~30d 后，叶片可分化出不定芽，再将叶片及分化的不定芽一起移至含 Kan 50mg/L 和 Cef 250mg/L 的培养基上继续进行筛选，会发现有部分不定芽逐渐变黄，甚至变白死亡。这样可提前去除一定量的假阳性不定芽，提高转化效率。

本试验从得到的 96 个卡那抗性芽中，去除了 31 株假阳性不定芽，筛选出 65 个抗性芽进行二次筛选。

不定芽是否可以在含 Kan 培养基上生根是鉴定植株是否成功转化的一个标志(王关林和方宏筠，2002)。二次筛选后，待不定芽长至 2~3cm 时，即可接种到含 Kan 50mg/L 和 Cef 150mg/L 的生根培养基上，此时会观察到仍有部分不定芽不能生根，甚至变白而死亡，又可进一步筛除假阳性植株。由于抗生素 Kan 和 Cef 等对植物生长和发育有一定的抑制作用(王关林和方宏筠，2002；王成等，2009；王瑶等，1999)，转化不定芽生根较正常植株推迟 5~15d。本研究一次筛选得到 65 个抗性芽，通过二次筛选去除了 22 个假阳性芽，只有 43 个抗性芽经生根长成完整的抗性植株。

取获得的 43 株卡那霉素抗性植株，提取植物基因组 DNA，以表达 *MdSPDS*1 基因的质粒作阳性对照(PC)，以未转基因植株作为阴性对照(WT)，进行 PCR 扩增，其中 9 株呈阳性，占抗性植株总数的 20.9%，阳性植株转化率达到 9.38%。阳性植株的 PCR 结果如图 1 所示，转化植株均扩增出与质粒 PC (*MdSPDS*1 基因，大小约 1200bp)大小一致的条带，而未转化植株没有得到相应条带，初步说明 *MdSPDS*1 基因已整合到毛白杨基因组中。

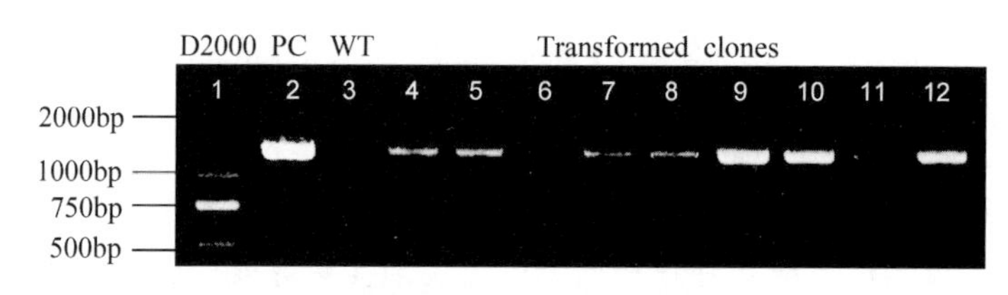

图 1　9 株阳性植株 PCR 扩增产物电泳结果

泳道 1：标准分子量 D2000，泳道 2：阳性对照 PC，泳道 3：阴性对照 WT，泳道 4~12：阳性植株。

3　结论与讨论

3.1　菌液浓度、侵染时间对遗传转化的影响

侵染过程中，适宜的菌液浓度和充分的侵染时间有利于农杆菌充分吸附到外植体表面，选择恰当的农杆菌浓度和侵染时间，对转化效率至关重要。不同菌液浓度对胡杨(*Populus euphratica*)转 *GUS* 基因有极显著的影响，OD_{600} 值为 0.4~0.6 的农杆菌菌液浓度较为合适(范源伟等，2009)。在以杂交杨(毛新杨×毛白杨)((*Populus tomentosa* × *P. bolleana*) × *P. tomentosa*)为材料的遗传转化体系中发现：当菌液的 OD_{600} 为 0.2 或 0.4，侵染时间 10~20min 时，Kan 抗性芽的产生频率较高；当菌液浓度过高或过低、侵染时间过长或过短时，抗性芽的产生频率明显降低(郝贵霞等，1999)。在三倍体毛白杨的遗传转化中，当菌液浓度 OD_{600} 值为 0.3~0.5 侵染时间为 15~20min 时，外植体转化频率最高(赵华燕等，2001)。本研究得出用 $OD_{600}=0.4$ 的菌液侵染 7min 是较合适的侵染方法。

根癌农杆菌感染植物过程中，农杆菌在植物细胞表面的附着是实现 T-DNA 转移和整合的前提，因此菌液活性和纯度直接影响根癌农杆菌菌体在植物细胞表面的附着状态。以

前的学者在研究中多采用二次摇菌的方法制备工程菌液(郝贵霞等, 1999; 樊军锋等, 2002; 陶晶等, 2001)。本研究也借鉴前人经验, 二次摇菌后当菌液浓度达到 $OD_{600}=0.4$ 时, 农杆菌处于生长对数期, 活性最强, 以确保菌液能充分侵染切口细胞。另外, 二次摇菌时不加 Kan, 可以使菌液保持最佳的活性, 提高侵染效率, 同时也降低了 Kan 在共培养阶段对细胞再生的抑制作用。

3.2 预培养和共培养时间对遗传转化的影响

研究表明, 预培养有利于杨树的转化(Confalonieri et al., 1994)。农杆菌介导杨树 NL-80106 遗传转化中发现, 预培养 1~2d 既可提高抗性芽诱导率, 又能减少假阳性植株的产生(饶红宇等, 2000)。此外, 农杆菌和外植体的共培养, 在整个遗传转化过程中是非常重要的环节, 由于农杆菌附着, T-DNA 的转移及整合都在共培养时期内完成。共培养时间的长短直接影响目的基因的整合及转化细胞的数量。若共培养时间过短, 农杆菌不能充分侵染切口细胞, 转化率低; 共培养时间过长, 则农杆菌生长过多, 会导致转化外植体死亡。因此共培养技术的掌握是植物遗传转化的关键, 是影响转化率的重要因素。不同植物、不同外植体对不同农杆菌菌株的敏感程度并不一样, 最适共培养时间也不同。共培养时间应以培养皿中转化叶片与培养基接触处出现肉眼可见菌落为宜(樊军锋等, 2002)。有研究指出, 不能以共培养时间长短来判断共培养是否成功, 如果共培养 2d 后仍未出现菌落, 则此转化的外植体一般不会产生抗性芽(周冀明等, 1996)。本实验中以共培养 3d, 观察到叶片边缘有明显的肉眼可见的菌落时为最佳共培养时间。

3.3 选择压对转化频率的影响

在农杆菌介导的杨树转基因试验中, 多将卡那霉素作为转化后的愈伤组织和植株的选择抗生素。抗生素对植株再生和生根有很大的抑制作用(Fillatti et al., 1987; Brasileiro, 1991; 赵世民等, 1999), 用适当浓度的 Kan 和抑菌素筛选获得抗性芽是很关键的一步。Kan 浓度过高, 对转化细胞造成强烈的毒害作用, 致使大量外植体在选择培养基上很快死亡, 从而降低了转化率; 浓度过低, 对抗性芽不能进行严格筛选, 会产生假抗性芽和大量嵌合体(Confalonierim et al., 2000)。抑菌抗生素在抑制农杆菌生长繁殖的同时, 对植物细胞同样也会产生一定的伤害作用, 抑菌素浓度过高或过低, 均不能恰当控制根癌农杆菌生长, 不能有效得到抗性芽。用根癌农杆菌转化植物细胞时, 过量 Cef 易引起伤口细胞褐化、死亡, 抑制芽的分化(姚娜等, 2007; 王树耀等, 2005)。本研究中借鉴了相关研究中采用的卡那霉素二次筛选的方法(樊军锋等, 2002), 先以较低浓度的 Kan 进行筛选, 防止某些转基因苗因为抗生素浓度过高而被杀死, 之后再经过 50mg/L Kan 的严格筛选, 抑制或杀死了大量非转化不定芽, 避免了大量非转化植株的产生, 减少了后继 PCR 检测工作量, 既提高了叶片的分化效率, 又提高了转化频率。

A Study on the Efficient Protocol for Transforming *MdSPDS*1 Gene into *Populus tomentosa* Carr.

Abstract　In order to establish the high efficiency genetic transformation system of *Populus tomentosa* Carr. by *Agrobacterium tumefaciens*, and to obtain mass-produced transgenic *P. tomentosa* with *MdSPDS*1 gene, several key factors of *P. tomentosa* transferring system were optimized during the *MdSPDS*1 transformation. The optimized condition for transformation of *P. tomentosa* was obtained as follows: the explant was pre-cultivated for 3 days and infected with *Agrobacterium* at the concentration of $OD_{600}=0.4$ for about 7min. The foliages were co-cultured with *Agrobacterium* for 3 days. The prophase and anaphase selecting methods were applied under the selection pressure of Kan 20mg/L and 50mg/L, respectively. At last, 9 out of 43(20.9%) Kan resistant plantlets were positive by PCR identification, suggesting that *MdSPDS*1 gene has been integrated into *P. tomentosa* genome. And the transformation frequency was increased to 9.38%. This work laid a foundation for appraisal of the gene function and the analysis of the salt-tolerant capability of the transgenic plants in the future.

Keywords　*MdSPDS*1, transformation, *Populus tomentosa*, *Agrobacterium tumefaciens*

第五部分

杨树光合作用、水分生理及矿质营养吸收

三倍体毛白杨无性系光合特性的研究*

摘　要　该文对 14 个毛白杨三倍体无性系和 6 个二倍体无性系盆栽苗的光合垂直变化和功能叶净光合速率进行了测定，对光合指标与生长的相关性进行了分析。结果表明三倍体无性系和二倍体无性系的光合垂直变化基本趋势相似，但在光合速率的波动幅度上存在着差异，生长快的三倍体无性系下部叶片能维持较高的光合速率。功能叶净光合速率在无性系间存在显著的差异，各光合指标与苗高和地径的相关均达到显著或极显著的水平，其中以净光合速率×总叶面积与生长的相关最紧密，相关系数在 0.80 以上。
关键词　毛白杨，三倍体，二倍体，光合，变异，相关

光合作用是树木生长的物质基础，叶片光合速率的高低直接影响林木的生长和产量。对杨树不同无性系间光合特性的变异已进行了许多的研究（Isebrands et al., 1988; Ceulemans et al., 1983, 1987; Nelson et al., 1988）。由北京林业大学培育出来的毛白杨优良三倍体无性系在生长上明显优于正常的二倍体无性系。但是对这些优良三倍体无性系的光合生理和光合特性遗传变异的研究报道还很少(杨敏生等，1991)。本文对 14 个毛白杨三倍体和 6 个二倍体无性系的光合特性进行对比研究，分析光合特性与生长的相关，从光合生理角度为揭示毛白杨三倍体无性系的速生性提供理论依据。

1　材料与方法

1.1　材料

供试材料包括 14 个毛白杨三倍体无性系和 6 个二倍体无性系。其中三倍体无性系是北京林业大学毛白杨课题组利用人工方法培育出来的杂种三倍体无性系，二倍体无性系为选种无性系。试验材料来源于北京林业大学毛白杨科研基地河北邯郸市峰峰矿区苗圃场。1998 年 3 月每个无性系挑选 9 个芽饱满度一致的嫁接条扦插于规格为 24cm×32cm×35cm 的砂盆中，盆土的组成和深度保持一致。无性系的放置采用完全随机区组排列，分 3 个区组，3 株小区。定期定量灌水。

1.2　方法

1.2.1　光合速率垂直变化测定

1998 年 8 月初对 3 个毛白杨三倍体无性系 B330，B301，B304 和 2 个二倍体无性系

---

* 本文原载《北京林业大学学报》，2000，22(6)：12-15，与李静怡合作发表。

BM33 和 BM86 的光合垂直变化进行测定。每个无性系在 3 个小区内各随机抽取一株，从顶端完全展开的第一片幼叶开始，依次测定主径上每片叶片的净光合速率。测定仪器为美国生产的 Li-6200 型便携式光合分析系统，利用镝灯作为人工光源。同时测定每片叶片的叶面积，测定方法采用硫酸纸称重法。

1.2.2　功能叶净光合速率的测定

1998 年 8 月 15 日和 10 月 12 日分别测定了 20 个无性系的净光合速率。每个无性系分别在 3 个小区内各随机抽取一株，所测定的叶片为苗木的第 7~8 片叶片。测定仪器为美国生产的 Li-6200 型便携式光合分析系统，利用镝灯作为人工光源。在测定前日对苗木进行灌溉，然后移至阴棚。在测定前进行预照处理，每片叶片的测定重复 2 次。采用抽样调查计算单株总叶面积。

对所测定的净光合速率和净光合速率×单株总叶面积进行方差分析。以这 3 个指标对苗高和地径分别进行一元线性回归分析。

2　结果与分析

2.1　光合速率垂直变化

图 1 为 3 个三倍体无性系 B330、B301、B304 和 2 个二倍体无性系 BM33、BM86 的光合速率垂直变化趋势。5 个无性系光合速率垂直变化的基本趋势是一致的，顶端第一片刚展平的幼叶其光合速率均为负值(图上未表示出来)，从第二片叶开始，光合速率变为正值，往下逐渐增大，在第 7~11 片叶光合速率达到最大值，然后向苗木的基部逐渐下降。二倍体无性系 BM86 和 BM33 的光合速率在第 7~9 片叶达到最大值；三倍体无性系 B330、B301、B304 分别在第 11，10 和 8 片叶达到最大值。

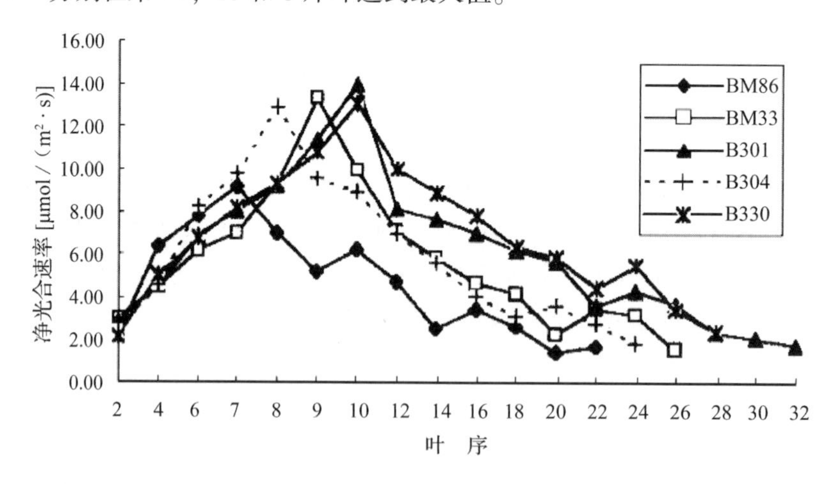

图 1　毛白杨无性系光合速率垂直变化

虽然各无性系的光合速率垂直变化在一般趋势上有相似之处，但在光合速率的波动幅度上存在着差异。生长差的二倍体无性系 BM86 的总体光合速率低，在第 7 片叶达到最大值后，光合速率迅速下降，其第 10 片以后的叶片光合速率很低，基本在 1.00~4.00μmol/

(m^2 · s)之间；二倍体无性系中生长最好的无性系 BM33 和三倍体无性系 B304 的光合速率变化趋势相似，两个无性系的光合速率在第 8~9 片叶达到最大值后，下降不像 BM86 那样迅速，在第 16~17 片叶还能保持在 4.00~5.00μmol/(m^2 · s)左右；三倍体无性系 B301 和 B330 下部叶片的光合速率要比其它 3 个无性系高，B330 第 24 片叶片的光合速率还保持在 5.63μmol/(m^2 · s)，B301 第 20 片叶片的光合速率为 5.76μmol/(m^2 · s)。以光合速率大于 5.00μmol/(m^2 · s)的叶片为标准，BM86 有 8 片，BM33 有 11 片，B304 有 12 片，B330 有 20 片，B301 有 16 片。因此 B330 和 B301 具有一个较大的由光合速率较高的叶片组成的垂直带。

由于叶片的光合潜力还依赖叶片的大小，本次实验测定了苗木从上到下所有叶片的叶面积，分析了 5 个无性系单叶光合（单叶面积×净光合速率，umol/s）的垂直变化趋势，结果如图 2 所示。

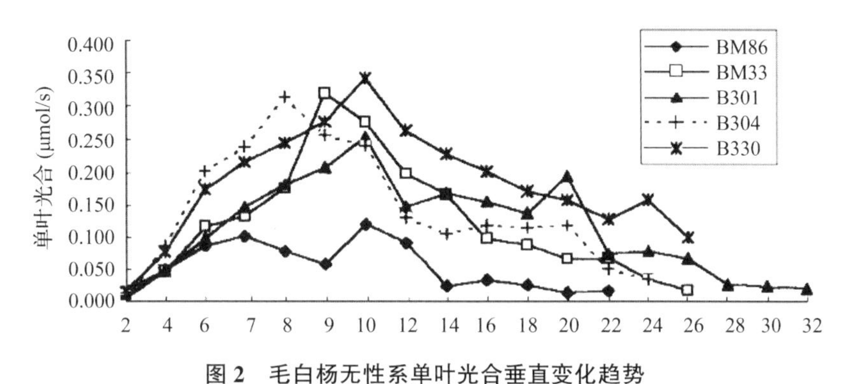

图 2　毛白杨无性系单叶光合垂直变化趋势

从图 2 可看出，单叶光合的变化趋势与净光合速率的变化趋势相似。顶部的叶片由于叶面积和光合速率均小，单叶光合低，随着叶片的发育和光合速率的增大，单叶光合不断增大，在达到最大值后，由于叶片光合能力的下降，单叶光合逐渐下降。但是，由于考虑了叶面积的大小，5 个无性系在单叶光合上的差异要比光合速率的差异更加明显。二倍体无性系 BM86 由于光合速率和单叶面积均低，其单叶光合远远低于其它 4 个无性系。三倍体无性系 B304 第 2~8 片叶片的单叶光合要高于其它无性系。三倍体无性系 B330 第 10 片以后的叶片的单叶光合明显高于其它无性系相对应部位叶片的单叶光合。

2.2　净光合速率的变异

各无性系在 8 月和 10 月所测定的净光合速率如图 3 所示。由于 8 月苗木正处于旺盛生长阶段，叶片的光合能力强，光合速率要高于 10 月苗木叶片的光合速率。

对各无性系的净光合速率（*Pn*）和净光合速率×总叶面积（*WTP*）进行方差分析，发现净光合速率和净光合速率×总叶面积在无性系间均存在极显著的差异(表 1)，其中以 WTP8 在无性系间的差异最显著。4 个光合指标的无性系重复力在 0.7561~0.8305 之间，表明无性系的净光合速率和净光合速率×总叶面积受强的遗传控制。

为了进一步探讨这些光合指标与生长的相关关系，利用上述 4 个光合指标分别与苗高和地径作一元线性回归，结果见表 2。4 个指标与苗高和地径的相关系数均达到了极显著的水平，其中以 WTP8 和 WTP10 与苗高和地径的相关最紧密，相关系数在 0.80 以上。

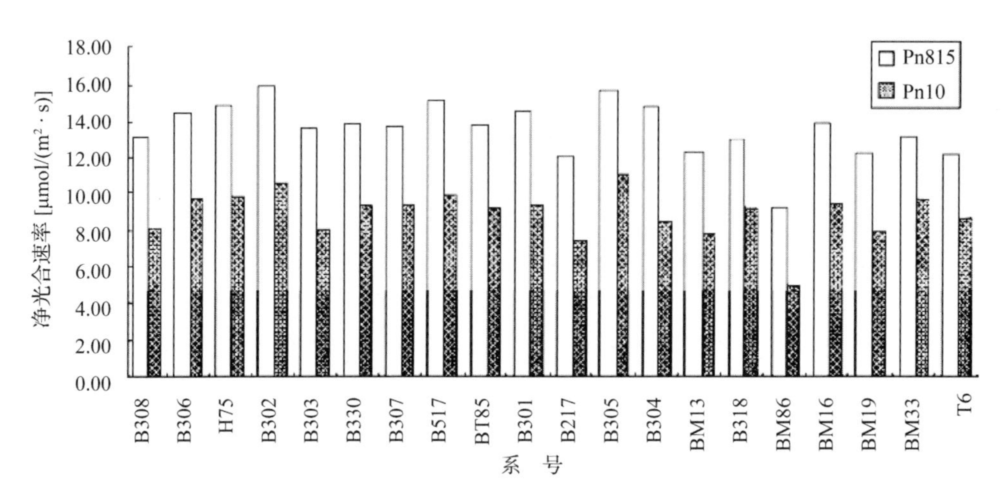

图 3　毛白杨无性系在不同时期测定的净光合速率

表 1　毛白杨无性系光合指标方差分析

指标	*SS*	*MS*	*F*	无性系重复力 *R*
Pn8	76. 787	4. 041	4. 10**	0. 7561
Pn10	96. 837	5. 097	4. 80**	0. 7917
WTP8	148. 169	7. 798	5. 90**	0. 8305
WTP10	69. 745	3. 671	4. 10**	0. 7561

表 2　光合指标与苗高和地径的相关性

指标	回归方程	相关系数	*F* 值	*P* 值
Pn8	$Y_h = 0.240 + 0.149X$	0. 754	23. 652	0. 000
	$Y_d = 0.370 + 0.096X$	0. 754	23. 652	0. 001
Pn10	$Y_h = 1.114 + 0.128X$	0. 612	10. 786	0. 004
	$Y_d = 0.908 + 0.085X$	0. 605	10. 371	0. 005
WTP8	$Y_h = 1.428 + 0.142X$	0. 836	41. 634	0. 000
	$Y_d = 1.122 + 0.093X$	0. 817	36. 161	0. 000
WTP10	$Y_h = 1.596 + 1.217X$	0. 806	33. 267	0. 000
	$Y_d = 1.217 + 0.135X$	0. 815	35. 643	0. 000

3　讨论

杨树叶片光合系统的发展与 RuBP 羧化加氧酶和希尔反应相关（Dickmann, 1971），而且叶片气孔的成熟度也影响光合活性（Michael et al. , 1990）。尹伟伦（1983）和杨敏生等（1991）研究认为光合速率的垂直变化与苗木的生长表现有关。生长势较好的树种最大光合速率出现的叶片位置靠下，并且下降得比较缓慢。本实验研究的 3 个三倍体无性系和 2 个二倍体无性系的最大光合速率出现在苗木的第 7~11 片叶片，没有很大的差别。但是生长快的三倍体无性系 B330 和 B301 具有较大的由较高光合速率的叶片组成的垂直带，特别是下部叶片能维持较高的光合速率。对三倍体无性系 B330 的叶片进行观察发现，不同叶龄的叶片从上到下在颜色和质地上看不出明显的变化。因此，三倍体无性系下部叶片光合能力衰退缓慢有可能是三倍体无性系生长快的重要生理原因之一。

在杨树中有许多关于净光合速率与生长呈正相关的报道。Gatherum 等（1967）报道单位面积的净光合速率与植物干重相关；Ceulemans 等（1983）研究认为，最大光合速率与无性系第一、第二年的苗高呈正相关，净光合速率可以作为一个早期的选择标准；Ceulemans 等（1987）的研究表明，一年生容器苗光饱和时的净光合速率与苗高显著相关。在本研究中，无性系在光合速率上存在极显著的差异。苗木 8 月和 10 月的净光合速率与生长均存在极显著的正相关。由于净光合速率×总叶面积考虑了叶面积的差异，因此更接近于苗木的光合潜力。一些研究表明净光合速率×总叶面积与生长或生物量紧密相关（Isebrands et al.，1988；张志毅等，1992；刘雅荣等，1983）。本研究中以净光合速率×总叶面积表示的 WTP 在无性系间存在极显著的差异，并且与生长的相关要高于净光合速率与生长的相关，相关系数在 0.80 以上（$P=0.000$）。因此，结合净光合速率和叶面积对毛白杨三倍体和二倍体无性系进行评价是比较可靠和理想的。

Genetic Variation in Photosynthetic Traits of Triploid Clones of *Populus tomentosa*

Abstract The within tree photosynthesis and the net photosynthetic rate were studied in 20 *Populus tomentosa* clones, including 14 triploid clones and 6 diploid clones. Similarity in the within tree photosynthesis patterns was found in 3 triploid clones: B330, B301, B304, and 2 diploid clones: BM33 and BM86, but higher photosynthetic rate was found in the old leaves of clones B330 and B301. Clonal differences in the photosynthetic traits were significant and repeatabilities of the photosynthetic traits ranged from 0.7561 to 0.8305. The photosynthetic traits were significantly correlated with stem height and basal diameter, and the whole tree photosynthesis showed a more significant correlation with growth ($r>0.80$).

Key words *Populus tomentosa*, triploid, diploid, photosynthesis, variation, correlation

美洲黑杨与大青杨杂种无性系苗期光合特性研究*

摘　要　对杨树光合特性的研究一直是杨树遗传改良的一个重要方面，为探讨美洲黑杨与大青杨杂种无性系苗期的光合特性和规律，本文以美洲黑杨与大青杨杂交所得 3 个生长及抗性等表现优良的杂种无性系 1 年生盆栽扦插苗为材料，采用 Li-6400 光合测定系统对 3 个无性系的光合作用指标进行测定。结果表明，在 7 月中旬，天气晴朗条件下，3 个无性系的净光合速率和蒸腾速率日变化均呈不对称的双峰曲线，气孔导度和胞间 CO_2浓度日变化分别呈单峰和倒双峰变化趋势；编号为 191 的无性系净光合速率日变化曲线两次峰值均高于 165 和 177，191 和 177 的低值高于 165；3 个无性系的净光合速率-光响应曲线相近，净光合速率均随着光合有效辐射的增加而增大，接近光饱和点时上升趋势平缓；191 的光饱和点高于 165 和 177，而光补偿点低于 165 和 177，表明 191 适应光强的能力强于 165 和 177。

关键词　美洲黑杨，大青杨，杂种，光合作用

美洲黑杨(*Populus deltoides* Bartr.) 原产于北美密西西比河沿岸，拥有生长期长、生长量大等优良特性，是北美重要的森林树种，又是美洲和欧洲的造林树种(赵天锡，1980)。

大青杨(*P. ussuriensis* Kom.)，又称憨大杨、哈达杨。其树高 30m 左右，胸高直径可达 2m，单株材积可达 $10m^3$。自然分布于我国东北的长白山、小兴安岭林区，材质优良，抗寒性强，是东北三省东部山区森林更新的主要树种之一(王冰等，2003)。

光合作用是绿色植物利用太阳光能同化二氧化碳和水，制造有机物质并释放氧气的过程，是绿色植物体内有机物质和能量的最终来源，在一定程度上决定着植物的生长(杨细明等，2008)。对杨树光合特性的研究一直是杨树遗传改良的一个重要方面，目前国内外对杨树的光合作用进行过大量的研究(Ceulemans et al.，1987；1989；胡新生等，1997；邓松录等，2006；张津林等，2006)，然而对于美洲黑杨与大青杨杂种的光合作用研究尚未见报导。本研究以美洲黑杨与大青杨杂交所得的 3 个生长及抗性等表现优良的杂种无性系一年生盆栽扦插苗为材料，通过对其光合日变化曲线及光响应规律进行研究比较，初步揭示了美洲黑杨与大青杨杂种无性系苗期的光合特性和规律，为深入探索美洲黑杨与大青杨杂种无性系光合特性与生长的关系以及为无性系的进一步选育提供科学依据。

1　材料与方法

1.1　试验材料

试验于 2008 年 7 月在北京林业大学苗圃院内进行。试验材料为 2008 年 3 月在河北省

* 本文原载《北京林业大学学报》，2009，31(5)：151-154，与江锡兵、李博、马开峰、何占国和刘承友合作发表。

平泉县种苗站美洲黑杨与大青杨杂种无性系人工测定林中选取的生长及抗性等表现优良的3个杂种无性系，编号分别为165、177和191。采集其一年生萌条，2008年4月在北京林业大学苗圃院内采用不催根扦插方法扦插于大盆中，大盆规格为30cm×25cm，每个无性系扦插15~25株。

1.2 试验方法

每个无性系选取标准株3株，选取饱满的枝条及同一部位受光量一致的完整功能叶片，采用美国Li-COR公司生产的Li-6400光合测定系统对3个无性系的光合作用指标进行测定。

光合作用-光响应曲线的测定：测定时间为2008年7月9日08:30~11:00。采用Li-6400-02B红蓝光源，光合有效辐射(*PAR*)在0~2000μmol/(m^2·s)范围内设定14个梯度，从高到低分别为2000、1800、1600、1400、1200、1000、800、600、400、200、100、50、20、0μmol/(m^2·s)，通过系统自动测量程序测定相应的净光合速率(*Pn*)，每个无性系作3次重复。采用二次多项式形式的最小二乘法进行光响应曲线拟合，求出光饱和点(*LSP*)，并将光合有效辐射(*PAR*)在0~200μmol/(m^2·s)范围内的测定值进行直线回归分析，求出光补偿点(*LCP*)和表观量子效率(*AQY*)。

光合作用日变化的测定：2008年7月12日(晴)，06:00~18:00，每2h测定1次叶片的净光合速率(*Pn*)，同时测定蒸腾速率(*Tr*)、气孔导度(*Gs*)、胞间CO_2浓度(*Ci*)及光合有效辐射(*PAR*)等生理参数，叶片每次按相同顺序测量，每1标准株测3叶片，每1叶片测10个数据。取平均值进行分析。

1.3 数据统计分析方法

应用SPSS统计分析软件进行数据分析处理，应用Excel软件进行图形和表格处理。

2 结果与分析

2.1 3个无性系苗期光合作用日变化

2.1.1 净光合速率(*Pn*)日变化

3个无性系*Pn*日变化均呈不对称的双峰曲线，如图1a所示。08:00时，3个无性系同时出现第一个高峰，也是一天之中最高值，分别为6.58、6.12和8.96μmol/(m^2·s)。随后开始迅速下降，出现光合“午休”现象，在14:00时，3个无性系同时达到低谷。光合“午休”后3个无性系的*Pn*均有小幅回升，且趋势相近，至16:00时，同时出现第二个高峰，分别为3.04、3.93和4.21μmol/(m^2·s)。在日变化进程中，191的峰值明显高于165和177，165的低值明显低于177和191。

2.1.2 蒸腾速率(*Tr*)日变化

如图1b所示，3个无性系的*Tr*日变化均呈双峰曲线，变化趋势与*Pn*日变化一致。在日变化进程中，3个无性系165、177、191的两次峰值分别为3.53、4.38、5.32mmol/(m^2·s)和2.77、2.91、4.12mmol/(m^2·s)，191的两次峰值明显高于165和177。

2.1.3　气孔导度(*Gs*)日变化

如图 1c 所示，06:00~08:00 时，3 个无性系 *Gs* 逐渐上升，至 08:00 时，同时达到全天中的最大值，分别为 0.13、0.12 和 0.26mol/(m^2·s)，191 的 *Gs* 值明显高于 165 和 177，均是它们的 2 倍左右。随着大气温度的不断上升，*Gs* 逐渐下降，在 16:00 时，191 又出现一个较明显的小峰，而 165 和 177 呈低值。

2.1.4　胞间 CO_2 浓度(*Ci*)日变化

如图 1d 所示，3 个无性系 *Ci* 日变化均呈倒双峰趋势。06:00 时，3 个无性系 *Ci* 值均为全天最大值，165、177 和 191 分别为 338、351 和 379mol/mol。随着大气 CO_2 浓度下降和气孔逐渐关闭，*Ci* 不断下降，165 和 177 于 10:00 时出现低谷，而 191 延迟至 12:00 时出现低谷。随后均呈上升趋势，至 16:00 时，3 个无性系又同时出现另一个低值，之后逐渐上升。

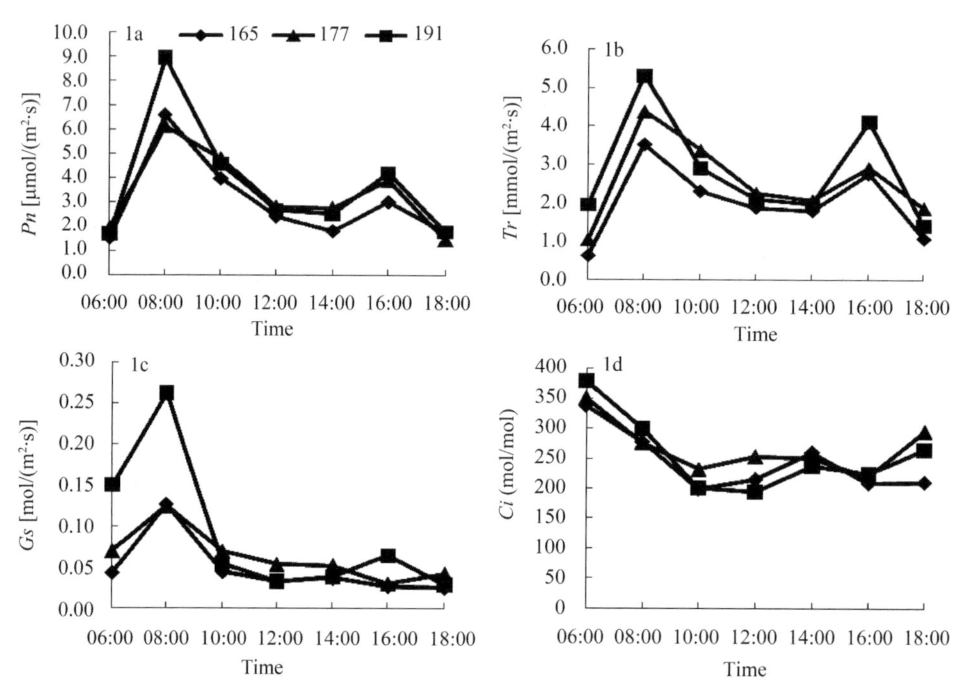

图 1　净光合速率及其他主要参数日变化

2.2　3 个无性系苗期光合作用—光响应规律

3 个无性系净光合速率(*Pn*)对光合有效辐射响应曲线如图 2 所示。可以看出，3 个无性系的曲线变化趋势非常接近，*Pn* 在光合有效辐射为 0~800μmol/(m^2·s)范围内迅速上升，尤其在 0~400μmol/(m^2·s)范围内接近直线上升趋势，光合有效辐射超过 800μmol/(m^2·s)后上升趋于平缓，逐渐接近光饱和。从表 1 分析得出，191 光饱和点(*LSP*)最高，165 次之，177 最低；165 的光补偿点(*LCP*)明显高于 177 和 191，三者的表观量子效率(*AQY*)相近。

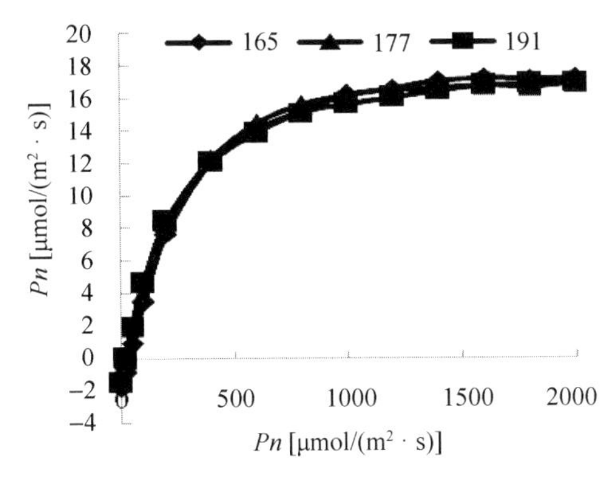

图2 3个无性系 *Pn*-光响应曲线

表1 3个无性系主要光合参数的差异

无性系	*LSP* (μmol/(m²·s))	*LCP* (μmol/(m²·s))	*AQY*
165	1389b	32. 50a	0. 046a
177	1333c	10. 22b	0. 046a
191	1438a	10. 02c	0. 045b

注：表中字母表示多重比较结果，不同小写字母表示无性系间差异达到 0. 05 显著水平。

3 小结与讨论

在 7 月中旬，天气晴朗的条件下，165、177 和 191 三个无性系的净光合速率日变化均为不对称的双峰曲线。最高峰值均出现在 08:00 时左右，然而此时光合有效辐射远没有达到一天的最大值，可能原因是由于此时大气温度适宜，光照充足，气孔开启程度较大，光合反应物 CO_2 浓度比较高，植物体内水分也比较充足，即影响 *Pn* 的杨树本身生理因子和环境因子相结合达到了全天的最佳或较佳组合状态，这与张津林等(2006)的研究结果相一致。3 个无性系 *Pn* 日变化进程均有光合“午休”现象，午间 *Pn* 降低时，*Ci* 却呈上升趋势，参照 Farquhar 等(1982)观点，只有在 *Ci* 降低而气孔限制值增大的前提下，气孔限制才是光合速率降低的主要原因；反之，如果光合速率降低时伴随着 *Ci* 的提高，那么非气孔因素即叶肉细胞的光合活性便是光合作用的主要限制因素。判定气孔限制和非气孔限制的重要指标和依据即 *Ci* 和气孔限制值。根据以上表明，日变化进程中 3 个无性系 *Pn* 午间降低主要是由非气孔限制因素引起的叶片光合能力降低造成的。*Pn* 日变化进程中，191 的两次峰值均高于 165 和 177，说明 191 在本身生理因子和环境因子相结合达到最佳或较佳状态时，相比于 165 和 177，能更好地利用光能，产生更多的净光合产物；165 在 14:00 时的低值明显低于 177 和 191，说明 177 和 191 在午间强光照射和高温的条件下适应性强于 165。

光饱和点和光补偿点是植物利用光强能力的重要指标(曹军胜，2005)，光饱和点高的植物被认为能更有效地利用强光(朱万泽，2004)，光补偿点低的植物被认为能更有效地利用弱光。从表 1 分析得知，191 的光饱和点高于 165 和 177，且差异显著，说明 191 利用强光的能力高于 165 和 177；191 的光补偿点低于 165 和 177，且差异显著，说明 191 在光强较弱的情况下能更有效地利用弱光。分析说明 191 对光强的适应能力强于 165 和 177。

此外，在日变化进程中 *Pn* 全天最大值为 8. 96μmol/(m²·s)，而在测定 *Pn* 对光合有效辐射响应曲线时 *Pn* 最大值却达到了 17. 2μmol/(m²·s)，两个过程中 *Pn* 最大值差别较大的原因可能与影响光合作用的诸多内部因子及环境因子有关，最可能原因是由于在测定日变化进程当天大气相对湿度普遍较低，叶片与大气水气压差增大导致部分气孔关闭，从而使得光合作用能力降低。具体原因需要进行进一步的研究论证。

Photosynthetic Characteristics of Hybrid Clones of *Populus deltoides* Bartr. and *P. ussuriensis* Kom.

Abstract　Photosynthetic characteristic is an important indication of genetic improvement of poplar. In order to study the photosynthetic characteristics of hybrid clones of *Populus deltoides* and *P. ussuriensis*, one-year-old potted cuttings of three hybrid clones(#165, #177, #191), which had better growth traits and resistance, were used as experimental materials for measurement of photosynthesis using Li-6400. The results from the measurement on a sunny day in the middle of July showed that there was a two-peak curve in the diurnal pattern of both the net photosynthetic rate and the transpiration rate of the three hybrids, while a one-peak curve and an inverse two-peak curve for stomata conductance and intercellular carbon dioxide concentration, respectively. Two peak values of daily changing curve of net photosynthetic rate of the sample of #191 had higher values in net photosynthetic rates than #165 and #177, and the vale values of #191 and #177 were higher than that in #165. The light response curves of net photosynthetic rate of three clones were similar to each other, and the net photosynthetic rate increased with the growth of photosynthetically active radiation, and then kept steady when reaching light saturation point. Light saturation point of #191 was higher than that of #165 and #177, suggesting that the ability of adapting radiation intensity of #191 is better than that of #165 and #177.

Key words　*Populus deltoides* Bartr., *P. ussuriensis* Kom., hybrid, photosynthesis

3 年生毛白杨无性系光合特性的比较研究*

摘　要　利用 Lico-6400 便携式光合测定仪，研究了 30 个毛白杨无性系光合指标的变化，并对环境因子与光合指标相关关系进行了探讨。结果表明：毛白杨无性系的净光合速率(*Pn*)、气孔导度(*Gs*)、蒸腾速率(*Tr*)日变化均呈典型双峰曲线，气孔限制是出现“午休”现象的主要调节因素，胞间 CO_2 浓度(*Ci*)日变化曲线呈典型的“V”字形。5 个毛白杨无性系的光合-光强(Pn-Par)响应曲线、光合-二氧化碳响应曲线(*Pn-Ca*)均呈“S”形，且符合 2 次曲线模型，饱和光强下 5 个毛白杨无性系瞬时 *Pn* 的排序为：BL204(24.64) > BL206(23.76) > BL30(21.50) > BL207(19.54) >BL63(18.64μmol·m^{-2}·s^{-1})。饱和 Par 和 Ca 条件下，无性系 BL206 瞬时 *Pn* 最大，达 30.15μmol·m^{-2}·s^{-1}，无性系 BL63 最小，只有 20.34μmol·m^{-2}·s^{-1}，5 个毛白杨无性系光补偿点(*lcp*)和二氧化碳补偿点(*ccp*)差异显著，*lcp* 变化范围为 33.08~81.17μmol·m^{-2}·s^{-1}，*ccp* 变化范围 74.03~93.35μmol·mol^{-1}。30 个毛白杨无性系间瞬时光合指标 *Pn*、*Gs*、*Ci* 和蒸腾速率(*Tr*)的差异均显著(Sig. <0.000)，各光合指标变异系数 CV 为 8.94%~23.22%，GCV 为 8.78%~22.79，重复力高，遗传因素在表型变异中起主要作用；毛白杨光合指标之间存在极显著相关，最高相关系数达 0.731。

关键词　毛白杨，无性系，净光合速率，变异

毛白杨(*Populus tomentosa*. Carr)是我国特有的杨属白杨派树种，具有生长迅速、材质优良、树干高大通直等特性，主要分布于我国黄淮海流域约 100 万 km^2 的范围内。在我国北方，尤其在黄河中下游林业生产和生态环境建设中占有重要地位。近些年，随着各种育种方法的不断成熟，生理、生化、分子标记等手段在杨树育种中被广泛应用，尤其是光合指标，在杨树抗性育种中起到迅速、便捷的指示剂作用。植物光合作用过程中，光合指标时刻受到环境因子的影响，国内外对杨树光合特性的报道有很多(Ceulenans et al., 1989; Bassman et al., 1991；吴瑞云，1999；苏东凯，2006；邓松录，2006；郑彩霞，2006；周永斌，2007；张守仁，2000；邱箭，2005；房用，2006)，但对自然环境下光合指标与环境因子相互作用的研究较少，同时利用光合指标对无性系进行立地选择的研究也不多(李静怡，2000)，本试验主要对自然条件下 30 个毛白杨无性系的瞬时光合指标进行测定分析，同时对光合指标与环境因子的相关性进行探讨，以期为了解毛白杨光合指标与环境因子的相互作用提供理论基础，为毛白杨无性系立地筛选提供新的手段和方法。

1　试验地概况

2006 年 4 月，在河北省邯郸市峰峰矿区苗圃场(E 114°03′40″, N 36°20′448″)营建毛

* 本文原载《林业科学研究》，2011，24(3)：370-378，与赵曦阳、马开峰、张明、边金亮、焦文燕合作发表。

白杨无性系试验林，该区年平均气温 13.5℃，最冷月份平均气温-2.3℃，极端最低气温-19℃，最热月份平均气温 26.9℃，极端最高气温 42.5℃，年平均降水量 589mm，无霜期约 200d。

2 试验材料与研究方法

2.1 试验材料

试验材料共 30 个毛白杨无性系，其中包括 29 个毛白杨杂交无性系(李善文，2005)(BL20、BL22、BL23、BL26、BL28、BL30、BL42、BL46、BL49、BL50、BL53、BL63、BL64、BL67、BL69、BL76、BL77、BL78、BL83、BL85、BL87、BL88、BL98、BL99、BL101、BL103、BL204、BL206、BL207)和无性系 LM50，试验林采用随机区组设计，4 株小区，单行排列，4 次重复(16 株/无性系)，株行距为 3m×4m，设 2 行保护行。数据调查于 2008 年 7 月 20 日进行。

2.2 研究方法

采用美国 Lico 公司生产的 Lico-6400 光合作用分析系统进行测定，该仪器可以对植株的单叶光合速率(*Pn*，$\mu mol \cdot m^{-2} \cdot s^{-1}$)、蒸腾速率(*Tr*，$mol \cdot m^{-2} \cdot s^{-1}$)、胞间 CO_2 浓度(*Ci*，$\mu mol \cdot mol^{-1}$)和气孔导度(*Gs*，$mol \cdot m^{-2} \cdot s^{-1}$)等光合指标进行活体测定，同时记录光合有效辐射(*PAR*，$\mu mol \cdot m^{-2} \cdot s^{-1}$)、空气温度($T_{air}$,℃)、空气 CO_2 浓度(*Ca*，$\mu mol \cdot mol^{-1}$)和空气相对湿度(*RH*,%)等环境因子。

2.2.1 叶片瞬时 *Pn* 测定

以试验林第二个区组中生长正常的无性系 BL204 单株为试验对象，选择 3 枝主干高度为 5m 处的南侧侧枝，自枝条顶端叶片至底端的叶片，测定其瞬时 *Pn*，测定时光照强度设定为 1400$\mu mol \cdot m^{-2} \cdot s^{-1}$，二氧化碳浓度设定为 400$\mu mol \cdot mol^{-1}$，温度设定为叶片温度，对其它环境条件没有特殊控制，叶片 *Pn* 稳定叶序作为枝条的功能叶片。利用相同方法对同一单株其它三个方向同一高度枝条上的叶片进行瞬时 *Pn* 测定，分析东、南、西、北方向叶片瞬时 *Pn* 是否存在显著差异，对无性系 BL204 其它单株相同功能叶片进行瞬时 *Pn* 测定，分析相同无性系不同单株间瞬时 *Pn* 是否存在显著差异。

2.2.2 光合指标日进程曲线的测定

2008 年 8 月 2 日，天气晴朗，6:00—19:00 对 2 个毛白杨无性系(BL204、BL30)净光合速率(Pn)日变化进行测定，每隔 1h 测定 1 次。每个系号随机抽取 1 株，每株选取 3 个主干高度 5m 处的侧枝，每个侧枝上选择 3 片功能叶片进行测定，仪器同时记录蒸腾速率(*Tr*)、气孔导度(*Gs*)、细胞间 CO_2 浓度(*Ci*)、环境 CO_2 浓度(*Ca*)、叶片温度(*Tair*)、光照强度(*Par*)和环境相对湿度(*RH*)。

2.2.3 毛白杨无性系光合—光强(*Pn-Par*)响应曲线的测定

对无性系 BL30、BL63、BL204、BL206 和 BL207 进行光曲线的测定，每个无性系选择生长良好 3 个单株，每株选择南侧 5m 高枝条上 3 片功能叶进行测定。测定时二氧化碳浓度控制在 400$\mu mol \cdot mol^{-1}$，光照强度梯度设定为 2000、1800、1600、1400、1200、1000、

800、600、400、300、200、100、50、0μmol·m^{-2}·s^{-1}，对温度和相对湿度因子没有特别控制，测定后的结果利用2次曲线方程对Par-Pn曲线进行模拟，方程模式：$Y=b_0+b_1X+b_2X^2$，同时计算曲线方程的光饱和点lsp，光补偿点lcp和最大*Pn*。

2.2.4 毛白杨无性系光合—二氧化碳(*Pn-Ca*)曲线的测定

对无性系BL30、BL63、BL204、BL206和BL207进行二氧化碳曲线的测定。材料同光曲线测定相同，测定时把光有效辐射控制在光饱和点，控制二氧化碳的浓度为1400、1200、1000、800、600、400、200、100、80、40、0μmoll·m^{-2}·s^{-1}(利用调节旋钮调节二氧化碳全吸收设定二氧化碳浓度为0)，对温度和湿度没有特别的控制，测定结果利用二次曲线方程对Ca-Pn曲线进行方程模拟，方程模型：$Y=b_0+b_1X+b_2X^2$，同时计算方程的 二氧化碳饱和点csp，二氧化碳补偿点*ccp*和最大*Pn*。

2.2.5 毛白杨无性系瞬时光合指标的测定

对30个毛白杨无性系进行瞬时光合指标的测定，每个无性系选择3个单株，每个单株选择3片功能叶片，光照强度设定为1400μmol·m^{-2}·s^{-1}，二氧化碳浓度设定为400μmol·mol^{-1}，其它环境因子没有特别控制，测定叶片瞬时P_n、T_R、C_i、G_S，利用公式$Wue=Pn/Tr$计算瞬时水分利用效率Wue，同时记录空气温度(*Tn*,℃)、空气CO_2浓度(*Ca*，μmol·mol^{-1})和空气相对湿度(*RH*,%)等环境因子。

2.2.6 光合指标与环境因子相关性分析

利用日变化测得的数据对各光合指标与环境因子进行相关性分析，同时利用逐步回归分析方法，计算*Pn*、*Tr*、*Ci*、*Gs*的回归方程，探讨环境因子对光合指标的贡献率。

2.2.7 数据统计分析方法

所有数据利用spss软件进行分析。

(1)方差分析线性模型为：$x_{ij}=\mu+P_i+a_j+e_{ij}$

式中：μ为总体平均值；P_i为无性系效应；a_j为区组效应；e_{ij}为环境误差。

(2)根据续九如(2006)的方法估算无性系重复力：$R=1-1/F$

式中：*F*为方差分析的*F*值。

(3)表型变异系数：$CV=S/\overline{X}\times100$

式中：*LS*为表型标准差；$\overline{X}$为某一性状群体平均值。

(4)遗传变异系数(续九如，1988)：$GCV=\sqrt{\sigma_g^2}/\overline{X}\times100\%$

式中：σ_g为遗传方差；$\overline{X}$为某一个性状的平均值。

(5)表型相关分析采用公式(续九如，2006)：$r_{p_{12}}=\dfrac{Cov_{p_{12}}}{\sqrt{\sigma_{p_1}^2\cdot\sigma_{p_2}^2}}$

式中，$Cov_{p_{12}}$为2个性状的表型协方差；$\sigma_{p_1}^2$、$\sigma_{p_2}^2$分别为2性状的表型方差。

3 结果与分析

3.1 毛白杨无性系叶片*Pn*变化规律

无性系BL204相同枝条叶片瞬时*Pn*变化如图1所示，从顶端至底端叶片*Pn*的变化呈先上升后下降曲线，第1片叶至第3片叶由于处于新生状态，叶片功能发育不成熟，叶绿素含量低，净光合速率低，第5~15片叶处于成熟稳定的功能叶序，第16片以下的叶片光合速率随着叶片的逐步老化而降低，选择第5~15叶片作为功能叶片。对同一单株的4个方向测定结果表明不同方向瞬时 *Pn* 差异不显著(sig 0.351)，相同无性系不同单株瞬时 *Pn* 差异也不显著(Sig. 0.449)。

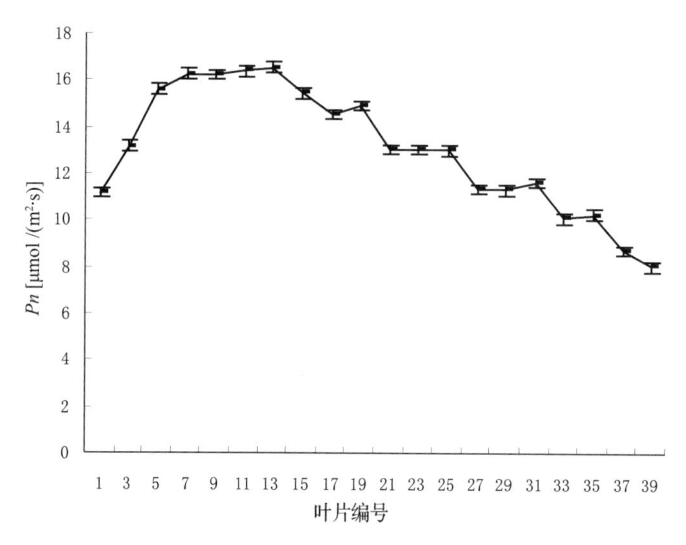

图1　同一枝条叶片净光合速率对比图

3.2　毛白杨光合指标日变化

由图2可知：净光合速率(*Pn*)的日变化呈典型双峰曲线，早晨6:00开始 *Pn* 迅速上升，无性系BL204第一个峰值出现在10:00(18.41μmol·m^{-2}·s^{-1})，之后下降，12:00(9.92μmol·m^{-2}·s^{-1})出现波谷，12:00后继续上升，16:00出现第2个波峰(18.66μmol·m^{-2}·s^{-1})。无性系BL30 *Pn* 第1个波峰出现在9:00(18.68μmol·m^{-2}·s^{-1})，之后迅速下降，波谷出现在12:00(12.07μmol·m^{-2}·s^{-1})，第二个峰值为14:00(19.76μmol·m^{-2}·s^{-1})，之后 *Pn* 缓慢下降，17:00仍保持较高的光合速率(16.69μmol·m^{-2}·s^{-1})，随着 *Par* 的变弱，18:00迅速下降至8.17μmol·m^{-2}·s^{-1}，19:00出现负值。2个毛白杨无性系的 *Gs* 与 *Tr* 日变化趋势基本与 *Pn* 相同，呈现双峰现象，中午12:00有波谷出现，说明2个毛白杨无性系的 *Pn* 与 *Gs*、*Tr* 有很强的相关性。2个无性系 *Ci* 的日变化趋势呈现典型的“V”字形曲线，从6:00开始下降，直到17:00出现波谷后稍有回升，这是植物利用二氧化碳进行光合作用而使环境中的二氧化碳浓度降低而导致的结果。环境因子的日变化趋势如图3所示，环境中 *Par* 与 *Tair* 呈单峰曲线，随着光照强度增加，温度上升，*Par* 于12:00出现最大值(1450.36μmol·m^{-2}·s^{-1})，*Tair* 于13:00出现最大值(34.29℃)。*Ca* 与 *RH* 呈先下降后上升的趋势。

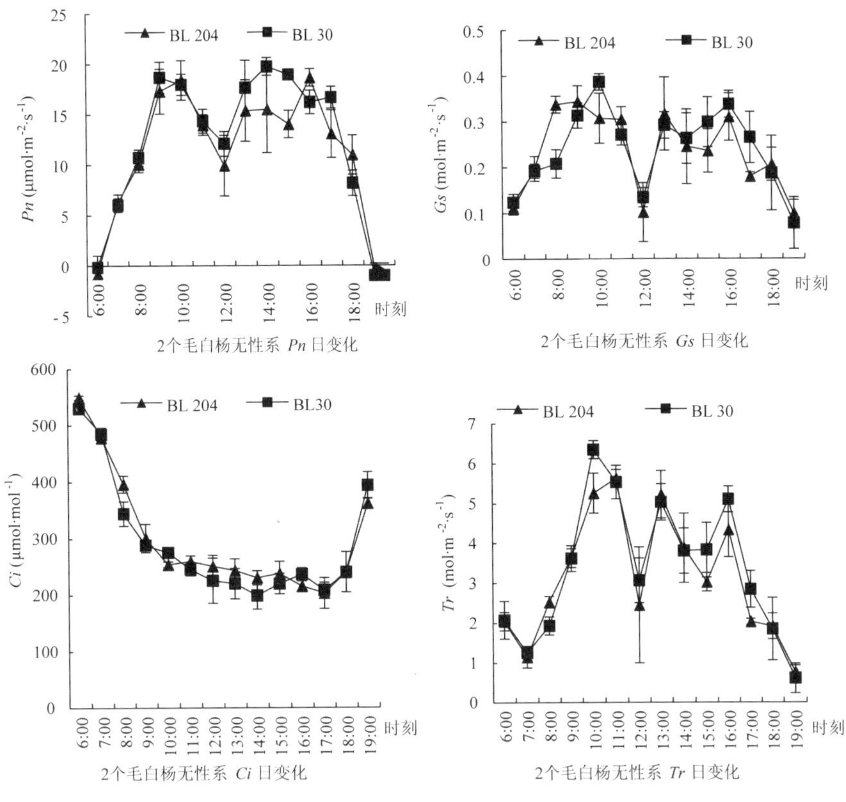

图2　2个毛白杨无性系光合指标日变化曲线

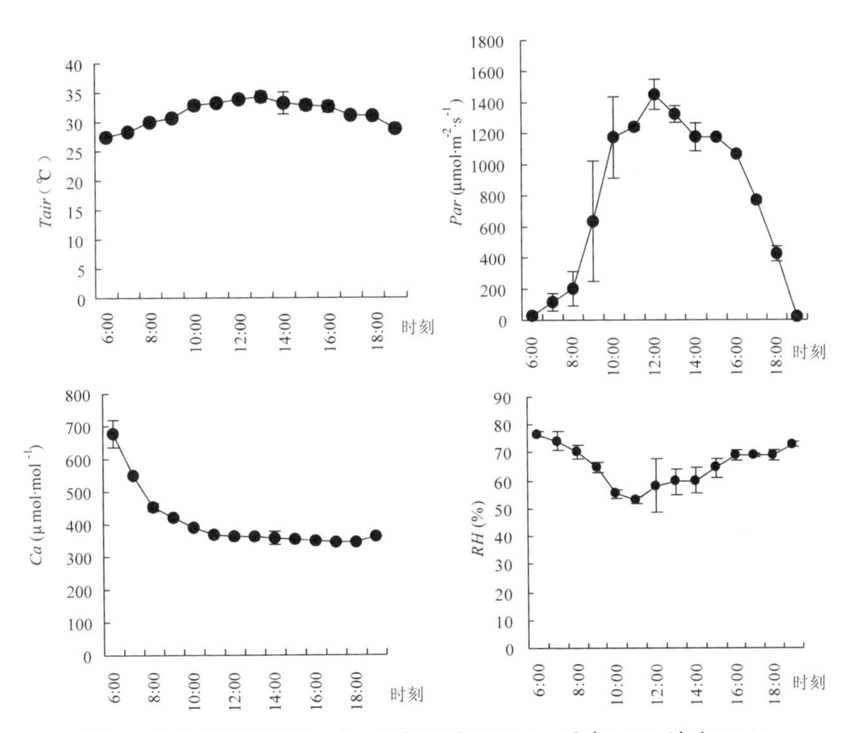

图3　光合有效辐射(Par)、空气温度(Tair)、空气 CO_2 浓度(Ca)和空气相对湿度(RH)的日变化曲线

3.3　5个毛白杨无性系的光合—光强(*Pn-Par*)响应曲线

5个毛白杨无性系的 *Pn-Par* 曲线(图4)呈S形，开始光照强度为0，5个毛白杨无性系的 *Pn* 均为负值，随着 *Par* 的增加，*Pn* 呈抛物线趋势上升，*Par* 增加到800μmol·m^{-2}·s^{-1}后，*Pn* 增加缓慢，当 *Par* 达到1400μmol·m^{-2}·s^{-1}，无性系瞬时 *Pn* 达到最大值，之后随着 *Par* 增加，*Pn* 不再增大，*Par* 达到2000μmol·m^{-2}·s^{-1}，无性系BL204出现轻微光抑制现象。毛白杨无性系的 *Pn-Par* 曲线模拟方程见表1，5个毛白杨无性系的lsp为1396.55~1469.86μmol·m^{-2}·s^{-1}，无性系BL63的lsp最高，但处于lsp的 *Pn* 只有18.64μmol·m^{-2}·s^{-1}。5个毛白杨无性系lcp差异显著，无性系BL63(81.17μmol·m^{-2}·s^{-1})最高，BL206(33.08μmol·m^{-2}·s^{-1})最低，说明无性系BL206在光照很弱的条件下就可以进行光合作用(冯岑等，2009)。5个毛白杨无性系的 *Gs-Par* 曲线和 *Tr-Par* 曲线均呈缓慢上升趋势，随着 *Par* 的增加，无性系BL30的 *Gs* 和 *Tr* 一直处于最大值，而BL207则一直处于最低值状态，*Par* 到达1800μmol·m^{-2}·s^{-1}，无性系 *Gs* 和 *Tr* 开始下降。5个毛白杨无性系的 *Ci* 均呈下降趋势，*Par* 达到600μmol·m^{-2}·s^{-1}以后，随着 *Par* 上升，*Ci* 处于平稳状态。

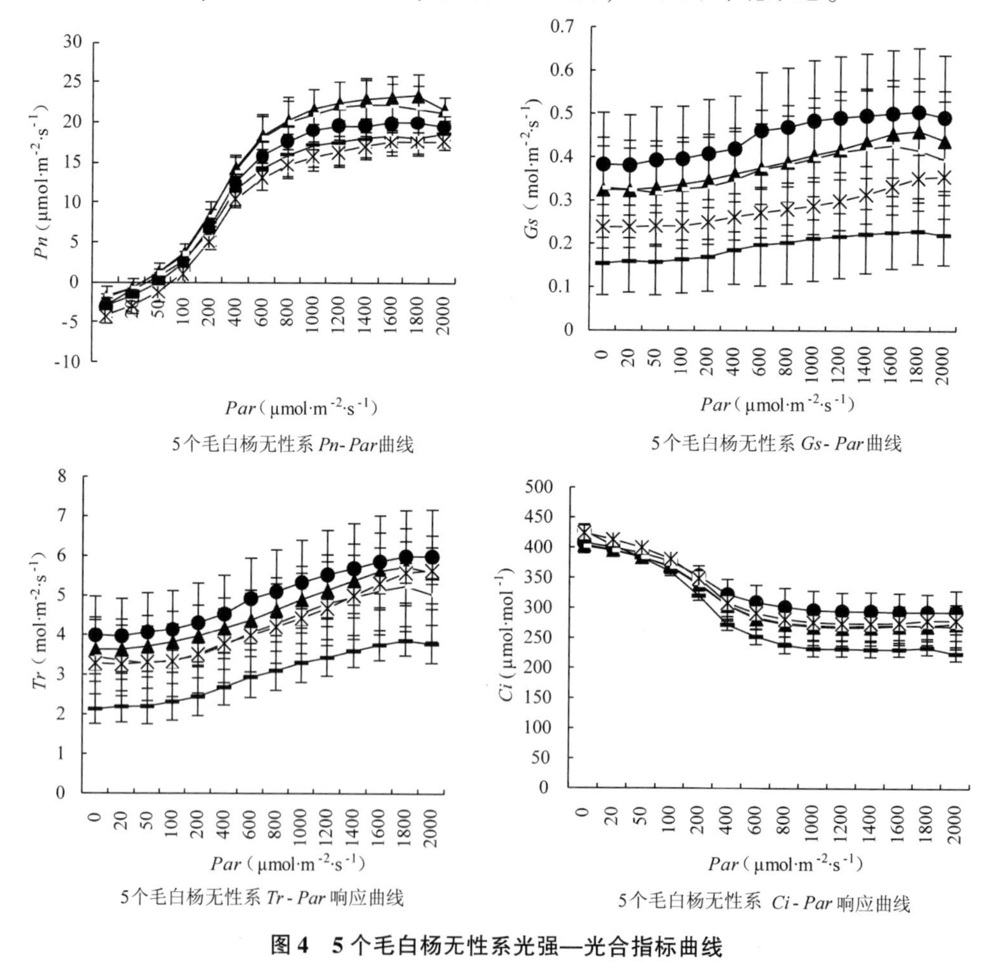

5个毛白杨无性系 *Pn-Par*曲线　　5个毛白杨无性系 *Gs-Par*曲线

5个毛白杨无性系 *Tr-Par* 响应曲线　　5个毛白杨无性系 *Ci-Par* 响应曲线

图4　5个毛白杨无性系光强—光合指标曲线

(注：—●— BL 30；—▬— BL 207；—▲— BL 204；—✕— BL 63；— — BL 206)

表 1 5 个毛白杨无性系 *Pn-Par* 曲线模拟方程以及最大 *Pn*、*lsp*、*lcp*

无性系	*Pn-Par* 曲线模拟方程	判定系数 R^2	最大 *Pn* ($\mu mol\cdot m^{-2}\cdot s^{-1}$)	*Lsp* ($\mu mol\cdot m^{-2}\cdot s^{-1}$)	*Lcp* ($\mu mol\cdot m^{-2}\cdot s^{-1}$)
BL30	$Y=-0.00001126\times Par^2+0.03189\times Par-1.0768$	0.972	21.50	1415.73	57.97
BL207	$Y=-0.00000964\times Par^2+0.02785\times Par-0.5732$	0.966	19.54	1444.61	54.18
BL204	$Y=-0.00001244\times Par^2+0.03517\times Par-0.2027$	0.979	24.64	1412.69	62.50
BL206	$Y=-0.00001216\times Par^2+0.03398\times Par-0.0353$	0.975	23.76	1396.55	33.08
BL63	$Y=-0.00000962\times Par^2+0.02829\times Par-2.1535$	0.966	18.64	1469.86	81.17

3.4 毛白杨无性系光合—二氧化碳(*Pn-Ca*)曲线

由图 5 可知：随着 *Ca* 增加，毛白杨无性系 *Pn* 增大，直到 Ca 达 800$\mu mol\cdot mol^{-1}$ 时，除 BL30 外，其他 4 个无性系的 *Pn* 均趋于平稳。由表 2 可知，无性系 csp 差异不显著，处

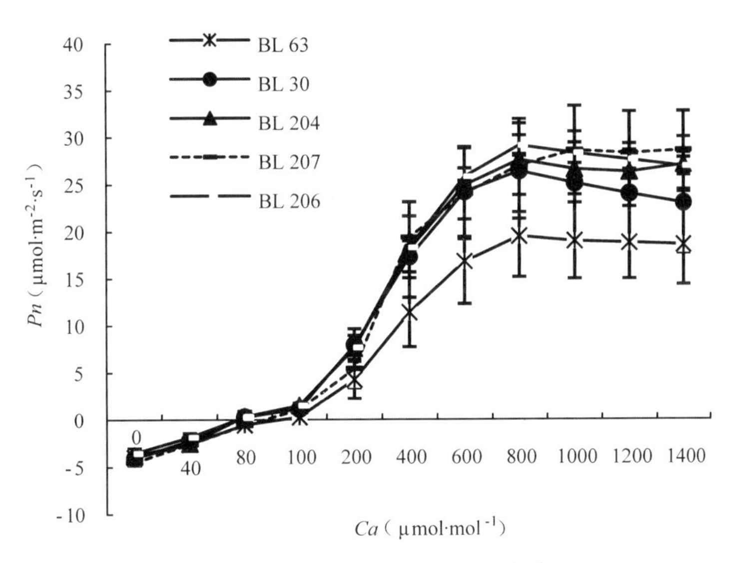

图 5 5 个毛白杨无性系 *Pn-Ca* 曲线

表 2 5 个毛白杨无性系 *Pn-Ca* 曲线模拟方程以及最大净光合速率(*Pn*)、二氧化碳饱和点(*csp*)、二氧化碳补偿点(*ccp*)

无性系	*Pn-Par* 曲线模拟方程	判定系数 R^2	最大 *Pn* ($\mu mol\cdot m^{-2}\cdot s^{-1}$)	*Csp* ($\mu mol\cdot mol^{-1}$)	*Ccp* ($\mu mol\cdot mol^{-1}$)
BL30	$Y=-0.00003320\times Par^2+0.06468\times Par-4.2576$	0.990	27.24	974.03	74.03
BL207	$Y=-0.00003000\times Par^2+0.06482\times Par-4.9111$	0.990	30.11	1080.50	80.68
BL204	$Y=-0.00003098\times Par^2+0.06416\times Par-4.1478$	0.988	29.07	1035.58	75.41
BL206	$Y=-0.00003306\times Par^2+0.06746\times Par-4.2544$	0.993	30.15	1020.11	74.94
BL63	$Y=-0.00002231\times Par^2+0.04668\times Par-4.0723$	0.995	20.34	1046.12	93.35

于 974. 03~1080. 50μmol · mol^{-1} 间，最大 *Pn* 和 *ccp* 差异显著，无性系 BL30 的 *ccp* 最小（74. 03μmol · mol^{-1}），无性系 BL63 的 *ccp* 最大（93. 35μmol · mol^{-1}）。饱和 CO_2、饱和 *Par* 条件下，无性系 BL206 的 *Pn*（30. 15μmol · m^{-2} · s^{-1}）最大，BL63 最小，只有 20. 34μmol · m^{-2} · s^{-1}。

3. 5　毛白杨无性系光合指标变异分析

光照强度与二氧化碳浓度固定的条件下，30 个毛白杨无性系的不同光合指标差异显著（表 3），平均 *Pn* 为 19. 82μmol · m^{-2} · s^{-1}，无性系 BL204 的 Pn 最大，为 24. 60，其次是 BL206 达 23. 17μmol · m^{-2} · s^{-1}，最小的是无性系 BL67，只有 15. 30μmol · m^{-2} · s^{-1}。Gs 平均值为 0. 37 mol · m^{-2} · s^{-1}，无性系 BL98（0. 518mol · m^{-2} · s^{-1}）最大，无性系 BL53（0. 196mol · m^{-2} · s^{-1}）最小；*Ci* 平均值为 263. 68μmol · mol^{-1}，最大值为无性系 BL26（309. 00μmol · mol^{-1}），最小值为无性系 BL23（212. 00μmol · mol^{-1}）；*Tr* 值最大的无性系为 BL98（5. 57 mol · m^{-2} · s^{-1}），最小值为 BL207（3. 16mol · m^{-2} · s^{-1}），Wue 是 *Pn* 与 *Tr* 的比值，30 个无性系中，无性系 BL206 的 Wue 最大（6. 76μmol · mol^{-1}），最小值是无性系 BL22（3. 34μmol · mol^{-1}）。30 个毛白杨无性系的 5 个光合指标的表型变异系数（*CV*）处于 8. 94%~23. 22%之间，遗传变异系数（GCV）处于 8. 78%~22. 79%之间，重复力均大于 0. 862，高变异、高重复力有利于无性系选择。

表 3　毛白杨无性系各光合指标遗传参数

性状	*df*	*mS*	*F*	Sig.	Mean	*S. D*	Min	Max	*CV*（%）	*GCV*（%）	R
Pn	29	13. 481	20. 279	0. 000	19. 82	2. 200	15. 30	24. 60	11. 10	10. 42	0. 9507
Gs	29	0. 0218	50. 519	0. 000	0. 37	0. 086	0. 196	0. 518	23. 22	22. 79	0. 9802
Ci	29	1640. 31	50. 871	0. 000	263. 68	23. 584	212. 00	309. 00	8. 94	8. 78	0. 9803
Tr	29	1. 1030	10. 103	0. 000	4. 38	0. 658	3. 16	5. 57	15. 02	13. 14	0. 9010
Wue	29	1. 189	7. 2489	0. 000	4. 60	0. 706	3. 34	6. 76	15. 34	12. 70	0. 8620

3. 6　光合指标与环境因子相关性分析及回归方程

植物的光合指标不仅受到生物结构特征的影响，同时与光照强度、温度、空气中的二氧化碳浓度和相对湿度等环境因子密切相关（温达志等，2000）。毛白杨无性系光合指标之间、光合指标与环境之间的相关系数见表 4，*Pn* 与 *Gs* 显著正相关，与 *Tr* 极显著正相关。环境因子 *Tair* 与 *Pn* 极显著正相关，*Ca*、*RH* 与 *Pn* 极显著负相关。*Gs* 与 *Ci*、*Tr* 和 *Ca* 极显著正相关。*Ci* 与 *Tr*、*Par* 和 *Tair* 极显著负相关，与 *Ca*、*RH* 极显著正相关。*Tr* 与所有环境因子相关均达显著水平。总之，光合指标之间、环境因子之间、光合指标与环境因子之间均存在着一定程度的相互作用关系。

为了探讨对光合指标影响最大的环境因子，进一步对各因子进行回归分析，采用逐步引入剔除法，对光合指标建立回归方程，模型的判定系数为 R^2，调整判定系数为 *Ra*，各个方程经过方差分析，得到 *F* 值与 *P*（Sig. ）值，检验方程的线性关系，结果见表 5，最先

引入 *Pn* 方程的变量是 *Par*，说明 *Par* 是影响叶片 *Pn* 的主导因子，*Ca* 也引入 *Pn* 回归方程，R^2=0.604，判定系数 *Ra*=0.601，Sig. 值 0.00，说明 *Pn* 与 *Par*、*Ca* 有显著线性关系。最先引入 *Ci* 回归方程的环境因子是 *Ca*，环境中的 *Ca* 对细胞间隙的二氧化碳浓度有最显著的影响，由日变化可以看出，2 个毛白杨无性系的 *Ci* 与 *Ca* 变化趋势相同，可以说明 *Ci* 与 *Ca* 具有一定的相关性。*Gs* 回归方程中最先引入的变量是 *Ca*，虽然判定系数只有 0.248，但是 Sig. 0.00，可以判定线性模型拟合显著。*Tr* 方程中最先引入的变量是 *RH*，其次是 *Par*。由表 4 还可以发现，4 个光合指标回归方程中，*Par* 均被引入，说明光照强度在不同程度上对 4 种光合指标起着制约和促进作用，在杨树光合作用中起重要作用。

表 4　光合指标与环境因子相关系数

因子	*Pn*	*Gs*	*Ci*	*Tr*	*Par*	*Tair*	*Ca*	*RH*
Pn	1	0.129*	−0.653**	0.731**	0.761**	0.691**	−0.555**	−0.536**
Gs		1	0.384**	0.430**	0.020	−0.096	0.407**	−0.038
Ci			1	−0.337**	−0.706**	−0.801**	0.895**	0.580**
Tr				1	0.745**	0.683**	−0.328**	−0.759**
Par					1	0.856**	−0.560**	−0.781**
Tair						1	−0.726**	−0.753**
Ca							1	0.488**
RH								1

注：* 代表 0.05 水平相关显著，** 代表 0.01 水平相关显著。

表 5　光合指标的回归方程

性状	光曲线方程	R^2	*Ra*	*F*	Sig.
Pn	$Pn=0.0071\times Par-0.0123\times Ca+10.2848$	0.604	0.601	189.815	0.00
Ci	$Ci=1.0777\times Ca-0.07271\times Par-72.9527$	0.862	0.861	779.826	0.00
Gs	$Gs=0.0008388\times Ca+0.00008156\times Par-0.1580$	0.254	0.248	42.4998	0.00
Tr	$Tr=-0.09639\times RH+0.001063\times Par+8.6450$	0.636	0.633	217.615	0.00

4　结论与讨论

杨树叶片从春季萌动、展叶、成熟到衰老的过程中，光合速率随着叶绿素含量的变化、组织成熟和衰老程度的变化而变化(陈冠喜等，2009)，幼叶叶绿素含量低，光合速率低，老化叶片功能下降，光合速率低，测定光合指标的功能叶片区域与杨树叶片的成熟区基本大体一致，从顶端至底端是低—高—低的过程。

毛白杨无性系 BL204 和 BL30 的 *Pn* 日变化呈典型的双峰曲线，中午由于环境中的 Par 过高导致光抑制现象发生，这与高健(2002)和邓松录(2006)对杨树无性系光合日进程研究的结果相似。气孔导度与光合速率同步化，表明是气孔导度降低限制了外界二氧化碳通过气孔进入细胞间隙，并进一步降低了光合速率，属于气孔调节现象(Arquhar et al.，1982)。2 个无性 BL204 和 BL30 的 *Pn* 波峰出现时间不同，表明不同基因型与对环境适应过程不同，上午的瞬时 *Pn* 没有下午高，可能是由于早晨环境相对湿度过高，导致叶片处于高度休眠状态，随着光照强度增加，温度上升，湿度下降，叶片逐渐打破休眠，下午出现高光合速率。

光可以参与调节酶的活性和气孔开度(李合生，2002)，植物出现光饱和点的实质是因为强光下暗反应跟不上光反应从而限制了光合速率随着光强的增加而增高，同时光饱和点是可以反映最大光能利用率重要的光合指标(Surabhi et al.，2009)。在光曲线的测定中，5个毛白杨无性系的最大 *Pn* 与 *lcp* 差异显著，在光照达到饱和状态下，无性系 BL204 与 BL206 显示了高光合速率，在 *Par* 上升过程中，这 2 个无性系也显示了高 *Tr* 速率，表明这 2 个无性系适合在光照强度与水分充足的条件下栽培。无性系 BL206 的 *lcp* 最低，可以看出 BL206 在低光照强度下就可以进行光合作用。无性系 BL207 的最大 *Pn* 只有 19.54μmol $\cdot m^{-2} \cdot s^{-1}$，但 *Tr* 一直是 5 个无性系中最低的一个，说明 BL207 无性系水分利用效率高，在水分条件受制约的情况下，可以选择栽培 BL207 无性系。

光合作用的过程是极复杂的过程，既受植物自身结构的调节，也受外界环境因子的影响(曹雪丹等，2008；尤扬等，2009；Calfapietra et al.，2005；Li et al.，2004)，毛白杨无性系的 *Pn* 与 *Gs*、*Ci* 和 *Tr* 相关均达到显著水平，*Pn* 增大，*Gs* 增大而导致 *Tr* 增加。叶片光合作用强，消耗的 CO_2 多，而气态的 CO_2 在液相的胞间或胞内的扩散阻力导致胞间 CO_2 浓度得不到迅速补充，其胞内 CO_2 浓度就会下降，所以出现 *Ci* 与 *Pn*、*Tr* 显著负相关的结果。杨树 *Tr* 是指水蒸气进入细胞间隙之后扩散到叶片外环境的过程，因此 *Tr* 与气孔的开放程度有密切的关系，而 *Gs* 和气孔的开闭一致，从日变化可以看出，*Gs* 与 *Tr* 的变化趋势基本一致，从相关性也可以看出 *Gs* 与 *Tr* 相关达到极显著水平。对蒸腾速率进行环境因子回归方程中发现最先进入方程的环境因子是 *RH*，说明由于空气的湿度大，水蒸气向环境中扩散的速度减低，所以 *RH* 成了影响 *Tr* 变化的主导因子。

影响植物光合作用的环境因子很多，有些因子直接参与光合作用，如光和 CO_2，另一些因子如水分和温度等，则主要是间接起作用(赵天锡，1994)，比如温度(Kim et al.，2007)的一部分效应就是增加叶片气孔内外蒸汽压的梯度以及光合作用中的酶活性，从而增加光合作用的进程(彭振华，2002)。在 *Par*-*Pn* 的光合曲线不同阶段，影响光合速率的主要因子不同，弱光下，*Par* 是控制 *Pn* 的主要因素，随着 *Par* 的提高，叶片吸收光能增多，光化学反应速度加快，CO_2 固定速率加快，从而 *Ca* 成了限制 *Pn* 的主导因子。从相关性来说，*Par*、*Tair* 与 *Pn* 极显著正相关说明了环境因子对 *Pn* 有重要影响，*Pn* 与环境因子回归方程的建立进一步说明环境中的 *Par* 与 *Ca* 是限制 *Pn* 的主导因子。由于 CO_2 是光合作用的主要原料，CO_2 浓度的高低是决定 *Ci* 大小的关键，*Ci* 对环境因子回归方程中起主导作用的因子是 *Ca*，也进一步证明了这一点。

本试验通过对 30 个毛白杨无性系瞬时光合指标测定分析，初步筛选出 2 个瞬时 *Pn* 和 *Tr* 均高的无性系 BL204 和 BL206，这 2 个无性系可以在水分和光照强度充分的条件下栽培，而 BL207 无性系则显示出高光合、低蒸腾的特性，可以在水分受制约的条件下栽培，其它无性系也显示了较高的瞬时光合速率和水分利用效率，还有待于进一步测定分析。

Comparative analysis of the photosynthetic characteristics of clones in three years old *Populus tomentosa*

Abstract Photosynthetic characteristics of 30 *Populus tomenntosa* clones were measured with lico-6400 photosynthetic instrument, and the relationship between the environmental factors and physiological indicators was studied by correlation analysis and stepwise regression equation. The results showed that the dirunal variation of *Pn* of *Populus tomentosa* clones presented a typical double-peak curve, and that stomatal limitation was a major regulatory factors of decreased photosynthesis. *Pn-Par*、*Pn-Ca* curves of five *Populus tomenntosa* clones shaped like "S", and also accorded with quadratic equation. Under saturated conditions of the *Ca* and luminous intensity, *Pn*、*lcp*、*csp* of the five clones showed significant differences. In addition, with the luminous intensity saturated, rankings for *Pn* were: BL204 (24. 64) >BL206(23. 76) >BL30(21. 50) >BL207(19. 54) >BL63(18. 64μmol · m^{-2} · s^{-1}). Whereas with the Ca saturated, BL206 had the maximum value: 30. 15μmol · m^{-2} · s^{-1}, and BL63had the minimum value: 20. 34μmol · m^{-2} · s^{-1}. Meanwhile, for the five clones, lcp ranged from 33. 08μmol · m^{-2} · s^{-1} to 81. 17μmol · m^{-2} · s^{-1}, and csp ranged from 74. 03μmol · mol^{-1} to 93. 35μmol · mol^{-1}. The values of Pn, Gs, Ci and Tr of 30 poplar clones were significantly different (Sig. <0. 000). The results also presented that the phenotype variation coefficients of photosynthetic factors changed between 8. 94% ~ 23. 22% , and genetic variation coefficients of photosynthetic factors were between 8. 78% ~ 22. 79%. And the high repeatability revealed that genetic factors played a major role for phenotypic variation. Furthermore, the photosynthetic factors of the poplar clones correlated extremely significantly, the highest correlation coefficient being 0. 731. We concluded that the photosynthetic indices could be used to evaluate poplar clones, providing theoretical basis for clone selection.

Keywords *Populus tomentosa*, clone, photosynthetic rate, variation

应用^{15}N示踪研究毛白杨苗木对不同形态氮素的吸收及分配*

摘要 以毛白杨(*Populus tomentosa*)新无性系83号插条苗为试材，应用^{15}N示踪技术研究在相同施氮量下毛白杨苗木对硝态氮($NO_3-^{15}N$)和铵态氮($NH_4-^{15}N$)的吸收、分配及利用特性。结果表明：施肥后不同时期毛白杨苗木对两种氮肥的吸收量和利用率存在极显著差异($P<0.01$)。施肥后28d，苗木对两种氮肥吸收利用达到最大值，其中标记$NO_3-^{15}N$肥吸收量为0.36g·株$^{-1}$，利用率达到35.98%；标记$NH_4-^{15}N$肥吸收量为0.15g·株$^{-1}$，利用率为14.53%。苗木$NO_3-^{15}N$肥平均利用率(19.75%)约为$NH_4-^{15}N$肥平均利用率(7.95%)的2.5倍。施肥后各个时期，全株的$NO_3-^{15}N$肥Ndff%值均显著大于$NH_4-^{15}N$肥Ndff%。各器官对$NO_3-^{15}N$肥的征调能力明显高于$NH_4-^{15}N$肥，茎对肥料竞争能力最强，其次为叶和根。氮素分配率在各器官中差异显著($P<0.05$)，总体趋势为叶>根>茎。叶中$NO_3-^{15}N$的分配率均高于$NH_4-^{15}N$。根中储存的氮素主要供地上部分生长所需，总体呈现逐渐下降的趋势。茎是^{15}N贮藏的"临时库"，苗木主要通过茎将吸收的氮素输送到叶等旺盛生长的部位。

关键词 毛白杨，硝态氮，铵态氮，吸收量，氮肥利用率，Ndff%，分配率

氮素既是植物最重要的结构物质，又是生理代谢中最活跃、无处不在的重要物质——酶的主要成分(赵平等，1998)，所以氮素对植物生理代谢和生长有重要作用。植物吸收氮素的多少对其生长、发育、开花、结果都有很大影响，尤其是对农作物的产量和品质有很大影响(陈龙池等，2002)。植物所利用的主要氮素形式是硝态氮(NO_3-N)和铵态氮(NH_4-N)。两种形态氮素在可选择的条件下，不同植物或同一植物的不同生育阶段，其相对吸收量则有明显差异(胡霭堂，2003)。有大量报道认为不同形态氮素能够影响植物内源激素的变化，从而影响植物的是生长发育，但迄今有关植物对不同氮素吸收的研究，多以农作物为研究对象(肖凯等，2000；王娜等，2002；Gariglio et al.，2000；Neuweiler et al.，1997)。

毛白杨(*Populus tomentosa*)作为华北地区速生丰产用材林主要树种(薛崇伯，1981)，氮肥有明显促进其生长的作用(刘寿坡等，1989；孙时轩等，1995；姜岳忠等，2004)。目前我国对毛白杨施肥的相关研究，主要集中于氮磷钾养分的配比，不同施肥量的增产效益和经济效益等方面，有关毛白杨对不同形态氮素的吸收、运转、分配及利用等方面，则未见报道。为此，本实验应用^{15}N示踪技术研究在相同施氮量下毛白杨苗木对不同形态氮素的吸收、分配及利用特性，以期为毛白杨的氮素管理，氮肥的合理施用以及营造毛白杨速生丰产林提供科学依据。

* 原载《北京林业大学学报》，2009，31(4)：97-101，与董雯怡、聂立水、韦安泰、李吉跃和沈应柏合作发表。

1 材料与方法

1.1 苗圃概况和试验材料

盆栽安置于北京林业大学苗圃，土壤容重：1.45g·cm^{-3}，盆栽土土壤养分状况为：全氮1.0g·kg^{-1}，有效磷1.5mg·kg^{-1}，速效钾75mg·kg^{-1}。供试苗木为三倍体毛白杨新无性系83号插条苗，于2007年3月盆插成苗，挑选健壮、生长状况基本一致的盆栽苗用作试验材料。核素标记肥料：^{15}N双标记硝酸铵，丰度10.19%，由上海化工研究院提供。

1.2 试验设计

2007年3月底将毛白杨小苗移栽到330mm×300mm的白色塑料花盆中(每盆1株)，放入温室进行培养。6月10日对苗木进行施肥。处理1：每盆施入纯氮2g，相当于NH_4NO_3 5.7142g，对NH_4NO_3中的NO_3-N做标记，记为标记NO_3-^{15}N肥，同时施入硝化抑制剂2g。处理2：每盆施入纯氮2g，相当于NH_4NO_3 5.7142g，对NH_4NO_3中的NH_4-N做标记，记为标记NH_4-^{15}N肥，同时施入硝化抑制剂2g。每个处理12个重复，共计24盆苗木。

1.3 取样及测定

采样：施肥后每隔7、14、28和56d进行采样。每次每个处理取样3株，共计6株。单株解析为：根、茎、叶三部位。样品用自来水洗净后，蒸馏水冲洗3次，晾干后于105~110℃下杀青30min后于65℃烘48h，称重。同时用不锈钢电磨粉碎，过0.25mm筛，分析根、茎、叶三部位全氮含量。全N含量用凯氏定N法测定，各部位^{15}N丰度在中国农业科学院原子能利用研究所用MAT-251质谱仪测定。器官吸收标记氮肥量、氮肥利用率、Ndff%和各器官吸收^{15}N分配率计算参照赵登超等(2006)的方法。

1.4 数据统计分析

采用Excel2003软件完成全部数据处理和作图，SPSS15.0统计软件进行ANOVA分析，并用Duncan检验法进行多重比较，字母法对其进行标记。

2 结果与分析

2.1 毛白杨苗木对NO_3-^{15}N和NH_4-^{15}N的吸收及利用

由表1~2可知，毛白杨苗木施肥后不同时期对两种氮肥的吸收量和利用率存在极显著差异，均呈先上升后下降的趋势。施肥后7 d和施肥后14d，苗木对两种肥料吸收差异较小($P>0.05$)，其吸收量和利用率均处于较低水平。施肥后28d，毛白杨苗木对两种氮肥吸收利用均达到最大值，明显高于施肥后其它时期($P<0.01$)。其中标记NO_3-^{15}N肥吸收量为0.36g·株$^{-1}$，利用率达到35.98%；标记NH_4-^{15}N肥吸收量为0.15g·株$^{-1}$，利用率为14.53%。至施肥后56d，苗木对标记NO_3-^{15}N肥的吸收和利用较28d时有所降低($P<$

表 1　毛白杨苗木对不同标记氮肥的吸收

处理	苗木吸收标记氮肥量($g\cdot 株^{-1}$)			
	7d	14d	28d	56d
标记 $NO_3-{}^{15}N$ 肥	0.06±0.01Bc	0.11±0.01Bc	0.36±0.03Aa	0.26±0.01Ab
标记 $NH_4-{}^{15}N$ 肥	0.02±0.00Cb	0.04±0.01BCb	0.15±0.03Aa	0.12±0.02ABa

注：表中数据为平均值±标准误；同一行上标相同字母为差异不显著；小写为 $P=0.05$ 水平，大写为 $P=0.01$ 水平；下同。

0.05)，而对标记 $NH_4-{}^{15}N$ 肥的吸收利用整体下降幅度不大，仍维持在较高水平。

毛白杨苗木在施肥后不同时期，吸收 $NO_3-{}^{15}N$ 肥量都高于吸收标记 $NH_4-{}^{15}N$ 肥量，$NO_3-{}^{15}N$ 肥平均利用率(不同时期氮肥利用率平均值)为 19.75%，$NH_4-{}^{15}N$ 肥利用率仅 7.95%，$NO_3-{}^{15}N$ 肥利用率约为 $NH_4-{}^{15}N$ 肥利用率的 2.5 倍。

表 2　毛白杨苗木对不同标记氮肥的利用　%

处理	氮肥利用率				平均氮肥利用率
	7d	14d	28d	56d	
标记 $NO_3-{}^{15}N$ 肥	6.20±0.48Dd	10.87±0.75Cc	35.98±0.72Aa	25.94±1.15Bb	19.75
标记 $NH_4-{}^{15}N$ 肥	1.67±0.44Bb	3.99±0.78Bb	14.53±1.73Aa	11.58±0.60Aa	7.95

2.2　毛白杨苗木各器官 $NO_3-{}^{15}N$ 肥和 $NH_4-{}^{15}N$ 肥的 Ndff%

器官的 Ndff%是指植株器官从肥料氮中吸收分配到的氮量对该器官全氮量的贡献率，它反映了植株器官对肥料氮的吸收竞争能力(徐季娥等，1993)。本试验中，毛白杨苗木施肥后不同时期各器官 Ndff%差异极显著(表 3)。对于 $NO_3-{}^{15}N$ 肥，苗木根的 Ndff%在施肥后各个时期均处于较低水平；施肥后 7d 至 28d，苗木茎的 Ndff%明显高于叶和根($P<0.01$)，表明茎对 $NO_3-{}^{15}N$ 肥有较强的征调能力；施肥后 28d，茎和叶的 Ndff%增至最大值，分别为 37.57%和 32.34%；56d 时，苗木叶对 $NO_3-{}^{15}N$ 肥的竞争能力增强，Ndff%明显高于茎和根($P<0.01$)。对于 $NH_4-{}^{15}N$ 肥，苗木根的 Ndff%呈先上升后下降的趋势，28d 时达到最大值(18.35%)，且明显高于同期茎、叶的 Ndff%（$P<0.05$)，表明施肥后 28d 是根对 $NH_4-{}^{15}N$ 肥征调能力最强的时期。苗木叶的 Ndff%随时间推移逐渐增加，至 56d 时，Ndff%达到 20.69%。施肥后各个时期苗木茎对 $NH_4-{}^{15}N$ 肥一直保持较强的竞争能力，其 Ndff%明显高于叶和根($P<0.01$)。总体来说，各器官对 $NO_3-{}^{15}N$ 肥的征调能力明显高于 $NH_4-{}^{15}N$ 肥，毛白杨苗木茎对肥料竞争能力最强，其次为叶和根。

表 3　施肥后不同时期毛白杨苗木各器官 Ndff%

处理	器官	7d	14d	28d	56d
标记 NO_3-^{15}N 肥	根	16.43±0.42Cc	22.33±0.59Bb	20.55±0.53Cc	19.92±0.53Cc
	茎	25.55±0.34Aa	32.60±0.68Aa	37.57±0.54Aa	25.39±0.54Bb
	叶	20.31±0.40Bb	30.60±0.60Aa	32.34±0.88Bb	31.02±0.74Aa
	全株	62.29	85.53	90.46	76.33
标记 NH_4-^{15}N 肥	根	9.00±0.48Bb	12.93±0.34Bb	18.35±0.54Aa	16.30±0.44Cc
	茎	13.96±0.42Aa	15.6±0.43Aa	17.02±0.66Aab	30.79±0.58Aa
	叶	7.74±0.22Bb	12.83±0.45Bb	15.71±0.43Ab	20.69±0.35Bb
	全株	30.71	41.44	51.08	67.78

从表 3 进一步看出，施肥后各个时期，全株的 NO_3-^{15}N 肥 Ndff%值均大于 NH_4-^{15}N 肥 Ndff%。对于 NO_3-^{15}N 肥，全株 Ndff%值呈先上升后下降的规律，施肥后 28d 达到最大值(90.46%)。而对于 NH_4-^{15}N 肥，全株 Ndff%值随时间推移呈逐渐上升的趋势，56d 时 Ndff%增至最高值(67.78%)。

2.3　^{15}N 在毛白杨各器官的分配率

各器官中 ^{15}N 占全株 ^{15}N 的百分率反应了肥料氮在树体内的分布及其各器官间运移的规律。由图 1，2，3 可看出，苗木吸收的肥料氮在各部位分配率差异较大，总体来看，在叶中的分配率最大，根茎中的分配率相对较小。施肥后不同时期 ^{15}N 在各器官中分配不同规律不同，这显然与毛白杨各器官的生长、代谢特点密切相关。

如图 1 所示，^{15}N 在根中的分配随时间推移呈现逐渐下降的趋势。施肥后 7d，苗木根系对 ^{15}N 的累积量较大，NO_3-^{15}N 的分配率为 25.49%；NH_4-^{15}N 的分配率为 29.13%。之后伴随着地上茎叶旺盛生长，根系活力降低，^{15}N 累积量也随之降低，至施肥后 56d 时，苗木根 NO_3-^{15}N 的分配率降至 13.33%；NH_4-^{15}N 的分配率降至 22.94%。

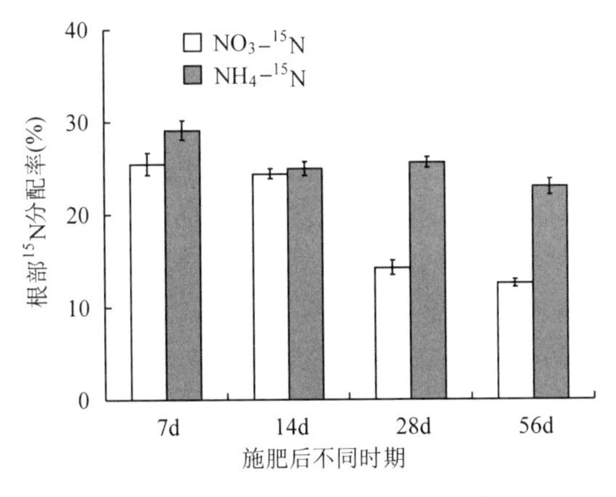

图 1　^{15}N 在毛白杨根中的分配率

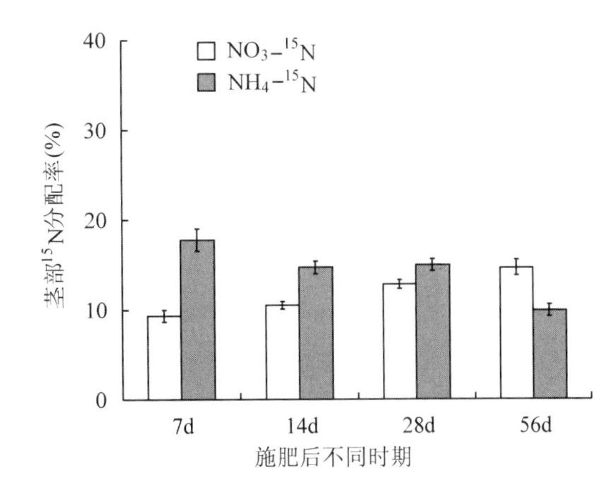

图 2　^{15}N 在毛白杨茎中的分配率

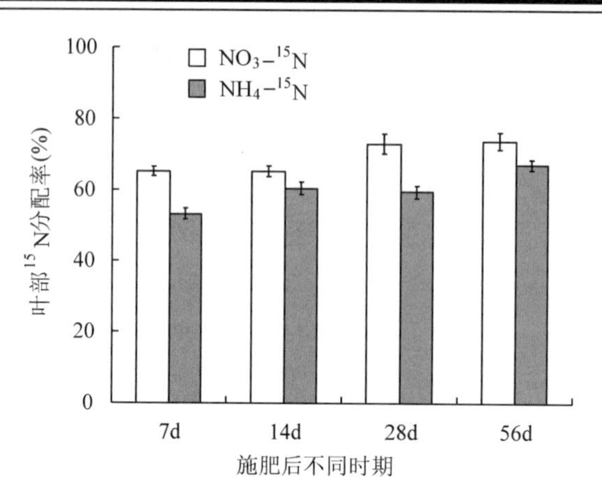

图 3　^{15}N 在毛白杨叶中的分配率

由图 2 可看出，$NO_3-^{15}N$ 和 $NH_4-^{15}N$ 在茎中的分配表现出不同的规律。在整个试验期，苗木茎中的 $NO_3-^{15}N$ 约呈现上升的趋势，从施肥后 7d 时分配率的 9.33%增至 56d 的 14.66%；而植株茎部的 $NH_4-^{15}N$ 运移和 $NO_3-^{15}N$ 有所不同，分配率随时间呈逐渐下降的趋势。施肥后 7d，$NH_4-^{15}N$ 在植株茎中的分配率为 17.76%，其后随着植株的生长，$NH_4-^{15}N$ 运移到其他器官，茎的分配率在施肥后 28d 至 56d 间下降较快，从 14.79%降至 9.90%，在茎中的残留较少。$NO_3-^{15}N$ 和 $NH_4-^{15}N$ 在茎中的分配率相对于其他器官较低，苗木主要通过茎将吸收的氮素输送到叶等旺盛生长的部位，茎中储存的氮素较少。

氮素在植物体内的分布，一般集中于生命活动最活跃的部分。在毛白杨各个生长器官中，叶是生长及代谢最旺盛的部位，因此，从图 3 可看出，叶中 $NO_3-^{15}N$ 和 $NH_4-^{15}N$ 的分配率都远远高于其他器官且 $NO_3-^{15}N$ 的分配率显著高于 $NH_4-^{15}N$。$NO_3-^{15}N$ 施肥后 7d 叶中分配率为 65.18%，随着叶片旺盛生长，分配率逐渐增加，在施肥后 56d 时叶中 $NO_3-^{15}N$ 累积达到全株的 72.02%。叶中 $NH_4-^{15}N$ 的分配率随着时间的推移及植株的生长呈现上升的规律，其值分别从施肥后 7d 的 53.11%升至 56d 时 67.16%。叶对 $NH_4-^{15}N$ 累积速率明显高于 $NO_3-^{15}N$，这可能是由于 $NO_3-^{15}N$ 在植株体内很容易在木质部移动有关(奚振邦，2003)。

3　结论与讨论

NO_3-N 和 NH_4-N 都是植物良好的氮源，但二者所带电荷不同，营养特点也不同。两种形态氮素在可选择的条件下，不同植物或同一植物的不同生育阶段，其相对吸收量则有明显差异。相关研究认为大田作物中，一般烟草(*Nicotiana tabacum*)、棉花(*Gossypium* spp.)等旱作物对硝态氮的反应较好，水稻(*Oryza sativa*)则多吸收铵态氮(胡霭堂，2003)。顾曼如等(1981)试验表明苹果(*Malus pumila*)植株吸收的^{15}N 量，无论是春施还是夏施，均是 NH_4-N 多于 NO_3-N，一般认为果树是喜硝植物。本研究表明，毛白杨苗木对 NO_3-N 的反应较好，吸收量最大的时期，标记 $NO_3-^{15}N$ 肥吸收量为 0.36g · 株$^{-1}$，利用率达到

35.98%；吸收标记 NH_4-^{15}N 肥 0.15g · 株$^{-1}$，利用率为 14.53%。NO_3-^{15}N 肥平均利用率为 19.75%，NH_4-^{15}N 肥利用率仅 7.95%，NO_3-^{15}N 肥利用率约为 NH_4-^{15}N 肥利用率的 2.5 倍。这可能是由于 NH_4^+ 易被土壤胶体吸附，部分进入黏土矿物的晶层被固定，不利于植株吸收，而且，在碱性环境中氨易挥发损失，尤其是挥发性氮肥本身就易挥发，与碱性物质接触会加剧氨的挥发损失。本研究主要用毛白杨苗木进行试验，对成林没有开展研究，有一定的局限性，要确定毛白杨植株对两种氮源吸收的问题，还需开展更多更广泛的试验。

毛白杨各器官对两种肥料氮的征调能力不同，对 NO_3-^{15}N 肥的征调能力明显高于 NH_4-^{15}N 肥，施肥后各个时期全株的 NO_3-^{15}N 肥 Ndff%值均大于 NH_4-^{15}N 肥 Ndff%。对于 NO_3-^{15}N 肥，全株 Ndff%值呈先上升后下降的规律，施肥后 28d 达到最大值(90.46%)。而对于 NH_4-^{15}N 肥，全株 Ndff%值随时间推移呈逐渐上升的趋势，56d 时 Ndff%增至最高值(67.78%)。

同种肥料各个器官 Ndff%值差异达到极显著水平。Sanchez 等(1992)的试验结果表明，梨树(*Pyrus sorotina*)春施^{15}N 肥料，施肥当年叶片、新梢及果实的 Ndff%值最高，表明这些部位的 N 含量比其它部位更依赖当年果树的吸氮量。本研究中，茎对肥料的竞争能力最强，吸收肥料氮对全氮贡献较大，其次为叶和根。施肥后 28d，茎和叶的 Ndff%增至最大值，分别为 37.57%和 32.34%；而此时根对 NH_4-^{15}N 肥的竞争能力也最强，Ndff%达到最大值(18.35%)。因此在这之前补充氮素，加施氮肥，以满足植株各器官旺盛生长的需要，促进根系和叶片的生长。

植株吸收的氮素在各器官中分配差异较大，肥料氮向各器官的分配与各时期生长中心关系密切。顾曼如等(1985)在苹果上的研究表明，土壤追施氮肥，^{15}N 的吸收分配随着生长中心转移而转移，而以叶内积累较多。叶是毛白杨生长及代谢最旺盛的部位，随植株生长，氮素不断累积，NO_3-^{15}N 和 NH_4-^{15}N 的分配率都远远高于其他器官，且 NO_3-^{15}N 的分配率显著高于 NH_4-^{15}N。茎是^{15}N 贮藏的"临时库"，NO_3-^{15}N 和 NH_4-^{15}N 在茎中的分配率较低，苗木主要通过茎将吸收的氮素输送到叶等旺盛生长的部位。范志强等(2004)在水曲柳(*Fraxinus mandshurica*)上的研究表明根系为氮的主要贮藏器官。Sanchez 等(1992)也认为，梨树采收期施用氮肥大多贮藏于根部，只有小部分贮藏在地上部的花芽等器官。而本研究中，NO_3-^{15}N 和 NH_4-^{15}N 在根中的分配率相对于叶较低，储存的氮素主要供地上部分生长所需，总体呈现逐渐下降的趋势。总体说来各器官分配氮素趋势为叶 > 根 > 茎。氮素在毛白杨植株体内的分布，集中于生命活动最活跃的部位。因此，氮素供应充分与否和氮素营养的好坏，在很大程度上影响着毛白杨植株的生长发育状况，需给予极大重视。

Effects of Nitrogen Forms on the Absorption and Distribution of nitrogen in *Populus tomentosa* Seedlings by Using the Technique of ^{15}N Trace

Abstract The technique of stable isotope ^{15}N trace was used to study the absorption, distribution and utilization to Nitrate Nitrogen and Ammonia Nitrogen at the same N application level with clone 83 of *Populus tomentosa* seedlings as materials. The results indicate that there was extremely significant difference in absorption and nitrogen use efficiency (*NUE*) in *P. tomentosa* seedlings at each stage after fertilization($P<0.01$). After 28 days of fertilization, the absorption and *NUE* reached the peak. The maximum absorption and NUE of $NO_3-^{15}N$ fertilizer was 0.36g · $plant^{-1}$ and 35.98% respectively, and that of $NH_4-^{15}N$ fertilizer was 0.15g · $plant^{-1}$ and 14.53% respectively. The average *NUE* of $NO_3-^{15}N$ fertilizer (19.75%) was almost 2.5 times more than that of $NH_4-^{15}N$ fertilizer(7.95%). At each stage after fertilization, the percent of nitrogen derived from fertilizer (Ndff%) of $NO_3-^{15}N$ in total plant was higher obviously than that of $NH_4-^{15}N$ fertilizer. The transfer ability of $NO_3-^{15}N$ fertilizer in each organ was much stronger than that of $NH_4-^{15}N$ fertilizer. The stem of *P. tomentosa* seedlings, which was followed by the leaf and the root, was the most competitive organ for $NO_3-^{15}N$ fertilizer. Nitrogen distribution rate changed significantly in different organs($P<0.05$) and the trend of that was leaf > root > stem. In the leaf, the distribution of $NO_3-^{15}N$ was higher than that of $NH_4-^{15}N$. The nitrogen which accumulated in the root was mainly supplied for the growth of above-ground organs and declined generally. Through the stem which was called "temporary storehouse", *P. tomentosa* seedlings mainly transported nitrogen to leaf and other organs which grow vigorously.

Key words *Populus tomentosa*, nitrate nitrogen, ammonia nitrogen, absorption, nitrogen use efficiency, the percent of nitrogen derived from fertilizer, distribution

鲁西平原毛白杨造林地土壤速效磷和速效钾空间变异性研究*

摘　要　研究了位于鲁西平原冠县苗圃毛白杨 6.0hm² 造林地土壤速效养分分布，为毛白杨造林试验地布设、养分精准管理、确定合理采样数提供依据。试验将土壤在水平方向按 50m×50m 分为 24 块样地，每块样地分为 0~20cm，20~40cm，40~60cm 三层取样，进行三维空间采样和室内分析。结果表明，速效养分含量属于较低水平，且随土层加深逐渐降低。土壤速效磷含量是 1.32~1.89mg/kg，总体均值为(1.51±0.09)mg/kg，3 个层次含量分别为(1.66±0.15)mg/kg、(1.44±0.07)mg/kg、(1.44±0.06)mg/kg。速效钾含量是 23~78mg/kg，总体均值(46.95±8.33)mg/kg，3 个层次的含量分别为(63.29±9.69)mg/kg、(43.38±8.35)mg/kg、(34.17±6.96)mg/kg。3 个层次速效磷变异系数分别为 9%、5%、4%，属于弱变异水平，速效钾变异系数分别为 15%、19%、20%，属于中等变异水平，变异水平和取样精度直接影响合理采样数。土壤速效磷含量在水平和垂直方向上变化不大，土壤速效钾含量自西向东先增后减，自北向南升降交替、变化复杂。

关键词　毛白杨，造林地，速效磷，速效钾，空间变异

土壤养分是土地生产力的基础，是植物生长的必要条件，也是衡量土壤质量好坏的重要指标(连纲等，2008)，其含量受成土母质、地形、人类活动等多种因素的影响(徐茂等，2007)，因此土壤养分在时间和空间上都存在着较大的变异性，土壤性质的空间变异性是指在一定的景观内，在同一时间、不同地点的土壤性质存在明显的差异性和多样性(Gambardella C A et al.，1994)。对土壤的精确管理和应用环境模型都需要对土壤质量指标的空间变异性进行分析(李双异等，2006)，尤其是对土壤养分空间变异的充分了解，是管理好土壤养分和合理施肥的基础(Franzen DW et al.，1996)，因此土壤养分的空间变异性越来越受到人们的重视(姚丽贤等，2004)。从 20 世纪 70 年代开始，国外许多学者将地统计学理论用于土壤科学，成功地进行了土壤特性空间变异性规律的研究(陈彦等，2005；赵彦峰等，2006；Cahn MD et al.，1994)。一般来说，养分元素如全磷会随着土壤剖面深度的加深而下降(Silver WL et al.，1994)。Cameron 也证明，磷的变异系数会随深度的增加而增加，而平均值是随着深度的加大而下降的，钾没有明显的变化趋势，Piercem 则认为深层土壤的变异程度要低于表层土壤(Cameron D R et al.，1994；Piercem RJ et al.，1995)。国内一些学者将地统计学和地理信息系统结合起来(王秋兵等，2009；雷永雯等，2004；孙波等，2002；金继运等，2001)，在土壤属性的空间变异性方面做出了有益的探索，但在土壤养分的空间变异研究方面还处于初步阶段，尤其对林业土壤速效养分时空变

* 本文原载《中国农学通报》，2009，25(23)：262-267，与聂立水、张志毅、韦安泰、董雯怡、张有慧合作发表。

异的研究鲜有报道(周慧珍等, 1996)。

笔者采用网格法布点(50m×50m)取样, 通过对山东省国有冠县苗圃毛白杨造林地6.0hm^{2}试验地的土壤速效磷和速效钾含量的测定和分析, 旨在了解土壤速效养分总体状况和空间分布, 为毛白杨人工林营造、试验地布设、精准施肥以及合理确定采样数提供科学依据, 并建立相应立地土壤肥力档案。

1 材料与方法

1.1 试验时间与地点

田间试验于2007年3月在山东省国有冠县苗圃内进行, 室内试验在北京林业大学水土保持学院土壤重点实验室进行。

1.2 试验地概况

试验地地处北纬36°30′, 东经115°27′之间的鲁西平原黄河故道上, 属暖温带季风大陆性气候, 热量丰富, 降雨较多, 但雨量不均, 蒸发量亦大。年平均气温12.8~13.4℃, 极端最高气温41.9℃, 极端最低气温-22.7℃, 年均≥10℃的有效积温4404~4424℃, 无霜期193~201天。年均降水量580~600mm, 全年降水分配3~5月占12.8%~14.1%, 6~8月占61.5%~67.3%, 9~11月占17.1%~20.9%, 12月至翌年2月占2.6%~3.5%。研究区土壤属潮土类, 黄潮土亚类。土壤以通体砂壤质为主, 个别地段夹有黏壤质层。多年来该研究区土壤上以种植毛白杨苗木为主, 有的年份轮作玉米(*Zea mays* L.)、棉花(*Gossypium hirsutum* Linn.)等。

1.3 样地布设与土样采集

为了全面反映试验区的土壤速效养分状况和空间分异规律, 在6.0hm^{2}试验区域内采用网格法布点取样, 依照从北到南(300m)、从西到东(200m)连续、均匀布设50m×50m的样地24块, 分别用序号1~24表示(图1)。在各样地内分别按距地面0~20cm、20~40cm、40~60cm分三层取样, 每个样地中各层以S形设置10个取样点构成1个混合土样作为该样地的分析土样, 共采集混合分析土样72个。

系列	C1	C2	C3	C4	
L1	1	7	13	19	N
L2	2	8	14	20	↑
L3	3	9	15	21	
L4	4	10	16	22	
L5	5	11	17	23	
L6	6	12	18	24	

图1 取土样地编号及分布

为了准确表示各个样点, 试验设置10个系列, 具体表示如下: 系列C1代表区块1~6样点, 系列C2代表区块7~12样点, 系列C3代表区块13~18样点, 系列C4代表区块19~24样点; 系列L1代表区块1、7、13、19样点, 系列L2代表区块2、8、14、20样点，系列 L3 代表区块 3、9、15、21 样点，系列 L4 代表区块 4、10、16、22 样点，系列 L5 代表区块 5、11、17、23 样点，系列 L6 代表区块 6、12、18、24 样点。其中系列 C1、C2、C3、C4 用于分析从北到南速效养分含量的差异，系列 L1、L2、L3、L4、L5、L6 用于分析从西到东速效养分含量的差异。

1.4 土样分析

试验分析在北京林业大学水土保持学院土壤重点实验室进行。土壤速效磷测定采用碳酸氢钠法浸提，钼锑抗显色分光光度计测定；土壤速效钾测定采用醋酸铵浸提原子吸收法测定(鲍士旦，2000)。

1.5 数据统计方法

土壤中各层速效磷和速效钾含量的均值分别由各层数据之和除以样品数获得。0~60cm 土壤速效磷或速效钾含量根据各层数据加权平均计算。变异系数(%)计算根据公式：$CV = SD/M \times 100\%$；土壤速效磷和速效钾总体均值空间变异根据公式：上限 $\mathrm{Max} = M + 2.576 \times SD/\sqrt{N}$，下限 $\mathrm{Min} = M - 2.576 \times SD/\sqrt{N}$，$\alpha = 0.01$。构成混合土样的样点数 $N = (CV/m)^2$，式中：SD 为土壤肥力指标标准差；M 为土壤肥力指标平均值；N 为样品数，m 为试验所允许的最大误差(要求的精密度)，单位为百分率(%)。其他数据和图表采用 Microsoft Office Excel 2003 处理。养分分级按全国第二次土壤普查暂行规程的标准进行。

2 结果与分析

2.1 土壤速效养分总体状况

土壤速效磷和速效钾是土壤磷钾养分供应水平高低的指标，土壤磷钾含量高低在一定程度反映了土壤中磷素和钾素的贮量和供应能力，了解土壤中有效磷钾的供应状况，对于施肥有着直接的意义。0~60cm 各层土壤速效磷和速效钾含量见表 1。试验区 0~60cm 土壤速效磷含量的变化范围是 1.32~1.89mg/kg，总体均值 1.51mg/kg，0~60cm 土壤速效钾含量的变化范围是 23~78mg/kg，总体均值 46.95mg/kg；按全国第二次土壤普查暂行规程的标准划分养分等级，土壤速效磷含量低于 3.00mg/kg，土壤速效磷含量为 6 级，属于最低水平；土壤速效钾含量为 5 级，属于较低水平。该区土壤通体砂质，土层较薄，保水保肥能力较差，在降雨量多的季节容易诱发水土流失，同时该地区春季风沙较大，必然带走大量的土壤养分，导致该地区土壤养分贫瘠。

表 1　毛白杨造林地土壤速效磷、速效钾含量的空间变异

项目	土层 (cm)	样本数	变幅 (mg/kg)	平均值 (mg/kg)	标准差 (mg/kg)	变异系数 (%)
速效磷	0~20	24	1.42~1.89	1.66	0.15	9
	20~40	24	1.35~1.63	1.44	0.07	5
	40~60	24	1.32~1.56	1.44	0.06	4
速效钾	0~20	24	48~78	63.29	9.69	15
	20~40	24	27~61	43.38	8.35	19
	40~60	24	23~51	34.17	6.96	20

2.2　土壤速效养分水平分异

2.2.1　土壤速效磷水平分异

0~20cm 土层的 24 个样本中，土壤速效磷含量在 1.70~1.80mg/kg 的范围出现的频数最多，其次是 1.40~1.60mg/kg 和 1.80~1.90mg/kg，1.60~1.70mg/kg 出现的频数最少(见图 2(a))。从西到东，系列 L1、L2、L3、L4 的含量先减小后增加，分别在 13、14、15、16 区达最大值，在 19、20、21、22 区达最低值。系列 L5、L6 先增加后减小，分别在 11、12 区达最大值，在 23、24 区达最小值。从北到南，C1、C3 系列的变化趋势都是先增加后减小，然后又增加；分别在 2、14 区达最大值。C2、C4 系列先增加后减小再增加；分别在 12、24 区达最大值(见图 3(a))。

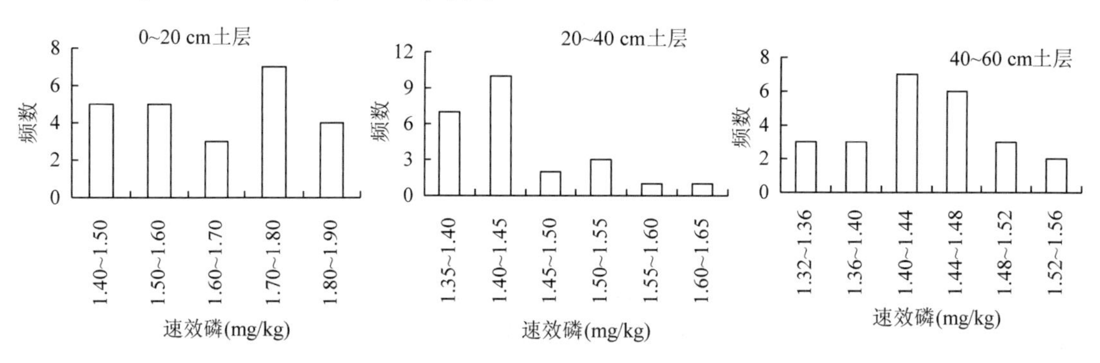

图 2　毛白杨造林地土壤速效磷频数分布图

20~40cm 土层的 24 个样本中，土壤速效磷含量在 1.35~1.45mg/kg 的范围出现的频数最多，在 1.45~1.65mg/kg 的范围出现的频数最少(见图 2(b))。从西到东，系列 L1、L2、L3、L4、L5、L6 土壤速效磷含量变化均是先增加后减小，6 个系列分别在 13、2、15、16、17、18 区达最大值。从北到南，系列 C1、C2、C3、C4 土壤速效磷含量变化基本一致，遵循先增加后减小再增加的趋势，分别在 4、8、17、22 区达最大值(见图 3(b))。

40~60cm 土层的 24 个样本中，土壤速效磷含量在 1.40~1.48mg/kg 的范围出现的频数最多，在 1.32~1.36mg/kg 和 1.48~1.56mg/kg 的范围出现的频数最少(见图 2(c))。从西到东，系列 L1、L2、L3、L4、L5、L6 均是先增加后减小；分别在 13、8、15、16、17、

12 区达最大值。从北向南，4 个系列速效磷含量的变化规律很复杂，但总的趋势是先增加后减小又增加；分别在 4、10、16、23 区达最大值(见图 3(c))。

由上可见，0~20cm 土层速效磷含量的变化趋势总体上是：从西到东是先减小后增加，从北到南是先增加后减小。20~40cm 和 40~60cm 土层从西到东的变化是先增加后减小，从北到南的变化是先增加后减小再增加。3 个层次速效磷含量的变异系数分别为 9%、5% 和 4%，可能与磷的移动性很小有关。若实验的精度为 90%，则 3 个层次分别取一个混合土样即可；若实验的精度为 95%，3 个层次可以取混合土样数分别为 4、1 和 1 个。

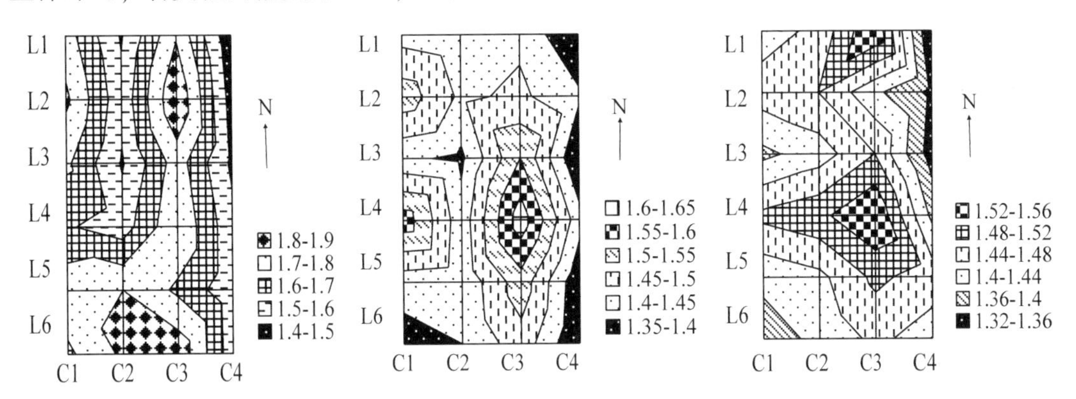

图 3　毛白杨造林地土壤速效磷含量(mg/kg)

(a)0~20cm 土层；(b)20~40cm 土层；(c)40~60cm 土层

2.2.2　土壤速效钾水平分异

0~20cm 土层的 24 个样本中，土壤速效磷含量没有明显差异，除在 60~66mg/kg 的范围出现的频数少一些，其他几个范围出现的频率相当(见图 4(a))。从西到东，系列 L1、L2、L3、L5、L6 的含量先增加后减小再增加，分别在 19、20、21、11、12 区达最大值，在 13、14、15、5、8 区达最低值；系列 L4 先增加后减小，分别在 16 区达最大值，在 22 区达最小值。从北到南，C1、C2、C3 系列的变化趋势都是先增加后减小，然后再增加减小，分别在 2、11、16 区达最大值；C4 系列先减小后增加，然后再减小增加，在 19 区达最大值，24 区达最小值(见图 5(a))。

20~40cm 土层的 24 个样本中，土壤速效磷含量在 34~48mg/kg 的范围出现的频数最多，其次是 55~62mg/kg 和 27~34mg/kg 范围，出现的频数最少的范围是 48~55mg/kg(见图 4(b))。从西到东，系列 L1、L2 土壤速效磷含量变化均是先增加后减小再增加，分别在 7、8 区达最大值，在 1、2 区达最小值；系列 L3、L4、L5、L6 变化规律是先增加后减小，分别在 9、10、17、18 区达最大值，在 21、22、23、24 区达最小值。从北到南，系列 C1、C2、C3、C4 土壤速效磷含量变化基本一致，遵循减小后增加的趋势，分别在 6、7、12、13 区达最大值，在 3、11、14、23 区达最小值(见图 5(b))。

40~60cm 土层的 24 个样本中，土壤速效磷含量在 27~37mg/kg 的范围出现的频数最多，在 37~52mg/kg 的范围出现的频数较少(见图 4(c))。从西到东，系列 L1、L2、L3、L4 均是先增加后减小；分别在 7、8、9、10 区达最大值，在 1、2、3、4 区达最小值；系列 L5、L6 先增加后减小再增加，分别在 11、12 区达最大值，在 5、18 区达最小值。从北向南，系列 C1、C4 先减小后增加，分别在 6、24 区达最大值，在 3、21 区达最小值；系列 C2、C3 先增加后减小，分别在 8、14 区达最大值，在 12、18 区达最小值(见图 5(c))。

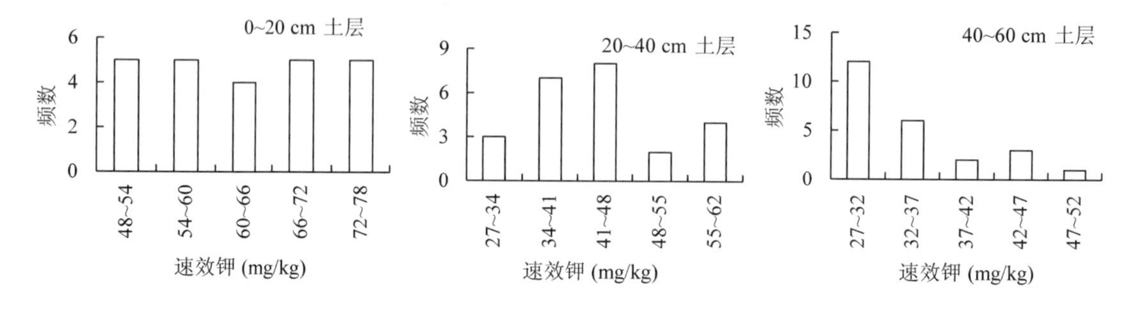

图 4　毛白杨造林地土壤速效钾频数分布图

由上可见，0~20cm 土层速效钾含量的变化趋势总体上是：从西到东是先增加后减小，从北到南变化复杂。20~40cm 和 40~60cm 土层从西到东的变化是先增加后减小，20~40cm 从北到南的变化是先减小后增加，40~60cm 从北到南的变化复杂。3 个层次速效钾含量的变异系数分别为 15%、19%和 20%，随着土层的加深，变异程度逐渐增加，但变化不明显。若实验的精度为 90%，则 3 个层次可取混合土样数分别为 3、4 和 4 个；若实验的精度为 95%，3 个层次可取混合土样数分别为 9、15 和 16 个。

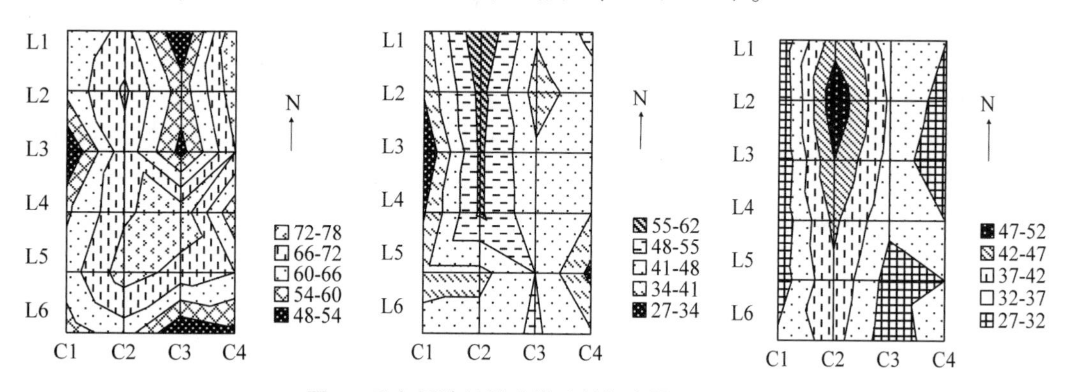

图 5　毛白杨造林地土壤速效钾含量(mg/kg)

(a)0~20cm 土层；(b)20~40cm 土层；(c)40~60cm 土层

2.3　土壤速效养分垂直分异

图 6 反映了土壤速效磷含量的垂直变化。可看出土壤速效磷含量均值表层最高，其它两层含量变化不明显。由于肥力指标主要取决于 0~20cm 土层物质的下移量和土壤本身的化学成分组成，故其空间分异程度相对较小。对速效磷含量进行方差分析，$F=36.70>F_{0.01}=4.98$，差异极显著，多重比较表明：0~20cm 层和其它两层差异极显著，20~40cm 和 40~60cm 土层速效磷含量差异不显著，在今后的取土中可只取 0~20cm、20~40cm 两层即可。图 7 反映了土壤速效钾含量的垂直变化。可看出土壤速效钾含量均值表层最高，随着土层的加深，速效钾含量逐渐减小，对速效钾含量进行方差分析，$F=75.24>F_{0.01}=$

4.98，差异极显著，多重比较表明3个层次速效钾含量差异极显著。速效钾的垂直变化大于速效磷的垂直变化，可能与土壤磷循环代谢有关，同时植物对钾的吸收对土壤本身含钾量影响不大。

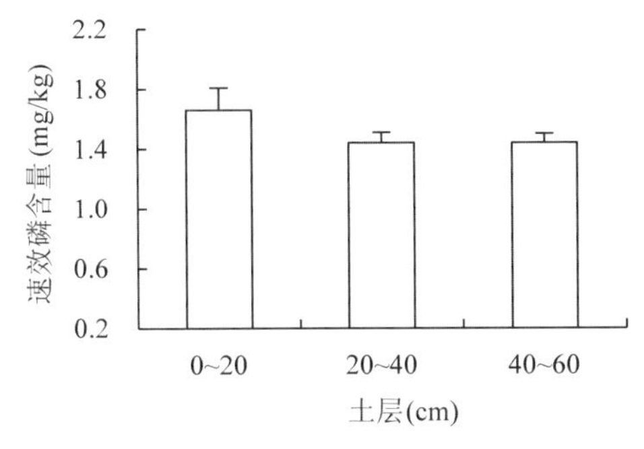

图6　土壤速效磷垂直变异

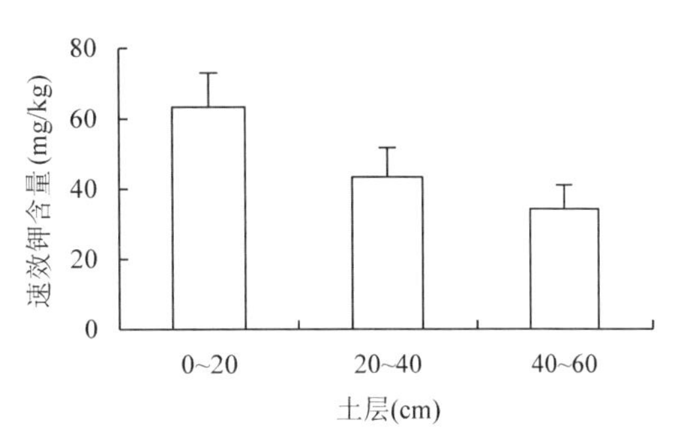

图7　土壤速效钾垂直变异

3　结论

(1)试验区土壤速效磷含量的范围是1.32~1.89mg/kg，总体均值为1.51mg/kg，速效钾含量范围是23~78mg/kg，总体均值46.95mg/kg，按全国第二次土壤普查暂行规程的标准划分养分等级，土壤速效磷含量为6级，属于最低水平；土壤速效钾含量为5级，属于较低水平。试验地可通过施肥来增加土壤中磷和钾的水平。

(2)0~20cm、20~40cm、40~60cm三层土壤速效磷和速效钾含量的变异系数分别为9%、5%、4%和15%、19%、20%。速效钾属于弱变异程度，速效磷属于中等变异程度，相同土壤层土壤肥力指标变异大小顺序为：速效钾>速效磷；随着土层加深，速效磷分异程度逐渐降低，速效钾分异程度逐渐增大。

(3)根据试验区土壤速效养分分布规律，若所能达到的试验精度分别为90%和95%，则0~20cm、20~40cm、40~60cm三层土壤速效磷可以取混合土样数分别为1、1、1个和4、1、1个，速效钾可以取混合土样数分别为3、4、4个和9、15、16个。

(4)土壤速效磷含量在水平和垂直变化不大，土壤速效钾含量自西向东先增后减，自北向南升降交替、变化复杂。试验区速效养分分布规律可能与母质、地理因素、人为活动、土地利用方式有关，有待进一步研究。

Study on the Spatial Variability of Soil Available Phosphorus and Available Potassium in *Populus Tomentosa* Planted Land in Western Plain of Shandong Province

Abstract　Analyze of the soil available phosphorus and available potassium in a 6.0 hectare *Populus tomentosa* planted land in Guanxian county of Shandong province was aimed to benefit test arrangement of *Populus tomentosa* planted land, the site specific nutrient management and determining rational sampling number. To have a better understanding the available phosphorus and available potassium status grid sampling method with grid size (50m×50m) was used. Each grid was subdivided into three layers (0~20cm, 20~40cm, 40~60cm). The results showed: available nutrient content was a lower level, which was decreased with the soil depth, and the range of available phosphorus was 1. 32~1. 89mg/kg, the population mean was (1. 51±0. 09) mg/kg, and the contents of three layers were (1. 66±0. 15) mg/kg, (1. 44±0. 07) mg/kg, (1. 44±0. 06) mg/kg, respectively. The ranges of available potassium was 23−78mg/kg, the population mean was (46. 95±8. 33) mg/kg, and the contents of three layers was (63. 29±9. 69) mg/kg, (43. 38±8. 35) mg/kg, (34. 17±6. 96) mg/kg, respectively. The Coefficient of Variance of the available phosphorus was 9%, 5%, 4% in three layers, respectively, which were weak variability. And available potassium was 15%, 19%, 20% in three layers, respectively, which were moderate variability. Rational sampling number was related to coefficient of variation and sampling precision. The value of available phosphorus changed little both in the horizontal direction and vertical direction, and the coefficient of variance of available potassium increased at first and then decreased from west to east, but ascended and descended, varied complicatedly from north to south.

Key words　*Populus tomentosa*, planted land, available phosphorus, available potassium, space variability.

毛白杨杂种无性系稳定碳同位素值的特征及其水分利用效率*

摘　要　2008年在北京林业大学对9个毛白杨(*Populus tomentosa*)杂种无性系苗木生长不同时期的叶片、小枝、根的碳同位素 $\delta^{13}C$ 和瞬时水分利用效率 WUEi 的差异进行研究，分析不同无性系间 $\delta^{13}C$ 与瞬时水分利用效率 WUEi 的相互关系，目的在于探求 $\delta^{13}C$ 在筛选高水分利用效率毛白杨杂种无性系中的应用价值。结果表明：不同生长时期叶片 $\delta^{13}C$ 表现为7月<8月<10月<9月，小枝 $\delta^{13}C$ 值8月<9月<10月，叶片 $\delta^{13}C$ 值、小枝 $\delta^{13}C$ 值在不同时期和不同无性系间的差异均达显著水平，无性系间的差异是引起 $\delta^{13}C$ 值变化的主要因素。不同部位碳同位素比值表现为叶片<小枝<根，毛白杨杂种无性系 $\delta^{13}C$ 值在不同部位和无性系间差异均达显著水平，部位间的差异是引起 $\delta^{13}C$ 值变化的主要因素。$\delta^{13}C$ 值较高的无性系30、83、BL5的 WUEi 也较高，$\delta^{13}C$ 值较低的无性系42、26、BT17的WUEi也较低，且不同时期 $\delta^{13}C$ 和WUEi呈较强的正相关，相关系数分别为0.766、0.872、0.675，高 $\delta^{13}C$ 可以作为筛选高WUEi毛白杨的有效指标，且在苗木生长旺盛时期选育能得到更为可靠的结果。

关键词　毛白杨，杂种无性系，碳同位素，水分利用效率

在自然条件下，碳有两种稳定同位素，其自然丰度 ^{12}C 占98.89‰，^{13}C 占1.11‰。植物叶片 $\delta^{13}C$ 可以用来作为一段时间内水分利用和损失的标准被广泛应用(Todd et al., 2002)。这主要是由于光合作用时，大气 CO_2 经气孔向叶内的扩散过程，CO_2 在叶中的溶解过程，以及羧化酶对 CO_2 的同化过程，均存在显著的碳同位素效应(Farquhar et al., 1989)。Farquhar等认为，植物组织的稳定碳同位素比率 $\delta^{13}C$ 和稳定碳同位素分辨力△与 C_3 植物的水分利用效率(Water Use Efficiency，WUE)具有很强的相关性，可以作为植物长期WUE的间接测定指标(Ehleringer et al., 1988；Ehleringer et al., 1993；苏波等，2000)，并在小麦(*Triticum aestivum* Linn.)的研究上得到证实(Farquhar et al., 1984)。目前，国内外学者，在草本植物如花生(*Arachis hypogaea*)(Hubock et al., 1989)、棉花(*Gossypium* spp.)(Hubock et al., 1987)、大麦(*Hordeum vulgare* Linn.)(Hubock et al., 1989)、甜菜(*Beta vulgaris* Linn.)(Monti et al., 1989)，木本植物如白云杉(*Picea glauca* Voss.)(Sun Z J et al., 1996)、桉树(*Eucalyptus radiate*)(Ernst et al., 2006)和黑杨(*Populus nigra* Linn.)无性系(赵凤君，2006)以及藤本植物葡萄(*Vitis vinifera* L.)(Claudia et al., 1989)等方面进行了 $\delta^{13}C$ 或△与长期WUE的相关性研究，大部分研究结果显示 $\delta^{13}C$ 与长期WUE呈正相关，△与长期WUE呈负相关。

毛白杨(*Populus tomentosa* Carr.)是我国特有乡土树种，由于其优良特性已被广泛地作

---

* 本文原载《生态环境学报》，2009，18(6)：2267-2271，与方晓娟、李吉跃、聂立水、沈应柏合作发表。

为短期轮伐的造林树种，在解决木材短缺方面占有重要位置。在水资源缺乏的现状下，快速、高效地选育速生且高水分利用效率(WUE)的无性系已成为迫切需求。

本文研究的主要目的是弄清毛白杨杂种无性系稳定碳同位素的组成特点，并探求 $\delta^{13}C$ 在筛选高水分利用效率毛白杨杂种无性系中的应用价值。

1 材料与方法

1.1 试验材料及培养

所研究的植物材料为 9 个 $1_{(2)}$-0 型毛白杨(*Populus tomentosa* Carr.)杂种无性系，分别为：S86、1316、BT17、BL5、26、42、30、50、83，为林木育种国家工程实验室培育保存。2007 年 3 月底将长约 15cm 的嫁接苗栽植到 33cm(高)×30cm(内径)的棕色塑料花盆中进行培养，盆栽土壤为砂壤土，土壤容重(1.411±0.02)$g \cdot cm^{-3}$，田间持水量(26.02±0.77)%。各无性系嫁接苗的大小和质量尽可能保持一致，每盆栽一株，苗木在充分供水条件下培养。2007 年年底平茬。2008 年 4 月测得盆栽土壤容重为(1.48±0.04)$g \cdot cm^{-3}$，田间持水量为(15.21±0.64)%。试验地点为北京林业大学。

1.2 研究方法

1.2.1 毛白杨 $\delta^{13}C$ 组成

分别于 2008 年 7 月 17 日、8 月 19 日、9 月 16 日、10 月 15 日选每无性系生长中等的苗木三株，每株取一片功能叶(第 6~9 片)，后三次加取小枝，最后一次加取根，将样品放入烘箱中 105℃杀青 30mins，置于 70℃恒温烘箱内烘干至恒重(约 72h)，称其干重，研磨过 100 目筛制成备用样品。在中国科学院植物研究所质谱仪分析室进行稳定碳同位素分析(测定精度为 0.2‰)，具体为取处理好的样品 3~5mg 封入真空的燃烧管，并加入催化剂和氧化剂，燃烧产生的 CO_2 经结晶纯化后，用 Delta plus XP 同位素比例质谱仪(Thermo Finnigan 公司，德国)测定碳同位素的比率，以 PDB(Pee Dee Belemnite)为标准，根据下面公式进行计算：

$$\delta^{13}C(‰)=\left\{\left[({}^{13}C/{}^{12}C)_{sample}-({}^{13}C/{}^{12}C)_{standard}\right]/({}^{13}C/{}^{12}C)_{standard}\right\}\times 1000$$

1.2.2 瞬时水分利用效率

分别于 2008 年 7 月 16 日、8 月 18 日、9 月 15 日用 Lico-6400 便携式光合作用分析系统(LICOR 公司，美国)进行测定。所有测定均在上午 9:30~11:30 之间完成。瞬时水分利用效率(*WUEi*，由 *Pn*/*Tr* 计算得出)，每个无性系 3 个重复。

1.2.3 数据处理

用 EXCEL 作图；用 SPSS13.0 进行双因素方差分析、S-N-K 多重比较和相关性分析。

2 结果与分析

2.1 叶片碳同位素组成

从表1可以看出：9个毛白杨杂种无性系叶片碳同位素比值在-28.549‰~-31.359‰之间变化，就均值来看，叶片 $\delta^{13}C$ 值7月<8月<10月<9月；就不同无性系的变化来看，BT17、26叶片 $\delta^{13}C$ 值较低，分别为-30.756±0.691‰和-30.330±0.322‰，而无性系30、83叶片 $\delta^{13}C$ 值较高，分别为-29.116±0.710‰和-29.262±0.604‰。方差分析显示(表4)，毛白杨杂种无性系叶片 $\delta^{13}C$ 值在不同时期和无性系间差异均达极显著水平，无性系间 $F=17.074$ 远大于时期间 $F=6.764$，说明无性系间的差异是引起叶片 $\delta^{13}C$ 值变化的主要因素。

表1 不同时期毛白杨杂种无性系叶片碳同位素比值($\delta^{13}C$‰)

无性系	测定时间(月.日)				
	7.17	8.19	9.16	10.15	均值
42	-30.856±0.035	-29.502±0.021	-29.825±0.025	-29.484±0.014	-29.917±0.645bcd
26	-30.802±0.033	-30.080±0.019	-30.201±0.017	-30.238±0.024	-30.330±0.322ab
S86	-30.530±0.032	-30.521±0.054	-28.857±0.061	-29.518±0.049	-29.857±0.818bcd
1316	-30.831±0.048	-29.318±0.018	-29.101±0.033	-29.261±0.030	-29.628±0.807bcd
BT17	-31.359±0.015	-29.902±0.021	-30.489±0.035	-31.275±0.042	-30.756±0.691a
83	-30.020±0.052	-29.317±0.036	-29.161±0.026	-28.549±0.034	-29.262±0.604cd
30	-30.164±0.025	-28.630±0.032	-28.733±0.036	-28.938±0.013	-29.116±0.710d
BL5	-30.726±0.036	-30.238±0.023	-29.305±0.052	-29.846±0.041	-30.029±0.602abc
50	-30.800±0.042	-30.123±0.056	-29.597±0.043	-29.812±0.051	-30.083±0.525abc
均值	-30.676±0.399a	-29.737±0.591b	-29.474±0.601b	-29.658±0.788b	-29.886±0.508

2.2 小枝碳同位素组成

从表2可以看出：9个毛白杨杂种无性系小枝碳同位素比值在-29.768‰~-28.236‰之间变化，就均值来看，小枝 $\delta^{13}C$ 值8月<9月<10月；就不同无性系的变化来看，42、26小枝 $\delta^{13}C$ 值较低，分别为-29.555±0.691‰和-29.447±0.048‰，而无性系30、83小枝 $\delta^{13}C$ 值较高，分别为-28.088±0.189‰和-28.610±0.232‰。方差分析显示(表4)，毛白杨杂种无性系小枝 $\delta^{13}C$ 值在不同时期差异不显著，无性系间差异达极显著水平，说明无性系间的差异是引起小枝 $\delta^{13}C$ 值变化的主要因素。

表 2　不同时期毛白杨杂种无性系小枝碳同位素比值($\delta^{13}C$‰)

无性系	测定时间(月·日)			
	8. 19	9. 16	10. 15	均值
42	-29. 768±0. 014	-29. 468±0. 021	-29. 428±0. 032	-29. 555±0. 186a
26	-29. 399±0. 025	-29. 494±0. 036	-29. 448±0. 027	-29. 447±0. 048a
S86	-29. 373±0. 012	-28. 494±0. 023	-28. 616±0. 035	-28. 828±0. 476ab
1316	-28. 236±0. 032	-29. 287±0. 026	-28. 395±0. 019	-28. 639±0. 567ab
BT17	-28. 938±0. 017	-29. 662±0. 026	-28. 706±0. 037	-29. 102±0. 499ab
83	-28. 709±0. 028	-28. 775±0. 016	-28. 345±0. 054	-28. 610±0. 232ab
30	-28. 112±0. 035	-28. 494±0. 043	-27. 658±0. 062	-28. 088±0. 419c
BL5	-29. 637±0. 046	-28. 433±0. 053	-28. 395±0. 023	-28. 822±0. 706ab
50	-29. 012±0. 039	-28. 976±0. 047	-28. 876±0. 058	-28. 955±0. 070ab
均值	-29. 020±0. 586a	-29. 009±0. 483a	-28. 652±0. 560a	-28. 894±0. 446

2. 3　碳同位素组成的器官差异

从表 3 可以看出：10 月 9 个毛白杨杂种无性系不同部位碳同位素比值表现为叶片<小枝<根，叶片 $\delta^{13}C$ 值比小枝偏负 1. 006‰，比根偏负 1. 632‰；就各无性系不同部位的平均值来看，26、BT17 的 $\delta^{13}C$ 值较低，分别为(-29. 609±0. 566‰)和(-29. 320±1. 731‰)，而无性系 30、83 的 $\delta^{13}C$ 值较高，分别为(-27. 940±0. 892‰)和(-28. 184±0. 466‰)。方差分析显示(表 4)，毛白杨杂种无性系 $\delta^{13}C$ 值在不同部位和无性系间差异均达极显著水平，部位间 $F=25.489$ 远大于时期间 $F=3.975$，说明部位间的差异是引起 $\delta^{13}C$ 值变化的主要因素。

表 3　毛白杨杂种无性系 10 月叶、小枝、根碳同位素比值($\delta^{13}C$‰)

无性系	叶	小枝	根	均值
42	-29. 484±0. 014	-29. 428±0. 032	-28. 971±0. 025	-29. 294±0. 281ab
26	-30. 238±0. 024	-29. 448±0. 027	-29. 140±0. 041	-29. 609±0. 566a
S86	-29. 518±0. 049	-28. 616±0. 035	-27. 593±0. 037	-28. 576±0. 963ab
1316	-29. 261±0. 030	-28. 395±0. 019	-28. 114±0. 042	-28. 590±0. 598ab
BT17	-31. 275±0. 042	-28. 706±0. 037	-27. 980±0. 031	-29. 320±1. 731ab
83	-28. 549±0. 034	-28. 345±0. 054	-27. 659±0. 028	-28. 184±0. 466b
30	-28. 938±0. 013	-27. 658±0. 062	-27. 223±0. 065	-27. 940±0. 892b
BL5	-29. 846±0. 041	-28. 395±0. 023	-27. 170±0. 029	-28. 470±1. 340ab
50	-29. 812±0. 051	-28. 876±0. 058	-28. 383±0. 063	-29. 024±0. 0726ab
均值	-29. 658±0. 788a	-28. 652±0. 560b	-28. 026±0. 705c	-28. 779±0. 563

2.4 瞬时水分利用效率

从图 1 可以看出：9 个毛白杨杂种无性系瞬时水分利用效率在 2.76～4.41μmol · mmol^{-1} 之间变化，就均值来看，*WUEi* 值大小为 7 月<8 月<9 月；就不同无性系的变化来看，26、42 的 WUEi 值较低，分别为(3.27±0.42)μmol · mmol^{-1} 和(3.62±0.21)μmol · mmol^{-1}，而无性系 30、50 的 *WUE* 值较高，分别为(3.97±0.71)μmol · mmol^{-1} 和(3.83±0.12)μmol · mmol^{-1}。方差分析显示(表 4)，毛白杨杂种无性系瞬时水分利用效率在不同时期和无性系间差异均达极显著水平。

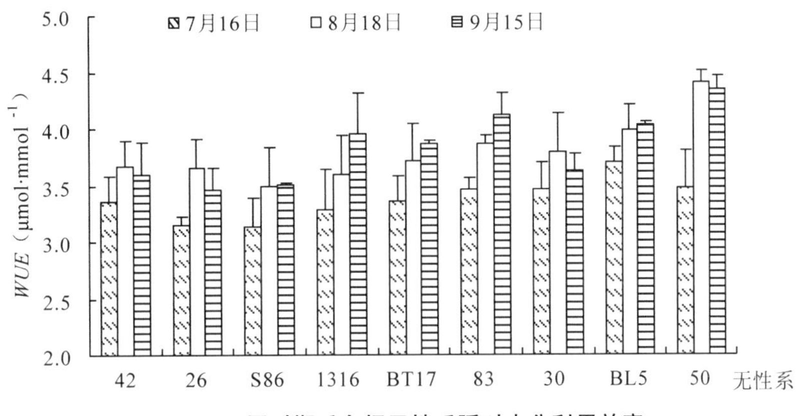

图 1　不同时期毛白杨无性系瞬时水分利用效率

2.5 叶片碳同位素组成与瞬时水分利用效率的相关性分析

植物叶片的稳定碳同位素比(δ^{13}C 值)能够很好地反映与植物光合、蒸腾作用相关联的水分利用效率，植物对 $^{13}CO_2$ 判别能力的大小是评价植物水分利用效率的有效指标。对九个毛白杨杂种无性系叶片的碳同位素组成和瞬时水分利用效率进行相关性分析得出二者呈显著正相关关系，不同时期的相关系数分别为：$R_7=0.766$，$P=0.016$；$R_8=0.772$，$P=0.002$；$R_9=0.675$，$P=0.023$。

表 4　双因素方差分析表

碳同位素组成	差异源	平方和	自由度	均方	*F* 值
叶片	时期	7.817	3	2.606	6.764**
	无性系	8.258	8	1.032	17.074**
	误差	3.663	24	0.153	
	总计	19.738	35		
小枝	时期	0.791	2	0.395	2.703
	无性系	4.783	8	0.598	4.086**
	误差	2.341	16	0.146	
	总计	7.914	26		

（续）

碳同位素组成	差异源	平方和	自由度	均方	F 值
不同部位	部位	12.202	2	6.101	25.489**
	无性系	7.612	8	0.951	3.975**
	误差	3.830	16	0.239	
	总计	23.644	26		
水分利用效率	时期	1.523	2	0.761	28.324**
	无性系	1.485	8	0.186	6.906**
	误差	0.430	16	0.027	
	总计	3.438	26		

注：* 表示在 0.05 水平上差异显著，** 表示在 0.01 水平上差异显著

3　结论与讨论

3.1　毛白杨杂种无性系碳同位素组成特点

9 个毛白杨杂种无性系碳同位素比值在-31.359‰~-27.170‰之间变化，叶片 $\delta^{13}C$ 值在-28.549‰~-31.359‰之间变化，表现为：7 月<8 月<10 月<9 月，小枝碳同位素比值在-29.768‰~-28.236‰之间变化，表现为：8 月<9 月<10 月，小枝 $\delta^{13}C$ 值变化范围低于叶片，方差分析结果显示：无性系间的差异是引起叶片、小枝 $\delta^{13}C$ 值变化的主要因素。不同部位碳同位素比值变化表现为叶片<小枝<根，Oliver Brendel 等(2003)对欧洲赤松(*Pinus sylvestris* L.)的研究结果表明针叶的 $\delta^{13}C$ 值约比小枝的 $\delta^{13}C$ 值偏负 1.3‰。刘海燕(2008)对油松(*Pinus tabulaeformis*)的研究表明小枝和针叶 $\delta^{13}C$ 值的差异要远大于 1.3‰，达到 2.10‰。祁连圆柏(*Sabina przewalskii* Kom.)树叶 $\delta^{13}C$ 值(-26.24‰)比树木 $\delta^{13}C$ 值(-22.94‰)偏负约 3.30‰(陈拓等，2004)。Carol & Winner 报道小麦(*Triticum aestivum* Linn.)根中 $\delta^{13}C$ 值高于叶片；冯虎元等(2001)对大豆(*Glycine max* Merr.)10 个品种的 $\delta^{13}C$ 值研究表明 $\delta^{13}C$ 根>$\delta^{13}C$ 茎>$\delta^{13}C$ 种子>$\delta^{13}C$ 叶；严昌荣等(2002)的研究结果也显示树干、根、小枝的 $\delta^{13}C$ 值一般高于叶片。Hall 等对豇豆(*Vigna unguiculata*)的分析发现籽粒的平均 $\Delta^{13}C$ 值明显低于叶片，且叶片和籽粒的 $\Delta^{13}C$ 之间高度相关，即籽粒的 $\delta^{13}C$ 值高于叶片的 $\delta^{13}C$ 值。尹伟伦等对杨树的研究发现 $\delta^{13}C$ 树干 > $\delta^{13}C$ 枝 > $\delta^{13}C$ 叶。可见，多数的研究结果为枝条的 $\delta^{13}C$ 值要高于叶片，而根茎等其他器官的 $\delta^{13}C$ 值大小则因研究物种不同而存在差异。

这种分异可能来自两个方面，一是植物不同器官的生物化学成分不同，生理生态特性差距较大。由于在合成这些物质时的途径不一样会出现不同的碳同位素的二次分馏，使各种化学成分具有各自的稳定碳同位素比率，如树干和根中木质素、粗纤维素占的比重大，叶片中蛋白质含量高，可溶性糖的 $\delta^{13}C$ 值高于纤维素 $\delta^{13}C$ 值，从而导致不同植物 $\delta^{13}C$ 值器官间差异较大(Francey et al., 1985)。二是不同器官的呼吸特性具有很大差异，从理论上讲，植物器官在呼吸时优先利用含 ^{12}C 的物质而使 ^{13}C 在组织中富集(Leavitt et al., 1985)。根据对北京山区几种植物非光合器官呼吸速率的测定，不同种和同一种的不同器

官具有不同的呼吸速率，这也从另一个侧面表明不同器官存在不同碳同位素比率的原因（韩兴国等，2000）。因而，在对不同植物的 $\delta^{13}C$ 值进行比较时，应根据具体植物的不同器官进行具体分析。

3.2 毛白杨杂种无性系碳同位素组成与瞬时水分利用效率

水分是树木生长发育所必需的，树木水分散失主要是由叶片蒸腾造成的。叶片光合作用与蒸腾作用是两个同时进行的气体交换过程，气孔作为气体交换的门户，其行为调节和控制光合与蒸腾。蒸腾作用与光合作用同步进行，两者的比值决定了植物叶片水平上水分利用效率的大小（李俊等，1997；张岁岐等，2002）。叶片水分利用效率作为植物生理活动过程中消耗水分形成有机物质的基本效率，成为确定植物体生长发育所需要的最佳水分供应的重要指标之一（王会肖等，2003）。

叶片 $\delta^{13}C$ 能很好地反映与植物光合作用和蒸腾作用相关的 WUEi，可以用来间接指示植物的长期水分利用效率（WUE_L）（刘光琇等，2004）。诸多研究认为，$\delta^{13}C$ 对植物 WUE_L 有较好的指示作用（赵凤君等，2006），但并不能很好地反映 *WUEi*，因为 Pn、Tr、Gs 和 Ci 等多种因子随环境条件变化较大，从而引起 *WUEi* 的变化，而 $\delta^{13}C$ 具有遗传稳定性。而本研究中，不同时期毛白杨杂种无性系叶片 $\delta^{13}C$ 值和瞬时水分利用效率 WUEi 呈显著正相关，相关系数分别达 0.766（7 月）、0.872（8 月）、0.675（9 月），说明利用 $\delta^{13}C$ 判断毛白杨 WUEi 具有一定的可行性，且在苗木生长旺盛时期（8 月）选育能得到更为可靠的结果。

The Characteristics of Stable Carbon Isotope and Water Use Efficiency for *Populus tomentosa* Hybrid Clones

Abstract Our objective was to explore the possibility of stable carbon isotope ($\delta^{13}C$) on selection of *Populus tomentosa* hybrid clones with high water use efficiency (WUEi). We measured $\delta^{13}C$ values and WUEi in leaves, branches, and roots from seedlings of nine different clones, and analyzed the relationship between $\delta^{13}C$ and WUEi at Beijing Forestry University in 2008. The results showed that $\delta^{13}C$ values of leaves were significantly different between clones at different times, so were $\delta^{13}C$ values of branches. $\delta^{13}C$ values of leaves were highest in September followed by October, August, and July, while $\delta^{13}C$ values of branches were highest in October followed by September and August, $\delta^{13}C$ values showed significant differences between different *Populus tomentosa* hybrid clones. It implied that the difference between clones was a major factor to cause the changes of $\delta^{13}C$ values. Carbon isotope ratios were also significantly different between different parts of clones. Ratios were highest in roots followed by branches and leaves, The differences between parts in the same period caused changes of $\delta^{13}C$ value as the main factor. The clones 30, 83, and BL5 with higher $\delta^{13}C$ had higher WUEi, meanwhile, the clones 42, 26, and BT17 with lower $\delta^{13}C$ had lower WUEi. It showed positive correlation between $\delta^{13}C$ and WUEi ($R=0.766$, $P=0.016$, $R=0.872$, $P=0.002$, $R=0.675$, $P=0.023$). $\delta^{13}C$ can be used as an effective indicator to select clones of *Populus tomentosa* with high water use efficiency, especially for the selection during vigorous seedling growth.

Key words *Populus tomentosa*, hybrid clones, carbon isotope, water use efficiency.

不同施肥处理对毛白杨人工林生长及营养状况的影响*

摘　要　为寻求毛白杨(*Populus tomentosa* Carr.)人工林施肥的最佳效果，该文采用$L_9(3^4)$正交试验设计研究了不同配比施肥和不同时期施肥对毛白杨人工林生长及营养状况的影响，通过测量分析其生长指标(胸径、树高、冠幅和一级分枝数)和叶片中营养元素含量的变化，结果表明：氮肥是影响毛白杨人工林生长的主要肥料，其次是钾肥、磷肥，而最佳施肥配比是纯N 240g/株、P_2O_5 120g/株、K_2O 80g/株，之比约为3:1.5:1；胸径、树高和冠幅生长主要集中在6月占整个生长季的34.23%、31.74%和36.39%，而施肥处理加快胸径、树高、冠幅生长主要集中在7月，分别比对照增加147.06%、23.26%、45.71%，施肥时期应选速生前期施肥即6月之前，再次追肥时间应选择7月期间；施肥影响叶片中氮、磷、钾含量的变化，其中，氮含量变化比磷、钾要明显，促进氮、磷、钾吸收最好的处理分别是处理7(N 240g/株、$P_2O_5$0g/株、K_2O 80g/株)、处理6(N 120g/株、P_2O_5 120g/株、K_2O 0g/株)和处理5(N 120g/株、P_2O_5 60g/株、K_2O 80g/株)。因此，选择合理的施肥配比和施肥时间能显著提高毛白杨人工林生长。

关键词　毛白杨人工林，正交试验设计，施肥，营养状况

毛白杨(*Populus tomentosa* Carr.)是中国特有的速生优质乡土树种之一，由于它具有生长快、材质优良、抗逆性强和适应性广等特点(刘超等，2008)，已在国内进行大面积速生丰产的营造，并取得了较好的经济效应，在中国的国民经济中起着重要作用(石金铭，1990)，但由于缺乏科学施肥理论的指导，盲目施肥导致肥料利用率降低、木材产量低、品质劣，因此，国内外许多学者对毛白杨人工林施肥开展初步研究(刘勇等，2000；孙时轩，1995；袁玉欣，1990；Brinkman J A et al.，1994)。其中，刘寿坡等(1988)对毛白杨幼林内进行施肥量和肥料配比试验，结果表明，施肥当年对生长的影响未达显著程度，第2、3年效果显著。N肥有明显促进生长的作用，P肥影响甚微，K肥有影响但未达显著程度。使用较大数量的N肥并配以少量均等的P肥和K肥效果较好；李鸣(2007)通过回归方程得出一年生3930杨苗氮、磷、钾最佳配方浓度范围分别是：136.72~158.67、61.81~79.57、59.43~81.46mg/L，施用比例总体约为2∶1∶1；胡亚利等(2008)通过对Ⅰ-107杨的施肥效应对比试验表明：施用SV杨树专用肥对1~2年生的杨树株高、地径的促增长效果显著；孙利涛(2008)提出尿素(N肥)对苗高生长的促进作用明显，但是当用量达到9g/盆时，表现出对生长的抑制作用。对于人工林施肥试验来说，不同地理位置和

* 本文原载《中国农学通报》，2010，26(9)：115-121，与胡磊、李吉跃、尚富华、辛颖、宋莲君和沈应柏合作发表。

土壤性状差别，可能使得研究结论不同。为找出适合本地区毛白杨人工林生长最佳施肥时期和施肥配方，此文采用 $L_9(3^4)$ 正交设计开展施肥处理对人工林生长和营养元素含量影响的研究。同时，也为毛白杨人工速生丰产林建设提供配方施肥的理论依据。

1 材料与方法

1.1 试验材料

试验品种为毛白杨无性系 S86，试验林为 3 年生人工林，位于河北省威县苗圃场，栽植密度为：3m×4m，平均胸径为 6.27cm，平均树高为 5.72m，威县苗圃场施肥试验前土壤的基本理化性质指标测定结果见表 1。试验用肥为常用的商业性肥料：尿素(含 N 量 46%)、过磷酸钙(含 P_2O_5 16%)、硫酸钾(含 K_2O 50%)。

表 1 试验地土壤的基本理化性质

土层(cm)	pH	全氮(g/kg)	速磷(mg/kg)	速钾(mg/kg)	有机质(g/kg)
0~20	8.60±0.35	0.698±0.09	1.17±0.12	140±5.23	9.8±0.25
20~40	8.82±0.13	0.602±0.16	2.26±0.10	130±4.65	9.0±0.19
40~60	8.44±0.22	0.451±0.06	1.17±0.07	90±3.39	7.0±0.24

1.2 试验方法

N、P、K 三因素 3 水平施肥试验按 $L_9(3^4)$ 正交表(续九如，1995)进行设计，共 9 个处理，重复 3 次，共 27 个小区，完全随机排列，每个小区共 16 株，小区间设有保护行。施肥 3 水平处理试验设计见表 2。

采用根外追肥施肥方法既距离树基部 50~60cm 处对称挖 2 个施肥穴(深度 20~25cm，宽度 25~30cm)。分别于 6 月 2 日(速生前期)和 8 月 5 日(速生期)2 次将肥料平均施入施肥穴中。

表 2 施肥处理的试验设计 g/株

水平	纯 N	P_2O_5	K_2O
1	0	0	0
2	120	60	40
3	240	120	80

1.3 指标测定

在 2008 年 5~10 月期间，每月初对胸径、树高、冠幅等生长指标进行测定。从 5 月开始的 5 个月内，每月下旬采集叶片进行叶片营养元素含量测定。所有处理的样叶均在一天内采完，并带回实验室后洗净，放入调温烘箱内，先在 105℃下杀青 30min，再调至 70℃左右烘至恒重为止，取出冷却至室温，粉碎后过筛(ϕ 为 0.25mm)。测定方法：氮采用半微量凯氏定氮法测定；磷采用钼锑抗比色法测定；钾采用火焰光度计法测定(鲍士旦，2005)。

生长综合指标(Overall desirability，OD)(李乃伟等，2008)：同一处理的各生长指标测定值与对照测定值的比值之和即为该处理的生长综合指标值。采用综合评定法对实验结果进行评价和分析，即用生长综合指标值进行直观分析和方差分析。

2　结果与分析

2.1　施肥对毛白杨人工林生长的影响

2.1.1　施肥对毛白杨人工林连年生长量的影响

在不同施肥条件下，毛白杨人工林各生长指标的生长量见表3。

表3　不同施肥处理对毛白杨生长的影响

处理号	因素与水平(g/株)			胸径连年生长量(cm)	树高连年生长量(m)	冠幅连年生长量(m)	一级分枝数(个)	综合指标
	N	P_2O_5	K_2O					
1(CK)	0	0	0	0.90±0.18	0.73±0.07	1.59±0.42	23.51±6.23	4.00
2	0	60	40	1.49±0.02	0.66±0.06	1.52±0.25	19.45±3.51	4.34
3	0	120	80	1.25±0.25	0.92±0.24	1.55±0.28	25.94±4.28	4.71
4	120	0	40	1.36±0.13	1.04±0.18	1.57±0.23	24.81±5.28	4.97
5	120	60	80	1.60±0.06	1.08±0.42	2.25±0.39	28.55±6.51	5.89
6	120	120	0	1.77±0.48	1.20±0.38	1.61±0.28	26.91±5.17	5.75
7	240	0	80	1.94±0.05	1.70±0.07	2.18±0.16	32.27±6.55	7.25
8	240	60	0	2.25±0.09	1.00±0.15	1.93±0.26	30.44±3.25	6.39
9	240	120	40	1.93±0.16	1.34±0.26	1.99±0.44	29.65±3.41	6.49
X_1	4.17	5.41	5.38					
X_2	5.54	5.54	5.27					
X_3	6.71	5.65	5.95					
R	2.54	0.24	0.68					

由表3可以看出，毛白杨人工林胸径、树高、冠幅生长和一级分枝数在不同施肥处理间差异显著项($P<0.05$)。其中，每株施240g氮肥的毛白杨幼树胸径连年生长量较高，其中，尤以配合施磷肥(60g/株)的处理胸径连年生长量最高，达2.25cm，为对照的250.00%。相反，在氮磷配合(氮肥：240g/株；磷肥：60g/株)施肥条件下，树高连年生长量未达最高值，而树高连年生长量最高的是氮钾配合施肥，其中，氮肥施240g/株，钾肥80g/株，最高树高达1.70m，为对照的232.88%。

由表3分析可得，不同施肥处理对毛白杨树冠生长、分枝数数量影响程度不同，处理间变异较大。其中，以氮、磷、钾的施用量分别为240、0和80g/株对树冠连年生长量较快是对照137.11%，最快的处理是对照141.51%，而一级分枝树生长达到最多为32个，是对照的139.12%。可见，适合的施肥比例对毛白杨树叶、树枝生长至关重要。

对生长综合指标的直观分析，可以看出以氮、磷、钾的施用量分别为 240、0 和 80g/株，最适合毛白杨生长。但从表 1 中的 R 值可以得出，三种肥料对毛白杨生长影响顺序为：氮>钾>磷。从平均值看，氮、磷、钾三因素中都是第 3 水平最好，通过正交试验找到了适合毛白杨生长的最佳施肥配比：氮、磷、钾的施用量分别为 240、120 和 80g/株，施肥用的氮、磷、钾之比约为 3:1.5:1。通过对生长综合指标的方差分析，结果表示：氮肥对毛白杨生长的影响达到 95%显著水平，而磷肥、钾肥的影响均不显著。因此，氮、磷、钾 3 种营养元素对毛白杨生长的效应不同，在实际生产中，对毛白杨施肥应该更注重氮肥的施用量。

2.1.2 施肥对毛白杨人工林生长动态的影响

在不同生长期，林木生长的速度不一样，而胸径、树高和树冠的动态变化是反映和分析毛白杨不同时期生长快慢的重要指标。在整个生长季，不同施肥处理对毛白杨胸径、树高和树冠生长动态变化如图 1、图 2、图 3 所示。

(1) 不同施肥处理毛白杨胸径生长动态分析。由图 1 可以看出，毛白杨胸径增长量在不同月份差异很大，经方差分析，不同月份毛白杨胸径增长量差异极显著($P<0.01$)。胸径增长量最快的是 6 月，其次是 8 月、7 月，平均增长 0.55、0.43、0.39cm，相对生长较慢的是 5 月，最慢是 9 月，胸径平均增长为 0.13、0.11cm。其中，6 月胸径增长量占整个生长季胸径增长量的 34.23%，8 月占 26.70%，7 月占 23.98%，5 月和 9 月分别只占 8.30%和 6.86%，表明在整个生长季过程中，毛白杨胸径生长主要集中在 6 月、7 月和 8 月，共占整个胸径增长量的 84.91%，而毛白杨胸径生长经历两个生长高峰期，分别是 6 月和 8 月，其中 6 月>8 月，共占整个胸径增长量的 60.93%。

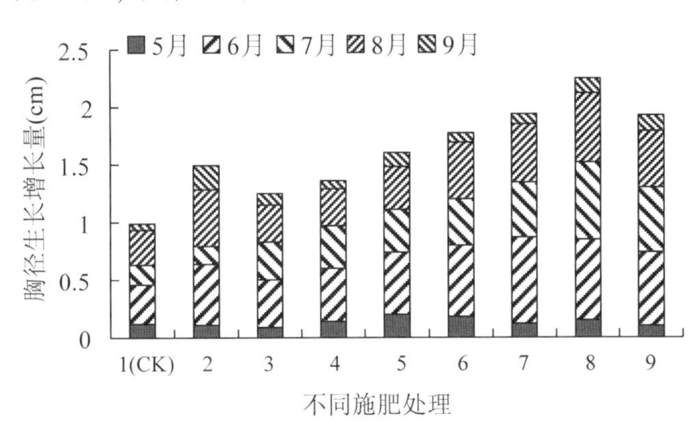

图 1 不同施肥处理毛白杨胸径生长量变化

由图 1 分析可知，5 月和 9 月胸径生长缓慢，处理之间没有差异；6 月、7 月和 8 月胸径生长较快，经方差分析不同施肥处理间对毛白杨胸径生长影响有显著差异($P<0.05$)。与对照相比，施肥处理 6 月胸径平均增长 70.59%，7 月胸径平均增长为 147.06%，8 月胸径平均增长为 50%。因此，施肥处理加快胸径生长，主要集中 6 月、7 月、8 月，其中加快生长最快的是 7 月，其次是 6 月和 8 月。而不同施肥处理对毛白杨胸径生长影响有明显差异($P<0.05$)。与对照相比，施肥处理后毛白杨胸径生长量都有增加，其中胸径生长明显加快的是处理 7、处理 8、处理 9，分别比对照增长 95.96%、127.27%、94.85%，其中

处理 8 最快。三个处理都是氮的最高水平，表明氮营养元素对毛白杨胸径生长有明显的促进作用。

(2)不同施肥处理毛白杨树高生长动态分析。由图 2 可以看出，毛白杨树高生长主要在 5 月、6 月和 7 月，树高增长量在不同月差异很大，经方差分析，不同月毛白杨树高增长量差异极显著($P<0.01$)。树高增长量最快的是 6 月，其次是 7 月、5 月，平均增长 0.67、0.53、0.0.42m，生长较慢的是 8 月，最慢是 9 月，树高平均增长为 0.35、0.14m。在整个生长季过程中，树高生长 6 月增长量占 31.74%，7 月增长量占 25.17%，5 月增长量占 19.68%，8 月和 9 月增长量分别占 16.64%和 6.77%，表明在整个生长季过程中，毛白杨树高生长主要在 6 月、其次 7 月，这 2 个月生长共占整个树高增长量的 56.91%。

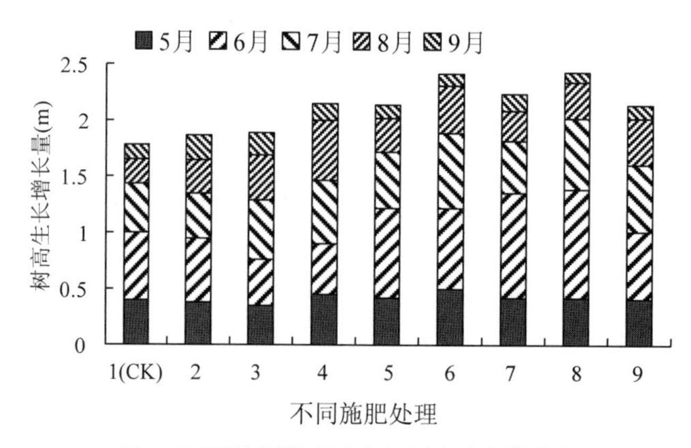

图 2　不同施肥处理毛白杨树高生长量变化

由图 2 分析可得，与对照相比，施肥处理促进毛白杨树高生长。8 月和 9 月树高生长缓慢，施肥处理之间差异较小。5 月、6 月和 7 月施肥处理树高生长较快，与对照相比，施肥处理后树高平均增长分别是 5.00%、11.67%、23.26%。因此，施肥处理加快树高生长，主要集中 5 月、6 月、7 月，其中加快树高生长最快的是 7 月，其次是 6 月和 5 月。而不同施肥处理对毛白杨树高生长影响有明显差异($P<0.05$)。与对照相比，施肥处理后毛白杨树高生长量都有增加，其中树高生长明显加快的是处理 6 和处理 8，分别比对照增长 33.33%、36.11%，其中处理 8 对毛白杨生长树高生长最快。

(3)不同施肥处理毛白杨冠幅生长动态分析。由图 3 可以看出，毛白杨冠幅增长量在不同月之间差异很大，经方差分析，不同月毛白杨冠幅增长量差异极显著($P<0.01$)。冠幅生长主要集中在 6 月和 7 月，平均增长 0.66，0.51m。在整个生长季过程中，毛白杨冠幅生长经历 1 个生长高峰期，为 6 月。生长较慢的是 5 月，8 月，最慢是 9 月，冠幅平均增长为 0.28、0.25、0.12m。其中，6 月冠幅增长量占整个生长季冠幅增长量的 36.39%，7 月占 28.21%，5 月占 15.26%，8 月和 9 月分别只占 13.64%和 6.41%，说明在整个生长季过程中，毛白杨冠幅生长主要集中在 6 月，其次是 7 月。

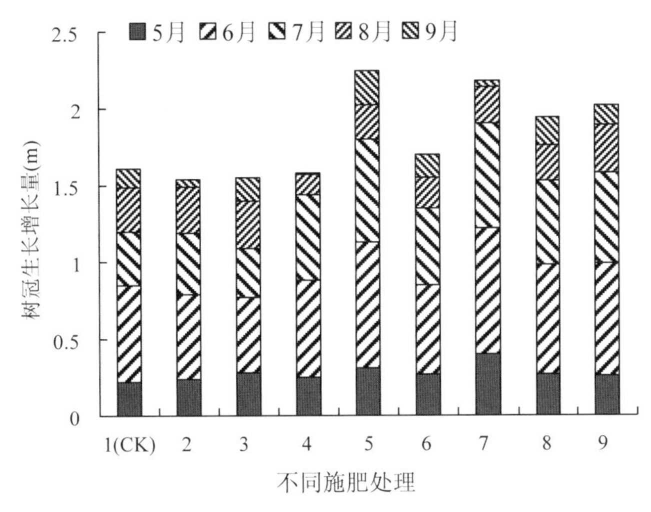

图3 不同施肥处理毛白杨树冠生长量变化

与对照相比，施肥处理6月冠幅平均增长4.76%，7月冠幅平均增长为45.71%,。因此，施肥处理加快冠幅生长，最快的是7月。经方差分析，不同施肥处理对毛白杨树冠冠幅生长影响不显著。与对照相比，不同施肥处理5和处理7冠幅生长较快，比对照增加39.75%、35.40%。其次是处理9和处理8是对照的25.47%、20.50%，表明施肥可以有效地促进毛白杨树冠冠幅生长，其中冠幅生长最快的施肥处理是5，其次是处理8。

2.2 施肥对毛白杨人工林叶片营养状况的影响

2.2.1 对氮营养元素的影响

从图4可以看出，不同施肥处理对叶片中氮含量影响不同，经方差分析，不同施肥处理对氮含量影响显著($P<0.05$)。其中处理7影响最大(表3)，含氮量29.30g/kg，其次是处理8和9，含氮量28.67、28.08g/kg，比对照增加了10.61%、8.23%和6.04%。随着两次肥料施入，叶片中氮的含量增加。不同施肥处理，叶片吸收氮含量增加不同，在6月20号的测定中，氮含量是处理7>处理8>处理6，比对照分别提高24.35%，22.32%和19.83%，在8月25号测定中，处理8>处理7>处理2，氮的含量居于前3位，比对照分别提高15.64%，7.53%，7.38%。可见，施用氮素肥料能明显增加叶片中氮的含量，施用氮含量为240g/株的处理，叶片中氮含量均较高，并且6月20号测定氮含量均比8月25号测定含量高，说明速生前期施肥比速生期施肥提高叶片氮吸收要明显，有可能施肥使得生长速度加快，生长量加大，需要氮元素较多，而导致后期生长叶片中氮含量有所减少。在9月29号测定，叶片中氮的含量都减小，不同施肥处理氮的含量相差不大，说明在秋季施肥对叶片中氮含量影响很小，不适合施肥。

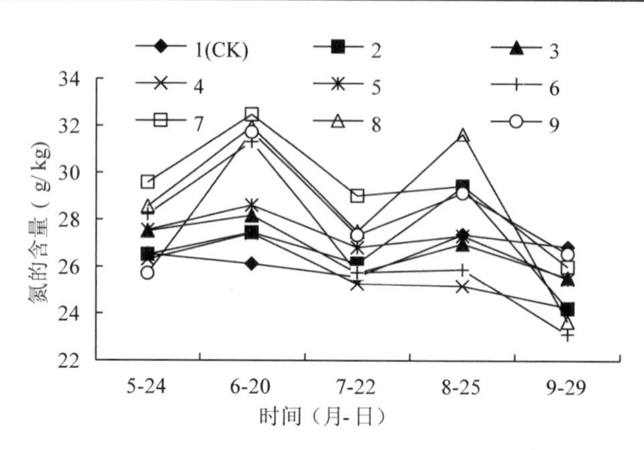

图 4　不同施肥处理对叶片中氮元素含量影响

2.2.2　对磷营养元素的影响

从图 5 可以看出，不同施肥处理对叶片中磷含量影响不同，经方差分析，不同施肥处理对磷含量影响显著($P<0.05$)。其中处理 6 影响最大，其次是处理 9，分别含磷量 1.14g/kg 和 1.04g/kg，比对照增加了 35.71%、23.81%。随着速生前期(6 月 2 号)和速生期(8 月 5 号)两次肥料施入，叶片中磷的含量增加。在 6 月 20 号的测定中，磷含量最大是处理 6，其次是处理 3，再次是处理 9，比对照分别提高 44.33%，39.18%和 37.11%，在 8 月 25 号测定中，磷含量有所增加，但不明显，最大是处理 6 其次是处理 9，比对照分别提高 36.90%，28.57%。生长后期(9 月 29 日)测定表明各处理叶片中磷的含量都减小。可见，施用磷素肥料能增加叶片中磷的含量，在速生前期施用比较好。在不同施肥处理中，促进磷吸收最快的是处理 6，其次是处理 9，说明施肥处理 6 和处理 9 对林木生长有较好的促进作用。

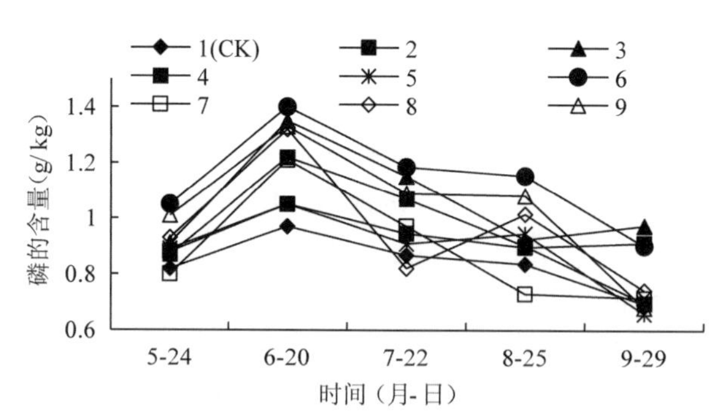

图 5　不同施肥处理对叶片中磷元素含量影响

2.2.3　对钾营养元素的影响

从图 6 可以看出，随着毛白杨生长，叶片中钾的含量逐渐减小，施肥并没有提高叶中钾的含量，相反施肥后叶中钾的含量在减小。不同施肥处理对叶片中钾含量影响不同，经方差分析，不同施肥处理对钾含量影响不显著。施肥处理中处理 5 含量最大，含钾量 10.84g/kg，其次是处理 3 和 9，含氮量 10.72、10.48g/kg，比对照增加了 16.43%、15.15%和 12.57%。随着 2 次肥料施入，叶片中钾的含量增加不明显。在 6 月 20 号的测定中，处理5、处理3、处理9含钾量分别为11.20、12.00、11.10g/kg，在8月25号测定中，处理5、处理3、处理9含钾量分别为11.00、10.00、10.10g/kg，分别下降1.82%，20.00%，9.90%。在5月24号生长初期，林木生长缓慢，叶片钾含量较高。随着林木生长，到9月29号测定中处理6、处理8叶片中钾的含量减小相对较快，相比5月24号减小了4.2、3.1 g/kg。而处理1、处理3钾的含量减小较慢，这是由于处理6、处理8毛白杨生长较快需要钾营养元素较多，从而使得叶片中钾的含量减小。处理1、处理3生长较慢，叶片中积累的钾含量较多。

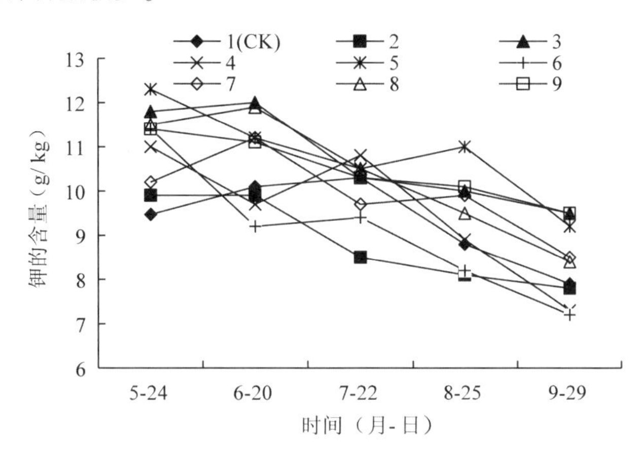

图6　不同施肥处理对叶片中钾元素含量影响

3　结论与讨论

毛白杨对土壤的营养条件很敏感，施肥能显著增加毛白杨的生长(孙时轩，1995；袁玉欣，1990)。氮、磷、钾对毛白杨生长的作用不同，相关研究认为毛白杨生长首先需要氮肥，并对促进毛白杨生长特别显著，且有持续性(曹邦华，2004；段树生，2001)。姜岳忠等研究认为，毛白杨苗木对养分的吸收以氮为主，磷、钾次之，苗木对氮、钾、磷的需比例为7.5∶3∶1。在此研究中，配比施肥能显著提高毛白杨生长，而其中最佳施肥配比为3∶1.5∶1(即，N 240g/株、P_2O_5 120g/株、K_2O 80g/株)，氮对毛白杨生长的影响达到95%显著水平，而磷、钾的影响均不显著。说明施肥配比非常重要，如果施肥单一或配比不合理，可能提高不了毛白杨生长。因此，在实际生产中应该注重配比施肥和氮肥的施用量。各施肥研究所得的最佳施肥配比不同，有可能与土壤肥力性质有关，再加上生长环境的不同。

在整个生长发育阶段，毛白杨对养分的需求是不同的，选择最佳时间进行科学施肥是丰产的关键措施(田库等，1989)。罗治建等(2005)的研究结果表明：每年6月1日4点穴施对杨树树高、胸径和材积生长的促进作用最明显。该研究综合不同施肥处理毛白杨生长动态分析结果显示：毛白杨胸径生长主要集中6月、8月，施肥加快胸径生长最快的是7月；树高生长主要集中在6月、7月，施肥加快树高生长最快的是7月；树冠冠幅生长主要在6月，施肥加快冠幅生长最快的是7月。因此，施肥时期应选择毛白杨胸径、树高、冠幅速生高峰期之前即 6 月 2 日(速生前期)。如果再进行第二次追肥，从研究毛白杨生长动态结果表明施肥加快毛白杨生长最快的是 7 月，所以，在这个时间段可以进行再追肥，不仅为速生高峰期生长提供足够的营养元素，还提高幼林的生长速度、延长速生期，增加幼林生长量。

叶片中营养元素的变化一定程度上反映出植物的生长状况(安国英，1997)' 对施肥后毛白杨叶片营养元素进行诊断，得出施肥能显著提高叶片中氮、磷、钾的含量，不同施肥处理叶片吸收氮、磷、钾含量不同，其中促使氮、磷、钾吸收最好的处理分别是处理 7(N 240g/株、P_2O_5 0g/株、K_2O 80g/株)、处理 6(N 120g/株、P_2O_5 120g/株、K_2O 0g/株)和处理 5(N 120g/株、P_2O_5 60g/株、K_2O 80g/株)。并且在速生前期施肥比在速生期施肥，叶片中营养元素含量要高，叶片中氮含量变化比磷、钾要明显。可能随着幼林的生长，叶片中氮、磷、钾含量逐渐减小，说明氮是幼林生长吸收的主要元素，其次是磷和钾。施肥促进幼林对氮、磷、钾的吸收，并促进幼林生长速度过快。林木营养状况，除受营养元素的影响外，还受环境因素如光、热、水、地形、土壤性质、微生物以及树木自身代谢等影响(刘军徽，2006)。

Effects of Different Fertilization Treatments on Growth and Nutritional Status of *Populus tomentosa* Plantation

Abstract In order to search the best effects of plantation fertilizing treatment of *Populus tomentosa* Carr. , the experimental studied the effects of plantation growth and nutritional status with different fertilizing time and regime by the orthogonal experiment design $L_9(3^4)$. By measuring the changes of its growth indicators (diameter height, crown width and a number of branches) and the nutrient content in leaves. The results showed that N was Mauritian. the effects of fertilization were in the order of N > K > P in the growth period, and the best N : P_2O_5 : K_2O fertilizing treatment recipe was 3:1. 5:1. Diameter, height and crown width concentrated in June to 34. 23%, 31. 74% and 36. 39% of total growing season . Fertilization treatments speeded up the diameter, tree height, crown width growth mainly concentrating in July . Compared with the control, the diameter, tree height, crown width growth was 147. 06%, 23. 26%, 45. 71%. Fertilizer should be elected during the fast-growing pre-fertilization that is prior to June . If top-dressing, we should choose the period July . Fertilization affected the change of nitrogen, phosphorus and potassium in leafs, where the changes of nitrogen was obvious than the phosphorus and potassium. The promotion absorption of nitrogen, phosphorus, and potassium were the best deal with Treatment 7 (N 240g/strains, P_2O_5 0g/strain, K_2O 80g/strain), treatment 6 (N 120g/strains, P_2O_5 120g/strains, K_2O 0g/strain, ,) and treatment 5 (N 120g/strains, P_2O_5 60g/plant, K_2O 80g/strain). Therefore, choosing a reasonable fertilization ratio and a reasonable period of fertilization can significantly increase the growth of *Populus tomentosa* plantations.

Key words *Populus tomentosa*, plantation, orthogonal experimental design, fertilization, nutritional status

修枝对毛白杨无性系生长、净光合速率和蒸腾速率的影响*

摘　要　以4年生毛白杨(*Populus tomentosa* Carr.)人工林为试验对象，在河北省威县苗圃(试验地I)和河北省霸州市万达林场(试验地II)开展修枝梯度为轻度修枝P1，中度修枝P2，强度修枝P3，以不修枝为对照的修枝试验，调查不同修枝强度毛白杨生长指标和不同冠层阳面叶片光合指标的变化，从生理机制解释修枝对毛白杨人工林生长的影响。结果表明：①修枝第一年会影响胸径生长，随修枝强度增大影响增强，但轻度修枝和中度修枝与对照差异不显著，试验地I强度修枝与对照差异显著，试验地II强度修枝与对照差异不显著($P>0.05$)；②修枝对毛白杨树高生长有一定促进作用，且轻度修枝和中度修枝与对照差异显著($P<0.05$)；③修枝前后叶面积指数(*LAI*)变化较大，叶量随时间变化增多，在对照叶面积指数下降时修枝处理叶面积指数仍在增加，修枝能够延长树木生长期；④修枝能提高毛白杨中上部叶片净光合速率(*Pn*)和蒸腾速率(*E*)，有利于提高光合和物质合成能力，而对照的下部叶片净光合速率和蒸腾速率都处于比较低的水平。因此从生长和光合生理综合分析，毛白杨人工林郁闭后树冠下部需要适度修枝。

关键词　毛白杨，修枝，生长，叶面积指数，净光合速率，蒸腾速率

毛白杨(*Populus tomentosa* Carr.)是中国北方重要的乡土树种，生长快，适应性强。毛白杨人工林栽植4年已达到郁闭，林分郁闭后，树冠下层的光照条件变差，下层枝变为消耗枝，应逐步修除，同时修去一些竞争枝和密集枝，能改善林内通风和光照条件，防止病虫害发生，并能减少水分蒸腾和降低养分消耗，对培育无节大径阶良材具有重要意义。

国外对林木修枝的研究较多，研究对象丰富，主要包括修枝对植株生长指标包括树高胸径生长、生物量生产力、生物量再分配和干形生长形态指标等方面的影响，以及修枝在农林复合系统中对林内光环境改善作用的研究。国内对林木修枝的研究开展较晚，多集中于用材林树种的研究，主要有修枝对减少泡桐丛枝，提高泡桐的生长和光合特性的影响研究，修枝对提高杉木土壤肥力及修枝后杉木林内多样性变化研究，修枝对提高欧美杨与农粮间作中粮食产量和修枝对改善欧美杨木材性质的研究，此外，近年来对冠层结构的研究有利于发现修枝对林分的影响机理。毛白杨修枝方面的研究主要集中在修枝强度、修枝季节等抚育技术的研究上，而修枝对毛白杨生长影响机制及修枝后毛白杨光合生理特性变化上的研究尚未见报道。笔者通过研究两块试验地不同修枝强度对毛白杨生长、形态和光合生理特性的影响，寻求4年生毛白杨的最佳修枝强度，并为修枝促进毛白杨人工林生长从

* 本文原载《中国农学通报》，2010，26(23)：134-139，与尚富华、李吉跃、胡磊、宋莲君、支恩波和沈应柏合作发表。

生理机制上提供理论依据。

1　材料与方法

1.1　试验地概况

试验地 I 设在河北省威县林木良种繁育基地，位于北纬 37.04°，东经 114.30°，地处华北平原南部，属冀中平原南部，自然环境优越，地势平坦，气候四季分明，为暖温带大陆性半干旱季风气候，年平均气温 13℃，年平均降水量 584mm，全年日照时数 2574.8h。

试验地 II 设在河北省霸州市万达林场，位于北纬 39.06°，东经 116.24°，地处华北平原北部，属冀中平原东部，地势平坦，土质较好，气候四季分明，为温带大陆性气候，年平均气温 11.5℃，年平均降水量 543.6mm，全年日照时数 2662h。

1.2　试验材料和方法

试验材料为毛白杨选种无性系 1316。根据杨树修枝强度相关研究并便于实际操作，试验分四个修枝强度(图 1)，分别为轻度修枝 P1(修至树高 5m)、中度修枝 P2(修至 6m)、强度修枝 P3(修至树高 7 m)，并以不修枝作对照 CK，于生长季前集中修除一定高度下的枝条，培养直立强壮的主干，并修除树干基部的萌条。

试验地 I 毛白杨栽植密度为 880 株/hm^2，株行距 3m×4m，南北走向，2004 年春季造林，苗木规格 1(年生)干 2(年生)根，无间作。于 2008 年 5 月初对毛白杨开展修枝试验，每个处理选取 48 株毛白杨，重复 3 次。

试验地 II 毛白杨栽植密度为 880 株/hm^2，株行距为 6m×2m，南北走向，2005 年春季造林，苗木规格 1(年生)干 2(年生)根，无间作。于 2009 年 5 月初对毛白杨开展修枝试验，每个处理选取 100 株毛白杨，重复 4 次。

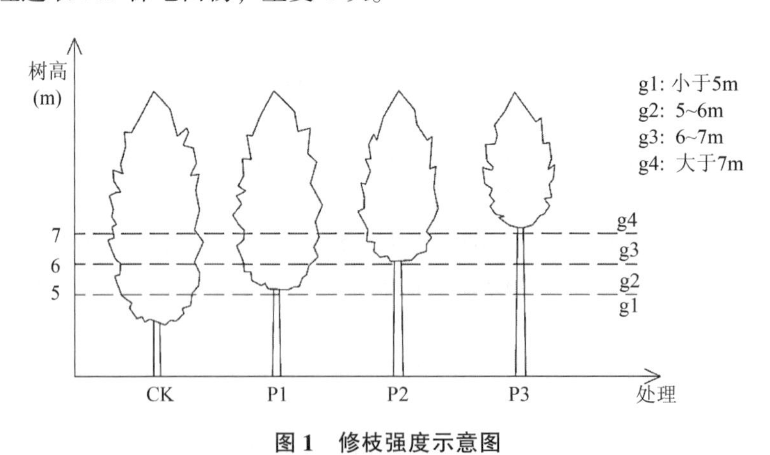

图 1　修枝强度示意图

修枝后分别于 2008 年和 2009 年 5 月初和 9 月初测定毛白杨胸径(*DBH*)、树高(*H*)等生长指标；利用鱼眼照相技术在试验地 I 采集 2008 年 5~9 月，在试验地 II 采集 2009 年 6 月、8 月和 9 月冠层内的半球图像通过系统 WinSCANOPY 分析得到叶面积指数(LAI)；使用 Li-6400 于 2008 年 8 月初选取数天在晴朗天气下测定试验地 I 毛白杨人工林不同冠层阳面叶片的净光合速率(Pn)和蒸腾速率(E)，于 2009 年 8 月初选取数天在晴朗天气下测定试验地 II 毛白杨人工林不同冠层阳面叶片的净光合速率和蒸腾速率，测定时间为 9:00～15:00，每 2 小时测定 1 次，18:00 测定 1 次，每个处理选取 3 株优势木，每株 3 次重复，叶室垂直面向太阳在自然光源下测定。

实验数据采用 Excel 2003 和 SPSS 13.0 进行方差分析(ANOVA)并进行 LSD 检验。

2 结果与分析

2.1 修枝后毛白杨胸径与树高变化分析

对两片试验林修枝后一个生长期的生长指标进行测量，得到不同修枝强度毛白杨胸径与树高生长增量变化情况(表 1)。

表 1 胸径与树高增量表

试验地	测定指标	修枝强度			
		对照	修枝至 5m/P1	修枝至 6m/P2	修枝至 7m/P3
I	胸径(cm)	0.66±0.075 a*	0.54±0.113 ab	0.47±0.064ab	0.30±0.048b
	树高(m)	0.99±0.075b	1.19±0.112a	1.21±0.087 a	1.12±0.048a
II	胸径(cm)	0.53±0.072a	0.51±0.048a	0.43±0.061a	0.40±0.082a
	树高(m)	1.10±0.100b	1.38±0.153 a	1.37±0.057 a	1.19±0.137 b

*不同小写字母表示在 0.05 水平上差异显著($P<0.05$)。

试验地 I，随修枝强度增大，胸径生长增量变小，CK、P1、P2 之间差异不显著，CK 与 P3 差异显著($P<0.05$)，P1、P2、P3 差异不显著。试验地 II，随修枝强度增大胸径生长增量变小，各处理 CK、P1、P2、P3 之间差异不显著。

试验地 I，修枝处理树高生长增量均大于对照，P1、P2、P3 与 CK 差异显著($P<0.05$)，P1、P2、P3 之间差异不显著，树高生长增量 P2>P1>P3>CK。试验地 II，树高生长增量随修枝强度增大而减小，修枝处理树高生长增量均大于对照，其中 P1、P2 与 CK、P3 差异显著($P<0.05$)，P1 与 P2 差异不显著，P3 与 CK 差异不显著，树高生长增量 P1>P2>P3>CK。两片试验地对照和 P3 之间树高生长增量比较接近。

2.2 修枝后毛白杨叶面积指数变化分析

记录 2 片试验林修枝后每月叶面积指数，得到不同修枝强度毛白杨人工林叶面积指数的变化情况(表 2)。

表 2　试验地叶面积指数变化表

试验地	测量日期	叶面积指数			
		对照	修枝至 5m/P1	修枝至 6m/P2	修枝至 7m/P3
I	2008-05-04①	1.283±0.031	1.317±0.033	1.321±0.043	1.302±0.058
	2008-06-03	1.414±0.042	0.969±0.046	0.891±0.052	0.739±0.077
	2008-07-06	1.627±0.069	1.116±0.057	1.076±0.048	0.927±0.069
	2008-08-07	1.428±0.052	1.211±0.028	1.145±0.059	1.078±0.037
	2008-09-10	1.281±0.039	1.176±0.073	1.097±0.073	1.006±0.064
II	2009-06-15②	1.397±0.029	1.016±0.033	0.825±0.065	0.731±0.041
	2009-08-13	1.713±0.043	1.368±0.063	1.328±0.058	1.209±0.036
	2009-09-15	1.482±0.071	1.212±0.056	1.219±0.070	1.017±0.042

注：①叶面积指数为试验地 I 修枝前的数值；②叶面积指数为试验地 II 修枝一个月后的数值。

试验地 I 中，对照 CK 的叶面积指数 5~7 月先增大，叶量逐渐增多，达到峰值后叶面积指数变小，叶量开始减少，但仍能保持较高的水平；试验处理 P1、P2、P3 修枝去除大量枝叶，叶面积指数减小明显，之后随时间变化叶面积指数增大，在 8 月达到高峰后叶面积指数逐渐变少，减少量小于未修枝减少量。8 月 P1、P2 和 P3 叶面积指数达到修枝后最大值，但都小于修枝前的水平。

试验地 II 中，叶面积指数随着修枝强度的增大变小，相邻修枝强度叶面积指数差值随修枝强度增大而减小；不同试验处理叶面积指数随时间变化先增大后减小，其中叶面积指数增大量 P2>P3>P1>CK，而叶面积指数减少量 CK>P3>P1>P2，修枝后叶量较对照增大快而减少慢，叶片凋落较对照晚。

2.3　修枝后毛白杨光合特性变化分析

2.3.1　净光合速率日变化规律

分析试验地 I 不同处理不同冠层阳面叶片的净光合速率日变化情况(图 2)，各处理不同冠层单叶的净光合速率变化规律呈现单峰曲线，均在 13:00 达到一天的峰值然后下降，最大值为 P1g4 的 8.74molCO_2 · m^{-2} · s^{-1}，最小值为 CKg1 的 6.04molCO_2 · m^{-2} · s^{-1}；各处理取样冠层越高，净光合速率越大，即净光合速率 g4>g3>g2>g1。在 9:00 至 13:00 之间，CKg1、CKg2 叶片净光合速率增长缓慢，其他处理的中上部叶片净光合增长较快。13:00 至 17:00，各处理不同冠层净光合速率不断降低，各时刻冠层上部 g4 叶片净光合速率强于中部 g3、g2 强于下部 g1。毛白杨不同冠层净光合速率分为三个层次，第一层次为 P3g4、P2g4 和 P1g4，即修枝后树冠上部，其叶片光合能力最强，净光合速率较高，第二层次是为 CKg4、CKg3、CKg2、P1g3、P1g2 和 P2g3，即对照的中上部(g2、g3、g4)叶片和修枝处理的中部(g2、g3)叶片，第三层次为 CKg1、CKg2，即对照的下部叶片，其处于阴生状态，净光合速率全天处于较低的水平。

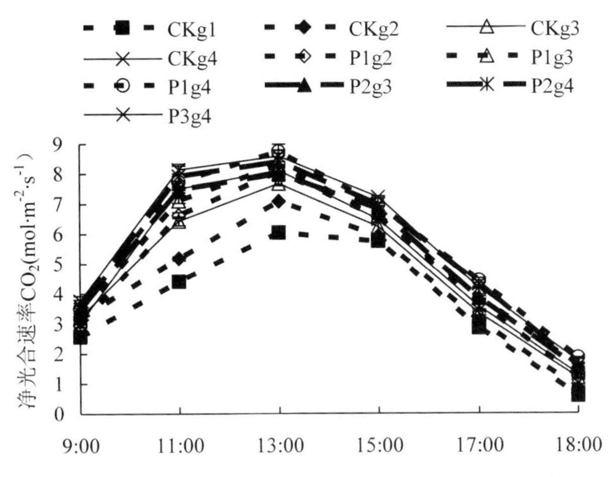

图 2　试验地 I 不同修枝强度对不同冠层净光合速率日变化影响

分析试验地 II 不同处理不同冠层阳面叶片的净光合速率日变化情况(图 3)，各处理不同冠层单叶净光合速率变化规律呈现先增长后降低的单峰曲线，其中 CKg1、CKg2、CKg3 在 13:00 达到峰值，三者中最大值为 CKg3 的 12. 3 $molCO_2 \cdot m^{-2} \cdot s^{-1}$，但小于各时刻各处理中最大值 P2g4 的 15. 95$molCO_2 \cdot m^{-2} \cdot s^{-1}$；其余处理不同冠层净光合速率在 11:00 达到峰值，最大值为 P2g4 的 15. 95$molCO_2 \cdot m^{-2} \cdot s^{-1}$。在试验地 II，净光合速率呈现上部冠层(g4)的叶片强于中部叶片(g2、g3)强于下部叶片(g1)的规律。各处理净光合速率至 18:00 基本降到同一水平。

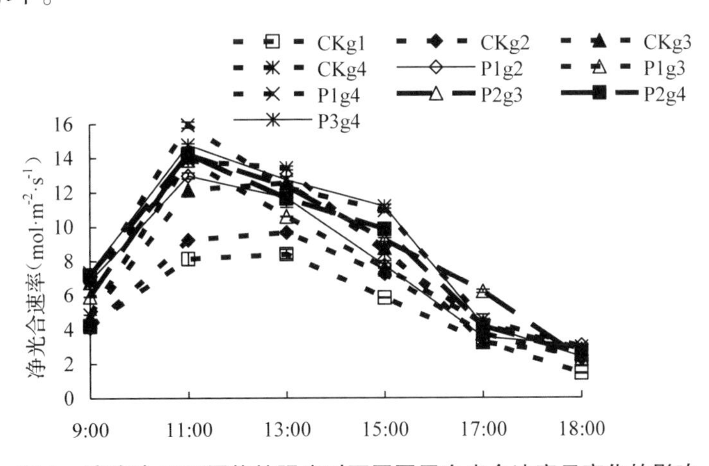

图 3　试验地 II 不同修枝强度对不同冠层净光合速率日变化的影响

2. 3. 2　蒸腾速率变化

根据试验度 I 不同处理不同冠层叶片的蒸腾速率日变化曲线(图 4)，各处理蒸腾速率变化规律呈现单峰曲线，均在 13:00 达到峰值，其中 P2g4 蒸腾速率最大为 5. 16$mmolH_2O \cdot m^{-2} \cdot s^{-1}$，之后各处理蒸腾速率开始下降，到 17:00 时，各处理蒸腾速率下降至同一水平。在各个时刻，修枝处理的蒸腾速率均大于对照，而各冠层蒸腾速率强弱关系为上部(g4)强于中部(g3、g2)强于下部(g1)。

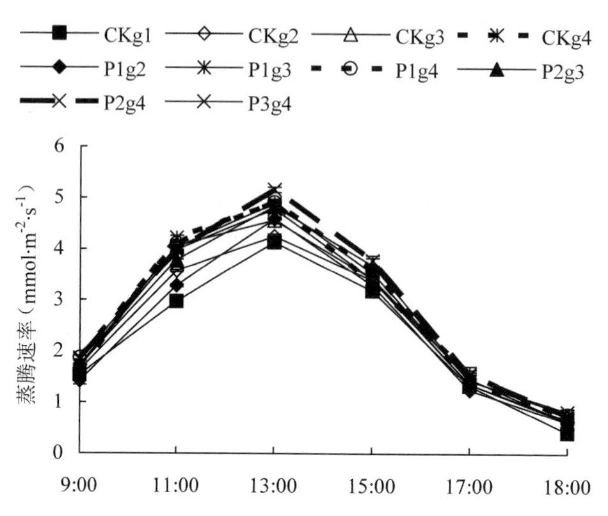

图 4　试验地 I 不同修枝强度对不同冠层蒸腾速率日变化的影响

试验地 II 不同处理不同冠层的叶片蒸腾速率日变化曲线(图 5)，其中处理 CKg2，CKg3 蒸腾速率变化呈现双峰曲线，峰值分别出现在 11:00 和 15:00，CKg3 在各时刻的蒸腾速率大于 CKg2；其余处理 CKg1，CKg4，P1g2，P1g3，P1g4，P2g3，P2g4 和 P3g4 蒸腾速率变化规律呈现单峰曲线，其中 CKg4，P1g2，P1g3，P1g4，P2g3，P2g4 在 11:00 达到峰值，CKg1，P3g4 在 13:00 达到峰值。各处理各冠层的最大值为 P1g4 在 11:00 的 6.98mmolH_2O · m^{-2} · s^{-1}。

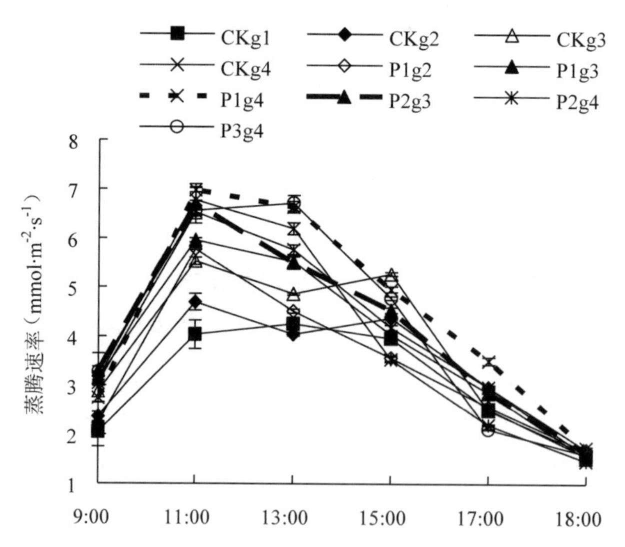

图 5　试验地 II 不同修枝强度对不同冠层蒸腾速率日变化的影响

3 讨论

3.1 修枝对毛白杨生长的影响

两试验林毛白杨品系相同，气候水分等环境条件相似，修枝后胸径、树高和叶面积变化规律有一致性，试验发现：

(1)修枝对毛白杨胸径生长有抑制作用，且随修枝强度增大，影响作用增强，这可能与修枝去除树干下部较多枝叶有关，树干下部的部分物质积累来源于临近枝叶的光合作用，两试验林中胸径生长增量变化规律相似，各处理差异均不显著，说明修枝对胸径生长影响不明显。

(2)试验还表明轻度和中度修枝对树高生长有一定的促进作用，而重度修枝树高生长量虽大于不修枝，对树高生长却没有明显促进作用。树高生长的产物主要来源于中上部叶片的光合作用，适度修枝促进树高生长，但强度太大也会影响树高的生长。

(3)叶面积指数指地上植株叶片的总面积与占地面积的比值，是反映植株叶量多少的重要指标，试验中叶面积指数(*LAI*)差值随修枝强度增大而减小，说明越靠近树冠下层叶量越大，重度修枝对叶量减少最明显。不同修枝处理的叶面积指数在不同月份开始减小，修枝强度越大，其叶面积指数变小的月份越晚，而且减少量越小，修枝可能刺激树干中上部新叶的不断生长，延长毛白杨生长季，有利于物质的合成和积累。

修枝第一年，中上部的光合产物可能用于切口的愈合、中上部新生枝叶的生长和树干直径和高度的生长，而对胸径生长物质合成供给减弱，有研究发现树干下部半径生长量随修枝强度增大而减小，而树干上部直径生长量修枝大于不修枝，因此修枝第一年虽然抑制胸径生长，但树干中上部直径生长增大，并且适度修枝对树高生长有促进作用，有利于树形调整和大径材培育，因此毛白杨第一年修枝强度不宜过大。

3.2 修枝对毛白杨光合生理的影响

试验中不同冠层光合能力的差异为研究修枝对毛白杨生理影响提供了依据，结合不同修枝强度和对照的生长和生理指标的变化，修枝第一年，下部枝叶光合产物主要用于树干的横向生长物质合成，故对照的胸径生长量均大于修枝处理，修枝处理的中上部叶片光合能力大于未修枝，其合成产物的去向还有待研究。

(1)在郁闭后的毛白杨人工林中，上部叶片净光合速率强于中下部，修枝能够提高毛白杨各冠层单叶净光合速率，且修枝后中上部叶片成为物质合成的重要部位，有利于林分树干直径生长的物质供给，对树高生长也有促进作用，这也解释毛白杨修枝后树高生长量大于对照的现象。

(2)蒸腾速率变化规律总体与净光合速率变化一致，但变化幅度较缓，修枝可能在一定程度上提高了毛白杨利用水分的能力，有利于物质合成过程中原料供给。在试验地 II 中 CKg2、CKg3 叶片在 15:00 出现第二个峰值，可能与太阳高度角有关系，下午太阳光斜射入林内，中下部叶片受光量增大，蒸腾速率增大，而且未修枝的 g2、g3 冠层的净光合速率在 15:00 至 17:00 的减少量小于其他处理相同冠层的净光合速率减少量。

根据两片试验林修枝第一年毛白杨生长指标、形态指标和光合能力的比较，实际生产中，第一年修枝至5m比较适宜，对于毛白杨的长期生长有促进作用。

3.3　对研究修枝影响机制的建议

修枝能够提高不同冠层叶片的光合能力，促进毛白杨林分生长，可以分为外因和内因：外因是修枝能够优化林内光照、水分和风速等环境因素，进而提高光能利用效率和水分利用效率；不同的冠层结构其叶面积指数和叶倾角不同，其光合作用面积、效能不同，会导致森林生产力的差异。内因认为树木修枝存在一种补偿机制，即林木在修枝后，由于下部枝叶缺失对物质积累造成的不利影响在一定条件下通过提高剩余枝叶的光合速率等途径补偿。结合植物光调节研究，植物耐阴性和植物激素调节等方面的研究，修枝对树木的影响还需要综合修枝后树木不同高度物质合成与分配，林内整体和局部环境变化规律，树木光合和水分生理，不同方位叶片的生理结构改变以及修枝对植株根系生长的影响等多方面研究，找到修枝阈值，为修枝抚育提供全面的理论依据和实践指导。

Effects of Pruning on Growth, Net Photosynthesis Rate and Transpiration Rate of *Populus tomentosa* in Plantation

Abstract　Pruning on 4-year-old plantations of *Populus tomentosa* with three degrees in Weixian (site I) and Bazhou (site II) of Hebei Province was studied in this paper. The three pruning degrees were low-grade P1, medium-grade P2, strength-grade P3 and the unpruned for control. Tree height, diameter at breast height, physiological parameters including net photosynthesis rate and transpiration rate in different canopies were measured, in order to explain the impact of pruning from the physiological mechanism. The results were: 1) Pruning inhibited the growth of diameter, in proportion to the increasing intensity of pruning. Difference in P1 and P2 were not significant with CK, while, P3 was significantly different with CK in site I, whereas not significant with CK in site II; 2) Pruning speeded the height growth, and P1, P2 were conspicuously with CK ($P<0.05$); 3) The leaf area index (LAI) changed significantly after pruning, and leaf density of pruned *Populous* stand increased while the LAI of CK decreased. Thus, pruning can extend growing stage of trees; 4) Pruning improved net photosynthesis rate and transpiration rate of leaves in the middle and upper of crown, but these indices in the bottom crown were low at the same time. Considering from the growth, photosynthetic and physiological indices of *Populus tomentosa*, branches in the bottom of canopy should be pruned after canopy closing.
Keywords　*Populus tomentosa* Carr., pruning; growth, leaf area index, net photosynthesis rate, transpiration rate

第六部分

杨树材性及木材加工

Study on Inheritance and Variation of Wood Fiber Length of *Populus tomentosa* Clones①

Abstract Inheritance and variation in wood fiber length (WFL) of the Chinese white poplar clones at 4 sites were investigated. The results show that there are significant differences in WFL on the site and the clone level. The clonal repeatability of WFL is estimated to be 0. 79, which indicates that the wood property character is under strong genetic control and can be improved by genetic means. Within-tree variation of WFL along the direct ion of tree height was also included. By correlation analysis of WFL and other factors, it is found the significant positive correlation between WFL and growth characteristics (volume, height, diameter at breast height and bole straightness) and the significant negative correlation between WFL and wood basic density. Results suggested select ion based on growth characteristics, wood properties or a combination of these two sides.

Keywords *P. tomentosa*, wood fiber length, inheritance and variation, genetic correlation

1 Introduction

Wood is the most stable and economic raw fiber material. As the fiber industry develops, the supply of fiber material is getting tighter and tighter, which forces people to count on resources of fast-growing tree species (Rogers, 1992). The consumption of paper is tremendous in China with its 120 million of population. The present situation is that the supply falls short of demand, with original paper material imported every year, which values nearly MYM 160~170 million. In order to quickly bring about a radical change in the passive situation of importing large quantities of paper and pulp and to realize self-sufficiency of paper and save unnecessary expenses, wood properties essential for quantity and quality of paper and pulp should be extensively studied and superior clones for pulpwood should be selected and cultured to quicken the construction of pulpwood plantation base.

Chinese white poplar, *Populus tomentosa* Carr. , is a very good tree species in the section *Leuce* under Populus in Salicaceae. Its wood has close texture and high whiteness from heart wood to sapwood. It was reported that the whiteness of its natural-color-pulp without bleaching reach as high as 50% that is close to the whiteness of general newsprint. When bleached with only 2% hy-

① 本文原载《北京林业大学学报》(*Journal of Beijing Forestry University*), 1997, 6(2): 21-34, 与宋婉、续九如和李新国合作发表。

drogen peroxide (H_2O_2) solution, it gets to 70% that comes up to the standard of the offset printing paper. Therefore both the paper-making cost and pollution to t he environment will be much decreased by utilizing the wood of Chinese white poplar. However, researches around the inheritance and variation of wood properties of the species are very few that is not able to keep pace with the imperative necessity of high-quality and consistent wood material in the present market.

Wood fiber properties are the important factors affecting the quantity and quality of fiber products, among which the fiber length is the most essential one. Long fiber length not only increases the tearing strength of paper but also helps to improve the ultimate tensile strength, bursting strength and bending endurance. Wood fiber length is believed to be the most important factor to determine high or low grade of pulpwood and has very significant function on increasing the quantity and quality of paper and pulp.

The objective of the study is to explore the inheritance and variation of wood fiber length of *Populus tomentosa* on the inter-site, interclonal and intraclonal levels through test and analysis of wood fiber length of the clones at 4 sites. Besides, the relationship between wood properties and tree growth factors is also studied to lay a foundation for further research around wood proper ties and to provide certain basis for scientific and efficient utilization of the wood of *Populus tomentosa*.

2 Materials and methods

2.1 Materials

In 1982, selection of superior trees of *P. tomentosa* was done on the basis of climatic zone distribution result by the ten-province-cooperation-group through-out China, considering the rules of exsitu selection conservation issued by the FAO. A total of 1047 superior trees were selected from the distribution area that crosses 1.0 million km^2. Then clonal arboretum was established by utilizing the juvenile materials. In 1987, 250 clones were selected by the nursery test and planted at the chosen sites in the 10 provinces. All the trees in the test plantations have grown to 8 years old when the investigation and sampling of this study were done.

In the winter of 1995, 24~30 clones (including two common control clones: *P. tomentosa* var. *yixian* and var. *baotou*) were chosen randomly from the test plantations at Yuanshi in Hebei province, Guanxian in Shandong province, Wenxian in Henan province and Tianshui in Gansu province. Among them, 22 clones were common for the 4 sites. In each clone, 3 ramets were randomly sampled at the breast height from the east direction with the 0.5cm growth borer to determine wood fiber length.

2.2 Variables measured

The 8th growth ring section of each wood sample was separated for WFL measurement. Tens of 0.1~0.2mm round chips were cut and then soaped into distilled water till saturation and dissociated at oven of 60 with dissociation resolution (made by mixing equal amount of 30% solution hydrogen peroxide (H_2O_2) and iced acetic acid (HAc)). When the wood samples were softened and turned white, the dissociation resolution was discarded and the wood samples were washed with distilled water until no acidity is tested. After dying with ethanol (50%) and safranin O resolution, wood fiber length (WFL) was determined by 50 measurements per sample under microscope (60×) using Taylor's method (Taylor, 1975).

Values of tree height (H), diameter at breast height (DBH) and bole straightness (BS) were got from the reports of 1995 by the ten-province cooperation group, and tree volume (V) was determined by formula 1 (Xu Weiying, 1988). Clonal wood basic density (WBD) used in the following correlation analyses is from the previous study made by the authors (Song Wan et al. 1996).

$$V = 0.5134H^{0.826956}(D/100)^{1.995375} \quad (1)$$

2.3 Data analyses

Two-factor variance and covariance analyses were done on the clone and the site level. Simple correlation analyses were carried out between wood properties and tree growth factors for the clones at each site and the significance was determined. The models for variance and covariance analysis were shown as Table 1.

Formulas used in estimation of various genetic parameters were:

$$R_c = V_c/(V_c + V_e) \quad (2)$$

Where R_c is the clonal repeatability, V_c is the variance of clones and V_e is the variance of environment.

$$r_g = COV_g(x, y)/\sqrt{\sigma_{gx}^2 \times \sigma_{gy}^2} \quad (3)$$

$$r_e = COV_e(x, y)/\sqrt{\sigma_{ex}^2 \times \sigma_{ey}^2} \quad (4)$$

$$r_p = COV_p(x, y)/\sqrt{\sigma_{px}^2 \times \sigma_{py}^2} \quad (5)$$

Where r_g, r_e, and r_p is respectively the genetic, environmental and phenogenic coefficient; $COV_c(x, y)$, $COV_e(x, y)$ and $COV_p(x, y)$ is the genetic, environmental and phenogenic covariance between x and y, respectively.

$\sigma_{gx}^2(y)$, $\sigma_{ex}^2(y)$, $\sigma_{px}^2(y)$ is the genetic, environmental and phenogenic variance of x or y, respectively.

Table 1 Analyses of variance and covariance format for WFL

Factor	Source of variance	*Df*	*EMS*	*ESP*
One	Clone	$C-1$	$V_e + BV_c$	$COV_e + BCOV_c$
	Error	$(C-1)(B-1)$	V_e	COV_e
	Total	$CB-1$		

(续)

Factor	Source of variance	Df	EMS	ESP
Two	Site	$S-1$	$V_e+BV_e+BCV_e$	$COV_e+BCOV_c+BCCOV_s$
	Clone	$S(C-1)$	V_e+BV_e	COV_e+BCOV_e
	Error	$SC(B-1)$	V_e	COV_e
	Total	$SCB-1$		

3 Results and analyses

3.1 Inheritance and variation of WFL

3.1.1 Among sites and clones

Unite variance analyses were done through using the tested fiber length data of the common 22 clones at 4 sites to study the variance on the site and the clone levels. The results were shown as Table 2.

Table 2 Variance analyses of WFL of *P. tomentosa* clones at 4 sites

Source of variation	Df	SS	MS	F
Site	3	0.5865	0.1955	15.73**
Chine	21	1.1580	0.0627	4.74**
Error	63	0.8333	0.0132	
Total	107	3.1559		

** Significant at 1% level of probability.

From the results, the differences of WFL among both sites and clones get to the great significant level, and the former are greater than the latter. Multiple comparisons of WFL at each site were performed to know the specific differences further. The result shows that WFL at Yuanshi is very significantly ($P<0.01$) different from that at Guanxian and Tianshui and is significantly ($P<0.05$) different from that at Wenxian. However, there are no significant differences among the latter three sites.

WFL of *P. tomentosa* averaged 1.221mm at 4 sites. It varied among 4 sites as shown in Table 3.

Table 3 Variation of WFL of *P. tomentosa* clones at 4 sites (unit: mm)

Site	Clone number	Average	Maximum	Minimum	Standard difference	Variation coefficien
Yuanshi	30	1.3316	1. 589	1.028	0.18112	0.135
Guanxian	24	1.1747	1.418	1.001	0.1300	0.111
Tianshui	25	1.1682	1.455	0.850	0.1512	0.129
Wenxian	25	1.1866	1.446	0.882	0.1559	0.131

From Table 3, the clones at Yuanshi have generally the longer average fiber length than those at Guanxian, Tianshui and Wenxian. The average variance coefficient of WFL at 4 sites is 12.7%.

In order to show directly the variance situation of WFL at 4 sites, the common clones were taken to make Figure 1 for the variation of clones at 4 sites.

The values of WFL of clones at Yuanshi are mostly greater than those of the other sites. Clones like B0067, B0101, B1211, B1414, B1913, B3515, B4104, B5079 and B5101, etc. have the greatest WFL values at Yuanshi; clones like B1715 and B4204 have the greatest values at Guanxian. However, the clones at Wenxian generally have the smallest WFL values.

Clonal repeatability was estimated for WFL of *P. tomentosa* clones from the variance analyses in Table 2 as 0.79. The value is similar to that of WBD and is also under great genetic control.

3.1.2 Within-tree variation along the tree height

Three clones (B1715, B1807 and B4104) were randomly sampled for the study on the variation of WFL along tree height. One ramet from each clone was tested along different tree height (0.15, 1.3, 2.3, 3.3, 4.3, 5.3, 6.3, 7.3, 8.3, 9.3m). The variation was shown as Fig. 2 in which the WFL is the averaged value of east and west directions of the wood.

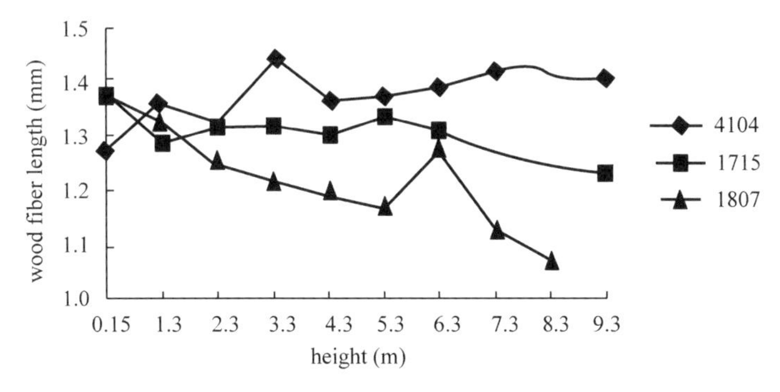

Fig. 1 Variation of WFL along tree height of three clones of *P. tomentosa*

From Fig. 1, the fiber length of the wood section with the same age (8th years) varied in a certain degree along the tree height. WFL of 1715 and 1807 has a trend of decrease as the tree height rises. Consistent with most study results, the closer the wood part to the top of the tree, the shorter its fiber length will be. However, the WFL of 4104 has a different variance trend from

1715 and 1807, it increased slightly in a certain range of the tree height. The result showed t hat variation trend of WFL varied from one clone to another.

3.2 Correlation among WFL, WBD and tree growth characteristics

The common 22 clones were taken to make the unite genetic correlation analyses among the 6 factors (WFL, WBD, BS, H, DBH and V). From the results shown in Table 4, great significantly negative genetic correlation exists between WFL and WBD. Furthermore, WFL has great significantly positive correlation with the four tree growth characteristics; while WBD has great significantly negative correlation with them. Especially, the correlation coefficient between tree straightness and other characters are mostly more than 1.000, which alludes that the grading standard and measuring method may be not appropriate and should be adjusted.

Table 4 Correlation analyses of the 7 characters of *P. tomentosa* at 4 sites

Trait	WBD	BS	H	DBH	V	WFL
WBD		−0.562**	−0.867**	−0.716**	−0.751**	−0.529**
BS	−0.356		>1**	>1**	>1**	>1**
H	−0.276	−0.002		0.986**	>1**	0.642**
DBH	−0.319	0.100	0.869		0.989**	0.806**
V	−0.317	0.074	0.898	0.970		0.714**
WFL	−0.332	0.260	0.379	0.425	0.395	

* The right upper part in the table is genetic correlation, and the lower is phonogenic correlation.

Table 5 Simple correlation analyses of 7 characters of *P. tomentosa* at 4 sites

(A: Yuanshi)

Trait	BS	DBH	H	V	WFL
WBD	0.1184	−0.1326	−0.3377	−0.3348	−0.1547
BS		0.4181*	0.6295**	0.608**	−0.2234
DBH			0.7375**	0.7959**	−0.0867
H				0.9866**	−0.0499
V					−0.0724

(B: Guanxian)

Trait	BS	DBH	H	V	WFL
WBD	0.0694	−0.3025	−0.1265	−0.2019	−0.2337
BS		0.4048	0.6640**	0.6075**	0.2974
DBH			0.8604**	0.9005**	0.7535**
H				0.9718**	0.6746**
V					0.6735**

（续）

Trait	BS	DBH	H	V	WFL
(C: Tianshui)					
WBD	−0.4120*	−0.4088*	−0.1863	−0.4105*	−0.3545
BS		0.6542**	0.5617**	0.6520**	0.3167
DBH			0.8486**	0.9855**	0.2081
H				0.8874**	0.1005
V					0.1782
(D: Wenxian)					
WBD	0.1196	−0.4234*	−0.3316	−0.3146	−0.5255**
BS		0.1629	0.1299	0.1214	−0.2373
DBH			0.8387	0.8825	0.4715*
H				0.9762**	0.4860*
V					0.4854**

* Significant at 5% level of probability.

In order to bypass the site effects on wood properties, simple correlation analyses were done for the characters at each site. Table 5 showed the results according to the data of 30 clones at Yuanshi, 22 at Guanxian, 25 at Tianshui and 25 at Wenxian, respectively.

From Table 5, at Guanxian, there are great significantly positive correlation between WFL and DBH, H and V with the coefficient being 0.7535, 0.6746 and 0.6735, respectively. At Wenxian, the correlation coefficients of the above characters arrive at the significant level, being 0.47~0.49. While at the other two sites, the correlation coefficients are not high enough to be significant, but they are still positive at Tianshui. The results of the simple correlation analyses are consistent with the trend got by the unite analyses of all 4 sites. Such positive correlation between WFL and tree growth factors illustrates that it is not contradictory to select simultaneously for longer WFL and larger V, and united selection of the two sides can be carried out.

Also from Table 5, WFL of the tested clones has great significantly negative correlation with wood basic density at Wenxian, with the coefficient being −0.5255. Though the correlation coefficients are not high enough to be significant, a certain degree of negative correlation between the two characters is also observed at the other three sites. It indicates that the two characters can be selected independently at the three sites to breed for superior clones with higher wood basic density and longer fiber length and faster growth rate. While at Wenian, as a result of the great significantly negative correlation, focus should be put on selection of clones with both longer fiber length and higher basic density. A scattered dot map (Fig. 2) was made for the clones at Wenxian, on which the ordinate represents WFL, the abscissa WBD, and the two dotted lines respectively t he average value of WFL and WBD.

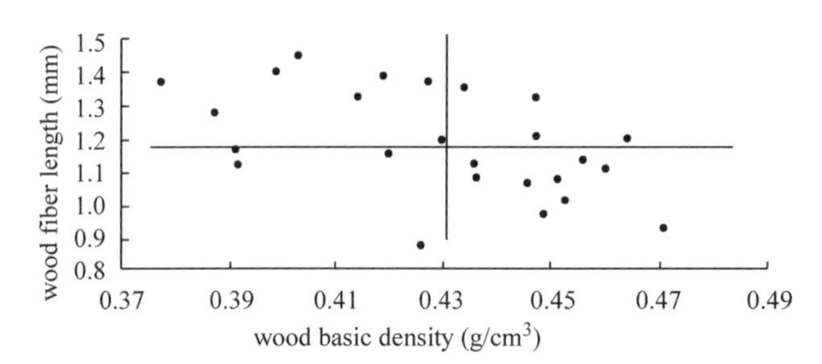

Fig. 2　Scattered dot map of wood basic density and wood fiber length of 25 *P. tomentosa* clones at Wenxian

There are still several clones with superior performances in both the characters though only 25 clones were analyzed.

4　Results and discussion

4.1　Inheritance and variation of WFL of *P. tomentosa* clones

On inter-site variation of WFL, most results of the former studies suggested that there exists certain site effect in WFL of poplars. The average WFL of *P. tremula* was believed to have significant variation among sites (Tank et al., 1987). Zeng Qiyun et al. (1990) compared the wood fibers of *P. tomentosa* var. *yixian* planted at different sites and drew the conclusion that WFL is significantly longer at sites with better soil, water and fertile conditions than at that with poorer condition. This study now got the results that significant variations exist for WFL of *P. tomentosa* at different sites. It may result from the different soil and management conditions at different sites. Compared with the other three sites, the poplar clones exhibited the best per formance at Yuanshi in terms of the growth rate, with largest diameter growth, highest tree height that reflected indirectly that the site condition at Yuanshi is the best. As a result, the WFL of clones at Yuanshi is the longest and it has significant or great significant differences from the other three sites.

Researches around the inter-clonal variation of WFL are numerous. For example, Yanchuk et al. (1983) studied the wood fiber proper ties of *P. tremuloides* and discovered that significant inter-clonal differences in WFL exist among just three clones. The later studies of his (1984) on 15 *P. tremuloides* clones showed that there is significant variation among the clones that occupied about 35% of the total variation. The studies of Cheng et al. (1979) believed that great significant differences exist among the tested 6 poplar clones on WFL. As to *P. tomentosa*, the studies of Wang Qi (1989) suggested great inter-clonal differences in WFL. In section Angeiros, the studies of Wang Mingxiu et al. (1989) concluded that there is broad genetic basis in WFL at seedling stage and indicated the possibility of selection for longer WFL during the stage. The hybrid poplars show great significant differences among clones with the longest being 20% ~ 106% times longer than the shortest. Besides, the studies of Wang Kesheng et al. (1995) on 18 improved clones and

25 hybrids in section Angeiros, and the studies of Jiang et al. , (1994) on 36 clones of *P. tremuloides* both showed fiber characters as fiber length has great significantly inter-clonal variation. Furthermore, most of the studies also showed consistently the strong genetic control of WFL. The broad heritability (H_B^2) of 30-aged WFL of *P. tremula* clones was estimated to be and that of the average fibers was 0.50. The clonal repeatability (R_c) of 33 hybrids in section Angeiros in WFL was 0.684(Liu et al. 1994). In this study the conclusion is that great significant differences in WFL exist among clones and sites; the R_c of WFL of *P. tomentosa* is 0.789. These results show that WFL is under strong genetic control and can be improved by genetic means, but different clone selection strategy should be carried out according to the change of plantation sites.

Most researches believed that there exists in-tree variation of WFL along the direction of tree height, though the variation degree is generally low and it has no great effect on wood utilization ratio. Wang Mingxiu et al. (1989) studied the clones in section Angeiros and pointed out that WFL of the same growth ring decreased gradually with the increase of tree height. The studies on *P. Yunnanensis* revealed that WFL has an undulate variation at the certain height at the tree base and trends to get shorter when closer to the top of the tree (Wang et al. , 1991). The results of a variety in *P. alba* showed, however, that WFL decreases slowly with the increase of the tree height from the base to the top (Xu Congjian, 1986). This study on *P. tomentosa* that there is a certain extent of variance on WFL in the vertical direction of the tree, and WFL shows a trend of decrease with the increase of the tree height but there are differences among different clones.

4.2 Correlation of WFL and tree growth

Posey et al. (1969) discovered that wood fiber length is positively correlated to growth rate and any other environmental factors which are helpful to improve the tree growth. While the studies of Yanchuk et al. (1984) on *P. tremuloides* showed that there are only very weak positive correlation between WFL and growth rate. Guo Xiangsheng (1988) discovered that poplar trees with faster growth rate tended to have shorter WFL, but he did not believe such differences in trees with slower growth rate are significant. Also, Zhu Xiangyu et al. (1993) supposed that the correlation between WFL and growth rate was not significant on basis of their studies on some poplar hybrids.

As to the correlation between WFL and WBD, there are different conclusions from different tree species. In poplars, most results showed that there was only faint correlation between them. For example, wood property researches on *P. deltoides* (Zhu Xiangyu et al. , 1993) showed that WFL was negatively correlated with growth rate but positively correlated with wood basic density to prove that WFL and basic density may be independently inherited and are controlled by different genetic mechanisms. Wang Mingxiu et al. (1989) reported on the clones in section Angeiros that correlation coefficient between wood fiber length and basic density has not arrived at the significantly level and there is no stable variation trend for the two characters in different growth rings of 16-year trees. Thus they concluded that correlation is prone to be affected by environmental conditions and the two characters may be independently inherited.

From the correlation analyses on wood properties and tree growth factors of 8-year-old *P. tomentosa* clones at 4 sites in this study, WFL shows stronger positive correlation with all the main growth characters (*H*, *DBH*, *V*) and medium negative correlation with WBD. On account of that, the oriented breeding for pulpwood should focus on the selection for the clones with faster growth rate generally possess longer WFL.

Recently Pang Guigan et al. (1995) made comprehensive evaluation on adaptability for BCMP (Bursting Chemical Mechanical Pulp) of ten short-rotation tree species. Among the tested samples, poplar woods are the best for their comprehensive performances in BCMP. T he wood of *P. tomentosa* possesses the lowest CMP electricity-consumption, higher paper whiteness value and tearing strength, tensile strength, bursting strength in all the species. It is believed to be economic and reasonable material for high-whiteness BCMP. It again proved the superiority of the wood of *P. tomentosa*.

The conclusions obtained from the paper cannot be taken as the final and general model, further research on genetics in wood proper ties of *P. tomentosa* is in progress. By comparing different studies, from different sites, we will be able to learn more about this native species whose importance for forestry practice is increasing.

Study on Inheritance and Variation of Wood Basic Density of *Populus tomentosa* Carr. Clones*

Abstract The wood basic density of about 100 *Populus tomentosa* Carr. clones grown in 4 clone testing plantations located in Hebei, Henan, Shandong and Gansu provinces was tested and its inheritance and variation among clones and sites were studied. The result shows that the average wood basic density of the studied clones is 0.4363g/cm^3 and there are significant differences in wood basic density among clones and sites. The high value of clonal repeat ability (0.82~0.91) means that the wood basic density is under strong genetic control. Correlation analysis shows general negative correlation exists between growth traits (including diameter at breast height, height, and volume) and wood basic density, although the correlation coefficients vary in a certain degree from one site to another. Studies on wood properties and relationships between tree growth and wood quality are of importance in such fields as wood properties breeding and combined improvement of both growth and wood quality.

Key words *Populus tomentosa* Carr., wood basic density, inheritance and variation

1 Introduction

Poplar wood has been playing more and more important roles in both solid wood and fiber product industries in the society. Many developed countries have attached importance to studies on inheritance and variation of proper wood properties. Systematic studies on *Populus tremuloides* Michx (Einspahr et al., 1972; Reddy, 1983; Yanchuk et al., 1983; 1984), *P. deltoides* March (Farmer et al., 1968; Posey et al., 1969; Olson et al., 1985) have been done. Genetic studies of poplar wood properties started late in China. Some researches on *P. davidiana* (Zhang et al., 1993; Yang et al., 1994; Gu et al., 1994) and a few species and hybrids in the Section Aigeiros and Tacamachacae (Wang et al., 1989; Zhu et al., 1993; Liu et al., 1994) have been started and some preliminary results have been got. However, much of the work about *P. tomentosa* is focused on growth improvement, and few studies have been carried out on wood properties (Zhu, 1992). As *Populus tomentosa* is a native tree species in China with wide distribution and varieties of uses, it is specially important to carry out studies on wood properties of the precious species.

This research is focused on the studies of wood basic density (WBD) the most determinant

* 本文原载《北京林业大学学报》(英文版)(*Journal of Beijing Forestry University*, English Ed.), 1997, 6(1): 8-18, 与宋婉、续九如合作发表。

factor affecting wood quality and quantity. Totally about 100 clones of *P. tomentosa* from 4 different sites were tested. The purpose is to explore inheritance and the further improvement of the species.

2 Materials and methods

The experiment materials are 100 clones of *P. tomentosa* from 4 clone testing plantation blocks growing at Yuanshi in Hebei, Wenxian in Henan, Guanxian in Shandong, and Tianshui in Gansu province. Those plantations were established at the same year (1987) and have the same field design using the common comparison clones method. There are altogether 250 clones at each plantation. The clones for this wood study were chosen randomly from the 250 clones in each plantation. Among the chosen clones at four sites, 70 clones are in common. The two common controls, *P. tomentosa* var. *yixian* (CKl) and *P. tomentosa* var. *baotou* (CK2) at 4 sites were taken as controls in the experiment.

The field investigation work was done from January to March in 1995. Three trees that had the average growth in each chosen clone were selected as sample trees. Two 5mm breast-high increment cores from east and west sides of the tree were taken from each tree and then used in obtaining wood basic density. An attempt was made to sample only clear wood by avoiding obvious knots or wounds. Cores containing branch wood or exhibiting an unusual amount of warping were discarded.

Wood basic density for each core was determined by the Maximum Moisture Content Method described by Smith (1954, 1955). Variance of basic density at clone and site levels was analysed. Correlation analyses among wood basic density, diameter at breast height, height, and volume were also done.

3 Results and analyses

3.1 Interclonal variation of *WBD*

The result of variance analysis of *WBD* among clones at each site is presented in Table 1.

Table 1 The single element(clone) variation analysis *of WBD* of poplar clones at 4 sites

Site	Yuanshi	Guanxian	Tianshui	Wenxian
df	112	97	100	95
F	10.70	5.60	6.70	6.60
Significance	* *	* *	* *	* *

* * significant at the1% level.

At each test site, great significant differences in *WBD* exist among clones of *P. tomentosa*. The result of a further evaluation of clonal repeatability of *WBD* at each site is shown in Table 2.

Table 2 The clonal repeatability of *WBD* of poplar clones at each test site

Site	*WBD* Repeatability(Re)
Yuanshi	0.9065
Guanxian	0.8214
Tianshui	0.8507
Wenxian	0.8485

The average value of *WBD* repeatability at the four test sites is 0.86 (from Table 2), which means the *WBD* should be controlled by genetic effect mostly and could be effectively improved through clonal selection.

3.2 Site effect on *WBD*

In order to recognize the site effect on clone *WBD*, the data of the 70 common clones in the four test sites are computed using combined variance analysis, and the result is shown in Table 3.

Table 3 Combined variance analysis of *WBD* of the common 70 clones at 4 sites

Source	*df*	*SS*	*MS*	*F*
Site	3	0.2152	0.0717	7.45**
Error	264	1.0942	0.0264	
Total	267	1.3094		

** significant at the1% level.

The result in Table 3 illustrates great significant differences in clone *WBD* existed at different sites. It means that different sites have great effect on *WBD*. The result of the further multiple comparison among four sites is shown in Table 4.

Table 4 The multiple comparison of *WBD* of poplar clones among four test sites (g/cm^3)

	Y_2 = 0.4678	Y_1 = 0.4429	Y_4 = 0.4417
Y_3 = 0.3934	0.0744**	0.0495**	0.0483**
Y_4 = 0.4417	0.0261**	0.0012	
Y_1 = 0.4429	0.0249**		

Y_1, Y_2, Y_3, Y_4 respectively represent the average *WBD* of poplar clones at Yuanshi, Guanxian, Tianshui and Wenxian.

There are great significant differences in wood basic density between every two sites except Yuanshi and Wenxian. Detailed variation information of clone *WBD* at four sites is shown in Table 5.

Table 5　Variation of clone *WBD* at four sites(g/cm^3)

Site	Clones	Average	Maximum(done)	Minimum(clone)	*SD*	*VC*
Yuanshi	113	0. 4429	0. 4974(4333)	0. 3628(1715)	0. 0341	0. 077
Guanxian	98	0. 4678	0. 4997(1808)	0. 4063(5101)	0. 0248	0. 053
Tianshui	101	0. 3934	0. 4494(3708)	0. 3349(5044)	0. 0308	0. 078
Wenxian	96	0. 4412	0. 5445(0096)	0. 3642(1210)	0. 0390	0. 088

SD is standard deviation, *VC* is variance coefficient.

The average *WBD* of all the studied clones at four sites is 0. 4363g/cm^3. The clones growing at Guanxian have the highest *WBD* (0. 4678g/cm^3), and those at Tianshui have the lowest *WBD* (0. 3934g/cm^3), with a difference of 0. 0744g/cm^3. The average variation coefficient at four sites is 7. 4%. Clones with the highest and lowest *WBD* values are different at different sites. CK1, CK2 have different values at different sites too.

3. 3　Correlation between growth traits and *WBD*

Clone growth data of 1994 at each test site were used in sample correlation analyses which were done among tree growth traits including diameter at breast height (*D*), height (*H*), volume (*V*) and *WBD* . Table 6 shows the results.

Table 6　Simple correlation coefficients between *WBD* and *D*, *H*, *V* of poplarclones at 4 sites

Sites	*n*	*D-WBD*	*H-WBD*	*V-WBD*
Yuanshi	113	- 0. 5194**	- 0. 2844**	-0. 5091**
Guanxian	98	- 0. 0797	0. 1379	-0. 1035
Tianshui	101	- 0. 0188	-0. 2598**	-0. 2695**
Wenxian	96	0. 1316	0. 2063*	-0. 0705

** significant at the 1% level, * significant at the 5% level.

At Yuanshi, great significant negative correlation exists between *WBD* and *D*, *H*, *V*. At Tianshui, *WBD* has great significant negative correlation with *H* and *V*, while with *D*, the negative correlation is not significant at the 5% level. At Wenxian, however, significant positive correlation exists between *H* and *WBD*, but not between *D*, *V* and *WBD*. At Guanxian, there are negative correlation between *WBD* and *D*, *V*, but both are not significant. From the above analyses, the correlation between clone *WBD* and growth traits is complicated, with different minus or plus values at different sites; but in general, there is a negative correlation between *WBD* and the growth traits.

Scattered dot diagrams were drawn for different sites to make the correlation clearer. The abscissa in each diagram represents the average volume (*V*) of each clone at each site and the ordinate shows the average *WBD* of each corresponding clone. The two lines on each diagram are respectively the average V and *WBD* of all the clones at each site.

There is a general trend on each diagram that *WBD* decreases with the increase of clone bole, which are at Yuanshi, with the average 0. 0834m^3comparing with the value 0. 0448m^3at Guanxian, 0. 0485m^3at Wenxian and 0. 0280m^3at Tianshui. In other words, the average volume of the clones at Yuanshi is nearly double that at Guanxian and Wenxian, and three times more than that at Tianshui. So the order of *V* from the biggest to the smallest is Yuanshi>Wenxian>Guanxian>Tianshui.

As to the order of *WBD* from the highest to the lowest, it is Guanxian>Yuanshi>Wenxian>Tianshui. Clones at Yuanshi have generally the biggest volume and relatively higher *WBD*; clones at Guanxian have generally the highest *WBD* and relatively bigger volume and clones at Tianshui have the smallest volume and also the lowest *WBD*. It indicates again that it is not necessarily true that the increase of the clone volume would bring with the decrease of *WBD* . The growth process of trees is so complex that growth rate is certainly not the only simple factor that affects *WBD* .

4 Conclusions and discussion

4.1 Inheritance and variation of *WBD*

Interclonal variations of wood properties are included in most researches on inheritance and variation of various poplar species. Most results are consistent that the significant differences exist among clones (Yanchuk et al. , 1983; 1984; Nepveu 1986; Wang et al. , 1989; Liu et al. , 1994). Most believe that there should be medium or strong genetic control over wood density (specific gravity). According to Farmer et al. (1968) and Olson et al. (1985), moderately higher broad sense inheritability in specific gravity exists in *Populus deltoides* (respectively h^2 = 0. 70 and h^2 = 0. 62). Similar results were reported by Liu et al. (1994) in some hybrid poplars (clonal repeatability R=0. 789), Wang et al. (1989) in some clones of the Section Aigeiros and Tacamachacae (h^2 = 0. 84), Gu et al. (1994) in *P. davidiana* (population repeatability R = 0. 541, individual repeatability R = 0. 471) .

The result in this study reveals that the clonal repeatability of *WBD* of P. *tomentosa* is 0. 82~0. 91, and the value is a little higher than most of the above mentioned results. It may depict a higher genetic control over *WBD* in *P. tomentosa*. The great differences among clones and strong genetic control in *WBD* mean a great hope of success in breeding efforts on improvement of *WBD*.

It is very important to understand the influence of different site conditions on poplar wood properties. It offers information about adaptability at any site and extending limits of superior clones and has a significant guiding meaning in culturing practice of poplar industry timber plantation. Most of the researches indicate site effect on wood properties in various poplar species (Zobel et al. , 1989). However, Gruss et al. (1993) reported no significant effect of site on wood properties in some selected poplar clones. Zobel et al. (1989) generalized on the basis of many studies that small site or soil difference has little influence on wood properties within a large geographic and climate region, however, real extreme differences such as between thick sand and clay, or mineral soil and organic soil might produce wood with completely different quality.

The result of this study is that great significant differences in *WBD* exist among different test sites except that between Yuanshi and Wenxian. Although the four studied sites spread along the range of natural distribution zone of *P. tomentosa*, different climate, soil, fertilizer and water conditions (Huang, 1992) and varied fostering and management methods at the four sites may result in the great site effect on *WBD*. Further researches are needed to explore the true reasons of the site effect on poplar wood properties. Also, the great site influence on *WBD* illustrates that in order to obtain higher gain from improving *WBD*, clonal selection program should take site effect into consideration.

4.2 Correlation between *WBD* and growth traits

The influence of growth rate on wood properties has been a hot question for decades, however, no consistent conclusion has been drawn until now. Various results including negative correlation, no correlation and positive correlation between wood properties and growth traits have been reported about poplars. Farmer et al. (1986) discovered from their researches on *P. deltoides* that the wood density decreases in trees with faster growth speed. Olson et al. (1985) reported that in *P. deltoides* it is impossible to improve both tree volume and wood specific gravity simultaneously using a selection index because of their negative correlation. Other researches done by Posey et al. (1969) on *P. deltoides*, Zhu Xiangyu (1993) on some hybrids in Section Aigeiros and Tacamachacae got the similar results. However, Guiher (1968) believed that the number of growth ring per inch can not represent big or small value of wood specific gravity. Wang (1989) held that the selection of growth traits will not affect tree's performance in wood specific gravity; and independent selection of the two characters can be carried out successfully. The research of Kennedy et al. (1959) reported that poplar wood specific gravity rises as the increase of growth rate, but few research results support the conclusion.

According to the correlation analyses of three growth traits and *WBD* at four test sites in this paper, a general negative correlation exists betweeen the two sides at three sites except at Guanxian. We can still expect clones with excellent performances in both volume and *WBD* . Comparing the clones from the 4 sites, clones at Yuanshi generally have the greatest volume and higher *WBD*; clones at Guanxian have the highest *WBD* and relatively greater volume and clones at Tianshui have both the smallest tree volume and the lowest wood basic density. So the conclusion is that the negative correlation depicts a general trend of *P. tomentosa* populations, and there exist individuals with both higher volume and wood basic density. Furthermore, the trees for wood samples are eight years old, only about half of the rotation term for wood utilization; the inconsistency of the results of the correlation between *WBD* and growth traits indicates that it is necessary to keep on studies as the trees grow into 10, 15, 20 years and thereafter.

Since the industry timber is concerned, wood density is the most significant factor that would affect wood quality and little fluctuation in it will result in great change in dry mass production per unit of area. Moreover, it also influences many properties of fiber products. This paper concluded some preliminary results of inheritance and variation of wood basic density in *P. tomentosa*. Many fields shall be further studied in the mode of inheritance and variation of wood properities in order to make full use of them for industrial uses, in selection of various woods such as pulp and paper wood, plywood and construction wood.

三倍体毛白杨无性系木材密度遗传变异研究*

摘　要　该文对三倍体毛白杨无性系气干密度的无性系间和株内径向、纵向的变异进行了研究，结果表明：无性系全年轮密度、早材密度和晚材密度平均值分别为 0.444，0.428 和 0.511g/cm³。气干密度在无性系间和年轮间存在显著差异；株内径向变异随树龄增加呈波动性的上升趋势；株内纵向变异趋势在取样范围内不一致。全年轮密度、早材密度和晚材密度的无性系重复力分别为 0.89，0.92 和 0.86，说明木材气干密度受到强度遗传控制。

关键词　毛白杨，三倍体无性系，木材密度，遗传变异

木材密度是影响木材质量的最重要因素(成俊卿，1985)，它的微小变化会导致单位面积上干物质产量发生巨大的变化，其大小直接影响到纸浆产量和木制品的质量和利用(Zobel et al.，1978；1989)，所以木材密度是林木品质育种和材性遗传改良中首先要考虑研究的重要指标。过去对木材密度的研究只是得出了密度的平均值，而不能反映木材的密度组成。X 射线木材密度计是快速高效测定木材密度组成的先进手段。阮锡根等(1995)利用 X 射线密度计对 17 个树种进行了测定，潘惠新等（1996；1998）、骆秀琴等（1997）都利用此方法对不同杨树树种木材密度进行了测定，结果表明木材密度在无性系间存在差异。毛白杨作为我国特有的乡土树种，Song Wan 等（1997）、李新国等(1999)曾对其木材密度进行了研究，但利用 X 射线密度计对三倍体毛白杨木材密度的研究还未见报道。本文以三倍体毛白杨无性系木材为材料，利用 X 射线密度仪对其木材气干密度的无性系间和株内径向、纵向的变异进行了研究，以期为三倍体毛白杨木材品质性状的选择提供技术指标。

1　材料与方法

1.1　材料来源

本研究材料采自定植于河北晋州国营苗圃的三倍体毛白杨无性系测定林。测定林采用完全随机区组设计，4 个区组，造林密度为 2m×2m，小区株数均为 10 株。测定林经过两次间伐后，到采样时小区株数为 4 株，树龄为 9 年生，无性系测定林在经过生长性状初步评定后，初选出生长量大、干形较好的无性系 L1，L2，L3，L4，L5，L6，L7，L8，L9，L10 用于木材气干密度测定，对照(CK)为二倍体毛白杨无性系。

1.2　研究方法

对每一无性系每隔 2m 高度沿树干纵向均截取 5cm 厚的圆盘，2 次重复，对各圆盘分

* 本文原载《北京林业大学学报》，2000，22(6)：16-20，与邢新婷合作发表。

东西南北四个方向取样，利用 MWMY 木材微密度测定系统，对全年轮木材平均密度、早材密度、晚材密度、年轮密度最大值、最小值等木材气干密度组成指标进行了测定。单位为 g/cm^3，精确至 0.001g/cm^3。该实验是在中国林业科学研究院木材工业研究所完成。

2 结果与分析

2.1 无性系木材气干密度变异

本试验沿树干每隔 2m 高度截取 5cm 厚的圆盘，利用 X 射线密度仪对气干材木材密度组成进行了测定，其结果见表 1。三倍体毛白杨无性系气干材木材密度组成 9 年的平均值、

表 1 三倍体毛白杨无性系木材气干密度组成平均值、变异幅度、标准差和变异系数（g/cm^3）

无性系	年轮密度(D_y)				早材密度(D_e)			
	平均值	变异幅度	标准差	变异系数	平均值	变异幅度	标准差	变异系数
L1	0.444	0.369~0.497	0.038	8.485	0.403	0.317~0.458	0.044	10.911
L2	0.470	0.416~0.519	0.030	6.451	0.434	0.346~0.494	0.044	10.237
L3	0.376	0.328~0.409	0.032	8.479	0.352	0.298~0.395	0.035	9.807
L4	0.463	0.403~0.519	0.035	7.529	0.444	0.393~0.496	0.037	8.311
L5	0.474	0.434~0.537	0.033	7.038	0.446	0.389~0.510	0.038	8.570
L6	0.471	0.378~0.515	0.038	8.069	0.444	0.357~0.492	0.039	8.819
L7	0.424	0.327~0.496	0.045	10.635	0.388	0.309~0.467	0.051	13.191
L8	0.465	0.418~0.514	0.029	6.319	0.435	0.376~0.489	0.032	7.421
L9	0.433	0.392~0.473	0.024	5.485	0.407	0.389~0.441	0.019	4.567
CK	0.417	0.376~0.437	0.019	4.567	0.525	0.486~0.581	0.028	5.413
无性系	晚材密度(D_L)				早晚材比(D_L/D_e)			
	平均值	变异幅度	标准差	变异系数	平均值	变异幅度	标准差	变异系数
L1	0.504	0.403~0.547	0.042	8.403	1.255	1.112~1.434	0.086	6.825
L2	0.542	0.458~0.616	0.048	8.926	1.256	1.105~1.481	0.109	8.684
L3	0.448	0.392~0.527	0.041	9.226	1.274	1.175~1.419	0.087	6.853
L4	0.546	0.469~0.615	0.050	9.077	1.229	1.150~1.311	0.060	4.881
L5	0.548	0.520~0.606	0.031	5.580	1.233	1.127~1.347	0.077	6.217
L6	0.533	0.450~0.570	0.037	6.890	1.206	1.109~1.291	0.059	4.901
L7	0.492	0.427~0.535	0.033	6.798	1.280	1.125~1.482	0.118	9.241
L8	0.538	0.459~0.619	0.058	10.694	1.235	1.168~1.385	0.081	6.554
L9	0.509	0.455~0.585	0.041	8.144	1.248	1.114~1.362	0.071	5.682
CK	0.450	0.435~0.471	0.012	2.643	0.858	0.799~0.903	0.029	3.406

（续）

无性系	最小密度(D_{min})				最大密度(D_{max})			
	平均值	变异幅度	标准差	变异系数	平均值	变异幅度	标准差	变异系数
L1	0.316	0.229~0.376	0.051	16.038	0.591	0.491~0.642	0.042	7.076
L2	0.331	0.264~0.374	0.038	11.440	0.641	0.532~0.745	0.066	10.227
L3	0.269	0.228~0.309	0.030	11.039	0.529	0.429~0.599	0.053	10.051
L4	0.329	0.269~0.392	0.041	12.393	0.604	0.547~0.706	0.056	9.219
L5	0.345	0.268~0.404	0.045	12.964	0.631	0.590~0.695	0.033	5.195
L6	0.341	0.232~0.390	0.043	12.637	0.624	0.529~0.674	0.042	6.741
L7	0.280	0.198~0.371	0.053	18.727	0.589	0.517~0.652	0.039	6.540
L8	0.347	0.309~0.384	0.022	6.383	0.631	0.549~0.711	0.060	9.504
L9	0.318	0.260~0.350	0.030	9.360	0.606	0.541~0.708	0.056	9.279
CK	0.341	0.297~0.369	0.023	6.722	0.598	0.543~0.678	0.041	6.825

变异幅度和变异系数存在着明显差异，这与潘惠新等（1996）的研究结果一致。三倍体毛白杨无性系的生长量都高于对照无性系，无性系 L5 的平均单株材积为 0.302m^3，对照无性系的平均单株材积为 0.112m^3，表 1 中无性系 L5 年轮平均密度最大为 0.474g/cm^3，平均高出对照无性系 13.69%，说明在无性系水平上有可能选育出木材生长量大且木材密度也较大的优良无性系。无性系 L3 各木材密度值都是最小的，其年轮平均密度为 0.376 g/cm^3，说明该无性系生长迅速但其木材密度却较低。无性系年轮密度平均值为 0.444g/cm^3，高出对照无性系 6.47%。早材密度对照无性系最大为 0.525g/cm^3，无性系 L3 最小为 0.352g/cm^3，无性系早材密度平均值为 0.428g/cm^3，平均低于对照无性系 18.48%，对照无性系早材密度值明显较高，其原因可能是因为木材密度受多种因素影响。晚材密度无性系 L5 最大为 0.548g/cm^3，无性系 L3 最小为 0.448g/cm^3，无性系晚材密度均值为 0.511g/cm^3，平均高出对照无性系 15.56%。年轮最大密度平均值为 0.604g/cm^3，平均高出对照无性系 1.17%，年轮最小密度平均值为 0.322g/cm^3，平均低于对照无性系 6.16%。各无性系晚材密度高于早材密度，这对木材加工是有利的。早晚材密度比对照无性系为最低，说明对照无性系早材密度占较大比重，但早材密度一般较低，影响其木材基本密度的大小。无性系 L7 的年轮密度平均值与对照无性系相差不大，但其年轮密度和早材密度的变异幅度、变异系数明显高于对照无性系；无性系 L4，L6 早材密度平均值相同，无性系 L1 与 L8 晚材密度值相近并超过对照无性系；无性系 L8 的年轮密度和早晚材比的均值、变异系数都大大超过对照。原因是由于年轮中早晚材密度所占比重不同，早材和晚材密度的变异大小影响到年轮平均密度的变化。

2.2 年轮平均密度、早材密度和晚材密度无性系间的变异

对三倍体毛白杨无性系的年轮平均密度、早材密度和晚材密度做无性系和年轮的双因素方差分析，结果见表 2。从表 2 中可以看到在无性系和年轮间，年轮平均密度、早材密度及晚材密度都存在极显著差异，无性系×年轮的交互作用不显著。年轮平均密度、早材密度及晚材密度无性系重复力分别为 0.89，0.92 和 0.86，说明年轮平均密度、早材密度

及晚材密度这三个年轮密度组成受较强遗传控制。这与顾万春等（1998）对毛白杨和潘惠新等（1996）对黑杨派杨树的测定结果一致。

表 2　三倍体毛白杨无性系木材气干密度方差分析

变异来源	自由度	年轮平均密度(D_y)(g/cm^3)			早材密度(D_e)(g/cm^3)			晚材密度(D_L)(g/cm^3)		
		均方	方差比	重复力	均方	方差比	重复力	均方	方差比	重复力
无性系	9	0.01117	9.387**	0.89	0.0167	12.846**	0.92	0.01478	7.005**	0.86
年轮	9	0.02001	16.815**		0.0414	31.846**		0.02872	13.611**	
无性系×年轮	81	0.00119	<1		0.0013	<1		0.00211	<1	
误差	100	0.00166			0.0017			0.00215		

2.3　木材气干密度组成株内径向变异和株内纵向变异

不同的树种木材密度变异的趋势也不尽相同，木材密度受很多因子的影响，因而比较复杂。木材密度的径向变异及其在不同高度、圆周不同方位上的变异，构成了树干内变异的全景。

2.3.1　木材气干密度株内径向变异

根据生长量及干形指标，对生长量较大且干形通直的无性系 L1，L2，L7，L8，L9 作图。以图 1 为例，各无性系木材气干密度组成径向变异大体呈上升趋势。在 1993 年时，无性系 L2 年轮密度值最大为 0.520g/cm^3，其次无性系 L7 为 0.500g/cm^3，无性系 L8 和 L9 相同为 0.460g/cm^3，而只有无性系 L1 年轮密度却降低为 0.420g/cm^3。在生长的早期，年轮平均密度变化较不稳定，年轮密度值增加较快，根据木材学理论这主要与细胞中物质含量多少以及所占空间有关。在 1993 年以后，年轮平均密度变化较小，比较平缓稳定。通过图 1~4 可以看到，木材气干密度组成的径向变异大体上是一致的，木材气干密度组成随着树龄增加呈波动性变化，1993 年以前木材气干密度组成变化较大，1993 年以后木材气干密度组成变化较缓慢一致。因为成熟材木材气干密度比较稳定，因此三倍体毛白杨无性系生长的前 4 年是否可以作为划分幼龄期和成熟期的界限还有待于深入研究。

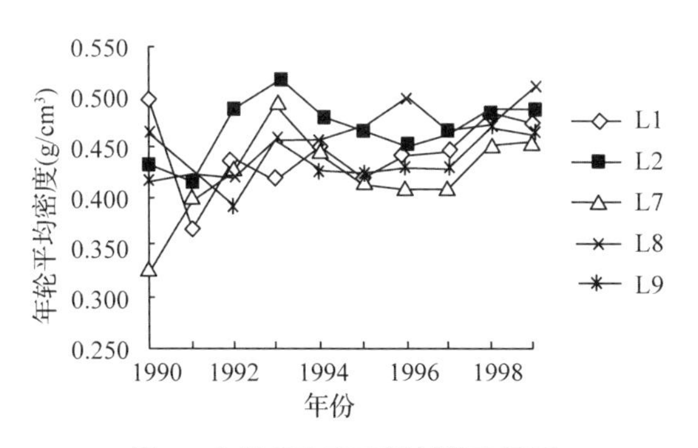

图 1　年轮平均密度随树龄变异图

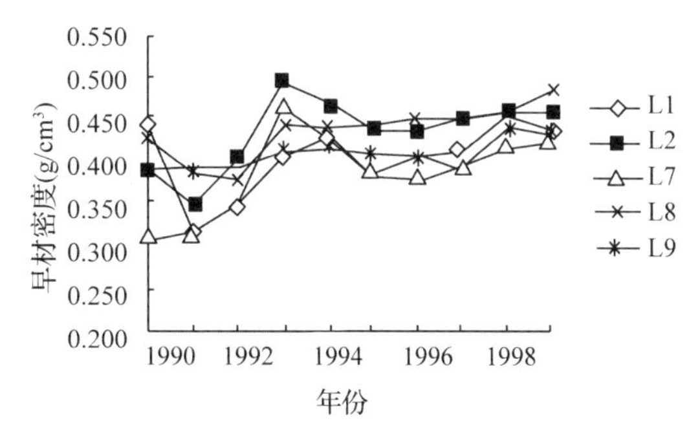

图 2　早材密度随树龄变异图

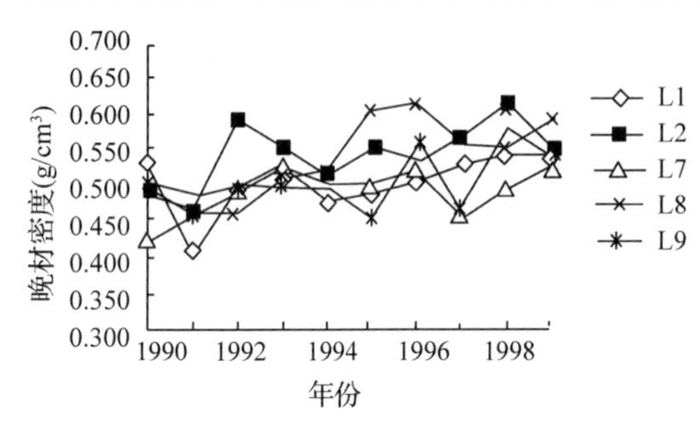

图 3　晚材密度随树龄变异图

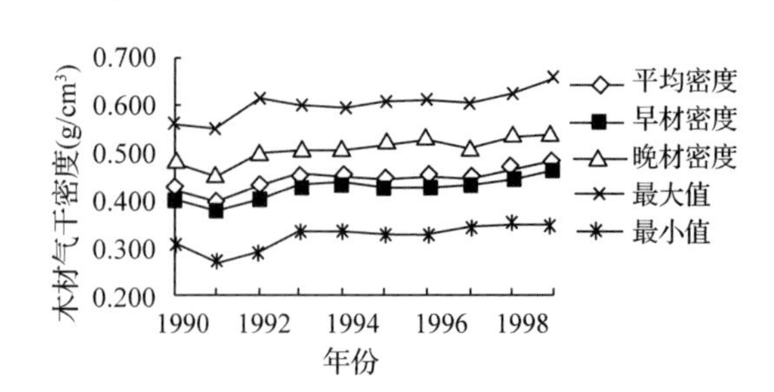

图 4　无性系木材气干密度组成随林龄变异图

为了了解三倍体毛白杨无性系群体的径向变异规律，对木材气干密度组成按无性系平均值分年轮作图。从图 4 中可看出，木材气干密度组成径向变异趋势是随着树龄增加，各密度值呈波动性的上升趋势。在 1991 年时，各木材气干密度组成值都有所降低，可能与木材内部细胞壁构造有关或者是因为气象因子造成。在 1993～1997 年各密度组成变化较缓慢、均匀，且木材气干密度值在增大，早材密度与年轮平均密度值相接近，说明三倍体毛白杨无性系年轮密度主要与早材气干密度有关。

2.3.2　木材气干密度的株内纵向变异

木材密度在株内纵向变异模式不一致，一般报导有两种变异模式：一种是木材密度在树干基部高，随树高先下降后升高；另一种是木材密度随树高增加而增加。多数研究认为木材密度在树干基部附近较高，向上随树高而降低，然后再升高，在三倍体毛白杨无性系中也发现了这两种变异规律。还发现另外一种变异模式如图 5 所示，三倍体毛白杨无性系木材气干密度组成在树干基部较低，树干中部上升，近树干顶部时木材密度又降低．因为木材密度与木材构造、浸提物等直接有关，而木材构造和浸提物又受到树龄、树干部位、立地条件及其它因子的影响，因此出现这种变异趋势。从图 6 中可以看出无性系 L9 其木材气干密度组成的株内纵向变异基本与多数研究结果相一致，这与国外学者 Yanchuk et al.(1984)和国内徐从建(1986)对杨树木材密度的研究也相符合。

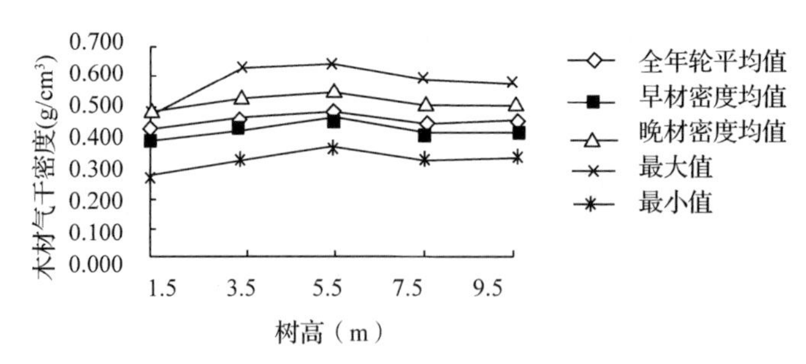

图 5　无性系 L7 木材气干密度组成株内纵向变异

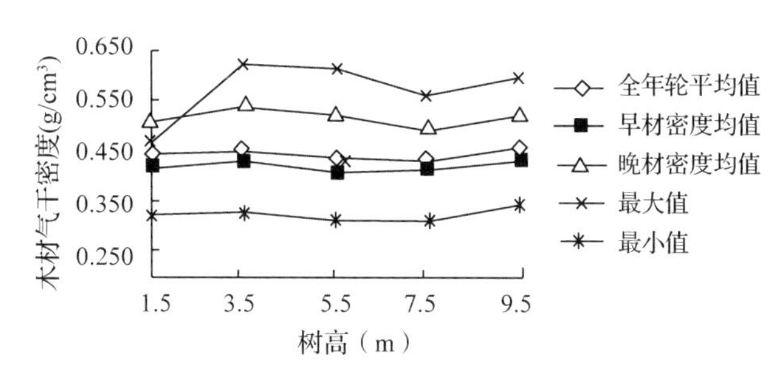

图6　无性系L9木材气干密度组成株内纵向变异

图7是无性系L1的木材气干密度组成纵向变异图。其木材气干密度组成株内纵向变异趋势属于第二种，木材气干密度从树基部到顶部升高，这与王明庥等(1989)对黑杨派新无性系的研究结果相一致。

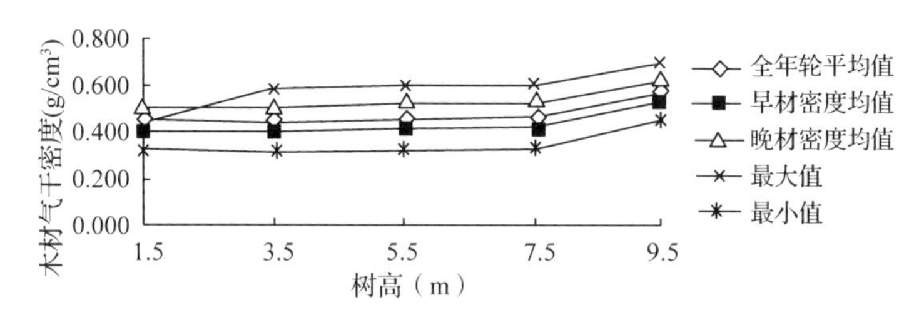

图7　无性系L1木材气干密度组成株内纵向变异

从图5至图7可以看出，三个无性系L1，L7，L9气干材年轮平均密度、早材密度、晚材密度在树高7.5m以下时除无性系L7相对变化较大外，无性系L1，L9变化速度较缓慢、均匀，超过树高7.5m变化较快。木材平均密度、早材密度、晚材密度株内纵向变异规律是在树干7.5m以下相对较稳定，接近树干顶部时木材密度变化较大。一般在木材应用时，多考虑的是全年轮密度的平均值，而全年轮密度平均值在树干中下部变异较小，所以不会影响到木材的应用。

3　结论

(1)三倍体毛白杨无性系气干材年轮密度平均值、早材密度平均值、晚材密度平均值、年轮最大密度平均值和年轮最小密度平均值分别为0.444，0.428，0.511，0.604和0.322g/cm^3。方差分析结果表明木材气干密度组成中年轮平均密度、早材密度和晚材密度在无性系间存在极显著差异，无性系×年轮交互作用不显著。遗传参数估算无性系木材密度重复力分别为0.89，0.92和0.86，说明年轮平均密度、早材密度和晚材密度具有较高遗传水平，受到强度遗传控制，因此根据木材密度对三倍体毛白杨进一步进行选择改良是有潜力的。

(2)三倍体毛白杨无性系木材气干密度的径向变异在株内和无性系总体上相一致。其木材气干密度组成随树龄的变异趋势是：从髓心到树皮木材气干密度组成呈波动性的上升趋势。

(3)三倍体毛白杨无性系木材气干密度在株内纵向上变异趋势不一致。其中有的无性系株内纵向变异趋势表现为木材气干密度组成在树干基部较低，在树干中部升到最大，近树干顶部时又降低。大多数无性系株内变异趋势与多数研究结果相一致，在树干基部较高，向上随树高增加而降低，然后又升高。还有的无性系株内纵向变异趋势表现为木材气干密度组成随树高增加而增大。

Genetic Variation in Wood Density of Triploid Clones of *Populus tomentosa*

Abstract The genetic variation in wood densities of triploid clones of *Populus tomentosa* Carr. were investigated and the variations in wood densities among clones, radial direction and vertical direction were studied in this paper. The average values of growth ring densities, early wood densities, late wood densities were 0. 444, 0. 428, 0. 511g/cm^3 respectitively. There were significant differences among clones and among growth rings in wood densities. The radial direction variation in wood densities were unstably accending, the vertical variation patterns were not identical. Clonal repeatabilities of growth ring densities, early wood densities, late wood densities of air dry timber were 0. 89, 0. 92 and 0. 86 respectitively. This indicated wood densities were under strong genetic controls.

Keywords *Populus tomentosa* Carr. , triploid clones, wood density, genetic variation

三倍体毛白杨无性系木材热学性质变异初探*

摘　要　该文对三倍体毛白杨无性系木材的热学性质进行了研究，结果表明热学性质在无性系间存在显著差异。比热系数、导热系数和导温系数的无性系重复力分别为 0.59，0.77 和 0.62，表明三倍体毛白杨无性系木材的热学性质受到较强的遗传控制。热学性质与木材密度呈正相关，导热系数与木材密度达到极显著相关，木材密度与导热系数、导温系数和比热系数的相关系数分别为 0.8100，0.5778 和 0.0938。

关键词　毛白杨，三倍体无性系，热学性质，导热系数，导温系数，比热系数

木材在制作单板前的蒸煮、单板的干燥及热压胶合过程中均涉及木材的热学性质。木材热学性质的主要指标有比热系数、导热系数、导温系数和蓄热系数等。木材的比热系数、导热系数和导温系数是木材热处理（如木材干燥、木材防腐处理、单板用木段的蒸煮等）中的重要工艺参数，因此了解和掌握三倍体毛白杨无性系木材的热学性质特点及其遗传变异规律具有重要的实用价值。一些学者（成俊卿，1983；杨庆贤，1992；马瑞堂等，1985）曾研究木材的热学性质，覃道春（1994）对马尾松树种的热学性质进行了研究，而有关毛白杨的热学性质尚未见报导。徐纬英（1988）在《杨树》中报道了沙兰杨和小叶杨的比热系数为 1632.85J/(kg·℃)，导热系数介于 305.64~431.24J/(m·h·℃)之间，杨木的导温系数介于 0.000423~0.000663m^2/h，但并未对热学性质的遗传变异规律进行研究。因此本文以三倍体毛白杨无性系为材料，对其热学性质在无性系间及株内纵向的变异规律进行了初步探索，以期为毛白杨热学性质的深入研究提供借鉴。

1　材料与方法

1.1　材料来源

本研究材料来自定植于河北晋州苗圃的三倍体毛白杨无性系测定林。测定林均采用完全随机区组设计，4 次重复，造林密度为 2m×2m，小区株数均为 10 株。测定林经过两次间伐后，到采样时小区株数为 4 株，树龄为 9 年生，无性系测定林在经过生长性状初步评定后，初选出生长量大、干形较好的无性系 L2，L3，L4，L5，L6，L7，L8，L9，L10 用于木材热学性质测定，对照（CK）为无性系二倍体毛白杨。

* 本文原载《北京林业大学学报》，2000，22(6)：21-23，与邢新婷和张文杰合作发表。

1.2　实验方法

根据中国林业科学研究院张文庆等制定出的木材热物理性质测定方法（草案），对各无性系基部木段按顺纹方向分别制作 120mm(L)×120mm(R)×12~40mm(T)的试件，每一无性系 6 块试件，每一试件重复测定 3 次，根据生长量选取无性系 L9 沿树干纵向每隔 2m 制作试件 6 块，每 1 试件重复测定 3 次，以测定树干纵向热学性质。同时测定试件质量计算其木材密度，使用 DRM—I 型导热系数测定仪在室温的条件下对试件进行顺纹热学性质测定。

2　结果与分析

2.1　三倍体毛白杨无性系木材顺纹热学性质测定结果

三倍体毛白杨各无性系木材顺纹热学性质测定结果见表 1。三倍体毛白杨木材顺纹热学性质在无性系间差异较大。比热系数无性系 L9 最大，其值为 1643.74J/(kg·℃)，对照无性系最小为 141.93J/(kg·℃)，平均值为 1196.59J/(kg·℃)，三倍体毛白杨无性系比热系数平均比对照无性系高出 743.08%。导热系数无性系 L8 最大为 370.11J/(m·h·℃)，无性系 L9 最小为 290.15J/(m·h·℃)，对照无性系为 321.96J/(m·h·℃)，平均值为 329.92J/(m·h·℃)，平均比对照无性系高出 2.47%。导温系数无性系 L6 最大为 0.000667m^2/h，无性系 L9 最小为 0.000520m^2/h，平均值为 0.000619m^2/h，平均比对照无性系高出 1.64%。无性系三倍体毛白杨的比热系数介于 141.93~1643.74J/(kg·℃)，导热系数介于 290.15~370.11J/(m·h·℃)，导温系数介于 0.000520~0.000667m^2/h，与徐纬英(1988)在《杨树》中报道的结果相近。从表中还可以看出比热系数数值较大而导温系数较小，反映三倍体毛白杨受热之后木材本身升温较快，而热量传播较慢，这样会加快单板材的含水率变化，这对三倍体毛白杨的单板胶合是有利的。

表 1　三倍体毛白杨无性系热学性质测定结果

无性系	木材密度 (kg/m^3)	含水率 (%)	比热系数 [J/(kg·℃)]	导热系数 [J/(m·h·℃)]	导温系数 (m^2/h)
L2	400.23	7.50	1327.22	307.31	0.000583
L3	367.19	7.41	1380.39	322.80	0.000640
L4	410.47	7.59	1312.56	347.92	0.000648
L5	405.28	7.62	1325.96	330.76	0.000631
L6	448.35	7.63	1157.65	344.57	0.000667
L7	419.79	7.73	1265.25	334.11	0.000635
L8	461.86	7.48	1214.59	370.11	0.000642
L9	371.08	7.42	1643.74	290.15	0.000520
CK	374.49	7.61	141.93	321.96	0.000609

从图1，2也可看出三倍体毛白杨无性系木材热学性质的3个系数在无性系间存在差异。

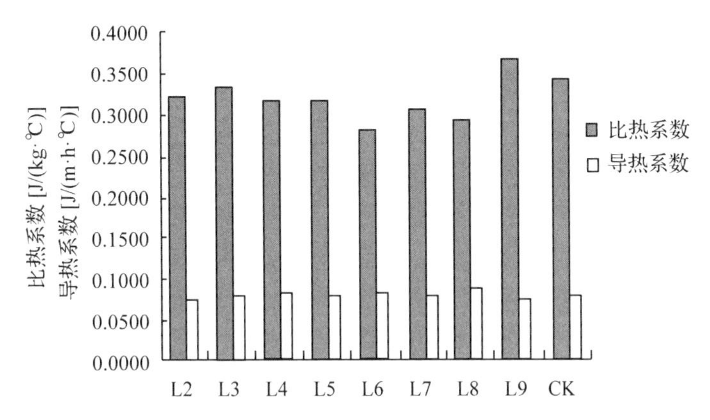

图1 三倍体毛白杨无性系木材比热系数和导热系数比较图

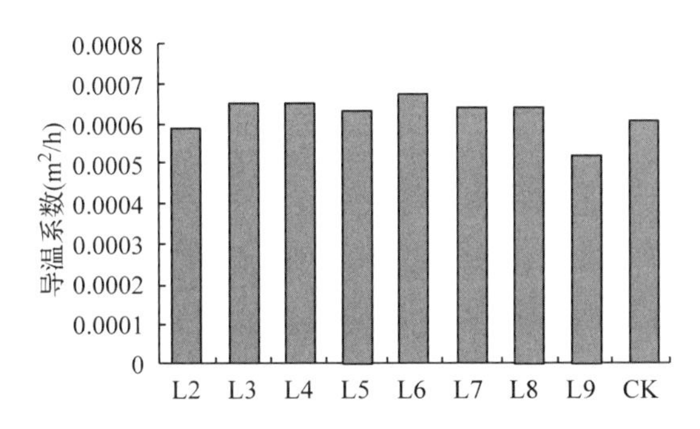

图2 三倍体毛白杨无性系木材导温系数比较图

2.2 三倍体毛白杨无性系间木材热学性质的变异

经方差分析得出各无性系间木材热学性质存在显著差异(表2)，比热系数、导热系数和导温系数的重复力分别为0.59，0.77和0.62，表明三倍体毛白杨无性系木材的热学性质受到较强的遗传控制，在无性系水平上进行选择会获得较高的遗传增益。

表2 三倍体毛白杨无性系热学性质方差分析结果

变异来源	自由度	比热系数			导热系数			导温系数		
		均方	均方比	重复力	均方	均方比	重复力	均方	均方比	重复力
无性系	9	0.001999	5.3449*	0.59	9.34E-05	10.9268*	0.77	5.92E-09	5.8646*	0.62
误 差	18	0.000374			8.55E-06			1.01E-09		

2.3 三倍体毛白杨木材热学性质株内纵向变异

毛白杨木材热学性质在株内也存在变异，表3中比热系数、导热系数随树高增加变异系数较大，且变异趋势不一致，而导温系数变异较小；随树干高度增加，导热系数和导温系数的均值在取样范围内由低到高，而比热系数随树干高度增加在取样范围内呈降低的变化趋势。由于所取材料来源于同一个无性系，因此所得结果只能说明该无性系的大致株内纵向变异趋势，要想得到无性系总体变异趋势，就要对每一无性系进行测定，但由于受制作试件要求的限制，所以只能对其变异趋势进行初步摸索。

表 3　三倍体毛白杨无性系热学性质株内纵向变异

热学系数	单 位	树干高度	试件数	均 值	标准差	变异系数
比热系数	[J/(kg·℃)]	1.5m	18	0.3626	0.0401	11.05
		5.5m	18	0.3403	0.0307	9.03
		9.5m	18	0.3262	0.0312	9.58
导热系数	[J/(m·h·℃)]	1.5m	18	0.0693	0.0068	9.83
		5.5m	18	0.0727	0.0086	11.80
		9.5m	18	0.0754	0.0083	11.00
导温系数	(m^2/h)	1.5m	18	0.000519	0.000047	9.07
		5.5m	18	0.000565	0.000054	9.63
		9.5m	18	0.000582	0.000054	9.33

2.4　热学性质与木材密度的回归分析

许多木材性质与木材密度都存在着密切关系。木材密度与各热学性质的回归分析结果列于表 4。从表中可以看出各热学系数与木材密度都呈正相关，其中导热系数与木材密度达到极显著相关，导温系数与木材密度相关也较紧密，只有比热系数与木材密度呈微弱正相关。可能原因是单个树种分析中，由于无性系内试件数较少，密度范围也较窄，使得相关性的表现受到限制。

表 4　木材密度与比热系数、导热系数和导温系数的回归分析

回归关系	相关系数	回归方程
木材密度与比热系数	0.09384	
木材密度与导热系数	0.80997**	$Y=0.0001X+0.0236$
木材密度与导温系数	0.57778	$Y=-4\times10^{-9}X^2+4\times10^{-6}X-0.0004$

注：Y 表示热学系数，X 表示木材密度.

3　结论

（1）在热学性质方面，三倍体毛白杨无性系间存在显著差异，比热系数、导温系数和导热系数均值分别为 1196.59J/(kg·℃)、0.000619m^2/h 和 329.92J/(m·h·℃)。无性系 L9 的比热系数比沙兰杨和小叶杨的要大，其它无性系低于这两个树种，且其导热系数和导温系数较低，这对木材热加工时含水率的改变是有利的。比热系数、导热系数和导温系数的重复力分别为 0.59，0.77 和 0.62，表明三倍体毛白杨无性系的热学性质受到较强的遗传控制，在无性系水平上进行选择会得到较高的遗传增益。

(2)三倍体毛白杨无性系热学性质在株内也存在变异，比热系数、导热系数随树高增加变异系数较大，且变异趋势不一致，这不利于三倍体毛白杨无性系单板的热加工，因此要通过育种手段进行选择，实现对木材热学性质的改良和控制。导温系数变异则较小，随树干高度增加，导热系数和导温系数的均值由低到高，而比热系数随树干高度增加呈降低的变化趋势。

(3)三倍体毛白杨无性系木材热学性质与木材密度呈正相关，导热系数与木材密度达到极显著相关，导温系数与木材密度相关也较紧密，只有比热系数与木材密度呈微弱正相关，表明在对三倍体毛白杨无性系进行木材密度的改良时也会对木材的热学性质进行改良。导热系数与木材密度呈极显著相关，对胶合板制造过程中的胶合是有利的。

Study on the Wood Thermal Properties of Triploid Clones of *P. tomentosa* Carr.

Abstract Wood thermal properties of triploid clones of *P. tomentosa* Carr. were reported in this paper. There were significant difference among triploid clones of *P. tomentosa* Carr. Clonal repeatabilities of specific heat coefficient, thermal conductivity coefficient and warm-transfer coefficient were 0. 59, 0. 77 and 0. 62 respectively, this indicated wood thermal properties were under moderate or strong genetic controls. Positive correlation existed between wood thermal property and wood density. There was significant correlation between thermal conductivity coefficient and wood density. The correlation coefficient between wood density and specific heat coefficient, thermal conductivity coefficient, warm-transfer coefficient were respectively 0. 09384, 0. 8099, 0. 57778.

Key words *P. tomentosa* Carr. , triploid clones, wood thermal property, specific heat coefficient, thermal conductivity coefficient, warm-transfer coefficient

Genetic Control of Air-dried Wood Density, Mechanical Properties and Its Implication for Veneer Timber Breeding of New Triploid Clones in *Populus tomentosa* Carr. *

Abstract The wood samples of 9 triploid clones of *Populus tomentosa* Carr. taken from a 9-year-old clonal test site were analyzed in order to investigate the genetic variation of wood properties, including air-dried wood density and some mechanical properties. The results showed that significant or extremely significant difference in air-dried wood density and the mechanical properties existed among the clones, this means these wood properties were under moderate or strong genetic controls and could be improved by genetic manipulations. The radial and vertical variation patterns of air-dried wood density were also studied and the results were found to coordinate with other previous research results. The vertical variation patterns of most mechanical properties within the individual tree also conformed to the general wood theories except the modulus of elasticity and cross section hardness. Among the mechanical properties, modulus of elasticity (MOE) and tangent section hardness were under strong genetic control, with the clonal repeatabilities being 0.90 and 0.80, respectively. However, the clonal repeatabilities of other mechanical properties under study were a little lower than above two indexes. Genetic correlation analysis indicated that super clonal selection and breeding for veneer timber could be realized through indirect selection of wood density and form indexes.

Keywords *Populus tomentosa* Carr., triploid clones, air-dried wood density, mechanical properties, genetic variation

1 Introduction

Many researches have confirmed that there is abundant genetic variation in most wood characters on different genetic control levels, and that the major wood characters are under strong genetic control and sensitive to genetic manipulations (Zobel et al., 1989). Although faster growth rate, better stem form and stronger adaptation to various site conditions have remained to be the main targets of tree breeding, desirable wood properties have been essential part of forest tree breeding programs. However, most timber breeding researches involved only few of wood characters such as wood density and wood fiber length, and studies of other important wood properties are few and far between. Although modulus of rupture (MOR) and modulus of elasticity (MOE) are important elements influencing ply-wood manufacture and bonding properties, their genetic patterns are poorly

* 本文原载《林业研究》(*Forestry Studies in China*), 2002, 4(2): 52-60, 与邢新婷合作发表。

known until today partly because of the high expenditures on test and evaluation procedures. With the great advancement of genetic improvement research of *Populus* especially in western Europe and north America, studies on genetic variations in wood properties have been strengthened, and more and more tree breeding programs have incorporated the comprehensive tree improvement plans including not only growth and resistance breeding but also wood quality improvement as well.

Populus tomentosa Carr. is one of the indigenous tree species in China in the section *Leuce* under genus *Populus* in Salicaceae. It is mainly distributed in the middle and lower reaches of the Yellow River and occupies about one million km^2 (according to Department of Science and Technology of State Forestry Administration, 1998). It has a number of desirable characteristics such as rapid growth rate, strong adaptation to harsh site conditions, excellent resistance, super wood quality, tall and straight trunk and handsome tree form, and it has been considered to be as an important tree species for establishing fast-growth and high-yield plantations in the vast plain areas of the Yellow River. As an indigenous tree species, a lot of genetic improvement researches have been carried out around growth-yield in the last several decades (Zhu, 1992). In recent years, researches in wood density and wood fiber length of *P. tomentosa* have obtained great progress (Gu et al., 1998; Xing et al., 2000; Song et al., 2000; Yao et al., 1998; Pu et al., 2002), but studies on the genetic variation in other important mechanical properties are few (Wang, 2001).

The purpose of this paper is to investigate the genetic variation of air-dried wood density and some important mechanical properties including modulus of rupture (MOR), modulus of elasticity (MOE), compression of strength (COS) and wood hardness by analyzing wood samples of triploid clones of *P. tomentosa*. Through this research, we hope to speed up the course of wood property studies of triploid clones of *P. tomentosa*, at the same time provide scientific basis for making veneer timber tree breeding plans.

2 Materials and methods

2.1 Experiment materials

The wood samples were taken from a 9-year-old *P. tomentosa* triploid clone test site located in Jinzhou City, Hebei Province. This testing plantation adopted complete random block design, with four times of replications, 10 individual trees in each block, and with the interplant space being 2m×2m. At the time of sample collection in March 1999, 4 individual trees were retained in each block after two intermediate cuttings, and the height and the diameter at breast height (DBH) of the whole test plantation were investigated. The individual tree volume was calculated by using the formula "$V=g_{1.3}\times h\times f_{1.3}$" (Guan, 1994). After preliminary evaluations of growth traits, 9 triploid clones with high growth rate and good stem form (labeled asL1, L2, L3, L4, L5, L6, L7, L8 and L9 respectively) were selected for further wood property test, with the control clone being T34, as labeled as CK.

The destructive breakdown sampling method was employed, and 2 trees were sampled in each clone and thus altogether 20 individual trees were tested.

2.2 Wood sample treatment and analysis of wood characters

Branch trimming was done at the time of tree falling. The trees were cut every 1.5m at the basal stem and every 2 m at the above stem for each clone. At the same time, 5cm thick wood discs were cut out every 2 m along the stem. Air-dried wood density, MOR, MOE, COS and wood hardness were tested on the disc samples, air-dried wood density was measured by MWMY micro-density test instrument, and the others were tested according to GB1928-91, 1936-91, 1935-91, 1941-91 and GB9846·8-10, respectively.

2.3 Data analysis

Block means for each stem were calculated and then the analysis of variance was computed using SYSTAT software. The linear model for variance analysis is "$Y=\mu+C_I+y_j+C_{ij}+e_{ijk}$", with Y being the test value, C_I the clonal effect value, y_j the site effect value of growth ring or tree height, C_{ij} the interaction effect between clone and growth ring or tree height, e_{ijk} the random error. Estimation of genetic parameters of clonal repeatability and genetic correlation was calculated according to Falconer (1981).

3 Results and analysis

3.1 Genetic variation of air-dried wood density among clones

The results of air-dried wood density and stem volume of 9 triploid clones in *P. tomentosa* were presented in Table 1. The L5 had the highest value of growth ring wood density as 0.474g · cm^{-3}, and its stem volume is 0.302m^3. This shows the wood density is not necessarily decreased with greater stem volume, and it is possible to select clones with both fast-growing and high wood density through selective breeding. L3 had the lowest wood density of 0.376g · cm^{-3}, indicating a fast growth rate but decreased wood density. The average value of growth ring wood density of the 9 clones was 0.444g · cm^{-3}, which meets the requirement of wood density for veneer timber. The wood densities of the triploid clones were all higher than CK except L3.

From the results in Table 1, the late wood densities of all triploid clones were higher than that of the early wood densities except CK, and this is desirable in wood processing. The results of variance analysis (Table 2) showed that great significant clonal effect in air-dried wood density existed among clones, implying a great selection potential. These results agreed with previous studies (Luo et al., 1997; Yanchuk et al., 1983; Nepveu et al., 1986).

Table 1 Average values of stem volume, the components of growth ring wood densities of 9 triploid clones in *P. tomentosa*

clone	Stem volume (m^3)	The average values of air-dried wood density (g/cm^3)				
		Annual ring density (D_y)	Early wood density (D_e)	Late wood density (D_L)	Max values of annual ring density (D_{max})	Min values of annual ring density (D_{min})
L1	0. 278	0. 444	0. 403	0. 504	0. 591	0. 316
L2	0. 279	0. 470	0. 434	0. 542	0. 641	0. 331
L3	0. 215	0. 376	0. 352	0. 448	0. 529	0. 269
L4	0. 268	0. 463	0. 444	0. 546	0. 604	0. 329
L5	0. 302	0. 474	0. 446	0. 548	0. 631	0. 345
L6	0. 222	0. 471	0. 444	0. 533	0. 624	0. 341
L7	0. 302	0. 424	0. 388	0. 492	0. 589	0. 280
L8	0. 371	0. 465	0. 435	0. 538	0. 631	0. 347
L9	0. 356	0. 433	0. 407	0. 509	0. 606	0. 318
CK	0. 112	0. 417	0. 525	0. 450	0. 598	0. 341

Table 2 Analysis of variation for air-dried wood densities components of triploid clones in *P. tomentosa*

Air-dried wood density	Clone	Growth ring	Clone×growth ring	Error
Annual ring density (Dy)	0. 01117** (9)	0. 02001** (9)	0. 00119(81)	0. 00166(100)
Early wood density (De)	0. 01670** (9)	0. 04140** (9)	0. 00130(81)	0. 00170(100)
late wood density (D_L)	0. 01478** (9)	0. 02872** (9)	0. 00211(81)	0. 00215(100)
Max. of annual ring density (D_{max})	0. 02059** (9)	0. 02047** (9)	0. 00322(81)	0. 00270(100)
Min. of annual ring density (D_{min})	0. 01462** (9)	0. 01552** (9)	0. 00163(81)	0. 00170(100)

Notes: The degree of freedom is indicated in each bracket, * * indicates a significant difference at the confidence level of 0. 99.

3. 2 Genetic variation of mechanical properties among clones

The test results of the mechanical properties of the clones were listed in Table 3. MOR of triploid clones in *P. tomentosa* ranged from 50. 3 to 60. 4MPa and MOE ranged from 7453 to 8478MPa. The minimum value of COS was 33. 0MPa, and the maximum was 40. 2MPa. Cross section hardness was higher than radial section hardness and tangent section hardness in all clones. The maximum value of cross section hardness was 3393N and the minimum was 2952N. Radial section hardness and tangent section hardness ranged from1887 to 2393N and 1937 to 2483N, respectively. The indexes of mechanical properties of the control clone were relatively lower than the triploid clones, indicating that wood properties of the triploid clones were generally superior to those of CK.

Table 3 Test results of modulus of rupture (MOR), modulus of elasticity (MOE), compression of strength (COS) and wood hardness

Clones	Stem volume (m^3)	MOR (MPa)	MOE (MPa)	COS (MPa)	Hardness in cross section (N)	Hardness in radial section (N)	Hardness in tangent section (N)
L1	0.278	56.3	8028	35.7	3393	2169	2276
L2	0.279	54.3	8255	37.3	3076	2173	2189
L3	0.215	50.3	7688	33.0	3245	1887	1937
L4	0.268	55.7	7731	37.9	3298	2160	2246
L5	0.302	52.2	8478	38.6	3315	2045	2168
L6	0.222	58.4	8474	36.6	3326	2287	2383
L7	0.302	52.2	7705	34.7	3097	2280	2293
L8	0.371	60.4	8474	40.2	3257	2393	2483
L9	0.356	52.4	7453	35.3	2952	1925	2003
CK	0.112	50.2	7898	-	3131	1937	1956

Table 4 Analysis of variation of the mechanical properties in triploid clones of *P. tomentosa*

Traits	Clone	Sample height	Clone×Sample height	Error
MOR(Mpa)	2.632* (9)	15.066** (2)	2.256* (18)	42.999 (210)
MOE(Mpa)	9.692* (9)	4.328* (2)	0.916(18)	559576.3(210)
COS(Mpa)	3.329* (9)	2.236(2)	13.55**(18)	4.29 (420)
Hardness in cross section(N)	2.026(9)	3.504(2)	4.720* (18)	54510.05(180)
Hardness in radial section(N)	3.503* (9)	3.722* (2)	4.240* (18)	54968.36(180)
Hardness in tangent section(N)	4.979* (9)	2.456(2)	3.520* (18)	43044.09 (180)

Notes: The degree of freedom is indicated in each bracket, * indicates a significant difference at the confidence level of 0.95 and ** indicates a significant difference at the confidence level of 0.99

Variance analysis was done employing the tested mechanical properties data to study the genetic variance at the clonal level. The results from the ANOVA were shown in Table 4. The differences of the mechanical properties among the clones were greatly significant except the cross section hardness, showing that it is possible to obtain superior triploid clones by selection for high quality veneer timber.

3.3 Wood property variations within trees in radial and tangent directions

Variations of wood properties exist not only among clones but also within trees in radial direction and at different tree heights. Variations within trees at different heights and circumferences constitute a panorama within tree variations.

The radial variation patterns of wood density were found to be very similar. The wood density components fluctuated with the increase of the annual ring; however there was a general trend of increase in the vertical direction.

3.3.1 Variations in radial and tangent directions of air-dried wood density

The wood density components fluctuated with the increase of the annual ring; however there

Table 5 Test results of mechanical properties at different heights of the trees

Clone	MOR(MPa)			MOE(MPa)			COS(MPa)			Hardness in cross section(N)			Hardness in radial section(N)			Hardness in tangent section(N)		
	1. 5m	5. 5m	9. 5m	1. 5m	5. 5m	9. 5m	1. 5m	5. 5m	9. 5m	1. 5m	5. 5m	9. 5m	1. 5m	5. 5m	9. 5m	1. 5m	5. 5m	9. 5m
L1	59. 5	56. 1	53. 3	7837	8121	8125	35. 1	36. 1	36. 0	3358	3413	3407	2269	2115	2122	2437	2170	2221
L2	60. 4	49. 1	53. 5	8611	7990	8165	38. 6	35. 9	37. 5	3000	2964	3263	2287	2043	2190	2154	2188	2226
L3	53. 1	50. 0	47. 7	7573	7544	7946	28. 8	34. 6	35. 6	3407	3098	3229	2174	1718	1770	2156	1763	1891
L4	58. 9	55. 5	52. 6	7868	7498	7826	39. 8	37. 4	36. 6	3404	3340	3151	2519	1987	1976	2475	2194	2068
L5	56. 2	48. 4	51. 9	8602	8499	8334	41. 1	36. 4	38. 4	3362	3229	3305	2245	1782	2109	2138	2146	2219
L6	63. 3	59. 3	52. 6	8640	8059	8723	34. 7	37. 7	37. 3	3376	3249	3353	2462	2282	2118	2668	2199	2282
L7	58. 5	49. 2	48. 9	7755	7397	7962	35. 9	34. 8	33. 4	2939	-	3254	2521	-	2039	2588	-	1998
L8	65. 6	58. 0	57. 5	8625	8450	8347	41. 2	40. 2	39. 1	3268	3235	3268	2770	2155	2255	2540	2323	2587
L9	57. 3	50. 7	49. 1	8118	6399	7843	38. 0	34. 0	33. 8	2787	2941	3127	1817	1893	2065	2015	1998	1996
CK	59. 1	53. 2	-	8050	7745	-	37. 8	35. 2	-	3174	2988	3232	1955	1819	2036	1794	1880	2193

Note: 1. 5, 5. 5 and 9. 5m in the Table referred to the tree height from the base of the trees

was a general trend of increase in the vertical direction.

In general, two vertical variation patterns within trees exist for wood density. One shows increase-decrease-increase again pattern from basal stem to the top, and the other demonstrates straight increase of the wood density with the increase of the tree height. Most of the precious researches on *Populus* have obtained the similar variation pattern of wood density as the former one. In this experiment both the variation patterns were found in triploid clones of *P. tomentosa*. In addition, a new variation trend described as "decrease-increase-again decrease" was disclosed.

The variations of air-dried wood density were relatively stable at the tree height ranging from 3.5m to 7.5m, but the extension of the variations enlarged near the top of the trees. We know that when manufacturing veneer, usually only the lower and middle parts of the log were used, so such variations of wood density would have little effect on timber utilization.

3.3.2　Within-tree vertical variations of the mechanical properties

MOR and MOE are important mechanical properties in veneer processing. The test results of these two indexes at different tree heights of different clones were listed in Table 5. The average MOR ranged from 47.7 to 65.6MPa, and the average MOE ranged from 6399 to 8723MPa. The wood hardness in cross, radial and tangent directions ranged from 2787 to 3407N, from 1718 to 2770N and from 1763 to 2668N, respectively. The general variation patterns of the mechanical properties within trees were illustrated as in Table 6. The variation trend of MOR in most of individual trees decreased with the increase of the tree height, which conformed to the established theory of wood science (Cheng, 1985). The variation trend of MOE, however, was different from that of the theory. The reasons are not only related to wood density, but also to the arrangement direction of cellulose and hemicelluloses in cell wall. The trends of COS, hardness of radial section and hardness of tangent section showed the same trend as MOR, but the hardness of cross section displayed an opposite pattern and the reasons invite further investigations in the future.

Table 6　Trends of longitudinal variation within trees from the base to the top of the trees of several mechanical indexes

Clone	MOR (MPa)	MOE (MPa)	COS (MPa)	Hardness in cross section (N)	Hardness in radial section (N)	Hardness in tangent section (N)
L1	↘	↗	↗	∧ total ↗	∨ total ↘	∨ total ↘
L2	∨ total ↘	∨ total ↘	∨ total ↘	∨ total ↗	∨ total ↘	↗
L3	↘	∨ total ↗	↗	∨ total ↘	∨ total ↘	∨ total ↘
L4	↘	∨ total ↘	↘	↘	↘	↘
L5	∨ total ↘	↘	∨ total ↘	∨ total ↘	∨ total ↘	↗
L6	↘	∧ total ↗	∧ total ↗	∨ total ↘	↘	∨ total ↘
L7	↘	∨ total ↗	↘	↗	↘	↘
L8	↘	↘	↘	∨ total ↘	↘	∨ total ↗
L9	↘	∨ total ↘	↘	↗	↗	↘
CK	↘	↘	↘	∨ total ↗	∨ total ↗	↗

Notes: The symbol '∨' referred to a variation pattern described as first decrease-then increase, '∧' referred to first increase-then decrease, ' ' referred to increase and ' ' referred to descend

3.4 Genetic control of air-dried wood density and the mechanical properties

Estimated genetic parameters such as air-dried wood density and the tested mechanical properties were listed in Table 7. All the wood characters were found to be under medium or strong genetic controls. Among the 6 indexes, MOE and hardness of tangent section were strongly genetically controlled, with the clonal repeatabilities being 0.90 and 0.80, respectively. MOR, COS and hardness in radial section were also under strong genetic controls since the clonal repeatabilities were all over 0.6. However the clonal repeatability of hardness in cross section was 0.51, indicating that the trait was under medium genetic control.

Table 7 Estimated genetic parameters for growth, air-dried wood density and the mechanical indexes

Traits	Mean	Range	Variance between clone	*R*
Tree height (m)	18.78	16.10~21.10	4.5147	0.55
DBH (cm)	19.27	13.55~22.15	11.6634	0.69
Stem volume (m^3)	0.271	0.112~0.371	0.0111	0.71
Air-dried density (g/cm^3)	0.444	0.376~0.474	0.0112	0.89
MOR (MPa)	54.8	50.3~60.4	255.3300	0.62
MOE (MPa)	8018	7453~8478	4966781	0.90
COS (MPa)	36.6	33.0~40.2	193.5540	0.70
Cross section hardness (N)	3209	2952~3393	520770.42	0.51
Radical section hardness (N)	2126	1887~2393	816076.96	0.71
Tangent section hardness (N)	2193	1937~2483	754415.90	0.80

Note: R * represents the clonal repeatability of the traits under study

3.5 Genetic correlations between air-dried wood density and the mechanical properties

Most previous studies have suggested that there exist very close relationships between wood density and other wood characters. Wood density was supposed to be the best index to evaluate the wood strength. On the other hand, since testing of other mechanical properties could be burdensome and expensive, it should be of great value if a strong relationship could be found between wood density and the important mechanical properties. The phenotypic and genetic correlations between air-dried wood density and the mechanical properties were as shown in Table 8.

Significantly positive phenotypic correlation was observed between air-dried wood density and tangent section hardness, while no significant correlation between wood density and other indexes was observed. This result generally indicated a non-significant phenotypic correlation between air-dried wood density and the investigated mechanical properties. However, from the results of genetic coefficients, great positive genetic correlation existed between air-dried wood density and MOE, cross section hardness, and radial section hardness, with the genetic correlation coefficients being 0. 6538, 0. 452 and 0. 7281 respectively. This suggests that these three mechanical indexes could be genetically manipulated by the indirect selection of air-dried wood density.

Table 8 Phenotypic and genetic correlations between air-dried wood density and mechanical properties

Correlation coefficient	Annual ring density (Dy)	MOR (MPa)	MOE (MPa)	COS (MPa)	Hardness in cross section(N)	Hardness in radial section(N)	Hardness in tangent section(N)
Annual ring density (Dy)	1	-0. 2651	0. 6538**	0. 2476	0. 452*	0. 7281**	0. 1573
MOR	0. 0467	1	0. 8286**	0. 8805**	0. 4423	0. 4898*	0. 3251
MOE	0. 3826	0. 5689**	1	0. 9941**	0. 8402**	-0. 1602	0. 4860*
COS	0. 3183	0. 6833**	0. 4612*	1	-0. 0142	0. 5101**	0. 7088**
Hardness in cross section	0. 2195	0. 3568	0. 2842	0. 1753	1	0. 9872**	0. 8451**
Hardness in radial section	0. 3734	0. 3233	0. 1054	0. 3024	0. 6088*	1	0. 9037**
Hardness in tangent section	0. 5681*	0. 1495	0. 2436	0. 3979	0. 5250*	0. 8512**	1

Notes: Genetic coefficients are listed above the diagonal, the phenotypic coefficients are under the diagonal, * indicates a significant correlation at $\alpha=0.05$ and ** indicates an extremely significant correlation at $\alpha=0.01$.

4 Conclusions and discussion

For *P. tomentosa*, studies on wood mechanical properties are much fewer than those on wood density and fiber morphology (Song et al., 2000; Pu et al., 2002; Xing et al., 2000). From previous researches (Gu et al., 1998; Wang et al., 2001; Cown et al., 1999). It is generally accepted that there exists certain degree of differences among clones in wood density and mechanical properties, but the within-tree variations of the wood properties have not been thoroughly investigated. In this paper, the within-tree air-dried wood density showed an increase with a certain degree of fluctuation in radial direction, and the variation trend was not identical along the tree height. The within-tree variation patterns of the most mechanical properties conformed to the established theories of wood science except MOE and cross section hardness. One reason of this discrepancy might be that the mechanical properties are not only greatly related to wood structure, but also to wood defects and faults, wood density, environmental temperature and humidity and so on. So when these mechanical indexes are tested, the correspondent national standards for test sample making should be strictly followed, and the tests should be carried out in the same time as possible, thus delimitating the effect of wood structure.

Air-dried wood density and mechanical properties were found to have great significant or significant inter-clonal variations, and they were under strong genetic controls. Most of studies also showed consistently the strong genetic control of wood characters. Compared with within-tree vertical variations, the inter-clonal variations have much greater significance to timber woods breeding. From the results of 9 triploid clones, there are differences in air-dried wood density and mechanical properties among the clones, and all these properties would have effect on the value of the timber. Thus in order to screen out the superior clones of *P. tomentosa*, during stock breeding and extending, wood quality factors which are mostly related with the specific breeding target should be systematically tested besides the considerations of growth rate, adaptation and resistance to diseases and insects.

Although there exist close genetic correlations between air-dried wood density and some indexes of mechanical properties, it is more difficult to directly select these mechanical indexes and it is easier to indirectly select wood density, growth yield, stem straightness and branch size, etc. Cown et al. (1999) considered the wood of *P. tomentosa* displayed better mechanical properties including higher values of COS, ultimate strength, MOE and tangent hardness, however moderate tension wood exists in the log. Because the timber with tension wood has gelatinous fibers, which cause wooly grain on the sawn wood surface, and sometimes cause serious saw pinch and heat, and even bowing, shake and shrinkage during wood seasoning, it has been an essential index to evaluate the wood quality whether the timber includes tension wood or not. Therefore it is useful to make testing samples and exact measurement of wood property indexes through selecting clone with straight stem, small sized branch, and few scars. When triploid clones of *P. tomentosa* are to be considered to serve as veneer timber, the principal morphological traits including stem straightness and branch traits should not be neglected and they should be considered together with the biomass per unit area as the preferred objectives of genetic improvement practice.

Being one of the major fast-growing tree species in China, poplar has versatile usages. It has been mainly used to manufacture plywood, chipboard and block board, while at the same time it serves as the desirable materials for fiberboard and pulping. Different assortments have their special requirements for selecting new poplar cultivars. Wood property breeding is the key to resolve the problems in industrial timber oriented breeding of poplar, and the core of the wood property breeding would be early test of wood properties. Wang et al. (1995) thoroughly studied the relationship among log peeling, veneer quality and wood properties of 7 poplar species. Their results showed that the peeling and veneer quality of the wood of 7 poplar species were not only related with wood density and wood hardness, but also were closely related with bending strength to grain—the bigger the tangent hardness value and the MOE value, the higher the veneer crack ratio. We suggest that superior logs fell in the medium and big diameter classes be firstly selected, and at the same time of yield breeding, according to the specific requirements of the veneer industrial timber, researches for highly efficient breeding technologies be then carried out. Thus the theories and methods of yield breeding, wood quality breeding and tree form quality breeding will be constantly improved and perfected, pushing forward the course of veneer timber breeding.

毛白杨无性系湿心材比例的遗传分析①

摘　要　对毛白杨无性系测定林分内25个无性系150棵单株木芯试样的湿心材比例进行测定和遗传分析。结果表明：无性系间湿心材比例差异达到5%显著水平，湿心材比例的无性系重复力高达0.749控制，在无性系间进行可获得良好效果；湿心材比例与木质素含量及与木材基本密度之间无相关性，对这3个性状可以进行单独选择；湿心材比例与胸径之间在1%水平上呈极显著的负表型相关性和显著的负遗传相关性，可联选择；在所研究的25个无性系中，可选择1232#无性系，它的湿心材比例较低、木质素含量较低、材色较白，且胸径较大。

关键词　毛白杨，无性系，湿心材比例，遗传分析

湿心材是树木生长中的一种反常现象，是树木加工利用中的一大难题(祖勃荪，2000)。Ward(1986)总结称，21种针叶材中13种有湿心材，50种阔叶材中38种有湿心材。国外对杨树湿心材的系统研究始于20世纪50年代，包括杨树湿心材的形成，湿心材形成与细菌、气体和水分的关系，湿心材的酸度，材色，湿心材的力学强度及渗透性等。结果表明：与正常材相比，由于湿心材含水率高、颜色深、抽提物多以及pH值偏酸或偏碱性，使木材物理、化学性质发生变化，导致干燥时易发生皱缩及沿年轮开裂、刨削难，呈碱性的湿心材很难用脲醛树脂胶合以及因变色导致的商品价值大幅下降等加工问题，直接影响到杨木木材的干燥、制材和胶合板的质量(Clausen et al.，1952；Kemp，1959；Haygreen et al.，1966)。

我国湿心材的研究较晚。杨树是我国制浆造纸业的主要树种，由于湿心材现象的普遍发生，致使湿心材的综纤维素含量和纤维素含量比正常材低，直接影响到出浆率以及印刷光泽度，这对制浆造纸业极为不利(安培均，1992；姜笑梅等，1993；诸葛强等，1997；晁龙军等，1996；1997；1998)。本文研究毛白杨(*Populus tomentosa*)湿心材所占比例，并对其进行遗传分析，为选育出无湿心材或湿心材比例低和木质素含量低的优良纸浆材毛白杨无性系提供科学依据。

1　材料与方法

1.1　材料

试材取自于1987年定植于山东冠县苗圃的纸浆材毛白杨无性系测定林。测定林中的无性系是北京林业大学毛白杨协作组从毛白杨分布区范围内选优得到的(朱之悌，1992)。

---

①　本文原载《林业科学》，2005，41(4)：140-144，与张冬梅、鲍甫成和黄荣凤合作发表。

株行距 4m×5m，8 次重复。

1.2 方法

选取测定林中的 25 个无性系各 6 株，用 9mm 生长锥在树木胸径处分南北向钻取木芯（弃去有疤木、应拉木及其他木材缺陷的木芯）。量取湿心材长度占整个木芯的比例，并测量取样单株的树高、胸径和材积。

截取不同无性系单株木芯上第 9 年轮的木材试样，采用最大含水量方法对木材试样的基本密度进行测定。将每个无性系 6 个单株的的木芯混合（包括湿心材和非湿心材部分）、干燥、粉碎、制成木粉，测定不同无性系间木质素含量。木质素含量测定方法依照 GB 2677.8—1981 进行。

木材白度利用 MSC-P 多光源分光测色仪测定，属于甘茨白度值。每个无性系 6 个单株的试样各测 6 个点，然后取平均值。计算公式采用我国用于评定纺织品白度方法（GB 8425—1987）中规定用的 D_{65} 光源、196410° 标准观察者数据，公式为：$W_{10} = Y_{10} + 800(0.3138 - x_{10}) + 1700(0.3310 - y_{10})$。

2 结果、分析与讨论

2.1 毛白杨无性系湿心材的遗传变异

改良木材的一条有效途径就是培育出具有所希望木材性质的树木品系。大多数木材的性质具有中等至较高的遗传性，因此通过遗传手段可以使其向我们需要的方向变化，木材的湿心材也不例外。

杨树湿心材的遗传变异较为复杂，从 150 个木芯试样的湿心材形态看，无性系间以及无性系内单株间的湿心材比例表现不一。从表 1 可清楚地看出，毛白杨无性系间湿心材百分比存在着明显的差异。湿心材比例较高的无性系有 40#、1210#、1808#、11#、3332#和 1005#无性系，它们的湿心材比例分别达 32.9%、27.3%、26.8%、26.1%、25.6% 和 24.1%；而有的较低，如：41#、27#、3-2-1#、60# 和 5101#无性系的湿心材比例分别为 5.5%、7.5%、8.2%、8.8%和 9.8%。无性系内单株间湿心材比例的变异幅度和变异系数也表现出很大的差异，如 34#、3-1-27#、5101#、60#、3-23-32#、27#无性系内单株间变异幅度较大，变异系数分别为 122.68%、119.69%、100.64%、110.39%、90.33%、和 89.40%；而 1808#、11#、1232#和 66#无性系的湿心材比例在单株间的变异较小，变异系数分别为 12.52%、18.01%、20.95%和 29.55%。

对湿心材比例进行方差分析见表 2。无性系南北木芯方向间的湿心材比例差异不显著，无性系间湿心材比例差异达到 5%显著水平，无性系内湿心材比例在 5%水平上差异不显著。变异是选择的基础，毛白杨无性系间湿心材比例差异的显著性为选择性育种奠定了基础。对毛白杨无性系木材湿心材比例的遗传参数估计结果（表 2）表明，无性系间湿心材比例的无性系重复力高达 0.749，受中度遗传控制；无性系内的湿心材比例个体重复力为 0.323，受弱度遗传控制。从而可推测不同杨树无性系的湿心材产生主要是受遗传控制的，有望通过无性系选择获得湿心材比例低的无性系。美洲黑杨（*Populus deltoides*）新无性系的湿心材遗传力高达0.6891，表明美洲黑杨湿心材性状变异也主要由无性系和基因型控制的（诸葛强等，1997）。有研究报道湿心材的形成分2种：原发性湿心材和继发性湿心材。原发性湿心材是指细菌在幼苗繁殖时已经潜伏于苗木中，随着树木的生长，细菌开始活动，逐渐形成湿心材。继发性湿心材则是由于苗木起初没有细菌，在生长过程中因受到生物或非生物的伤害，细菌由伤口进入树木形成层等诱发湿心材发生（祖勃荪，2000）。从本研究结果看，毛白杨木材无性系间的湿心材含量受中度的遗传控制，说明该性状与杨树本身固有的生理因素及结构有关，受树木生长后期的环境因子的影响相对较小。推测毛白杨木材的湿心材形成属于原发性湿心材，可能是由寄生菌造成的病原后果，原因也是毛白杨本身的生理特征及结构适合于细菌潜伏于繁殖材料中，随着树木的生长，细菌开始活动，逐渐形成湿心材。对于毛白杨木材的湿心材的形成还需对病菌进行分离、纯化，导入健康植株等进一步的研究。

表1　毛白杨无性系胸径、木质素含量湿心材比例的平均值、变幅和变异系数

无性系	单株	湿心材比例			胸径	木质量含量
		变幅	平均	变异系数(%)	(cm)	(%)
41	6	0.016~0.087	0.055	42.33	19.43	20.61
27	6	0.012~0.177	0.075	89.4	17.98	19.97
3-2-1	6	0.018~0.165	0.082	75.27	20.58	22.28
60	6	0.005~0.235	0.088	110.39	18.968	21.51
5101	6	0.014~0.190	0.098	100.64	16.97	22.05
22	6	0.060~0.160	0.100	51.47	15.92	18.69
5009	6	0.016~0.224	0.109	71.21	15.68	18.65
34	6	0.024~0.317	0.116	122.68	17.46	22.42
53	6	0.055~0.230	0.122	61.33	17.62	22.42
1232	6	0.011~0.285	0.125	20.95	18.23	19.68
5003	6	0.015~0.246	0.129	74.65	14.72	20.6
3-23-32	6	0.040~0.293	0.142	90.33	16.92	19.89
67	6	0.055~0.239	0.147	61.60	16.63	20.67
3-1-27	6	0.110~0.183	0.153	119.69	19.62	19.41
4337	6	0.018~0.302	0.172	70.14	17.1	20.63
66	6	0.033~0.283	0.197	29.55	19.53	19.66
1807	6	0.075~0.365	0.223	51.97	18.48	19.63
3-48-2	6	0.06~0.384	0.229	44.62	16.52	21.91
1272	6	0.069~0.341	0.233	38.21	17.42	20.39
1005	6	0.011~0.380	0.241	50.19	16.03	22.10
3332	6	0.010~0.360	0.256	34.24	14.23	21.42
11	6	0.216~0.336	0.261	18.01	18.3	20.74
1808	6	0.221~0.319	0.268	12.52	20.82	20.65

（续）

无性系	单株	湿心材比例			胸径	木质量含量
		变幅	平均	变异系数(%)	(cm)	(%)
1210	6	0.190~0.360	0.273	30.66	15.43	23.28
40	6	0.166~0.431	0.329	39.7	18.5	19.52
平均			0.169		17.56	20.75

表 2 毛白杨无性系湿心材百分比方差分析及遗传参数估计*

变异来源	自由度	平方和	均方	*F* 值	重复力	*F* 值
无性系间	24	1.6971	0.0707	4.4855	0.749	1.5610
南北向木芯间	1	0.0041	0.0041	0.2601		3.8789
交互作用	24	0.1209	0.0050	0.3194		1.5610
无性系内	250	3.9411	0.0158		0.323	
总和	299	5.7632				

* 差异显著水平 0.05。

2.2 毛白杨无性系湿心材比例与木材白度的关系

木材白度即木材本身的光散射和反射性能，又称木材色度。木材本身的颜色是影响漂白后纸浆白度的重要因子。研究发现，纸浆白度与所用木材白度有良好的相关性，而且纸浆可漂性与木材初始白度亦有一定关系(房桂干等，1995)。姚光裕(1997)认为高白度杨木化学浆是生产高级纸张的一种重要配比浆原料。通过对毛白杨无性系木芯试样颜色特征(人为分为白和偏红两类)的观察和白度的测定发现，25 个无性系中 9 个无性系的木芯试样呈现较白的外部特征(表 3)，其中 8 个无性系的白度测定值大于平均数，即 3-48-2#、3-2-1#、4337#、41#、22#、5101#、66#和 1232#无性系。而这些无性系中湿心材比例低于平均水平的有 41#、3-2-1#、5101#、22#和 1232#无性系。显然，在湿心材比例和木材白度这 2 个性状上可选择这些无性系。然而，选择时应结合每个无性系湿心材比例的变异系数(表 1)综合分析，在这几个无性系中 1232# 的单株变异系数较小为 20.95%，说明该无性系在湿心材比例这个性状上受外界因素的影响较小，容易遗传控制，显然，1232# 是材质较白湿心材比例较低且能稳定遗传的无性系。41#无性系的变异系数为 42.33%，也可以考虑。尽管 3-2-1#、22#和 5101#无性系的湿心材比例较低，但这些无性系单株的湿心材比例变异较大，受遗传外的因素影响大些，显然不是最佳选择。

2.3 毛白杨无性系湿心材比例与木质素含量的关系

根据文献记载，湿心材的形成与细菌有关，由于细菌的活动使湿心材的构造、化学、物理力学性质发生变化(Schink et al.，1981；Rink et al.，1987；1989；Yanig et al.，1989；Wilkins et al.，1990)。也有报导认为湿心材的形成和树木细胞产生的多种酶如多酚氧化酶、过氧化物酶、纤维素酶、淀粉酶、苯丙氨酸酶等生理活动相联系(Loewas，1976；Yanchuk et al.，1986)。胡景江等(1999)研究发现，木质素是植物体内的一种重要的物理抗菌物质，它能与羟脯氨酸糖蛋白(HRGP)一起作为结构屏障物，起强固细胞壁的作用，认为HRGP和木质素与杨树对溃疡病的抗性有关。本文将25个无性系的每个无性系的6个单株湿心材比例的平均数和相对应的6个单株的无性系木质素含量平均数进行了相关分析，结果表明，二者相关系数为0.1991($F_{0.05}$= 0.381)，未达到相关显著水平。显然，对湿心材的比例和木质素含量这2个性状可以独立进行选择。在木质素含量性状上，低于25个无性系总平均水平(20.75)的有16个无性系(表1)，可供选择。结合木芯试样白度测定值(表3)，16个无性系中4337#、41#、22#、5101#和1232#无性系的白度测定值高于平均数，因此，欲选择木质素含量低、色较白的树种时可选择这些无性系。

本次试验的试材是木芯，而木质素分析需要的木粉量较大，所以木质素含量只有利用每个无性系6个单株的木芯混合物测定，从而造成湿心材比例和木质素含量的相关分析是无性系的单株平均数间的结果。毛白杨树木个体内，正常材木质素含量和湿心材木质素含量的变异情况以及湿心材比例与湿心材处木质素含量的关系，还需进一步探讨。

表3　毛白杨无性系木芯试样形态特征与白度

无性系	木芯外观	白度值	无性系	木芯外观	白度值
3-48-2	白	30.93	1210	偏红	19.69
3-2-1	白	27.98	40	偏红	19.69
4337	白	27.81	1807	偏红	19.19
41	白	25.69	3332	偏红	18.55
22	白	23.66	60	白	18.17
5101	白	23.42	5009	偏红	17.34
66	白	23.65	5003	偏红	17.34
1232	白	22.00	53	偏红	17.24
27	偏红	21.36	1272	偏红	17.19
3-23-32	偏红	20.17	1005	偏红	16.76
67	偏红	20.14	34	偏红	13.68
1808	偏红	19.34	3-1-27	偏红	10.15
11	偏红	19.72	平均		20.46

2.4　毛白杨无性系湿心材比例与密度及木材生长性状关系

为了进一步揭示毛白杨无性系湿心材比例与木材密度及生长性状的关系，将毛白杨无性系测定林中的25个无性系的150单株的基本密度(p)、胸径(D)、树高(H)和材积(V)指标，与无性系湿心材比例之间做简单相关分析，并作散点图(图1A~D)。图中横坐标为各无性系的湿心材比例(W)，纵坐标分别为无性系的基本密度、胸径、树高及材积。毛白杨无性系湿心材比例分别与木材的基本密度、胸径、树高及材积呈负相关关系，表型相关系数分别为-0.0199、-0.3743**($F_{0.01}$=0.208)、-0.1558和-0.1202，其中湿心材比例与胸径在1%的水平上呈现显著的负表型相关性。

生物性状的变异来源于2个方面，一是遗传因素决定的，称遗传变异；二是受外界条件影响而产生的，称环境变异。遗传因素和环境因素的联合作用决定了生物性状的表型变异。对林木育种来说，性状间的遗传相关对遗传改良意义更大。通过对毛白杨无性系湿心材比例和木材的密度、胸径、树高及材积的遗传相关分析，遗传相关系数分别为：0.0000、-0.4258**（$F_{0.01}=0.208$）、-0.1234 和-0.1602*（$F_{0.05}=0.159$）。结果表明湿心材比例分别与木材基本密度以及树高之间的遗传相关性都很微弱，说明湿心材比例和木材基本密度、树高分别受相互独立的遗传基因控制，它们之间可以独立选择；湿心材比例与材积、与木材胸径之间分别在5%和1%的水平上呈显著的负遗传相关。推测毛白杨无性系湿心材的形成有可能与生长过程中胸径、材积的增加受到连锁基因的控制作用。

有研究认为有湿心材的树可以继续活几十年，对木材强度也不造成严重损害(祖勃荪，2000)。本研究结果显示湿心材比例与木材胸径以及湿心材比例与材积的遗传负相关达到显著水平，所以在湿心材比例低于平均水平且胸径较大的41#、27#、3-2-1#、60#、53#、1232#无性系中，可选择湿心材比例变异系数较小的41#和1232#无性系。

3 结论

毛白杨无性系间湿心材比例差异达到5%显著水平，无性系重复力高达0.749，湿心材的产生主要是受遗传控制，无性系间选择有效，有望获得无湿心材或湿心材比例较低的优良毛白杨无性系。毛白杨湿心材比例与木质素含量及与木材基本密度之间的相关系数分别为0.1991 和-0.0199，相关不显著，这3个性状可以独立进行选择；湿心材比例与胸径在1%水平上达到极显著负表型相关和极显著负遗传相关，可联合选择。在所研究的25个毛白杨无性系中，可选择1232#无性系，它的湿心材比例较低、木质素含量较低、材色较白，且胸径较大。

Genetic Analysis of Wetwood Proportion on Clone Test Stand of *Populus tomentosa*

Abstract 25 clones 6 trees from each clone were sampled (aged 15 years) from clone test stand of *Populus tomentosa*. Wetwood proportion of every tree was calculated, their inheritance and variation among clones were studied. The high value of clones repeatability of wetwood proportion (0.749) indicated that the character was under genetic control, and selection could be conducted effectively between clones. Whereas correlation analysis showed that no significantly correlation between wetwood proportion and lignin content and height, general significantly negative correlation between wetwood proportion and growth traits (diameter at breast height and volume) were discovered.

Keywords *Populus tomentosa*, clones, wetwood proportion, inheritance analysis

QTL Analysis of Growth and Wood Chemical Content Traits in an Interspecific Backcross Family of White Poplar (*Populus tomentosa* × *P. bolleana*) ×*P. tomentosa* *

Abstract The genetic control of tree growth and wood chemical traits was studied using interspecific backcross progenies between clone "TB01" (*Populus tomentosa* ×*P. bolleana*) and clone "LM50" (*P. tomentosa*). A total of 247 and 146 AFLP markers from genetic maps previously constructed in backcross parents LM50 and TB01 were used for the QTL analyses. These markers were distributed among 19 linkage groups and covered 3265cm and 1992cm in the backcross parents, respectively. A total of 32 putative QTLs, associated with 5 growth and chemical traits including sylleptic branch number, sylleptic branch angle, stem volume, wood cellulose content and wood lignin content, was detected. These QTLs were dispersed among 16 linkage groups in the parent LM50 and 10 groups in the parent TB01. The phenotypic variance explained by each QTL ranged from 7.0% to 14.6%. QTLs controlling sylleptic branch number and stem volume were co-localized in two linkage groups TLG6 and TLG8, respectively. The favorable alleles were mostly from *P. tomentosa*, which is phenotypically superior over *P. bolleana* for sylleptic branch angle, stem volume and wood chemical traits. The favorable alleles for the sylleptic branch number was from *P. bolleana*. These AFLP markers that were associated with the QTLs have potential use in *P. tomentosa* breeding program.

Keywords *Populus tomentosa* Carr., QTL mapping, tree growth, wood chemical traits

1 Introduction

The Chinese white poplar (*Populus tomentosa* Carr.), indigenous to north China, is one of key commercial tree species for timber production and plays a significant role in ecological and environmental protection along the Yellow River (Zhu and Zhang, 1997). Over the past several decades, several breeding methods such as interspecific hybridization, creation of polyploidy, and clonal selection have been successfully used for genetic improvement of *P. tomentosa* (Zhu and Zhang, 1997). Many technical issues relating to improvement of propagation, wood production, and wood quality have not been adequately resolved due to the lack of knowledge on the genetic mechanism and architecture controlling these commercial important traits. Numerous studies have indicated that these economic traits were controlled by many quantitative trait loci (QTLs) (Brad-

---

* 本文原载《加拿大林业研究》(*Canadian Journal of Forest Research*), 2006, 36: 2015~2020, 与张德强和杨凯合作发表。

shaw and Stettler, 1995; Grattapaglia et al. , 1996; Marques et al. , 1999). The QTLs that are responsible for the variation of such traits could potentially be identified and mapped in a segregating population using molecular markers. Knowledge of the number, genomic region, and individual effects of genes controlling these traits can be used to enhance understanding of genetic architecture of these traits and assist in breeding program in forest trees.

During the past decade, the application of molecular markers to study genetic control of quantitative traits in forest trees has received extensive attention (Bradshaw and Stettler, 1995; Groover et al. , 1994; Wheeler et al. , 2005). Notable examples include QTL studies for tree growth, wood quality and vegetative propagation traits in *Eucalyptus* (Grattapaglia et al. , 1996; Byrne et al. , 1997; Verhaegen et al. , 1997; Marques et al. , 1999), and growth, adaptation, wood specific gravity, and wood physical and chemical properties in *Pinus* (Emeberi et al. , 1998; Hurme et al. , 2000; Kaya et al. , 1999; Sewell et al. , 2002). Using F_2 population of *P. trichocarpa* and *P. deltoides*, QTLs for growth, form, phenology and disease resistance, have been detected in poplars (Bradshaw and Stettler, 1995; Frewen et al. , 2000; Wu et al. , 1997; Villar et al. , 1996). However, QTLs controlling growth and wood chemical content traits have not been reported for *P. tomentosa* to date. In a previous study, parent-specific genetic maps using AFLP markers were constructed for two poplar parents (clones): parent LM50 (*P. tomentosa*) and parent TB01 (a hybrid progeny between *P. tomentosa* and *P. bolleana*) based on their 120 progenies (Zhang et al. , 2004). In this study, the AFLP markers from these genetic maps were examined for their linkage to QTLs that affect stem growth and wood chemical content traits. QTLs associated with five economically important traits: sylleptic branch number, sylleptic branch angle, stem volume, wood cellulose content and wood lignin content were identified in this study.

2 Materials and methods

2.1 Mapping population

A total of 120 progeny, a subset chosen randomly from an interspecific backcross population of 696 progeny was used in this research. These progenies were derived from a cross between two elite poplar parents TB01and LM50. The mapping pedigree was planted using a Randomized Complete Block Design with three replications, as previously described by Zhang et al. (2003).

2.2 Construction of linkage maps

Separate genetic linkage maps of the parents were previously constructed using AFLP markers and a pseudo-test-cross mapping strategy (Zhang et al. , 2004). The framework map for the parent LM50 consisted of 247 markers, and had 19 major linkage groups. The total map distance was about 3265cm with an average distance of 13. 2cm between adjacent markers. For the parent TB01, a total of 146 markers were mapped into similar 19 linkage groups, covered approximately 1992cm of the genome, with an average marker distance of 13. 6 cM. These parent-specific maps were used as a framework for QTL analysis for stem volume growth and wood chemical property traits.

2.3 Phenotypic trait assessment

2.3.1 Growth traits

Three growth traits: sylleptic branch numbers (SBN), sylleptic branch angle (SBA) and stem volume (SV) were measured using the materials of annual stem with biennial root. As for SBN, we counted the numbers of sylleptic branches on the tree stem. The angle of sylleptic branch (unit degree) in the main stem was measured by a protractor. Stem volume index (m^3) was computed as $\pi\times$ (basal diameter/2)$^2\times$height according to Bradshaw et al. (1995).

2.3.2 Wood chemical traits

Two wood chemical traits were measured. Nitric-ethanol wood cellulose content (WCC) was determined according to Borchardt and Piper (1970). Wood lignin content (WLC) was measured using the Klason lignin method (Effland, 1977), which measures residual lignin following hydrolysis of the sample tissue using concentrated sulfuric acid.

2.4 QTL analysis

QTL analysis was performed using a backcross model, based on the individual linkage maps for parents TB01 and LM50, respectively. A LOD score above 2.0 was used as the criterion for the presence of a linked QTL in a given chromosome interval. Additive genetic effects attributed to individual QTLs, and the percentage of phenotypic variation explained by each QTL, were estimated using MAPMAKER/QTL (Lander and Botstein, 1989).

3 Results

3.1 Phenotypic variation and correlations among growth and wood chemical traits

All five measured traits showed significant differences between the two original parent lines *P. tomentosa* and *P. bolleana* (data not shown). The F_1 hybrid TB01 (*P. tomentosa* and *P. bolleana*) showed intermediate values between the two parent means for SBN, SBA, and stem volume traits. The F_1 was higher than both parents in WCC, indicating positive heterosis. In contrast, the F_1 had lower values than their parents for WLC. A segregating population was obtained when the F_1 hybrid TB01 was backcrossed with the *P. tomentosa* parent LM50, and showed considerable variation in the five observed traits. The mean, minimum and maximum along with other statistics are shown for the five traits in Table 1. As expected, the frequency distribution for each trait with annual stem with biennial root measured approximately followed a normal distribution (Fig 1). The between-traits phenotypic correlations were also computed. SBN was positively correlated with SBA and stem volume. Among the wood chemical traits, a significant negative correlation was observed between WCC and WLC. Several significant positive correlations were also observed between growth and wood chemical traits, for example positive correlations between stem volume and WCC, and between SBN and WLC (Table 2).

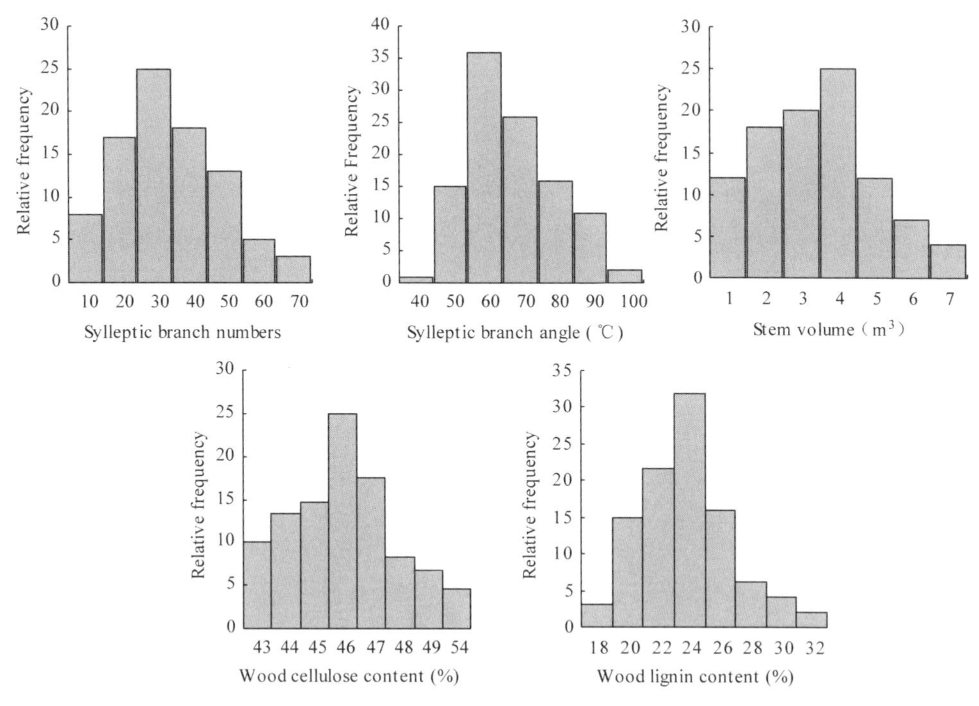

Fig. 1 Phenotypic frequency distributions for five growth and wood chemical contents traits in the BC_1 progeny.

Table 1 Statistics of five growth and wood chemical content traits in the mapping population.

Trait*	μ	Minimum	Maximum	σ	CV
SBN	27.73	2	84	17.86	0.64
SBA (℃)	61.04	36	81	9.56	0.16
SV (m^3)	2.53	0.19	6.98	1.59	0.63
WCC(%)	47.32	41.37	53.36	2.55	0.05
WLC(%)	24.29	18.95	31.93	3.36	0.14

Note: For a description of traits see materials and methods.

Table 2 Phenotypic correlations among the five traits in the interspecific backcross population.

Traits[1]	SBN	SBA	SV	WCC
SBN				
SBA (℃)	0.211 *			
SV(m^3)	0.248 * *	0.106 *		
WCC (%)	ns	ns	0.216 *	
WLC (%)	0.125 *	ns	ns	-0.236 *

Note: For a description of the traits see materials and methods. Only significant correlations are shown (* , $P< 0.05$ and * * , $P< 0.01$; ns, not significant)

Table 3 Putative QTLs detected for stem volume and wood chemical content traits by interval mapping via MapMaker/QTL in an interspecific backcross population of 120 progeny from clone "TB01" and clone "LM50" in poplar

Traits * /QTL	Descendance of QTL	Linkage group	Interval	Length ** (cM)	Position *** (cM)	LOD	Variance § (%)	Additive effect
SBA								
SBATL2	Clone" LM50"	TLG2	E60M34268-E63M39436	16. 0	2. 0	2. 74	11. 5	-6. 05
SBATL5	Clone" LM50"	TLG5	E35M52350r-E44M6097r	16. 7	16. 2	2. 0	7. 6	-4. 06
SBATL7	Clone" LM50"	TLG7	E63M32396-E44M60413	11. 2	3. 9	2. 02	7. 8	+5. 08
SBATL10	Clone" LM50"	TLG10	E44M40100r-E34M47391r	12. 1	12. 0	3. 1	12. 8	+6. 31
SBATL16	Clone" LM50"	TLG16	E61M41145-E61M41144	4. 7	1. 3	2. 0	7. 4	+4. 37
SBATBL14	Clone" TB01"	TBLG14	E33M48277-E65M3181r	12. 5	6. 3	2. 12	10. 6	-5. 82
Total							57. 7	
SBN								
SBNTL3	Clone" LM50"	TLG3	E61M4152r-E63M33357	11. 3	0. 1	2. 02	7. 3	+8. 76
SBNTL6	Clone" LM50"	TLG6	E33M62439-E61M3192	30. 1	0. 1	2. 05	7. 6	+9. 43
SBNTL8	Clone" LM50"	TLG8	E60M33270r-E33M44290r	12. 2	0. 1	2. 0	7. 2	+8. 06
SBNTL9	Clone" LM50"	TLG9	E63M36218-E63M39248	16. 9	15. 2	2. 15	9. 2	+9. 87
SBNTL14	Clone" LM50"	TLG14	E35M52154-E61M3170	19. 1	0. 2	2. 0	7. 3	+8. 76
SBNTL19	Clone" LM50"	TLG19	E60M33227r-E33M40307	20. 3	6. 0	2. 03	7. 5	+8. 89
SBNTBL1	Clone" TB01"	TBLG1	E33M44298-E34M44490r	11. 1	5. 4	2. 1	8. 6	-9. 55
SBNTBL17	Clone" TB01"	TBLG17	E63M52396-E60M3267	22. 0	2. 9	2. 0	7. 0	+8. 16
Total							61. 7	
SV								
SVTL1	Clone" LM50"	TLG1	E33M32273-E33M32258r	6. 2	3. 5	2. 0	8. 0	+0. 936
SVTL6	Clone" LM50"	TLG6	E33M62439-E61M3192	30. 1	0. 2	2. 16	11. 0	+0. 940
SVTL8	Clone" LM50"	TLG8	E60M33270r-E33M44290r	12. 2	1. 4	2. 04	8. 2	+1. 198
SVTBLG2	Clone" TB01"	TBLG2	E44M46517-E3342134r	33. 0	7. 2	2. 08	8. 6	+0. 950
Total							35. 8	

Continued

Traits * /QTL	Descendance of QTL	Linkage group	Interval	Length * * (cM)	Position * * * (cM)	LOD	Variance § (%)	Additive effect
WCC								
WCCTL1	Clone"LM50"	TLG1	E61M31154-E61M31157	8.0	0.7	2.05	9.0	+1.48
WCCTL5	Clone"LM50"	TLG5	E35M52350r-E44M6097r	16.7	6.8	2.15	9.8	+1.60
WCCTL12	Clone"LM50"	TLG12	E34M48492r-E44M44664	25.3	0.4	2.63	10.1	-1.63
WCCTL13	Clone"LM50"	TLG13	E63M36140r-E3338126r	36.7	12.6	2.02	8.7	-1.44
WCCTL18	Clone"LM50"	TLG18	E63M3396-E65M3474	17.1	9.1	2.92	12.9	+1.83
WCCTBL9	Clone"TB01"	TBLG9	E33M79222-E63M36283	28.0	4.0	2.10	9.4	-1.56
WCCTBL11	Clone"TB01"	TBLG11	E33M79660-E44M40293	40.6	9.2	3.20	14.6	+2.41
WCCTBL14	Clone"TB01"	TBLG14	E35M50220-E60M34338	26.7	23.1	2.08	9.2	+1.52
Total							83.7	
WLC								
WLCTL2	Clone"LM50"	TLG2	E63M39436-E35M52185	16.0	0.4	2.06	9.2	+1.52
WLCTL15	Clone"LM50"	TLG15	E33M32134-E63M6198	24.9	1.7	2.20	10.0	-1.64
WLCTBL3	Clone"TB01"	TBLG3	E63M3672-E61M31125	13.9	0.3	2.04	8.8	+1.46
WLCTBL5	Clone"TB01"	TBLG5	E33M82207-E44M60644	12.4	12.3	2.00	8.5	-1.60
WLCTBL6	Clone"TB01"	TBLG6	E33M40134-E65M34504	14.3	1.2	2.18	9.8	+1.44
WLCTBL8	Clone"TB01"	TBLG8	E33M4266-E33M41151	35.3	17.7	2.02	8.6	+1.42
Total							54.9	

* See Materials and methods for the description of the traits.

* * Interval between the two flanking markers (cM) where a QTL is located

* * * QTL position from the first marker (cM)

§ Phenotypic variation explained by each QTL

3.2 QTL analysis

Thirty-two putative QTLs were found associated with the five growth and wood chemical traits using the two parental genetic maps. The number of QTLs for each trait ranged from four to eight. The putative QTLs were distributed among 16 linkage groups of parent LM50 and 10 linkage groups of parent TB01, respectively. The proportion of phenotypic variation explained by each of the QTLs varied from 7.0% (e.g. TBLG17) to 14.6% (e.g. TBLG11) (Table 3). A detailed summary of the detected QTLs is shown in Table 3.

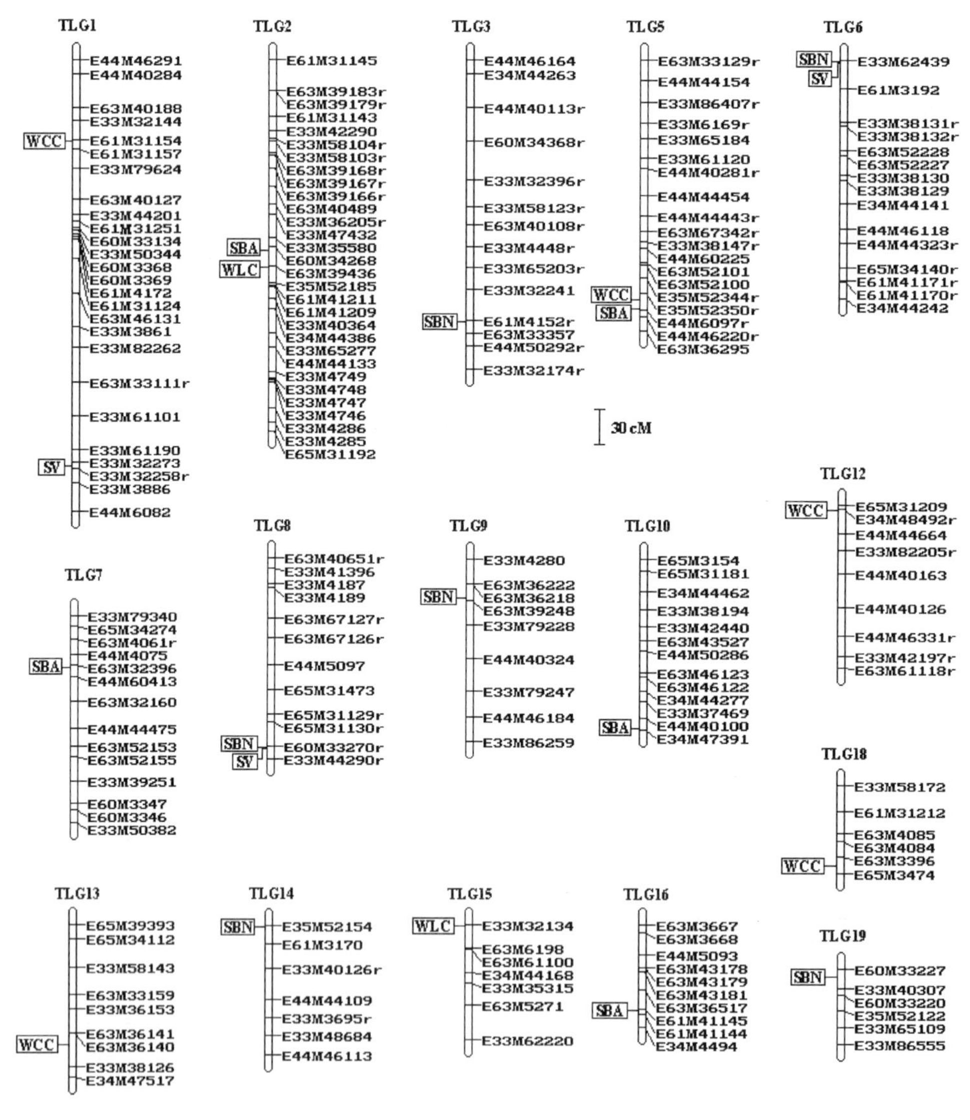

Fig. 2 Approximate locations of QTLs for growth and wood chemical contents in the genetic map of clone "LM50" (*P. tomentosa*).

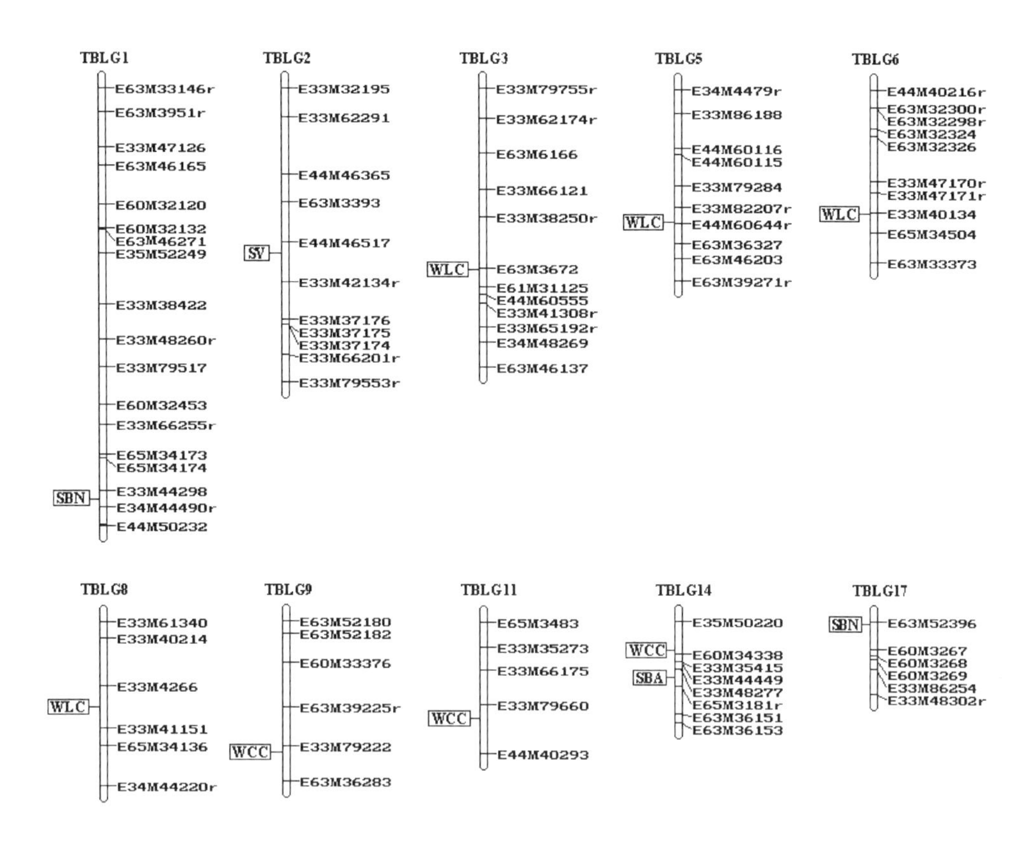

Fig. 3 Approximate locations of QTLs for growth and wood chemical contents in the genetic map of clone "TB01" (*P. tomentosa* and *P. bolleana*).

3.2.1 Sylleptic branch angle (SBA)

For SBA, the six QTLs were detected and distributed in six linkage groups (e.g. TLG2, TLG5, TLG7, TLG10, TLG16 and TBLG14, Table 3, Fig. 2 and 3). Three of the QTLs had a positive impact on phenotypic values and another 3 had negative effects. The percentage of phenotypic variance explained by each QTL ranged from 7.4% (e.g. TLG16) to 12.8% (e.g. TLG10). By addition, total phenotypic variation explained by each of six putative QTL was 57.7%. The largest effect was associated with maker SBATL10, which had additive effects of 6.31 and accounted for 12.8% of the phenotypic variance.

3.2.2 Sylleptic branch number (SBN)

QTLs were detected in eight linkage groups for SBN (Table 3, Fig. 2 and 3). Seven QTLs from both parents increased phenotypic value. The additive effects of alleles associated with markers SBNTL3, SBNTL6, SBNTL8, SBNTL9, SBNTL14, SBNTL19 and SBNTBL17 were estimated as 8.76, 9.43, 8.06, 9.87, 8.76, 8.89 and 8.16, respectively. The proportion of the total variance explained by a single QTL varied between 7.0% (e.g. TBLG17) and 9.2% (e.g. TLG9). The total phenotypic variation explained by these seven putative QTLs was added 61.7%

without considering co-linearity.

3.2.3　Stem volume

Four putative QTLs were associated with stem volume (Table 3, Fig. 2 and 3). The percentage of variation explained by each individual QTL ranged from 8.0% (e.g. TLG1) to 11.0% (e.g. TLG6). QTLs linked with these four markers increased stem volume from 0.936 to 1.198m^3. One QTL linked to marker SVTL6 on TLG 6 had the largest effect on stem volume, accounting for 11.0% of the phenotypic variance. These four QTLs successfully explained 35.8% of the total phenotypic variation without considering co-linearity. It was also observed that QTLs for SBN and stem volume were co-localized in two linkage groups TLG6 and TLG8, respectively.

3.2.4　Nitric-ethanol cellulose content (WCC)

Eight QTLs were observed associated with WCC and distributed on linkage group TLG1, TLG5, TLG12, TLG13, TLG18, TBLG9, TBLG11 and TBLG14, respectively (Table 3). Of the eight QTL detected, five were located on the genetic map of LM50 and the other 3 were mapped on the linkage map of TB01. These putative QTLs accounted for between 8.7% (e.g. TLG13) and 14.6% (e.g. TBLG11) of phenotypic variation and accounted for 83.7% of the total variation by addition. One of these QTLs was located on the same region of TLG5 (interval of E35M52350r–E44M6097r) where a QTL for SBN was found. The marker CCCTBL11 on linkage group TBLG11 had the largest additive effect on WCC and the allele in parent TB01 accounted for 2.41% increase in WCC.

3.2.5　Klason lignin content (WLC)

For WLC, six putative QTLs were identified (Table 3), on linkage groups TLG2 and TLG15 in LM50 (Fig. 2), and on linkage groups TBLG3, TBLG6, TBLG8 and TBLG5 in TB01 (Fig. 3), respectively. These six QTLs explained 54.9% of the phenotypic variation in total. QTLs for WLC accounted for variation from 8.5% (e.g. TBLG5) to 10.0% (e.g. TLG15) which were lower relative to QTLs associated with WCC. More QTLs were detected in the maternal parent TB01 than in the paternal parent LM50.

4　Discussion

4.1　Phenotypic variation of growth and wood chemical traits

Significant differences in the five growth and chemical traits were observed between parental clones *P. tomentosa* and *P. bolleana* (Zhang et al., 2003). The *P. tomentosa* parent contributed almost all the favorable alleles to the phenotypic values of four of the five traits studied (SBN is an exemption) in this interspecific backcross population. Significant variation and a normal distribution for the five traits suggest that this population is suitable for QTL analysis with interval mapping strategy. The transgressive segregation was observed in both directions for all traits, indicating that neither of the parents carried all positive or negative alleles, as commonly found for quantitative genetic traits in other backcross populations (Crouzillat et al., 1996; Butruille et al., 1999; Xiao et al. , 1998; Hurme et al. , 2000).

4.2 QTL identification in interspecific BC1 population

In *P. tomentosa*, progress in breeding for wood growth and chemical properties was impacted by inbreeding depression and a long generation interval, among others. In order to elucidate the genetic control of growth and wood chemical traits, backcrossing was used to generate a segregating population suitable for QTL mapping. With the construction of genetic maps using molecular markers, the genetic architecture of these traits could be studied. We identified 32 putative QTLs, related to growth and wood chemical traits in this study. The number of QTLs identified for each trait ranged from 4 to 8, with each QTL accounting for intermediate proportion of the phenotypic variation (e. g. 7.0%~14.6%). Major QTLs have been detected using F_2 progeny in *P. trichocarpa* and *P. deltoides*, with RFLP and RAPD markers, and a single QTL accounted for 24.4% and 33.4% of the phenotypic variation for a growth and development trait, respectively (Bradshaw and Stettler, 1995). QTLs in our study accounted for less phenotypic variation. For example, the QTL with largest effect only accounted for 14.6% phenotypic variation. The mapping population, the type and density of markers used in the genetic maps, and the QTL mapping software employed may explain the difference. The efficiency of QTL detection obtained from a backcross is usually less than from an F_2 population since only one, rather than two, recombinant gametes are sampled in plants. The limited population sizes used in this study may impact on estimation of QTL number, effects, and position (Utz et al. , 2000; Allison et al. , 2002; Schön et al. , 2004; Vales et al. , 2005). The small size of the mapping population may decrease the sensitivity of detecting QTLs so that only QTLs with large effect may be detected. When there are many QTLs associated with a trait, the individual effect of a given QTL may be small, so the power of QTL detection with a small population size would be low. Therefore, the QTL number may be underestimated and the QTL effects may be overestimated. Secondly, incorrect estimation of recombination frequency can occur when using dominant markers, and could lead to a lower LOD score (Staub and Serquen, 1996). Analysis with dominant markers can underestimate the magnitude of variation that otherwise could be fully accounted for by segregating multiple alleles in the offspring of out-cross species (Staub and Serquen, 1996). Furthermore, our medium-density parental maps still have gaps and might not cover the entire genome of *Populus*. As a result, some QTLs of large effect might have not been detected. Further investigation is needed in order to fully make use of genetic maps in QTL analysis.

The QTLs detected in this study have potential use in poplar breeding program. During the past two decades, QTLs mapping has been used as the key tool for identifying the genetic basis underlying quantitative traits and has contributed greatly to our understanding of complex trait architecture, such as number, magnitude of effect, and mode of action of QTLs (Sewell and Neale, 2000). If individuals with favorable alleles at these loci can be identified, this will increase efficiency of selection for breeding material. In this study, several QTLs with favorable alleles controlling stem volume and wood cellulose content have been detected. Among which, SVTL6 and WC-CTL1 have additive effect of 0.940m^3 for stem volume and 1.48% for cellulose content, respectively. These two QTLs have small distances with markers E33M62439 and E61M31154(less than 1.0cm). We may select desirable progeny with favorable QTL alleles linked with alleles of these two markers for backcrossing with parental *P. tomentosa* and incorporate these two favourable QTL alleles into a single line in BC_2 pedigree if there is no recombination between marker and QTL alleles. However, we need to validate identified QTLs in different environments and ages before incorporating them into breeding program.

Paterson et al. (1991) suggested that the studies conducted in a single environment are likely to underestimate the number of QTLs which can influence a certain trait. The present study identified several QTLs for each trait, which together could account for a majority (*PVP*>50.0%) of the total phenotypic variation, with the one exception being the stem volume trait. Accordingly, this is a first step in exploiting AFLPs in a *P. tomentosa* genome mapping program, and in the long-term for both MAS and positional cloning of genes controlling high-value economic traits.

4.3 Among-trait correlation and pleiotropic effects of QTLs

Quantitative genetics theory suggests that correlations among traits may due to pleiotropic effects of same genes or linkage disequilibrium from tight linkage of several genes or selection etc. Thus, it is not surprising that correlated traits may have QTLs mapping to the same genomic regions (Veldboom et al., 1994; Xiao et al., 1996). We observed a significant positive correlation between SBN and stem volume, and similarly, two QTLs (on linkage group TLG6 and TLG8, respectively) affecting both traits were mapped to the same genomic regions (E33M62439-E61M3192 and E60M33270r-E33M44290r), respectively. The presence of QTLs for both SBN and stem volume, detected in the same genomic regions, may be partially explain such correlation. In a previous study by Bradshaw and Stettler (1995), pleiotropy was suggested for QTLs on linkage group E that were simultaneously associated with spring flush, stem height, basal area and SBN. It was also observed that close trait correlations couldn't be explained either by QTLs mapped to the same genomic region, or by non-significant correlations found to map to this same position (Damerval et al., 1994; Wu et al., 1997). For example, in this study, the QTLs located on the same position (E35M52350r-E44M6097r on linkage group TLG5) were found to be associated with SBR and WCC although there was no significant correlation between these two traits.

4.4 Segregation distortion

Segregation distortion in molecular marker studies was often observed in progeny of F_2, BC_1 and F_1 hybrids derived from inter-and intraspecific crosses for crops (Bentolila et al., 1992; Graner et al., 1991; Causse et al., 1994), forest trees (Bradshaw and Stettler, 1994; Ceveral et al., 2001; Remington and O'Malley, 2000), and many other plant species. In our study, 36 AFLP markers mapped in the separated genetic maps between clone LM50 and clone TB01 were surveyed for distorted segregation ratios ($P < 0.01$). A whole linkage group had markers with distorted segregation ratios, but large regions of markers showing significantly skewed segregation were also detected on linkage groups TLG2, TLG4 and TLG6 of LM50. Of 32 putative QTLs identified in this BC_1 population, 7 (22%) had association with the markers exhibiting distorted segregation. Previous studies have described an association between deviating markers and genes related to embryonic viability and disease resistance in forest trees (Bradshaw and Stettler 1994; Remington and O' Malley, 2000). Bradshaw and Stettler reported that a recessive lethal allele in the F_2 population of *P. trichocarpa* and *P. deltoides* affecting embryonic development was tightly linked to the deviated markers (Bradshaw and Stettler, 1994). Remington and O' Malley (2000) described a genome-wide evaluation of embryonic viability loci in a selfed family of loblolly pine (*Pinus taeda* L.) using AFLP markers. Cervera et al. (2001) reported that markers co-segregating with the *Melampsora larici-populina* resistance gene showed a significant deviation in *P. deltoides* because of death of susceptible trees. Our results indicated that deviated markers were associated with QTLs affecting wood growth and chemical property traits, but not the embryonic development. This pattern may suggest that genetic mechanisms, rather than chance or error, account for the observed transmission ratio distortion. In our interspecific backcross population, alleles from the male parent LM50 were generally dominant over the female parent TB01. However it may be difficult to determine the genetic causes of segregration distortion in the genetic maps. Inbreeding depression may be a source of transmission ratio distortion in the BC_1 populations.

毛白杨无性系纤维特性及微纤丝角的遗传分析*

毛白杨(*Populus tomentosa*)是我国特有的白杨派珍贵乡土树种，早期的研究多集中在以生长量为改良目标的育种工作上(朱之悌, 1992)。由于造纸原料的短缺，发展杨树速生丰产纸浆林刻不容缓，对毛白杨木材性状的研究也日益增多(赵勇刚等, 1996；顾万春等, 1998；宋婉等, 2000；邢新婷等, 2000；2004)。毛白杨的重要用途之一就是作为制浆造纸的原料。影响纸浆性能的性状很多，这些性状的表型变异受遗传因素和环境因素两方面影响，遗传因子有时能造成木材材性在无性系间的差异，有时则不然，环境和竞争的共同影响常常掩盖真实遗传差异，所以需要对影响木材材性的遗传因子进行更深入的研究。目前对杨树多个纤维性状进行系统遗传分析研究还不多。本文通过对毛白杨无性系的木材纤维形态各指标以及纤维组织比量的遗传分析，弄清这些材性指标受遗传控制的强度及遗传改良潜力，对提高和加速我国纸浆材树种的遗传改良水平具有重要的理论和现实意义，同时可为定向培育优良纸浆材毛白杨新品种提供科学依据。

1 材料与方法

1.1 材料

试材取自1987年定植于山东省国营冠县苗圃的全国毛白杨无性系测定林。测定林中的无性系是北京林业大学毛白杨协作组从毛白杨分布区范围内选优得到的(朱之悌, 1992)。株行距4m×5m，8次重复。到本研究调查和采样时树龄均为15年。

1.2 方法

选取无性系测定林中25个无性系各6株，用9mm生长锥在树木胸径处分南北向钻取木芯。截取不同无性系单株木芯上第9年轮的木材试样，在Quantimat-570图像分析仪下，每个指标选取50个视野，测量木纤维直径、木纤维腔径、木纤维壁厚，并对木材纤维比量、导管比量、射线比量和胞壁率等指标做分析。

纤维长度和宽度采取离析测试法测定，纤维长度量取完整纤维端间距离，宽度在纤维中部最宽处量取。每个样品测定50根纤维的长度和宽度数据。

微纤丝角测定是将20um厚的弦切面切片，经脱木素、染色和固定处理，使用偏光显微镜测量2~3个切片，随机测取25根纤维的微纤丝角，取平均值，得出该试样的微纤丝角。

---

* 本文原载《林业科学》, 2007, 43(4): 129-133, 与张冬梅、黄荣凤合作发表。

1.3 数据统计分析

使用 SAS 软件中 ANOVA 模块进行完全随机区组的单因素方差分析。利用 SPQG (3.00)林木遗传育种估算性状间的表型相关、遗传相关和环境相关系数。根据单因素试验随机线性模型：$x_{ij}= u+ a_i+ e_{ij}$，x_{ij}表示第 i 个处理的第 j 个观测值，u 为平均值，ai 为样本的第 i 处理的效应，e_{ij} 为试验误差。以每个无性系 6 个单株调查值的平均值为单位进行方差分析，估算各性状的重复力。

2 结果与讨论

2.1 木纤维形态性状的遗传变异分析

木材纤维形态是断定木材是否适合于制浆造纸的主要因素，了解毛白杨无性系纤维形态特征的变异规律是对其材质改良的基础。木材纤维形态性状主要包括：木纤维长度、宽度、微纤丝角以及木材截面的形态学指标。

2.1.1 木纤维长度、宽度及长宽比

通过对毛白杨无性系 150 单株的木材纤维长度、宽度及长宽比的测定分析(表 1)，纤维长度平均值达到 1.145mm，150 单株的木材纤维长度均在 0.9~1.6mm 之间，符合国际木材解剖学会规定的中级长度标准(Bate，1937)。变异系数普遍较小，说明这些无性系在纤维长度这个性状上的表现较稳定，利于遗传控制。方差分析结果(表 1) 显示：无性系间纤维长度差异达到显著水平($F_{0.05}$= 1.605)。纤维长度的遗传参数估计结果(表 1)表明：无性系重复力高达 0.748，受强度遗传控制；个体重复力达 0.331，受弱度遗传控制。从这一结果可推测杨树无性系的纤维长度主要是受遗传控制的，在毛白杨无性系间选择可获得纤维较长且能稳定遗传的优良无性系。

纤维宽度是植物纤维的第二个主要特征性状。对这些毛白杨无性系纤维宽度的统计分析发现(表 1)，纤维宽度平均值达到 20.780um，纤维宽度变异系数较小，说明这些无性系在这个性状上的表现较稳定，利于遗传控制。同样纤维宽度的方差分析结果(表 1)显示：无性系间纤维宽度差异也达显著水平。遗传参数估计结果(表 1)表明：无性系重复力高达 0.820，受强度遗传控制；个体重复力达 0.440，受中度遗传控制。说明毛白杨无性系的纤维宽度也是受较强遗传控制的，在无性系水平上进行选择会得到较高的遗传增益。

纤维长宽比也是衡量造纸纤维质量的一项指标。本文根据对采样的 25 个毛白杨无性系 150 单株的纤维长宽比的分析，最低长宽比为 45.660，最大为 59.100，平均达到 55.140，是符合造纸原料要求的。从表 1 中可以看出，毛白杨长宽比在无性系间的差异达到了显著水平($F_{0.05}$= 1.605)。同时，对遗传参数的分析结果(表 1)表明：长宽比的无性系重复力为 0.653，达到中等遗传控制水平，个体重复力为 0.377，受弱度遗传控制。从这一点可以看出，该性状在无性系间选择是有潜力的。

对木材材质性状进行遗传分析是为育种服务的。通过性状遗传变异规律的研究，开展性状遗传改良。20 世纪 60 年代就已有许多文章报道细胞长度是受遗传控制的(Dinwoodie，1961；Van Buijtenen，1965)。在阔叶树上，对纤维长度遗传性的研究有限，大多数研究表

明该性状是受中等强度的遗传控制(Otegbeye et al., 1980; Mohrdiek, 1979; Bakulin, 1980)。在毛白杨上也有报道，纤维长度均受中至强度遗传控制(顾万春等，1998；宋婉等，2000；邢新婷等，2000；2004)。本次研究结果进一步印证了毛白杨纤维长度受遗传控制的事实，表明对其实行遗传改良是可以获得较高遗传增益的。

表 1　毛白杨无性系木纤维形态的平均值、变异系数及其方差分析和遗传参数评估

统计项	纤维长	纤维宽	长宽比
平均	1.145mm	20.780μm	55.140
变异系数(%)	5.060	5.110	
F 值	3.960*	3.960*	2.890*
无性系重力	0.748	0.820	0.653
个体重复力	0.331	0.440	0.377

①* 0.05 差异显著水平。下同。

对纤维长度进行遗传操作的研究早在 1958 年 Builtenen 等就做过尝试，通过对杨树(*Populus* spp.)的染色体加倍，发现纤维长度增加了 21% ~ 26%；在阔叶材中，人们已经证明了多倍体化对细胞大小的影响，所以通过人工加倍染色体来培育多倍体的杂种杨以增加纤维长度(Armstrong et al., 1979)。联邦德国黑森州林科所选育出来的三倍体欧洲山杨(*Populus tremula*)和美国威士康星州纸张化学学院培养的三倍体与四倍体美洲山杨(*Populus tremuloides*)，已成功地用于山杨丰产林中，在材质、抗性和生长量上得到了显著的改进。我国也成功地培育出三倍体毛白杨，试验证明无论从生长性状还是纤维长度明显优于二倍体毛白杨(朱之悌，1989；姚春丽等，1998；赵泾峰等，2001)。

2.1.2　*木纤维壁厚和弦向腔径的遗传变异分析*

用 Quant imat-570 图像分析仪对毛白杨无性系 150 单株木材横切面的木纤维直径、木纤维弦向腔径、木纤维壁厚进行测定(表 2)，平均细胞壁厚度为 7.255um 对其方差分析的结果(表 2)表明：该性状在无性系间的差异达到了显著水平($F_{0.05}$ = 1.605)。遗传参数的分析结果(表 2)表明：木纤维细胞壁厚的无性系重复力为 0.546，达到中等遗传控制水平；个体重复力为 0.167，为弱的遗传控制。

同样对木纤维弦向腔径进行统计(表 2)，平均值为 11.661um，方差分析结果(表 2)表明：该性状在无性间的差异达到了显著水平($F_{0.05}$ = 1.605)。遗传参数的分析结果(表 2)表明：木纤维弦向腔径的无性系重复力为 0.863，达到中等强度遗传控制水平；个体重复力为 0.013，受弱的遗传控制。说明这个性状在无性系间选择也是有潜力的。从各无性系木纤维内腔径的变异系数来看，无性系内单株的变化普遍较低，该性状能稳定遗传。

在纤维形态的各指标中，对纸浆材无性系的选育来说，纤维壁腔比和腔径比更能说明问题。一般认为壁腔比小于 1 者为上等原料，大于 1 者为低等原料。通过对毛白杨纤维壁腔比的测定(表 2)，平均值为 0.627，作为纤维原料适宜造纸。纤维腔径比一般情况下越大越好(戴邦潮等，2001)。对毛白杨纤维腔径比的测定(表 2)，平均值为 0.617，表明毛白杨符合阔叶材制浆原料要求。

表 2　毛白杨无性系的方差分析及遗传参数评估

统计项	纤维胞壁厚(双壁)	纤维弦向腔径	纤维壁腔比	纤维腔径比
平均	7.255μm	11.661μm	0.627	0.617
F 值	2.201*	7.319*	2.980*	3.015*
无性系重复力	0.546	0.863	0.664	0.668
个体重复力	0.167	0.013	0.248	0.251

本文对毛白杨纤维壁腔比及纤维腔径比的方差分析(表 2)表明：这 2 个性状在无性系间的差异达到了显著水平($F_{0.05}$= 1.605)。同时，对遗传参数的分析结果(表 2)表明：纤维壁腔比及纤维腔径比的无性系重复力分别为 0.664 和 0.668，都达到中等遗传控制水平；个体重复力分别为 0.248 和 0.251，均为弱遗传控制。从这一点可以看出，这 2 个性状在无性系间选择是有潜力的。

与纸品有关的纤维性状的遗传力报道极少，Mergen 等(1960) 证实了黑松(*Pinus thunbergii*)×赤松(*Pinus densiflora*) 杂交种的管胞壁厚是受遗传控制的，杂交种的平均管胞壁厚超过亲本黑松而低于赤松。Otegbeye 等(1980)的研究表明：幼龄多枝桉树(*Eucalyptus viminalis*) 纤维直径和纤维壁厚的家系遗传力在 0.82~0.94 间，说明木材细胞壁厚和细胞直径是受强度遗传控制的。然而对杨树木纤维直径、木纤维弦向腔径、木纤维壁厚的遗传分析还未见报道。通过对毛白杨无性系这些性状的分析，对其实行遗传改良具有重要的理论指导意义。

2.1.3　毛白杨无性系微纤丝角的遗传变异分析

微纤丝角是次生壁 S2 层中微纤丝排列方向与细胞轴向之间的夹角，是细胞壁的基本性质之一，也是评估木材材质和纸张强度的重要因子。微纤丝角小，木材和纸张强度大，纵向收缩小，所以，各国学者在进行树木木材品质性状遗传改良时，十分重视微纤丝角的研究(孙成志等，1987)。

对毛白杨 150 单株的微纤丝角分析(表 3) 表明：3-48-2 无性系的微纤丝角最大为 21.52°；5009 无性系的微纤丝角最小为 12.75°，平均为 16.18°。从这一结果看，毛白杨木材的微纤丝角普遍较小，说明纤维的弹性模量和强度较强，利用该原料制出理想纸品的潜力很大。无性系内单株的变异系数变化普遍较低(表 3)，因此说明该性状能稳定遗传。

微纤丝角方差分析结果(表 4)显示：微纤丝角在无性系间的差异达到了差异显著水平($F_{0.05}$= 1.605)。遗传参数的分析结果(表 4) 表明：微纤丝角的无性系重复力为 0.961，达到了极强遗传水平；个体重复力为 0.651，达到了中等遗传水平。所以，无论是在无性系间还是无性系内的个体间对微纤丝角进行选择都将是有效的。

有研究表明微纤丝角和管胞长度(Echols，1958)、细胞壁厚度是紧密相关的(Hiller，1964)，从而推测微纤丝角可能受中强度遗传控制。Mergen 等(1960)在对黑松×赤松杂交种的研究中证实了微纤丝角是受遗传控制的。在杨树上对微纤丝角的研究不多，柴修武等(1992)对 13 个杨树无性系微纤丝角的研究表明：微纤丝角在无性系间的差异很显著，但并未对其进行遗传分析。本次研究结果证实了对毛白杨无性系微纤丝实行遗传改良的可能性。

表 3 毛白杨无性系微纤丝角的平均值、变异幅度及变异系数

无性系	平均值(°)	变化幅度(°)	变异系数(%)	无性系	平均值(°)	变化幅度(°)	变异系数(%)
3-48-2	21.52	18.52~23.52	8.31	3-23-32	14.14	13.64~14.52	2.58
3-2-1	20.26	19.00~22.38	6.43	5009	12.75	11.32~13.80	6.76
0027	20.78	18.72~23.72	8.39	1807	13.11	12.48~14.04	4.26
0060	16.71	14.40~19.16	11.88	0022	13.41	11.84~14.64	7.23
1210	15.78	14.96~17.26	5.63	0053	13.89	13.12~15.00	4.79
0041	14.88	14.24~15.36	2.78	1272	15.17	14.80~16.00	3.03
4337	15.38	14.80~16.24	3.52	3332	16.44	14.80~18.40	7.15
1005	16.78	15.00~19.76	10.74	1808	17.44	15.92~19.36	8.77
0034	18.76	16.28~21.48	9.64	0040	15.77	14.20~17.24	7.32
0067	17.21	15.12~18.56	7.20	5003	16.89	15.96~17.52	3.82
3-1-27	15.12	13.48~16.36	7.17	0066	17.87	17.24~18.64	2.78
1232	13.82	12.96~14.68	4.10	0011	16.13	15.44~17.04	3.79
5101	14.62	13.40~15.36	4.98	平均	16.18		

表 4 毛白杨无性系微纤丝角的方差分析及遗传参数估计

变异来源	自由度	离差平和	均方	F 值	重复力
无性系间	24	775.941	32.331	25.297*	0.961
无性系间	125	159.757	1.278		0.651
总和	149	935.698			

2.2 木材组织比量的遗传变异分析

有报道称杨树木材作为纤维资源利用时，单位体积或单位质量的纤维数量和纤维质量，是决定杨树无性系品种材质优劣的关键(张景良，1991)。尽管本研究是从木材的横切面对木材的组织比量进行测定，但在同等条件下，仍能反映不同无性系间木纤维组织比量的差别。利用木材横切面对毛白杨无性系木材组织比量的测定结果(表 5)表明：导管占 25.73%，木射线占 14.25%，纤维占 60.02%。在毛白杨木材横切面中，纤维比量占组织的一半以上，表明毛白杨确实是很好的纤维用材。

组织比量的方差分析结果(表 5)显示：除了导管比例外，纤维比例和射线比例在无性系间的差异均达到了显著水平($F_{0.05}=1.605$)。遗传参数的分析结果(表 5)表明：导管比例和纤维比例的无性系重复力分别为 0.336 和 0.437，均属弱度遗传控制；个体重复力极低。射线比例的无性系重复力分别为 0.711，达到了中等遗传控制水平；个体重复力为弱度遗传控制，预示了对射线比例的选择在无性系间更有潜力。目前，对毛白杨无性系间组织比量(导管比例、射线比例以及纤维含量) 的研究还不多，对其遗传变异规律的研究还未见报道，然而这方面研究结果对优良纸浆材无性系的选育意义重大。

表 5 毛白杨无性系组织比量的方差分析及遗传参数估计

统计项	导管比量	射线比量	纤维比量
平均	25.73	14.25	60.02
F 值	1.507	3.460*	1.776*
无性系重力	0.336	0.711	0.437
个体重复力	0.078	0.291	0.115

3 结论

通过对毛白杨无性系木纤维形态各指标参数的分析，所有性状在无性系间均达到了差异显著水平。其中，纤维宽、木纤维腔径和微纤丝角的无性系重复力分别为0.820、0.863和0.961，均受强度遗传控制，纤维长、长宽比、木纤维壁厚、木纤维腔径和木纤维壁腔比5个性状无性系重复力分别为0.748、0.653、0.546、0.668、0.664，均受中度遗传控制，表明这些性状在无性系间进行选择很有潜力，揭示了通过一定的遗传育种措施和养护措施改变毛白杨木纤维宽度、调节木纤维腔径和微纤丝角，对这些性状实行定向遗传改良是切实可行的。

毛白杨无性系组织比量的分析结果表明：除导管比例外，纤维比例和射线比例在无性系间的差异均达到了显著水平。导管比例和纤维比例的无性系重复力较低，表明这2个性状在无性系间的选择潜力不大；射线比例的无性系重复力为0.711，受中度遗传控制。对毛白杨木材横切面的组织比量进行综合考虑，对这些性状遗传改良潜力不大，显然，毛白杨作为纤维资源利用时，更重要的还要考虑单位体积或单位重量的纤维数量和纤维质量，这些仍是决定毛白杨无性系品种材质优劣的关键。

Genetic Analysis of Fiber Traits and Microfibrillar Angles on Clones of *Populus tomentosa*

Abstract 25 clones (6 trees from each clone) were sampled (aged 15 years) from clone test stand of *Populus tomentosa*. The fiber forms (fiber length, fiber width, ratio of fiber length and fiber width, fiber wall thinckness, fiber lumen diameter, ratio of 2 fiber wall thinckness and fiber lumen diameter, ratio of fiber lumen diameter and cell diameter) were analysed. Variance analysis showed that these 7 traits were significantly different among clones at 0.05 level. 2 characters, Fiber width and fiber lumen diameter, were strongly inherited, and others 5 traits were under moderate genetic control. The selection could be conducted effectively among clones. The variation analysis of microfibrillar angles showed that there was significantly difference among clones at 0.05 level. The clones repeatability (0.961) and single repeatability (0.651) of microfibrillar angles indicated that the character was under genetic control. And selection could be conducted effectively among clones and within clones. The variance analysis of transverse section tissue percentage include vessel percentage, fiber percentage and wood ray percentage of *Populus tomentosa* were perfected. All traits except vessel percentage were observed significant clonal differences at 0. 05 level. The values of clone repeatability of vessel percentage and fiber percentage were 0. 437 and 0. 336 separately, and the values of clone repeatability of wood ray percentage was 0. 711. This showed that genetic control for wood ray percentage is feasible.

Keywords　*Populus tomentosa*, pulpwood, fiber forms, microfibrillar angles, genetic analysis

附　录

一、教材与国家标准的编著

名　称	出版社/出版日期	承担任务	备　注
林业生物技术	中国林业出版社/2008.05	主编	全国高等农林院校教材
林木遗传学基础	中国林业出版社/2012.06	主编	全国高等农林院校规划教材
植物新品种特异性、一致性、稳定性测试指南(杨属)	中华人民共和国国家质量监督检验检疫总局，中国国家标准化管理委员会发布/2015.12.31	参编	中华人民共和国国家标准 GB/T 32344—2015

二、研究生培养

1. 硕士生

姓　名	论 文 题 目	起始时间	合作导师	职　称
宋　婉	毛白杨无性系木材材性遗传变异研究	1993.09—1997.07	续九如	
高程达	毛白杨抗寒性研究及其优良无性系选择	1993.09—1997.07	朱之悌 马国华	教授
于雪松	三倍体毛白杨有性生殖能力的研究	1995.09—1998.07		研究员
李静怡	毛白杨新无性系苗期生长、光合、形态性状变异研究及相关分析	1996.09—2000.07		
张万科	毛白杨新无性系多点测定与造林密度试验	1997.09—2000.07	李新国	
邢新婷	三倍体毛白杨无性系木材品质性状变异及选择的研究	1997.09—2000.07	张文杰	副研究员
李雪平	低温锻炼和 ABA 处理对毛白杨膜稳定性及抗冻性的影响	1998.09—2001.07		副研究员
袁定昌	美洲黑杨杂种优势分子机理的研究	1999.09—2002.07	苏晓华	
肖基浒	引进国外白杨品种资源的繁殖技术研究	1999.09—2002.07		
黄权军	白杨及其杂种染色体加倍与三倍体培育的研究	1999.09—2002.07	康向阳	
张子辉	毛白杨新无性系对比试验与造林密度效应研究	2000.09—2003.07	李新国	
张　谦	转 *CpT* Ⅰ基因毛白杨的分子检测与虫试	2000.09—2003.07		

（续）

姓　名	论 文 题 目	起始时间	合作导师	职　称
于海武	毛白杨无性系耐盐性比较与转耐盐基因的研究	2001. 09—2004. 07		
张有慧	欧美杨新品种 L35 快速繁殖技术研究	2001. 09—2004. 07		教授级高工
于志水	白杨无性系茎尖培养及体细胞胚胎发生技术研究	2002. 09—2005. 07		高级教师
冯夏莲	滇杨遗传多样性与杨属派间遗传分化的研究	2003. 09—2006. 07		
张慕博	杨属已知品种数据库管理信息系统的研建	2003. 09—2006. 07		
姚　娜	不同基因型毛白杨离体再生及细胞悬浮培养的比较研究	2004. 09—2007. 07		
乔梦吉	柽柳离体培养体系的建立及耐盐性研究	2004. 09—2007. 07	林善枝	
陶凤杰	悬铃木再生体系的建立和开花关键基因转化的研究	2004. 09—2007. 07		
孙丰波	毛白杨无性系木材品质性状遗传变异研究	2004. 09—2007. 07		副研究员
徐华金	几种彩叶植物的引种栽培及适应性研究	2004. 09—2007. 07		
李　琰	*PtDRG*01 基因原核表达分析及遗传转化的初步研究	2005. 09—2008. 07		
刘婷婷	*MdSPDS*1 基因的功能分析及转化毛白杨的初步研究	2005. 09—2008. 07	庞晓明	
李海霞	转 *PtDRG*01 基因烟草的抗病试验与表达研究	2005. 09—2008. 07		
曹冠琳	*PtLF* 和 *PtAG* 不育结构转化烟草和毛白杨的研究	2006. 09—2009. 07	安新民	
龙　苹	毛白杨遗传转化体系优化和转 *MdSPDS*1 基因植株耐盐性分析	2006. 09—2009. 07	庞晓明	
于　来	*PtAP*3、*PtAG*、*PtAP*1 不育结构转化毛白杨和拟南芥的研究	2007. 09—2010. 07	安新民	
王　强	邛崃山系三种大熊猫主食竹种更新对比研究	2008. 09—2011. 07	李颖岳	
杜　鹃	毛白杨纤维素合酶基因 *PtoCesA*4 的克隆、表达及 SNP 分析	2008. 09—2011. 07	张德强	
陈　磊	杨树感染溃疡病后 microRNA 的鉴定与表达分析	2009. 09—2012. 06	徐吉臣 王延伟	

（续）

姓 名	论文题目	起始时间	合作导师	职 称
张译云	毛白杨低温胁迫相关 microRNAs 的鉴定及差异表达分析	2009. 09—2012. 06	徐吉臣 王延伟	
郭 斌	欧美杨×藏川杨杂交群体构建及其苗期相关性状 QTLs 初步分析	2009. 09—2012. 06	安新民	
季乐翔	毛白杨与河北杨失水过程基因表达规律的研究	2010. 09—2013. 06	安新民	
吴丽萍	‘冬枣’花药培养体系优化和再生植株鉴定研究	2010. 09—2013. 06	庞晓明	
王斯琪	‘冬枣’×‘映山红’遗传连锁图谱构建	2010. 09—2013. 06	庞晓明	

2. 博士生

姓 名	论文题目	起始时间	合作导师	职 称
李纪元	枫杨苗期性状遗传变异及其耐涝机理的研究	1997. 09—2000. 07	张志毅	研究员
林善枝	杨树抗冻性机理及分子生物学研究	1997. 09—2001. 07		教授
陈 敏	光肩星天牛内切葡聚糖酶的特性及其结合短肽的筛选	1998. 09—2003. 07	卢孟柱	副教授
张德强	毛白杨遗传连锁图谱的构建及重要性状的分子标记	1999. 09—2002. 07	杨 凯	教授
黄秦军	美洲黑杨×青杨连锁图构建及重要材性 QTLs 分析	2000. 09—2003. 07	苏晓华	研究员
李善文	杨树杂交亲本与子代遗传变异及其分子基础研究	2001. 09—2004. 07	李百炼	研究员
何承忠	毛白杨遗传多样性及起源研究	2002. 09—2005. 07		教授
林元震	甜杨葡萄糖-6-磷酸脱氢酶基因克隆及结构分析与功能鉴定	2001. 09—2006. 07 （硕博连读）		副教授
王冬梅	毛白杨花芽分化规律与开花调控的分子基础研究	2002. 09—2007. 07 （硕博连读）	安新民	
张 谦	毛白杨抗锈病基因筛选与 NBS 型抗病基因分析	2003. 09—2007. 07		研究员
周祥明	美洲黑杨雄性花芽 cDNA 文库构建及花发育相关基因的克隆与鉴定	2003. 09—2007. 07	苏晓华	
李义良	转基因杨树的分子检测及抗逆性评价	2003. 09—2008. 09 （硕博连读）	苏晓华	教授级高工

（续）

姓　名	论 文 题 目	起始时间	合作导师	职　称
樊永明	麦草甲酸法制浆脱木素化学及蒸煮历程的研究	2004.09—2007.07	谢益民	教授
李　博	毛白杨与毛新杨转录组图谱构建及若干性状的遗传学联合分析	2004.09—2009.07（硕博连读）		副研究员
杨爱珍	桃核发育的生理生化特性及基因表达差异研究	2006.09—2009.07	王有年	高级实验师
王泽亮	河北杨干旱胁迫响应基因的分离与鉴定	2006.09—2010.07		副研究员
江锡兵	美洲黑杨与大青杨杂种无性系遗传变异研究	2006.09—2011.07（硕博连读）	张金凤	副研究员
杨章旗	马尾松材性与产脂性状遗传改良研究	2006.09—2012.06		教授级高工，院士有效候选人
薄文浩	藏川杨遗传多样性及杂交子代遗传变异研究	2006.09—2011.12（硕博连读）	邬荣领	
郑会全	毛白杨 TIR-NBS 类抗病相关 *PtDrl*02 基因启动子的结构与功能研究	2007.09—2010.07		教授级高工
赵曦阳	白杨杂交试验与杂种无性系多性状综合评价	2007.09—2010.07		副教授
王利保	白杨优良无性系与栽培密度互作效应研究	2008.09—2013.06	康向阳	副教授
曾慧杰	金银花优株评价及培育技术研究	2009.09—2013.06	李　云	副研究员
雷　杨	毛白杨 WRKY 转录因子基因的结构、功能分析与调控表达研究	2009.09—2013.06	张德强	高级教师
杨晓霞	多阶段生长性状关联分析统计模型	2009.09—2013.06	邬荣领	副教授
任媛媛	毛白杨干旱、水涝、盐胁迫相关 micro RNA 的鉴定与表达分析	2007.09—2013.06（硕博连读）	康向阳 王延伟	
马开峰	毛白杨基因组 DNA 甲基化遗传变异及遗传效应	2007.09—2013.06	张德强	
宋跃朋	毛白杨花发育遗传调控研究	2008.09—2013.06	张德强	副教授

3. 博士后

姓　名	论 文 题 目	起始时间	合作导师	职　称
张冬梅	纸浆材毛白杨无性系材质性状的遗传分析	2001.07—2003.06	鲍甫成	研究员
安新民	毛白杨开花关键基因的分离及其功能分析	2002.07—2004.06		教授

三、主持和参与的科研项目

1. 国家“六五”、“七五”、“八五”科技攻关课题，“毛白杨良种选育”，主要参加人
2. 国家教委优秀年轻教师基金项目，“白杨染色体加倍技术研究及三倍体育种”，1989—1994，主持
3. 国家“八五”科技攻关加强专题，“毛白杨三倍体育种研究”，1995—1998，副主持
4. 国家“九五”科技攻关专题，“毛白杨单板类人造板材新品种选育”，1996—2000，主持
5. 国家“948”项目，“国外白杨品种资源和抗旱基因及转化技术引进(98-4-04)”，1998—2003，主持
6. 教育部高校青年骨干教师基金项目，“白杨 $2n$ 花粉诱导的细胞学机理研究”，2000—2003，主持
7. 国家转基因植物研究与产业化专项课题，“转抗虫基因毛白杨田间试验(J00-B-001-02)”2000—2005，主持
8. 国家自然科学基金项目，“毛白杨遗传图谱构建和重要性状的分子标记(30170780)”，2001—2003，主持
9. 国家自然科学基金项目，“甜杨抗冻基因的克隆及表达鉴定(30271093)”，2002—2005，主持
10. 国家“十五”科技攻关课题，“毛白杨优良新品种选育(2002BA515B0303)”，2001—2005，主持
11. 国家农业科技成果转化资金项目，“三倍体毛白杨新品种及繁育技术中试与示范(02EFN217101267)”，2002—2005，主持
12. 教育部博士点基金项目，“杨树杂交育种及杂种优势形成的分子机理研究(20020022004)”，2002—2005，主持
13. 国家“863”项目，“杨树纸浆材新品种选育研究(2001AA244031)(白杨)”，2002—2006，主持
14. 国家“863”项目，“抗旱节水白杨新品种筛选与利用(2002AA2Z4011)”，2002—2006，主持
15. 国家“973”项目课题，“分子改良树木的性状表达与鉴定(G1999016005)”，主要参加人
16. 国务院总理基金，“三倍体毛白杨区域化试验及产业化技术研究”，主要参加人
17. 国家自然科学基金项目，“杨树杂交育种中杂种优势分子机理的研究(30571516)”，2006—2008，主持
18. 国家林业局“948”项目，“杨树控花基因分离鉴定与转化技术引进(2006-4-72)”，2006—2010，主持
19. 国家林业局项目，“林木转基因研究现状调查及存在问题研究(LY-09-02)”，2009—2010，主持
20. 国家“十一五”科技支撑项目，“杨树速生丰产林培育关键技术研究与示范(2006BAD24B04)”，2006—2010，主持

21. 国家“十一五”科技支撑项目，“高产优质毛白杨速生材新品种选育(2006BAD01A1502)”，2006—2010，主持
22. 国家自然科学基金项目，“毛白杨叶锈病病原菌诱导表达 NBS 型基因的分子作用机制(30872043)”，2009—2011，主持
23. 国家“863”项目，“速生杨树品种抗虫、抗旱耐盐转基因育种研究(2009AA10Z107)”，2009—2011，主持
24. 林业公益性行业科研专项，“重要乡土树种核心种质评价及高效育种共性技术研究(201004009)”，2009—2012，主持

四、科研获奖

1. “毛白杨成年优树的复壮与快速繁殖技术研究”，1988 年获林业部科技进步二等奖，排名第二
2. “全国毛白杨基因资源的收集、保存和利用的研究”，1990 获林业部科技进步一等奖，1992 年获国家科技进步二等奖，排名第 13
3. “毛白杨无性系同工酶基因标记研究”，1993 年获北京市科协优秀论文三等奖，排名第一
4. “白杨染色体加倍技术研究及三倍体育种研究”，1995 年获林业部首届青年学术论文一等奖，排名第一
5. “毛白杨多圃配套系列育苗技术研究”，1996 年获林业部科技进步二等奖，1997 年获国家科技进步二等奖，排名第五
6. “三倍体毛白杨新品种(三毛杨 1~6 号)”，2000 年获国家植物新品种保护专利，排名第二
7. “三倍体毛白杨新品种选育研究”，2003 年获国家科技进步二等奖，排名第二

五、表彰和荣誉

1992 年　部级有突出贡献中青年专家
1993 年　享受国家政府特殊津贴
1996 年　获中国林业青年科技奖
1997 年　获中国青年科技奖
1997 年　部级跨世纪学科学术带头人
1998 年　获共青团北京市“五四”青年奖章
1998 年　获全国优秀教师宝钢教育基金奖
1998 年　获台湾刘业经教授基金奖
2000 年　获北京市优秀工会工作者称号
2001 年　获北京市经济技术创新标兵称号
2004 年　获学校师德标兵称号
2005 年　获全国优秀博士论文指导教师奖
2006 年　获北京市第二届教学名师称号

2006 年　获北京市教育创新标兵称号
2007 年　获全国林业产业突出贡献奖先进个人
2008 年　获联合国粮农组织（FAO）森林资源部国际森林年纪念奖章
2010 年　获全国生态建设突出贡献奖——林木种苗先进工作者

六、社会兼职

- 国家科学技术进步奖林业组评委
- 国家留学基金评审专家
- 国家留学回国人员基金评审专家
- 教育部精品课程评审专家
- 国家林木良种审定委员会委员
- 国家林业局林木基因安全技术专家组成员
- 国家农业综合开发项目专家组成员
- 中国太平洋经济合作全国委员会粮农资源开发委员会委员
- 中国林学会学术工作委员会委员
- 中国杨树委员会委员
- 中国林学会林木遗传育种分会委员、副秘书长
- 《北京林业大学学报》编委
- 《中国林业研究》(*Forestry Studies in China*)编委
- 国家林木品种审定委员会用材林专业委员会副主任
- 全国植物新品种测试标准化技术委员会委员
- 中国杨树委员会副主任
- 中国林学会林木遗传育种分会常委、副主任
- 中国生物化学与分子生物学会农学分会理事
- 中国植物生理学会生物质与生物能源专业委员会委员
- 北京林学会理事
- 北京市第十二届、第十三届人大代表，北京市人大常委会农村工作委员会委员
- 北京市青年联合会委员
- 北京市党外高级知识分子联谊会理事
- 北京市海淀区政协委员

后记

张志毅教授是我的博士后合作导师，在工作生活中亦师亦友。2011 年春节期间，突闻张老师不幸辞世，震惊错愕，瞬间泪水奔涌模糊了视线，陷入万分悲痛之中，顿感茫然失措。与张老师相识、相知、相悉的一幅幅画面不时从脑海中掠过，记忆中工作生活的点点滴滴常常浮现，数次梦中重现和张老师交谈的情景，惊醒后黯然神伤，思考着自己应该做一件事情。强烈的责任感促使我产生整理张老师学术论文的想法，这一想法得到了学科负责人康向阳教授的赞同和张老师所有弟子们的大力支持。

《杨树遗传改良研究》文集的编辑出版是在学科和张老师众多弟子的支持下完成的。曹冠琳、刘文凤主要参与了论文的收集和编辑，宋跃朋、龙萃、薄文浩、马开峰、赵曦阳等帮助收集整理了张老师所指导的研究生信息和影像资料等，李善文、何承忠、张德强、肖基浒参与了部分编辑工作，冯秀兰、康向阳、李悦、李云和张金凤老师作为顾问为文集顺利完成提供了指导，张老师的众多弟子均给予了积极的配合和支持，在此一并表示由衷的感谢！由于篇幅所限，文集没能全部收录张老师和弟子合作发表的所有文章，在此深表遗憾，敬请谅解。

承蒙山西大学裴雁曦教授为文集题写了新中国林业事业的奠基人梁希先生对新中国林人的寄语和北京林业大学校训，在此深表谢意！

北京林业大学原校长、中国工程院院士尹伟伦教授为文集做序，倍感荣幸。尹院士在百忙之中与我进行了多次沟通，不厌其烦，反复修改，几易其稿，在此向尹院士表示真诚的谢意！

在张老师离开我们十年之际，用《杨树遗传改良研究》文集来缅怀他，重温他的学术成果，传承和弘扬其“常规育种与现代分子育种融合创新”的学术思想，力争在国家生态文明建设中做出更大贡献，这是对张志毅老师最好的追忆。

安新民

2021 年 6 月 24 日

將河山裝成錦繡
把國土繪成丹青
知山知水
樹木樹人

成长岁月

世界地图

教书
育人

1. 张志毅教授携弟子在山东冠县苗圃基地视察
2. 张志毅教授与青年教师、研究生在实验室合影
3. 张志毅教授在指导研究生开展实验
4. 张志毅教授和毕业研究生在科研楼前合影
5. 张志毅教授和“翱翔计划”中学生在科研楼前合影
6. 张志毅教授在新中国林业事业的奠基人——梁希雕像前留影

实验室和基地建设

1. 2010 年 3 月张志毅教授在北方平原林业创新与示范实践基地
2. 2010 年 4 月曹部长视察实验室
3. 2009 年 3 月张志毅教授同专家组介绍林木育种国家工程实验室建设情况
4. 2009 年 7 月张志毅教授与林木育种国家工程实验室理事会参会委员合影
5. 2009 年 7 月张志毅教授主持林木育种国家工程实验室理事会暨学术委员会会议
6. 2007 年 8 月张志毅赴青海西藏考察藏川杨资源

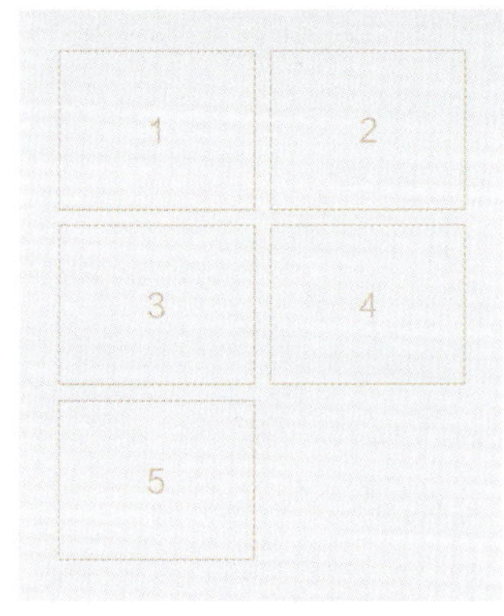

1. 白杨杂交子代优良无性系扩繁
2. 张志毅教授和团队的研究成果（白杨杂交子代群体）
3. 张志毅教授在山东冠县苗圃基地视察白杨杂种优良无性系
4. 黑杨杂交子代扩繁育苗
5. 美大青杨杂交子代扩繁育苗

1	2
3	4
5	6

1. 河北平泉美洲黑杨 × 大青杨杂交子代优良无性系试验林
2. 河南鄢陵黑杨杂交子代优良无性系试验林
3. 山东冠县白杨杂交子代优良无性系试验林
4. 山东冠县黑杨杂交子代优良无性系试验林
5. 山东临沂黑杨杂交子代优良无性系试验林
6. 山东宁阳白杨杂交子代优良无性系试验林

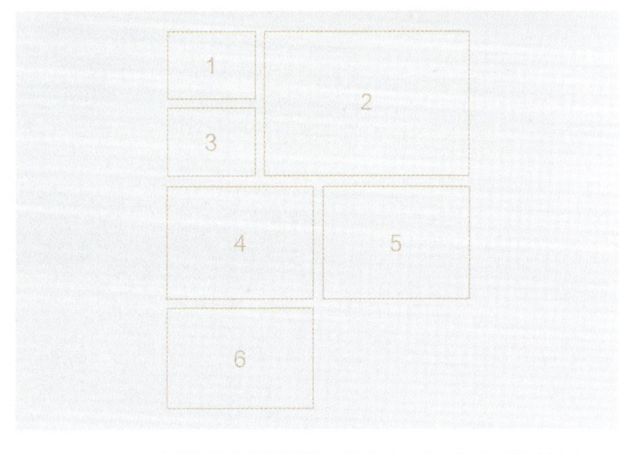

1. 独特的“蜘蛛网”栽植密度设计试验
2. 张志毅教授和“千人计划”邬荣领教授及山东冠县、河北威县苗圃领导等合影
3. 张志毅教授和“千人计划”邬荣领教授及山东冠县与威县苗圃领导一起考察白杨试验林
4. 张志毅教授在杨树育种研讨会上作报告
5. 张志毅教授主持统计和计算遗传学国家研讨会
6. 张志毅教授与美国宾夕法尼亚州立大学来访专家合影

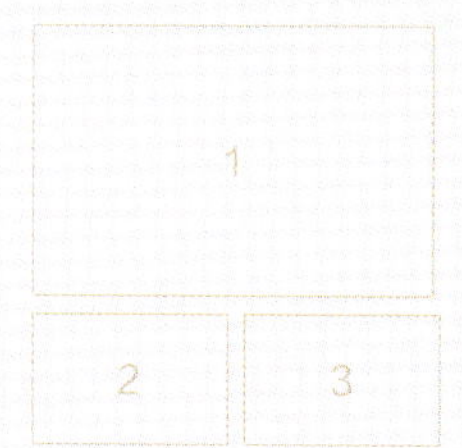

学术交流

1. 张志毅教授在“北方平原基地建设”毛白杨基因库现场讲解
2. 张志毅教授在第二届中国林业学术大会第二分会场主持会议
3. 张志毅教授在作“948”项目现场查定汇报

1	2
3	4
5	6

1. 张志毅教授出国考察进行学术交流
2. 张志毅教授出国考察和国外专家合影
3. 张志毅教授和“千人计划”学者邬荣领教授在实验基地合影
4. 张志毅教授和“长江学者” 李百炼教授及张德强教授合影
5. 张志毅教授和基因组所于军教授签署实验室合作协议
6. 张志毅教授和新加坡南洋理工大学杨远方教授等在实验楼前合影

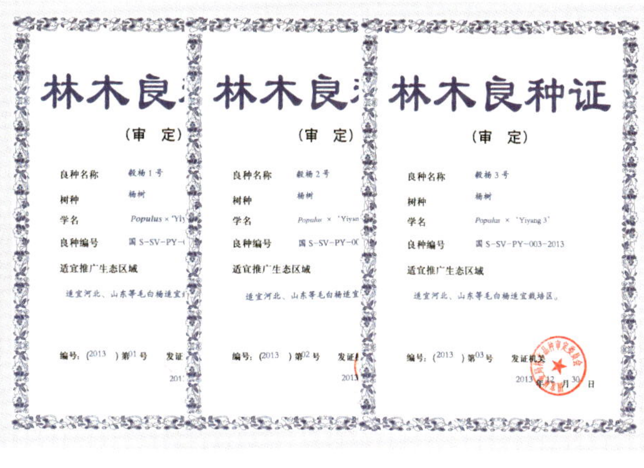

丰硕成果

1. 孜孜不倦求真知
2. 倾囊相授育新人
3. 张志毅教授在北京市第十三届人民代表大会第三次会议留影
4. 张志毅教授和朱之悌院士喜获国家科技进步二等奖
5. 张志毅教授获奖证书和荣誉证书
6. 良种证书